Industrial Air Pollution Handbook

Industrial Air Pollution Handbook

Editor: **Albert Parker**

McGRAW-HILL Book Company (UK) Limited

London · New York · St Louis · San Francisco · Auckland · Beirut
Bogotá · Düsseldorf · Johannesburg · Lisbon · Lucerne · Madrid · Mexico
Montreal · New Delhi · Panama · Paris · San Juan · São Paulo
Singapore · Sydney · Tokyo · Toronto

Published by
McGRAW-HILL Book Company (UK) Limited
MAIDENHEAD . BERKSHIRE . ENGLAND

Industrial air pollution handbook.
1. Air—Pollution—Great Britain
I. Parker, Albert
628.5'3'0941 TD883.7.G7 77-30126

ISBN 0-07-084486-0

1 2 3 4 5 WC&S 8 0 7 9 8

PRINTED AND BOUND IN GREAT BRITAIN

Preface

This book has been planned to provide information which will be of value to all who are concerned with surveys of general pollution of the air in urban and rural areas, and to those who are involved in reducing to a realistic minimum the emission of air pollutants from industrial processes.

It would be an advantage if there were internationally accepted methods and standards for the measurement of air pollution, so that a sound comparison of all results could be made. That is not the position at present, though there are moves in that direction; nevertheless, this book will have a greater relevance outside, as well as inside, the UK.

There is a comprehensive chapter on the effects of air pollution on human health, on plants and animals, and on materials, while another chapter deals with the existing legislation. However, the major part of the book deals with methods of reduction of the emission of pollutants, so far as is practicable on the basis of present knowledge and experience. As it is impossible in one volume to deal with the total range of industries, it has been necessary to be selective, but we have ensured that all important methods of emission reduction have been covered.

The editor has written three chapters and parts of three others, and the remainder have been written by authors who are experts in the subjects they cover. The editor expresses his appreciation of their work in preparing their contributions and thanks them for their cooperation throughout the life of the project.

ALBERT PARKER
November 1977

The editor

Albert Parker, CBE, BSc(Hons), MSc (Manchester), DSc (Birmingham), FRIC, FIChemE, FIWater Pollution Control, Hon FIGasE, Hon FInst Fuel, Hon FIPubHlthEng, Hon Mbr Coke Oven Managers' Assoc, Life Vice-President RSHlth, Hon FAmerPubHealth Assoc.

The editor became Graduate Scholar and Beyer Research Fellow of Manchester University. He was then a Lecturer at Birmingham University, and subsequently officer in charge of a team on development work and research for a joint committee of the Institution of Gas Engineers and the Fuel Department of Leeds University on methods of manufacture of gas and the treatment and disposal of gas liquor effluents. Afterwards, he became Assistant Director and later Director of Water Pollution Research in the former Government Department of Scientific and Industrial Research, and then Director of Fuel Research and Air Pollution Research in the same Government Department. On leaving Government Service, he established a practice as a Consultant Chemical Engineer on a variety of scientific and technical problems in Britain and several other countries.

During the Second World War he was Lieutenant-Colonel in charge of a group of British and American scientists, all in military uniform, advising the American Army during their advance through and beyond the Ruhr on certain facets of the German production of synthetic fuels for German aircraft, tanks, etc.

From 1951 to 1964, he was Honorary Secretary of British National Committee of World Energy Conference. He prepared and presented documents on world energy resources for the World Energy Conference in Melbourne, Australia, in 1962 and for the Conference in Moscow in 1968.

He was Honorary Editor IChemE 1937–45 and Vice-President 1941–3. From 1953 until 1955 he was President of the Fuel Luncheon Club, London, and from 1955 until 1956 Chairman of the Council of RSH, of which he is now Life Vice-President. He was Honorary Secretary of the British National Committee of the World Energy Conference 1951–64 and Chairman of the Committees guiding the production of the sixth and seventh editions of Technical Data on Fuel. He was President of the National Society for Clean Air from 1963 to 1965, and is now Honorary Vice-President.

He was awarded the Osborne Reynolds Medal IChemE, 1941, the Melchett Medal Inst Fuel, 1955, the Telford Premium, 1942, and the Chadwick Medal, 1955, of the Inst. Civil Eng. for papers on water pollution and on air pollution. He gave the Thomas Hawksley Lecture IMechE, 1949. He gave the Mitchell Memorial Lecture and was awarded the Gold Medal of the Stoke-on-Trent Assn of Engineers in 1956. He gave the Cantor Lectures of the Roy. Soc. Arts, 1960. He was Chairman of the Expert Committee of WHO on Air Pollution, 1958, and was given a Clean Air Award of the Nat. Soc. for Clean Air, 1973.

Contents

Contributors

1. The editor.
2. Part 1. Marjorie Clifton, MD, and Sir Charles Stuart-Harris, CBE, MD, DSc, FRCP, Professor at University of Sheffield Medical School.
   Part 2. L. H. P. Jones, PhD (Melbourne), BAgrSc, FRACI, FRIC, FIBiol, and D. W. Cowling, MIBiol, both of Grassland Research Institute, Hurley, Maidenhead, Berkshire.
   Part 3. The editor.
3. Part 1. The editor.
   Part 2. Pamela M. Bryant, BSc, FRIC, Senior Environmental Adviser to the National Radiological Protection Board, Harwell. President, UK Society for Radiological Protection 1977–78.
4. S. R. Craxford, ISO, MA(Oxon), DPhil, Consultant.
5. The late G. Spurr, BSc(Hons), FInstP, FRMetS, was on the staff of the Environmental and General Studies Section, Central Electricity Generating Board, London.
6. K. P. Grubb, BSc(Eng), MICE, Senior Assistant Civil Engineer, Central Electricity Generating Board, London.
7. J. Bettelheim, CEng, DiplEng, CSc(Prague), FIChemE, Research Officer, B. H. M. Billinge, PhD, BSc, Group Leader, and W. S. Kyte, CEng, MA(Cantab), PhD, FIChemE, Research Officer, all of Research Laboratories, Central Electricity Generating Board, Leatherhead, Surrey.
8. K. Darby, Managing Director of Lodge-Cottrell Ltd, Birmingham.
9. J. Bettelheim, CEng, DiplEng, CSc(Prague), FIChemE, Research Officer, and B. H. M. Billinge, PhD, BSc, Group Leader, both of Research Laboratories, Central Electricity Generating Board, Leatherhead, Surrey.
10. A. A. Barrett, CEng, FIMechE, Plant Engineering Department of the Generation Development of the Construction Division of the Central Electricity Generating Board, Cheltenham, Gloucestershire.
11. Pamela M. Bryant, BSc, FRIC, Senior Environmental Adviser to the National Radiological Protection Board, Harwell.

12. Part 1. G. E. Hall, CChem, MSc, MRIC, Senior Technical Officer, Head of Member Services, British Carbonization Research Association, Wingerworth, Chesterfield, Derbyshire.
    Part 2. The editor.
13. M. T. Westaway, PhD, FInstPet, Environmental Studies, and G. J. Brockis, BSc, FInstPet, Head of Environmental Sudies, both of BP Research Centre, Sunbury on Thames, Middlesex.
14. G. E. Speight, MBE, BSc, FRIC, AssocMet, FIM, formerly Manager, Environmental Pollution Control, Head Office, British Steel Corporation, London.
15. A. K. Barbour, CChem, PhD, BSc, FRIC, of R.T.Z. Services Ltd, Bristol, J. F. Castle, CEng, FIChemE, MInstF, of R.T.Z. Consultants, London, and S. E. Woods, DPhil, FRS, of R.T.Z. Services Ltd, Bristol,
16. B. C. Spall, MA, DPhil(Oxon), Chief Chemist, Levington Research, Levington, Ipswich, Suffolk of the Fertilizer Division of Fisons Ltd.
17. R. J. Gates, CEng, MA, MIMechE, Chief Engineer, Rugby Portland Cement Co. Ltd, Rugby.
18. E. Rowden, MBE, BSc, ARCS, DIC, FRIC, FICeram, Consultant on Air Pollution Problems to the British Ceramic Research Association, Penkhull, Stoke-on-Trent.
19. The editor.
20. The late H. B. Johnson, BSc, FIEE, was a Partner in Merz and McLellan, Consulting Engineers, and J. M. Burnett, BA, MIEE, an Associate in Merz and McLellan, London and Newcastle upon Tyne.
21. The editor.
22. R. A. Haslett, PhD, BSc, Ricardo, Consulting Engineers, Shoreham by Sea, Sussex.

Units and conversion factors

TABLE 1

Basic SI units

Quantity	Name of unit	Unit symbol
length	metre	m
mass	kilogram	kg
time	second	s
electric current	ampère	A
thermodynamic temperature	degree Kelvin	K
luminous intensity	candela	cd

TABLE 2

Some derived SI units with special names

Physical quantity	SI unit	Unit symbol
force	newton	N $= kg\ m\ s^{-2}$
work, energy, quantity of heat	joule	J $= N\ m$
power	watt	W $= J\ s^{-1}$
electric charge	coulomb	C $= A\ s$
electrical potential	volt	V $= W\ A^{-1}$
electric capacitance	farad	F $= A\ s\ V^{-1}$
electric resistance	ohm	Ω $= V\ A^{-1}$
frequency	hertz	Hz $= s^{-1}$
magnetic flux	weber	Wb $= V\ s$
magnetic flux density	tesla	T $= Wb\ m^{-2}$
inductance	henry	H $= V\ s\ A^{-1}$
luminous flux	lumen	lm $= cd\ sr$
illumination	lux	lx $= lm\ m^{-2}$

TABLE 3

Some additional derived SI units

Physical quantity	SI unit	Unit symbol
area	square metre	m^2
volume	cubic metre	m^3
mass density	kilogramme per cubic metre	$kg\ m^{-3}$
velocity	metre per second	$m\ s^{-1}$
angular velocity	radian per second	$rad\ s^{-1}$
acceleration	metre per second squared	$m\ s^{-2}$
angular acceleration	radian per second squared	$rad\ s^{-2}$
pressure	newton per square metre	$N\ m^{-2}$
surface tension	newton per metre	$N\ m^{-1}$
dynamic viscosity	newton second per metre squared	$Ns\ m^{-2}$
kinematic viscosity diffusion coefficient	metre squared per second	$m^2\ s^{-1}$
thermal conductivity	watt per metre degree Kelvin	$W\ m^{-1}\ K^{-1}$
electric field strength	volt per metre	$V\ m^{-1}$
magnetic field strength	ampere per metre	$A\ m^{-1}$
luminance	candela per square metre	$cd\ m^{-2}$

TABLE 4

Multiples and sub-multiples of units

Factor by which the unit is multiplied	Prefix	Symbol
$1\ 000\ 000\ 000\ 000 = 10^{12}$	tera	T
$1\ 000\ 000\ 000 = 10^{9}$	giga	G
$1\ 000\ 000 = 10^{6}$	mega	M
$1\ 000 = 10^{3}$	kilo	k
$100 = 10^{2}$	hecto	h
$10 = 10^{1}$	deca	da
$0{\cdot}1 = 10^{-1}$	deci	d
$0{\cdot}01 = 10^{-2}$	centi	c
$0{\cdot}001 = 10^{-3}$	milli	m
$0{\cdot}000\ 001 = 10^{-6}$	micro	μ
$0{\cdot}000\ 000\ 001 = 10^{-9}$	nano	n
$0{\cdot}000\ 000\ 000\ 001 = 10^{-12}$	pico	p
$0{\cdot}000\ 000\ 000\ 000\ 001 = 10^{-15}$	femto	f
$0{\cdot}000\ 000\ 000\ 000\ 000\ 001 = 10^{-18}$	atto	a

Much of the useful technical literature available, particularly that published in the UK and the USA, gives quantitative data in units that are neither metric nor SI. Eventually, technical literature will use metric and SI units. In some instances it will mean modifications in units used in certain items of legislation. There are some differences in the units that have been and are still used in the UK and the US. In Tables 5–17 conversion factors are given likely to be useful to readers of this book.

TABLE 5

Comparison of some UK and US units

UK	US
1 yard = 0·914398 m	1 yard = 0·914402 m
1 lb = 0·453 592 338 kg	1 lb = 0·453 592 427 7 kg
1 lb = 16 oz = 7000 grain	1 lb = 16 oz = 7000 grain
1 cwt = 112 lb	1 short cwt = 100 lb
1 ton = 20 cwt = 2240 lb	1 short ton = 20 short cwt = 2000 lb
1 gal is the volume occupied by 10 lb of distilled water at 62°F, i.e. 4·545 96 l	1 gal is 3·785 33 l which = 0·832 680 UK gallon

TABLE 6

Miscellaneous relationships between units

1 mi = 1760 yd = 5280 ft = 1·609 344 km
1 in^2 = 6·4516 cm^2
1 acre = 4840 yd^2
1 lb = 7000 grain 1 US barrel = 42 US gallons
1 mb = 1000 dyn cm^{-2}
1 hp = 550 ft lbf s^{-1}
1 cheval vapeur = 75 kgf m s^{-1}
1 therm = 10^5 British thermal units
1 thermie = 10^6 15° C calories

TABLE 7

Length

	cm	m	in	ft	yd
1 cm	1	0·01	0·393 701	0·032 808 4	0·010 936 1
1 m	100	1	39·370 1	3·280 84	1·093 61
1 in	2·54	0·025 4	1	0·083 333 3	0·027 777 8
1 ft	30·48	0·304 8	12	1	0·333 333
1 yd	91·44	0·914 4	36	3	1

1 km = 0·621 371 mi
1 micron (μ) = 1 micrometre (μm) = 10^{-6} m = 10^{-3} mm
1 angstrom (Å) = 0·000 1 micron = 10^{-7} mm

TABLE 8

Area

	m^2	in^2	ft^2	yd^2
1 m^2	1	1550·00	10·763 9	1·195 99
1 in^2	0·000 645 16	1	0·006 944 44	0·000 771 605
1 ft^2	0·092 903 0	144	1	0·111 111
1 yd^2	0·836 127	1296	9	1

1 hectare = 10 000 m^2 = 2·471 05 acres
1 mi^2 = 640 acres = 258·999 hectares

TABLE 9

Volume

	l	in³	UK gal	US gal
1 l	1	61·025 5	0·219 976	0·264 178
1 in³	0·016 386 6	1	0·003 604 65	0·004 328 98
1 UK gal	4·545 96	277·42	1	1·200 94
1 US gal	3·785 33	231·00	0·832 680 1	1

	m³	yd³	UK gal	US gal	US barrel
1 m³	1	1·307 95	219·969	264·170	6·289 76
1 yd³	0·764 555	1	168·179	201·973	4·808 88
1 UK gal	0·004 546 09	0·005 946 07	1	1·200 94	0·028 593 8
1 US gal	0·003 785 43	0·004 951 15	0·832 680	1	0·023 809 5
1 US barrel	0·158 988	0·207 948	34·972 6	42	1

TABLE 10

Mass

	kg	grain	oz	lb
1 kg	1	15 432·4	35·274 0	2·204 62
1 grain	0·000 064 798 9	1	0·002 285 71	0·000 142 857
1 oz (avoirdupois)	0·028 349 5	437·5	1	0·062 5
1 lb	0·453 592 37	7000	16	1

	kg	tonne	ton	sh tn
1 kg	1	0·001	0·000 984 207	0·001 102 31
1 tonne (1000 kg)	1000	1	0·984 207	1·102 31
1 ton (2240 lb)	1016·05	1·016 05	1	1·12
1 short ton (2000 lb)	907·185	0·907 185	0·892 857	1

TABLE 11

Force

	dyn	gf	pdl	lbf
1 dyn	1	0·001 019 72	$72{\cdot}3301 \times 10^{-6}$	$2{\cdot}248\ 09 \times 10^{-6}$
1 gf	980·665	1	0·070 931 6	0·002 204 62
1 pdl	13 825·5	14·098 1	1	0·031 081 0
1 lbf	444 822	453·592	32·174 0	1

TABLE 12

Pressure

	mb	kgf cm^{-2}	pdl ft^{-2}
1 mb	1	1·019 72 × 10^{-3}	67·196 9
1 kgf cm^{-2}	980·665	1	65 897·6
1 pdl ft^{-2}	0·014 881 6	15·175 0 × 10^{-6}	1
1 lbf ft^{-2}	0·478 803	0·488 243 × 10^{-3}	32·174 0
1 tonf ft^{-2}	1 072·52	1·093 66	72 069·9

	lbf ft^{2}	tonf ft^{-2}
1 mb	2·008 54	0·932 385 × 10^{-3}
1 kgf cm^{-2}	2048·16	0·914 358
1 pdl ft^{-2}	0·031 081 0	13·875 4 × 10^{-6}
1 lbf ft^{-2}	1	0·446 429 × 10^{-3}
1 tonf ft^{-2}	2240	1

TABLE 13

Pressure

	mb	atm	kgf cm^{-2}
1 mb	1	0·986 923 × 10^{-3}	1·019 72 × 10^{-3}
1 standard atmosphere	1 013·25	1	1·033 23
1 kgf cm^{-2}	980·665	0·967 841	1
1 torr	1·333 22	1·315 79 × 10^{-3}	1·359 51 × 10^{-3}
1 barometric inch of mercury	33·863 9	0·033 421 1	0·034 531 6

	torr	in Hg
1 mb	0·750 062	0·029 530
1 standard atmosphere	760	29·921 3
1 kgf cm^{-2}	735·559	28·959
1 torr	1	0·039 370 1
1 barometric inch of mercury	25·4	1

TABLE 14

Pressure

	mb	kgf m^{-2}	lbf ft^{-2}	mmHg
1 mb	1	10·197 2	2·088 54	0·750 062
1 kgf m^{-2}	0·098 066 5	1	0·204 816	0·073 555 9
1 lbf ft^{-2}	0·478 803	4·882 43	1	0·359 131
1 barometric millimetre of mercury	1·333 22	13·595 1	2·784 5	1

TABLE 15

Power

	W	kgf m s^{-1}	ft lbf s^{-1}	hp
1 W	1	0·101 972	0·737 562	1·341 02 × 10^{-3}
1 kgf m s^{-1}	9·806 65	1	7·233 01	0·013 150 9
1 ft lbf s^{-1}	1·355 82	0·138 255	1	1·818 18 × 10^{-3}
1 hp	745·700	76·040 2	550	1

1 W = 3·412 14 Btu h^{-1}; 1 Btu h^{-1} = 0·293 071 W
1 kW = 0·947 817 Btu s^{-1}; 1 Btu s^{-1} = 1·055 06 kW

TABLE 16

Heat, work and energy

	j	kcal	Chu
1 J = 1 W s = 10^{7} erg	1	0·238 846 × 10^{-3}	0·526 565 × 10^{-3}
1 kcal	4 186·8	1	2·204 62
1 Centigrade heat unit (Celsius heat unit)	1 899·10	0·453 592	1
1 British thermal unit	1 055·06	0·251 996	0·555 556
1 kgf m	9·806 65	2·342 28 × 10^{-3}	5·163 84 × 10^{-3}
1 ft lbf	1·355 82	0·323 832 × 10^{-3}	0·713 928 × 10^{-3}

	Btu	kgf m	ft lbf
1 J = 1 W s = 10^{7} erg	0·957 817 × 10^{-3}	0·101 972	0·737 562
1 kcal	3·968 32	426·935	3 088·03
1 Centigrade heat unit (Celsius heat unit)	1·8	193·655	1 400 71
1 British thermal unit	1	107·586	778·169
1 kgf m	9·294 91 × 10^{-3}	1	7·233 01
1 ft lbf	1·285 07 × 10^{-3}	0·138 255	1

TABLE 17

Heat, work and energy

	kJ	kcal	Btu
1 kJ	1	0·238 846	0·947 817
1 kcal	4·186 8	1	3·968 32
1 British thermal unit	1·055 06	0·251 996	1
1 therm	1·055 06 × 10^{5}	0·251 996 × 10^{5}	10^{5}
1 kWh	3 600	859·845	3 412·14
1 hp h	2 684·52	641·186	2 544·43

	therm	kWh	hp h
1 kJ	0·947 817 × 10^{-5}	2·777 78 × 10^{-4}	3·725 06 × 10^{-4}
1 kcal	3·968 32 × 10^{-5}	0·001 163	1·559 61 × 10^{-3}
1 British thermal unit	10^{-5}	2·930 71 × 10^{-4}	3·930 15 × 10^{-4}
1 therm	1	29·3071	39·301 5
1 kWh	0·034 121 4	1	1·341 02
1 hp h	0·025 444 3	0·745 700	1

1 Air pollution from the use of fuels

ALBERT PARKER

On average we each breathe in a day 14 to 18 kg of air while we consume only 1·5 to 2·0 kg of water in one form or another and no more than 0·7 kg of dry solid matter as food. Man cannot survive for more than a few minutes without air whereas he can be kept alive for days without drinking water and for weeks without food. Though we have long insisted on the need for uncontaminated drinking water and food, it is only in recent years that we began seriously to recognize the importance to health of clean air.

United Kingdom

The UK was the first country to develop industry on a considerable scale to enable the country to support an increasing population at a rising standard of living. This industrialization was based during the eighteenth and nineteenth centuries on the use of indigenous coal as the major source of heat and power. It was mainly from the use—or rather the misuse—of coal that the air of the towns became heavily polluted by smoke, grit and dust, and oxides of sulphur. During the past fifty years there has been an increasing amount of gaseous pollution caused by the greater use of petroleum oils for furnaces and for road transport. In many instances, there has also been serious local pollution of the air in certain areas due to discharges from the manufacturing processes of a variety of materials. Methods of reducing air pollution from a selection of important industries are considered in subsequent chapters.

Coal has been used in the UK for more than a thousand years. The earliest recorded use was in 852[1] when the population of the country was probably below one million. Over the centuries coal, at first slowly and then more rapidly, replaced wood as fuel and became the major source of heat for houses and of heat and power for industry. Even in the days when the use of coal was very small—it had reached only about 3 million tonnes in the year 1700—there were periodic complaints of smoke nuisance. As early as 1257, Eleanor, Queen of Henry III, left Nottingham for Tutbury because of smoke. In 1273 the use of coal was prohibited in London as it was considered that the clouds of smoke were prejudicial to health. A Royal Proclamation in 1306 in the reign of Edward I prohibited the use of coal in furnaces in London while Parliament was in session; it is recorded that one offender was executed. There was a similar proclamation in the reign of Elizabeth I.

In 1595 Thomas Owen brought Welsh coal and anthracite to London to demonstrate its relative smokelessness. Londoners petitioned Parliament in 1648 to prohibit the supply of coal from Newcastle upon Tyne because of the injury it caused. In 1661 John Evelyn wrote his famous pamphlet entitled 'Fumifugium, or the Inconvenience of the Aer and Smoake of London Dissipated, together with some Remedies'.[2] On the initiative of the Kyrle Society and the Public Health Society a Smoke Abatement Committee was formed in 1881 and in the same year organized a smoke abatement exhibition in Kensington. In the following year a National Smoke Abatement Institution was established. Led by Sir William Richard, RA and Dr H. A. Des Voeux, the Coal Smoke Abatement Society, mainly concerned with London, was formed in 1899. At a smoke abatement exhibition in Sheffield in 1909, the Smoke Abatement League of Great Britain was formed, interested mainly in smoke abatement in the provinces; centres were later established in Manchester and Glasgow. Up to that time, however, the leaders of the various voluntary organizations concerned with smoke abatement were generally considered to be 'cranks', most people being of the opinion that prevention of smoke emission from the use of coal for heating in homes and factories was technically and economically impracticable.

In 1929 the Coal Smoke Abatement Society and the Smoke Abatement League of Great Britain amalgamated to form the National Smoke Abatement Society, which began the publication of a journal *Clean Air*. In 1958 the Society was incorporated under the present name of the National Society for Clean Air, since it had now become concerned with all forms of air pollution and not merely with pollution by smoke. Over the years there have been various legislative measures, which are considered in chapter 3, but none of them was very effective in limiting the emission of smoke and other pollutants into the air until the twentieth century.

It required several major 'smog' disasters to convince people that serious efforts should be made to reduce pollution of the air by smoke. The word smog was first used by Dr Des Voeux to describe the combinations of smoke and natural fog, that occurred periodically in thickly populated areas such as London, Manchester, Sheffield and Glasgow during the winter, and reduced visibility to only a few metres. Smogs cause a significant rise in the death rates from respiratory diseases and have a deleterious effect on the heart and circulatory systems.

Natural fogs are caused by the formation of layers of temperature inversion in which the air from the ground level, upwards to a height which may vary from time to time and place to place, is colder than the air above; the velocity of the wind must also be low, not more than about one metre per second. The moisture in the air then condenses to fine particles to form a mist or fog, which is not moved by upward circulation. As a result, discharges of smoke and other pollutants, at relatively low temperature and velocity coming from low chimneys, collect in the inversion layer. In consequence the concentration of pollutants increases and can reach ten or more times the concentration found under more normal weather conditions, when the temperature of the air decreases with height above ground level. In the populated areas of the UK, inversion layers from ground level upwards rarely extend to a height of more than about 130 m. In some parts of the world they extend on occasion to much greater heights. It is during periods of temperature inversion, lasting continuously for one or more days, that the smog disasters have occurred.

Major smog disasters

It is only during recent years that the authorities concerned have kept records of the rise in the concentrations of pollutants in the air during smogs and of the effects on health, and even these records have not always been sufficiently detailed. Of the recorded smogs, the following are worthy of mention. Smogs in London in 1880 and in 1891 caused increases in the number of deaths from bronchitis in one week of 692 and 572 respectively. The worst of the recorded smogs in London lasted from 5 to 8 December 1952 and caused an estimated total of 4000 additional deaths in the Greater London area.[3] Smogs in London in January 1956 and in December 1962 caused additional deaths estimated at 1000 and 750 respectively.

A scheme for the investigation of air pollution in the UK arose from a conference of delegates of municipal authorities and other organizations held in association with the international smoke abatement exhibition in London in 1912. After several changes as the scheme was developed, the responsibility for the investigation between 1945 and 1958 was vested in the Fuel Research Organization of the then Department of Scientific and Industrial Research. Part of the investigation was the measurement of air pollution in various parts of the country. The measurements were made by local authorities and others by agreed methods and the results were sent to the Fuel Research Station for correlation and publication. At the time of the smog in 1952, there were nine sites in the London area which had instruments for measuring the daily average concentrations of smoke and eight sites which had instruments for measuring the daily average concentrations of oxides of sulphur expressed as sulphur dioxide.

Over the years 1950 to 1954 the average concentration of smoke in the London area during the summer months of April to September was about 115 $\mu g\ m^{-3}$ (microgrammes per cubic metre) and in the winter months of October to March about 320 $\mu g\ m^{-3}$; but 4460 $\mu g\ m^{-3}$ was obtained at one site on one day during a smog in December. The average concentration of sulphur dioxide in the summer months was 183 $\mu g\ m^{-3}$ and in the winter months 422 $\mu g\ m^{-3}$; a daily result of 3832 $\mu g\ m^{-3}$ on one day at one site was obtained in December. Later experience has shown that much higher concentrations of both smoke and sulphur dioxide must have occurred over some periods much shorter than 24 hours. The development of the national survey of air pollution in the UK is described on pages 7 to 12.

The smog of December 1952 stirred the Government to action. In July 1953 it appointed a Committee on Air Pollution, with the late Sir Hugh Beaver as Chairman, 'To examine the nature, causes and effects of air pollution and the efficacy of preventive measures; to consider what further preventive measures are practicable; and to make recommendations'. The Committee worked with great thoroughness and expedition and presented two valuable reports, an interim one in November 1953[4] and a final one in November 1954.[5] These reports led to the Clean Air Act 1956[6] enabling local authorities to designate smoke control areas in which the burning of bituminous coal in open domestic grates would be prohibited. In addition, the emission of smoke from industrial chimneys in all parts of the country was to be strictly limited. As a result of this Act and the effects of other factors it is certain that the UK will never again have smogs of the severity of that in December 1952.

Changes in the amounts and uses of sources of energy

As pollution of the air by smoke has been caused almost entirely by the inefficient burning of bituminous coal with incomplete combustion in domestic grates and industrial and other furnaces, and pollution by oxides of sulphur mainly by the uses of coal and petroleum oils as fuels, it is useful to consider the changes in the UK in the annual amounts of coal and other sources of energy used over a number of years.

In 1700 the population of the country was about 6 million and the consumption of coal about 3 million tonnes. The population rose to 10 million in 1800, 20 million in 1850 and 37 million in 1900, while the inland use of coal increased to 12 million tonnes in 1800, 55 million tonnes in 1850 and 170 million tonnes in 1900. Before 1900 the quantity of petroleum oil used as fuel was negligible; it reached 3 million tonnes in 1920 and 6 million tonnes in 1930. The coal equivalents of the primary forms of energy used in the UK in 1938, 1956 and each alternate year from 1960 to 1974 are given in Table 1.1.[7] The following factors were used in calculating the coal equivalents of the other forms of primary energy.[7]

1 tonne of petroleum oil = 1·7 tonnes of coal

276 therms (8076 kWh) of natural gas = 1 tonne of coal

Nuclear and hydroelectricity were converted to the amount of coal needed to produce the same amount of electricity at the efficiency of contemporary steam power stations.

The figures in Table 1.1 show that the coal equivalent of the primary sources of energy used in the UK increased by about 75 per cent from 191·8 million tonnes in 1938 to 336·2 million tonnes in 1974. Over that period the population rose by about 18·5 per cent from 47·5 million to 56·3 million, so that the coal equivalent *per capita* increased by 48 per cent from 4·03 to 5·97 tonnes. The proportion of the coal equivalent provided by coal decreased from 93 per cent in 1938 to 35 per cent in 1974, while the percentage provided by oil rose from 6·6 in 1938 to 48 in 1972 and then decreased to 45 per cent in 1974. Throughout the period 1938 to 1974 hydroelectricity provided

TABLE 1.1

Coal equivalents of the primary forms of energy used in the UK in 1938, 1956, and alternate years from 1960 to 1974 (Quantities in million metric tonnes)

Year	Coal	Oil	Hydro-electricity	Nuclear electricity	Natural gas	Total	*Per capita* tonnes
1938	178·5	12·6	0·7	—	—	191·8	4·03
1956	217·4	39·1	1·3	—	—	256·8	5·03
1960	198·6	66·6	1·7	0·9	0·1	267·9	5·09
1962	194·0	79·9	2·1	1·5	0·1	277·6	5·18
1964	189·6	94·8	1·9	3·2	0·3	289·8	5·35
1966	176·7	113·5	2·4	7·9	1·1	301·6	5·49
1968	167·1	127·9	2·2	10·3	4·5	312·0	5·64
1970	156·9	147·9	2·6	9·6	16·2	333·2	5·98
1972	122·8	160·1	2·0	10·7	37·3	332·9	5·96
1974	117·7	151·4	2·1	12·1	52·9	336·2	5·97

less than 1 per cent of the total coal equivalent. Nuclear electricity was less than 1 per cent from 1960 to 1963 and then increased over the years to reach 3·6 per cent in 1974. Natural gas provided less than 1 per cent from 1960 to 1967 and then rose each year to reach nearly 16 per cent in 1974. As a result, of the coal equivalent of the total consumption of primary forms of energy in 1974, the percentages were oil 45, coal 35, natural gas 15·7, nuclear electricity 3·6 and hydroelectricity less than 1 per cent.

The burning of bituminous coal in domestic open grates leads to the emission of more smoke per ton of coal than any of the other uses of coal. The quantities of coal used in the UK in 1938, 1956 and each alternate year from 1960 to 1974 in grates or furnaces that may cause smoke emission are given in Table 1.2.[7]

The domestic use of coal, excluding miners' coal, fell from 47·3 million tonnes in 1938 to 9·8 million tonnes in 1974. The main reasons for this great decrease in the

TABLE 1.2

Quantities of coal used in the UK in 1938, 1956 and alternate years from 1960 to 1974 in grates or furnaces that may case smoke emission (million metric tonnes)

Year	Domestic excluding miners'	Miners' coal	Railways	Power stations	Industrial† miscellaneous and collieries
1938	47·3	4·7	13·6	15·3	61·1
1956	31·1	5·4	12·3	46·3	63·6
1960	29·6	5·1	9·1	53·5	48·3
1962	27·9	4·8	6·2	61·4	43·9
1964	23·4	4·4	3·9	68·5	40·5
1966	21·2	3·9	1·7	69·0	36·6
1968	18·4	3·3	0·2	75·6	29·7
1970	15·4	2·7	0·1	77·2	25·6
1972	10·6	2·2	0·1	66·7	15·9
1974	9·8	2·1	0·1	66·0	14·5

† Excluding coke ovens, gas supply industry and plants for making solid smokeless fuels.

domestic use of bituminous coal are, first, that during the Second World War coal and coke for domestic use were rationed and this rationing was continued for some time after the war. In consequence, householders began to use more electricity and gas for heating and continued to do so perhaps because they were more convenient and cleaner than heating by bituminous coal in open fires. Second, as we have seen, the Clean Air Act 1956 enabled local authorities to establish smoke control areas in which the burning of bituminous coal in open domestic grates was prohibited. An appreciable number of smoke control areas have now been established, and progress in that direction has continued but not so rapidly as desirable. Third, since 1945 there has been an increase in the number of housewives in the UK taking employment; this has led to an increasing demand for the convenience of gas and electricity for heating and to some extent for heating by oil. It is difficult to estimate the relative effects of each of the three factors.

It is interesting to note that for many years household heating by open fires has been proportionately much less prevalent in Western Europe, the US and Canada than in the UK. The reduction in the amount of concessionary coal to miners in the UK, who largely burn coal in open fires, from 5·4 million tonnes in 1956 to 2·1 million tonnes in 1974 was due to a substantial reduction in the number of wage earners engaged in the deep mining of coal. There is a system whereby coal miners can elect to receive a somewhat smaller quantity of solid smokeless fuel in place of concessionary coal, but most miners have not chosen to make this change, which, in the opinion of the author of this chapter, is disappointing.

From 1956 to 1968 steam-driven railway locomotives, which emitted smoke, were steadily replaced by diesel engines and electric traction so that the consumption of bituminous coal by the railways in the UK fell from 12·3 million tonnes in 1956 to 0·2 million tonne in 1968 with a further decrease to somewhat less than 0·1 million tonnes in 1974. The coal consumption by power stations increased from 15·3 million tonnes in 1938 to 77·2 million tonnes in 1970 and then declined to 66·0 million tonnes in 1974. Power stations also consumed considerable quantities of oil, 5·56 million tonnes equivalent to 9·45 million tonnes of coal in 1960, rising to 18·87 million tonnes equivalent to 32·1 million tonnes of coal in 1972 and then falling to 17·2 million tonnes equivalent to 29·2 million tonnes of coal in 1974. Industrial undertakings, including collieries, reduced their consumption of coal from 63·6 million tonnes in 1956 to 14·6 million tonnes in 1974, coal being replaced by oil, electricity from the public supply, and, to some extent during 1970 to 1974, by natural gas.

Emissions of smoke and oxides of sulphur in the UK from the uses of fuels

Estimates are given in Table 1.3[8] of the quantities of smoke and oxides of sulphur discharged into the air of the UK from the uses of fuels in 1938, 1956 and alternate years from 1960 to 1974.

The observations on the figures in Table 1.2 give the reasons for the great reduction in the emission of smoke from domestic open fires. The percentage reduction in smoke emission from the uses of coal by industry for raising steam and heating furnaces was greater than the percentage reduction in the amount of coal used. This was partly due to

TABLE 1.3

Discharges of smoke and oxides of sulphur into the air of the UK from the uses of coal, coke and other solid smokeless fuels, and oil, over the years 1938 to 1974 (million metric tonnes)

	Smoke		Oxides of Sulphur			
Year	Domestic	Total	Coal	Coke†	Oil	Total
1938	1·74	2·75	3·89	0·24	0·06	4·19
1956	1·28	2·29	4·52	0·36	0·54	5·42
1960	1·21	1·47	4·37	0·36	1·26	5·99
1962	1·18	1·42	4·32	0·36	1·51	6·19
1964	0·98	1·14	4·34	0·31	1·74	6·39
1966	0·84	0·98	4·03	0·29	2·05	6·37
1968	0·75	0·84	3·72	0·29	2·13	6·14
1970	0·64	0·72	3·40	0·17	2·50	6·07
1972	0·45	0·50	2·34	0·13	2·73	5·20
1974	0·42	0·47	2·63	0·15	2·64	5·42

† Including other solid smokeless fuels.

the Clean Air Act 1956 which indirectly also stimulated efforts to improve the efficiency of the use of coal by industry. The emission of smoke from the use of coal by industry and the railways has been reduced over the years 1956 to 1974 by 95 per cent from 1·01 to 0·05 million tonnes. Over the same period the emission of smoke from domestic fires was reduced by 67 per cent from 1·28 to 0·42 million tonnes. In 1956, of the 2·29 million tonnes of smoke 56 per cent was from domestic fires and 44 per cent from the uses of coal by industry and the railways. In 1974, of the 0·47 million tonnes of smoke emitted 89 per cent was from domestic fires. It is clear that further effort should be made to reduce smoke emission from domestic chimneys, particularly as the heights of domestic chimneys and the exit velocities of the gases are much lower than the heights and exit velocities of industrial chimneys.

The estimated emissions of oxides of sulphur from the uses of coal, coke and oil in the UK rose from 4·2 million tonnes in 1938 to 6·0 million tonnes in 1960, remained at somewhat more than 6 million tonnes from 1962 to 1970, and then fell to 5·2 in 1972 and 5·4 million tonnes in 1974. The author of this chapter has found that many who use figures such as those in Table 1.3 assume that the figres are more accurate than is justifiable. The statistics for the quantities of each type of fuel sold may be reasonably accurate but there are no accurate estimates of the amounts held in stock at any particular time by users and merchants. It is not justifiable, therefore, to draw conclusions from differences in the estimates of less than 10 per cent from one year to another.

Motor vehicles

There have been greater increases over the years 1938 to 1974 in most countries in the consumption of gasolene (petrol) and derv fuel (diesel oil) by vehicles for road transport. The great increases in the UK are shown by the figures in Table 1.4.[8]

TABLE 1.4

Consumption of gasolene and derv fuel for road transport in the UK over the years 1938 to 1974 (million metric tonnes)

Year	Gasolene	Derv fuel
1938	4·91	0·39
1950	5·28	1·05
1955	6·34	1·63
1960	7·75	2·62
1965	10·91	3·91
1970	14·23	5·04
1974	16·48	5·52

In the year 1974 the consumption of gasolene at 16·48 million tonnes was more than three times the amount of 5·28 million tonnes in 1950 and the consumption of derv fuel at 5·52 million tonnes in 1974 was more than five times that of 1·05 million tonnes in 1950. The pollutants emitted in the exhaust gases from both types of engine include carbon monoxide, hydrocarbons, aldehydes, oxides of nitrogen and oxides of sulphur. Estimates of the quantities of these pollutants discharged in the UK from road vehicles in 1974 are given in Table 1.5.[8]

TABLE 1.5

Estimates of pollutants from road vehicles in the UK in the year 1974 (consumption of gasolene 16·48 million tonnes, consumption of derv fuel 5·52 million tonnes) (million metric tonnes)

Pollutant	Gasolene engines	Diesel engines
Carbon monoxide	7·8	0·12
Hydrocarbons	0·38	0·023
Aldehydes	0·01	0·003
Oxides of nitrogen	0·24	0·07
Oxides of sulphur	0·013	0·04

The average amount of carbon monoxide emitted at ground level from motor vehicles using gasolene is equal in weight to about 47 per cent of the weight of gasolene used and the concentration in the moist waste gases is about 3 per cent by volume, which is about 3·5 per cent of the dry waste gas by volume. These are overall figures but there are considerable variations under the different conditions of operation of idling, acceleration and deceleration and much depends on the setting of the carburettor.[9] Combustion is more complete with a well-adjusted diesel engine, with the result that the average amount of carbon monoxide in the exhaust gases is only about 0·2 per cent by weight of the fuel used. It is worthy of mention that the estimated quantity of carbon monoxide discharged into the air of the UK in 1974 in the gases from the industrial and domestic uses of fuels was about 8 million tonnes including about 3 million tonnes from domestic heating appliances; but these discharges were from chimneys well above ground level whereas the discharges from road vehicles are at ground level.

The total amount of lead in the lead alkyl compounds added to the 16·48 million tonnes of motor spirit used in the UK in 1974 was about 11 660 tonnes, or an average of 0·52 g of lead per litre. The lead is converted to a variety of compounds and about one-third is retained in the lubricating oil and in the exhaust system. This means that the amount of lead in the compounds discharged in the exhaust gases from petrol-driven vehicles in the UK in 1974 was about 7800 tonnes.

Survey of air pollution in the United Kingdom

The first steps towards systematic measurement of the pollution of the air of the country by solid matter, smoke and oxides of sulphur was taken at the beginning of the twentieth century by those enthusiasts who established the Coal Smoke Abatement Society in 1899 and in 1912 formed a committee, including representatives of several local authorities, for the investigation of air pollution. Their report for the year 1914–15 recorded observations using 38 instruments placed in different areas. In 1917 the committee became an advisory committee to the Meteorological Office. In 1927 the task of coordinating the local observations and undertaking research was transferred to the then Government Department of Scientific and Industrial Research. The committee was then re-constituted as an Atmospheric Pollution Research Committee and a Standing Conference of cooperating bodies was established to ensure close liaison between the department and the local authorities and others making measurements and taking part in investigations. The Standing Conference meets at intervals of six months. At the meetings, which are well attended, the progress of the survey of pollution is discussed, and special investigations are considered.[10] In 1945 the investigations and coordination of the methods and results of measurements become the responsibility of the Fuel Research Station of the Department of Scientific and Industrial Research and the committee came under the Advisory Fuel Research Board. This was a wise move as by 1945 the Fuel Research Station, which was built at Greenwich in 1918, had a large staff of scientists and engineers and was well equipped with large scale and pilot plant and laboratories to undertake intensive research on the causes and mitigation of pollution and on methods of measurement.[11] In addition, the organization had coal survey laboratories operating in all the principal coalfields of the country.

In 1957 a newly formed Research Council of the Department of Scientific and Industrial Research decided that with the growth of research activities on fuel problems by the nationalized fuel industries and by certain industrial Research Associations it was not necessary to continue much of the work at the Fuel Research Station and that the work on air pollution should be transferred to a new research station, the Warren Spring Laboratory, to be built at Stevenage, Hertfordshire. This new laboratory would continue the work on air pollution and would investigate a number of problems of interest to the country as they arose. The transfer of staff from the Fuel Research Station to the Warren Spring Laboratory was completed in 1959 when the Fuel Research Station was closed. In 1947 the Coal Survey Laboratories had been taken over by the National Coal Board.

The progress of interest and activity in daily measurements of smoke and sulphur dioxide in the air near ground level by local authorities and other organizations in many parts of the UK is shown by the number of instruments in regular operation in 1945, 1958 and 1975 as given in Table 1.6.[12] In addition there have been numerous monthly measurements and analyses of deposited solid matter and a number of special investigations.

TABLE 1.6.

Numbers of measuring instruments maintained by cooperating bodies in the UK in 1945, 1958 and 1975

Year	1945	1958	1975
Smoke (daily)	10	303	1186
Sulphur oxides (daily)	9	250	1186
Deposited solid matter (monthly)	102	1167	505

There are available equipments that automatically draw samples of air and determine the concentrations of sulphur oxides and smoke each day for a number of days, and require attention only once each seven or eight days. There are also methods of measuring the concentrations of sulphur oxides in the air over periods as short as a minute or less. Determinations of many elements in air have been undertaken. For example, during the years 1953–6, the Fuel Research Station worked with the British Empire Cancer Campaign in an investigation of the possible effects of air pollution on the incidence of cancer in parts of Lancashire, Cheshire and North Wales. The work included measurements of smoke and sulphur dioxide in the air and the weight and composition of deposited solid matter. Collections of smoke stains on filters and of samples of selected coals and coal ash were examined for the determination of so-called trace elements. The minimum and maximum concentrations of the elements determined in the air are given in ng m^{-3} (1 ng = 1 g divided by one thousand million) in the following list. Antimony (Sb) 1–20, arsenic (As) 5–100, beryllium (Be) 0·1–1, chromium (Cr) 1–10, cobalt (Co) 0·2–2, copper (Cu) 5–70, lead (Pb) 70–1000, manganese (Mn) 5–50, molybdenum (Mo) 0·3–5, nickel (Ni) 1–100, titanium (Ti) 10–100, vanadium (V) 1–50, zinc (Zn) 40–500. At a conference of the National Society for Clean Air in 1974, Mr N. J. Pattenden of the Environmental and Medical Science Division of the Atomic Energy Research Establishment at Harwell, gave a paper on the determination of various elements in air. The teams with which Mr Pattenden has been associated have made measurements of elements in the air at about twelve places in England, Scotland including the Shetland Islands, and Wales. The paper presented in 1974 dealt mainly with the results obtained at Trebanos, about 10 km north of Swansea, and at Kidwelly, about 27 km north west of Swansea. The minimum and maximum values in ng m^{-3} for the two sites considered together for a selected number of elements are given in the following list. Antimony less than 0·03–14, arsenic less than 0·12–30, caesium less than 0·01–3, cerium less than 0·01–1·6, chromium 0·10–60, cobalt less than 0·01–90, copper less than 1·1–300, lead 8–900, manganese less than 0·3–70, nickel less than 3–400, scandium 0·001–0·5, selenium 0·12–8, vanadium 0·10–40, zinc 0·6–2000.

Smoke

The Warren Spring Laboratory, now part of the Department of Trade and Industry, in reports on the National Survey of Air Pollution[12] have given annual averages over the years 1958–9 to 1972–3 of the daily measurements of the concentrations of smoke and oxides of sulphur in the urban areas of the UK. More than 80 per cent of the population live in towns and most industry is associated with the towns. Over the years 1958–9 to 1972–3 the overall average annual concentration of smoke in the urban areas fell from about 180 $\mu g\ m^{-3}$ to 52 $\mu g\ m^{-3}$, which is a decrease of about 70 per cent. The averages of the daily measurements over the winter months of October to March were about 40–50 per cent greater than the annual averages and the averages for the months April to September were about one half of the annual averages. The average concentrations of smoke in the air of the urban areas in the northern half of the country were appreciably higher than in the southern half. At times of temperature inversion with fog persisting for a day or more, usually in November, December or January, the daily concentration of smoke sometimes reached values ten or more times the annual average. In the London area, the annual average concentration of smoke fell from about 175 $\mu g\ m^{-3}$ in 1958–9 to about 36 $\mu g\ m^{-3}$ in 1972–3, which is a decrease of nearly 80 per cent.

Oxides of sulphur

The overall annual average of oxides of sulphur, expressed as sulphur dioxide, in the air near ground level in the urban areas of the UK fell by about 46 per cent from 177 $\mu g\ m^{-3}$ in 1958–9 to 95 $\mu g\ m^{-3}$ in 1972–3. There were higher average concentrations over the months of October to March and lower average concentrations over the months April to September. As in the case of smoke, the concentrations of oxides of sulphur at times of temperature inversion with persistent fog were very much higher than the winter averages. In general, apart from the London area, the average concentrations of oxides of sulphur in the southern half of the country over the years considered were less than in the northern half. In the London area, however, the average annual concentration was relatively high at 233 $\mu g\ m^{-3}$ in 1958–9 and about 120 $\mu g\ m^{-3}$ in 1972–3, which is a reduction over the period of about 52 per cent. There are arrangements in the City of London whereby the sulphur content of fuel oils used in the City can be restricted.[13]

Carbon monoxide and lead in the air from road vehicles

Pollution of the air at or near ground level by carbon monoxide and lead are of concern in the areas of dense motor traffic. In surveys of the concentrations of carbon monoxide in busy London streets in 1955 and 1956, the Fuel Research Station found considerable variations from hour to hour near ground level. In streets where the traffic was dense and congested, concentrations as high as 50 to 80 parts per million by volume were found for very short periods at a height of 1·5 m above the pavement. In Blackwall Tunnel, when the ventilation was very poor, concentrations as high as 500 parts per million were found for short periods at times of exceptionally heavy traffic. In Hyde Park, away from motor traffic, the concentration of carbon monoxide was less than one part per million.[14] The Warren Spring Laboratory in 1967 made measurements of carbon monoxide at a height of 2·7 m on the kerbside in a number of towns.[15] At one site in one town the average concentrations over periods of 15 minutes ranged from zero to 1·2 parts per million during the hours from midnight to 6.00 hours, fluctuated between 2 and 13 parts over the period 6.00 hours to 21.00 hours and between 5 and 1·5 parts per million between 21.00 hours and midnight. Continuous measurements were made at one site for a year. During the year the concentration of carbon monoxide was above 20 parts per million for an overall total of 4 h with short-term high peaks, including a total of about 40 min with concentrations in the range of 40–60 parts and shorter periods in the ranges of 60–80 parts and of over 80 parts per million. The possible effects of carbon monoxide from motor vehicles on the health of residents near roads with periodic high density of motor traffic have been considered in the UK but so far no detrimental effects have been found. It is worthy of note that the effect of carbon monoxide from motor vehicles on the amount of carboxy haemoglobin in the blood is generally much less than that from medium and heavy smoking of cigarettes.

The average amount of lead in the form of alkyl compounds added to gasolene used in the UK in 1974, as already mentioned, was 0·52 g of lead per litre. This means that the amount of lead in the undiluted and moist exhaust gas discharged from vehicles using gasoline is on average about 0·037 g m^{-3}. The exhaust gas is rapidly diluted in the air. Recent investigations have shown that the lead content in the blood of children living close to a road in the Midlands with heavy motor traffic is somewhat higher than in children living well away from such traffic, but so far the levels of lead have not been considered detrimental to health. There are a number of other sources of lead in blood. The concentrations of lead found in the air of busy streets with dense motor traffic, for example 3·2 $\mu g\ m^{-3}$ in the air of Fleet Street, London and 5·9 $\mu g\ m^{-3}$ in the High Street, Warwick, are very small compared with the permissible threshold limit value of 200 $\mu g\ m^{-3}$ for workers in industry.

Cost of damage by air pollution

In an appendix to their report published in 1954,[5] the Committee on Air Pollution appointed by the UK Government in 1953 referred to previous estimates of the costs of damage by air pollution in certain cities in the UK and in the US. The estimates in £ per capita per annum, at the then current rates of exchange for the US dollar were as follows. £4 for Pittsburg in 1913, £1 for Manchester in 1919, £1·2 for London and £1·5 for Manchester in 1925, between £2 and £3 for Liverpool in 1939, £1 for Great Britain in 1945 and about £3 in 1952, £5·75 in Greater New York and £7 for Chicago in 1952. As the Committee on Air Pollution said in their report, even the best estimate of the cost of damage by air pollution cannot be more than a reasoned guess. The Committee then proceeded to make their own estimate of the cost of the damage per annum for the UK and gave the following figures in £ million: laundry 25, painting and decorating 30, cleaning and depreciation of buildings other than houses 20, corrosion of metals 25, damage to textiles and other goods 52·5, damage to agriculture plus loss of efficiency of workers in industry and losses in transport 100, making a total for these items of about £250 million equivalent to about £5 per head of the population, including about £10 per head per annum in the so-called black areas. This estimate did not include loss due to the inefficient combustion of coal which the Committee estimated at between £25 million and £50 million a year.

In November 1972 the Programmes and Analysis Unit of the British Department of Trade and Industry and the UK Atomic Energy Authority working in cooperation issued a report[15] on 'An Economic and Technical Appraisal of Air Pollution in the United Kingdom'. That report gave the direct annual cost of the damage due to air pollution at about £410 million per annum, including £5 million for window cleaning, £42 million for corrosion and protection of metals, £33 million for damage to textiles, paper, etc., £195 million for damage to agricultural produce and £130 million for damage to health. In addition, the report estimated a total of about £780 million a year for social or disamenity cost, including £510 million for health and £100 million for amenity, so that the overall total of these estimates was in the region of £1190 million a year, which is equivalent to about £21 per capita.

Estimates of this kind can be no more than rough guesses. For example, in a damp climate such as that in the UK, with air containing carbon dioxide and other gases from natural sources, and with material blown from the ground by the wind, there would be the corrosion of metals and damage to other materials without any pollution from the various activities of man. Air pollution from industrial and other activities certainly increases the rates of corrosion and damage to materials but it is difficult to ascertain the amounts of these increases. There is also difficulty in assigning a monetary figure for damage to health and agriculture with all the vagaries of climate and other factors. The difficulties in estimating costs of damage by air pollution must not detract, however, from the desirability and the need to reduce air pollution so far as is practicable.

Costs of reducing air pollution

It is difficult also to obtain trustworthy total figures for the costs borne in the first place by industry in installing additional or modified plant and in changing processes with the main object of reducing air pollution. In the end the community pays any extra costs in the prices that have to be paid for the finished products. A good beginning in making estimates was made by the Chief Alkali Inspector for England and Wales in 1968.[16] In his statement in the 105th Annual Report on Alkali, etc., Works for 1958 he gave the figures reproduced in Table 1.7 for the expenditure incurred over the years 1958–68 for each of ten classes of works operated in England and Wales using processes scheduled under the Alkali Act.

TABLE 1.7

Cost of air pollution control for scheduled processes in 1958–68 in £millions

Works	Capital	Research and development	Working costs	
			10-year	Latest year
Electricity	75·731	0·856	126·691	15·300
Cement	6·216	0·301	6·442	1·000
Petroleum	6·822	0·536	11·667	1·788
Gas	2·839	—	4·474	0·350
Coke ovens	2·909	0·242	6·126	0·710
Lime	0·976	0·004	0·707	0·118
Ceramics	2·090	0·163	3·011	0·382
Iron and steel	26·430	1·235	93·351	10·364
Non-ferrous metals	5·762	0·656	16·449	2·262
Chemical	20·527	0·952	55·516	6·782
Totals	150·302	4·945	324·434	39·056

The Chief Alkali Inspector for England and Wales stated that the figures for expenditure on research and development included only amounts returned by research associations, nationalized industries and scheduled works for work on air pollution control; they did not include money spent by plant manufacturers. Working costs included research and development, depreciation, interest on capital, costs of operation, maintenance and repairs, and miscellaneous and overheads insofar as they were related directly to air pollution control. Many works, especially those scheduled before 1958, had installed and were operating control equipment prior to 1958; for such works the capital costs, depreciation and overheads were not included in the table, though operation and maintenance were included for works which had done their accountancy properly. Costs for scheduled works in Scotland and Northern Ireland presumably were not included. There are also numerous works not scheduled under the Alkali Act that have to adopt some measure of pollution control. The costs for the whole of industry in the UK, therefore, must be appreciably greater than the amounts given in Table 1.7. There is no doubt that costs have risen appreciably since 1968 and they are likely to rise further during the next few years.

Air pollution in other countries

In broad terms the problems of mitigation of air pollution in most highly industrialized countries with areas of dense population are similar to those in the UK. There is now close consultation between the EEC countries, the UK, the US and some other countries in efforts to keep air pollution to the practicable minimum. No other country has made continuous measurements of general air pollution systematically over so many years as the UK in measurements of pollution by smoke, oxides of sulphur and deposited solid matter in most parts of the country.

It is of interest to consider some of the measurements and estimates of general pollution from the use of fuels in the US where the consumption of fuels is twice as great per capita as in the UK. In such consideration it must be appreciated that the land area of the US is about 9·4 million km^2, which is about 39 times the land area of 241 000 km^2 of the UK. The population of the US in 1975 was about 214 million so that the overall average density in that year was 23 persons per km^2, whereas the population of the UK at nearly 56 million was equivalent to an overall average density of about 230 persons per km^2. To give the same coverage of systematic daily measurements of the concentrations of oxides of sulphur, for example, in the US as in the UK with its measuring instruments at about 1200 sites would require about 4000 instruments in regular daily use in relation to size of population or 40 000 in relation to the area of the country.

It is not usual in the US to measure smoke in the air near ground level by the method used in the UK. Measurements are made in certain areas of the US of 'smoke shade', often called coefficient of haze, which is a measure of the brightness or darkness of a spot on a filter paper of a diameter of 1 cm that has had 304·8 m of air drawn through it. The results for smoke shade are not directly comparable with UK measurements of smoke in $\mu g\ m^{-3}$.

From regular measurements of the concentration of sulphur dioxide in the air at 38 sites in New York City the average in 1969 was 7·6 parts per hundred million by volume (217 $\mu g\ m^{-3}$). There was afterwards a reduction of the sulphur content of the fuels allowed to be used in the City, with the result that the average of the concentrations of sulphur dioxide in 1971 as measured at the same 38 sites was 3·9 parts per hundred million[17] (112 $\mu g\ m^{-3}$).

Scrubbers to remove sulphur dioxide from flue gases of power stations

In an article by A. V. Slack of the SAS Corporation, Wilson Lake Shores, Sheffield, Alabama in the *Electrical World* of December 15, 1975, under the title '1975 Scrubber Report—Technology Still Undergoing Labor Pains', it is stated that scrubbers using limestone, lime, magnesia, or sodium salt are in operation in 24 units at electricity power stations in the USA to reduce the quantity of sulphur dioxide in the flue gases. It is also stated that 23 units are under construction and further units are planned. Scrubbers to remove sulphur dioxide from the flue gases were in operation for a few years before the outbreak of war in 1939 at the Battersea and Fulham power stations in London, UK; both scrubbing systems were put out of action during the war because of the visibility of the plumes of vapour to aircraft. The scrubber unit at the Fulham power station has not since been operated; the unit at Battersea is still in operation to some extent but will probably cease in the near future. A unit is in operation at the Bankside power station in London; it was installed after the war and removes 97 per cent of the sulphur dioxide in the flue gases. Bankside is directly opposite St Paul's Cathedral and it was decided on the basis of general architectural considerations that a very high chimney should not be built opposite the Cathedral. The opinion of experts in the UK is that in relation to air pollution near ground level it is better to discharge the untreated hot flue gases at a suitable velocity rather than to scrub the gases by any system so far devised to remove part of the sulphur dioxide and then discharge the treated flue gases at a lower temperature.

If there is a power station at a site where there is little discharge of sulphur dioxide from other sources it would be useful if the advocates of scrubbing the flue gases would arrange for series of measurements to be made on the effects of discharging the hot untreated gases from the high chimney of the power station on the concentration of sulphur dioxide in the air near ground level at points on circles at several appropriate distances from the power station. The dispersion of the plume of gases in the air can also be studied by laser beams projected into the plume. Both methods of investigation have been used in the UK and have led to the conclusion that there is no real benefit from scrubbing the waste gases to remove about 75 per cent of the sulphur dioxide and then discharging the gases at relatively low temperature. There are advocates in certain European EEC countries of the system of scrubbing the flue gases from power stations by the systems so far developed; but in the opinion of the author of this chapter it is unlikely that the electricity supply industry in the UK will accept such a proposal. There is no technical objection to limiting the sulphur content of the fuels, particularly oil fuel, to say 1 per cent used at power stations, as is done in some areas, though it increases the cost of fuel and thus increases the cost of producing electricity.

Pollution from motor vehicles

Almost the whole of road transport in the US is by vehicles using gasolene; there are relatively few vehicles driven by diesel engines. The latest publication by the United Nations, issued in September, 1975, gives statistics for the years 1970 to 1973. According to that document[18] the amount of motor gasolene used in the US in 1973 was 289 million tonnes, equivalent to 1372 kg per capita as compared with 310 kg per capita in that year in the UK. This difference is due partly to larger engines and partly to greater distances travelled by vehicles in the US. The factors used in calculating the emissions of pollutants from motor vehicles from the amount of gasolene used are not the same in the US as the factors used in obtaining the estimates for the UK given in Table 1.5.

From figures by Zupan[17] for earlier years it can be estimated that from the amount of gasolene used in the US in 1973 the total emissions from motor vehicles included about 115 million tonnes of carbon monoxide, 8·3 million tonnes of hydrocarbons and 4·9 million tonnes of oxides of nitrogen. If the stringent limits of emissions of pollutants from motor vehicles that have been proposed in the US for the future are achieved, there would be much smaller quantities of carbon monoxide and hydrocarbons discharged for the same quantity of gasolene used; but it is not possible to predict in present circumstances what will be the consumption of gasolene in each country in the future. The estimated quantity of carbon monoxide discharged from other uses of fuels in the US was between 50 and 60 million tonnes in 1969.

The consumption of gasolene in million tonnes in 1970 and 1973 and the amounts in kg per capita are given in Table 1.8 for those countries using the largest amounts per capita.

TABLE 1.8

Consumption of gasolene in countries using the largest quantities per capita in 1970 and 1973

Country	Million tonnes 1970	1973	Kilogrammes 1970	Per Capita 1973
United States	251	289	1224	1372
Canada	19·8	23·5	924	1061
Australia	7·55	8·96	603	682
Denmark	2·00	2·23	422	444
Switzerland	2·17	2·60	350	402
Sweden	2·66	3·26	330	400
Austria	1·94	2·52	262	335
Norway	1·00	1·24	259	312
German F.R.	15·6	19·3	257	312
UK	14·5	17·4	259	310
Japan	15·2	20·5	167	187

With the growing general interest in the quality of the environment, including the practicable limitation of pollution of the air, there will gradually be accumulated much more information in various countries of the concentrations of the more common air pollutants. It is important that agreed, standardized methods of measurement, to be modified periodically as improved methods are developed, should be universally used, otherwise comparisons will be difficult. There is a move towards standardization in the OEEC countries in which the UK is taking a prominent part.

References

1. 'History of air pollution in Great Britain', *Clean Air Year Books* 1975 and 1976, National Society for Clean Air, Brighton, England.
2. John Evelyn's pamphlet issued in 1661. Copies obtainable from the National Society for Clean Air, Brighton, England.
3. Wilkins, E. T. 'Air pollution and the London fog of December, 1952', *J. Roy. Sanitary Institute* (now the Royal Society of Health), **74**, 1, 1954.
4. Committee on Air Pollution, interim report, November 1953, HMSO, London.
5. Committee on Air Pollution, final report, November 1954, HMSO, London.

6. Clean Air Act 1956, HMSO, London.
7. *Ministry of Power Statistical Digest*, 1967, and *Digest of Energy Statistics*, 1968 and 1969; *Ministry of Technology Digest of Energy Statistics*, 1970; *Department of Trade and Industry Digest of Energy Statistics*, 1971, 1972 and 1973; *Department of Energy Digest of UK Energy Statistics*, 1974 and 1975, HMSO, London.
8. Parker, A., *Clean Air Journal*, **4**, (14), 1974; *Clean Air Year Books* 1975 and 1976. National Society for Clean Air.
9. *Air pollution from road vehicles*, Report by Technical Committee, National Society for Clean Air.
10. *Atmospheric Pollution in Leicester*—a scientific survey studied the movement of pollution produced in and around Leicester. HMSO, London, 1945.
11. Reports by the Fuel Research Board and the Director of Fuel Research covering the periods 1939–46, 1946–9, 1949–50 and then each year from 1951 to 1958 gave accounts of investigations with the object of reducing pollution from the use of fuels. HMSO, London.
12. Over a long period many reports have been issued on the investigations of atmospheric pollution in the UK. The 27th report published in 1955 included the observations made in the 10 years to 31 March 1954. The 28th–31st reports covered the years ended 31 March 1955, 1956, 1957 and 1958 when the work was taken over by the Warren Spring Laboratory who have since issued reports. The Warren Spring Laboratory also issue special reports and issue annually to the organizations cooperating in making measurements a document giving the results of observations for each year to 31 March. The published reports are obtainable from HMSO, London.
13. *Sulphur dioxide—an air pollutant*, National Society for Clean Air.
14. Lawther, P. J. and B. T. Commins, 'Pollution from road vehicles and health,' Clean Air Conference, 1970, National Society for Clean Air.
15. *An economic and technical appraisal of air pollution in the United Kingdom*, Programmes and Analysis Unit of the Department of Trade and Industry and the UK Atomic Energy Authority, HMSO, London, 1972.
16. 105th Annual Report on Alkali, etc., Works by the Chief Inspectors, HMSO, London, 1968.
17. Zupan, J. M., *The distribution of air quality in the New York region*, Resources for the Future, Inc., Washington DC, Johns Hopkins University Press, Baltimore and London.
18. World Energy Supplies 1970–1973, *Statistical papers*, Series J, No. 18, United Nations, New York, 1975.

2 Effects of air pollution

Part 1 Air pollution and health

MARJORIE CLIFTON and SIR CHARLES STUART-HARRIS

2.1.1. Historical considerations

For many centuries, polluted air has been considered to be a hazard to health. In Britain the burning of coal instead of wood or charcoal led to many complaints of the offensive nature of smoke. Queen Eleanor's illness in the thirteenth century, when she sought refuge at Tutbury from the smoke of Nottingham's artisans, may have been the first record of ill-health due to pollution of the air. Queen Elizabeth I forbade the use of coal near the palace at Westminster and in 1661 a diatribe was published by John Evelyn[1] on the 'Smoake of London'. He proposed that factories should be moved away from London further down the Thames valley and a green belt be formed around the heart of the city. In the nineteenth century, the English sanitarians Southwood Smith and William Farr pleaded for the maintenance of clean air and attributed disorders of the lung to pollution. Sir John Simon[2] published reports on health in the City of London and in 1854 forecast that smoke from domestic chimneys would ultimately be controlled. Yet it was not until a century later that a massive incidence of pollution during a fog in London led to a public outcry and legislation to control smoke in Britain.

Pollution episodes, however, had occurred before this event in the Meuse Valley in Belgium in 1930[3] and the Donora Valley, Pennsylvania in 1948.[4] In both instances, a steep-sided valley became filled with air polluted by smoke from coal-burning domestic appliances and industrial furnaces and some deaths were attributed to exposure to the polluted atmosphere. In December 1952, a dense fog occurred in London during inversion weather and lasted for four days. During this time the measured concentrations of smoke particles and of sulphur dioxide rose to more than ten times the previous levels. Healthy cattle in London for the Annual Smithfield Show died acutely and damage occurred to plants at Kew Gardens and to clothing. It was estimated that 4000 excess deaths occurred during and after this episode and though most were in elderly persons with chronic heart or lung disease, some were in infants and previously healthy adults.[5,6]

This episode led to the setting-up of the Beaver Committee on Air Pollution,[7] which reported in 1954 and was followed by the Clean Air Act of 1956.[8] Yet even at this time efforts to identify a specific atmospheric constituent in polluted air on which the blame for the event in 1952 could be placed had failed and smoke particles, sulphur dioxide and even sulphur trioxide were generally held responsible by reason of their high concentrations for the hazard to health. Enforcement of the Clean Air Act during the next few years by local authorities led to limitation of the discharge of black smoke by industry and to the use of smokeless fuel for domestic heating. Smoke concentrations in London and other cities began to decline to very much lower levels from those in the late 1950s and are still declining. Concentrations of sulphur dioxide, however, have been less reduced because of the sulphur content of solid fuels including solid smokeless fuels and the sulphur present in oil fuels. Thus far, no specific benefit to health has been clearly discerned but the lesser frequency of pollution episodes has been obvious.[9] As will be discussed later, the exposure of children and adults to a chronic level of smoke pollution much lower than that during fog episodes is not blameless but may produce effects which are difficult to discern by means of ordinary health statistics. Such subtle effects of air pollution require sophisticated epidemiological enquiries.

Meanwhile, other clinical effects attributable to polluted air have become recognized in other countries. The notorious Los Angeles 'smog' causes a lachrymatory effect because of intense irritation of the eyes. It has been attributed to chemicals resulting from bright sunlight acting upon the exhaust fumes of automobiles. In Japan, particularly in the Yokohama area, nocturnal attacks of breathlessness resembling asthma have been reported during the last twenty years and blamed upon pollution derived from industrialization.[10] No specific hazard has been identified, though higher concentrations of pollutants may develop during the night than during the day as a consequence of climatic factors.

The world-wide effects on health of air pollution are at present imprecisely known, and though Britain has been a pioneer in its study, it is likely that different hazards exist in different countries. The particular fuels that are used for industrial and domestic purposes, the chemicals and gases expelled by industrial concerns and methods used for their limitation and climatic differences cause variations in the atmosphere and very little is yet known concerning the relative effects of these factors upon health.

2.1.2. Mortality and air pollution

Studies have been made retrospectively in Britain using mortality statistics and correlating these with available measured concentrations of smoke particles and sulphur dioxide. Gore and Shaddick[11] made a detailed analysis of mortality during two years in the County of London and estimated the amount of increased pollution required in London to give rise to an immediate excess in mortality. They found considerable difficulties in arriving at a definite critical level either of smoke or sulphur dioxide which increased mortality but stated that 2000 $\mu g\ m^{-3}$ of smoke and 1144 $\mu g\ m^{-3}$ of sulphur dioxide constituted such a level. Martin[12] also analysed the excess deaths in the London area during a number of winters and showed that high levels of pollution by smoke and sulphur dioxide were correlated with high overall mortality, but thought that illness rates (morbidity) might be a more sensitive indicator of the effect of pollution. Both these studies have led to the conclusion that a much smaller increase in smoke pollution than that which occurred in 1952 is hazardous to health and that there are those in the population of a city who are by reason of previous chronic disease particularly susceptible to such increased pollution. Excess mortality is thus largely derived from persons with chronic bronchitis or emphysema of the lungs and those with heart disease ascribed to disease of the coronary arteries. The lack of understanding that such personal susceptibility exists is probably one reason why the lethal effects of pollution went unrecognized for so long. After all, London had suffered from yellow fogs for many years and an excess mortality was noted in the winters of 1873, 1880, 1892 as well as 1948.

The decline in smoke accompanied as it was by a smaller change in sulphur dioxide pollution in London was expected to reveal which of these two constituents is the more hazardous to health. The evidence derived from mortality statistics is suggestive but not final. Thus in 1962 a four-day fog in London caused the concentration of sulphur dioxide to rise to as high a level as that in 1952. The lower excess mortality of 700[13] could be ascribed to the smaller smoke pollution but the recognition of the hazard of smog to health and the warnings given to chronically ill persons to stay indoors could equally well have been responsible. It thus seems misleading to attempt to discern a critical level of smoke or of sulphur dioxide concentrations either alone or together from mortality statistics alone. This is particularly so when comparing mortality statistics from different areas with different experiences of pollution, because so many other factors are at work in determining death.

2.1.3. Respiratory illness and air pollution

The realization that many of those suffering from acute illness after exposure to polluted fogs were patients with chronic bronchitis and emphysema was quickly followed by attempts to study such persons prospectively. It has, of course, long been known that patients say that their chronic cough and breathlessness is worse in adverse weather and particularly during fogs. However, increased breathlessness and production of sputum also follow acute respiratory infections such as common colds and influenza. All such illnesses are termed 'exacerbations' and they range in severity from a single day of worsening of symptoms to episodes of acute distress requiring admission to hospital. Many are treated at home but if patients are away from work their sickness certificates are usually reported as 'bronchitis'. Clifton *et al.*[14] used the daily totals of bronchitis certificates from the working population in Sheffield as an index of incapacity with which to attempt correlation with changes in measured concentrations of smoke and sulphur dioxide. They found that relatively small increases in pollution during brief smog episodes were accompanied or followed by a doubling of numbers of certificates of bronchitis for a day or so.

However, information based on the bronchitis patient's own appraisal of his state of health on the relatively crude basis of 'better' or 'worse' has in the hands of Lawther *et al.*[15] in London proved to be a relatively sensitive index of changes in atmospheric pollution rather than weather. During the winters of 1955, 1957 and 1959 the diaries of 100 or more patients were analysed and compared with measured pollution. Lawther *et al.*[15] concluded that the minimum daily concentration leading to a change in health was 500 $\mu g\ m^{-3}$ of sulphur dioxide accompanied by 250 $\mu g\ m^{-3}$ of smoke. Lawther also thought that during such 24-hour averages of pollution, relatively brief periods of much higher concentrations are probably responsible for the clinical effects, but of this there is no proof. Nor was it possible to distinguish the separate effects of smoke from those of sulphur dioxide; during the last year of study, 1967–8, when smoke concentrations were very low, only one day occurred in which sulphur dioxide exceeded 500 $\mu g\ m^{-3}$, and it was not possible to relate minor fluctuations in illness to the lesser peaks in pollution.

Apart from these British studies, few definite correlations have been shown by other workers between pollution levels and bronchitic illnesses. In Edinburgh, Fisher *et al.*[16] found that infection rather than air pollution appeared to be the chief factor in causing exacerbations; and Burrows *et al.*[17] in Chicago found seasonal fluctuations in symptoms of patients with chronic bronchitis which seemed to be related to mean temperature. Brief exacerbations were correlated, however, with high levels of sulphur dioxide, and Carnow *et al.*,[18] also in Chicago, found a good correlation between increases in sulphur dioxide and acute illnesses in elderly bronchitis patients.

Apart altogether from bronchitis, important results relating the numbers of acute illnesses of the respiratory tract in children to broad differences in exposure to atmospheric pollution were obtained by Douglas and Waller.[19] In their prospective study of a cohort of children born in Britain in 1946, colds and other upper respiratory illnesses occurred to the same extent in children up to seven years of life who lived in areas of high as in those living in areas with very low air pollution as judged by coal consumption. Attacks of bronchitis and pneumonia during the first five years of life, however, were three times as frequent in children living in areas of high pollution compared with those exposed to very low pollution. Social factors did not appear to account for the observed differences. Such large differences strongly suggest the serious effect of chronically raised levels of pollution by smoke and sulphur dioxide quite apart from those due to briefer more intense pollution episodes. As this study was carried out before the Clean Air Act, the levels of pollution to which the children were exposed can be assumed to be very much higher than those encountered today.

A study made in 1966 by Colley and Reid[20] again showed that children born in different parts of Britain still exhibit differences in the number of attacks of acute respiratory illnesses. In this study, there were clinical examinations and a survey of symptoms in 10 000 children aged 6 to 10 living in contrasting areas of England and Wales. A history of past chest illness was commoner in children living in areas with high levels of air pollution than in those in less polluted districts. By the time that adult life is reached the effects of cigarette smoking appear to blur the past history. Thus, Colley *et al.*[21] found that smokers at the age of 20 suffered a chronic cough two and a half times as often as non-smokers, and their childhood exposure to atmospheric pollution had a lesser effect.

Differences undoubtedly exist in the prevalence of chronic bronchitis in adults living in towns, conurbations or rural districts (Holland[22]). City dwellers have the highest and country dwellers the lowest rates of bronchitis. Occupational and social differences and the added factor of cigarette smoking render interpretation of these differences difficult. When persons in a single occupation category of men living in socially comparable areas of England and the US were compared, the British men exhibited a greater incidence of respiratory symptoms and as a group had a poorer lung function than the Americans (Holland *et al.*[23]). Smoking habits did not account for these differences, which were attributed to a greater exposure in England to smoke and sulphur dioxide. The ventilatory function of the lungs of Kent school children was studied in boys of 5–8 by Holland *et al.*[24] Highly significant differences were found between boys from different geographical areas. Social circumstances such as housing seemed important in these differences but pollution by smoke and sulphur dioxide was worse in areas where the children's lungs performed least well.

These studies indicate the great complexity of the factors which cause a deterioration of health and the great difficulty in arriving at firm conclusions regarding a single environmental circumstance such as pollution. Lawther *et al.*[25] studied the lung function of four non-smoking healthy adults daily over a five-year period and for both short- and long-term changes. Though an inverse relation existed in one person between the one-second forced expiratory volume and measured concentrations of smoke and sulphur dioxide on the same day, this was not so in the other persons, whose mid-expiratory volumes were sensitive to pollutants. The authors were reluctant to conclude, however, that pollution directly accounted for the observed differences largely because season and respiratory infections also caused effects. In any case, there is no certainty that smoke and sulphur dioxide, which have been chiefly measured, are the only constituents harmful to health. Sulphuric acid and oxides of nitrogen for instance cannot be exonerated. One curious anomally exists in the case of South Wales, where there is a higher rate of bronchitis than would be expected from the pollution data, Colley and Reid[20] found this to be true also for acute respiratory illnesses in Welsh children.

2.1.4. Experimental studies on healthy men and animals

Many ways have been devised of studying the effect of inhaled particles and of gases on the lungs of experimental animals. These have varied from observation of mortality and changes in the lung structure or the lining epithelium of the air passages to physiological measurements of the air-way resistance to air flow. Both acute and prolonged exposures have been used. The most commonly used materials have been gases such as sulphur dioxide and oxides of nitrogen and particulate matter such as carbon. Mice, rats, guinea pigs, monkeys and dogs have been the subjects of such investigations. The results obtained are difficult to summarize. In the case of sulphur dioxide it has been shown that rats, guinea pigs and mice tolerate exposure to concentrations well above the maxima encountered during smog conditions and suffer remarkably few ill effects. Reid[26] found it necessary to use 300–400 ppm (parts per million by volume (one part of SO_2 per million by volume = 2860 μg m^{-3})) in periods of five hours a day, five days a week for six weeks to produce increase in mucus secretion in the epithelium of the air passages of rats analogous to that found in human chronic bronchitis. Guinea pigs exposed chronically to concentrations varying from 0·1 to 5·7 ppm (286–12 302 μg m^{-3}) of sulphur dioxide for 12 months showed no ill effects according to Alarie *et al.*[27] even though sophisticated measurements of pulmonary function were made. On the other hand other workers have observed an increase in resistance of the lungs to air flow after short-term exposure to sulphur dioxide. In dogs similar increases in resistance to air flow suggesting narrowing of the air passages were produced by exposure to sulphur dioxide in concentrations of 7–61 ppm (20 020–174 460 μg m^{-3}),[28] values far in excess of those found in polluted atmosphere where the sulphur rarely exceeds 1 ppm (2860 μg m^{-3}).

In contrast to animals it appears that man's respiratory tract is affected by lower concentrations of sulphur dioxide. The effect produced is one of constriction of the air passages causing an increase in airway resistance to flow. There is, however, considerable individual variation, some persons experiencing ill effects with as little as 1–2 ppm (2860–5720 μg m^{-3}) of sulphur dioxide. However, Frank *et al.*[29] found no significant increase in air flow resistance in 10 of 11 healthy volunteers given 1 ppm, but the remaining subject exhibited a small increase with this concentration. Amounts of 5 and 13 ppm (14 300 and 371 800 μg m^{-3}) caused the air flow resistance regularly to increase in the first ten minutes after exposure in all persons. The possibility exists that sulphur dioxide absorbed on carbon particles might cause clinical effects at lower concentrations than those quoted, but no proof of this has been obtained experimentally.

Sulphuric acid, which is derived from sulphur trioxide (SO_3) in polluted foggy air, has also been tested experimentally. A concentration of 5·2 ppm (18 580 μg m^{-3}) was observed by Sim and Pattle[30] to increase air-way resistance in man. Guinea pigs are much more sensitive to sulphuric acid (H_2SO_4) mist than are other animals and develop broncho-constriction and even death.

Nitrogen dioxide has been found to cause injury and destructive changes in the lung of exposed rats. Both acute effects after high concentrations and more subtle chronic effects including experimental emphysema have been observed after lesser concentrations (Freeman *et al.* [31]). In man, acute effects follow exposure to very high concentrations of 500 ppm (1 027 800 μg m^{-3}). Acute pulmonary oedema (water-logging) or the clinical picture of acute inflammation of the air passages result. The effect in man of chronic exposure to lower concentrations is not clearly known.

Some workers have attempted to experiment with polluted atmospheric air. In Los Angeles acute increase in air-way resistance after exposure of guinea pigs to 'smog' was noted by Swann and Balchum,[32] but chronic exposure of guinea pigs to Los Angeles air for two years caused no significant differences (Swann *et al.*[33]). Nor were acute effects on lung function of school children in Los Angeles discernible after natural exposure to air with different levels of nitrogen dioxide (McMillan *et al.*[34]).

Ozone in concentrations of 0·7–0·9 ppm (1501 to 1930 $\mu g\ m^{-3}$), similar to the amounts found on occasions in Los Angeles air, has been shown to cause a reduction in the ability of the lung to take up oxygen, presumably due to fluid exudation in the walls of the air sacs (Young *et al.*[35]). In other experiments, 0·7 ppm seemed to irritate the airways and inhibit the taking of a deep breath (Bates *et al.* [36]). Higher concentrations narrow the airways.

The chief lesson to be drawn from these experimental approaches is that healthy animals and men fail to respond acutely to exposure to concentrations of individual pollutants such as are found in mixtures in polluted air. Higher concentrations, for example of sulphur dioxide, cause narrowing of the airways in healthy subjects and it is probable that a person with existing chronic disease of the lower airways reacts to the lower concentrations found in ambient air. Experimental attempts to produce changes in the lungs by chronic exposure to low levels of pollution have generally failed in animals and cannot of course be used on man. Particulate matter alone appears to produce effects if inhaled in relatively high concentrations; it is possible that a synergic action between particles and gases may occur. It is also likely that infection may combine in a subtle way to add to the effect of chemical pollutants. It has, for instance, been shown that cigarette smoke causes a delay in recovery from acute virus infections of the respiratory tract, perhaps because it reduces ciliary action required to cleanse the airways (Dalhamn[37]).

2.1.5. Carbon monoxide poisoning

The ability of carbon monoxide (CO) to kill depends on its high affinity for haemoglobin and the stability of carboxyhaemoglobin, which prevents oxygen carriage by the blood. Exposure to non-lethal concentrations leads to some accumulation of COHb in the blood, and this is probably why the drivers of automobiles and even heavy cigarette smokers have higher concentrations of COHb in the blood than are found normally. Clinical effects, which are produced in the nervous system, are unlikely with COHb less than 5 per cent, but drivers may have higher levels (Moureu[38]). Policemen on point duty in heavy traffic in narrow streets may similarly be affected. Exposure of volunteers to carbon monoxide to produce levels of 10 to 25 per cent COHb has been found to cause dizzyness, fatigue and headache (Post-Lingen[39]). The lower concentration of COHb in these experiments could only be detected by partly uncontrolled tests of nervous function (flicker fusion).

2.1.6. Air pollution and lung cancer

The view that lung cancer may be related to atmospheric pollution is based firstly on the demonstration of low concentrations of carcinogenic compounds, particularly benzpyrene, in the smoke of town air (Commins[40]). Secondly, epidemiological evidence exists relating the mortality from lung cancer and smoke pollution records, such that the highest death rates were found in the most polluted areas (Stocks[41]). Other studies (Ashley[42]) have shown less significant correlation between mortality from lung cancer and pollution. The correlation is certainly much less than between bronchitis mortality and pollution. Probably the effect of cigarette smoking, which is strikingly correlated with lung cancer, may mask the differences between persons living in polluted and less polluted environments.

2.1.7. Health and air quality criteria

The development of control measures to prevent or limit air pollution is now a matter of concern in all countries and not merely in those with highly industrialized communities. The enunciation of clean air standards necessary to safeguard health and legislation to enforce pollution control vary from one country to another. Reference to Hall *et al.*,[43] who have published a profile study of air pollution control in various countries, makes this clear. It is a fact that the system of monitoring pollution differs in different countries. That adopted in the UK enables unusually wide coverage to be maintained because of the existence of a reasonably adequate body of trained workers employed by the Government and by local authorities. Unless an adequate recording system exists, it is impossible to apply standards for air based on maximum permissible concentrations and amounts of pollutants.

Although maximum concentrations such as of ozone, particulate matter, sulphur and nitrogen oxides have been recommended in several countries including the western European countries, the US and USSR, there are limitations to the assumption frequently made that they are based on health considerations. This can be seen by comparing the standards with the minimum concentrations, for example of sulphur dioxide, found to exert an effect in man. Lawther *et al.*[15] found that an average concentration of 500 $\mu g\ m^{-3}$ of sulphur dioxide over 24 hours, when accompanied by 250 $\mu g\ m^{-3}$ of smoke particles, was hazardous to patients. These concentrations are much lower than those found by earlier workers, who depended on correlation with morbidity or mortality statistics and who, like Lawther, could not separate the effect of sulphur dioxide alone from those of smoke and sulphur dioxide together. Direct experiment on healthy persons, however, indicates that it may require as much as 14 300 $\mu g\ m^{-3}$ of sulphur dioxide in a 30 min exposure in the absence of smoke to cause a physiological reaction, though some susceptible persons react after exposure to 2860 $\mu g\ m^{-3}$ (Frank *et al.*[29]).

The published maximum permissible concentrations of sulphur dioxide in the air in the US (Air Quality Criteria[44]) as recommended are 0·08 ppm (229 $\mu g\ m^{-3}$) average over 24 hours and for exposure for one hour 0·5 ppm (1430 $\mu g\ m^{-3}$). In the USSR, the limits are 0·058 ppm (166 $\mu g\ m^{-3}$) over 24 hours and 0·19 ppm (543 $\mu g\ m^{-3}$) over a period of 20 min. There is also a recommended limit in the US for the concentration of smoke particles in the air.[45] Such levels are based on extrapolation from observed effects on man or on damage to materials or plants (Goldsmith[46]).

Much thought has been given to air pollution from motor vehicles, which can be reduced to some extent by changes in engine design, composition of the fuel used, and automatic treatment of the exhaust gases before discharge to the atmosphere. Limitations to the composition of the exhaust gases from new vehicles are already in operation in the US, Western Europe, Japan and some other countries, and there are moves to make the limits more stringent with improvements in engine design, subject to reasonable cost. Standards have been formulated for oxidants based on experimental and epidemiological association of eye irritation or damage to plants. Air quality standards for carbon monoxide, such as an average concentration of 30 ppm (37538 μg m^{-3}) for 8 hours are based on the belief that this will produce not more than 5 per cent of carbon monoxide haemoglobin in the blood of persons thus exposed.

In conclusion, caution is necessary in basing arguments on a hazard to health from exposure to any particular concentration of any air pollutant. Long-term effects from chronic low-level pollution are not certainly known, and in man there are interactions between pollution, infection of the respiratory tract and personal habits such as cigarette smoking. Even if the air that is breathed became as pure as that in primeval times, there would still be disease of the lungs and of the nose and throat from other causes. In many ways present efforts to limit pollution should be regarded as attempts to provide a more pleasant environment and prevent a rapidly worsening position with more industry to maintain a world population that is increasing far too rapidly.

References

1. Evelyn, J., *Fumifugium or the inconvenience of the aer and smoake of London dissipated*, 1661. Reprints obtainable from the National Society for Clean Air, Brighton, England.
2. Simon, J., *Public health reports of John Simon*, Vol. 1, Sanitary Inst. of Great Britain, J. and A. Churchill, London, 1887.
3. Firket, J., 'Fog along the Meuse valley', *Trans. Faraday Soc.*, **32**, 1192–97, 1936.
4. Schrenk, H. H., H. Heimann, G. D. Clayton, H. Wexler and W. M. Gafafer, 'Air pollution in Donora, Pa., USA', *US Pub. Hlth. Bull.*, **306**, Washington, DC, 1949.
5. Logan, W. P. D., 'Mortality in the London fog incident, 1952', *Lancet*, **1**, 336–8. London, 1953.
6. 'Mortality and morbidity during the London fog of December, 1952'; *Ministry of Health Report No. 95 on Public Health and Medical Subjects*, HMSO, London, 1954.
7. *Committee on Air Pollution, Report, Cmd. 9322*, HMSO, London, 1954.
8. Clean Air Act 1956, HMSO, London.
9. *Air pollution and health*, Report of Royal College of Physicians of London. Pitman Med. and Sci. Pub. Co., London, 1970.

10. Smith, R. B. W., E. J. Kolb, H. W. Phelps, H. A. Weiss and A. B. Hollinden, 'Tokyo–Yokohama asthma', *Arch. Environ. Hlth.*, **8**, 805–17, 1964.
11. Gore, A. T. and C. W. Shaddick, 'Atmospheric pollution and mortality in the County of London', *Brit. J. Prev. Soc. Med.*, **12**, 104–13, 1958.
12. Martin, A. E., 'Mortality and morbidity statistics and air pollution', *Proc. Roy. Soc. Med.*, **57**, 969–75, 1964.
13. Scott, J. A., 'The London fog of December, 1962', *The Med. Officer*, **109**, 250–2, 1963.
14. Clifton, M., D. Kerridge, J. Pemberton, W. Moulds and J. K. Donoghue, 'Morbidity and mortality from bronchitis in Sheffield in four periods of severe air pollution', *Proc. Conf. Natl. Soc. for Clean Air*, 189–192, 1959.
15. Lawther, P. J., R. E. Waller and M. Henderson, 'Air pollution and exacerbations of bronchitis', *Thorax*, **25**, 525–39, 1970.
16. Fisher, M., A. J. Akhtar, M. A. Calder, M. A. J. Moffatt, S. M. Stewart, H. Zealley and J. W. Crofton, 'Pilot study of factors associated with exacerbations in chronic bronchitis', *Brit. Med. J.*, **4**, 187–92, 1969.
17. Burrows, B., A. L. Kellogg and J. Buskey, 'Relationship of symptoms of chronic bronchitis and emphysema to weather and air pollution', *Arch. Environ. Hlth.*, **16**, 406–13, 1968.
18. Carnow, B. W., M. H. Lepper, R. B. Shekelle and J. Stamler, 'Chicago air pollution study, SO_2 levels and acute illness in patients with chronic bronchopulmonary disease', *Arch. Environ. Hlth.*, **18**, 768–76, 1969.
19. Douglas, J. W. B. and R. E. Waller, 'Air pollution and respiratory infection in children', *Brit. J. Prev. Soc. Med.*, **20**, 1–8, 1966.
20. Colley, J. R. T. and D. D. Reid, 'Urban and social orgins of childhood bronchitis in England and Wales', *Brit. Med. J.*, **2**, 213–17, 1970.
21. Colley, J. R. T., J. W. B. Douglas and D. D. Reid, 'Respiratory disease in young adults: influence of early childhood lower respiratory tract illness, social class, air pollution and smoking', *Brit. Med. J.*, **3**, 195–8, 1973.
22. Holland, W. W. 'The study of geographic differences in the prevalence of chronic bronchitis', *The Statistician*, **16**, 5–22, 1966.
23. Holland, W. W., D. D. Reid, R. Seltser, and R. W. Stone, 'Respiratory disease in England and the United States', *Arch. Environ. Hlth.*, **10**, 338–43, 1965.
24. Holland, W. W., T. Halil, A. E. Bennett and A. Elliott, 'Factors influencing the onset of chronic respiratory disease', *Brit. Med. J.*, **2**, 205–8, 1969.
25. Lawther, P. J., A. G. F. Brooks, P. W. Lord and R. E. Waller, 'Day-to-day changes in ventilatory function in relation to the environment', *Environmental Research*, **7**, 27–53, 1974.
26. Reid, L., 'An experimental study of hypersecretion of mucus in the bronchial tree', *Brit. J. Exp. Path.*, **44**, 437–45, 1963.
27. Alarie, Y., C. E. Ulrich, W. M. Busey, H. E. Swann, Jr. and H. N. Macfarland, 'Long-term continuous exposure of guinea pigs to sulphur dioxide', *Arch. Environ. Hlth.*, **21**, 769–77, 1970.

28. Frank, N. R. and F. E. Speizer, 'SO_2 effects on the respiratory system in dogs', *Arch. Environ. Hlth.*, **11**, 624–34, 1965.
29. Frank, N. R., M. O. Amdur, J. Worcester and J. L. Whittenberger, 'Effects of acute controlled exposure to SO_2 on respiratory mechanics in healthy male adults', *J. Appl. Physiol.*, **17**, 252–8, 1962.
30. Sim, V. M., and R. E. Pattle, 'Effect of possible smog irritants on human subjects', *J. Amer. Med. Assn.*, **165**, 1908–13, 1957.
31. Freeman, G., S. C. Crane, R. J. Stephens and N. J. Furiosi, 'Environmental factors in emphysema and a model system with NO_2', *Yale J. of Biol. and Med.*, **40**, 566–75, 1968.
32. Swann, H. E., Jr. and O. J. Balchum, 'Biological effects of urban air pollution. IV. Effects of acute smog episodes on respiration of guinea pigs', *Arch. Environ. Hlth.*, **12**, 698–704, 1966.
33. Swann, H. E., Jr., D. Brunol, L. G. Wayne and O. J. Balchum, 'Biological effects of urban air pollution. II. Chronic exposure of guinea pigs', *Arch. Environ. Hlth.*, **11**, 765–9, 1965.
34. McMillan, R. S., D. H. Wiseman, B. Hanes and P. F. Wehrle, 'Effects of oxidant air pollution on peak expiratory flow rates in Los Angeles school children', *Arch. Environ. Hlth.*, **18**, 941–9, 1969.
35. Young, W. A., D. B. Shaw, and D. V. Bates, 'Effect of low concentrations of ozone on pulmonary function in man', *J. Appl. Physiol.*, **19**, 765–8, 1964.
36. Bates, D. V., G. Bell, C. Burnham, M. Hazuvha, J. Mantha, L. D. Pengelly, and F. Silverman, 'Problems in studies of human exposure to air pollutants', *Canad. Med. Assn. J.*, **103**, 833–837, 1970.
37. Dalhamn, T., 'Effect of cigarette smoke on ciliary activity', *Amer. Rev. Resp. Dis.*, **93**(3), 108–14, 1966.
38. Moureu, H., 'Carbon monoxide as a test for air pollution in Paris due to motor-vehicle traffic', *Proc. Roy. Soc. Med.*, **57**, 1015–20, 1964.
39. Von Post-Lingen, M–L., 'The significance of exposure to small concentrations of carbon monoxide', *Proc. Roy. Soc. Med.*, **57**, 1021–9, 1964.
40. Commins, B. T., 'Polycyclic hydrocarbons in rural and urban air', *Int. J. Air Polln.*, **1**, 14–17, 1958.
41. Stocks, P., 'Cancer and bronchitis mortality in relation to atmospheric deposit and smoke', *Brit. Med. J.*, **1**, 74–9, 1959.
42. Ashley, D. J. B., 'The distribution of lung cancer and bronchitis in England and Wales', *Brit. J. Cancer*, **21**, 243–59, 1967.
43. Hall, H. J., H. I. Fuller and A. C. Stern, Profile study of air pollution control activities in foreign countries—first year study. *US Dept. of Health, Education and Welfare, National Air Pollution Control Administration Publication APTD. 0601*, N. Carolina, 1970.
44. *Air quality criteria for sulphur oxides*, US Dept. of Health, Education and Welfare, Washington, DC, 1969.

45. *Air quality criteria for particulate matter*, US Dept. of Health, Education and Welfare, Washington, DC, 1969.
46. Goldsmith, J. R. 'Uses of medical and epidemiological research in the control and prevention of air pollution', *Proc. Roy. Soc. Med.*, **57**, 1034–40, 1964.

Further reading

Air pollution and health, Report of Roy. College of Physicians, Pitman Med. and Sci. Publ. Co., London, 1970.

Medical and epidemiological aspects of air pollution, Symposium No. 6, Proc. Roy. Soc. Medicine, 1964.

Mortality

McCarroll, J. and W. Bradley, 'Excess mortality as an indicator of health effects of air pollution', *Amer. J. Pub. Hlth.*, **56**, 1933–42, 1966.

Hickey, R. J., E. P. S. G. Hoff and R. C. Clelland, 'Relationship between air pollution and certain chronic diseases', *Arch. Environ. Hlth.*, **15**, 728–38, 1967.

Gardner, M. J., M. D. Crawford and J. N. Morris, 'Patterns of Mortality in middle and early old age in the county boroughs of England and Wales', *Brit. J. Prev. Soc. Med.*, **23**, 133–40, 1969.

Illness

Anderson, D. O., B. G. Ferris, Jr. and R. Zickmantel, 'Levels of air pollution and respiratory disease in Berlin, New Hampshire', *Amer. Rev. Resp. Dis.*, **90**, 877–87, 1964.

Sterling, T. D., J. J. Phair, S. V. Pollack, D. A. Schumsky and I. DeGroot, 'Urban morbidity and air pollution', *Arch. Environ. Hlth.*, **13**, 158–70, 1966.

Anderson, D. O., 'The effects of air contamination on health, I, II, and III', *Canad. Med. Ass. J.*, **97**, 528–36, 585–93, 802–6, 1967.

Lunn, J. E., J. Knowelden and A. J. Handyside, 'Patterns of respiratory illness in Sheffield infant school children', *Brit. J. Prev. Soc. Med.*, **21**, 7–16, 1967.

Winkelstein, W., Jr. and S. Kantor, 'Respiratory symptoms of air pollution in an urban population of North Eastern U.S.', *Arch Environ. Hlth.*, **18**, 760–7, 1969.

Winkelstein, W., Jr., S. Kantor, E. W. Davis, C. S. Maneri and W. E. Mosher, 'The relationship of air pollution and economic status to total mortality and selected respiratory system mortality in men. II. Oxides of sulphur', *Arch. Environ. Hlth.*, **16**, 401–5, 1968.

Battigelli, M. C., 'Sulphur dioxide and acute effects of air pollution', *J. Occup. Med.*, **10**, 500–10, 1968.

Snell, R. E. and P. C. Luchsinger, 'Effects of sulphur dioxide on expiratory flow rates and total respiratory resistance in normal human subjects', *Arch. Environ. Hlth.*, **18**, 693–8, 1969.

Experimental and toxicological

Pattle, R. E., and H. Cullumbine, 'Toxicity of some atmospheric pollutants', *Brit. Med. J.*, **2**, 913–16, 1956.

Pattle, R. E. and F. Burgess, 'Toxic effects of mixtures of sulphur dioxide and smoke with air', *J. Path. and Bact.*, **73**, 411–19, 1957.

Aviado, D. M. and H. Salem, 'Acute effects of air pollutants on the lungs', *Arch. Environ. Hlth.*, **16**, 903–7, 1968.

Amdur, M. O., 'Toxicologic appraisal of particulate matter, oxides of sulphur and sulphuric acid', *J. Air Polln. Control Assoc.*, **19**, 638–46, 1969.

2 Effects of air pollution

Part 2 Air pollution and plants and farm animals

L. H. P. JONES and D. W. COWLING

Air pollution from man's industrial and domestic activities has long been known to have significant and occasionally devastating effects on plants, under both cultivated and natural conditions, and on animals. Economic losses through damage to agricultural and horticultural crops and to forest trees has stimulated considerable research on the effects of individual air pollutants on plants, whereas much of the research on animals has developed because of concern about the effects on human health. The principal air pollutants responsible for damage to plants are sulphur dioxide, fluorides, smoke, soot, dusts and the components of photochemical smog; most harm to farm animals results from fluorides and metallic dusts. Generally, plants are more susceptible to damage by gaseous pollutants, whereas animals seem to be more susceptible to damage either directly or indirectly by particulate pollutants.

In this part the effects of air pollution on plants and animals are discussed in three sections. The first deals with general principles and the second with the effects of individual air pollutants. The list of these is not complete and radionuclides are among several pollutants that are omitted. In the final section the influences of agriculture, horticulture and forestry on air quality are briefly discussed.

Since the publication in 1903 of the handbook by Hasselhoff and Lindau[1] there has been an ever-increasing stream of research reports, reviews and symposia publications concerned with the subject of air pollution in relation to plants. Among reviews that are helpful we list, in particular, the following, which are grouped according to topic: comprehensive reviews dealing with the effects of most air pollutants,[2–10] especially on forest trees,[11–15] effects of sulphur dioxide,[16,17] of oxidant pollutants,[18–19] of ozone,[20] and of fluorides,[21,22] and shorter general reviews.[23–26] Among books concerned wholly or in substantial part with the effects of pollution on plants are those by Garber[27] and by Treshow.[28] Two pictorial atlases are available illustrating the effects on plants of most pollutants[29] or of sulphur dioxide only.[30]

There are fewer publications dealing with the effects of air pollution on animals. More attention has been given to pets in cities with pollution problems than to farm livestock, and reports of animal deaths in major air-pollution episodes are fragmentary and inconclusive.[31] Useful sources of reference to the effects of many air pollutants on animals[32–35] have been augmented recently in a comprehensive review by Lillie.[36] The effects on farm animals of fluorides have been studied extensively; in addition to the classic monograph by Roholm,[37] there are several more recent reviews dealing with industrial fluorine intoxication or fluorosis.[38–40] Burns and Allcroft[41] surveyed the problem in the UK and Allcroft *et al.*[42] presented the case history for an experimental farm. A comprehensive account of fluorine as a trace element in human and animal nutrition is given by Underwood.[43]

2.2.1. General principles

Injury, or damage, by air pollutants is discussed under three main headings: the receptor, either plant or animal; the pollutant itself, particularly its physical and chemical form; the environment in which the receptor and pollutant co-exist. This division should not obscure the fact that injury or damage is the result of interactions between the receptor, the pollutant and the environment. The situation is analogous to that described by Gäumann[44] for plant diseases caused by fungi; disease only develops when a specific relationship exists between pathogen, plant and environment.

It may be noted that the terms injury and damage have sometimes been used synonymously in descriptions of the effects of an air pollutant on plants. Recently, however, injury has been defined as including all responses ranging from leaf necrosis and abscission to reduced photosynthesis and interference with other biochemical processes. Damage includes only those injuries and functional disturbances that interfere with the desired use of the plant.[45] Thus, leaf necrosis on turnip is described as injury, whereas similar necrosis on spinach is described as damage. This practice is followed here, although it becomes difficult to apply in some circumstances.

2.2.1.1. The receptor of pollutants

The plant. The aerial parts of plants occupy the lowest layers of the atmosphere and their subterranean parts occupy the upper layers of the soil. Plants therefore occupy a central position in the soil–plant–atmosphere continuum, deriving from the soil the essential mineral nutrients and water necessary for growth and development and from the atmosphere the carbon dioxide necessary for photosynthesis. This process uses energy from sunlight to fix the carbon dioxide that enters the leaves through small pores or stomata in their surfaces. Through these pores there is normally an exchange of oxygen and water and, sometimes, fungi and substances such as pollutants may also gain entry into the leaf from the atmosphere.

In addition to providing a pathway of entry for gaseous pollutants the leaf is important as a surface for deposition of particulate matter. Very often the leaf is the part of the plant which first and most obviously displays visible symptoms of injury, although injury may also be observed on other above-ground parts, notably the flowers and fruit. Symptoms of injury due to airborne pollutants first appear in the leaf tissues at a microscopical level, and the particular cells or tissues affected often indicate the causative agent. It is therefore useful at this point to consider the anatomy of a typical leaf.

In transverse section (Fig. 2.2.1) the leaf is bounded by an upper and lower epidermis, a single layer of cells whose outer walls become thickened to form a relatively impermeable continuous 'cuticle'. Scattered throughout this layer are the stomata, each consisting of a pair of guard cells surrounding an elliptical pore; changes in the turgor of the guard cells in response to external or internal stimuli control the opening and closing of the pore. There are tens of thousands of stomata per square centimetre of leaf surface and the pore size varies among species; when open it ranges from 9 to 52 μm in length and from 2 to 10 μm wide. The internal tissues of the leaf are (a) the palisade tissue, an upper layer comprising one or more tiers of elongated, chloroplast-rich cells, separated by narrow intercellular spaces, (b) the spongy mesophyll, a lower layer of irregularly shaped cells, which contain few chloroplasts and which are separated by conspicuous intercellular spaces, and (c) the conducting tissues, or veins, which carry water, salts and metabolites through the leaf.

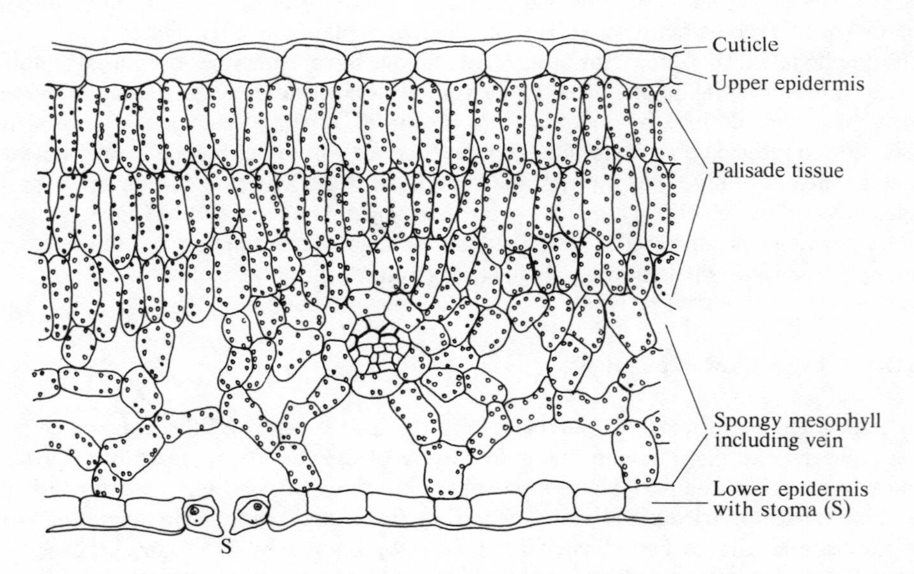

Fig. 2.2.1. Anatomy of typical leaf in transverse section. (From Eames and MacDaniels, *Introduction to Plant Anatomy*, 2nd edn., McGraw-Hill, 1947)

The tissues of leaves that have recently completed their growth by expansion are, in general, more susceptible to injury than those of younger or older leaves. Exceptions to this are found with fluorides and hydrogen sulphide, which are most injurious to the youngest leaves.[46]

As well as the varying susceptibility of plant tissues and parts, there are variations in susceptibility to injury by any one pollutant according to species and genotype, and any one species may be more or less susceptible to different pollutants. For example, barley and lucerne are more susceptible to sulphur dioxide and less susceptible to hydrogen fluoride than maize and gladiolus.[5] Variations due to genotype have been found in tobacco[47] and onion;[48] in white pine they are sufficiently marked to enable selected clones to be used as indicators of pollutants.[12]

The animal. Air-borne pollutants may enter animals through two main pathways, namely ingestion of contaminated plant material, and inhalation.[36] Other less obvious pathways include ingestion of pollutants through licking or grooming the coat or feathers, and direct deposition causing irritation of the eyes or skin.

Pollutants entering through inhalation may cause responses in various parts of the respiratory tract. Sulphur dioxide causes constriction of the air passages which results in coughing, whilst nitrogen dioxide and ozone produce changes in the pulmonary capillaries which reduce absorption. Other pollutants may influence the bronchial blood vessels so that absorption through the mucosa is reduced, or they may affect the circulatory blood system more generally so that the distribution of an absorbed compound is retarded. Interference with the function of macrophage cells, which are responsible for clearing the alveoli and other lung tissue of foreign matter, may be another important effect of inhaled pollutants.[36]

Toxic effects in animals may show as chemical, physiological and morphological changes in tissues and organs, but there may also be sensory effects and overt behavioural signs which are as obvious as visual symptoms of injury in plants. However, with animals it is more difficult to diagnose the cause of such symptoms, because there have been few definitive experiments involving exposure to specific pollutants. There are differences in susceptibility to injury both between and within species of animals. These may sometimes be due to differences in mobility and powers of selecting food. On the other hand, housed animals may suffer from air pollution caused by themselves, for example, ammonia as a pollutant in chicken houses.[49]

2.2.1.2. The pollutant

Pollutants may be classified according to their physical and chemical properties. They may be gaseous, e.g., nitrogen oxides and sulphur dioxide; or solid, e.g., soot, dust and sodium aluminium fluoride; or present in aqueous solution, as for example, sulphuric acid in rain. Some may have reducing capacity, e.g., sulphur dioxide, while others have oxidizing capacity, e.g., ozone; some may be acidic, e.g., hydrogen fluoride, while others are alkaline, e.g., cement dusts.

Interactions between two or more air pollutants may increase or decrease their potential effects as individual pollutants. Where there is an increased effect, the phenomenon is known as synergism. One such reaction was found with tobacco which was injured by a mixture of 64 $\mu g\ m^{-3}$ (0·03 ppm) ozone and 858 $\mu g\ m^{-3}$ (0·30 ppm) sulphur dioxide, whereas the same exposure to the individual pollutants produced no injury.[50] Synergism had been found also with tobacco between nitrogen dioxide and sulphur dioxide.[51] Sometimes chemical reactions between pollutants may influence toxicity to plants and animals. Thus, sulphur dioxide may react with ammonia to produce the relatively harmless salt, ammonium sulphate.[52]

The size of particulate pollutants may influence their rate of deposition on plant leaves and their passage through both stomata and the respiratory tract of animals. Generally, the physical state of the pollutant will have a considerable effect on its dispersal and dilution in the atmosphere.

The pollutant and the plant. The most important factors relating to a pollutant are its concentration and the duration of its existence at that concentration, which when combined give the time–concentration product. This relationship is relevant both to accidental exposure of plants and to exposures that are apparently or practically unavoidable. O'Gara[53] drew attention to the time–concentration relationship as a determinant of injury following studies of the effects of sulphur dioxide on about 200 species of plants. He proposed the 'Law of gas action on the plant cell', as expressed by the following equation:

$$t(L - l) = C, \qquad (2.2.1)$$

where t is the time of exposure, L is the concentration of the gas during the exposure, l, which varies with the species, is the concentration that just fails to injure the plant, and C is a constant. Later studies showed that this equation was not entirely adequate because doubling the concentration (L) was more injurious than doubling the time (t) for a given constant (C).

For air pollutants in general, Guderian *et al.*[45] modified Eq. (2.2.1) and developed the concept of an *irritation threshold concentration* for a given species, which is the concentration that may be tolerated throughout its life without causing injury: it is more or less equivalent to l in Eq. (2.2.1). For the most sensitive species this concentration forms the *first tolerance limit*. Above the irritation threshold concentration the time–concentration product is used to define a *second tolerance limit* above which injury will occur. From experimental data, Guderian *et al.*[45] established the following equation:

$$t = ke^{-a(C-C_r)} \qquad (2.2.2)$$

where t is the time of exposure, C is the concentration of the pollutant during the exposure, C_r is the concentration at the irritation threshold, k is a constant for the species, and a is a constant which accounts for the deviation from an inverse relationship between t and C. A comparison of irritation threshold curves of different species allows their relative susceptibilities to be defined.

Such mathematical approaches are complex and have limitations mainly because the constants in the equations are dependent on environmental conditions. Nevertheless, they stimulate thought and provide a basis for further experimentation; for these reasons they deserve more attention.

The pollutant and the animal. A rather less sophisticated approach has been taken in studies of the effects of air pollutants upon animals, and much of the work has been concerned with small animals such as rats, mice and rabbits. Laboratory experiments in which such animals are exposed to substantial concentrations of specific air pollutants have served to indicate their relative toxicities. The concentrations of the pollutant are frequently high in order to determine acute toxicity, the most common expression of this being the lethal dose (*LD*) or concentration (*LC*) that kills 50 or 100 per cent of the population, e.g., LD_{50} or LD_{100}. A time factor is sometimes included in the toxicity study to take account of time–concentration effects.[54] Although studies of acute toxicity are valuable, the most urgent need is to examine more closely chronic injury that might result from prolonged exposure to lower concentrations of pollutants, similar to those encountered in polluted atmospheres.

2.2.1.3. The environment

Many environmental factors affect the reaction of plants and animals to air pollutants, most often through their effects on the receptor, although they may also affect the physical or chemical state of the pollutant after its emission.

The environment and the plant. Environmental factors have been examined by including them as variables in controlled experiments in which a plant is exposed to known concentrations of a pollutant. Light, both in terms of intensity and photoperiod, controls many aspects of growth and development and it also greatly influences the sensitivity of plants to air pollutants.[6,55,56] The effect of temperature is not easily separated from that of light because of the strong correlation between the two factors. However, plants usually become more susceptible to injury as temperature increases over the range 5–30°C.[16] Low humidity and high concentration of carbon dioxide in the air have been found to reduce the susceptibility of some plants to injury by particular pollutants.[5,57]

Soil physical and chemical characteristics influence the susceptibility of plants to injury by air pollutants. Thus, Katz[16] found that lucerne suffered less injury from sulphur dioxide when it grew with poor water supply than when it was grown with adequate water. It has been suggested that plants with good supplies of mineral nutrients are more susceptible to injury than those with poor supplies, but there is insufficient definitive information to support this generalization.

The environment and the animal. Animals are protected from environmental hazards by a system of interrelated mechanisms and only show symptoms of stress when the capacity of these mechanisms is exceeded. Additionally, they may suffer stress because of poor husbandry, poor feeding, including deficiency or excess of mineral nutrients in their feed, metabolic disorders, endemic diseases and internal or external parasites. An air pollutant may add to the stress that farm animals experience because of these factors, and any symptoms that develop are difficult to distinguish. The lack of studies dealing with the influence of environmental factors on the response of farm animals to pollutants makes further comment difficult.

2.2.1.4. Hidden injury in plants

The idea that air pollutants may injure plants without producing visible symptoms appears to have arisen at the end of the nineteenth century,[58] since when such injury has been referred to as 'hidden' or invisible'. In 1923, Stoklasa[59] suggested that invisible injury was associated with decreased growth, accelerated ageing and reduced photosynthetic rate. However, the concept of hidden injury received little support from the results of experiments with lucerne and sulphur dioxide in which yield and photosynthesis were not reduced until symptoms of injury appeared on the leaves.[16,60] In reviewing the effects of sulphur dioxide on plants, Katz[16] rejected the concept and concluded: 'It is hoped that the "invisible injury" theory has now been disposed of, once and for all time, and will not be resurrected again in problems involving sulphur dioxide damage to plants.' Nevertheless, invisible injury to citrus resulting from continuous and prolonged exposure to sulphur dioxide has been reported by Matsushima and Harada.[61]

Thomas[6] admitted the possibility of invisible injury in the case of fluorides and redefined it as 'the reduction of photosynthesis below the level expected from the amount of leaf destruction'. Studies of the effects of photochemical pollutants have led to differing conclusions: some authors[62,63] accept that they cause invisible injury while others[64] do not.

Clearly, the definition of hidden injury is unsatisfactory and will remain difficult. The electron microscope and modern analytical techniques now make it possible to detect injury at levels that were previously hidden, and future definitions might profitably take this into account.

2.2.2. Specific pollutants

2.2.2.1. Sulphur dioxide

Sulphur dioxide is present in air as the result both of natural processes and the activities of man; in industrial countries it is derived mainly from the combustion of fuels that contain sulphur and from the smelting of non-ferrous metals. After emission, sulphur dioxide disperses in the air and reacts with other substances; eventually it reaches a 'sink' which may be land surfaces, the oceans or a receptor, such as plants or man. Along with the process of dispersion there is one of dilution and thus concentrations in the air vary within wide limits.

Effects on plants. Plants may be injured by sulphur dioxide in the air if its concentration and the duration of exposure exceed certain limits. Much information about effects on plants has come from studies in areas around particular emission sources such as smelters. Thus, Holmes *et al.* produced the Selby Report[65] about damage in the San Fransisco area, and O'Gara and his contemporaries[53] examined damage to plants in Utah resulting from the activities of the American Smelting and Mining Corporation. Investigations in Utah were continued by Thomas and Hill[60,66] who also reported some of O'Gara's studies.[67] In the 'thirties Katz *et al.*[68] examined damage around the Trail smelting works in British Columbia.

The symptoms of acute injury, which are generally more pronounced on recently fully expanded leaves, are initially seen as water-logged blotches. Later there is a bleaching of marginal and interveinal areas of the leaves, although the veins themselves often remain green. The bleached areas become brown to red-brown with eventual necrosis of the tissues. As acute symptoms develop, leaves may be shed, but in some species the necrotic areas may fall out giving the appearance of 'shot-holes'. Chronic injury resulting from long-term exposure to slightly lower concentrations of sulphur dioxide is seen as interveinal chlorosis or yellowing and, on conifer needles, as pale green bands which may, in time, become yellow to red-brown. A directional effect due to a prevailing wind carrying sulphur dioxide may give rise to mis-shaped trees, the leaves on the windward side suffering most damage. At the microscopical level, the chlorophyll-containing cells may show dehydration whilst the chloroplasts themselves are reduced by disintegration. The guard cells, particularly in conifer needles, may become red. It is generally accepted that the injury due to sulphur dioxide arises through its reducing capacity rather than through an acidifying effect.

Although sulphur dioxide, after absorption through the stomata, may satisfy part of the plant's requirement for sulphur,[69] there is a limit to the rate at which it can be metabolized by the plant and thus rendered useful rather than harmful. Plants are able to tolerate relatively high levels of sulphate in their tissues, but sulphite, a possible intermediate in the oxidation of the pollutant, is thirty times as toxic as sulphate. Photosynthesis is decreased and respiration is increased by non-injurious levels of sulphur dioxide but the effects are reversible.[60]

Plants are more sensitive to injury when environmental factors encourage the stomata to open.[16,60] Thus, injury often accompanies high growth rates and is more likely to occur in the light than in the dark, at high humidity than at low, at adequate moisture supply than at inadequate, and at moderately high temperatures. Whilst injury to plants is to a large extent dependent on stomatal behaviour, this alone does not account for all the variation among plants in their susceptibility. For example, the stomata of potato are open through day and night but the plant is relatively resistant, whereas maize, whose stomata are normally closed, is a relatively susceptible species.[6] As a general rule, the susceptibility to sulphur dioxide increases as the day progresses; however, it decreases in the afternoon, possibly because of the accumulation of sugars.[16]

O'Gara's data have been used to rank plants within three groups according to their susceptibility to injury.[53] Over 200 species were exposed to controlled concentrations of sulphur dioxide in an experimental enclosure, and their resistance to injury was related to that of lucerne which, as the most susceptible plant, was given a value of 1. Many

TABLE 2.2.1

Tolerance limits of various plants to sulphur dioxide (after Zahn[70])

Resistance group	Plants	Tolerance limit† $\mu g\ SO_2\ m^{-3}$ air
I	Clovers	430–570
II	Wheat, leafy vegetables (excluding cabbage), bean, strawberry, roses	570–860
III	Root crops, rape, cabbage	860–1140

† Tolerance limit: The concentration which must be exceeded before injury occurs.

common agricultural plants, e.g., barley (value 1), oat (1·3), wheat (1·5), clover (1·4), and turnip (1·5) are in the susceptible group; rye (2·3) and many cruciferous plants (1·6–2·3) are intermediate; potato (3·0), citrus (6·5–6·9), conifers (7·0–15·0) and oak (14·0) are in the resistant group. Zahn[70] considered that because some sulphur dioxide may be adsorbed on the surfaces of the experimental enclosure, it is difficult to determine precisely the time–concentration product that would cause injury. Accordingly, he proposed three groups within which plants were not further ranked (Table 2.2.1).

During the 'thirties a large number of other experiments were conducted in which barley, lucerne and conifers were exposed to sulphur dioxide in enclosures in the field.[16,60,68] The results showed that lucerne was not injured by long-term exposure to air containing 430 $\mu g\ SO_2\ m^{-3}$ (0·15 ppm), and that yield was unaffected by this concentration. Short-term exposures at higher concentrations often caused yield reductions which could be related to the amount of leaf damage.[66] This held with lucerne regardless of the stage of growth. With cereals, reductions in grain yield were greater if injury occurred at ear emergence than if it occurred earlier because the plants were then able to compensate by tillering.[68]

A concentration of 572 $\mu g\ SO_2\ m^{-3}$ (0·20 ppm) has become widely accepted as a critical limit below which most plant species would not suffer injury even after prolonged exposure. Recently, however, Wentzel[71] suggested that this limit should be reduced to 57 $\mu g\ m^{-3}$ (0·02 ppm), following an examination of the results of large-scale experiments conducted under field conditions in Germany.[72–74] The suggestion is of considerable significance because the official American Air Quality Criteria[75] accept that injury to plants has occurred when the average concentration in the air exceeds 85 $\mu g\ SO_2\ m^{-3}$ (0·03 ppm). At the present time the average concentration over rural areas in Britain is reported[76] to be about 50 $\mu g\ m^{-3}$ (0·02 ppm), a value approaching the limit suggested by Wentzel.[71]

Many lichens—plants formed by the association of a fungus and an alga—are unable to tolerate exposure to sulphur dioxide in concentrations above Wentzel's limit[77] and there are many reports of their disappearance around industrial areas. It is frequently suggested that certain species of lichens,[78,79] and indeed of higher plants, might serve as 'indicator' plants of sulphur dioxide concentrations in the air. The topic of 'indicator' plants for this and other air pollutants has been fully reviewed by Heck.[80]

Rather than basing critical limits for injury on average concentrations of sulphur dioxide in the air, Guderian *et al.*[45] have integrated the product of time and concentration during periods when the concentration is about a certain minimum level. The concept of 'recovery' periods in which low concentrations are interspersed with periods of high concentration has been used to develop a mathematical function that would predict injury to plants, and to establish a new critical limit.[81,82]

In contrast to the considerable amount of information about the effects of atmospheric sulphur dioxide on plants, there is very little about its effects on soils. In various investigations[68,72,83] it has been found that the pH of the surface horizon of soils is lowered by sulphur dioxide. The effect was greatest in areas close to the emission source, where concentrations of the gas in the air were highest. The decreases in pH were associated with striking increases in soluble[68,83] and total sulphur content[83] of the soils.

Effects on animals. High levels of sulphur dioxide in the air have not been positively identified as adversely affecting the health of farm livestock in the field, but they have been implicated in causing injury when associated with smoke and fog. Much of the research on this subject has been conducted with small mammals and because of its relevance to human health, it is dealt with in Part 1 of this chapter.

Laboratory studies with swine have shown that eye irritation occurs following an eight-hour exposure to 14·3 mg SO_2 m^{-3} (5·0 ppm) and more severe symptoms appear at higher concentrations.[84] However, all the concentrations chosen were in excess of those likely to be encountered under usual conditions of air pollution.

Cattle fed with forage that had 25 per cent of its leaves damaged by sulphur dioxide in the air showed no ill effects and performed no differently from animals fed with uncontaminated forage.[85]

2.2.2.2. Fluorides

Damage to plants and animals by fluorides is usually localized in areas affected by emissons of fluoride-containing compounds, principally from brick, ceramic and other industries which use fluoride-containing clays or which, as in the case of aluminium production, employ cryolite as a flux in their processes.

Effects on plants. The symptoms of acute fluoride injury to plants vary between monocotyledons (plants with long narrow leaves, e.g., gladiolus) and dicotyledons (plants with broad leaves, e.g., fruit trees).[6] In monocotyledons, the symptoms appear at the leaf tips, which become brown and necrotic, and there is frequently a clearly defined red-brown margin between the injured and healthy part of the leaf. In dicotyledons the necrosis is initially confined to the margins of the leaf and the affected areas may later become detached leaving a notched margin. In both groups of plants the younger leaves are affected first. With more severe injury the necrosis spreads further down or into the leaf and abscission frequently occurs. Symptoms of chronic injury resulting from exposure to lower concentrations of fluorides appear as an interveinal chlorosis. Injury to fruit, particularly peach, has been reported, the symptoms being a softened area on the suture of the fruit.[86]

Most plants have a limited capacity to absorb fluoride from the soil even when fluoride-containing fertilizers are applied;[87] if they have not been subjected to contamination from air-borne fluorides their aerial parts commonly contain between 1 and 15 ppm fluorine on a dry-matter basis. Certain species, notably the camellias, are exceptional in this respect and may contain as much as 1900 ppm fluorine.[6] In most species the translocation of fluoride from the roots is restricted, so that these normally have a higher content than the aerial parts. In contrast, plants growing in fluoride-polluted atmospheres have a much higher content in the aerial parts than in the roots, some of the fluoride being adsorbed on the leaf surface.[88,89] The fluoride that enters the leaf through the stomata is translocated to the leaf extremities (i.e., the tips or margins), where it may cause the collapse of cells and the visual symptoms described earlier. The fluorine content of plants at which symptoms of injury appear differs considerably among species. Gladiolus shows symptoms with 50 ppm fluorine in its tissues, whereas cotton may contain 5000 ppm without showing symptoms.[6] Maize and tomato exposed under similar conditions to fluorides in the air were found to contain 70 and 200 ppm fluorine, respectively, in their leaves but only the maize showed symptoms of injury[90] The reason for such differences is not completely understood but they may be due, at least in part, to (a) the extent of external adsorption on the leaf and (b) the extent of distribution within the plant.

Several workers[5,91,92] have classified plant species on the basis of their susceptibility to injury by fluorides. Gladiolus, maize and peach are among the most susceptible species, barley, oat and apple are intermediate, whilst tomato, carrot and brassica are among the most resistant. Injury is likely to occur in susceptible species if the concentration in the air exceeds $0{\cdot}9\ \mu g\ F\ m^{-3}$ (0·001 ppm) for periods of 7 to 21 days and, in general, plants are more susceptible under dark than under light conditions.

Because of the link between afforested areas and hydro-electricity and, in turn, aluminium smelting, a considerable amount of information is recorded about fluoride damage to natural and planted coniferous forests. In Norway, emissions from smelters of 8 to 30 kg fluoride per hour have caused visible injury to conifers up to a distance of 32 km.[93] Pine was generally more resistant to injury than spruce although the latter showed a wide range in the severity of injury; European larch and Western Hemlock spruce were relatively resistant.

An unusual association between fluoride damage and fertilizer practice has been found with apricot trees in France.[94] In some orchards there was little or no damage to fruit or leaves whilst in adjacent orchards it was severe; apparently, the only factor that differed was the formulation of the potassium–phosphorus fertilizer.

Effects on animals. The ingestion of toxic or potentially toxic amounts of fluorine by farm animals may occur (a) in areas where the drinking water is naturally high in fluoride, (b) when the animal's diet is injudiciously supplemented with fluoride-containing minerals, and (c) in restricted areas where industrial emissions of fluoride fumes or dusts contaminate the herbage and water consumed by the animals. Although the animal disorder arising from the latter is sometimes referred to as industrial fluorine intoxication, the term fluorosis is used here for the sake of brevity. In areas of industrial pollution plants may accumulate fluoride to an extent which, whilst not injurious to themselves, may produce fluorosis when ingested by animals over a long period. The alternative pathway of entry into animals, namely through inhalation, is unlikely to be important; for example, it has been calculated that in a fluoride-polluted atmosphere this pathway would account for only one-hundredth of the total intake.[33]

Initially, the ingestion of large but sub-lethal amounts of fluoride produces no visible ill-effects because the animal is protected by two mechanisms, namely, a rise in urinary excretion and a deposition of the retained fluorine in the bones and teeth. As the ceilings of these mechanisms are approached, clinical signs of fluorosis, notably dental lesions and loss of appetite, usually appear.[43] Teeth that are erupting concurrently with high fluorine intakes develop poorly and are liable to rapid wear. Bone damage, which is similarly a characteristic of fluorosis, can occur in animals at any age. Exostoses (bone over-growth) may develop on the skeletal structure giving rise to stiff, laborious gait and lameness.[95] These disorders eventually lead to a loss of the animal's ability to maintain its food intake,[96] and its productivity is reduced, whether in terms of milk from cows or of reproductive capacity through cows or ewes.

Whilst no single criterion is diagnostic of fluorosis, the levels of fluorine in urine and bone are useful indicators when considered along with visible symptoms. Urinary fluoride levels in sheep and cattle that are not exposed to excess fluorine are usually about 5 ppm. With incipient fluorosis they range from 20 to 30 ppm, whilst with severe fluorosis they may be as high as 50 ppm.[96] In the bones of adult farm animals the level of fluorine rarely exceeds 1200 ppm (ash basis) and is usually in the range 300–600 ppm. In cows, levels in excess of 5500 ppm of fluorine are associated with fluorosis and levels above 10 000 ppm may occur in severely affected individuals.[97] The levels are somewhat lower in the bones of sheep suffering from fluorosis.[98]

Tolerance levels to dietary fluorine are influenced by the species and age of the animal, the solubility of the compound, and the duration and continuity of ingestion. Many experimental feeding studies have been conducted to define the tolerance levels for various species; the results from some of these, as summarized by Phillips *et al.*,[39] are given in Table 2.2.2. Soluble forms of fluorine such as sodium fluoride are more

TABLE 2.2.2

Safe levels of fluorine in the total ration (ppm F, dry-matter basis)
(After Phillips *et al.*[39])

Animal	NaF or other soluble fluoride	Phosphatic limestone or rock phosphates
Dairy cow	30–50	60–100
Beef cow	40–50	65–100
Sheep	70–100	100–200
Chickens	150–300	300–400
Turkeys	300–400	

toxic per unit of fluorine than less soluble forms such as rock phosphate and cryolite; calcium fluoride is highly insoluble and is even less toxic. Particle size also influences toxicity; thus, finely divided cryolite may be as toxic as sodium fluoride.[37]

Apart from controlling the emission of industrial fluorides, two main approaches have been suggested to minimize the incidence of fluorosis.[42] Improved grassland management, particularly increased yields of grass, will dilute the concentration of fluorine in the crop and thus its intake by grazing animals. Intake of fluorine may also be reduced by feeding defluorinated cereals as supplements in early spring and late autumn when the growth rate of grass is low and its fluorine content is relatively high. The incorporation of aluminium sulphate in feeds has been proposed as a means of rendering fluoride non-toxic, apparently by reducing its absorption from the alimentary tract.[99] However, Allcroft *et al.*,[42] who fed aluminium sulphate to dairy cows, concluded that it was of little practical value in the control of fluorosis. The second approach is to use herds which contain no home-reared stock, and in which the adult animal does not spend the whole of its life in the contaminated area.

2.2.2.3. Smoke, soot and dusts

Smoke emitted from the combustion of the major fuels comprises particles of soot, fine dust and ash in a mixture of gases. Other dust and ash may be blown into the air from sources such as ash heaps, coal dumps, cement factories and smelters of non-ferrous metals, but these are less important as sources of particulate pollutants in the air than the combustion of fuels. While dispersed in the air, such particles may affect plant growth by reducing the intensity of light and its quality; this effect is limited to the vicinity of towns or industrial sites. After deposition particulate pollutants may be injurious to plants and animals in various ways, but again they are unlikely to cause problems at any considerable distance from the point of origin. As solid particles are frequently deposited in areas where gaseous pollutants are present their effects may be modified and difficult to isolate.

Effects on plants. Solid particles, especially soot, adversely affect plant growth by reducing the light energy available for photosynthesis through a blackening of leaves and of glasshouse panes. The effect on glasshouses may be so serious as to make it necessary for horticulturalists to move their industry to cleaner, unaffected areas. Soot and fine dust may also interfere with gaseous exchange in plants by blocking the pores of stomata. In experiments on laurel it has been found that the removal of such material from the leaves increased their growth rate, but the results were not definitive.[100] Bleasdale[101] doubted that there was significant blocking of stomata, because they are usually more numerous on the under surfaces of leaves than on the upper surfaces where most deposits would accumulate.

There are conflicting reports on the effects of deposits of cement dust. Some indicate that they have no direct effect on the plant but that they indirectly affect growth through raising soil pH.[8,102] Others have found that cement dust causes direct injury.[103,104] This has been attributed to the dust on wet leaves forming alkaline droplets which saponify the protective cuticle, and penetrate and injure the underlying palisade cells.

Deposits of various solids on soft fruits and vegetables, such as lettuce and cauliflower, may be sufficiently obvious and objectionable to reduce their market value, or to require the grower to incur extra costs in preparing them for market.

Effects on animals. Dusts of an unspecified nature may cause respiratory irritation in farm animals, although often the dusts may be of farm rather than of industrial origin. Cement dusts have been implicated as the cause of adverse effects on cattle but the results of investigations were inconclusive.[105] Laboratory studies have indicated that toxicity may arise through the inhalation of dusts containing specific contaminants such as beryllium and certain heavy metals, but the studies have not involved farm animals.[36]

The ingestion by farm animals of heavy metals and other trace elements may be increased substantially as a result of the deposition of certain dusts on the aerial parts of plants. A few examples will illustrate this point. In a case of poisoning of cattle in an area adjacent to a zinc smelter in Norway, the grass available for grazing animals contained, on a dry-matter basis, 3100 ppm zinc, 700 ppm lead and 116 ppm copper compared with 38, 2 and 6 ppm of the elements, respectively, in grass unexposed to dust.[106] Recent studies[107] in Wales have shown large increases in the contents of zinc, lead, copper, cadmium and nickel in hay and grass grown downwind in an urban–industrial complex. These increases were attributed to air-borne dusts originating partly from contaminated wastes and soils and partly from present-day industrial activity. A horse which ingested the contaminated hay and grass suffered from lead toxicity. and examination *post-mortem* revealed that both the lead and cadmium contents of the kidney were almost certainly lethal. Lead in dusts has caused toxicity in animals in restricted areas elsewhere,[108,109] as also have nickel,[110] molybdenum,[111] vanadium and arsenic.[36]

The ingestion by animals of heavy metals and other trace elements in dusts on the leaves of plants warrants further investigation because it circumvents the soil-to-plant pathway in which there are barriers that restrict the movement of such elements in toxic amounts to the animal.[112–114]

2.2.2.4. Photochemical, or oxidant, smog

In contrast to London-type smog which contains smoke, fog and a high concentration of sulphur dioxide, the photochemical smog which is found in and around many cities in the US contains neither smoke nor fog and is characterized by a relatively high concentration of oxidants, chiefly ozone. This smog is often referred to as the Los Angeles type because it was first recognized there, in the 'forties, as a serious pollution problem by its effects on plants.[5,6] As a result of considerable research, it is now known that the photochemical smog complex results from reactions in the atmosphere between oxides of nitrogen and certain hydrocarbons, both derived from the incomplete combustion of liquid fuels by industry and motor vehicles. The phytotoxic products of photochemical reactions that have been identified to date are ozone, peroxyacyl nitrates (PAN's) and nitrogen dioxide.

It is generally thought that photochemical smog is unlikely to be important in Western Europe as an air pollutant or as a cause of injury to plants. However, there is evidence to suggest a need for vigilance in the future. First, injury to plants, resembling that caused by photochemical smog, has been observed near London and Paris.[5] Secondly, levels of oxidants and of ozone, recently measured in the Netherlands[115] and the UK,[116] have been found to exceed 0·1 ppm in the air. These compare with tolerance levels for plants of 215 $\mu g\ m^{-3}$ (0·1 ppm)[117] and 430 $\mu g\ m^{-3}$ (0·2 ppm)[77] ozone, and with levels in clean air of 43–65 $\mu g\ m^{-3}$ (0·02–0·03 ppm) ozone.

Effects on plants. The phytotoxicity of photochemical smog is very striking.[118] The characteristic lesions, which have been reproduced experimentally only by exposure to PAN, are a silvering or bronzing of the under surface of the leaf. The first symptom is a swelling of the guard cells giving a blistered appearance. This is soon followed by dehydration and shrinkage of these and of mesophyll cells surrounding the stomatal cavities. The death of the affected tissues produces the characteristic lesions. Recently expanded tissues of leaves are most susceptible to injury. Young leaves are more resistant because of their compact cells and poorly developed stomata, whilst older leaves are resistant, presumably because of the thickening of their cell walls.[118]

There is considerable variation among species in susceptibility to injury by photochemical smog, and several workers[5,51] have classified plants on this basis. Spinach, lucerne and oat are among the most susceptible species, radish, onion and potato are intermediate, whilst bean, carrot and brassica are among the most resistant.

Although ozone comprises up to 90 per cent of the total oxidants in photochemical smog[18] and its concentration may be as high as 2150 $\mu g\ m^{-3}$ (1·0 ppm) in polluted air, it causes symptoms of injury which are distinct from those described above. The upper surfaces of injured leaves show punctate lesions which are initially dark brown or black but later become milk-white.[20] Injury is at first restricted to the upper palisade cells just below the epidermis, and may extend to produce necrosis and sometimes a yellowing and shedding of leaves. Exposure to ozone has been shown to reduce photosynthesis, probably as the result of damage to the chloroplasts; it also increases respiration and lowers cell-wall permeability.[119,120] Although injury by cold weather may produce symptoms resembling those caused by ozone, the so-called 'weather fleck' of tobacco is now usually ascribed to injury by ozone[121] as also is 'tip-burn' of white pine.[122] Daines[123] lists 21 different species that are injured by ozone in New Jersey, including potato, spinach and cucumber.

In early investigations on photochemical smog it was concluded that the main role of nitrogen dioxide was as an initiator of photochemical reactions, and that its concentration in polluted air, commonly 410–1025 $\mu g\ m^{-3}$ (0·2–0·5 ppm), was too low to cause injury to plants.[6] However, recent investigations at the University of California at Riverside have shown that ambient concentrations of 615–1025 $\mu g\ m^{-3}$ (0·3–0·5 ppm) for 10 to 22 days caused reduced growth in bean and tomato,[124] and that prolonged exposures to concentrations exceeding 510 $\mu g\ m^{-3}$ (0·25 ppm) were damaging to navel oranges.[125] These observations indicate that the impact of photochemical smog on plants may sometimes be expressed in terms of reductions in growth and yield without obvious symptoms.

Ethylene, although not a photochemical product, is a component of the smog complex. It is produced with other hydrocarbons from the incomplete combustion of liquid fuels by industry and motor vehicles, and it is also a product of biosynthesis within plant tissues. Much of our knowledge about its effect on plants derives from investigations at the Boyce Thompson Institute of injury to plants in glasshouses caused by leakages of the ethylene-containing gas used for lighting.[2] Because ethylene is a physiologically active gas, it produces symptoms of injury at extremely low concentrations. These include early senescence, epinasty (distortion of leaves), abnormal directional effects on stem growth, reduced growth, and flower and fruit drop.[126,127] Flower drop has been observed with highly sensitive orchids at concentrations of ethylene as low as 2·5 $\mu g\ m^{-3}$ (0·002 ppm).[127]

Effects on animals. There is little information about the effects of photochemical smog upon farm animals, but veterinarians in Southern California occasionally attribute illness among animals to such air pollution.[34] The effects of several components of the smog complex have been studied with small animals. Jaffe,[128] in reviewing studies of ozone, suggested that exposure to concentrations of 108–430 $\mu g\ m^{-3}$ (0·05–0·2 ppm) may increase mortality in animals with respiratory infections, and cause irritation of the mucosa of the respiratory tract, whilst concentrations of 1290–1720 $\mu g\ m^{-3}$ (0·6–0·8 ppm) cause choking and coughing. The mortality rate of freshly hatched chicks increased considerably when they were exposed continuously to concentrations of 2·15–8·60 mg m^{-3} (1·0–4·0 ppm) ozone and for an LD_{50} exposure the time-concentration value was calculated as 54 ppm-hours.[129]

Nitrogen oxides of the photochemical smog complex have not been implicated in injury to farm animals. However, nitrogen dioxide, which is sometimes produced in silos during the fermentation of forage crops that are high in nitrate, has been found to be toxic to farm animals, especially cattle and swine.[36]

2.2.3. Rural industries and air quality

Most of the concern about air pollution and its effects on plants and animals has been directed toward pollutants resulting from industrial and domestic activities, and little attention has been given to the influences of agriculture, horticulture and forestry on air quality. These influences can be either adverse or beneficial and it is appropriate that this chapter concludes with a brief reference to some of them.

The use of pesticides and herbicides as dusts or sprays has become an essential feature of modern crop husbandry and forestry. Their effects are generally confined to relatively small areas containing the target organism, but they may drift a long way from their point of application and harm sensitive plant and animal life,[130] including crops[131,132] and occasionally farm animals.[133] The more persistent pesticides such as the chlorinated hydrocarbons, which have caused considerable public anxiety, are now widespread on the earth's surface, in the air and in rain.[134] They tend to accumulate in fatty tissues and to increase in concentration with each step in the food chain.[135] As a result, some wildlife species have been adversely affected, but there is no documented evidence that this build-up constitutes a threat to farm animals.[134]

Wind erosion of soils increases the dust content of the atmosphere. Although more common in dry climates, the blowing of soil particles can occur wherever there is incomplete vegetative cover and, inevitably, during tillage. The burning of many plant materials such as crop residues or stubble, and of forests themselves, contributes smoke, fine ash and oxides of nitrogen, as well as dust, to the air.

The beneficial influences of agriculture, horticulture and forestry on air quality far outweigh their adverse influences. Plants wherever they grow—in forests, in fields and in horticulture—remove carbon dioxide from the atmosphere and add oxygen, thus helping to maintain the balance between these two gases. In addition, trees can filter out particulate matter from moving air masses, thus improving the quality of the air near ground level.

It has already been noted that plants act as sinks for gaseous pollutants, but recent evidence indicates that soil must be considered as a major natural sink. With sulphur dioxide this role of plant and soil may sometimes be beneficial for plant growth.[136,137]

Although it is widely considered that carbon monoxide is almost entirely a man-made pollutant,[135] there is now evidence, from the Argonne National Laboratory, US, that natural processes, notably the oxidation of methane, contribute over 90 per cent of the total entering the atmosphere.[138] However, there is little, if any, accumulation on a world-wide basis, the ambient concentration usually being 126–630 $\mu g\ m^{-3}$ (0·1–0·5 ppm). Until recently the explanation for this was unknown. Both soil and plant have now been shown to have a remarkable ability to remove carbon monoxide from controlled, test atmospheres.[139,140] In soil this ability is positively correlated with the content of organic matter and is ascribed to the activity of micro-organisms. Extrapolating from the experimental results, it has been estimated that the total soil surface of the US has a capacity to absorb 570 million tonnes of carbon monoxide per annum, which is about 6·5 times the amount currently emitted as a result of man's activities in that country.[139] The absorption of carbon monoxide by plant leaves varies widely with species, but even in those of intermediate absorption capacity the rate, on a ground-area basis, approaches that found in soil.[140] Other experiments have shown that soils can remove ethylene by microbial degradation, and sulphur dioxide and nitrogen dioxide, predominantly by chemical mechanisms as yet not elucidated. Again it has been estimated that, on an annual basis, in the US the capacity of the soil is large enough to remove one-half of the ethylene and all of the sulphur dioxide and nitrogen dioxide released as air pollutants.[141]

It is apparent that agriculture, horticulture and forestry have a vital interest in maintaining air as a clean and healthy resource because they can suffer losses from air pollution. Scientists concerned with these rural industries have a responsibility for carrying out research that aims at alleviating injury to crop plants, trees and farm animals caused by air pollution, whatever its origin. Ultimately it may be more difficult to control emissions to the air from rural sources than from industrial and domestic activities.

References

1. Hasselhoff, E. and G. Lindau, *Die Beschadigung der Vegetation durch Rauch*, Bornträger, Leipzig, 1903.
2. Crocker, W., *Growth of plants*, Reinhold, New York, 1948.
3. Naegele, J. A., *Air pollution damage to vegetation*, Advances in Chemistry Series 122, American Chemical Society, Washington DC, 1973.
4. Thomas, M. D., 'Gas damage to plants', *Ann. Rev. Plant. Physiol.*, **2**, 293, 1951.
5. Thomas, M. D. and R. H. Hendricks, 'Effect of air pollution on plants', in P. L. Magill *et al.* (eds.), *Air pollution handbook*, Section 9, p. 1, McGraw-Hill, New York, 1956.
6. Thomas, M. D., 'Effects of air pollution on plants', in *Air pollution*, Monograph Series 46, 233, WHO, Geneva, 1961.
7. Garber, K., 'Recent literature on smoke, dust and waste-gas damage to plants, *Angewandte Botanik*, **36**, 127, 1962.
8. Webster, C. C., *The effects of air pollution on plants and soil*, Agricultural Research Council, London, 1967.
9. Brandt, C. S. and W. W. Heck, 'Effects of air pollutants on vegetation', in A. C. Stern (ed.), *Air pollution*, Vol. 2, p. 401, Academic Press, New York, 1968.
10. Linzon, S. N., 'Effects of air pollution on vegetation', in B. M. McCormac, (ed.), *Introduction to the scientific study of air pollution*, Reidel, Dordrecht, 1971.
11. Scurfield, G., 'Air pollution and tree growth', *Forestry Abstracts*, **21**, 339 and 517, 1960.
12. Hepting, G. H., 'Diseases of forest and tree crops caused by air pollutants', *Phytopathology*, **58**, 1098, 1968.
13. Keller, T., 'The effect of air pollution on forests—a review of some recent literature', *Schweizerische Seitschrift für Forstwesen*, **119**, 353, 1968.
14. Dochinger, L. S., *et al.*, 'Methods for the identification and evaluation of air pollutants injurious to forests' (Papers presented to Working Party on Fume Damage, Congress of International Union of Forestry Research Organization), *Mitteilungen aus der Forstlichen Bundesversuchsanstalt*, No. 92, 1971.
15. Commonwealth Forestry Bureau, *Atmospheric pollution*, Annotated bibliography, Commonwealth Forestry Bureau, Oxford, 1971.
16. Katz, M., 'Sulphur dioxide in the atmosphere and its relation to plant life', *Ind. Eng. Chem.*, **41**, 2450, 1949.
17. Guderian, R. and H. van Haut, 'Nachweis von Schwefeldioxid-Wirkungen an Pflanzen', *Staub*, **30**, 22, 1970.

18. Taylor, O. C., 'Effects of oxidant air pollutants', *J. Occup. Med.*, **10**, 485, 1968.
19. Dugger, W. M. and I. P. Ting, 'Air pollution oxidants—their effects on metabolic processes in plants'. *Ann. Rev. Plant Physiol.*, **21**, 215, 1970.
20. Rich, S., 'Ozone damage to plants', *Ann. Rev. Phytopath.*, **2**, 253, 1964.
21. McCune, D. C., *et al.*, 'A symposium—the technical significance of air quality standards. Fluoride criteria for vegetation reflect the diversity of plant kingdom', *Environmental Science and Technology*, **3**, 720, 1969.
22. Treshow, M., 'Fluorides as air pollutants affecting plants', *Ann. Rev. Phytopath.*, **9**, 21, 1971.
23. Daines, R. H., *et al.*, 'Air pollution and plant response in the northeastern United States', in N. C. Brady (ed.), *Agriculture and the quality of our environment*, p. 11, American Association for Advancement of Science, Washington DC, 1967.
24. Heggestad, H. E., 'Diseases of crops and ornamental plants incited by air pollutants', *Phytopathology*, **58**, 1089, 1968.
25. Daines, R. H., 'Air pollution and plant response', in J. E. Gunckel (ed.), *Current topics in plant science*, p. 436, Academic Press, New York, 1969.
26. Weinstein, L. H. and D. C. McCune, 'Implications of air pollution for plant life', *Proc. Amer. Phil. Soc.*, **114**, 18, 1970.
27. Garber, K., *Luftverunreinigung und ihre Wirkungen*, Borntraeger, Berlin, 1967.
28. Treshow, M., *Environment and plant response*, Part 4, p. 245, McGraw-Hill, New York, 1970.
29. Jacobson, J. S. and A. C. Hill, *Recognition of air pollution injury to vegetation; a pictorial atlas*, Air Pollution Control Association, Pittsburgh, 1970.
30. Haut, H. van and H. Stratmann, *Colour-plate atlas of the effects of sulphur dioxide on plants*, Girardet, Essen, 1970.
31. Perry, J., *Our polluted world; can man survive?*, Franklin Watts, New York, 1967.
32. Phillips, P. H., 'The effects of air pollutants on farm animals', in P. L. Magill *et al.* (eds.), *Air pollution handbook*, Section 8, McGraw-Hill, New York, 1956.
33. American Industrial Hygiene Association, 'Effects of air pollution on farm animals', in *Air pollution manual*, Part 1, p. 63, American Industrial Hygiene Association, Michigan, 1960.
34. Catcott, E. J., 'Effects of air pollution on animals', in *Air pollution*, Monograph Series 46, p. 221, WHO, Geneva, 1961.
35. Stockinger, H. E. and D. L. Coffin, 'Biological effects of air pollutants', in A. C. Stern (ed.) *Air Pollution*, Vol. 2, p. 445, Academic Press, New York, 1968.
36. Lillie, R. J., 'Air pollutants affecting the performance of domestic animals', *Agriculture Handbook 380, US Department of Agriculture*, Washington DC, 1970.
37. Roholm, K., *Fluorine intoxication*, Lewis, London, 1937.
38. Mitchell, H. H. and M. Edman, 'The fluorine problem in feeding livestock', *Nutrition Abstracts and Reviews*, **21**, 787, 1952.
39. Phillips, P. H., *et al.*, 'The fluorosis problem in livestock production' (Report of Committee on Animal Production, National Academy of Science), *National Research Council Publicn. No. 824*, Washington DC, 1960.

40. Suttie, J. W., 'Effects of inorganic fluorides on animals', *J. Air Pollution Control Assoc.*, **14**, 461, 1964.
41. Burns, K. N. and R. Allcroft, 'Fluorosis in cattle. Occurrence and effects in industrial areas of England and Wales, 1954–7', *Animal Disease Surveys, Report No. 2, Part I*, HMSO, London, 1964.
42. Allcroft, R., *et al.*, 'Fluorosis in cattle. Development and alleviation: experimental studies', *Animal Disease Surveys, Report No. 2, Part II*, HMSO, London, 1965.
43. Underwood, E. J., 'Fluorine', in *Trace elements in human and animal nutrition*', Chapter 13, Academic Press, New York, 1971.
44. Gäumann, E., Principles of plant infection, Translated W. B. Brierley, Crosby Lockwood, London, 1950.
45. Guderian, R., *et al.*, 'Probleme der Erfassung und Beurteilung von Wirkungen gasförmiger Luftverunreinigungen auf die Vegetation', *Zeitschrift für Pflanzenkrankheiten und Pflanzenschutz*, **67**, 257, 1960.
46. Zimmerman, P. W., 'Impurities in the air and their influence on plant life', *Proc. 1st Nat. Air Pollution Symp.* (Pasadena, California, 1949), 135, 1950.
47. Heggestad, H. E. and H. A. Menser, 'Leaf spot-sensitive tobacco strain Bel W.3, a biological indicator of the air pollutant ozone', *Phytopathology*, **52**, 735, 1962.
48. Engle, R. L., *et al.*, 'Tipburn, an ozone incited response in onion, *Allium cepa* L.', *Proc. Amer. Soc. of Hort. Sci.*, **86**, 468, 1965.
49. Valentine, H., 'A study of the effect of different ventilation rates on the ammonia concentration in the atmosphere of broiler houses', *British Poultry Sci.*, **5**, 149, 1964.
50. Menser, H. A. and H. E. Heggestad, 'Ozone and sulfur dioxide synergism: injury to tobacco plants', *Science*, **153**, 424, 1966.
51. Heck, W. W., 'Discussion of Dr. Taylor's paper', *J. Occup. Med.*, **10**, 496, 1968.
52. McKay, H. A. C., 'Ammonia and air pollution', *Chemistry and Industry*, **34**, 1162, 1969.
53. O'Gara, P. J., 'Sulphur dioxide and fume problems and their solutions', *Ind. Eng. Chem.*, **14**, 744, 1922.
54. Rosenholtz, M. J., *et al.*, 'A toxicopathological study in animals after brief single exposures to hydrogen fluoride', *J. Amer. Indust. Hyg. Assoc.*, **24**, 253, 1963.
55. Dugger, W. M., *et al.*, 'Stomatal action in plants as related to damage from photochemical oxidants', *Plant Physiol.*, **37**, 487, 1962.
56. Dugger, W. M., *et al.*, 'The effect of light on predisposing plants to ozone and PAN damage', *J. Air Pollution Control Assoc.*, **13**, 423, 1963.
57. Heck, W. W. and J. A. Dunning, 'The effects of ozone on tobacoo as conditioned by several ecological factors', *J. Air Pollution Control Assoc.*, **17**, 112, 1967.
58. Sorauer, P. and E. Ramann, 'Sogenannte unsichtbare Rauch beschädigungen', *Botanisches Zentralblatt*, **80**, 50, 1899.
59. Stoklasa, J., *Die Beschädigung der Vegetation durch Rauchgase und Fabriksexhalation*, Urban und Schwartzenburg, Berlin, 1923.
60. Thomas, M. D. and G. R. Hill, 'Relation of sulphur dioxide in the atmosphere to photosynthesis and respiration of alfalfa', *Plant Physiol.*, **12**, 309, 1937.

61. Matsushima, J. and M. Harada, 'Sulphur dioxide injury to fruit trees', *Bulletin of Faculty of Agriculture, Mie University, Tsu, Japan*, **30**, 11, 1964.
62. Middleton, J. T., *et al.*, 'Damage to vegetation from polluted atmospheres', *J. Air Pollution Control Assoc.*, **8**, 9, 1958.
63. Taylor, O. C., 'Air pollution with relation to agronomic crops. IV. Plant growth suppressed by exposure to air-borne oxidants (smog)', *Agronomy J.*, **50**, 556, 1958.
64. Darley, E. F. and J. T. Middleton, 'Problem of air pollution in plant pathology', *Ann. Rev. Phytopath.*, **4**, 103, 1966.
65. Holmes, J. A., *et al.*, 'Selby Report', *U.S. Bureau of Mines, Bulletin 98*, Washington DC, 1915.
66. Hill, G. R. and M. D. Thomas, 'Influence of leaf destruction by sulphur dioxide and by clipping on the yield of alfalfa', *Plant Physiol.*, **8**, 223, 1933.
67. Thomas, M. D. and G. R. Hill, 'Absorption of sulphur dioxide by alfalfa and its relation to leaf injury', *Plant Physiol.*, **10**, 291, 1935.
68. Katz, M., *et al.*, Effect of sulphur dioxide on vegetation, National Research Council of Canada, Ottawa, 1939.
69. Faller, N., *et al.*, 'Die Aufnahme von Schwefeldioxyd aus der Luft, I', *Plant and Soil*, **33**, 171, 1970.
70. Zahn, R., 'Wirkungen von Schwefeldioxyd auf die Vegetation, Ergebnisse aus Begasungsversuchen', *Staub*, **21**, 56, 1961.
71. Wentzel, K. F., 'Empfindlichkeit und Resistenzunterschiede der Pflanzen gegenüber Luftverunreinigung', *Proc. 1st European Congress on the influence of air pollution on plants and animals*, 357, Centre for Agricultural Publishing and Documentation, Wageningen, 1969.
72. Guderian, R. and H. Stratmann, 'Freilandversuche zur Ermittlung von Schwefeldioxidwirkungen auf die Vegetation, Teil I', *Forschungsberichte des Landes Nordrhein-Westfalen*, Nr. 1118, 1962.
73. Stratmann, H., 'Freilandversuche zur Ermittlung von Schwefeldioxidwirkungen auf die Vegetation, Teil II', *Forschungsberichte des Landes Nordrhein-Westfalen*, Nr. 1184, 1963.
74. Guderian, R. and H. Stratmann, 'Freilandversuche zur Ermittlung von Schwefeldioxidwirkungen auf die Vegetation, Teil III', *Forschungsberichte des Landes Nordrhein-Westfalen*, Nr. 1920, 1968.
75. National Air Pollution Control Association, 'Air quality criteria for sulfur oxides', *Publication No. AP-50, U.S. Department of Health, Education and Welfare*, Washington DC, 1969.
76. Craxford, S. R. and D. L. R. Bailey, 'The acidity of rain and the sulphur dioxide in the air in country districts in the U.K. and Ireland', *Report, Warren Spring Laboratory*, HMSO, London, 1970.
77. Eilers, H., 'Epilogue', *Proc. 1st European Congress on the influence of air pollution on plants and animals*, 401, Centre for Agricultural Publishing and Documentation, Wageningen, 1969.
78. Barkman, J. J., 'The influence of air pollution on bryophytes and lichens', *Proc.*

1st European Congress on the influence of air pollution on plants and animals, 197, Centre for Agricultural Publishing and Documentation, Wageningen, 1969.
79. Gilbert, O. L., 'The effect of SO_2 on lichens and bryophytes around Newcastle upon Tyne', *Proc. 1st European Congress on the influence of air pollution on plants and animals*, 233, Centre for Agricultural Publishing and Documentation, Wageningen, 1969.
80. Heck, W. W., 'The use of plants as indicators of air pollution', *Internat. J. Air and Water Pollution*, **10**, 99, 1966.
81. Zahn, R., 'Untersuchungen über die Bedeutung kontinuierlicher und intermittierender Schwefeldioxidwirkungen für die Pflanzenreaktion', *Staub*, **23**, 343, 1963.
82. Zahn, R., 'Über Schwefeldioxid-Wirkungsgrenzen für Pflanzen', *Staub*, **29**, 375, 1969.
83. Johannsen, O., 'On sulfur problems in Swedish agriculture', *Kungliga Lantbrukshogskolans Annaler*, **25**, 57, 1959.
84. O'Donoghue, J. G. and F. E. Graesser, 'Effects of sulphur dioxide on guinea pigs and swine', *Canad. J. Compar. Med. and Vet. Sci.*, **26**, 255, 1962.
85. Cunningham, O. C., *et al.*, 'Nutritive value for dairy cows of alfalfa hay injured by sulphur dioxide', *J. Agric. Res.*, **55**, 381, 1937.
86. Benson, N. R., 'Fluoride injury or soft suture and splitting of peaches', *Proc. Amer. Soc. Hortic. Sci.*, **74**, 184, 1959.
87. Hansen, E. D., *et al.*, 'Air pollution with relation to agronomic crops. VII Fluoride uptake from soils', *Agronomy J.*, **50**, 565, 1958.
88. Daines, R. H., *et al.*, 'The effect of fluorine on plants as determined by soil nutrition and fumigation studies', in L. D. McCabe (ed.), *Air pollution: Proceedings of United States Technical Conference on air pollution*' p. 97, McGraw-Hill, New York, 1952.
89. Jacobson, J. S., *et al.*, 'The accumulation of fluorine by plants', *J. Air Pollution Control Assoc.*, **16**, 412, 1966.
90. Leone, I. A., *et al.*, 'Atmospheric fluoride; its uptake and distribution in tomato and corn plants', *Plant Physiol.*, **31**, 329, 1956.
91. Tendron, G., 'Effects of air pollution on animals and plants', *Proc. European Conference on air pollution*, 25, Council of Europe, Strasbourg, 1964.
92. Buck, M., 'Untersuchungen über die Wirkung von Luftverunreinigungen auf Pflanzen', *Proc. 1st European Congress on the influence of air pollution on plants and animals*, 53, Centre for Agricultural Publishing and Documentation, Wageningen, 1969.
93. Robak, H., 'Aluminium plants and conifers in Norway', *Proc. 1st European Congress on the influence of air pollution on plants and animals*, 27, Centre for Agricultural Publishing and Documentation, Wageningen, 1969.
94. Bolay, A., *et al.*, 'Interaction between fertilization and the damaging effects of fluorides', *Proc. 1st European Congress on the influence of air pollution on plants and animals*, 143, Centre for Agricultural Publishing and Documentation, Wageningen, 1969.

95. Blakemore, F., *et al.*, 'Industrial fluorosis of farm animals in England, attributable to the manufacture of bricks, the calcining of ironstone, and to enamelling processes', *J. Compar. Path.*, **58**, 267, 1948.
96. Phillips, P. H. and J. W. Suttie, 'The significance of time in intoxication of domestic animals by fluoride', *Arch. Indust. Health*, **21**, 343, 1960.
97. Suttie, J. W., *et al.*, 'Studies of the effects of dietary sodium fluoride on dairy cows', *J. Nutrition*, **65**, 293, 1958.
98. Jackson, D. and S. M. Weidmann, 'Fluorine in human bone related to age and the water supply of different regions', *J. Path. and Bacter.*, **76**, 451, 1958.
99. Becker, D. E., *et al.*, 'The alleviation of fluorine toxicosis by means of certain aluminium compounds', *J. Animal Sci.*, **9**, 647, 1950.
100. Cohen, J. B. and A. G. Ruston, *Smoke: a study of town air*, Arnold, London, 1925.
101. Bleasdale, J. K. A., *Atmospheric pollution and plant growth*, Ph.D. Thesis, University of Manchester. 1952.
102. Scheffer, F., *et al.*, 'Untersuchungen über den Einfluss von Cementefen-flugstaub auf boden und pflanze', *Staub*, **21**, 251, 1961.
103. Czaja, A. T., 'Zementstaubwirkungen auf Pflanzen: Die enstehung der Zement Krusten', *Qualitas Plantarum et Materiae Vegetabiles*, **8**, 201, 1961.
104. Czaja, A. T., 'Über das Problem der Zementstaubwirkung auf Pflanzen', *Staub*, **22**, 228, 1962.
105. Schurmann, E., 'The effect of exhaust gas dust from cement works on the health of cattle', *Zement-Kalk-Gips*, **51**, 5, 1962.
106. Ender, F., 'The effect of air pollution on animals', *Proc. 1st European Congress on the influence of air pollution on plants and animals*, 245, Centre for Agricultural Publishing and Documentation, Wageningen, 1969.
107. Goodman, G. T. and T. M. Roberts, 'Plants and soils as indicators of metals in the air', *Nature*, **231**, 287, 1971.
108. Kradel, D. C., *et al.*, 'Lead poisoning and eosinophilic meningoencephalitis in cattle', *Vet. Med. and Small Animal Clinician*, **60**, 1045, 1965.
109. Rains, D. W., 'Lead accumulation by wild oats (*Avena fatua*) in a contaminated area', *Nature*, **233**, 210, 1971.
110. Ashton, W. M., 'Nickel pollution', *Nature*, **237**, 46, 1972.
111. Verwey, J. H. P., 'Industrial molybdenosis in grazing cattle', *Proc. 1st European Congress on the influence of air pollution on plants and animals*, 269, Centre for Agricultural Publishing and Documentation, Wageningen, 1969.
112. Allaway, W. H., 'Agronomic controls over the environmental cycling of trace elements', *Adv. Agron.*, **20**, 235, 1968.
113. Jones, L. H. P and C. R. Clement, 'Lead uptake by plants and its significance for animals', in P. Hepple (ed.), *Lead in the environment*, p. 29, Institute of Petroleum, London, 1972.
114. Jones, L. H. P., *et al.*, 'Lead uptake from solution by perennial ryegrass and its transport from roots to shoots', *Plant and Soil*, **38**, 403, 1973.

115. Galbally, I., *et al.*, 'Preliminary discussion on some oxidant measurements at Vlaardingen, The Netherlands', *Atmospheric Environment*, **5**, 187, 1971.
116. Atkins, D. H. F., *et al.*, 'Photochemical ozone and sulphuric acid aerosol formation in the atmosphere over southern England', *Nature*, **235**, 372, 1972.
117. Darley, E. F., 'The role of photochemical air pollution on vegetation', *Proc. 1st European Congress on the influence of air pollution on plants and animals*, 137, Centre for Agricultural Publishing and Documentation, Wageningen, 1969.
118. Bobrov, R. A., 'The leaf structure of *Poa annua* with observations on its smog sensitivity in Los Angeles County', *Amer. J. Bot.*, **42**, 467, 1955.
119. Dugger, W. M. and I. P. Ting, 'Physiological and biochemical effects of air pollution oxidants on plants', *Recent Adv. in Phytochem.*, **3**, 31, 1970.
120. Dugger, W. M., *et al.*, 'Physiological and biochemical effects of atmospheric oxidants on plants', *J. Air Pollution Control Assoc.*, **16**, 467, 1966.
121. Heggestad, H. E. and J. T. Middleton, 'Ozone in high concentration as a cause of tobacco leaf injury', *Science*, **129**, 208, 1959.
122. Berry, C. R. and L. A. Ripperton, 'Ozone, a possible cause of white pine emergence tip-burn', *Phytopathology*, **53**, 552, 1963.
123. Daines, R. H., 'Effects of air pollution on crops and livestock', *Proc. National Conference on Air Pollution*, 179, US Department of Health, Education and Welfare, Washington DC, 1963.
124. Taylor, O. C. and F. M. Eaton, 'Suppression of plant growth by nitrogen dioxide', *Plant Physiol.*, **41**, 132, 1966.
125. Thompson, C. R., *et al.*, 'Effects of continuous exposure of navel oranges to nitrogen dioxide', *Atmospheric Environment*, **4**, 349, 1970.
126. Knight, L. I. and W. Crocker, 'Toxicity of smoke', *Botanical Gazette*, **55**, 337, 1913.
127. Crocker, W., *et al.*, 'Ethylene induced epinasty and the relation of gravity to it', *Contributions from the Boyce Thompson Institute, New York*, **4**, 177, 1932.
128. Jaffe, L. S., 'The biological effects of ozone on man and animals', *J. Amer. Indust. Hyg. Assoc.*, **28**, 267, 1967.
129. Quilligan, J. J., *et al.*, 'The toxicity of ozone for young chicks', *Arch. Indust. Health*, **18**, 16, 1958.
130. Walker, K. C., 'Agricultural practices influencing air quality', in N. C. Brady (ed.), *Agriculture and the quality of our environment*, p. 105, American Association for Advancement of Science, Washington DC, 1967.
131. Middleton, J. T., 'Air—an essential resource for agriculture', in N. C. Brady (ed.), *Agriculture and the quality of our environment*, p. 3, American Association for Advancement of Science, Washington DC, 1967.
132. Mukammal, E. I., *et al.*, 'Air pollutants, meteorology and plant injury', *Technical Note 96*, World Meteorological Organization, Geneva, 1968.
133. Quinby, G. E. and G. M. Doornink, 'Tetraethyl pyrophosphate poisoning following airplane dusting', *J. Amer. Med. Assoc.*, **191**, 1, 1965.
134. Agricultural Research Council, *Third report of Research Council on toxic chemicals*', HMSO, London, 1970.

135. American Chemical Society, *Cleaning our environment: the chemical basis for action*, Washington DC, 1969.
136. Cowling, D. W. and L. H. P. Jones, 'A deficiency in soil sulfur supplies for perennial ryegrass in England', *Soil Science*, **110**, 346, 1970.
137. Jones, L. H. P., *et al.*, 'Plant-available and extractable sulfur in some soils of England and Wales', *Soil Science*, **114**, 104, 1972.
138. Anonymous, 'Isotopic study confirms CO sources', *Chemical and Engineering News*, **50**(27), 2, 1972.
139. Inman, R. E., *et al.*, 'Soil: a natural sink for carbon monoxide', *Science*, **172**, 1229, 1971.
140. Bidwell, R. G. S. and D. E. Fraser, 'Carbon monoxide uptake and metabolism by leaves', *Canad. J. Bot.*, **50**, 1435, 1972.
141. Abeles, F. B., *et al.*, 'Fate of air pollutants: removal of ethylene, sulfur dioxide and nitrogen dioxide by soil', *Science*, **173**, 914, 1971.

2 Effects of air pollution

Part 3 Air pollution and materials

ALBERT PARKER

The damaging effects of air pollution on materials were clearly shown in the UK before the passing of the Clean Air Act 1956[1] and before the limits of emission from industrial undertakings registered under the Alkali Works Etc. Regulation Act, 1906 and subsequent Alkali Orders were made more stringent to meet the recent requirements of the Alkali Inspectorate.

2.3.1. Soiling by grit, dirt and smoke

The grit and dust of the kind collected in gauges used for the measurement of deposited solid matter are deposited on the ground, on the roofs, window sills and other ledges of buildings and on other structures mostly within a few kilometres of the points of discharge into the air. The larger and heavier particles fall near the sources of emission and the smaller particles are carried farther before being deposited; the distances travelled are dependent on the heights of the points of discharge, the heights and other dimensions of buildings in the area, the local topography and the meteorological conditions including the velocity and turbulence of the wind. The smaller particles penetrate houses and other buildings through open windows, crevices and ventilators and are deposited on floors, walls and ceilings, furniture, curtains and other materials. These deposits increase the labour and cost of cleaning and decorations. They frequently contain sharp, gritty matter which causes scratches and other damage. If the grit and dust impinge on freshly painted surfaces they stick to and damage the paintwork and reduce the period before repainting is necessary.

The Clean Air Act 1956 greatly reduced the emission of smoke from industry and it arranged for the establishment of smoke control areas in which the burning of bituminous coal in domestic open grates is prohibited. Before this, the grit, dust and smoke in thickly populated industrial areas were heavily contaminated with sticky soot and tarry matter. Such particles are often damp or wet with atmospheric moisture or rain; they readily stick to surfaces with which they come into contact, and are not easily removed. When the particles are moist or wet they absorb sulphur oxide from the contaminated air and the oxide is converted to sulphurous and sulphuric acids.

As smoke consists of very fine particles of soot and tarry matter, some of it can be carried distances of 50 km or more before being deposited on buildings and other structures or being washed down by rain. Smoke from the burning of bituminous coal in the traditional open domestic fire has a greater detrimental effect than that from industrial chimneys, because it is discharged at lower levels and velocities and contains a much larger proportion of tarry matter. Many of the buildings in the Mall and Pall Mall in London are painted cream or pale buff. Observations before 1960 showed that when repainted, even before the paint had dried, the surfaces collected numerous black spots. Within two years the buildings had lost their exterior freshness and were distinctly soiled. A typical example of the soiling effect of the polluted air in thickly populated industrial areas is shown by the photograph (Fig. 2.3.1) of St George's Hall, Liverpool, that was taken during cleaning of the building in the 1960's. When deposition of soot, grime and rain on a structure is not uniform, because of projections or uneven exposure of surfaces to wind and rain, ugly effects can be produced as shown by the photograph

Fig. 2.3.1. St George's Hall, Liverpool, during cleaning. (Acknowledgement to National Society for Clean Air, Brighton, England.)

Fig. 2.3.2. Epstein's sculpture on London Transport Offices, Broadway, taken only a few years after installation. (Acknowledgement to National Society for Clean Air.)

(Fig. 2.3.2) of Epstein's sculpture on the London Transport offices in Broadway, Westminster, which has since been cleaned.

One common soiling effect of suspended dirt in the air is the marking of walls and ceilings in buildings heated by convection systems such as hot-water radiators and heated wall panels. This is caused by thermal precipitation. The warmed air travelling upwards carries the dirt particles, which are moved by differential molecular bombardment to cooler layers of air against the colder walls and ceilings on which the particles are deposited. Patterns of light and shade on a lath and plaster ceiling are due to preferential deposition of dust particles on those cooler parts of the ceiling not backed by beams or laths. Such soiling causes unsightly staining and the expense of more frequent decoration.

There are instances in which soiling by dirt particles is increased by electrostatic precipitation, for example the 'fog-marking' of nylon and acetate rayon fabrics during weaving. The exposed yarns in the loom become electrified by friction. Owing to their high insulating properties the yarns retain the electrical charge for some time after the loom is stopped, and attract from the air particles that are neutral or of opposite charge. The effect is shown by a streaky pattern of soiling on the fabric when weaving is continued after an overnight stoppage unless precautions are taken to prevent the deposition.

2.3.2. Damage by acids

Air that is not polluted by the products of combustion of fuels or by discharges from industrial processes contains carbon dioxide from the natural respiration of animals and plants. This gas dissolves to some extent in water to produce a solution of carbonic acid. Rainwater contains carbonic acid and is corrosive in character but much less corrosive than solutions of stronger acids such as sulphurous, sulphuric, hydrochloric and nitric acids. The proportion of carbon dioxide in the air of thickly populated industrial areas is often more than 1·5 times as great, especially at times of fog, as in air unpolluted by the combustion of fuels and by industry.

2.3.2.1. Stone and similar building materials

Building stones can be classified broadly into two main groups, those that consist of or contain significant amounts of calcium or magnesium carbonates and those that are practically free from carbonates. Limestones, ranging in hardness and density from chalk to marble, belong to the first group and are all susceptible to attack by acid in the atmosphere. Sandstones consist essentially of grains of sand cemented together to form a coherent mass, which may be fine or coarse in texture. In some sandstones the cementitious material is predominantly carbonate, which is attacked by acids; in others it is of acid-resistant material such as silica. Extreme varieties of sandstone in the UK are the soft calcareous stones from the North Downs (mainly in Kent) and the hard siliceous stone (Craigleith stone of the Edinburgh area). Granite and similar igneous rocks are generally resistant to acid atmospheres; they sometimes contain calcium carbonate in the form of calcite as an accessory mineral, but in such small amounts as to have little detrimental effect on their durability.

Rainfall containing carbonic acid slowly attacks limestone to give a solution of calcium bicarbonate. On evaporation by exposure to the air, the solution loses some carbon dioxide and deposits calcium carbonate. It is by the same process that stalactites and stalagmites are formed in caves. It is clear, therefore, that even in the absence of pollution from an industrial area there would be some attack on stone of the limestone type in damp climates.

The action of oxides of sulphur in polluted air in the presence of moisture is to convert the calcium and magnesium carbonates of stone to calcium and magnesium sulphates. Calcium sulphate is slightly soluble in water and magnesium sulphate is very soluble. On drying the solutions, the calcium and magnesium sulphates form hydrated crystals of greater volume than the original carbonates, with the result that incrustations and friable scale are formed to produce effects known as blistering, flaking and exfoliation. The presence of grit and dust seems to accelerate the corrosion. It may be that these particles encourage oxidation of the sulphur dioxide and sulphurous acid to the stronger sulphuric acid. The photograph (Fig. 2.3.3) shows the exfoliation of Headington stone (a limestone formerly much used for buildings in Oxford) as a result partly of attack by carbonic acid and mainly of attack by sulphur acids. During repairs in 1922 to the Victoria Tower of the Houses of Parliament in London it was found that some of the stone and material of fixing had been so badly damaged by acid that the stone fell away when touched.

One example of the destructive action of a damp, acid atmosphere on stone is the deterioration in the condition of the Cleopatra's Needle that was moved in 1878 from

Fig. 2.3.3. Exfoliation of a soft limestone (Headington) of an Oxford College. (Acknowledgement to Building Research Station, Watford, England.)

Alexandria to the Victoria Embankment in London. That obelisk was originally erected at Heliopolis, then re-erected in Alexandria by Rameses II in the thirteenth century BC. Though the age of the obelisk is at least 3000 years, it has suffered more deterioration in about 90 years in London than in the whole of its previous exposure in Egypt.

Slates containing carbonates, and the calcareous sandstones often used as roofing materials are attacked by polluted, acid atmospheres. Damage usually occurs mainly on the undersides, especially between the laps where water is held for some time as a thin film. Though brickwork can soon become badly discoloured by soot and grime, most types are fairly resistant to damage by sulphur oxides in the air, unless the bricks absorb sulphates produced by the action of the oxides of sulphur on adjacent mortar and limestone.

2.3.2.2. Iron, steel and other metals

A considerable amount of work over a long period of years has been undertaken by many organizations and individuals on the mechanism of corrosion of metals and on the factors, including the pollutants in the atmosphere, affecting the rate of corrosion. In relation to air pollution, the factor of greatest importance is the extent of pollution by oxides of sulphur, especially in damp climates. In one series of experiments, for example, specimens of a copper-steel and of zinc were exposed to the air at sixteen places in Great Britain, ranging in character from a relatively unpolluted country area to a heavily polluted industrial town. The extent of pollution by oxides of sulphur at each site was measured over a period of one year. It was found that the amount of corrosion of the metals increased considerably with increase in the extent of air pollution by sulphur oxides. With the copper-steel, the corrosion was more than four times as great; and with zinc it was about ten times as great in the highly polluted area as in the relatively unpolluted country area. Investigations with other corrodible metals have given similar results. It has been found that the rails of railway tracks wear more rapidly for the same traffic in industrial than in rural atmospheres and more rapidly in tunnels traversed by steam locomotives than in tunnels of electrified lines.

In addition to the loss of metal and damage to metallic structures, the corrosion can have serious effects on associated materials. The corrosion products have volumes greater than those of the original metals, so that intense local pressures can be produced in adjacent rigidly fixed materials. Serious disintegration of stone, caused largely by the expansion of iron by corrosion, is shown by the photograph (Fig. 2.3.4) taken in 1921 of a pinnacle on one of the towers of the Houses of Parliament. The iron tie rod to secure the stonework of the finial and weather vane above had corroded and expanded.

2.3.2.3. Leather

It has long been known that oxides of sulphur in the air have a destructive effect on leather. In 1843 Faraday, as a result of investigation, concluded that the rotting of the

Fig. 2.3.4. Photograph taken in 1921 of a pinnacle on one of the towers of the Houses of Parliament, London, during repairs. An iron tie rod had corroded and expanded and thus assisted in disintegrating the stone. (Acknowledgement to *Memorandum on the Defective Condition of the Stonework of the Houses of Parliament*, HMSO, London, 1926.)

leather upholstery of the chairs in the Athenaeum Club in London was largely caused by the sulphur compounds in the air. This general conclusion has been confirmed, particularly with vegetable-tanned leather, by more recent investigations by several authorities. The deterioration can in time be so great that the leather can be reduced to a powder by gentle rubbing. The leather bindings of books in libraries in relatively unpolluted atmospheres remain in reasonably good condition much longer than when similarly exposed in libraries in thickly populated industrial areas. In polluted atmospheres, the rate of deterioration of leather bindings of books on open shelves is greater in rooms through which there is a high circulation of air from outside than in rooms in which the rate of air circulation is small. Deterioration is more rapid on open shelves than in glass-fronted book cases. This is shown by the photograph (Fig. 2.3.5) of the backs of similarly bound books which had been kept on open shelves or behind glass doors for a number of years. The sides of books kept in closely packed rows so that air circulation is restricted remain in good condition for a much longer period than the exposed backs of the books.

Fig. 2.3.5. Leather backs of books after being kept (a) behind glass doors and (b) on open shelves for a considerable period of time. (Acknowledgement to British Leather Manufacturers' Research Association, Milton Park, Egham, Surrey, England.)

Leather safety belts exposed to polluted air can be so attacked and weakened that after some years they break under a light load and are useless for the purpose for which they were made.

There is no doubt that the attack of polluted air on leather is due to the oxides of sulphur, which are absorbed by the leather to sulphates. Analyses of leather that has rotted under such conditions have on occasions shown amounts of sulphate as high as 6 per cent of the weight of the leather.

2.3.2.4. Paper

There are many documents, books and other records of value that should be preserved, but many of these records are gradually being destroyed by polluted air. As with leather, paper absorbs oxides of sulphur, which are converted to sulphuric acid and this acid discolours the paper and renders it brittle and fragile. The rate of production of acid depends to some extent on the nature and amount of metallic salts in the paper as some of the salts catalyse the chemical reactions involved. As would be expected, the edges of the paper suffer greater damage than the middle parts of the pages. Analyses of paper from old books have shown amounts of sulphate in the edges ranging up to 1·5 per cent with differences between the amounts in the edges and the centre of a page sometimes approaching 1 per cent.

2.3.2.5. Textiles

Though there is a lack of precise information on the damage to textiles by acid oxides of sulphur in the air, it is certain that curtains and other fabrics exposed to frequent changes in atmospheric humidity become fragile and rotted more quickly in polluted than in unpolluted air. As with leather and paper, they absorb sulphur oxides, particularly in humid air, and the oxides are converted to sulphuric acid. The damage is accelerated by dust and smoke particles deposited on the fabric. There were many complaints of an unusual amount of damage to nylon hose during the London smog of December, 1952. It has also been said that 'runs' in nylon have been started by the exhaust gases from diesel-driven buses on humid days, probably by the sulphur oxide in the gases. The finer low-denier stockings are more susceptible to this damage than those of the coarser higher-denier thread.

2.3.2.6. Other materials

Valuable pictures can be damaged by exposure in the polluted atmospheres of many areas. The damage may include deterioration of the canvas, paper and paint. In the air of many districts there are minute quantities of hydrogen sulphide, which is emitted when coal is heated and is not completely burned; this gas is also evolved from highly polluted rivers and streams as a result of biological decomposition. Hydrogen sulphide blackens paints with a lead base by converting the lead compound to black lead sulphide; it also tarnishes silver by forming black silver sulphide.

Acknowledgements

In preparing the section on the effects of air pollution on materials, the author has drawn on the work of the former Fuel Research Organization and of the Chemical Research Laboratory of the former HM Department of Scientific and Industrial Research. He has also used the work of many other organizations and individuals, including particularly the Building Research Station of the Department of the Environment, the British Iron and Steel Research Association, the Research Department of the British Transport Commission, the British Leather Manufacturer's Research Association and the Cotton, Silk and Man-Made Fibres Research Association.

Reference

1. Parker, A., 'The destructive effect of air pollution on materials', National Smoke Abatement Society, now National Society for Clean Air, Brighton, England, 1955.

3 Legislative control

Air pollution

ALBERT PARKER and PAMELA M. BRYANT

There has been more uniform control of air pollution for the whole of the UK from the use of fuels and from industrial processes for a longer period than for the whole of any other country. This is not surprising, as Great Britain was the first to begin extensive development of industry based on coal as the source of heat and power. In this chapter, consideration will be given first to legislation aimed at the mitigation of pollution by smoke, grit and dust and oxides of sulphur from the use of fuels, and then to the limitation of pollution from certain specialized industrial processes.

Part 1 UK legislation: smoke, grit, dust and fume emission

ALBERT PARKER

Smoke emission

In chapter 1, there are references to complaints of serious pollution by smoke from the use of coal for heating in parts of England as early as the thirteenth, fourteenth and sixteenth centuries. In 1273, 1306 and during the reign of Queen Elizabeth I (1558–1603) there were periods when the use of coal in London was prohibited. In 1648, Londoners petitioned Parliament to prohibit the import of coal from Newcastle upon Tyne, whence it was brought by sea to London. During the seventeenth, eighteenth and nineteenth centuries and the first half of the twentieth century, there were various authoritative documents, exhibitions and other activities by enthusiasts stressing the need for legislation or improved legislation for a reduction of pollution by smoke, grit and dust and by various discharges from industrial processes. A list of the documents, including the reports of several committees appointed by Parliament, items of legislation, and exhibitions is given in chronological order in the *Clean Air Year Book* for 1975 issued by the National Society for Clean Air.[1] This list is kept up to date with each new issue of the *Clean Air Year Book*. These books also give in some detail the laws in the UK relating to air pollution as they are applied.

Legislation in the UK before 1956 in relation to reduction of pollution by smoke included the following:

1. Railway Clauses Consolidated Act 1845, requiring railway engines to consume their own smoke;
2. Town Improvement Clauses Act 1847, containing a section dealing with smoke from factories;
3. Smoke Abatement Acts 1853–6, for the London Metropolitan Area;
4. Sanitary Act 1866, which gave Sanitary Authorities the power to take action against smoke nuisances;
5. Public Health Act 1875, containing a section on smoke abatement;
6. Public Health (London) Act 1891;
7. Public Health (Smoke Abatement) Act 1926, in which the Acts of 1875 and 1891 were amended and extended;
8. Public Health Act 1936, including the provisions of the Public Health (Smoke Abatement) Act 1926.

None of these Acts was particularly effective in appreciably reducing pollution of the air by smoke, partly because many industrialists considered that prevention of smoke emission to any great extent was impracticable and partly because many householders wished to continue heating their houses by coal in traditional open grates, though they gave rise to considerable quantities of smoke. As a result of the activities of certain enthusiasts, however, under a Manchester Corporation Act 1946, that city had arrangements for a smokeless zone. The first smokeless zone was established in Coventry in 1951 and in 1955 the whole of the City of London was declared a smokeless zone. It was the great smog in London and other cities in 1952 that really stirred the people and Parliament to intensive action greatly to reduce smoke pollution. Following the smog in December 1952, the Government in July 1953 appointed a Committee on Air Pollution with the late Sir Hugh Beaver as chairman. That Committee worked diligently and fast and issued an interim report[2] in November 1953 and a final report[3] in November 1954. The reports of that committee led to the Clean Air Act 1956.

The main legislative enactments now in operation in the UK limiting the emissions of smoke are the Clean Air Acts of 1956 and 1968, combined with certain Regulations. In the Act of 1956 and the Regulations 'dark' smoke is defined as smoke that is as dark as or darker than shade 2 of the British Standard Ringelmann chart and 'black' smoke means smoke that is as dark as or darker than shade 4 of the Ringelmann chart. Descriptions and illustrations of the main and miniature Ringelmann charts and the use of the charts are given in chapter 4 of this book. The boiler furnace systems using coal or oil as fuel at numerous industrial undertakings in the UK are now equipped with devices that measure and indicate or record the density of the smoke in the gases rising through the chimney. Most of these devices include an alarm system that rings a loud bell when the density of the smoke reaches a value, generally set at the equivalent of between shades 1 and 2 of the Ringelmann chart.

Under the Clean Air Act, 1956 all new furnaces, except domestic boilers rated at less than 55,000 Btu h^{-1}(13·86 Mcal h^{-1} or 16·1 kW h^{-1}) are required to be capable of operation, so far as practicable, without emitting smoke when burning the type of fuel for which the furnace was designed. It is also required that the proposed installation of new furnaces (other than domestic boilers rated at less than 55,000 Btu h^{-1}) must be notified to the local authority. This notification, even if the local authority approve the erection of the proposed furnace, does not allow the discharge of smoke greater than the permissible emissions.

Allowable emissions of smoke from furnaces on land are included in the Dark Smoke (Permitted Periods) Regulations 1958 (Statutory Instrument 1958 No. 498). They are given in Table 3.1. If one boiler or unit of industrial plant is fired by more than one furnace all discharging into the same chimney, those furnaces are deemed to be one furnace. The continuous emission of dark smoke for a period greater than 4 min, caused otherwise than by soot blowing, is prohibited, and no emission of black smoke exceeding an aggregate of 2 min is allowed in any period of 30 min. The penalty for an offence under the Act and the Regulations is a fine not exceeding £100 for each day on which the offence occurs; it may be that the maximum of £100 for each day will later be increased.

Under Section 20 of the 1956 Act and the Dark Smoke (Permitted Periods) (Vessels) Regulations 1958 (Statutory Instrument 1958 No. 878), vessels in waters not navigable by sea-going ships and in certain waters within the seaward limits of the territorial waters of the UK must comply with the dark smoke provisions in Section 1 of the 1956 Act and with the Regulations in the Statutory Instrument 1958 No. 878. The permitted periods for the emission of dark smoke are given in Table 3.2.

A vessel is not under way when it is at anchor or made fast to the shore or bottom, but a vessel which is aground is deemed to be under way. There are provisos in the Regulations that the continuous emission of dark smoke caused otherwise than by soot

TABLE 3.1.

Dark Smoke (Permitted Periods) Regulations 1958

Number of furnaces served by the chimney	Permitted emission of dark smoke (Ringelmann 2) in any period of 8 h (min)	
	If soot not blowing during period	If soot blowing during period
1	10	14
2	18	25
3	24	34
4 or more	29	41

TABLE 3.2

Dark smoke (permitted periods) for vessels

Class of case	Permitted periods
1. Forced draught oil-fired boiler furnace, or an oil engine	10 min aggregate in any period of 2 h
2. Natural draught oil-fired boiler furnaces (except cases within class 4)	10 min aggregate in any period of 1 h
3. Coal-fired boiler furnace:	
(a) when vessel is not under way (except cases within class 4)	10 min aggregate in any period of 1 h
(b) when vessel is under way	20 min aggregate in any period of 1 h
4. Natural draught oil-fired boiler furnace or a coal-fired boiler furnace:	
(a) a vessel with funnels shortened for navigating the Manchester Ship Canal	20 min aggregate in any period of 1 h
(b) a tug not under way but preparing to get under way or supplying power to other vessels or to shore installations	
(c) a vessel not under way but using main power for dredging, lifting, pumping or other special operation for which the vessel is designed	
5. Emissions from any other source	5 min aggregate in any period of 1 h

blowing a water tube boiler must not exceed 4 min for classes 1 and 2, 10 min for natural draught oil-fired boiler furnaces in class 4 and in no case shall black smoke be emitted for an aggregate of 3 min in any period of 30 min.

Under Section 1 of the 1956 Act, there are the following defences that may be available if an occupier of premises is prosecuted for excessive dark or black smoke emission. They are as follows:

1. Emissions were solely due to lighting a furnace from cold and all practicable steps had been taken to minimize emissions;
2. Emissions were solely due to unavoidable failure of part of the plant and the contravention could not have been prevented after failure occurred;
3. Emissions were solely due to unavoidable use of unsuitable fuel, suitable fuel not being available, the best available fuel was used and all practicable steps were taken to minimize emissions;
4. Emissions were due to any combination of (a), (b) and (c).

These defences are only available if every practicable effort was made to avoid or minimize the emissions.

Dark smoke from bonfires, etc. Section 1 of the Clean Air Act 1968 prohibits, subject to certain exemptions, the emission of dark smoke from industrial or trade premises (as distinct from chimneys) by the indiscriminate burning of industrial or trade refuse in the open; it also controls smoke emissions from industrial processes not having a chimney. Industrial or trade premises includes not only premises normally used for industrial or trade purposes but also premises not normally used, but which at the time of the offence were being used for burning industrial or trade waste. Under the Clean Air (Emission of Dark Smoke) (Exemption) Regulations 1969 the materials exempted, subject to certain conditions, are those given in Table 3.3.

TABLE 3.3

Exempted matter under the Clean Air (Emission of Dark Smoke) (Exemption) Regulations 1969

Matter	Conditions
1. Timber and any other waste matter (other than natural or synthetic rubber or flock or feathers) which result from demolition of a building or clearance of a site for any building operation or work of engineering construction (within the meaning of section 176 of the Factories Act 1961)	A, B and C
2. Explosive (within the meaning of the Explosives Act 1875) which has become waste; and matter which has been contaminated by such explosive	A and C
3. Matter which is burnt in connection with: (a) research into cause or control of fire (b) training in fire fighting	C
4. Tar, pitch, asphalt and other matter which is burnt for the preparation and laying of any surface, or is burnt off, any surface in connection with resurfacing, together with any fuel used for any such purpose	C
5. Carcases of animals or poultry which: (a) have died, or are reasonably believed to have died because of disease; (b) have been slaughtered because of disease; or (c) have had to be slaughtered pursuant to the Diseases of Animals Act 1950	Conditions A and C unless the burning is done by or on behalf of an inspector (within the meaning of Section 84 of the Diseases of Animals Act 1950)
6. Containers which are contaminated by any pesticide or any toxic substance used for veterinary or agricultural purposes	A, B and C

Conditions

A. That there is no other reasonably safe and practicable method of disposing of the matter
B. That the burning is done in such manner as to minimize the emission of dark smoke
C. That the burning is carried out under the direct and continuous supervision of the occupier of the premises concerned or a person authorized to act on his behalf.

Smoke control areas. Sections 11 to 15 of the Clean Air Act 1956, Sections 8 to 10 of the 1968 Act and Section 95 of the Housing Act 1964 refer to smoke control areas, the administrative procedures and the enforcement of smoke control orders. Under Section 11 of the 1956 Act a local authority is allowed to make a smoke control order, which must be submitted to the Secretary of State for the Environment for confirmation. When the order is operative it is an offence for an occupier of premises in the area to allow smoke emission from a chimney, unless the smoke is caused by the use of an authorized fuel or the fireplace which the chimney serves is exempt from the order.

The various Clean Air (Authorized Fuel) Regulations issued during the years 1956 to 1971 now allow the use of the following fuels in smoke control areas: anthracite and low volatile steam coal, a named list of 15 manufactured solid smokeless fuels (though some of these solid fuels may cause light smoke emission for short periods) and gas and electricity. Other solid fuels, when they are introduced, may be submitted to the Department of the Environment for testing to ascertain whether or not they are suitable to be included in the list of authorized fuels for use in smoke control areas.

The Smoke Control Areas (Exempted Fireplaces) orders issued in 1970, 1971 and 1972 conditionally exempt the emission of smoke from chimneys serving certain fireplaces. The list given in Table 3.4 is a rearranged summary of the more detailed list given in the 1975 *Year Book* of the National Society for Clean Air.[1]

The 1968 Act extended the smoke control provisions of the 1956 Act and the Minister now has the power to require a local authority to make smoke control orders. It is an offence to acquire or sell by retail for delivery in a smoke control area any fuel

TABLE 3.4

Fireplaces from whose chimney emission of smoke is conditionally exempt under orders issued in 1970–72

Class of fireplace	Conditions
1. Any fireplace designed or adapted for combustion of liquid fuel	No fuel shall be used other than that for which the mechanical stoker was designed (applies to classes 1–3)
2. Any fireplace (other than one fired by pulverized fuel) constructed on or after 31 December 1956 and installed before 1 May 1970 and equipped with mechanical stokers or adapted between those dates for use with such stokers	
3. Any fireplace designed to burn coal (other than a fireplace fired by pulverized coal) with a heating capacity exceeding 150 000 Btu h^{-1} (37·8 Mcal h^{-1}) constructed and installed on or after 31 December 1965 and equipped with mechanical stokers or adapted on or after that date for use with such stokers	
4. The fireplace known as the Solid Fuel Ductair Unit	
5. Six named domestic heating appliances	Only selected washed coal under the name Housewarm by agreement with the National Coal Board shall be used
6. Three fireplaces for burning wood waste; the fireplaces are named	Only wood waste in a clean condition to be used and in one case waste of a size stated by the makers of the appliance

other than an authorized fuel, unless in each case the premises or fireplace are exempt. The financial provisions of the 1956 Act require the local authority to pay to owners or occupiers of dwelling houses a grant of seven-tenths of the reasonable costs of fireplace adaptions and they may repay all or any part of the remaining three-tenths. The local authority recovers four-sevenths of the grant from the Exchequer. A grant is not available in the case of a new dwelling the building of which was begun on or after 16 August 1964. There are maximum amounts for the grant towards the cost of each type of a number of appliances to enable householders in a smoke control area to install appliances for the use of authorized or smokeless fuels. A list of the appliances with the maximum costs on which the grants are based are given in Cost Lists for Various Appliances (*Circular 135/73*) reproduced in the *Clean Air Year Book* for 1975 of the National Society for Clean Air.[1] These costs are under revision by the Department of the Environment.

In addition to the procedure that can be taken against offences for excessive smoke emission and smoke emission in a smoke control area under the 1956 and 1968 Acts, it is possible to proceed under the nuisance provisions of the Public Health Act 1936 as extended by the Clean Air Act 1956 to include excessive smoke as a nuisance; but action under the Public Health Act is a cumbersome and lengthy procedure that is rarely used for smoke nuisances.

Control of chimney heights

Under Section 6 of the Clean Air Act 1968, the local authority must consider chimney height when

1. a new chimney is to be erected to serve a new or existing furnace or group of furnaces;
2. the capacity of an existing furnace or group of furnaces served by an existing chimney is to be enlarged; or
3. a furnace served by an existing chimney is replaced by a larger furnace, provided that in all three cases the furnace or group of furnaces is to burn
   (a) pulverized fuel or
   (b) any other solid matter at a rate of 100 lb h^{-1} (45·359 kg h^{-1}) or more, or
   (c) any liquid or gaseous matter at a rate equivalent to 1·25 m Btu h^{-1}.

The local authority considering any application that must be made must give a written decision within 28 days unless it is agreed in writing by both parties that a longer period may be allowed. The local authority must be satisfied that the proposed or agreed chimney height will be sufficient to prevent, so far as practicable, the smoke, grit, dust, gases or fumes emitted from becoming prejudicial to health or a nuisance, after taking into account the purpose of the chimney, the position and description of nearby buildings, the levels of neighbouring ground and any other matters requiring consideration in the circumstances. If the applicant is not satisfied with the decision of the local authority, he may within 28 days appeal against the decision to the Secretary of State for the Environment or the Secretary of State for Wales or Scotland as appropriate. The Secretary of State must give his decision in writing and his reasons for the decision. In the practical determination of the heights of chimneys for furnaces burning solid, liquid or gaseous fuels, account must be taken of the quantities of smoke, grit, dust, fume and gaseous substances, particularly oxides of sulphur, that the gases leaving the chimney may contain.

To assist local authorities, applicants and others in reaching decisions on appropriate chimney heights, the then Ministry of Housing and Local Government issued a *Memorandum on Chimney Heights* in 1963, which was revised in 1967. In that *Memorandum*, which includes charts in the form of nomograms, the factors taken into account are the mass emission of sulphur dioxide in the gases discharged and the type of locality. Five types of locality are considered:

1. An undeveloped area where background pollution is low and there is no development within half a mile (0·8 kilometres) of the new chimney;
2. A partially developed area with scattered houses, low background pollution and no other comparable industrial emissions within half a mile of the new chimney;
3. A built-up residential area with only moderate background pollution and without other comparable industrial emissions;
4. An urban area of mixed industrial and residential development, considerable background pollution and other comparable industrial emissions within a quarter of a mile of the new chimney, and
5. A large city or an urban area of mixed heavy industrial and dense residential development and severe background pollution.

Category (3) represents average conditions. A chimney in areas (1) or (2) need not be so high as in area (3) and chimneys in areas (4) and (5) need to be higher than in area (3) with similar topography.

From the sulphur content of the fuel used, the mass emission of sulphur dioxide per hour with the boiler or plant operated at its maximum continuous rating can readily be calculated for use with the charts. When coal or coke is the fuel allowance is made for the fact that on average 10 per cent of the sulphur remains in the ash. It is not always known what is the precise sulphur content. When it is not known, it is usual to assume for British coal that it is 1·6 per cent. For oils as fuel, the specification maximum for the particular type of oil is used in the calculation of the quantity of sulphur dioxide in the gases discharged.

On the basis of the calculated amount of sulphur dioxide discharged per hour the so-called uncorrected height of the chimney required can be obtained from one or other of the charts for emissions of sulphur dioxide ranging from 3 to 1800 lb h^{-1} (1·36 to 816 kg h^{-1}) and for each of the five types of area. The uncorrected height can then be adjusted to what is termed the final chimney height from charts that take into account the type of area and the height and length of the building and of nearby buildings. In some instances there should be further adjustments to the chimney height to allow for the effects of the local topography if there are appreciable changes in the height of the surrounding area.

While the physical height and effective dispersion height, including rise of plume, are important in relation to dispersion of the gases in the air, there are also other factors of importance. These include the shape of the chimney, adequate thermal insulation; and a top outlet sectional area capable of allowing a sufficient exit velocity of the gases discharged, if downflow of gases and air in the top of the chimney, acid condensation with smut emission and downwash on the lee side of the chimney are to be avoided.

The Clean Air (Heights of Chimneys) (Exemption) Regulations 1969 (Statutory Instrument 1969 No. 411) made by the Secretary of State provide that approval of chimney height is not required where a boiler or plant is used

1. as temporary replacement for a boiler or plant under inspection, maintenance, repair, rebuilding or being replaced by another permanent boiler or plant,
2. as a temporary source of heat or power for work of building or engineering construction,
3. as an auxiliary plant to bring other plant to an operating temperature, or
4. as a mobile or transportable source of heat or power for agricultural operations.

The Regulations, however, do not provide exemption from other relevant provisions of the Clean Air Acts and the Public Health Act 1936.

Control of grit, dust and fume emissions

In the Clean Air (Emission of Grit and Dust from Furnaces) Regulations 1971 grit is defined as particles exceeding 76 μm in diameter, and in the 1968 Act fume is defined as solid particulate matter smaller than dust, but there is no definition of dust. It is generally accepted, however, that dust is particulate matter of size between 1 and 76 μg, and fume consists of particles smaller than 1 μg.

The Clean Air (Emission of Grit and Dust from Furnaces) Regulations 1971 (Statutory Instrument 1971 No. 162) came into operation on 1 November 1971. They apply to

1. boilers,
2. indirect heating appliances in which the combustion gases are not in contact with the material being heated, and
3. furnaces in which the combustion gases are in contact with the material being heated but the material does not in itself contribute to the grit and dust in the combustion gases.

Until 1 January 1978, the Regulations apply only to plant for which an agreement for purchase or installation was entered into after 1 November 1971. There is thus an exemption for seven years for installations in existence on 1 November 1971. The

Regulations do not apply to incinerators burning refuse or waste matter, whether or not the resulting heat is used for any purpose.

Schedules 1 and 2 of the Regulations tabulate the quantities of grit and dust that may be emitted, and where the rating of the boiler or furnace is intermediate between tabulated values, the maximum permitted emission is obtained by interpolation.

Schedule 1 applies to furnaces rated by heat output and includes boilers and industrial heating appliances in which the material being heated is a gas or a liquid, e.g., an air heater. The range of heat outputs tabulated includes 23 items from 825 lb h^{-1} (373 kg h^{-1}) of steam, calculated at from and at 100°C or in thousands of Btu h^{-1} (252 kcal h^{-1}) to 475 000 lb h^{-1} of steam (252 kcal h^{-1}) to 475 000 lb h^{-1} of steam (215 tonnes h^{-1} or 215·5 Mcal h^{-1}). Schedule 1 gives the maximum permitted emissions of grit and dust in lb h^{-1} for each of the rates of flow of steam at maximum continuous rating.[1] Some of the items in the Schedule (with calculated equivalents in metric units) are given in Table 3.5.

Schedule 2 applies to indirect heating appliances and to furnaces in which the combustion gases are in contact with the material to be heated, but the material does not in itself contribute to the grit and dust in the gases discharged. It gives the maximum permitted emissions of grit and dust in 19 items for furnaces rated by heat inputs ranging from 1·25 million to 575 million Btu h^{-1} (0·315 to 145 Gcal h^{-1}). A selection of the items in the schedule is given in Table 3.6.

For Schedule 1 furnaces with maximum continuous rating not exceeding 16 800 lb of steam h^{-1} or 16·8 MBtu h^{-1} (7·62 tonnes h^{-1} or 4·23 Mcal h^{-1}) and for Schedule 2 furnaces with a designed input not exceeding 25 MBtu h^{-1} (6·3 Mcal h^{-1}) the proportion of grit in a sample of grit and dust must not exceed 33 per cent. In other cases, which are of higher ratings, the amount of grit in the grit and dust must not exceed 20 per cent. With multiflue chimneys each flue is taken as a separate chimney in the calculations of ac-

TABLE 3.5

Schedule 1: Furnaces rated by heat output; maximum permitted emissions of grit and dust

Maximum continuous rating steam per hour from and at 100°C		Quantity of grit and dust per hour			
		Furnaces burning solid matter		Furnaces burning liquid matter	
10^3 lb	Mg	lb	kg	lb	kg
1	0·454	1·33	0·60	0·28	0·13
5	2·27	6·67	3·03	1·4	0·64
10	4·54	10	4·54	2·8	1·27
25	11·3	20	9·1	7·0	3·18
50	22·7	37	16·8	12·5	5·67
100	45·4	66	29·9	18	8·2
200	90·7	122	55·3	29	13·2
300	136·1	172	78·0	41	18·6
400	181·4	217	98·4	50	22·7
475	215·5	250	113·4	57	25·8

TABLE 3.6

Schedule 2: Furnaces rated by heat input; maximum permitted emissions of grit and dust

Heat input per hour		Quantity of grit and dust per hour			
		Furnaces burning solid matter		Furnaces burning liquid matter	
MBtu	Gcal	lb	kg	lb	kg
1·25	0·315	1·1	0·50	0·28	0·13
5·0	1·26	4·3	1·95	1·1	0·50
10	2·52	7·6	3·45	2·2	0·99
25	6·3	14·1	6·40	5·5	2·49
50	12·6	25	11·3	10·9	4·94
100	25·2	45	20·4	16	7·3
200	50·4	90	41	26	11·8
300	75·6	132	60	35	15·9
500	126	218	99	54	24·4
575	145	250	113	57	25·5

ceptable emissions. If the emission of grit and dust from a chimney serving a furnace to which the Regulations apply, exceeds the permitted limit, it is a defence that the best practicable means have been used to prevent the excessive emission.

Until Regulations are made to apply to existing furnaces or further Regulations are made for furnaces where the material being heated contributes to the grit and dust emission, the occupier of the premises must use any practicable means to minimize the grit and dust emission. The definition of practicable in relation to the 1956 and 1968 Acts means reasonably practicable having regard to local conditions and circumstances, financial implications and the current state of technical knowledge, and includes the proper maintenance and use of the plant.

Under the Clean Air Act 1956 certain larger new furnaces, the installation of which was begun (or was subject to an agreement signed) between July 1956 and October 1969, should be equipped with approved grit and dust arrestment plant and the arrestors must be properly maintained and used. These provisions still apply to furnaces used to burn pulverized fuel; or to burn solid fuel in any other form or solid waste, provided that the plant was installed or an agreement was signed between July 1956 and October 1969. The 1968 Act provides that if new plant is installed after 1 October 1969 or is subject to an agreement after that date and is to be used to burn (a) pulverized fuel or (b) any other solid matter at a rate of 100 lb h^{-1} (45·36 kg h^{-1}) or any liquid or gaseous matter at a rate equivalent to 1·25 MBtu h^{-1} (0·315 Gcal h^{-1}) or more, it must, unless exemption has been obtained, be equipped with grit and dust arrestment plant, approved by the local authority. If approval is refused in writing by the local authority there is the right of appeal within 28 days to the Secretary of State.

Under Section 4 of the 1968 Act the Secretary of State has power to provide in Regulations that furnaces of any particular class, whilst used for prescribed purposes, should be exempted. The Clean Air (Arrestment Plant) (Exemption) Regulations 1969 (Statutory Instrument 1969 No. 1262) provides the following exemptions:

1. (a) Mobile or transportable furnaces providing a temporary source of heat or power during building operations, engineering construction or for investigation or research, and
   (b) Providing heat or power for agriculture;
2. Furnaces (other than those designed to burn solid matter at a rate of 1 tonne h^{-1} or more)
   (a) burning liquid matter or gas or both,
   (b) certain furnaces burning solid matter at a rate of not more than 25 lb h^{-1} (11·34 kg h^{-1}) for each 1 ft^2 (929 cm^2) of grate surface (these include hand-fired sectional furnaces, magazine type gravity-fed furnaces and those fitted with a chain grate stoker),
   (c) furnaces fitted with an underfeed stoker designed to burn solid matter at a rate not exceeding 251 lb h^{-1} (114 kg) for each 1 ft^2 (929 cm^2) of the plan area of the combustion chamber, and
   (d) furnaces fitted with a coking stoker designed to burn solid matter at a rate not exceeding 251 lb h^{-1} (114 kg) for each 1 ft^2 (929 cm^2) of area covered by the fire bars excluding the solid coking plate.

It is an offence to use an exempt furnace for a purpose other than that prescribed in the exemption.

Measurement of grit and dust emissions

Under the legal provisions in the Clean Air Acts of 1956 and 1968, in relation to the measurement of emissions of grit and dust, there are the Clean Air (Measurement of Grit and Dust) Regulations 1971 (Statutory Instrument 1971 No. 161). If a furnace or range of furnaces served by one chimney is to burn (a) pulverized fuel, (b) any other solid matter, at a rate of 100 lb h^{-1} (45·36 kg h^{-1}) or (c) any liquid or gaseous matter at a rate equivalent to 1·25 MBtu h^{-1} (0·315 Gcal h^{-1}) or more, the local authority may serve a notice on the owner of the plant or occupier of the building requiring emissions of grit and dust to be measured. The local authority must give at least six weeks' notice in writing, requiring the industrialist to make any necessary adaptations to the chimney or flues and to have the necessary equipment to enable the measurements to be made in accordance with the British Standard 3405 : 1961; this Standard is under review. Before making the measurements, the industrialist must give the local authority at least 48 hours' notice in writing of the date and time of the beginning of the measurements. The 1968 Act requires the industrialist to permit the local authority to be represented during the tests and must send the results to the local authority within 14 days of making the measurements. The report has to include the date of the test, the number of furnaces discharging into the chimney on that date, and with plant fired by solid fuel the percentage of grit in the samples of the emission. A notice from the local authority may require the making of measurements at stated intervals of not less than three months.

If the plant burns less than 1 tonne h^{-1} of solid matter, other than pulverized fuel, or less than 28 MBtu h^{-1} (7·06 Gcal h^{-1}) of liquid or gaseous matter, the industrialist may serve a counter notice requiring the local authority to make the measurement at its own expense, and the local authority is required to do so from time to time unless the counter notice is withdrawn. If a counter notice is served, the industrialist is required to make the necessary adaptations to the chimney or flues, including the provision of any necessary scaffolding and such facilities as electrical connections to enable the sampling equipment to be used. The local authority must provide the sampling equipment and conduct the test.

Miscellaneous provisions

In addition to the provisions already mentioned the following are of interest.

Fumes are defined in the Clean Air Act 1968 as any airborne solid matter smaller than dust. The Secretary of State has the power to make regulations limiting fume emissions, but such regulations have not so far been made under that Act. There are, however, provisions limiting the emissions of fumes from certain special processes controlled by the Alkali and Clean Air Inspectorate under the Alkali etc. Works Regulation Act 1906 and by Alkali Orders under that Act. The Alkali Act 1906 and Alkali Orders are considered later in this chapter.

Section 18 of the Clean Air Act 1956 provides that the owner of a coal mine or quarry with a spoilbank in use on 5 July 1956 or begun after that date must employ all practicable means to prevent spontaneous combustion and the emission of smoke and fumes. If the use of a spoilbank had ceased by 5 July 1956 and its control had passed from the owner of the mine or quarry, any emission of smoke or fumes comes within the provisions of Part III of the Public Health Act 1936 as amended.

The Building Regulations 1972 include a section M2 in relation to the prevention of smoke emission as follows. In any building (other than a building erected under former control) there shall not be installed for the purpose of heating or cooking in that or any other building any appliance which discharges the products of combustion into the atmosphere, unless that appliance is designed to burn as fuel either gas, coke, or anthracite. Provided that nothing in this regulation shall prohibit the installation of

1. a furnace which complies with Section 3 of the Clean Air Act 1956,
2. an appliance of a class exempted conditionally or unconditionally from the provisions of Section 11 of the Clean Air Act 1956 (which relates to smoke control areas) by any order for the time being in force under subsection (4) of that section, or
3. a solid fuel appliance with a bottom grate unsuitable for burning coke or anthracite but designed so as to be capable of use with an alternative bottom grate which is suitable for burning such fuel.

The Control of Pollution Act 1974, when fully operative, will

1. enable Ministers
   (a) after consultation with various bodies, to make Regulations limiting the composition of any motor fuel for use in the UK,

(b) after consultation, to limit the sulphur content of a wide range of fuel oils used in industry, commerce and in domestic boilers, etc.,

(c) require local authorities to monitor air pollution as directed by the Minister;

2. modify the Alkali Act 1906 and the Clean Air Act 1968;
3. substantially extend and amend the provisions of the Clean Air Acts in relation to research and publicity and to obtaining information about emissions to atmosphere;
4. raise the penalties in the Public Health Acts and Clean Air Acts for offences under Clean Air and related legislation.

Notices requiring information about air pollution are served by the local authority when it requires information about the emission of pollutants, and they require the occupier of the premises to provide the information within six weeks or such longer period as allowed. With regard to works registered under the Alkali Act, the local authority can require only information of a kind being supplied to HM Alkali and Clean Air Inspectorate. A person to whom such a notice is served may appeal to the Secretary of State on the grounds that disclosure would unreasonably prejudice some private interest, or be contrary to public interest, or that the information is not readily available and cannot be collected without incurring undue expenditure. The Secretary of State may make Regulations as to appeals including the timing of appeals and the circumstances in which all or any part of the appellant's case may be withheld from the local authority. If the appeal is upheld the Secretary of State may direct the local authority to withdraw or modify the notice or to take steps to ensure that the prejudicial information is not disclosed to the public.

Alkali etc. Works Regulation Acts and Alkali etc. Works Orders

In the eighteenth century sodium carbonate was made from the ashes of sea weed. In 1782, Collinson took out a patent in England for making sodium carbonate alkali from common salt, but so far as is known it was not operated on a commercial scale. By 1794, Leblanc in France had a process similar in broad principles in operation near Paris. The Leblanc process was first operated in England in 1814 on a very small scale at a site near the River Tyne in Northumberland. In 1823 James Muspratt erected and operated an alkali works near Liverpool; he used the principles of the Leblanc process but with a number of modifications, particularly in the production of the sulphuric acid required in the process. This was the beginning of the establishment of a growing alkali industry in England.

Briefly, the Leblanc process included the following stages. The sodium chloride was heated with sulphuric acid to produce sodium sulphate, known as salt cake, which was then heated to a high temperature with coal and limestone to form so-called black ash. The sodium carbonate in the black ash was extracted by warm water in which it dissolved. The solution was evaporated to give a solid, which was calcined to produce crude soda ash. This product was purified to provide anhydrous sodium carbonate. If washing soda ($Na_2CO_3 . 10H_2O$) was required, the anhydrous carbonate was dissolved in hot water from which the crystalline washing soda separated on cooling.

From alkali works using the Leblanc process, large quantities of hydrogen chloride were discharged into the air, and in the earlier stages of development of the industry little was done to remove the hydrogen chloride from the gases and vapours discharged. This led to serious and justifiable complaints from neighbouring residents of the detrimental effects of the discharges on health and materials. A Royal Commission was appointed to study the problem, and as a result of the Commission's report the first Alkali etc. Regulation Act was passed in 1863. The Act was extended on several later occasions and the various modifications and additions were consolidated in the Alkali etc. Regulation Act 1906. Further additions to the processes within the Act have been made by Alkali Orders issued at intervals during the period 1928–71. There is the Alkali Act 1906 with Alkali Works Orders 1966 and 1971 in operation in England and Wales, and in Scotland there are the Alkali etc. Works Regulation (Scotland) Acts 1906 and 1951 and an Alkali etc. Works (Scotland) Order 1972. The Acts and Orders for England and Scotland are basically similar, the differences being due to differences in the processes in operation in the two areas.

The Acts with the appropriate Orders are administered by Inspectors appointed by the Government Departments concerned, and the Inspectors employed have all had scientific and technical training followed by experience in industry. Each year the Chief Inspector issues a report on the work to the appropriate Minister or Ministers. During recent years the annual reports for England and Wales have been issued to the Secretary of State for the Environment and to the Secretary of State for Wales and the reports for Scotland have been issued to the Secretary of State for Scotland. The Annual Reports are published by HM Stationery Office, and each includes the reports by both the Chief Inspector for England and Wales and the Chief Inspector for Scotland. Anyone interested in how the system operates should study the last few available reports. The report for 1974 is entitled *111th Annual Report on Alkali etc. Works.*

In 1974 the number of works registered in England and Wales was 2147 of which only 3 were alkali works; in Scotland the number of works registered was 325, of which one was an alkali works. Each of the registered works must be registered annually to be allowed to operate. The general practice is to insist on the best practicable means being adopted to reduce the emission of polluting effluents to the practicable minimum. The use of the word 'practicable' means that the plant and equipment must be maintained in good order and efficient operation, and takes into account the additional cost necessarily incurred.

The following list of registrable works is an indication of the great variety of works now covered in England and Wales by the Alkali etc. Works Regulation Act 1906 and Alkali etc. Works Orders 1966 and 1971.

Registrable works in England and Wales under Alkali Act and Orders

A. *Non-scheduled works*

I. Alkali works
Works for:
(a) The manufacture of sulphate of soda or sulphate of potash, or
(b) The treatment of copper ores by common salt or other chlorides whereby any sulphate is formed, in which muriatic acid gas is evolved

B. *Scheduled works*

1. *Sulphuric acid works*
   Works in which the manufacture of sulphuric acid is carried on by the lead chamber process by which sulphurous acid is converted into sulphuric acid by the agency of oxides of nitrogen and by the use of a lead chamber or by any other process involving the use of oxides of nitrogen
2. *Sulphuric acid (class II) works*
   Works in which the manufacture of sulphuric acid is carried on by any process other than the lead chamber process, and works for the concentration or distillation of sulphuric acid
3. *Chemical manure works*
   Works in which the manufacture of chemical manure is carried on, and works in which any mineral phosphate is subjected to treatment involving chemical change through the application or use of any acid and works for the granulating of chemical manures involving the evolution of any noxious or offensive gas
4. *Gas liquor works*
   Works (not being sulphate of ammonia works or muriate of ammonia works) in which sulphuretted hydrogen or any other noxious or offensive gas is evolved by the use of ammoniacal liquor in any manufacturing process, and works in which any such liquor is desulphurized by the application of heat in any process connected with the purification of gas
5. *Nitric acid works*
   Works in which the manufacturer of nitric acid is carried on and works in which nitric acid is recovered from oxides of nitrogen and works where in the manufacture of any product any acid-forming oxide of nitrogen is evolved
6. *Sulphate of ammonia works, and muriate of ammonia works*
   Works in which the manufacture of sulphate of ammonia or of muriate of ammonia is carried on
7. *Chlorine works*
   Works in which chlorine is made or used in any manufacturing process

8. *Muriatic acid works*
   (a) Muriatic acid works (not being alkali works as defined in the Act) where muriatic acid gas is evolved either during the preparation of liquid muriatic acid or for use in any manufacturing process or as the result of the use of chlorides in a chemical process;
   (b) Tinplate flux works, works in which any residue or flux from tinplate works is calcined for the utilization of such residue or flux, and in which muriatic acid gas is evolved;
   (c) Salt works, works (not being works in which salt is produced by refining rock salt, otherwise than by the dissolution of rock salt at the place of deposit) in which the extraction of salt from brine is carried on, and in which muriatric gas is evolved
9. *Sulphide works*
   Works in which sulphuretted hydrogen is evolved by the decomposition of metallic sulphides, or in which sulphuretted hydrogen is used in the production of such sulphides, or in any works in which sulphuretted hydrogen is evolved as part of a chemical process
10. *Alkali waste works*
   Works in which alkali waste or the drainage therefrom is subjected to any chemical process for the recovery of sulphur or for the utilization of any constituent of such waste or drainage
11. *Venetian red works*
   Works for the manufacture of Venetian red, crocus or polishing powder, by heating sulphate or some other salt of iron
12. *Lead deposit works*
   Works in which the sulphate of lead deposit from sulphuric acid chambers is dried or smelted
13. *Arsenic works*
   Works for the preparation of arsenious acid, or where nitric acid or a nitrate is used in the manufacture of arsenic acid or an arseniate and works in which any volatile compound of arsenic is evolved in any manufacturing process and works in which arsenic is made
14. *Nitrate and chloride of iron works*
   Works in which nitric acid or a nitrate is used in the manufacture of nitrate or chloride of iron
15. *Bisulphide of carbon works*
   Works for the manufacture, use or recovery of bisulphide of carbon
16. *Sulphocyanide works*
   Works in which the manufacture of any sulphocyanide is carried on by the reaction of bisulphide of carbon upon ammonia or any of its compounds
17. *Picric acid works*
   Works in which nitric acid or a nitrate is used in the manufacture of picric acid
18. *Paraffin oil works*
   Works in which crude shale oil is refined

19. *Bisulphite works*
Works in which sulphurous acid is used in the manufacture of acid sulphites of the alkalis or alkaline earths, works for the manufacture of liquid sulphur dioxide or of sulphurous acid or of any sulphite, and works (not being smelting works as defined in section 8, sub-section (1) of the Act of 1906 or other works defined elsewhere in this schedule) in which oxides of sulphur are evolved in any chemical manufacturing process
20. *Tar works*
Works where gas tar or coal tar is distilled or is heated in any manufacturing process and works in which creosote or any other product of the distillation of gas tar or coal tar is distilled or is heated in any manufacturing operation involving the evolution of any noxious or offensive gas
21. *Zinc works*
Works in which, by the application of heat, zinc is extracted from the ore or from any residue containing that metal, and works in which compounds of zinc are made by dry processes giving rise to fume
22. *Benzene works*
Works (not being tar works as already defined) in which any wash oil used for the scrubbing of coal gas is distilled, or in which any crude benzol is distilled
23. *Pyridine works*
Works in which pyridine or picolines are recovered or made
24. *Bromine works*
Works in which bromine is made or is used in any manufacturing operation
25. *Hydrofluoric acid works*
Works in which hydrofluoric acid is evolved in the manufacture of liquid hydrofluoric acid or its compounds
26. *Cement production works*
Works in which argillaceous and calcareous materials are used in the production of cement clinker, and works in which cement clinker is ground or cement is packed
27. *Lead works*
Works (not being works for the recovery of lead from scrap by direct liquation) in which, by the application of heat, lead is extracted from any material containing lead or its compounds and works in which compounds of lead are manufactured from metallic lead or its compounds by dry processes which give rise to dust or fume
28. *Fluorine works*
Works in which fluorine or its compounds with other halogens are made or used in any manufacturing process, and works for the manufacture of fluorides, borofluorides or silicofluorides

29. *Acid sludge works*
Works in which acid sludge produced in the refining of coal tar, petroleum or other hydrocarbon derivatives is treated in such manner as to cause the evolution of any noxious or offensive gas
30. *Iron works and steel works*
Works in which
(a) iron or ferro-alloys are produced in a blast furnace and in which raw materials for use in blast furnaces are handled or prepared; or
(b) iron ores for use in blast furnaces are calcined or sintered; or
(c) iron or steel is melted in cupolas employing a heated air blast, or in electric arc furnaces,
(d) steel is produced, melted or refined in Bessemer, Tropenas, open hearth or electric arc furnaces; or
(e) oxygen or air enriched with oxygen is used for the refining of iron or for the production, shaping or finishing of steel; or
(f) ferro-alloys are made by processes giving rise to fume
31. *Copper works*
Works in which
(a) by the application of heat
(i) copper is extracted from any ore or concentrate or from any material containing copper or its compounds; or
(ii) molten copper is refined; or
(iii) copper or copper alloy swarf is degreased; or
(iv) copper alloys are recovered from scrap fabricated metal, swarf or residues by processes designed to reduce the zinc content; or
(b) copper or copper alloy is melted and cast in moulds the internal surfaces of which have been coated with grease-bound or oil-bound dressings. (But this paragraph does not apply to works in which the aggregate casting capacity does not exceed 10·165 tonnes (10 tons) per day.)
32. *Aluminium works*
Works in which
(a) aluminium swarf is degreased by the application of heat; or
(b) aluminium or aluminium alloys are recovered from aluminium or aluminium alloy scrap fabricated metal, swarf, skimmings, drosses or other residues by melting but not including works in which aluminium or aluminium alloys are separated from ferrous metals by liquation in sloping hearth furnaces; or
(c) aluminium is recovered from slag; or
(d) molten aluminium or aluminium alloys are treated by any process involving the evolution of chlorine or its compounds; or
(e) aluminium is extracted from any material containing aluminium by a process evolving any noxious or offensive gases, or
(f) oxide of aluminium is extracted from any ore and materials used in the above processes or the products are treated or handled by methods which cause noxious or offensive gases to be evolved

33. *Electricity works*
Works in which
(a) solid or liquid fuel is burned to raise steam for the generation of electricity for distribution to the general public or for purposes of public transport; or
(b) boilers having an aggregate maximum continuous rating of not less than 204 tonnes h^{-1} (450 000 lb h^{-1}) of steam and normally fired by solid or liquid fuel are used to produce steam for the generation of electricity for purposes other than those mentioned in the preceding paragraph

34. *Producer gas works*
Works in which producer gas is made from coal and in which raw producer gas is transmitted or used

35. *Gas and coke works*
Works (not being producer gas works) in which
(a) coal, oil or mixtures of coal or oil with other carbonaceous materials or products of petroleum refining or natural gas or methane from coalmines or gas derived from fermentation of carbonaceous materials are handled or prepared for carbonization or gasification or reforming and in which these materials are subsequently carbonized or gasified or reformed; or
(b) water gas is produced or purified; or
(c) coke or semi-coke is produced and quenched, cut, crushed or graded; or
(d) gases derived from any process mentioned in paragraph (a) are subjected to purification processes

36. *Ceramic works*
Works in which
(a) pottery products (including domestic earthenware and china, sanitary ware, electrical porcelain, glazed tiles and teapots) are made in intermittent kilns fired by coal or oil; or
(b) heavy clay or refractory goods are fired by coal or oil in
  (i) intermittent kilns; or
  (ii) continuous grate-fired kilns, not being tunnel kilns; or
  (iii) any kiln in which a reducing atmosphere is essential; or
(c) salt glazing of any earthenware or clay material is carried on

37. *Lime works*
Works in which calcium carbonate or calcium-magnesium carbonate is burnt through the agency of coal or oil.

38. *Sulphate reduction works*
Works in which metallic sulphates are reduced to the corresponding sulphides by heating with carbonaceous matter

39. *Caustic soda works*
Works in which
(a) either concentrated solutions of caustic soda or fused caustic soda are produced in vessels heated by coal; or
(b) black liquor produced in the manufacture of paper is calcined in the recovery of caustic soda

40. *Chemical incineration works*
Works for the destruction by burning of wastes produced in the course of organic chemical reactions which occur during the manufacture of materials for the fabrication of plastics and fibres, and works for the destruction by burning of chemical wastes containing combined chlorine, fluorine, nitrogen, phosphorus or sulphur

41. *Uranium works*
Works (not being works licensed under the Nuclear Installations (Licensing and Insurance) Act, 1959, and not being nuclear reactors or works involving the processing of irradiated fuel therefrom for the purpose of removing fission products) in which
(a) any ore or concentrate or any material containing uranium or its compounds is treated for the production of uranium or its alloys or its compounds; or
(b) any volatile compounds of uranium are manufactured or used; or
(c) uranium or its compounds are manufactured, fashioned or fabricated by any dry process giving rise to dust or fume

42. *Beryllium works*
Works in which
(a) any ore or concentrate or any material containing beryllium or its compounds is treated for the production of beryllium or its alloys or its compounds; or
(b) any material containing beryllium or its alloys or its compounds is treated, processed or fabricated in any manner giving rise to dust or fume

43. *Selenium works*
Works in which
(a) any ore or concentrate or any material containing selenium or its compounds is treated for the production of selenium or its alloys or its compounds; or
(b) any material containing selenium or its alloys or its compounds other than as colouring matter is treated, processed or fabricated in any manner giving rise to dust or fume

44. *Phosphorus works*
Works in which
(a) phosphorus is made; or
(b) yellow phosphorus is used in any chemical or metallurgical process

45. *Ammonia works*
Works in which ammonia is
(a) made; or
(b) used in the ammonia-soda process; or
(c) used in the manufacture of carbonate, nitrate or phosphate of ammonia or urea or nitriles

46. *Hydrogen cyanide works*
Works in which hydrogen cyanide is made or is used in any chemical manufacturing process
47. *Acetylene works*
Works in which acetylene is made and used in any chemical manufacturing process
48. *Amines works*
Works in which
(a) any methylamine or any ethylamine is made; or
(b) any methylamine or any ethylamine is used in any chemical process
49. *Calcium carbide works*
Works in which calcium carbide is made
50. *Aldehyde works*
Works in which formaldehyde, acetaldehyde or acrolein or the methyl, ethyl or propyl derivatives of acrolein are made
51. *Anhydride works*
Works in which acetic, maleic or phthalic anhydrides or the corresponding acids are made
52. *Chromium works*
Works in which any chrome ore or concentrate is treated for the production therefrom of chromium compounds or chromium metal is made by dry process giving rise to fume
53. *Magnesium works*
Works in which magnesium or any compound of magnesium is made by dry processes giving rise to fume
54. *Cadmium works*
Works in which metallic cadmium is recovered or cadmium alloys are made or any compound of cadmium is made by dry processes giving rise to fume
55. *Manganese works*
Works in which manganese or its alloys or any compound of manganese is made by dry processes giving rise to fume
56. *Metal recovery works*
Works in which metal is recovered from scrap cable by burning the insulation
57. *Petroleum works*
Works in which
(a) crude petroleum is handled or stored; or
(b) crude petroleum is refined by any operation carried out at petroleum refineries to convert crude petroleum into saleable products; or
(c) any product of such refining is subjected to further refining; or
(d) natural gas is refined; or

(e) any product of any of the foregoing operations is used, except as a solvent, in any subsequent chemical manufacturing process, not being a chemical manufacturing process described in any other scheduled work; or

(f) used lubricating oil is prepared for re-use by any thermal process

58. *Acrylates works*

Works in which acrylates are

(a) made; or

(b) purified; or

(c) made or purified and polymerized

59. *Di-isocyanates works*

Works in which di-isocyanates are made, or partly polymerized, or used in the manufacture of expanded plastics

60. *Mineral works*

Works in which metallurgical slags or pulverized fuel ash or minerals are subjected to any size reduction, grading or heating by processes giving rise to dust, including the subsequent handling of the products of any such process but not including works for the processing of coal, lime works, ceramic works or foundries except in so far as the process carried on is described in any other scheduled work.

61. *Smelting works*

Works in which sulphide ores, including regulus, are calcined or smelted.

List of noxious or offensive gases under the Alkali etc. Works Regulation Act and Alkali etc. Works Orders

---

Muriatic acid
Sulphuric acid and sulphuric anhydride
Sulphurous acid and sulphurous anhydride (except those arising solely from the combustion of coal)
Nitric acid and acid forming oxides of nitrogen
Chlorine and its acid compounds
Bromine and its acid compounds
Iodine and its acid compounds
Fluorine and its compounds
Arsenic and its compounds
Ammonia and its compounds
Cyanogen compounds
Pyridine
Bisulphide of carbon
Fumes containing copper, lead, antimony, arsenic, zinc, aluminium, iron, silicon, calcium, or their compounds
Fumes containing chlorine or its compounds
Smoke, grit and dust
Fumes containing uranium, beryllium, cadmium, selenium, sodium, potassium or their compounds
Carbon monoxide
Acetic anhydride and acetic acid
Acrylates
Aldehydes
Amines
Di-isocyanates
Fumes containing chromium, magnesium, manganese, molybdenum,

List of noxious or offensive gases under the Alkali etc. Works Regulation Act and Alkali etc. Works Orders—continued

Chloride of sulphur	phosphorus, titanium, tungsten, vanadium or their compounds
Acetylene	Maleic anhydride, maleic acid and fumaric acid
Sulphuretted hydrogen	Products containing hydrogen from the partial oxidation of hydrocarbons
Volatile organic sulphur compounds	Phthalic anhydride and phthalic acid
Fumes from benzene works	Picolines
Fumes from cement works	
Fumes from tar works	
Fumes from paraffin oil works	
Fumes from petroleum works	

As advances in science and technology lead to the development of practicable methods, taking costs into account, of reducing the emission of pollutants from works registered under the Alkali Acts and Alkali Orders, the Alkali Inspectorates insist on more stringent limits of the emissions. For example, with the improvements in the efficiency of electrostatic precipitators, the limit of emission of solid matter discharged in the gases leaving the chimneys of large modern cement works has been steadily reduced from $1{\cdot}15$ g m^{-3} ($0{\cdot}5$ grain ft^{-3}) as measured at ordinary temperature (15°C) to $0{\cdot}23$ g m^{-3} ($0{\cdot}1$ grain ft^{-3}).

Health and Safety at Work etc. Act 1974

Under the Factories Act 1961 all practicable steps have to be taken to minimize the risks to employees of accidents and of harmful exposure to dust, fume or any other pollutant of such a character and extent as to be likely to be injurious or offensive to the persons employed. Under the Ministry of Labour there were factory inspectors, who visited the various factories to ensure that the best practicable means were being used to prevent accidents and exposure of the employees to detrimental dust, fumes and other pollutants. In relation to gaseous pollutants and fumes, the inspectors were largely guided by a list of threshold limit values of about 400 vapours and fumes agreed at periodic meetings of the American Conference of Governmental Industrial Hygienists. The list is periodically revised; in effect it gives the limits of concentration of each of the various substances above which employees should not be exposed during the normal period of working hours per day. The Ministry of Labour issued a number of booklets, published by HM Stationery Office, on various aspects of safety, health and welfare in factories; these booklets were periodically revised and brought up to date.

The purpose of the Health and Safety at Work etc. Act 1974 is to provide a comprehensive and integrated legal and administrative system of ensuring that measures are taken in factories to avoid conditions detrimental to the safety, health and welfare of people at work and to avoid detriment to others from the activities of persons at work. To achieve this integration, the separate inspectorates and agencies involved in these various activities are brought together in a Health and Safety Executive under a Health and Safety Commission. The Commission was established on 1 October 1974 and the Executive on 1 January 1975. Initially these bodies will be responsible for enforcing the existing legislation, including the Alkali etc. Works Regulation Acts and the Factories Acts. This legislation will be progressively repealed and replaced by provisions in relation to the Health and Safety at Work etc. Act 1974. The lists of registrable works and of noxious or offensive gases under the Alkali etc. Works Regulation Acts and Orders will remain the same under the 1974 Act, with revisions from time to time with changes and developments in industrial processes. The practice of requiring the use of the best practicable means will be continued under the 1974 Act, probably in accordance with codes of practice approved by the Secretaries of State concerned. Local authorities will continue to be responsible for the control of pollution under the Clean Air Acts.

Legislation on pollution from motor vehicles

Though research has not so far shown that current levels of emissions from motor vehicles are detrimental to health, pollutants such as lead, carbon monoxide, hydrocarbons and smoke are undesirable and their effects are being studied in many areas. Existing legislation aims at preventing total emissions from rising significantly, even with an appreciable increase in the number of vehicles in use. In the ECE and EEC countries it has been agreed to aim at reducing the emissions from petrol-driven vehicles by 20 per cent for carbon monoxide and by 15 per cent for hydrocarbons. With the exception of small three-wheeled cars and some other models, petrol-driven cars and light commercial vehicles made on or after 20 September 1973 and first used on or after 10 November 1973 must comply with the limits prescribed for emissions of carbon monoxide and hydrocarbons in Regulation No. 15 of the United Nations' Economic Commission for Europe. The Regulation prescribes that the vehicles shall be tested on a chassis dynamometer under a cycle of conditions representative of driving in a congested urban area after a start from cold. The maximum permissible amounts of carbon monoxide and of hydrocarbons discharged in relation to the weight of the vehicle tested are given in a table. A further test is the measurement, with a hot engine running at idling speed, of the amount of carbon monoxide in the exhaust gases, which should not exceed 4·5 per cent by volume. Reductions of up to 30 per cent in carbon monoxide and 35 per cent in hydrocarbons as compared with the amounts in the exhaust gases of uncontrolled vehicles should result. In the UK, the Secretary of State for the Environment in 1972 announced a programme of phased reductions in the permissible lead content of petrol. In November 1974, the permitted content was reduced to $0{\cdot}55\ g\ l^{-1}$ as compared with $0{\cdot}84\ g\ l^{-1}$ in 1972.

In the UK, there have for some years been regulations in relation to motor vehicles, the latest being the Motor Vehicles (Construction and Use) Regulations 1973. All diesel engine vehicles must be so constructed that they do not emit avoidable smoke or visible vapours, and each new diesel engine for use in a road vehicle must be of a type in accordance with British Standard A.U.141a, which prescribes maximum allowable smoke emission in exhaust gases over a wide range of conditions of operation of the eingine. Excess fuel devices (chokes) must be so fitted that they cannot be used while the vehicle is in motion, if such use would increase smoke emission. Inspectors, in cooperation with the police, stop diesel vehicles emitting excessive quantities of smoke and in bad cases order that the vehicles must not be used again until the engine has been serviced or replaced to ensure satisfactory emission from the exhaust. Most commercial vehicles must be fitted by the manufacturers with a plate giving particulars of weight, etc., and the figures on the plate must show that the power/weight ratio is not below a specified minimum. No vehicle in use may emit smoke, vapour, dust, grit or other substances so as to endanger other road users, and engines must not be altered in a way which increases smoke emission.

Control in some other countries

In 1958 there was issued a report[4] on air pollution by an expert committee of the World Health Organization. Arrangements were then for a WHO document[5] dealing with certain aspects in more detail. That document, published in 1961, included a chapter on Legislation: Standards and Enforcements, in 14 Countries. Since that time there have been improvements in legislation and the standards required in most of the countries considered. It is not possible in the space allotted to give an account of legislative and other controls of air pollution in other countries in so much detail as given for the UK. All that can be done is to draw attention to the main differences in the control systems in a few other countries as compared with systems in the UK.

United States

It must be recognized that the character and amounts of pollutants discharged into the air in the US vary greatly from one state to another and in different parts of the same state. Some areas are highly industrialized and have high densities of population while in many large areas there is little industry and the population density is very low. The overall average population density in the US is only about 22 km^{-2} as compared with 325 km^{-2} in England and Wales and 227 km^{-2} for the whole of the UK. It is not surprising, therefore, that there is not the same uniformity of control over the whole of the US as in the UK.

Prior to the Federal Air Quality Act 1967,[6] which has since been modified or more or less replaced on several occasions, control of air pollution was left entirely to the separate states and local governments. As a result there have been widely varying levels of control in different areas, from virtually no control to strict control. For some years the main activity of the Federal Government, through the Department of Health, Education and Welfare, has been to undertake and assist research, local investigations, surveys, training of personnel, providing scientific and technical assistance, and finance by grants in aid. The objects, as in other countries, are to reduce to the practicable minimum the discharges of pollutants that may have a detrimental effect on health and welfare, agriculture, farm animals and property.

The earliest definite move to reduce air pollution was by cities limiting the allowable emissions of smoke from the use of fuels. The great improvement in Pittsburg many years ago in this respect is well known. By 1912, 23 cities, each with a population of more than 200 000, had smoke-control programmes, many of which also included the limitation of emissions of grit and dust from the chimneys. As in the UK, the Ringelmann chart has been used to measure and assist in controlling the density of the smoke discharged. The amount of grit and dust emitted has been limited to a specified weight in a measured volume of the gases discharged from the chimney. There are also in some towns restrictions on the use of solid fuels to those containing not more than specified contents of ash, sulphur and volatile matter, and limitations of the sulphur content of oil fuels used.

A problem that has arisen with state and local government control is that discharges from industrial activities in certain states have led to objectionable pollution in adjacent states, and activities in local government areas have caused undue pollution in neighbouring areas. To overcome these difficulties, the Federal Government has encouraged and advised on joint arrangements for pollution control by adjacent states and areas. There are still great differences, however, in the extent and efficiency of control in different states and local authority areas. It seems probable that the Federal Government will gradually obtain further powers whereby more efficient control will have to be exercised by states and local authorities.

Los Angeles is the area that seems to have the most stringent methods of control of man-made pollution in the US. This is not surprising in view of the somewhat unusual climatic conditions. This area is subject to serious temperature inversions that are caused by air moving from the Pacific during the daytime and then reaching the hills and mountains to the east of Los Angeles. This air is cooled over the hills and mountains during the night, and afterwards travels back down into Los Angeles during the following morning and early afternoon. This causes a fog, of the nature of a mist, which often restricts visibility to as little as a kilometre; it is quite unlike the black smoky fog that occurred in London in December 1952 and reduced visibility to no more than a few metres. The mist or smog in Los Angeles is usually at its worst in warm weather; the writer has experience of it on days when the temperature was in the region of 38°C (100°F). It can be highly lachrymatory and unpleasant in character. Some years ago it was thought that the smog was intensified by the smoke and fumes from the burning of household refuse in the yards or gardens of houses. Arrangements were then made to collect the refuse for centralized disposal; this did not prevent the formation of the lachrymatory smog.

Intensive investigations by Haagen-Smit and Wayne[7] and by others have explained the mechanism of the formation of the Los Angeles smog. Briefly, the mechanism is a photochemical reaction in bright sunlight between hydrocarbons and nitrogen oxide from the exhausts of motor vehicles with ozone in the atmosphere. The lachrymatory material in this photochemical smog may be a mixture of several irritants including peroxyacyl nitrates.

In relation to pollution from motor vehicles,[6] under Federal legislation the Secretary of Health, Education and Welfare is authorized to issue, by Regulations, standards of emission of any substance discharged from new motor vehicles and their engines. There is provision for testing new vehicles and engines in accordance with a schedule and for those passing the tests, certificates that are valid for not less than a year are issued. Vehicles and engines made to the same detailed specification are assumed to satisfy the tests. It is unlawful to introduce into interstate commerce new motor vehicles or engines which do not comply with the applicable standards. States and local governments in general are prohibited from imposing different requirements for new vehicles and engines, but this does not affect their power to impose requirements on used vehicles and engines. There is a proviso for waiver of the prohibition of state action for new vehicles and engines for states which had already met the requirements prescribed in the Regulations under Federal Law. Only California was eligible for the waiver; that state already had standards higher than those of the Regulations under the Federal Law. In view of the conditions during the smogs of Los Angeles, this waiver is not surprising.

Canada

The only Federal legislation in Canada is that under the Canada Shipping Act, which includes Regulations controlling the emission of smoke from ships. There is legislation in each of several provinces, including Alberta, British Columbia, Manitoba, Nova Scotia, Ontario and Saskatchewan.

In Ontario, which has a larger population and more industry than any of the other provinces, there is provincial legislation limiting air pollution by grit and dust, liquids, gases and odours of a type that may cause damage to health and safety, cause discomfort, may cause injury or damage to plant and animal life or property, or may interfere with visibility and the normal conduct of transport or business. Certificates of approval for methods and equipment for controlling air pollution have to be obtained from the Minister of Health before construction of factories and other buildings likely to lead to the discharge of pollutants. There are limitations on emissions of smoke, grit and dust, sulphur dioxide and other pollutants. There are also Regulations for the control of pollution from motor vehicles. In the other provinces mentioned, there are broadly similar regulations, with control either by appropriate departments of the province or by municipalities. In Quebec, air pollution investigations and recommendations for control of air pollution are made by the Industrial Hygiene Division of the Department of Health; but in Quebec the air pollution problems are not sufficient to need legislative control or air pollution staff.

Australia

Australia has a total area of about 7·7 Mkm^2 and a population of about 13·5 millions, so the overall average density of population is less than 2 km^{-2}. The states with the largest populations and some important industrial areas are on the eastern side of the country. They are New South Wales with a population of 4·8 millions and an average population density of about 6·25 km^{-2}, Victoria with a population of 3·7 millions and a density of 16·4 km^{-2} and Queensland with a population of 1·9 millions and a density of only 1·2 km^{-2}.

There is no nation-wide legislation in Australia in relation to air pollution. Since about 1955, New South Wales, Victoria and Queensland have gradually introduced legislation and control measures largely on the lines of those in operation in the UK, from whose experts they have received general guidance.

South Africa

South Africa has a total area of 1·22 Mkm^2 and a population of about 25 millions, so that the overall average density of population is 20 km^{-2}. In relation to the economy of the country, mineral production, including gold, diamonds, coal, copper, tin, silver and asbestos is of the greatest importance. The principal agricultural products are maize, wheat, ground nuts, oats and barley. The problems of mitigation of air pollution are not so widespread and varied as in the more highly industrialized countries. Attention has been directed during recent years to the areas of the cities with the largest populations, more particularly in the Transvaal. They include the following, the approximate populations of which in millions are given in brackets: in the Transvaal, Johannesburg (1·5) and Pretoria (0·53), in Cape Province, Cape Town (0·68) and Port Elizabeth (0·4) and in Natal, Durban (0·74). The greatest progress has been made in recent years in the Transvaal, where the initial steps taken during recent years have been on the lines of the action in Great Britain.

European Common Market countries

The areas, populations and population densities of the Common Market countries that have considerable industrial activity are given in Table 3.7. The densities of population are a rough guide to the relative proportions of the populations engaged in industry in the countries mentioned.

TABLE 3.7

Areas, populations and population densities in 1975 of the Common Market countries

Country	Population (millions)	Area (thousand km^2)	Density (km^2)
Belgium	9·71	30·51	318
Denmark	5·10	43·07	118
France	53·10	547·00	97
Germany Federal Republic	62·15	248·56	250
Eire (Ireland)	3·07	70·28	44
Italy	55·56	301·22	184
Netherlands	13·76	40·84	337
UK	56·46	244·05	231

The legislative provisions in relation to air pollution, the regulations under them and the procedures for enforcement and for inspection of plant and premises differ considerably in the countries of the European common market. The legislation and regulations in the UK have been described in detail in this chapter. Before 1937, when the Republic of Ireland (Eire) became a separate country, it was part of the UK and had the same air pollution legislation. For example, Alkali Inspectors based in England periodically inspected plants in that part of Ireland now known as Eire. The same provisions are now under the control of Eire officials. There is also more recent legislation combined with regulations in relation to emissions of smoke, particulate matter and polluting gases; these do not seem yet to have reached the stage of general enforcement. Eire is not a highly industrialized country.

There is effective anti-pollution legislation and enforcement in Belgium, France, the Federal Republic of Germany and the Netherlands. With gradual changes in legislation, regulations and inspection procedure, further reduction in air pollution will be achieved. It is worthy of note that there are frequent meetings of experts from each of the industrialized common market countries to discuss legislative and technical matters, including methods of measurement and surveys of general pollution. For example, agreement has been reached on methods of testing diesel engines for use in road vehicles, with a limit of smoke emission in the tests, with the object of ensuring that there is not excessive smoke emission from such heavy vehicles in use on the roads. There is also general agreement on the maximum allowable content of lead in gasolene. In addition, there is close liaison with work in Sweden on the question of the extent to which air pollution in that country, particularly by oxides of sulphur, is carried by the wind over appreciable distances from other countries of Europe.

References

Legislative Enactments, Regulations and other associated documents for the control of air pollution in the UK the titles of which are given in this chapter are published by HMSO, London.

1. *Clean air year book, 1975*, National Society for Clean Air, Brighton, England, 1975.
2. *Interim Report of UK Government Committee on Air Pollution*, HMSO, London, 1953.
3. *Final Report of UK Government Committee on Air Pollution*, HMSO, London, 1954.
4. *Air pollution*, Fifth Report of Expert Committee on Environmental Sanitation, Technical Report Series No. 157, World Health Organization, Geneva, 1958.
5. Parker, A., *Air pollution legislation: standards and enforcement*, Monograph Series No. 46, Air Pollution, pp. 365–80, World Health Organization, Geneva, 1961.
6. Edelman, S., in A. C. Stern (ed.), *Air Pollution*, 2nd edn, Vol, 3, chapter 50, Academic Press, New York, 1971.
7. Haagen-Smit, A. J. and L. G. Wayne, in A. C. Stern (ed.), *Air Pollution*, 2nd edn, Vol. 1, chapter 6, Academic Press, New York, 1968.

Part 2 Legislation in the UK on radioactive wastes

PAMELA M. BRYANT

In the UK there is no single general statute concerned with nuclear energy and radiological protection. A number of separate statutes and non-statutory codes of practice involving several Government Departments are directed to specific aspects of the use of radioactive materials and ionizing radiation.

The Radioactive Substances Act 1960 applies to premises used for the purpose of an undertaking and regulates the keeping and use of radioactive materials from the point of view of waste control. Under Section 1 of this Act, persons are required to be registered in respect of premises unless they are specifically exempted or the radioactive material consists of mobile radioactive apparatus. The UK Atomic Energy Authority (UKAEA), occupiers of sites licensed under the Nuclear Installations Act, 1965 ('nuclear installations' consisting broadly of reactor and fuel processing sites), National Health Service hospitals (except in Northern Ireland) and Crown premises do not require registration. Separate arrangements apply to hospitals and Crown premises; they are obliged to notify by letter the use of radioactive materials. Under Section 3, persons are also required to be registered in respect of mobile radioactive apparatus unless specifically exempted. The exemptions include the UKAEA and Government Departments.

Under the Radioactive Substances Act, authorizations are required for the disposal of waste unless exemptions are made. Such exemptions may be made either absolutely or subject to limitations or conditions. The UKAEA and nuclear installations require joint authorization by the Minister of Agriculture, Fisheries and Food and the Secretary of State for the Environment.* Authorizations for other premises are normally issued by the Secretary of State for the Environment* alone, though there is an administrative arrangement between departments with regard to premises of the Ministry of Defence when the Minister of Agriculture, Fisheries and Food has an interest in the disposals. These authorizations specify numerical limits in the case of liquid and solid effluents. For discharges to the atmosphere they are more general and normally require the use of the best practicable means to reduce the discharge, having due regard to local circumstances and to ensure that agriculture and public health are not endangered. Inspectors appointed under the Act have the power to enter and inspect any premises and, inter alia, inspect records of waste discharges. Under the Act, authorizations are

* In Scotland the Secretary of State for Scotland is responsible, operating through the Scottish Development Department; in Wales it is the Secretary of State for Wales, operating through the Welsh Office, and in Northern Ireland it is the Department of the Environment. It is normal practice for the occupier to establish working limits for discharges to air, specific to the site and radioactive materials involved, and these are used as guidance in the control of discharges. Such working limits are derived from the accepted radiological protection criteria.

also required for accumulation of waste, unless exemptions are made. The UKAEA, nuclear installations and Crown premises are exempted.

The radiological protection policy associated with waste disposal was set out in 1959 in a White Paper.[1] It is

1. to ensure, irrespective of cost, that no member of the public shall receive a radiation dose exceeding the Dose Limit recommended by the International Commission on Radiological Protection;
2. to ensure, irrespective of cost, that the whole population of the country shall not receive an average dose of more than 1 rad per person in 30 years, and
3. to do what is reasonably practicable, having regard to cost, convenience and the national importance of this subject, to reduce the doses far below these levels.

Where the potential hazard to the public is minimal, exemptions from the registration and authorization requirements are granted under the Radioactive Substances Act. For example, the Act itself exempts radioactive clocks and watches from both requirements except where they are manufactured or repaired by processes involving the use of luminous material. Other exemptions by Order have been made as the occasion arose. For example, fire detectors containing small quantities of radioactive material have been exempted by a specific exemption order from the registration and authorization requirements subject to certain conditions such as return of the detector to the manufacturer for disposal.

Reference

1. *The control of radioactive wastes*, Cmd. 884, HMSO, London, 1954.

4 Methods of measurement and survey

S. R. CRAXFORD

The methods in common use for the assessment of air pollution have been described in the many text books and instruction manuals covering the field. It is considered that little useful purpose would be served by making this chapter yet another descriptive compendium of these methods. There is, however, little appreciation of what a survey of air pollution should achieve and of what the methods available really measure. A short discussion of the material on these lines should be valuable, particularly if it induces potential users to give careful thought to their requirements and the possibilities of meeting them.

Two general points are conveniently made here, though they arise from subsequent discussion of individual methods. The first is that while many methods purport to give a measure of pollution by a specific pollutant, closer examination shows that the result obtained is dependent to some extent on the method of measurement chosen and the possible effects of the presence of other pollutants on the result. No misunderstanding occurs if the investigator realizes the limited accuracy of a method chosen for its simplicity and convenience. The second point concerns the role of the British Standards Institution and the corresponding bodies in other countries. The methods specified, which form the main concern of this chapter must, if they are to be generally useful, be simple and robust enough to be applied by the factory, on the ground area of the factory site or in the surrounding area by the staff of the factory, by public health inspectors or others concerned.

Mitigation of air pollution is an international problem, as the wind carries polluted air from one area to another. This is particularly important in a region such as western Europe, where several countries are highly industrialized and have geographical boundaries adjacent to one or more other countries. If comparable results of measurements are to be obtained it is important that precisely the same methods of measurement of individual pollutants should be used in each country. In this respect the British Standards Institution and the corresponding organizations in other countries are doing useful work in specifying in detail the equipment and methods of operation. These standards are devised by committees of experts in each country, and the move towards using the same equipment and methods in the various countries is encouraged by the International Standards Organization whose committees include representatives of the various countries concerned. Much has been done in this direction, but very much more needs to be achieved. The necessary negotiations often seem to occupy far too much time before agreement is reached. As improved equipment and methods of measurement are devised, new standards can be prepared and published to replace the earlier standards. The need for agreement on automatic systems available for making instantaneous measurements and continuous measurements with chart recorders is also a subject that should be studied; makers of such equipment provide details of the operation of their equipment.

Grit and dust

Measurement of emissions

The basic principles of measurement of solids in flue gases were defined by Hawksley *et al.*[1] in 1961 and suitable apparatus was described. An official publication[2] in 1967 gave advice on the approximations that might be allowed while still retaining sufficient validity of the results. The corresponding British Standard Specification[3] was derived from these documents.

The main difficulty of measuring the amount of grit and dust in a stream of gas arises from the fact that owing to their weight and momentum the particles do not uniformly follow the flow of gas where the speed and direction of the gas are changing. This affects the choice of sampling points and the method of sampling. Ideally, sampling should be done near the top of the chimney, but as this is not usually practicable, the samples have to be taken at the most suitable part of the system that is accessible and where two or more sampling holes each of about 100 mm diameter can be drilled into the chimney or flue and where a platform can be provided for the manipulation of the sampling probe and ancillary apparatus. The most suitable places are in the lower part of the chimney, as far as practicable from the point of entry of the flue gas, or in a straight vertical length of flue as far as possible from bends in the flue, particularly bends in the flue leading to the points of sampling and away from cyclone cleaning equipment. As the distribution of solids in the gas stream over the cross-section of the flue is rarely uniform, at least four points of sampling over the cross-section should be used. To avoid disturbing the flow of gas and solids in the flue at the points of sampling, the rate of withdrawing the sample at each point must be such that the gas and solids enter the sampling probe at the same velocity as that at which they travel along the flue.

Immediately before each measurement at a sampling point, a standard Pitot tube[4] is inserted into the flue to measure the velocity of the gas at the particular sampling point. From a calibration chart provided with the apparatus, the pressure drop required across the sampling probe to ensure sampling at the same velocity as in the flue is obtained. The Pitot tube is then removed and replaced by the sampling probe, to which suction is applied from a pump; the suction is regulated by a valve to give the correct pressure drop. The sampling probe is in effect a small cyclone. The inlet for the dust-laden gas points upstream in the flue and the gas entering it passes tangentially into the cyclone. The resulting swirl spins the grit and dust into the hopper part of the sampler and the dust-free gas is removed through the suction pump. At the end of the timed sampling period—say 10 min for the flue gas from coal-fired furnaces, much longer for the relatively clean gas from oil firing—the probe is removed and the material in the hopper is weighed. It can be separated into grit and dust by a 75 μm British Standard sieve.

If sampling is done at four points in the sampling plane as suggested, the average weight of grit and dust collected at the four points, multiplied by the ratio of the cross-sectional area of the flue at the sampling plane to the area of the inlet of the sampling probe gives the total weight of grit and dust emitted through the flue for discharge through the chimney during the period of sampling. With four sampling points and one test at each, as described, the accuracy should be within 25 per cent; by repeating the test at each point and averaging the results the accuracy should be improved to within about 20 per cent; by using eight sampling points with one test at each the accuracy should be within 15 per cent. To strain for greater precision makes the determination long and laborious.

For the larger coal-fired plants burning solid matter, the weight of grit plus dust allowed to be discharged from the chimney is less in relation to the quantity of fuel burned than for the smaller plants; and the amounts allowed for discharge from oil-fired boilers are much less than with furnaces burning solid fuel under the 1971 Regulations in the UK (see chapter 3).

Grit and dust deposition

The measurement of the rate of deposition of grit and dust from the air cannot be a precise determination. If a bowl is exposed on land the material falling into it includes not only the grit and dust discharged from chimneys in the area but also the grit and dust blown by the wind from heaps of coal and of various raw materials used by industries of various kinds. In addition there is the grit and dust of various kinds that have fallen onto roadways, roofs of buildings, etc. and later been entrained by the wind. The amounts of these falling materials that enter the bowl are affected by disturbances to the air flow caused by turbulence at the rim of the bowl; and some of the material deposited in the bowl may later be blown out again by the wind. The various types of deposit gauge in use differ in design and height above the ground because of compromises necessary to meet different and sometimes conflicting requirements. As a result, the measurements made with the different types of deposit gauge cannot be expected to give results in agreement with one another; even gauges of the same design give widely varying results when placed in adjacent positions.

American standard deposit gauge. This gauge,[5] often used in the US, is of simple design. It is an open-top polythene cylinder of diameter 150 to 200 mm and vertical sides of a height of not less than 275 mm. It is exposed so that the top of the gauge is not less than 2·5 m above the ground or not less than 1·25 m above any other surface, such as a flat roof, on which it may be placed. It must be placed so far away from higher objects such as walls or trees that a line from the gauge to the top of such objects is not inclined at more than 30° to the horizontal. The gauge is exposed for a month; the contents are then separated into material insoluble in water and material soluble in water and the amounts of each are determined. Further detailed analyses required for any more detailed surveys are also made. The advantage of this gauge is its low cost.

Fig. 4.1. British Standard deposit gauge. (Published by permission of the British Standards Institution. Photograph by the British Carbonization Research Association.)

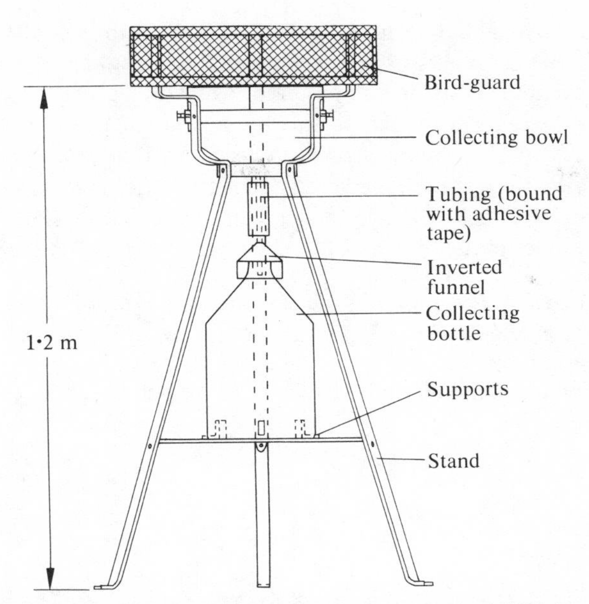

Fig. 4.2. British Standard deposit gauge. (Line drawing from British Standard 1747: Part 1: 1969 by permission of the British Standards Institution.)

British standard deposit gauge. The extensive surveys of grit and dust fall that have been made in the UK during the last 50 years have used this gauge,[6] which is likely to remain the UK standard for many more years. The collection bowl (Figs. 4.1 and 4.2) is a glass or polythene funnel of a diameter of 315 mm and with vertical sides of a height of 105 mm. It is fixed in a stand so that the mouth of the funnel at the top is 1·2 m above ground level. The aerodynamics of the equipment are affected to some extent by an open cylindrical, wire-mesh bird guard which projects about 100 mm above the bowl to prevent birds perching on the top edge of the funnel. The funnel-shaped bowl drains into a collecting bottle. In dry weather the deposit remains in the bowl but when it rains a greater or less proportion is washed down by the rain into the collecting bottle. The conditions of exposure are much the same as for the American gauge. At the end of a

TABLE 4.1

Insoluble matter collected by four pairs of adjacent British deposit gauges
(Average results over 24 months)

Site	Mean rate of deposit (mg m^{-2} day^{-1})	Mean percentage difference between the two gauges
1	404	6·0
2	171	6·5
3	56	36·0
4	23	5·5

month any dry material in the funnel is washed down into the collecting bottle with distilled water and the contents of the bottle are analysed for soluble matter, insoluble matter, and for any other constituents required. Table 4:1 gives the results of four tests[7] each made with two gauges standing side by side at each of the four sites. The results give some indication of the reproducibility of the data that can be expected.

In the UK the average annual rate of deposit of insoluble material, as measured by the British Standard deposit gauge ranges from about 200 mg m^{-2} day^{-1} in industrial areas down to values in the region of 60 mg m^{-2} day^{-1} in the country.

VDI (German) sticky plate gauge. A sheet of aluminium foil, about 40 mm by 80 mm, is exposed horizontally at about 1·5 m above the ground on the top of a stake.[8] A bird guard consisting of ten vertical, stout wires, each 120 mm long, surrounds the foil. The foil is covered with a thin layer of petroleum jelly, dried in a desiccator, and weighed. It

Fig. 4.3. Directional dust gauge by D. H. Lucas and D. J. Moore, of the Central Electricity Research Laboratory of Leatherhead, England. (Photograph by the British Carbonization Research Association, Chesterfield, England.)

is then exposed for a sufficient period, redried and weighed. The amount of deposit collected should not exceed 30 mg or the sticky nature of the surface is impaired. This gauge is in widespread use on the continent of Europe.

Directional dust gauge. The nuisance caused by dust in the air may be more closely related to the amount carried horizontally, or at a low angle to the horizontal, by the wind than to the amount falling vertically or near vertically.[9] A directional gauge was therefore designed. It is a vertical plastic cylinder with a vertical slit, 45 mm wide and 350 mm high, as inlet. When the wind is from the direction towards which the slit is facing, dust borne by it enters the slit and falls into the collecting bottle. By mounting four of these units on a stake (Fig. 4.3) with the slits facing the four points of the compass, a directional gauge is obtained that can be useful in determining the relative severity of the pollution by dust from individual areas or sources. After exposure for a month any material remaining in a cylinder is washed down into the collecting bottle with distilled water and the contents of the bottle are analysed in the usual way. Alternatively, an arbitrary measure may be obtained by transferring the contents to a special absorptiometer cell and measuring the obscuration of light as compared with that caused by the same amount of water in the cell.

Suspended particulate matter: smoke

When the particle size is very low, not more than a few micrometres, solid or liquid impurities in the air do not fall out readily under the influence of gravity but remain suspended for long periods of time. The most widespread of these pollutants is smoke, which includes brown tarry droplets produced by the destructive distillation of coal in the lower temperature zones of a fire, and black solid particles of impure carbon which can arise from the cracking at high temperatures of primary distillation products. Smoke, of whatever kind, is usually accompanied by very fine particles of dust which may be of any colour—white from cement works, red from steel works, pale brown from boiler plant fired with pulverized fuel, etc., according to the source.

Emissions of fine particulate matter

The only way in which particulate matter, fine enough to remain in suspension, can be determined in absolute terms is by filtration of a sample of the flue gas from which the coarser material has been removed, and weighing the fraction collected on the filter. This is provided for in the British Standard specification[3] for the determination of solids in flue gas, already discussed, by the provision of a filter immediately following the cyclone head of the probe. This is slow and laborious and various simpler methods are commonly used for special purposes where it is unnecessary to obtain absolute values of concentrations of suspended matter.

The Ringelmann chart

This rough, approximate method is designed to enable an estimate to be made of the darkness of the smoke issuing from a chimney. Because of its simplicity it was chosen as one of the industrial provisions of the British Clean Air Act 1956. A Ringelmann chart is a rectangular white card divided into five 101·6 mm (4 inch) squares of which one is white and the others (Fig. 4.4) have black cross-hatching so that 20, 40, 60 and 80 per cent of the white background is obscured.[10] When held at a sufficient distance from the observer, usually more than 15 m, the black hatchings appear to merge with the white background to give various shades of grey. The chart is mounted on a pole so as to fall in the line of sight of the observer looking at the smoke rising from the chimney; he decides which of the Ringelmann shades most nearly matches the smoke. Dark smoke is defined in the Clean Air Act 1956 as darker than Ringelmann No. 2 shade (40 per cent black hatching).

As this chart, mounted on a pole, is somewhat inconvenient in use, a miniature chart has been produced by the British Standards Institution. This miniature chart includes five 25 mm squares, showing five shades of grey corresponding with the large Ringelmann chart. The miniature chart (Fig. 4.5) can be held on a light telescopic rod by the observer himself at 1·5 m to 2·0 m from his eye.

Obscuration gauges

Provision was also made in the Clean Air Act 1956 for smoke meters or indicators to be fitted, as might be required, in the flues of boilers or furnaces; many such instruments are available. Most of them operate by projecting a beam of light across the flue on to a photoelectric cell on the other side. The output from the cell is fed to a recorder or indicator in such a way as to show the loss of intensity of the light due to the presence of particulate matter in the flue gas. Some instruments also operate a sound alarm when the loss of intensity of the light reaches an amount that can be altered as required by a key adjustment. Usually the alarm is set to operate at a proportion of smoke slightly less than the maximum allowed. Precautions have to be taken to prevent the deposition of smoke from the flue gas on the windows of the instrument on either side of the flue; otherwise a spuriously high value for the obscuration will be obtained. Such precautions usually include the supply of a continuous flow of clean air across the inner surfaces of the windows to prevent flue gas from coming into contact with the windows. The windows may also be protected by an aluminium honeycomb, within the narrow cells of which turbulence is zero so that the access of even the finest dust or of condensible gases to the window is prevented.[11]

For a given type of plant, calibration curves may be plotted connecting the reading of the gauge either with the Ringelmann number of the plume of smoke or fine particulate matter issuing from the chimney, or with the concentration of the solids in the flue gas.[11,12] It must be emphasized that this calibration will only be valid for a given type of smoke or other suspended particulate matter.

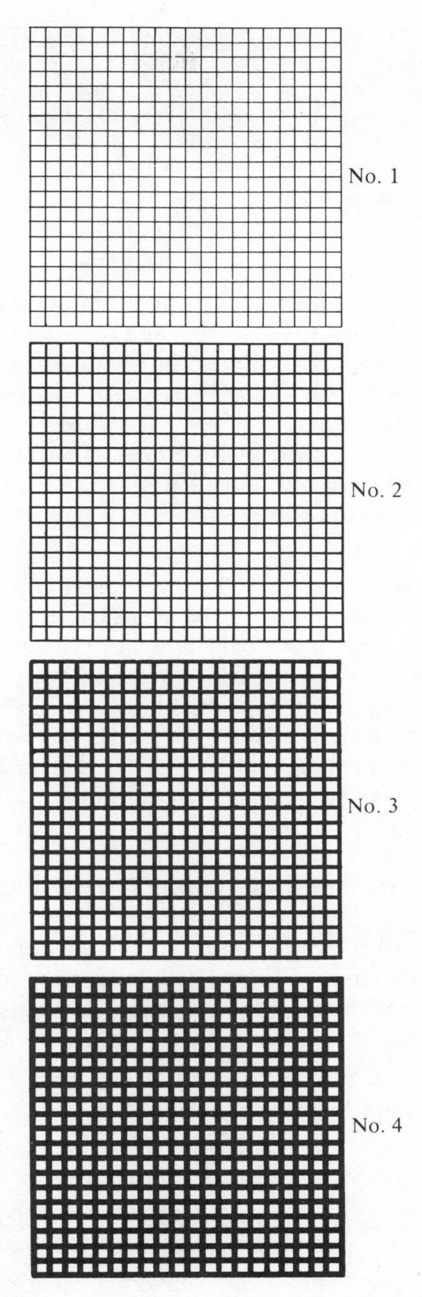

Fig. 4.4. British Standard Ringelmann chart. (By permission of the British Standards Institution. Block and print provided by the National Society for Clean Air, Brighton, England.)

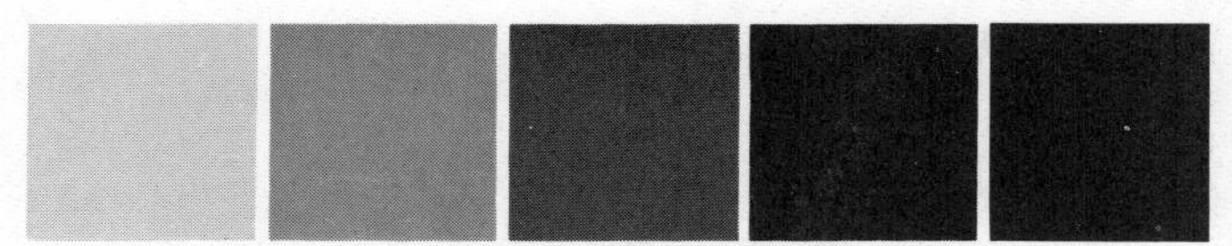

A half-tone reproduction of the miniature smoke chart

Fig. 4.5. British Standard miniature smoke chart. (By permission of the British Standards Institution. Block and print provided by the National Society for Clean Air.)

The Hartridge smoke meter is another type of obscuration gauge; this smoke meter is accepted for use in the British Standard method of testing exhaust gases from diesel engines.[13] A continuous sample of gas is taken from the exhaust pipe of the engine and passed through a cylinder 55 mm diameter and 407 mm long with a lamp bulb at one end and a photoelectric cell at the other end. The decrease in the intensity of the light caused by the solids in the exhaust gas is obtained by comparing the output of the photoelectric cell when the cell and the lamp are swung into corresponding positions across an identical cylinder through which pure air is passed.

The Bosch–Dunedin smoke meter is also accepted for use in the British Standard method of testing the exhaust gases from diesel engines; this meter depends on the staining of white filter paper when a specified amount of exhaust gas is passed through it. A pump, operated by hand, takes one-third of a litre of gas from the exhaust pipe of the engine and draws it slowly through a Whatman No. 4 filter paper, which is then removed from the apparatus. The darkness of the smoke stain is assessed with a photoelectric reflectometer. A calibration curve has been determined for diesel smoke[13] relating concentration of smoke in g m^{-3} to the darkness of the stain.

Measurement of smoke and total suspended particulate matter in the air

The only method of measuring the concentration of suspended matter in the air in absolute terms is to filter a sample of air and determine the weight of material collected on the filter. This procedure is circumstantial and time-consuming and for many surveys is only used to calibrate simpler methods depending on the staining of filter papers through which samples of air have been passed.

Gravimetric methods

The American high volume sampler is the simplest of the direct weighing methods and has been adopted by the US Environmental Protection Agency as their official reference method.[14] A glass-fibre filter, supported on a coarse wire gauze, is held horizontally in a filter holder so as to leave an area of 406·5 cm^2 facing upwards exposed to the air under a standard shield to prevent rain and adventitious matter reaching the surface. Air is drawn through at a rate of 1·13 to 1·70 $m^3\ min^{-1}$, which is calculated to prevent particles coarser than about 100 μm gaining access to the filter. Sampling is usually continued for 24 h. If, for example, the air contains 100 $\mu g\ m^{-3}$ of suspended matter—a common concentration in urban areas in the more polluted parts of the UK—the weight of material collected on the filter will be in the region of 200 mg, which can be obtained from the increase in weight of the filter without special precautions. Minor changes in the design of the apparatus could lead to improvement; the blast of filtered air, with the present arrangement, is directed to the ground where it can stir up dust and contaminate the air being sampled.

The corresponding British Standard method[15] uses a sampling rate some 20 times less than the American high volume sampler, so that with air containing 100 $\mu g\ m^{-3}$ of suspended matter, a sample taken over the same time interval would weigh only about 10 mg; more care must therefore be taken in drying and weighing the sample under standard conditions. This method was designed to collect material of the same size range as is collected by the British Standard method for smoke and uses the same type of apparatus.

Smoke stain methods

The British Standard method for the estimation of smoke in the air in terms of equivalent standard smoke is in widespread use in the UK in the National Survey of Air Pollution.[15,16] It is also used in other countries and is, in effect, the method adopted by OECD for such surveys.[17]

The air to be sampled is drawn through the apparatus (Fig. 4.6) at a rate of about 2 $m^3\ day^{-1}$ (24 h) by a suction pump. It enters through an inverted funnel placed outside the building in which the apparatus is housed and far enough from the wall not to be affected too much by changes induced in the general air flow by the presence of the building. The mouth of the funnel is 40 mm in diameter, a size chosen so that, with the flow rate used, falling material coarser than about 10 to 20 μm should not be drawn up into it. The air then passes upwards through a sheet of Whatman No. 1 filter paper, held as

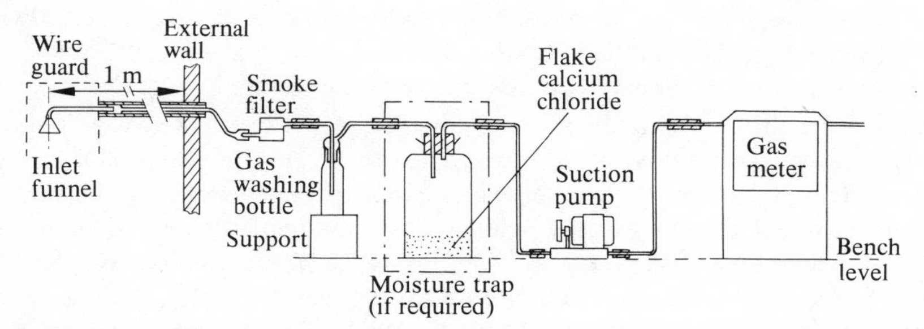

Fig. 4.6. British Standard 1747: Parts 2 and 3: 1969 on determination of suspended matter and of sulphur dioxide. (By permission of the British Standards Institution.)

shown, in a brass filter clamp, the two halves of which are clamped tightly together. A bubbler containing dilute hydrogen peroxide for the determination of sulphur dioxide (described later) follows, and then a vessel containing calcium chloride to remove moisture, followed by a suction pump and a gas meter. After sampling for 24 hours the filter paper is removed and the darkness of the smoke stain on it is determined with a photoelectric reflectometer. The concentration of smoke in the sample of air can then be read off from a calibration curve relating the darkness of the stain to the concentration of 'equivalent standard smoke' in microgrammes per cubic metre. This calibration needs some explanation. At a number of sites curves were drawn relating the darkness of the smoke stain to the concentration of particulate matter measured by what was in effect the British Standard gravimetric method already described. Different curves were obtained in different places according to the darkness of the smoke stain per unit weight of smoke. For example, smoke from road traffic is some four times as dark as an average town smoke and smoke in the area of cement works is much lighter than this average. The important point, however, is that the curves all belonged to the same family and could be converted one to another by multiplication by an appropriate factor. This enabled the curve for average town smoke to be taken as the standard and allowed concentrations of whatever type of smoke, darker or lighter than average, to be expressed in terms of equivalent standard smoke.

While it may be impracticable on the grounds of manpower and cost to conduct a survey on more than a very modest scale using a gravimetric method, the smoke-stain method enables surveys of much wider scope to be undertaken. It must be emphasized, however, that if the results are to be considered in terms of actuai weight concentrations of smoke, frequent calibration against a gravimetric method is essential. It must be borne in mind that the intrinsic darkness of smoke may vary not only from place to place, but at any one site with the time of day, the day of the week and the season of the year and progressively over a run of years. All these factors affect the proportions of the various components that go to make an average smoke. It has been argued that from the point of view of health and amenity it is the dark carbonaceous products of the imperfect combustion of fuels that are important and that the light coloured components are of minor interest. On this view it is the darkness index, as expressed by the concentration of equivalent standard smoke, that is important and not the actual weight concentration of total suspended particulate matter.

An experimental study by Lee *et al.*[18] of results obtained at a set of sites with the various methods available for the measurement of airborne particulate matter has provided valuable data on the limits of comparability of the results.

In considering the results of measurements with the daily smoke filter, it should be noted that the concentrations to be measured in terms of equivalent standard smoke vary from a few microgrammes per cubic metre in a country area in summer to perhaps the region of 400 $\mu g\ m^{-3}$ on a highly polluted day in an industrial town in winter. Concentrations of the order of 2000 $\mu g\ m^{-3}$ were observed in the London smog in December 1962. The annual average urban concentration varies from about 30 $\mu g\ m^{-3}$ in the south of England to nearly 90 $\mu g\ m^{-3}$ in the north. The determination itself shows reasonably good reproducibility. Using sets of observations with two trains of apparatus each with its own inlet funnel, the coefficient of variation (standard deviation/mean) was found to be about 6 per cent.[19]

Semi-automatic methods—smoke stains

In conducting surveys using smoke stain methods, labour can be saved by preparing eight trains of apparatus at the beginning of a week and arranging for automatic switching arrangements to divert the stream of air to be sampled from one train to the next every 24 hours. Several different types of apparatus are available for this purpose. In some instances it is useful to know how the amount of smoke in the air varies from hour to hour. This can be found conveniently using a ribbon of white filter paper clamped tightly between the two halves of an ordinary brass filter clamp so that air can be drawn through a circular disc of the paper in the usual way. At the end of each hour an automatic mechanism opens the clamp, moves the paper ribbon along 20–30 mm and recloses the clamp. In this way a circular smoke stain is obtained for each hour to be assessed with either a direct reading or an automatic reflectometer. The springs holding the two parts of the filter clamp must be sufficiently strong to make an air-tight seal on the filter paper. These pieces of apparatus are reliable and need little attention, but the average smoke concentration obtained from the 24 stains collected in a day should be checked occasionally against the standard daily instrument.

Sulphur oxides

All fossil fuels contain sulphur, which appears largely as sulphur dioxide in the flue gases when the fuel is burnt. The concentration in the resulting flue gas varies with the sulphur content of the fuel and with the conditions of combustion. As an example, coal with 1·5 per cent sulphur by weight when burnt in a power station gives a flue gas containing between 2 and 3 g m^{-3} of sulphur oxides or in the region of 0·1 per cent by volume. Of the sulphur oxides, more than 95 per cent by volume is sulphur dioxide and less than 5 per cent is sulphur trioxide. These concentrations present no difficulties in measurement.

The problem is very different when the chimney gases have been greatly diluted in the atmosphere and it is a question of determining the concentrations of oxides of sulphur in town and country air. Concentrations of sulphur dioxide in the air range from a few microgrammes per cubic metre on a clean day in summer in the country to 1000 $\mu g\ m^{-3}$ or more on a badly polluted day in an industrial town in winter. In the London smog in December 1962, values exceeding 4000 $\mu g\ m^{-3}$ were recorded. Annual average figures in urban air[20] vary from about 50 $\mu g\ m^{-3}$ in the south of England to about 100 $\mu g\ m^{-3}$ in the north. London has the high average of about 130 $\mu g\ m^{-3}$. The difficulties of precise measurement arise not so much from the low concentrations as from the presence in the air of other substances at concentrations at which they interfere to some extent.

Emissions of sulphur oxides

The British Standard method for the determination of both oxides simultaneously,[21] which should be read in conjunction with the corresponding specification on methods of sampling flue gas,[22] is still the best available. Suspended solids are first removed from the hot gas by a filter and the gas is then saturated with cold 80 per cent isopropanol to promote the formation of sulphuric acid mist from the sulphur trioxide and to inhibit the oxidation of sulphur dioxide. The sulphuric acid mist is collected by passing the gas through a sintered glass disc and is retained, together with excess isopropanol, in a collecting vessel. Some sulphur dioxide is dissolved at this stage, but most passes on with the gas stream and is absorbed in standard iodine solution. The flow of gas is continued until the iodine is just decolourized, which allows the amount of sulphur dioxide absorbed from the gas to be calculated. The sulphuric acid in the isopropanol in the collecting vessel is determined by titrating an aliquot part with barium perchlorate using thorin as indicator. A second aliquot part is titrated with standard iodine to give the small amount of sulphur dioxide retained in the isopropanol.

The accuracy of this method of determining the concentration of sulphur dioxide in the flue gases to be discharged from a chimney is within about 5 per cent with a concentration in the region of 200 ppm by volume and within 2 per cent for concentrations of about 2000 ppm. For sulphur trioxide the accuracy is within 10 per cent at 5 ppm by volume and within 5 per cent at 20–50 ppm.

Sulphur dioxide in the atmosphere

The basic choice is between a method that measures concentrations specifically but is inconvenient for use as a regular routine, and a much simpler method that is subject to some interference from other substances that may be present in the air. It seems reasonable, as in the selection of a method for the determination of smoke, to adopt the simpler method for general survey work but to check it frequently against the reference method to be certain that under the conditions of use for the particular purposes it gives sufficiently accurate results.

The pararosaniline method has been adopted as the reference method by the US Environmental Protection Agency under the US Clean Air Act. It is described in detail in the Federal Register.[14] In principle, sulphur dioxide is absorbed from a stream of air by a solution of potassium tetrachloromercurate. A dichlorosulphitomercurate complex is formed which resists oxidation by the oxygen of the air. Once formed, the complex is stable to strong oxidants such as ozone and oxides of nitrogen. At the end of the sampling period the complex is treated with pararosaniline and formaldehyde, when the intensely coloured pararosaniline methyl sulphonic acid is formed. The method is not an absolute determination in the sense that would require the concentration of sulphur dioxide to be calculated from the readings by the rules of chemistry and physics. The absorbance of the intensely coloured product is measured in a spectrophotometer. For each sample of the reagents and for each spectrophotometer a calibration curve has to be constructed relating concentration of sulphur dioxide to absorbance; fortunately the relationship is rectilinear. The calibration may be done by starting with solutions of tetrachlormercurate and adding known amounts of sodium sulphite instead of bubbling air containing sulphur dioxide through them. Alternatively, air containing known amounts of sulphur dioxide may be used. This is prepared by passing purified air at known rates over standard permeation tubes which release sulphur dioxide at a known constant and very slow rate. These tubes can be obtained from the US Bureau of Standards.

For sampling in the field, 30 min samples are taken using 10 ml of tetrachlormercurate solution in a midget impinger, sampling at the rate of 1 lb min^{-1}. For longer periods larger amounts of liquid and correspondingly slower flow rates are used. There is some doubt as to whether samples should be taken for periods longer than 6 h, particularly on hot summer days. The lowest limit of detection for a 30 min sample is 25 $\mu g\ m^{-3}$, but by using longer sampling times much smaller concentrations can be measured, with checking against atmospheres made up with a permeation tube.

The hydrogen peroxide method is the much simpler procedure adopted as the British Standard method[16,23] and also adopted by OECD.[17] It uses the same apparatus as for the British Standard method for smoke (Fig. 4.6). The sample of air to be analysed is passed at a rate of 2 $m^3\ day^{-1}$ (24 h) first through a smoke filter and then bubbled through 0·3 per cent (w/v) hydrogen peroxide solution (also known as one-volume hydrogen peroxide) that has been adjusted initially to pH 4·5. At the end of the 24 h the solution is titrated with 0·01 N sodium carbonate solution to an end point of pH 4·5. The titration may be done with a mixed indicator or with a pH meter. The end point of pH 4·5 is chosen so that the large amount of carbonic acid present from the solution of carbon dioxide from the air cannot interfere. The results are recorded as microgrammes of sulphur dioxide per cubic metre on the convention that the increase in acidity of the hydrogen peroxide is due solely to the oxidation of sulphur dioxide to sulphuric acid. For concentrations up to 500 $\mu g\ m^{-3}$, replicate determinations should be within 20 $\mu g\ m^{-3}$ of the mean and for higher concentrations within 4 per cent.

The method is not specific for sulphur dioxide, but really measures net gaseous acidity. Nitrogen dioxide contributes a little to the acidity, but the main interference is from ammonia which neutralizes a part of the sulphur dioxide. In addition, in humid weather, sulphur dioxide may be absorbed by the smoke filter to a greater extent than normally occurs. While for ordinary urban atmospheres net gaseous acidity probably does not differ from the true sulphur dioxide concentration by more than about 10 per cent, the errors are likely to be greater when dealing with country air, where the concentration of sulphur dioxide is low and that of ammonia can be relatively high. For such conditions, the simple titration with alkali must be replaced by titration with barium perchlorate, using thorin as indicator. This makes the method specific for sulphur dioxide, but it is time-consuming as, for each titration, a curve showing intensity of colour against amount of reagent added must be constructed in order to obtain an accurate end-point. For very low concentrations of sulphur dioxide, the sampling is better continued for two weeks instead of for one day, but a somewhat different type of apparatus is required to minimize evaporation of the hydrogen peroxide solution.

Sulphur dioxide recorders are available for use when continuous records are required for various purposes concerned with the magnitude of the peak concentrations concealed in the daily averages and the reasons for their occurrence. One of the best types of recorder available uses the flame photometric method.[24] Pure hydrogen is burnt in a stream of the air to be analysed and the sulphur compounds in the air produce a strong luminescence between 300 and 423 nm. By arranging a photomultiplier tube to view the region above the flame through an optical filter transmitting at 394 ± 5 nm, the method is specific for sulphur and can be used for concentrations above about 25 $\mu g\ m^{-3}$.

This type of apparatus is modern and expensive, and most recorders now in use belong to an earlier generation of instrument. Many depend on bringing the stream of air to be analysed into contact with a slow counter-flow of reagent, changes in which are then measured continuously in some way that gives an electrical response that can appear on a recorder chart or be fed into data processing equipment. This is easy with the pararosaniline method, as the output of the spectrophotometer is an electrical one. It is achieved for the hydrogen peroxide method by measuring the acid in the exposed solution by means of its electrical conductivity. Many different types of this basic equipment are available; all of them can probably be made to work satisfactorily. On average, it probably requires one good technician, with support as required from an instrument workshop, to keep five such recorders operating properly. Much of the dissatisfaction with these recorders, frequently expressed, has probably arisen from a failure to appreciate the need for skilled attention and servicing to avert breakdown.

One of the simpler of the recording systems and probably requiring less maintenance to keep it in working order, is that using the coulometric method. The sample of air to be analysed is drawn through a coulometric titration cell containing a solution of buffered halide. Sulphur dioxide present in the sample of air reacts with the halide and increases the resistance to the flow of electricity through the cell. The current from the reference electrode needed to return the cell to a steady state is directly related to the concentration of sulphur dioxide. After amplification, the current from the reference electrode is fed to a recorder or a data processing system. The method is suitable for measuring concentrations of sulphur dioxide above about 60 $\mu g\ m^{-3}$.

Sulphuric acid mist in the atmosphere

The method normally used for determining the amount of sulphuric acid mist in air is that described by Commins.[25] The sample of air is filtered through Whatman No. 1 paper. The acid and soluble salts are removed by washing with distilled water and the free acid is titrated with standard sodium tetraborate solution to pH 7·0. The amount of acid indicated in this way is often an underestimate, as some of the acid may react with water-insoluble bases such as calcium carbonate that may be present on the filter. This error is avoided by adding excess sodium tetraborate initially and back-titrating. The method can detect less than 1 $\mu g\ m^{-3}$ with an air sample of about 10 m^3 and a rate of flow in the region of 10 l min^{-1}.

At best, this procedure should give net particulate acidity, but there is no means of determining what relationship this bears to the concentration of sulphuric acid mist. Recently, doubt has been cast on the validity of the titration, because weak electrolytes are usually present. If, instead, a Gran electrometric titration is done, the analysis can be refined to give the net particulate strong acidity. In this modification, for each addition of reagent the pH is measured with a glass electrode, and a type of titration curve is plotted from which the end point for the neutralization of strong acid can be obtained.[26]

Carbon monoxide

Carbon monoxide, whether in the exhaust gases from an engine or in the atmosphere, is now almost always measured by the non-dispersive, infra-red absorption method.[14] It is non-dispersive in the sense that there is no need to disperse the incoming radiation by wave length as in a spectrophotometer. It depends on the strong absorption by carbon monoxide in the infra-red around 4·6 μm. Radiation from a source emitting a wide range of wavelengths in the infra-red is split into two parallel beams which are directed through two tubes, one containing the air to be analysed and the other pure air. A detector is placed at the end of each tube remote from the source of radiation; it consists of a diaphragm assembly sealed in an atmosphere of pure carbon monoxide. Radiation of a wavelength of about 4·6 μm reaching these detectors is absorbed by the carbon monoxide in them and causes heating. This in turn increases the pressure and displaces the diaphragm. By arranging the diaphragm as part of a condenser system the displacement is detected electronically and is amplified to provide an output signal. The instrument requires calibration with a series of standard gas mixtures, which are readily available.

Since water vapour absorbs in the same region of the infra-red it is essential to dry the sample of air or other gas before it is introduced into the analyser. This can be done either by refrigeration or by one of the usual drying agents such as silica gel. The analyser may be used in the field by passing the air to be analysed through it continuously or samples may be taken in plastic bags and analysed by transferring to the instrument in the laboratory.

For urban air the range of the analyser is usually chosen as 0–60 mg m^{-3}; the sensitivity is then about 0·6 mg m^{-3}, and the accuracy within 1 per cent of the full scale reading. Values in city streets may be in the region of 5–50 mg m^{-3} with peak values reaching 300 mg m^{-3} or more. For exhaust gases from motor vehicles, sampled out of the exhaust pipe, the analyser is set to read from 0 to 5 or 0 to 10 per cent by volume, which is 0 to 60 or 0 to 120 g m^{-3}. The accuracy should be within 5 per cent of the full scale reading.

Hydrocarbons

When hydrocarbons are introduced into a hydrogen flame they cause ionization and the current flowing between electrodes with a potential difference in the region of 200 V between them gives a measure of the rate at which hydrocarbons are being added to the flame. A number of commercial instruments are available working on this principle. They can be set to measure the comparatively high concentrations found in motor engine exhausts[27] or the relatively low concentrations in the ambient air.[14] They all require calibration over the range of concentration concerned, and this is done with standard propane mixtures. If the apparatus is combined with a gas chromatograph, the methane in the sample may be separated and the remainder determined to obtain a value for hydrocarbons other than methane, as is required under the US Clean Air Act.

Oxides of nitrogen

The validity of the methods commonly used for the measurement of oxides of nitrogen in flue gas and in the air has been uncertain until recent years, when the emergence of the chemiluminescent method, which is specific for nitric oxide, has enabled the position in relation to the other methods to be clarified.

Nitric oxide

The chemiluminescent method depends on the fact that when nitric oxide reacts with ozone to give nitrogen dioxide about 10 per cent of the nitrogen dioxide is electronically excited and its transition to the ground state yields a detectable light emission with wavelength in the region of 590 to 630 nm. The emission increases with reduction in pressure and its intensity is proportional to the mass flow rate of nitric oxide into the reactor. The light emitted is passed through optical filters to remove unwanted wavelengths and is measured with a photo-multiplier tube and its associated electronics.[14] In the commercially available instruments, ozone in the correct concentrations is produced by a built-in ozonizer. Calibration is easy as dilute nitric oxide of the necessary purity is readily available in cylinders. The accuracy is within about 2 per cent of the full scale reading. The sensitivity varies with the pressure in the reactor, ranging from $1{\cdot}2\ \mu g\ m^{-3}$ at 1 torr to $6\ \mu g\ m^{-3}$ at atmospheric pressure. The pressure is usually maintained at a value somewhere between the two.

Chemical methods for the determination of nitric oxide which depend on its oxidation, for example with permanganate, to nitrogen dioxide, which then reacts with various reagents to give highly coloured products, are not recommended as the oxidation stage is not reliable.

Nitrogen dioxide

Of the methods available the best are probably the chemiluminescence method and the older Saltzman chemical procedure.

Chemiluminescence method. The simplest method is to decompose the nitrogen dioxide to nitric oxide, which is then determined as already described. The decomposition can be done by passing the sample of air through a stainless steel tube at 800°C when the decomposition occurs quantitatively. However, any ammonia in the sample is at least partially converted to nitric oxide and thus interferes with the determination. This can be avoided by using a much lower temperature, below 250°C. The favoured method at present is to use a quartz tube packed with fine strands of an inert mesh, such as gold or glass wool, coated with carbon and maintained at 220°C. If unstable nitrogen compounds of the type of peroxyacetyl nitrate are present they are decomposed and augment the recorded concentration of nitrogen dioxide.

Saltzman method.[28,29] A sample of the air is drawn through a filter at a rate of 0·6 l min^{-1} and then through two bubblers in series, each containing a solution of sulphanilic acid and naphthyl ethylenediamine to give an azo dye with an intense magenta colour. The intensity of the colour is measured with a spectrophotometer or a simpler colorimeter.

The method is not an absolute determination as it depends on a calibration that has to be carried out for each set of apparatus. This can be done by plotting intensity of colour against known concentrations of nitrogen dioxide in air, obtained by use of a permeation tube, when it is found that the relationship is strictly rectilinear. Instead of using known concentrations of nitrogen dioxide in air it is simpler to add known amounts of sodium nitrite to the Saltzman reagent. The relationship is again rectilinear, but for some unknown reason the colour intensity produced by x moles of nitrogen dioxide is the same as that produced by $0{\cdot}74x$ moles of sodium nitrite. In spite of this second arbitrary feature in the determination it is considered to be satisfactory. The method is sensitive; in a one hour sample the minimum concentration that can be detected is 10 $\mu g\ m^{-3}$. Sulphur dioxide, ozone and peroxyacetyl nitrate interfere, but in town air the extent of the interference is normally small.

Ozone and oxidants

Until recent years it has been impossible to make a measurement of ozone in air without interference from other oxidizing substances; this can now be done with the introduction of the chemiluminescence technique. It should be mentioned that the simpler but much less precise neutral, buffered potassium iodide method that has been in use for many years gives a useful indication of the pollution of the air by oxidizing substances.

Ozone

The chemiluminescence method, adopted by the US Clean Air Administration as the reference method,[14] depends on the reaction of the ozone in the sample of air with ethylene. The sample flows through the apparatus at a rate of about 1 l min^{-1} with an excess of ethylene. A short-lived ozonide is formed which decomposes to formaldehyde with the emission of light. This light is measured by a photomultiplier tube, the output of which is recorded and gives a measure of the ozone originally present. The instrument is calibrated with purified air from a cylinder, which is passed through an ozonizer to give a series of concentrations of ozone, determined by neutral potassium iodide.

The method is applicable to concentrations from 10 to 2000 $\mu g\ m^{-3}$ and is accurate to within about 7 per cent. It may be noted that concentrations in the region of 50 $\mu g\ m^{-3}$ occur in nature. The US Clean Air Act has set a limiting target of 160 $\mu g\ m^{-3}$ for an average one hour concentration that should not be exceeded more than once a year. Concentrations in the region of 100 $\mu g\ m^{-3}$ have been observed in London.

Oxidants

When air containing ozone and other oxidants is bubbled through neutral, buffered potassium iodide solution, iodine is liberated. The ozone reacts quantitatively. Some 10 per cent of nitrogen dioxide and some 20 per cent of any peroxyacetyl nitrate that may be in the sample of air also react. The most serious interference is from sulphur dioxide, which reacts quantitatively in the opposite direction to that of ozone. The sulphur dioxide must therefore be removed before the sample of air reaches the bubbler containing potassium iodide. This is done by inserting a large U-tube packed with glass fibre paper impregnated with chromium trioxide; the chromium trioxide also converts any nitric oxide present to nitrogen dioxide, which must be taken into account when assessing the oxidants.

The bubbler, a glass midget impinger, contains one per cent potassium iodide solution, buffered with the requisite mixture of potassium dihydrogen phosphate and disodium hydrogen phosphate. The amount of iodine formed is estimated colorimetrically at 352 nm.

Surveys

Many of the methods described in this chapter are for the measurement of concentrations of pollutants in the atmosphere and are used for surveys of pollution. It is considered useful to conclude with some comments on the design of surveys, particularly surveys to be made for the assessment of pollution in relation to public health and amenity and legal requirements and other obligations. This section is not concerned with specialist surveys made by research organizations in investigations of subjects such as microclimate. The points to be considered can perhaps best be discussed in relation to three examples of surveys of pollution by gases and aerosols.

Surveys in country areas

The simplest case is the survey of an area of mainly agricultural land in relation to possible damage from pollution blowing in from an industrial installation outside the area or arising from such an installation within it. This type of survey often has to be undertaken before planning new housing estates on the outskirts of towns. It is a simple type of survey because several serious sources of pollution are absent and because there are no great complications caused by turbulent air flow around buildings.

The measuring instruments are arranged either in a square grid pattern or in a grid formed by concentric circles drawn round the source of pollution and on radial lines drawn from it. If apparatus worked by batteries is chosen, there is no difficulty in siting the items of equipment according to plan in small huts that can be placed where required. Sufficient flexibility must be allowed in siting the instruments to ensure that the inlet is not in a hedge or in a small wood where the damp foliage would absorb pollutants and lead to spuriously low results. If it is essential to have a supply of mains electricity to operate the instruments, considerable divergencies from the planned grid have to be tolerated to avoid unwarranted expense in installing temporary power lines of undue length. There are no definite rules. The best arrangement may be to install the instruments as near to the planned sites as is possible at reasonable expense and ascertain whether the pattern of pollution is sufficiently precise for the purpose of the particular survey. Only if the answer is firmly in the negative should the extra cost of improving the placing of instruments be incurred.

There are no rules governing the closeness of the grid. It seems reasonable to start with a coarse grid and judge from the results whether the pattern of pollution is being defined sufficiently accurately for the purposes of the particular survey; only if it is not sufficient should the grid spacing be decreased. It is always worth giving thought to avoid striving after an unnecessary degree of precision and so to avoid squandering resources. In many instances it is the financial and manpower resources that determine the density of the grid.

Basically, the measurements required at each point of the grid should be daily (24 h) measurements which can be obtained with comparatively simple and relatively cheap apparatus. If variation of pollution during the day is important, sufficient supplementary information can usually be obtained by the use of one automatic continuous recorder, which can be moved from site to site as required. It is expensive to have continuous recorders at all sites. The analysis of the data involves, at some stage, knowledge of changes in wind direction and speed. Data that apply sufficiently well to the area concerned can usually be obtained from the nearest meteorological station.

The data coming in from the survey should be scrutinized daily or at least weekly for unexpected or unreasonable results. These may be due to instrumental failure or error in analysis and their early detection may avoid wasting long runs of expensive data. On the other hand the odd results may be genuine and their immediate investigation may reveal unsuspected escapes of pollution from neighbouring works caused by changes in plant operation or by plant failure.

Surveys in urban areas

The next more complicated example is broadly similar to the last, but it is assumed that the area is largely covered by housing estates. This introduces a number of extra problems. The first concerns the siting of instruments to enable the pollution pattern to be established with sufficient precision, as very wide divergencies from any systematic grid have to be tolerated. In practice, almost the only sites where the equipment can be protected from vandals and will be accessible to the survey staff are sites in public buildings. Schools, clinics, hospitals, town halls, police stations, etc., have to serve as sites for most of the instruments. The success in choosing them can only be judged from the pattern of pollution obtained, which, if necessary, serves as a guide in further efforts towards better siting.

Another complication arises from the fact that the houses and buildings in which the instruments are sited are themselves emitting pollutants, as are also the houses in the immediate vicinity. In an urban area turbulent air flow in the streets generally ensures that the air below chimney level is usually well mixed so that the height above ground of the inlets of the measuring instruments is not critical. Ideally, perhaps, it should be at breathing level, about 1·8 m above the ground, but it is almost impossible to protect an inlet at this height from vandals; a height in the region of 3 m is much better. It is usually stated that the inlet should be at least 1 m clear from the surface of the wall of the building in which the instrument is housed. The object is to avoid air currents closely associated with the building, which might, for example, bring chimney gases down the wall and increase the concentration of pollutants in the air being sampled. It also avoids making the sample unrepresentative in the opposite sense, as could come about by absorption of pollutants by the wet wall of the building. The ideas behind all this are probably sound but there is probably no quantitative foundation for the 1 m rule. It is certainly more important to avoid placing the inlet where it might be affected either by a plume of gases from any nearby chimney or by puffs of gas that are so often brought down in the turbulent wake of the building itself or of a neighbouring taller building.

If the survey is concerned with pollution from road traffic, the inlets of the instruments must be placed above the pavements of the streets concerned, which must be those with heavy congested traffic, as concentrations of pollutants emitted by traffic fall off rapidly on leaving such a street and going only a little way down a side street. On the other hand, if the survey is concerned with the general pollution in the urban area care must be taken to avoid siting any of the instruments in a street carrying heavy traffic or the general pattern of pollution will be greatly distorted.

Surveys of a town

Many factors determine the pollution of any constituent part of a town. One of the important factors is the rate of emission of the pollutants concerned per unit area. With this in mind, it seems reasonable in the first place to divide a town into four types of district:

1. commercial centre with central heating and high emissions of sulphur dioxide but little smoke;

2. inner housing area—old-fashioned congested housing with high density emissions of smoke due to the high density housing, but less sulphur dioxide;
3. outer local authority housing estates—low density smoke emissions corresponding to the lower density housing and possibly in smoke control areas;
4. industrial areas—large individual emitters of sulphur dioxide and possibly small quantities of some other pollutants, but relatively little smoke.

The simplest type of survey could then be made by siting one instrument in each type of district. If the survey were concerned with the approximate average pollution of the town, each instrument would be sited in what would appear by subjective assessment to be an average part of each type of district. If, however, the object were to find out whether or not abatement measures should be taken, the instrument in each type of district would be sited in what appeared to be the most polluted part of the district. If, as a result, it were found that pollution was undesirably high, a modest extension of the survey would serve to define the areas where action should be taken.

References

1. Hawksley, P. G. W., S. Badzioch and J. H. Blackett, *Measurement of solids in flue gases*, British Coal Utilization Research Association, now obtainable from the Institute of Fuel, London, 1961.
2. *Grit and dust—the measurements of emissions from boiler and furnace chimneys. Standard levels of emission,* Code 75-187, HMSO, London, 1967.
3. *Simplified methods for measurement of grit and dust emissions from chimneys,* BS 3405, British Standards Institution, London, 1971.
4. *Code for flow measurement,* BS 1042, British Standards Institution, London, 1946.
5. 'TR-2-Air pollution measurements committee. Recommended standard method for continuing dust fall survey (APM-1 Revision 1)', *J. Air Poll. Contr. Assoc.,* **16**, 372, 1966.
6. *Methods for the Measurement of Air Pollution, Part 1, Deposit Gauges,* BS 1747: Part 1. British Standards Institution, London, 1969.
7. Craxford, S. R., D. W. Slimming and E. T. Wilkins, 'The measurement of atmospheric pollution: the accuracy of the instruments and the significance of the results', *Proc. Harrogate Conf. Nat. Soc. Clean Air,* Brighton, England, 1960.
8. V. D. I. Richtlinien 2119, Staubniederschlagsmessungen Gerätebeschreibungen u. Gebrauchsanweisungen, 1962.
9. Lucas, D. H. and D. J. Moore, 'The measurement in the field of pollution by dust', *Int. J. Air and Wat. Poll.,* **8**, 441, Pergamon Press, London, 1964.
10. *The use of the Ringelmann and miniature smoke charts,* BS 2742, British Standards Institution, 1969.

11. Lucas, D. H. and W. L. Snowsill, 'Routine dust measurement methods developed at cen. elec. res. lab.', *Proc. Harrogate Conf.*, Nat. Soc. Clean Air, Brighton, 1960.
12. Hurley, T. F. and D. L. R. Bailey, 'The correlation of optical density with the concentration and composition of the smoke emitted from a Lancashire boiler', *J. Inst. Fuel,* **31**, 534, 1958.
13. *The performance of diesel engines for road vehicles,* BS Au 141. British Standards Institution, 1967.
14. National Primary and Secondary Ambient Air Quality Standards. US Environmental Protection Agency, Federal Register, **36**, 3825, 22384. Washington, 1971.
15. *Methods for the measurement of air pollution.* Determination of concentrations of suspended matter, BS 1747: Part 2, British Standards Institution, 1969.
16. *National survey of smoke and sulphur dioxide. Instruction Manual,* Warren Spring Laboratory, Stevenage, 1966.
17. *Methods of measuring air pollution,* OECD, Paris, 1964.
18. Lee, R. E., J. S. Caldwell and G. B. Morgan, 'The evaluation of methods for measuring suspended particulates in air', *Atmos. Envir.*, **6,** 593, Pergaman Press, London, 1972.
19. Weatherley, M.-L. P. M., *Measurement of atmospheric smoke and sulphur dioxide,* Warren Spring Laboratory, Stevenage, Hertfordshire, Report RR/AP/70, 1962.
20. Craxford, S. R., M.-L. P. M. Weatherley and B. D. Gooriak, *National survey of air pollution 1961–71, Vol. 1, Part 2, The United Kingdom, a summary,* HMSO, London, 1972.
21. *Methods for the sampling and analysis of flue gases. Part 4, miscellaneous analyses,* BS 1756: Part 4, British Standards Institution, 1965.
22. *Methods for the sampling and analysis of flue gases. Part 1, Methods of sampling,* BS 1756: Part 1, British Standards Institution, 1971.
23. *Methods for the measurement of air pollution. Part 3, Determination of sulphur dioxide,* BS 1747: Part 3, British Standards Institution, 1969.
24. Stevens, R. K. and A. E. O'Keeffe, 'Modern aspects of air pollution monitoring', *Anal. Chem.,* **42**, 143A, 1970. (See also R. K. Stevens, T. A. Clark, C. E. Decker and L. F. Ballard, Paper to the APCA meeting, Miami, 1972.)
25. Commins, B. T., 'Determining particulate acid in town air', *Analyst,* **88**, 364, 1963.
26. Gran, G., 'Determination of equivalence point in potentiometric titration', *Analyst,* **77**, 661, 1952.
27. 'Continuous hydrocarbon analysis of diesel emissions', *Soc. Automotive Eng. J.,* 215, 1970.
28. Saltzman, B. E., 'Colorimetric microdetermination of nitrogen dioxide in the atmosphere', *Anal. Chem.,* **26**, 1949, 1954.
29. Byers, D. H. and B. E. Saltzman, *Am. Ind. Hyg. Assoc. J.* **19**, 251, 1958; Saltzman, B. E. and N. Gilbert, *Anal. Chem.* **31**, 1914, 1959; Saltzman, B. E. and A. F. Wartburg, *Anal. Chem.* **37**, 779, 1965.

5 Meteorological factors and dispersion

G. SPURR

Meteorology

The dissipation of waste material discharged into the atmosphere depends on the meteorological conditions prevailing in the earth's atmospheric boundary layer, a surface layer not more than 1000 m thick. In the broadest sense dispersion is controlled by the fluctuations of wind speed and direction. Other factors such as atmospheric temperature stability, topographical features, and surface roughness strongly influence the extent to which emissions are diluted by the wind.

Wind

The general flow of air over the earth's surface is produced by large-scale pressure systems, and wind strength and direction at any point are determined by the position, intensity and movement of these systems. Within this framework, there are important local factors such as the presence of sea or land, mountains or large conurbations which all influence the details of air movement.

The changes of wind direction and speed with time at a particular site can be presented diagrammatically in the form of a wind rose, showing climatological statistics of great importance in air pollution studies. Wind roses can be obtained from most meteorological stations. A wind rose or polar diagram consists of a series of lines emanating from the centre of a circle and pointing in the direction from which the wind blows. The length of each line represents the frequency of the wind from that particular direction. The speeds of the wind on the Beaufort scale, in the different directions, are

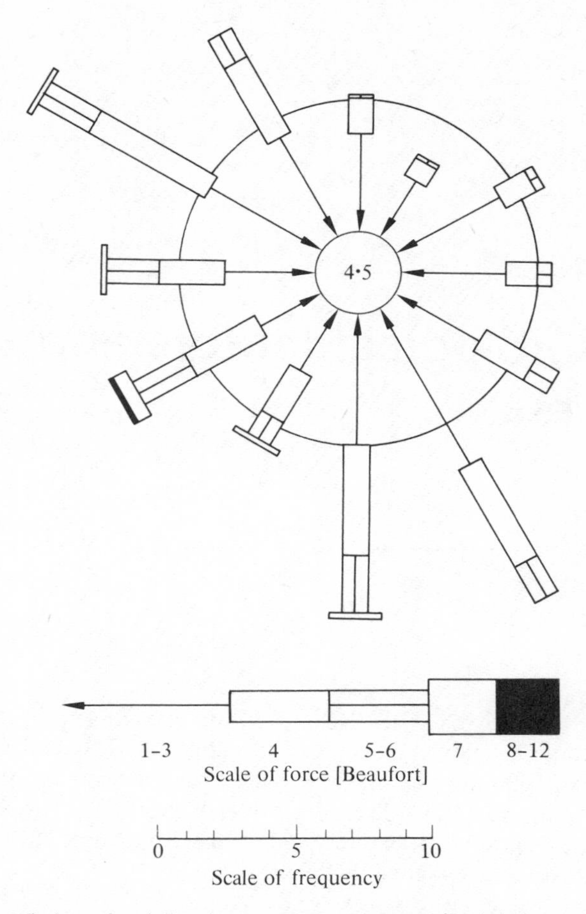

Fig. 5.1. Wind speed and direction recorded at Liverpool (Speke) airport 1957–61.

also indicated in the rose in Fig. 5.1. The circle marked 4·5 in the centre of the wind rose given as an example means that for 4·5 per cent of the time covered there were calms with very low wind velocities.

Scientists who specialize in calculating, using various formulae, the dispersion in the air of pollutants discharged in the gases from chimneys, usually use wind speeds expressed in metres per second. This method of expressing wind speeds is more precise than given by the number on the Beaufort scale. In Table 5.1 the ranges of wind speed for each number of the Beaufort scale are given in miles per hour and equivalent metres per second.

Most areas are subject to some prevailing wind, which means that the wind blows more frequently in one direction than in others. In the British Isles prevailing winds blow from the south-west although the presence of land blocks or deep valleys give rise to local variations.

TABLE 5.1

Beaufort scale of wind velocities

Force	Beaufort scale description	Velocity (mi h^{-1})	(m s^{-1})
0	Calm	0	0
1	Light	1–3	0·45–1·34
2	Light	4–7	1·8–3·1
3	Light	8–12	3·6–5·4
4	Moderate	13–18	5·8–8·0
5	Fresh	19–24	9·5–10·7
6	Strong	25–31	11–14
7	Strong	32–38	14–17
8	Fresh gale	39–46	17–21
9	Strong gale	47–54	21–24
10	Whole gale	55–63	25–28
11	Storm gale	64–72	29–32
12	Hurricane	73–82	33–37

Wind shear. At the top of the earth's atmospheric boundary layer (on average about 600 m) the wind flow is unaffected by the earth's surface and is almost parallel to the isobars. The speed of this 'gradient wind' is determined by the horizontal pressure gradient. Closer to the ground friction due to the surface roughness reduces the wind speed and changes its direction in addition to increasing the turbulence. The vertical wind speed profile within the turbulent boundary layer can be represented by a simple power law such as

$$\frac{u_2}{u_1}=\left(\frac{z}{z_1}\right)^{\alpha},$$

where u_1 is a reference velocity at height z_1 (usually the height of the boundary layer) and α varies between 0·14 and 0·40 depending on the roughness of the underlying terrain (Fig. 5.2) and on the temperature stability of the atmosphere (Davenport[1]).

Turbulence

Atmospheric turbulence can be convective or mechanical in origin. Convective turbulence is produced by variations in the solar heating of the surface and also depends on the thermal stability of the atmosphere. It is at a maximum on a clear sunny day in the afternoon and at a minimum at night or in the early morning. Mechanical turbulence arises from the movement of air over the earth's surface—the greater the surface roughness, the greater the turbulence. The standard deviation of wind fluctuations from a mean value taken over a period of, say, one hour is a useful quantitative means of describing turbulence intensity.

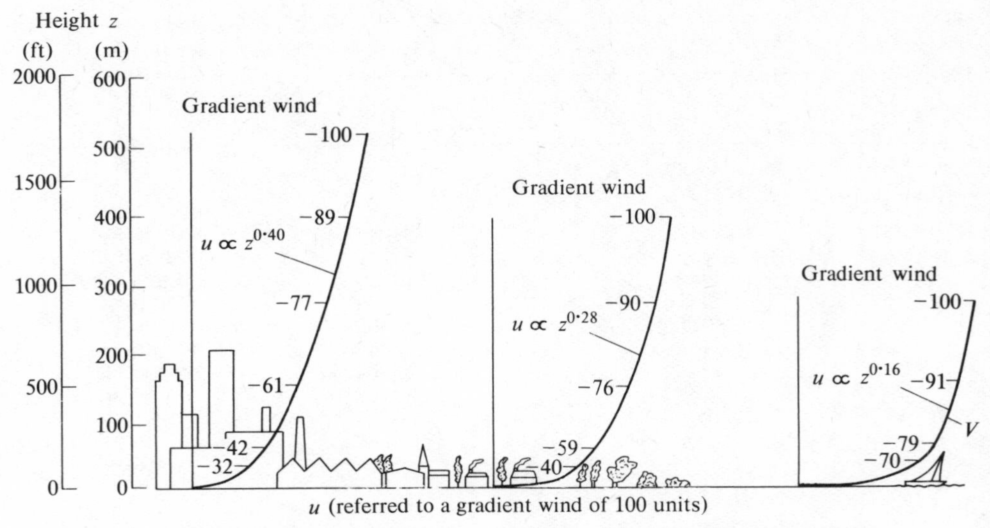

Fig. 5.2. **Profiles of mean wind velocity over level terrains of differing roughness.**

Stability

If a parcel of air is carried upwards in the atmosphere it expands and cools due to the reduction in air pressure. Assuming no exchange of heat with the surrounding air, the parcel cools at a rate of about 10°C km^{-1}, defined as the dry adiabatic lapse rate. Such a lapse rate of temperature rarely occurs in the atmosphere, since although mechanical

TABLE 5.2

Relation of turbulence types to weather conditions

A. Extremely unstable conditions
B. Moderately unstable conditions
C. Slightly unstable conditions
D. Neutral conditions†
E. Slightly stable conditions
F. Moderately stable conditions

Surface wind speed m/s^{-1}	Daytime insolation			Night-time conditions	
	Strong	Moderate	Slight	Thin overcast or $\geq \frac{4}{8}$ cloudiness†	$\leq \frac{3}{8}$ cloudiness
<2	A	A–B	B		
2	A–B	B	C	E	F
4	B	B–C	C	D	E
6	C	C–D	D	D	D
>6	C	D	D	D	D

† Applicable to heavy overcast, day or night.
‡ The degree of cloudiness is defined as that fraction of the sky above the local apparent horizon which is covered by clouds.

turbulence tends to maintain a dry adiabatic lapse rate, convection during the day and outgoing radiation at night destroy the neutrality of the atmosphere. Table 5.2 outlines the relation of turbulence types to weather conditions.

The environmental lapse rate (Fig. 5.3) varies in different air masses and at different times of day. On days with strong solar heating of the ground the decrease of air temperature with height often exceeds 10°C km^{-1}, although this superadiabatic condition is usually confined to the lowest 200 m of the atmosphere. A parcel of air rising in such an environment always finds itself in surroundings cooler than itself and continues to rise, favouring active convective instability.

An atmosphere in which the temperature decreases more gradually than 10°C km^{-1} or even increases with height is stable; that is, any parcel of air which is displaced to another level returns to its original position. In the case of a temperature 'inversion' all vertical motion and turbulence are suppressed and any air pollution trapped in an inversion cannot disperse. On the other hand emissions released above the inversion do not ordinarily reach the ground.

Two types of inversion are worth special mention. A surface inversion of temperature is normally formed at night by radiation cooling of the ground and subsequently of the air near the ground. Radiation inversions are rarely deeper than 150 m in the UK, but in stagnant winter conditions may persist for a few days. Elevated inversions at 500 m or above are formed in a number of ways. Below this inversion the lapse rate is frequently neutral, allowing emissions to disperse uniformly throughout the surface layer; the inversion aloft acts as a barrier between the mixing layer and the free air above.

Influence of location and terrain

The location of a country in relation to the major weather systems and to the oceans and continents is of great importance in determining the characteristic turbulence of the lower atmosphere. Large areas of North America and Central Europe are subject to persistent high pressure systems in winter-time, giving rise to days of stagnant air with a shallow mixing layer below an elevated inversion. The British Isles, on the other hand, are exposed to trains of low pressure systems from the Atlantic and are well ventilated by the dominant south-westerly winds. More locally, topographical features modify the general pattern of wind speed and direction. A plume discharged from a chimney in a valley may be confined within the valley sides if the main wind is bearing along the axis of the valley. If the wind is blowing across the valley the plume may be caught in the cavity created by the wake of the valley side, giving high ground-level concentrations.

At coastal sites, the disparity between temperature over land and over the sea surface creates localized winds extending a few kilometres out to sea and as much as 20 km inland. During the day an on-shore breeze is set up which is stable over the sea, but which quickly becomes unstable and turbulent over land. Measurements made at Chicago (Lyons and Olsson[2]) on the shore of Lake Michigan (Fig. 5.4) suggest that there is a closed circulation cell over the shore line. Chimney emissions released within the inflow

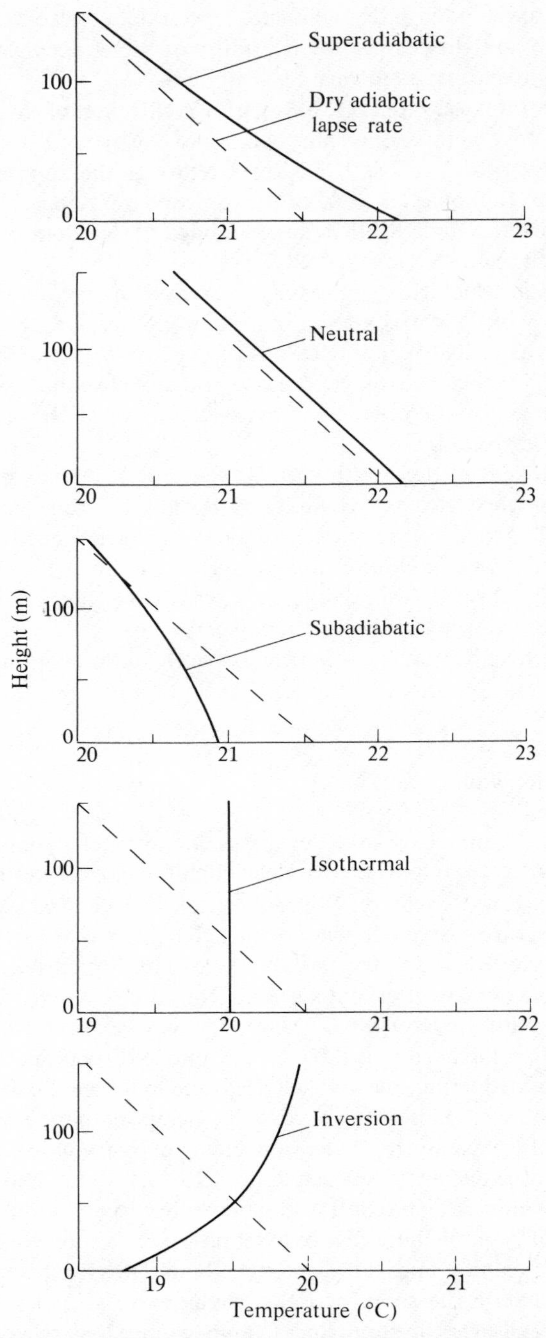

Fig. 5.3. Typical environmental lapse rates. Typical examples of vertical temperature profiles are shown in comparison with the dry adiabatic lapse rate (–1°C/100 m) which serves as a reference for distinguishing unstable from stable cases. The position of the dashed line representing the adiabatic lapse rate is not important; it is significant only as far as its slope is concerned.

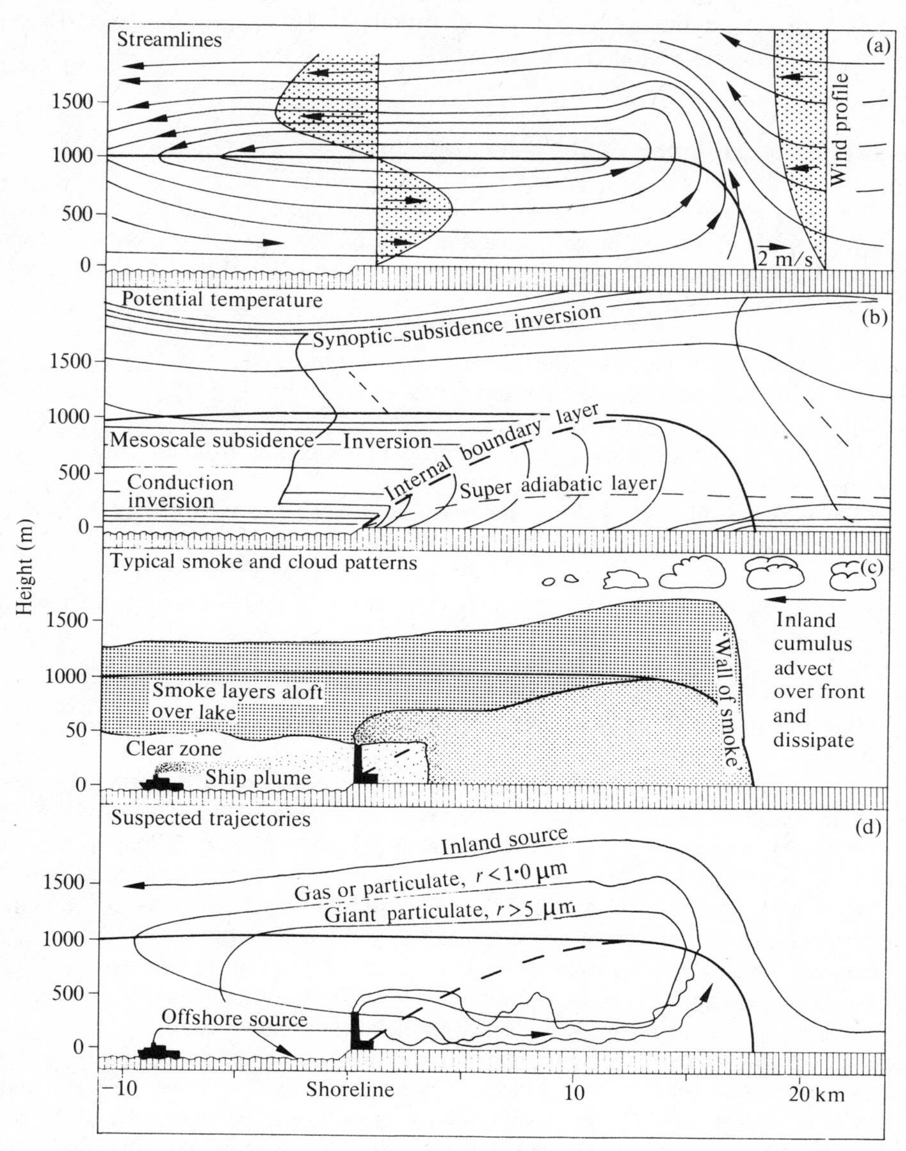

Fig. 5.4. (a) Typical streamline pattern in a well developed lake breeze cell, heavy line represents boundary of inflow. View is looking to south with lake on left, during mid-afternoon. (b) Schematic isopleths of potential temperature with packing indicating synoptic and mesoscale subsidence inversions, as well as the conduction inversion. Actual temperature soundings over land and lake are shown compared to the dry adiabatic (Γ_d). (c) General smoke pattern found, including fumigation of an elevated shoreline point source. (d) Probable trajectories of both small and large particulates from several sources.

layer move inland, rise in the convective updraught at the 'sea breeze front' and return in the layer aloft. Because of this return to the shore line, concentrations inland are appreciably higher than would otherwise be expected.

Dispersion and dilution of chimney emissions

On being discharged from a chimney both particulate and gaseous emissions rapidly mix with the surrounding atmosphere, and the dilution increases as the plume is carried downwind. Dispersion is most easily described by considering three separate phases of plume behaviour:

1. Shortly after release to the atmosphere a plume may be affected by the aerodynamic disturbances created by the stack, by adjacent buildings or by topographical irregularities.
2. Because of its buoyancy and initial vertical momentum, a plume rises relative to the mean motion of the ambient air.
3. The ultimate dilution of the plume depends on the degree of turbulence in the atmosphere.

All these effects occur at the same time in nature, but it is convenient and, in general, only possible to treat them separately.

Aerodynamic effects of buildings and irregular topography

If the efflux velocity of the stack gases is too low in relation to the horizontal motion of the air, the plume is dragged down into the low-pressure region in the wake of the stack. Wind tunnel studies have shown that this downwash is only slight as long as the efflux velocity is more than 1·5 times the wind speed; certainly downwash can be prevented if this velocity ratio is greater than 2. When a large building is situated near a chimney stack, the plume is distorted by the disturbed air stream which accommodates itself to the shape of the building. As long ago as 1932 a rule of thumb for stack design was proposed recommending that stacks should be at least $2\frac{1}{2}$ times the height of the surrounding buildings, as shown in Fig. 5.5. If the stack height is much lower than this, the plume is caught in the wake of the building, possibly bringing high concentrations of flue gases to the ground near to the building. Location of the stack relative to a high building is also important. Ideally the stack should be situated as near to the centre of the building as possible so that whatever the wind direction, the plume reaches the free air aloft before the wake has fully developed.

There are no rules of any consequence relating to the effects of irregular terrain. A plant located in a deep valley suffers serious plume downwash unless the chimney extends well above the rim of the valley. Again a plume affects an isolated hilltop in strong winds unless this is taken into account in the stack design.

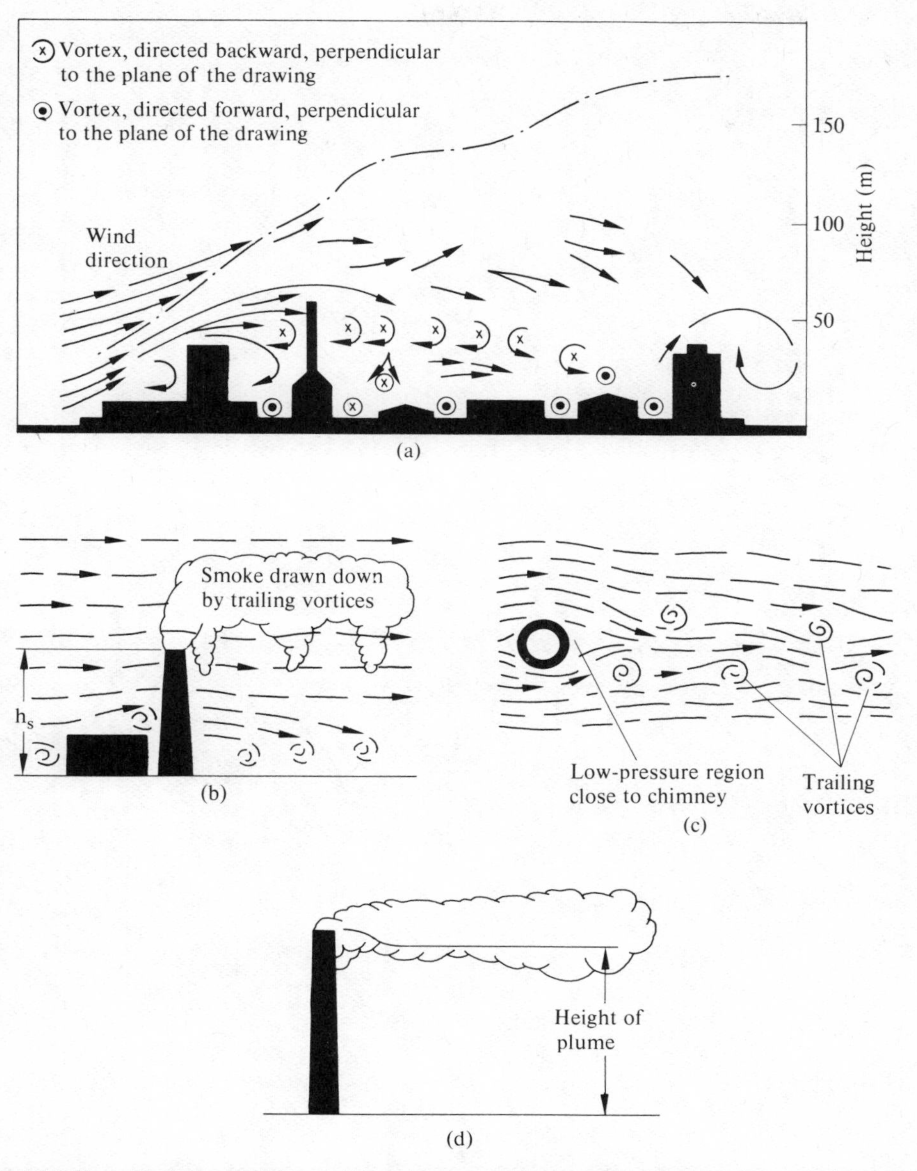

Fig. 5.5. Effect of buildings and chimneys on smoke plumes. (a) Pattern of flow over factory buildings. The flow is seriously disturbed below the broken line.[4] Chimney plumes are affected by buildings upstream as shown and downstream. (b) Chimney lower than $2\frac{1}{2}$ times height of adjacent buildings—effect on smoke. (c) Plan of (b). (d) Downwash pulling plume axis below chimney top.

It is difficult if not impossible to calculate plume behaviour in the presence of aerodynamic disturbances, and in these circumstances wind tunnel tests are usually made. In the past, such tests had serious limitations since the earlier tunnels were not designed specifically for studying atmospheric boundary layer problems. Nowadays tunnels with long working sections (Fig. 5.6) can simulate the turbulence characteristics and velocity profiles of the atmospheric boundary layer, largely by creating the correct turbulent conditions initially and then maintaining them by scaling the ground roughness upwind of the model under test (Counihan[3]).

Plume rise

As the turbulent gases leave the stack they mix with the ambient air, and the plume is bent over, quickly acquiring the mean wind speed of the air it has entrained. However, because of its initial vertical momentum and buoyancy, the plume continues to rise relative to the ambient air. Plume rise is probably the most important single factor affecting the eventual concentrations of effluent at ground level, and its prediction has been the subject of many treatments with over 30 formulae in existence (Briggs[4]). However, with the acquisition of more and more full-scale measurements it has become evident that a relatively simple formula can be used to derive the 'effective chimney height' with sufficient accuracy to enable the chimneys to be designed with confidence.

The height of plume rise depends on the heat content of the emissions, the initial momentum, wind speed at plume level, the distance downwind of the stack, the height of the source and the variation of atmospheric temperature with height. From extensive studies of plume behaviour at Tilbury Power Station, England (Fig. 5.7) the following summary statements (Lucas[5]) can be made, and will help the planning engineer to design a chimney height for a given plant:

1. More sophisticated methods of measurement such as LIDAR (Fig. 5.8), which extend beyond the distance where a plume may cease to be visible, have shown that plume rise can continue for a downwind distance of 1000 m or more (Hamilton[6]).
2. In all but light winds plume rise is little affected by the stability of the atmosphere.
3. With light winds plume rise is reduced by an inversion above the level of the chimney top.
4. Plume rise in neutral conditions is given by

   $$\Delta h = \alpha Q^{1/4}/u,$$

   where Δh is the plume rise, Q is the heat emission rate, u is the wind speed and α is a constant for the chimney and site. There is evidence to show that α is dependent on the height of the source. Turbulence in the atmosphere decreases with height so that a plume emitted at a greater height has a greater average rise. On the basis of

Fig. 5.6. In the CEGB wind tunnel at the research station at Marchwood. The atmospheric boundary layer is simulated by means of vortex generators at the air intake and maintained over the working section by means of sealed roughness elements. (a) shows a 1:300 scale model of an urban site, while (b) is a 1:1000 scale topographic model of a coastal site.

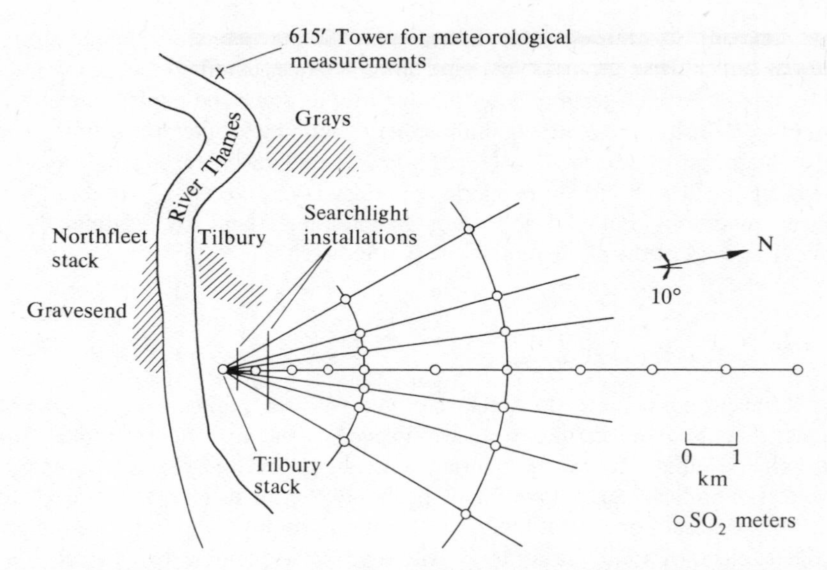

Fig. 5.7. Tilbury plume rise and dispersion study. Layout of sulphur dioxide recorders and meteorological instruments.

many experimental observations, an empirical expression has been derived to account for this dependence on stack height, giving

$$\Delta h = \left(\frac{408 + 0{\cdot}67\, h_c}{u}\right) Q^{1/4},$$

where Δh and h_c are measured in m, u in m s^{-1} and Q in mW.

Atmospheric dispersion

On being emitted from a chimney both particulate and gaseous effluents are rapidly mixed with the surrounding atmosphere and the degree of dilution increases as the plume is carried downwind. In the absence of any obvious alternative, the well-known mathematical dispersion formulae developed by Sir Graham Sutton[7] has been used for many years to estimate the downwind concentrations after dilution. Strictly speaking, these formulae apply only to a source operating for a short time over level ground with a steady wind direction and a neutral atmosphere. Extensions to other conditions have necessitated the empirical adjustment of the diffusion parameters.

In recent years dispersion data from a number of extensive field studies have been presented in the form of Sutton's dispersion formulae, but with some modifications (Gifford[8]).

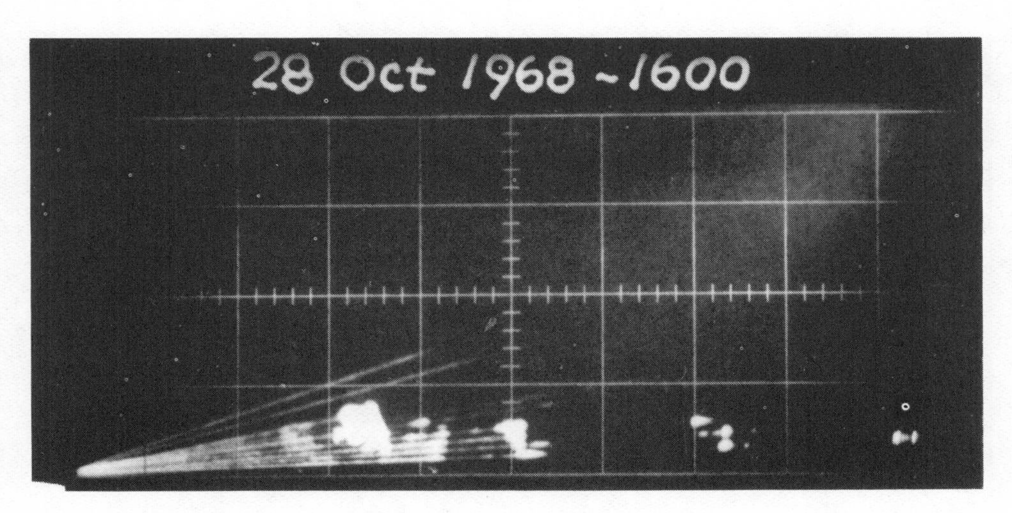

Fig. 5.8. (a) In LIDAR equipment a light beam from a ruby laser is reflected back to a telescope by particles in the atmosphere. (b) A LIDAR 'map' shows the position of a chimney plume by the strong reflections.

The simple Gaussian interpolation formula in current use has the form

$$C \equiv \frac{Q}{\pi \sigma_y \sigma_z u} \exp \left\{ -\tfrac{1}{2} \left(\frac{y^2}{\sigma_y^2} \frac{h^2}{\sigma_z^2} \right) \right\}, \tag{5.1}$$

where Q = source strength (g s^{-1})
u = average wind speed (m s^{-1})
h = height of the source (m)
y = crosswind distance from the plume axis (m)
C = ground level air concentration (g m^{-3}).

From experimental data, values can be assigned to the dispersion coefficients σ_y and σ_z (Figs. 5.9 and 5.10) for the different atmospheric stability categories suggested by

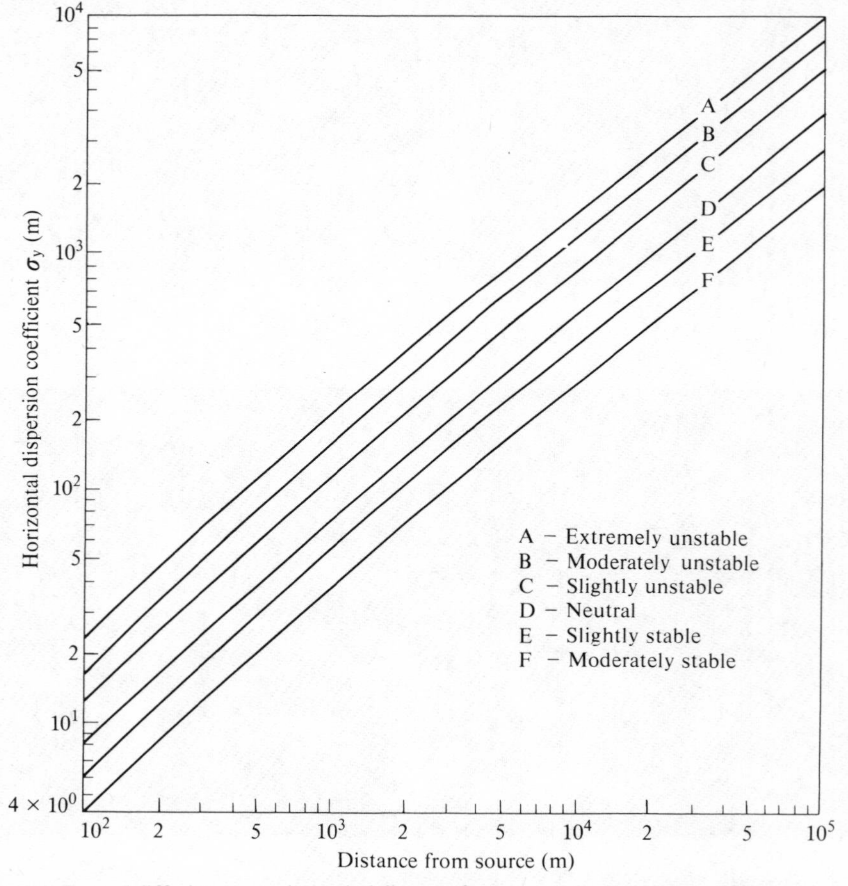

Fig. 5.9. Lateral diffusion, σ_y vs. downwind distance from source for Pasquill's turbulence types.

Pasquill,[9] see Table 5.1, and are functions of the downwind distance, representing the crosswind and vertical standard deviations of the plume concentration distribution (Gifford[10]).

In practical applications it is often convenient and sufficient to determine only the principal properties of a concentration distribution. These are the maximum ground-level concentration and its distance from the source.

The three-minute maximum ground-level concentration is given by

$$C_{\max} = \frac{2Q_s}{e\pi h^2 u} \frac{\sigma_z}{\sigma_y}, \tag{5.2}$$

occurring at the distance where $\sigma_z = h/\sqrt{2}$, which is about 15 chimney heights (or effective chimney heights) in a neutral atmosphere.

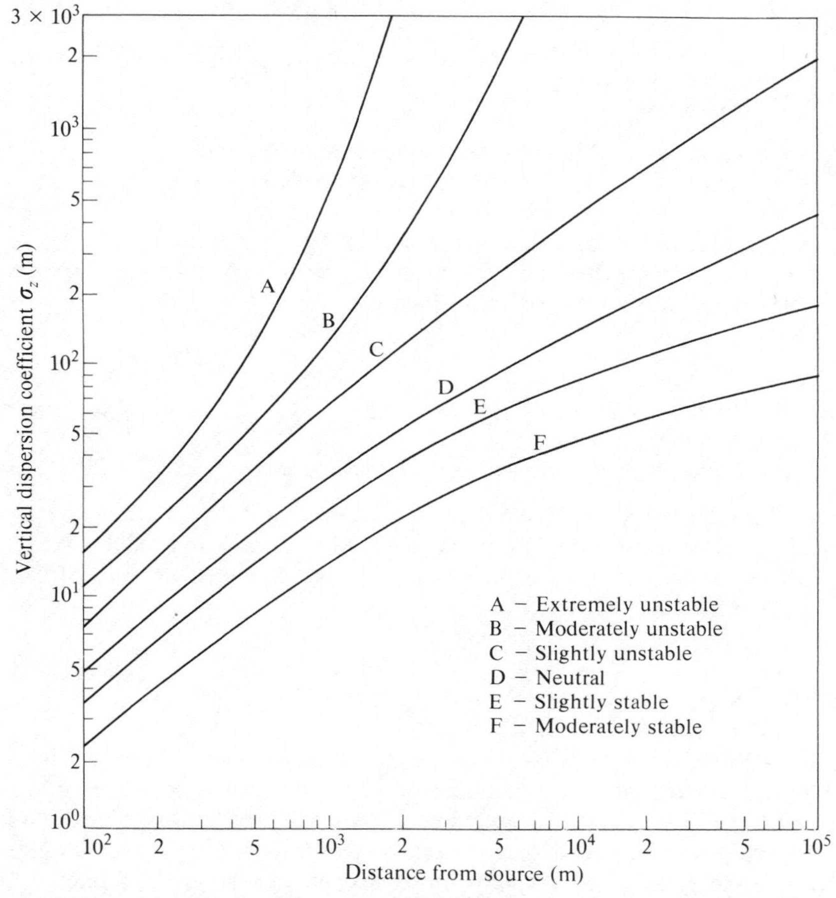

Fig. 5.10. Vertical diffusion, σ_z vs. downwind distance from source for Pasquill's turbulence types.

It has been found by observation round large power stations that the ground-level concentrations are in general less than or equal to the values calculated by the classical formulae described above. This appears to be due to the fact that the vertical and cross-wind spreads of the plume vary with distance from the source in different ways. Moore[11] has explained these differences by assuming that the size of the eddies transporting the material to the ground is generally smaller than the height of the plume. He derives the following expressions for predicting concentrations:

$$C_m = \left(\frac{Q_s}{H^2}\right)\left[\left(\frac{A}{H}\right) + \left(\frac{C}{u^2}\right)\right] \times 10^8 \tag{5.3}$$

for neutral or stably stratified conditions and

$$C_m = \left(\frac{Q_s}{H^2}\right)\left[\left(\frac{A}{H}\right)^2 + \left(\frac{C}{u^2}\right)^2\right]^{1/2} \times 10^8 \tag{5.4}$$

for unstable conditions,

where C_m = maximum hourly ground-level concentration (pphm)
Q_s = rate of sulphur dioxide emission ($m^3\ s^{-1}$ reduced to STP)
H = plume height above ground (m)
u = wind speed at plume level ($m\ s^{-1}$)
A and C are parameters whose values vary with vertical temperature gradient, but within stability groups are as follows:

	A	C
unstable	1·65	0·58
stable but no inversion	1·87	0·22
stable but turbulent below an elevated inversion	1·35	0·20.

The term containing A represents the effect of mechanical turbulence and the term containing C represents the effect of thermals on the turbulence in a stable environment.

The maximum hourly value calculated on this basis is about one third of the three-minute maximum value.

Volume sources

Equation (5.1) requires modification when considering a volume source which may be located in the lee or wake of a building or may represent a collection of multiple sources. At large distances the plume from a volume source approximates to a plume from a single point source. At distances near the position of the maximum ground-level concentration the form of the plume is more difficult to describe. A number of semi-empirical formulae have been proposed to calculate the effect of a building wake.

Fuquay, Munn and Cole[12] suggested an initial dilution factor $D_B = cA\bar{u}$ where A is the cross-sectional area of the building and c is a factor to be determined, probably about 2. The building factor can be combined with the atmospheric dilution giving

$$\Psi = \frac{Q\exp(-h^2/2\sigma_z^2)}{(\pi\sigma_y\sigma_z + cA)\bar{u}}, \qquad (5.5)$$

where C = ground level concentration (pphm)
Q = source strength ($m^3\ s^{-1}$)
h = source height in the mean wind speed (m)
σ_y and σ_z are the lateral and vertical diffusion coefficients.

As an alternative, a 'virtual source' model can be used where an equivalent ground-level point source is assumed, located at a distance upwind from the building such that the cross-sectional area of the plume is the same as the downwind face of the building.

Time average concentrations

'Instantaneous' ground-level concentrations downwind of an elevated source vary continuously and are difficult to measure. As a rule the time interval over which the concentration is measured depends on the time-constant of the instrument and is rarely less than three minutes. Concentrations averaged over short periods may be many times greater than the longer term mean, say taken over a day and it is important to be able to relate the two. Barry[13] has shown that observational data on short-term concentrations and the long-term mean fit a probability distribution function of the form

$$F(C) = F(0) \exp[-F(0)\, C/M] \qquad (5.6)$$

Where $F(C)$ is the frequency of concentrations, averaged over a short time (t_p), that exceed the value C, $F(0)$ is the frequency of concentrations exceeding zero, M is the concentration averaged over a longer time t_m. The relationship between the peak P and the mean concentration M is

$$F(P) = t_p/t_m = F(0) \exp[-F(0)\, P/M], \qquad (5.7)$$

from which

$$P/M = [\log F(0) + \log (t_m/t_p)]/F(0).$$

$F(0)$, the frequency of concentrations exceeding zero, is difficult to evaluate, and as an alternative approach empirical relationships between long-term averages and short-term peaks can be found, provided a large number of results are available. From a comprehensive sulphur dioxide survey made round High Marnham power station on the River Trent, several hundred thousand readings were collected over a period of three years. From these data it was possible to build up a set of factors (Table 5.3) (Martin and Barber[14]) relating averages over different periods of time.

TABLE 5.3

Time-mean relationships of SO_2 ground level concentrations

Sampling period	Mean value of concentration	Maximum concentration (less than 2 per cent of all occasions)
3 min	$C/5$	$1{\cdot}3\,C$
1 h	$C/13$	$C/1{\cdot}5$
1 day	$C/60$	$C/7{\cdot}5$
1 month	$C/230$	$C/45$

C is the maximum three-minute concentration, $C/5$ means one fifth of C, $C/13$ means one thirteenth, etc.

The factors in the table apply to measurements made around a circle of radius equal to the distance of maximum short-term concentration under average weather conditions. C is the maximum calculated three-minute concentration, using the formulae described earlier.

References

1. Davenport, A. G. 'Wind effects on buildings and structures', *National Physical Lab. Symposium,* HMSO, 1965.
2. Lyons, W. A. and L. E. Olsson, 'Mesoscale air pollution transport in the Chicago Lake breeze', *J. Air Poll. Control Assoc.*, **22,** 876, 1972.
3. Counihan, J. 'Simulation of an adiabatic urban boundary layer in a wind tunnel', *Atmospheric Environment,* **7**, 673, 1973.
4. Briggs, G. A., *Plume rise,* AEC Critical Review Series, US Atomic Energy Commission, 1969.
5. Lucas, D. H., 'Application and evaluation of results of the Tilbury rise and dispersion experiment, *Atmospheric Environment,* **1**, 421, 1967.
6. Hamilton, P. M., 'Use of LIDAR in the study of chimney plumes', *Phil. Trans. Roy. Soc.* **265**, 153, 1969.
7. Sutton, O. G., 'The theoretical distribution of airborne pollution from factory chimneys', *Q. J. Roy. Met. Soc.,* **73**, 426, 1947.
8. Gifford, F. A., Jr., 'Atmospheric dispersion', *Nuclear Safety,* **1**(3), 56, 1960.
9. Pasquill, F., *Atmospheric diffusion,* Van Nostrand, London, 1962.
10. Gifford, F. A., Jr., 'Use of routine meteorological observations for estimating atmospheric dispersion', **2** (4), 47, 1961.
11. Moore, D. J., 'Factors influencing plume rise and ground level concentrations', *Chimney Design Symposium,* Edinburgh University, 1973.

12. Fuquay Munn, R. E. and A. F. W. Cole, 'Turbulence and diffusion in the wake of a building', *First Canadian Conference on Micrometeorology,* Toronto, 1965.
13. Barry, P. J., 'Use of argon-41 to study the dispersion of stack effluents', *IAEA Symposium on the Use of Nuclear Techniques in the Measurement and Control of Environmental Pollution,* AECL Report No. 3731, 1970.
14. Martin, A. and F. R. Barber, 'Investigations of sulphur dioxide pollution around a modern power station', *J. Inst. Fuel,* **39**, 294, 1966.

6 Height and design of chimneys

K. P. GRUBB

A chimney is defined as a 'flue carrying off smoke or steam of fire, furnace, engine, etc.' This basic premise does not, however, define the true function of a chimney, which is to reduce the pollution of the air near ground level from oxides of sulphur and other gases and from particulate matter including dust, sooty particles and acid smuts, the products of combustion of modern fuels. In designing a chimney to fulfil this function the engineer must first concern himself with two basic parameters, H_e, the effective height of emission, and Q, the rate of emission.

If ground level concentrations of pollutants such as sulphur dioxide are to be kept within acceptable limits, then from the following expression, $C \propto Q/H_e^2$, it can be seen that C, the concentration of emitted pollutant, will be greatly affected by H_e, the effective height of emission and less affected by Q, the rate of emission. It is therefore in the national interest to provide as high a chimney as is practicable for any boiler or other installation, subject to the other basic requirements of visual amenity, structural design and air transport regulations.

In considering chimney design in more detail, the engineer must also take account of other factors which may affect or control to some degree the two basic parameters. To summarize, chimney design can be said to consist of the study of a number of independent and dependent variables:

1. height of chimney, H_c—subject to regulations and technical considerations;
2. flue gas quantity, Q—defined by boiler design and size;
3. flue gas temperature, T_g—defined by boiler design;
4. available draught, h—dependent upon H_c and T_g.
5. efflux velocity of flue gas, V—dependent upon h but minimum value generally defined.

6. lateral dimension of chimney, i.e., flue diameter D—dependent upon Q and controlled by V.

Each variable will be considered in more detail to see how it is affected or defined by other influences.

Height of chimney, H_c

The term *height* is used in various senses in chimney design and the engineer must distinguish between H_c the physical height of the chimney, and H_e, the effective height of emission.

Generally,

$$H_e = H_c + H_t,$$

where H_t is the 'thermal rise' and is the distance above H_c to which the buoyancy lifts the waste gases before dispersal commences. However, it must in turn depend upon other factors such as wind speed, atmospheric temperature and the chimney height H_c. In addition, the terms 'uncorrected' and 'final' chimney heights are used but their meanings will become evident in the text.

Where new chimneys are to be built to serve new or existing boiler plant or where existing plant is to be modified or enlarged, the determination of the chimney height is the subject of legislature, in the form of the Clean Air Acts,[1] first introduced in 1956 to reduce the nuisance of air pollution in Great Britain. The only plants exempt from these Acts are those belonging to the Electricity Supply Industry and certain other plants which are subject to the more stringent requirements of the Alkali Act and must be approved by the Alkali Inspectorate.

Guidance on the choice of chimney heights is given in the *Clean Air Act—Memorandum on Chimney Heights*,[2] published by HMSO, which assists in the determination of height by considering the maximum rate of emission of sulphur dioxide, the height of structures in the near vicinity and the existing levels of pollution in the surrounding district.

Such details will be considered by the local authority concerned who, as the responsible body under the Acts, will not approve the submitted chimney height unless it is satisfied that the discharge of pollutants from the plant will not become a nuisance or a danger to the health of the general populace. The procedure by which any industrial developer or other boiler owner makes application for the approval of a chimney design is laid down in a prescribed form[3] which is obtained from the relevant local authority.

The *Memorandum on Chimney Heights* gives simple guidelines for the chimney designer, based upon the following parameters:

1. Maximum rate of emission of sulphur dioxide in kg h^{-1}. This is calculated from the rate of burning of the fuel and its sulphur content.

2. Flue gas efflux velocity. This is required to be not less than 6 m s^{-1} for small installations and 15 m s^{-1} for plant with a capacity rated at 20 000 kg h^{-1} steam or the equivalent. The provision of forced-draught or induced-draught fans is considered and the efflux velocities are modified accordingly.
3. Topography and character of surrounding area. Five categories are given which distinguish between rural, residental and industrial environments.
4. Height of adjacent buildings.

The designer has first to determine an 'uncorrected' chimney height, based on the rate of sulphur dioxide emission, which is modified by a series of nomograms in the Memorandum drawn up to take into account any neighbouring sources of pollution, local background levels of pollution and the character of the surrounding district. Some allowance is also made for the type of fuel it is proposed to use. A 'final' chimney height is now obtained by adjusting the 'uncorrected' height by means of further nomograms which take into account the height of the attached building and/or neighbouring buildings. A final correction is made to ensure that the chimney height exceeds the height of the building roof by at least 3 m and that the 'final' height is greater than the 'uncorrected' chimney height. From this point onwards the 'final' height can be taken as being equal to H_c.

Flue gas quantity, Q

Total waste gas flow is normally specified by the boiler manufacturer, who also states the draught requirements for efficient combustion conditions.

Flue gas temperature, T_g

The flue gas temperature at the outlet from the boiler or other plant which the chimney serves is either specified by the manufacturer or is possibly a variable quantity. The latter is so in the case of incinerator furnaces, in which back-end temperatures must vary with the type of raw material being consumed.

The flue gas temperature governs the amount of draught available for the combustion process in a natural-draught system and affects the efflux velocity in all systems; in addition it plays a major part in the choice of flue lining materials.

Some cooling of the gases must take place within the chimney, but in concrete chimneys with brick linings, or well-lagged steel chimneys, no great inaccuracy results from assuming no loss of temperature. In unlagged chimneys an average value of flue gas temperature must be calculated, but in calculating the heat losses through the shell care must be taken in assessing suitable values of the external and internal film coefficients, and indeed the whole question of heat losses cannot be divorced from a more comprehensive consideration of the desirability of lagging and the avoidance of acid smuts, which are discussed in detail later.

Available draught, *h*

The buoyancy effect of the column of hot gases in a chimney must provide sufficient draught to overcome the total losses in the combustion system and retain enough kinetic energy to discharge the products of combustion into the atmosphere at the required efflux velocity. This concept is true for a natural-draught system, and with certain reservations can be said to apply also to the various mechanical-draught systems.

The buoyancy effect, or to be more specific, the pressure differential created by the chimney is the difference between the density of the gas column in the chimney and that of a column of air of the same height but at atmospheric pressure and temperature. In calculating the pressure differential, the assumption is made that the density of the hot gases in the chimney is approximately the same as that of air at the same temperature and pressure, and this assumption does not lead to great inaccuracy. The height of the column, H, does not equal the height of the chimney, H_c, since it is measured from the centre line of the chimney inlet to the outlet.

The pressure of each column equals the height of the column times its density. Therefore

$$\Delta P' = H(\rho_a - \rho_g),$$

where $\Delta P'$ = pressure differential (kgf m^{-2})
H = height of chimney above inlet (m)
ρ_a = density of air at ambient temperature (kg m^{-3})
ρ_g = density of flue gases at operating temperature (kg m^{-3}).

In SI units,

$$\Delta P = 9{\cdot}806H(\rho_a - \rho_g)\,\mathrm{N\,m^{-2}};$$

but

$$p = \Delta P/10^2 = 0{\cdot}098\,06H(\rho_a - \rho_g),$$

where p = pressure differential (millibars).

Assuming that the densities of air and of the flue gases at the same temperatures and pressure are approximately equal, then

$$\rho_g = \rho_a T_a/T_g,$$

where T_a and T_g are absolute values of temperature (°K);
hence

$$p = 0{\cdot}098\,06H\rho_a T_a\left(\frac{1}{T_a} - \frac{1}{T_g}\right).$$

From this expression it will be seen that p decreases as T_a increases. For design purposes, therefore, it is necessary to use the highest ambient temperature that the air may attain.

Assuming that the density of air at 15°C and 1013 mbar is 1·229 kg m^{-3}, then

$$p = 0{\cdot}098\ 06 \times 1{\cdot}229 \times 288H\left(\frac{1}{288} - \frac{1}{T_g}\right)$$

$$= 34{\cdot}7H\left(0{\cdot}0035 - \frac{1}{T_g}\right) \text{ mbar.} \qquad (6.1)$$

A more accurate value of p may be calculated if the density of the flue gas and the ambient temperature are known, but for most practical purposes, and in particular for a preliminary design, the above equation is sufficient. A curve (Fig. 6.1) may be drawn for a theoretical chimney height of 30 m showing the variation in pressure differential, or draught, with flue gas temperature, and from this curve the available draught for higher chimneys may be calculated.

From a consideration of the two factors affecting the natural draught produced by a chimney, namely flue gas temperature and height, it can be appreciated that on occasions there is insufficient draught produced to (1) overcome the losses in the system, and (2) provide sufficient velocity head at the chimney outlet to guarantee a reasonable value of efflux velocity, i.e., one which ensures that downwash is reduced to a minimum over the range of wind speeds normally encountered.

In a natural-draught system, ideally

$$p \geq p_e + p_f + p_b,$$

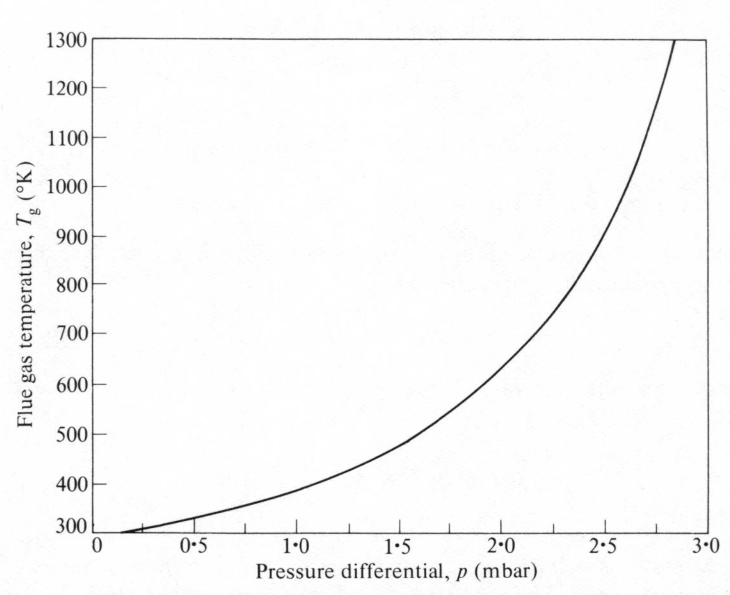

Fig. 6.1. **Pressure differential for a 30 m high chimney showing variation with flue gas temperature.**

where p_e = velocity head at outlet

p_f = friction losses

p_b = total losses due to bends and changes of section.

In considering the flow of the gases at the chimney outlet we have

velocity head, $h = V^2/2g$,

where V = efflux velocity (m s^{-1})

g = acceleration due to gravity (m s^{-2})

Expressed as a pressure

$$p_e = \rho_g h \text{ kg m}^{-3}$$

where ρ_g = gas density (kg m^{-3}).

That is,

$$p_e = \frac{9{\cdot}806\, \rho_g V^2}{2g} = \frac{\rho_g V^2}{2} \text{ N m}^{-2}.$$

Using basic chimney units, this becomes

$$p_e = \left(\frac{4Q}{\pi D^2}\right)^2 \times 1{\cdot}229 \times \frac{288}{T_g} \times \frac{1}{2},$$

where Q = gas flow (m^3 s^{-1})

D = diameter of chimney at outlet (m)

T_g = flue gas temperature (°K).

Hence

$$p_e = \frac{Q^2}{D^4 T_g} \frac{16 \times 1{\cdot}229 \times 288}{\pi^2 \times 2}$$

$$= \frac{286 Q^2}{D^4 T_g} \text{ N m}^{-2}$$

$$= \frac{2{\cdot}86 Q^2}{D^4 T_g} \text{ mbar.} \qquad (6.2)$$

Gases in motion are subjected to certain resistances which result in a reduction in the kinetic energy. Such resistances, which can be better expressed as a loss of velocity head, are caused by friction with the duct lining material and by bends and changes of section in the ducting and flues.

From a study of the mechanics of fluids, the friction loss in a straight length of ducting or flue can be expressed as

$$h_f = \frac{f l V_f^2}{2gM}$$

where $l =$ length of duct (m)

$f =$ friction factor

$$M = \frac{\text{perimeter of duct (m)}}{\text{cross sectional area (m}^2)}$$

$V_f =$ velocity of flow (m/s)

Alternatively, expressed as a pressure,

$$p_f = \frac{f\rho_g l V_f^2}{2M}$$

$$= \frac{\rho_a T_a}{T_g} \frac{flV_f^2}{2M} \text{ N m}^2.$$

Expressed in chimney units and taking $M = D/4$,

$$p_f = 2\rho_a T_a \times \frac{flV_f^2}{DT_g}$$

$$= 2 \times 1{\cdot}229 \times 288 \times \frac{flV_f^2}{DT_g}$$

$$= 707 \frac{flV_f^2}{DT_g} \text{ N m}^2$$

$$= 7{\cdot}07 \frac{flV_f^2}{DT_g} \text{ mbar.} \tag{6.3}$$

Energy losses due to causes such as bends and section changes are best expressed in terms of the velocity head modified by an energy loss coefficient, which varies with the shape of the bend or section. The main loss is at the right-angled bend at the base of the chimney, and if possible this should take the form of a quadrant of large radius. If construction problems preclude this form of bend, then splitter plates or a cascade should be provided to reduce the turbulence and consequent head loss associated with such bends.

From the basic expression for velocity head, expressed as a pressure loss,

$$p = \frac{\rho_g V_f^2}{2} \text{N m}^{-2};$$

hence bend and section loss

$$p_b = \frac{k\rho_g V_f^2}{2} \text{ N m}^{-2},$$

where k = energy loss coefficient for section, and in chimney units,

$$p_b = \frac{k\rho_a T_a}{T_g} \frac{V_f^2}{2}$$

$$= \frac{1 \cdot 229 \times 288}{2} \frac{kV_f^2}{T_g} \text{ N m}^{-2}$$

$$= 177 \frac{kV_f^2}{T_g} \text{ N m}^2$$

$$= 1 \cdot 77 \frac{kV_f^2}{T_g} \text{ mbar.}$$

Other losses may occur in the system, but these are generally negligible. In one form of lining construction, employing refractory brickwork, the bricks are laid on corbels at regular intervals up the chimney and forming a series of truncated cones. The total loss of head caused by the repetitive sudden enlargements of section is insignificant and can be ignored.

The estimation of the values of k and f for bend and friction losses has been the subject of much experimental work, but Tables 6.1 and 6.2 summarize values of these constants which cover most of the cases encountered in industrial chimney design. For larger flues, reference should be made to standard works on boiler design.[4,5]

Summarizing this section, we have:

1. pressure differential, or draught,

$$p = 34 \cdot 7H\left(0 \cdot 0035 - \frac{1}{T_g}\right) \text{ mbar,}$$

2. velocity head at outlet, (exit loss)

$$p_e = \frac{2 \cdot 86Q^2}{D^4 T_g} \text{ mbar,}$$

3. friction loss

$$p_f = 7 \cdot 07 \frac{fV_f^2}{DT_g} \text{ mbar,}$$

4. bend and section losses

$$p_b = 1 \cdot 77 \frac{kV_f^2}{T_g} \text{ mbar.}$$

As stated previously,

$$p \geqslant p_e + p_f + p_b,$$

TABLE 6.1
Energy loss coefficient, k

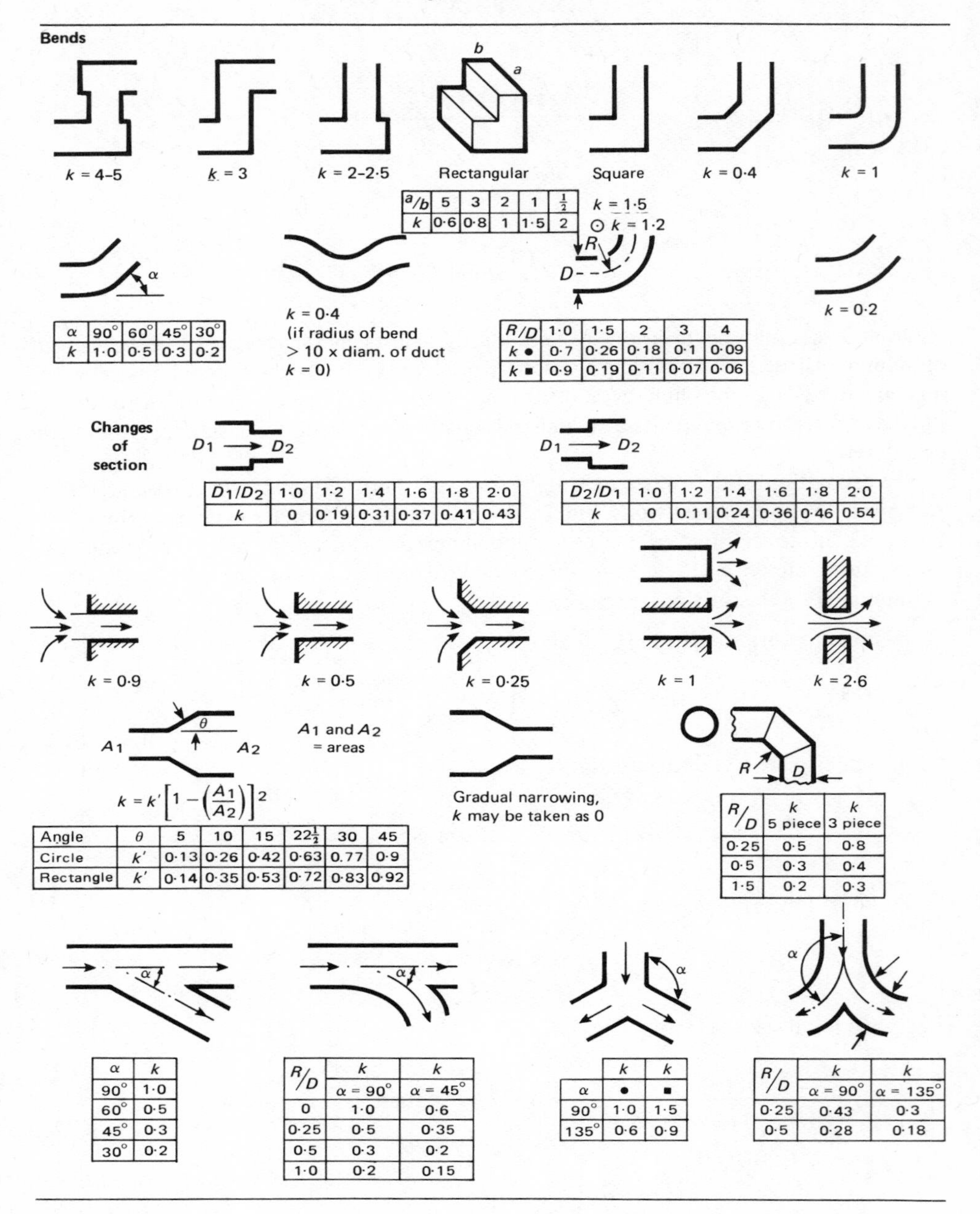

a/b	5	3	2	1	½
k	0·6	0·8	1	1·5	2

α	90°	60°	45°	30°
k	1·0	0·5	0·3	0·2

R/D	1·0	1·5	2	3	4
k ●	0·7	0·26	0·18	0·1	0·09
k ■	0·9	0·19	0·11	0·07	0·06

D_1/D_2	1·0	1·2	1·4	1·6	1·8	2·0
k	0	0·19	0·31	0·37	0·41	0·43

D_2/D_1	1·0	1·2	1·4	1·6	1·8	2·0
k	0	0.11	0·24	0·36	0·46	0·54

$$k = k'\left[1 - \left(\frac{A_1}{A_2}\right)\right]^2$$

Angle	θ	5	10	15	22½	30	45
Circle	k'	0·13	0·26	0·42	0·63	0.77	0·9
Rectangle	k'	0·14	0·35	0·53	0·72	0·83	0·92

R/D	k 5 piece	k 3 piece
0·25	0·5	0·8
0·5	0·3	0·4
1·5	0·2	0·3

α	k
90°	1·0
60°	0·5
45°	0·3
30°	0·2

R/D	k α = 90°	k α = 45°
0	1·0	0·6
0·25	0·5	0·35
0·5	0·3	0·2
1·0	0·2	0·15

α	k ●	k ■
90°	1·0	1·5
135°	0·6	0·9

R/D	k α = 90°	k α = 135°
0·25	0·43	0·3
0·5	0·28	0·18

TABLE 6.2

Friction factors, f, for duct, flue and chimney friction loss calculations (Mean gas temperatures 450–650°K)

Flue diameter (mm)	Mean gas velocity (m s^{-1})	Smooth concrete or welded steel	Riveted steel or smooth cement pargeting	Brick or rough cement pargeting
	1·5	0·0054	0·0069	0·0190
150	3·0	0·0048	0·0066	0·0187
	4·5	0·0045	0·0064	0·0185
	6·0 and over	0·0044	0·0063	0·0180
	1·5	0·0047	0·0058	0·0180
225	3·0	0·0043	0·0057	0·0176
	4·5	0·0040	0·0056	0·0171
	6·0 and over	0·0039	0·0055	0·0168
	1·5	0·0042	0·0054	0·0146
300	3·0	0·0039	0·0051	0·0145
	4·5	0·0037	0·0050	0·0144
	6·0 and over	0·0036	0·0049	0·0143
	1·5	0·0040	0·0050	0·0138
350	3·0	0·0037	0·0049	0·0135
	4·5	0·0035	0·0048	0·0130
	6·0 and over	0·0034	0·0046	0·0130
	1·5	0·0037	0·0047	0·0130
450	3·0	0·0035	0·0045	0·0128
	4·5	0·0033	0·0044	0·0127
	6·0 and over	0·0032	0·0043	0·0127
	1·5	0·0036	0·0043	0·0125
600	3·0	0·0033	0·0040	0·0123
	4·5	0·0031	0·0039	0·0120
	6·0 and over	0·0030	0·0038	0·0119
	1·5	0·0030	0·0035	0·0111
1200	3·0	0·0033	0·0040	0·0123
	4·5	0·0031	0·0039	0·0120
	6·0 and over	0·0030	0·0038	0·0119
	1·5	0·0030	0·0035	0·0111
1200	3·0	0·0027	0·0034	0·0110
	4·5	0·0026	0·0033	0·0109
	6·0 and over	0·0025	0·0032	0·0108
	1·5	0·0027	0·0031	0·0101
1800	3·0	0·0025	0·0029	0·0098
	4·5	0·0024	0·0028	0·0097
	6·0 and over	0·0023	0·0027	0·0095

Note: For other gas velocities and flue diameters, interpolate. For constructions other than those shown, select the figure on the basis of similarity of the interior finish of the flue.

so any shortcoming in the value of p results in loss of efficiency in the combustion zone and trouble with downwash due to low efflux velocity. In these circumstances a mechanical-draught system should be considered.

It should be noted that all the losses considered vary as velocity, but whereas the velocity at the outlet can be clearly defined ($= 4Q/\pi D^2$) the velocity of flow in the chimney would be better considered as an average value over the total length of the flue, in particular where the chimney has a tapering section resulting in significantly differing velocities at the upper and lower sections.

It should also be noted that the flue gas temperature, T_g, has a significant effect on both the pressure differential and each pressure loss. Again, an average value of flue gas temperature must be considered, except where temperature loss is insignificant, as in the case of a concrete chimney with an inner brick lining. The estimation of average gas temperatures is a doubtful pursuit, since although the heat transfer characteristics of most materials of construction are sufficiently accurately known, the values of the film coefficients of the thin stagnant films of gas and air at the inner and outer surfaces of the chimney vary with the temperature and pressure of the fluid considered and are also dependent upon the velocity of flow.

In most chimney calculations, heat is considered to be transferred by conduction and convection, and the effect of radiation is ignored. No distinction is normally made between the two prior effects, and the value of the film coefficient normally encompasses the conductivity of the stagnant film and the convection due to mixing for the remainder of the fluid.

Within the limitations of the temperature ranges shown in Figs. 6.2 and 6.3, these curves are suitable and convenient to use. They are predominantly drawn up for application to steel flues or linings, and for higher temperatures reference should be made to standard works on heat transfer.[6] Some information is also available on comparable film coefficients for brick flues.[7]

In the majority of cases the transfer of heat takes place through a composite structure consisting of a flue lining, sometimes with insulation or an air space, and then through the structural shell. Each component part has a resistance to heat flow which results in a reduction in temperature at the successive interfaces.

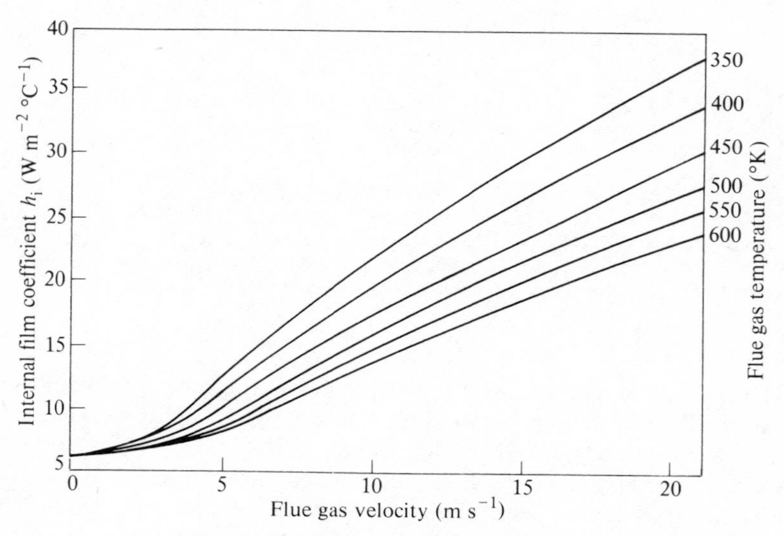

Fig. 6.2. Values of internal film coefficient h_i.

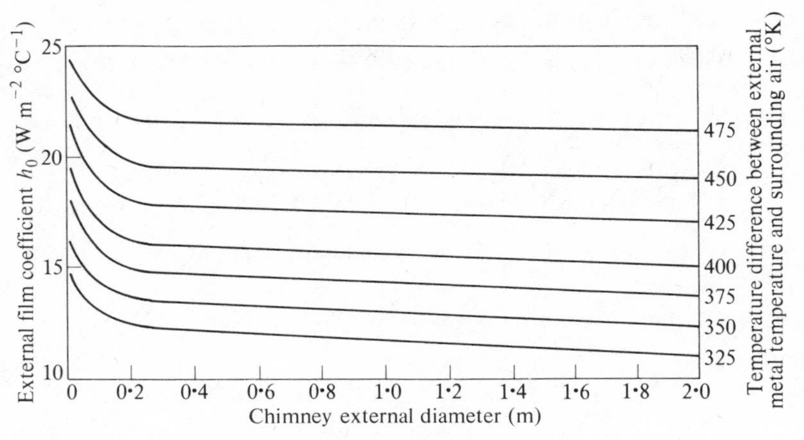

Fig. 6.3. Values of external film coefficient h_0.

Let a_i and a_0 be the film coefficients at the inner and outer surfaces (W m^{-2} °C^{-1}). Let λ_1, λ_2, and λ_3 be the thermal conductivities of the individual component materials (W m^{-1} °C^{-1}). Let d_1, d_2 and d_3 be the thicknesses of the individual component materials (m). Then

$$\frac{1}{U} = \frac{1}{a_i} + \frac{d_1}{\lambda_1} + \frac{d_2}{\lambda_2} + \frac{d_3}{\lambda_3} + \frac{1}{a_0}, \tag{6.4}$$

where U = overall heat transfer coefficient, and $1/U$ may be considered as the sum of the individual resistances of the various components to the flow of heat.

The total heat loss from chimney surface is given by

$$UA(T_g - T_a) = UA\,\Delta T \quad \text{W},$$

where

$$A = \text{surface area of flue (m}^2\text{)}. \tag{6.5}$$

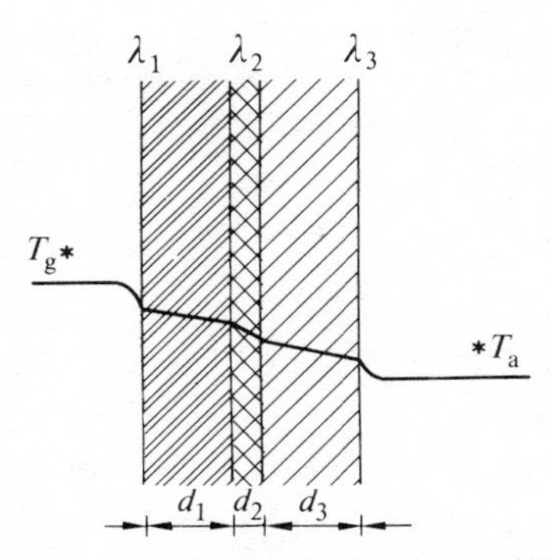

Fig. 6.4. Temperature drop in flue wall of composite construction.

From a consideration of the gas flow, also

$$\text{total heat loss from chimney surface} = \rho_g QC(T'_g - T''_g), \qquad (6.6)$$

where ρ_g = density of flue gases at T'_g (kg m^{-3})
Q = gas flow (m^3 s^{-1})
C = specific heat of waste gases at constant pressure (J kg^{-1} °K^{-1})
T'_g, T''_g = temperature of flue gases at inlet and outlet (°K).

The value of T'_g is known, but ΔT is best calculated from an average value of T_g, that is, $\frac{1}{2}(T'_g + T''_g)$.

Under equilibrium conditions, (6.5) and (6.6) can be equated; hence

$$UA\left(\frac{T'_g + T''_g}{2} - T_a\right) = \rho_g Qc(T'_g - T''_g),$$

from which T''_g can be calculated.

From this calculation, the maximum heat loss occurs at a low value of T_a and it should be assumed that the ambient temperature is at freezing point.

However, the values of the film coefficients a_i and a_0 must be those equivalent to a flue gas temperature of T'_g as a first approximation. An iterative process of not more than two steps then gives a sufficiently accurate value of the gas face and other interface temperatures.

To calculate the interface temperature, the total temperature drop through the flue walls is proportioned in the ratio of the thermal resistance of each component part to the total resistance of the composite structure. For example,

$$\text{temperature drop through gas face stagnant film} = \frac{U}{a_i}(T''_g - T_a) \qquad \text{(A)}$$

and

$$\text{temperature drop through component 1} = \frac{Ud_1}{\lambda_1}(T''_g - T_a), \qquad \text{(B)}$$

and so on. Hence the temperature at each interface can be calculated.

It is important to use the minimum value of the flue gas temperature in this calculation, since it is necessary to know the minimum value of the inner surface temperature which may occur so that the possibility of acid smut formation may be avoided.

Thermal shock

The term *thermal shock* is used to mean the stresses which a chimney lining experiences when a boiler or other heat producing plant is brought up on load quickly. The temperature gradients calculated previously are those which exist under steady state conditions, i.e., gradient is independent of time. However, on starting up and in particular on a quick run-up, a transient thermal gradient, or one changing with time, first occurs. As a practical example of this, panels of acid resisting bricks, 110 mm and 220 mm thick, laid in sodium silicate mortar were subjected to cycling tests[8] involving heating one face of each panel from room temperature to 300°C. This took 3 min and 4 min for each panel respectively. Thermocouples were installed in the mortar joints to determine temperature distribution. Figure 6.5 and 6.6 show the distribution of temperature through each panel, and from these curves it can be clearly seen that the temperature gradient exists in only 25 per cent of the total thickness of the brickwork. The distribution of stress through the panels, based on a theory[9] referring more to an isotropic uncracked material, was calculated and is shown in Figs. 6.7 and 6.8. The stresses calculated were not sufficient to cause cracking of the bricks initially, but were sufficient to crack the mortar. If such stressing were followed for some reason by rapid cooling of the hot face, high tensile stresses would be set up on the layers immediately adjacent to this face, since conditions would then be reversed. The panels were in fact heated and cooled in this manner through 12 cycles and extensive cracking did occur.

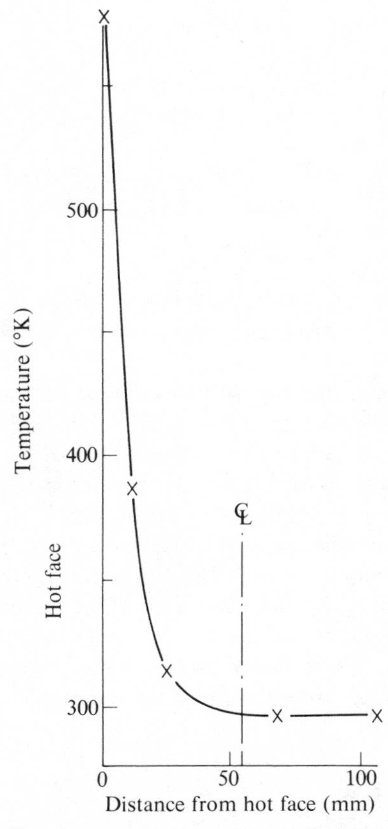

Fig. 6.5. Distribution of temperature due to thermal shock in a panel 110 mm thick.

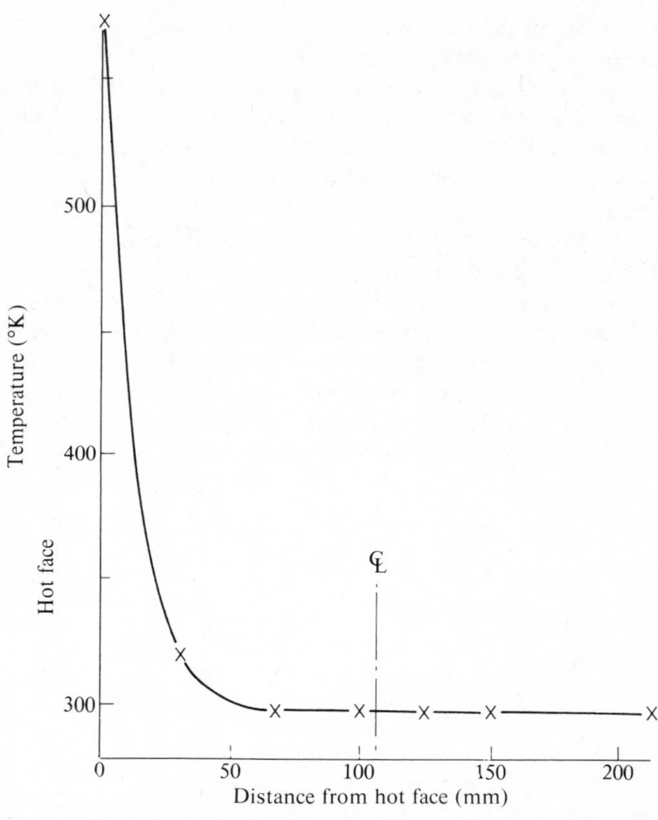

Fig. 6.6. Distribution of temperature due to thermal shock in a panel 220 mm thick.

The part that lagging plays in thermal shock should be considered. It is obvious from Figs. 6.5 and 6.6 that there was no rise of temperature on the outer face of both panels; hence lagging can make no difference to the magnitude of the thermal stresses under these conditions. However, during the steady state condition, lagging reduces the thermal gradient but raises the mean temperature of the lining, thus accentuating the difference in the coefficients of expansion of the brick and the mortar. This does not mean that lagging should be eliminated, since it has an important part to play in chimney design, but rather it emphasizes the need for a careful consideration of the operating regime which the chimney is likely to experience. An example can be taken from the mechanical engineering field, where boiler drums, turbine rotors and pressure vessels are heated or cooled gradually in order to reduce these high initial thermal gradients.

Efflux velocity, *V*, of flue gas

The choice of a suitable value of efflux velocity at full load conditions is important if

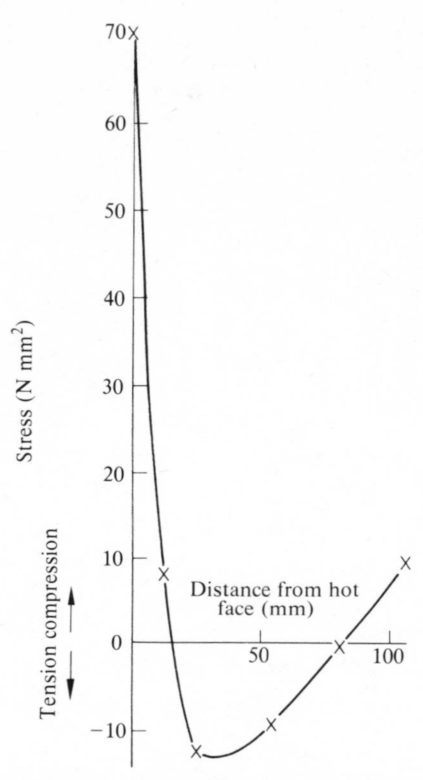

Fig. 6.7. Theoretical stress distribution at maximum face temperature in a panel 110 mm thick.

downwash is to be avoided in the lee of the chimney. Downwash causes unsightly stainings on the chimney surface, but of major importance is the resultant loss of buoyancy, which decreases in turn the rise or effective height of the plume. A secondary effect is that cold air may enter the chimney at high wind velocities causing undesirable cooling of the interior face of the flue.

It is difficult to recommend suitable values for efflux velocity because of the variety and heights of modern chimneys. At one end of the scale, large multi-flue chimneys have been built for power station complexes with heights of 200–260 m and efflux velocities of 25 m s^{-1}. The success of this policy of building high chimneys with high efflux velocities has been recognized and publicly acknowledged by several authorities in the air pollution field. However, an upper limit must be set, since with too high a velocity the rapid mixing induced with the atmosphere diminishes the subsequent thermal rise. It is sufficient to provide a velocity that ensures a clean emission from the chimney in all but exceptionally strong winds. Such idealisms are perhaps all very well for chimneys associated with some of the largest balanced-draught fan systems in the world, but at the other end of the scale a true natural-draught system on a small industrial boiler would probably have difficulty in creating an efflux velocity of 6 m s^{-1}.

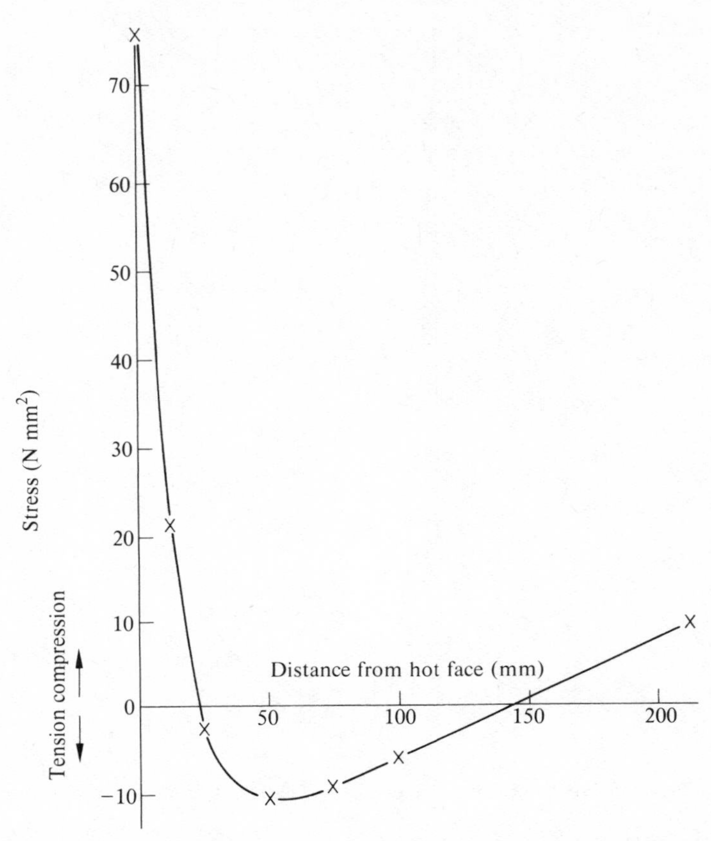

Fig. 6.8. Theoretical stress distribution at maximum face temperature in a panel 220 mm thick.

Since a general recommendation is desirable, however, the values given below will suffice:

1. Natural-draught systems — 6 m s^{-1}
2. Forced-draught systems
   - Chimneys up to 20 m high — 6 m s^{-1}
   - Chimneys from 20 to 45 m high — 9 m s^{-1}
   - Chimneys over 45 m high — 12 m s^{-1}
3. Induced-draught systems: velocities at full load conditions should be as high as fan power requirements and noise levels permit and should not be less than those suggested for forced-draught systems.

Similarly, the Memorandum on Chimney Heights[2] recommends that efflux velocities for boiler plant within its scope should be 6 m s^{-1} for boilers equipped with forced-draught fans and 7·5 m s^{-1} for boilers equipped with induced-draught fans.

Another authority has suggested that the efflux velocity of any chimney should be not less than 15 m s^{-1}. However, the critical factor affecting the velocity is the available draught, and where natural draught is inadequate, mechanical draught systems are necessary to ensure a sufficiently high value. To suggest a common minimum value for chimneys serving boilers of various duties is perhaps idealistic, but where designers can exceed the figures above, at no great cost penalty, then this is highly desirable if downwash is to be avoided. To attain velocities in excess of 15 m s^{-1}, however, requires forced- or induced-draught fans or a combination of both. The design of a fan-aided system is outside the scope of this chapter, and the reader is referred to standard reference works[4,5] on boiler design. Suffice it to say that induced-draught fans are generally installed to augment the natural draught when chimney heights are insufficient. A second advantage is that draught is available before ideal combustion conditions are established, hence bringing the system up to operating temperature more quickly. A forced-draught fan is normally installed when the manufacturer is seeking high efficiency, thus necessitating efficient mixing of air and fuel in the combustion chamber. Larger installations with variable loading conditions ideally have a mixed system employing both induced- and forced-draught fans.

Lateral dimension, *D*, of chimney

The aesthetic appearance of a chimney is more properly the province of the architect, but it should be the aim of every designer to produce chimney shapes of not over-generous proportions. The diameter of the structural shell is governed by the internal diameter, D, of the flue or flues, which in turn is decided by the gas flow, Q, and the efflux velocity, V. The diameter of the structural shell at the bottom of a single flue chimney is decided more from a consideration of structural stability than from the size of the flue at this level, although on very large boiler installations it is common practice to limit the gas flow velocity in the main ducting to reduce friction and bend losses. The same cannot be said of multi-flue chimneys, where the flue layout and sizes generally produce a structural shell with a cross-section larger than is required for stability against wind forces.

It is found that chimneys having an aspect ratio (height to average external diameter) of 15–20, are generally pleasing to the eye, but chimneys with aspect ratios of less than 15 appear short and stubby by comparison.

It is not structurally necessary to have a uniform taper on a chimney, since the design moments vary parabolically as with any cantilevered structure. The chimney therefore can very often be made straight sided over most of its height or given a slight taper or tapers and then in both cases flared out over the bottom 20–25 per cent of the total height to provide the required stability and to accommodate the inlet bends and turning vanes. The material of construction also affects to a great degree the final form of the chimney. For instance, the rise in popularity of the pre-cast concrete chimney with an integral refractory lining means that the shape of the structure should ensure that the individual sections are similar in order to reduce formwork costs by repetitive use of one shape. This is also true of chimneys in glass-reinforced plastics where tooling and mould costs are high. Concrete chimneys, where cast *in situ*, can have pronounced tapers, since most forms of climbing formwork are capable of being modified to suit the successive reduced perimeters as the height increases. However, the fairly recent innovation of the 'slipform' shutter is best used where parallel-sided chimneys are required, since with this method the formwork is jacked up at a slow continuous rate which does not easily allow the perimeter to be reduced.

Smut emission

The problem of acid smut emission can occur in chimneys where low temperature conditions exist on the gas face of the lining due either to poor insulation causing heat losses or to intermittent operation of the boiler plant. The problem is also exacerbated by the fact that advances in boiler design for large installations have resulted in low back-end temperatures due to the utilization of waste heat in air heaters and economizers. As an example, modern power stations have attained back-end temperatures of 110°C for coal-fired boilers and 150°C for oil-fired boilers.

With the exception of wood, there are no solid fuels which may be described as sulphur free. Some coals have very high sulphur contents, but generally the sulphur contents of most British coals are reasonably low. With liquid fuels, the sulphur content is more varied and depends in the main on the origin and subsequent processing. Heavy fuel oils contain about 3 per cent of sulphur, whilst distillate fuels such as gas oil and diesel oil may contain from 0·1–1 per cent. Natural gases are even more varied in their content, ranging from sulphur free to those with a very high content of hydrogen sulphide, sometimes as much as 15 per cent by volume.

During the combustion process, the sulphur in the fuel produces sulphur dioxide (SO_2) and sulphur trioxide (SO_3). Usually only a very small part, generally less than 5 per cent of the SO_2 is oxidized to SO_3. Higher conversion rates can occur with large amounts of excess air and suitable temperatures, but conversion rate is reduced with an increase in temperature above about 600°C.

A sulphur-free flue gas, if cooled to a sufficiently low temperature, becomes supersaturated with water vapour, and condensate or dew forms with further cooling. This *water dew point* will occur at temperatures of 40–50°C. The presence of SO_3 in flue gases means that sulphuric acid is present in vapour form and an *acid dew point* exists, which, similarly, is the point at which saturation of the flue gas with sulphuric acid vapour occurs. Cooling can then result in condensation or the formation of an acid dew on the flue lining at temperatures of about 130–160°C.

Further cooling results in a marked increase in the strength and rate of deposition, reaching a peak some 40–50°C below the acid dew point. The cooling effect is caused by the heat losses through the flue lining producing a 'cold' face on which the condensate can form. No condensation takes place in the main flow of gas unless the temperature is drastically reduced throughout the whole body of the gas, in which case a mist of acid particles forms. Such low temperatures are not normally reached in practice. The quantity and concentration of the sulphuric acid formed varies so much with the type of boiler and the method of operation that a detailed discussion is not possible. The problem has been the subject of much research and is well documented.[10]

During combustion, solids are produced in the form of soot or unburnt carbon which are trapped on the acidic condensate, if this is present, forming a messy agglomerate or smut. As the smuts build up into loose layers they are easily dislodged by changes in the gas flow and are then discharged into the atmosphere to cause unsightly damage wherever they may land. If the formation of the acid dew is drastically reduced then the combustion products are normally discharged as fine solids which are dispersed harmlessly over a wider area.

Reduction of heat losses

It is obvious from the foregoing that the deposition of acid, with attendant acid smuts and corrosion, may be reduced or largely avoided by ensuring that the condensing surface is at a temperature which is equal in value to or slightly greater than the acid dew point of the gas. This is an ideal solution which is not always easy to effect in practice. From expression (B) on p. 154 for the temperature drop through each component of a composite flue lining, it will be seen that the ideal insulating material should have a large resistance to heat flow resulting in a high temperature drop across the interfaces. Such a material should therefore be thick and have a low thermal conductivity. The insulator can be the flue lining material itself if it possesses suitable qualities, or it can be an external material which could probably not stand direct exposure to the flue gases.

The insulating effect or low heat conductivity of such materials is due primarily to their porous nature. Most good insulators contain voids or small air pockets in which there can be little or no air movement and hence no convection. However, the insulation value of the material is reduced if the spaces are enlarged, because convection currents are then possible. On becoming wet, an insulator has a higher thermal conductivity than when dry, because the voids are now saturated with a medium of a higher conductivity.

Choice of materials

The choice of a suitable material to be used as chimney insulation is usually governed by the structural form of the chimney. The passing of the Clean Air Act has brought about a radical change in the design of chimneys under 60 m in height, since it is now common practice to group several flues in a common windshield so that the minimum requirements for chimney height and efflux velocity can be satisfied.

For a multi-boiler installation, plume rise can therefore be maximized and this is not affected so much by variations in the boiler loading as it would be if the boilers were discharging through a single chimney or through a stack of individual chimneys.

Patent chimneys are now available which are quickly erected and incorporate a high degree of pre-casting in the case of concrete or of pre-fabrication in the case of steel.

Chimneys of height greater than 60 m follow more conventional lines and generally comprise a concrete windshield supporting brick flues on corbels or intermediate floors, or supporting steel linings in various ways.

Some insulating materials which are commercially available are summarized in Table 6.3. It should be borne in mind that some insulating materials have structural strength and may be used as a flue lining and that some structural materials possess thermal insulation properties, albeit very low.

Therefore the choice of an insulating material cannot be divorced from a consideration of a suitable flue lining material for the particular conditions prevailing, since

TABLE 6.3

Values of thermal conductivity

Type	Description	Thermal conductivity† (W m^{-10} C^{-1}
Loose fill		
Perlite	Acidic volcanic glass, commercially available as expanded granules of size, 60–80 mesh, bulk density 30–100 kg m^{-3}	0·025–0·144
Vermiculite	Hydrated laminated material, mainly silicates resembling mica in appearance, commercially available as exfoliated granules, bulk density 60–200 kg m^{-3} in varying grades	0·036–0·288
Lagging		
Mineral wool	Resilient mineral wool fibres, commercially available as semi-rigid mattress with wire netting, bulk density 80–160 kg m^{-3} in varying grades	0·037–0·120
Glass fibre	Long glass fibre insulation, commercially available as slabs, rolls and wired mattresses, density 64 kg mg^{-3} approximately	0·037–0·081
Foamed glass		
Blocks	Expanded glass blocks of closed cellular structure, inert and non-absorbent, commercially available as blocks and slabs, bulk density 144 kg m^{-3}	approx 0·055
Insulating bricks		
Moler	Diatomaceous earth bricks commonly known as *Moler*. Cast in varying shapes for flue linings, density 700 kg m^{-3}	0·144–0·173
Castable linings		
Aluminous cement and lightweight aggregate	High alumina cement and various proprietary aggregates, e.g., Lytag, Moler, etc. Good refractory and insulating properties, bulk density 1100 kg m^{-3}	Varies with aggregate and mix, typical range 0·199–0·236

† Thermal conductivity values vary with operating temperatures and reference should be made to manufacturers' catalogues for specific applications.

in most cases the latter make some contribution to the overall heat transfer coefficient, U, and should the acid conditions be mild enough to allow an insulating brick to be used as a flue lining, then the provision of extra insulation probably does not need to be considered.

Acid-resisting bricks

There are very dense, good quality engineering bricks, characterized by extreme hardness, with crushing strengths of 84 MN m^{-2} and vitrified over their entire cross-section. They are required to be in accordance with BS 3679 which specifies dimensional tolerances to ensure tight joints and also specifies limits for apparent porosity. They have been used for chimney linings over many years and are resistant to most forms and concentrations of acid, suffering little or no degradation in service. They can be laid in single brick thickness (115 mm) at a radius not exceeding 3 m, to form cylinders up to 10 m high which are structurally stable.

They need to be laid in an acid resisting mortar, and cracking can occur in either brick or mortar under high thermal gradients. The calculation of thermal stresses is made difficult by the fact that the brick/mortar combination cannot be analysed as a homogenous material, although for thin linings an approximate calculation can be based on the properties of the brick alone. Some authorities have postulated the concept of a 'composite' material having composite values of modulus of elasticity and coefficient of expansion, but since the expressions for these constants take into account the physical dimensions of a single brick and a single mortar joint the values obtained do not differ greatly from those for brick.

Because of the limitations on height and the high density of these bricks, their use is limited to lining reinforced concentrate chimneys, where they are built as discrete cylinders supported on corbels cast integral with the windshield. It has been the practice for many years to provide a narrow ventilated air space behind the lining, so that if cracking occurred leaking gases would be removed by the air flow. However, with present-day emphasis on conserving thermal energy in the gas stream and the fact that these bricks have a fairly high thermal conductivity, approximately 1·0096 W m^{-1} $°C^{-1}$, modern thinking is to insulate linings, built from these bricks, with lagging or foamed glass blocks.

Steel liners

Steel has been used for many years as a construction material for chimneys, either free-standing or guyed, and such chimneys have been a familiar sight on small power stations and other small boiler installations. These old steel chimneys are normally unlagged but have a thin lining, generally 50 mm thick, of refractory mortar which has been either cast or gunned onto the interior.

In recent years steel has been more commonly used as a lining material, being easily rolled and capable of a high degree of pre-fabrication. Its most common uses are summarized below:

1. as small-bore flues, within a high multi-flue concrete chimney;
2. as small bore flues within steel windshields;
3. as the inner lining of a double-skinned single flue chimney in which the outer skin is either steel or polished aluminium;
4. as a new lining to an older chimney, possibly brick, where the boiler being served has been uprated or replaced by a unit of higher efficiency, involving a change in fuel.

Steel cannot be employed as a lining without careful consideration of the insulation which must accompany it. Loose fill is a popular solution to the cases mentioned and serves a dual purpose in case 1 by protecting the concrete windshield against high thermal stresses. The material is normally provided in a thickness of not less than 150 mm, and is either poured into the interspaces or pumped by special equipment. It has the disadvantage of compacting to about 90 per cent of its volume in about the first year of service, and requires topping up.

Where larger interspaces occur in a multi-flue chimney and the use of loose fill is not practical, the steel flues can be insulated by lagging which is in blanket form and of varying thicknesses. It is normally reinforced on the outer face with wire netting or expanded metal, and is retained in place by narrow straps, the ends of which are machine cramped.

Some single flue chimneys of type 3 have been the subject of experimentation where an aluminium outer skin, chosen for its appearance and weathering properties, was placed around the steel lining. An air gap, less than 25 mm, was provided, but no other form of insulation. The reflectivity of the inner surface of the aluminium also contributed to the overall resistance to the heat flow. With such an arrangement the temperature on the inner face of the lining was increased by approximately 30°C over that of a bare steel chimney under the same operating conditions. There is an optimum distance which the air gap must not exceed, otherwise convection currents may be set up in this space. This effect can be reduced by providing asbestos spacers at intervals of, say, 1 m to split the air column into separate compartments.

Most liners of this kind are fabricated from mild steel plates, varying in thickness from 2 mm to 5 mm, but BS 4076[11] stipulates a maximum temperature of 480°C. In incinerator structures or in chimneys serving some gas turbine plants, higher temperatures may exist or a rapid run-up onto full load causing thermal shock may be common. In the former case a castable refractory lining may be satisfactory, but in the second case a boiler quality steel with better yield characteristics at elevated temperatures should be considered.

Insulating bricks

The most commonly used form of insulating brick is the moler brick, made from diatomaceous clay quarried in Denmark, and characterized by low density, high absorption and a low thermal conductivity. Because of the low density it has been the practice to build moler brick linings of substantial height, up to 30 m, with a radius of 1·5 m and a wall thickness of 200 mm.

However, a series of failures of such free-standing liners, characterized by severe cracking and partial or complete collapse, has led to recommendations on the limiting heights of free-standing moler linings (Table 6.4). Where moler bricks form the lining of a single flue chimney it is advisable to build the brickwork close to the surrounding windshield, although due account must be taken of the cantilever and ovalling deflections of the windshield so that the brickwork is not overstressed.

The high porosity of moler bricks makes them capable of absorbing a high percentage of water, whether from soaking as part of the laying process or from atmospheric conditions. Tests using strain gauges have shown that high local stresses can be set up by wetting small areas of linings. It is advisable, therefore, to curtail the brickwork for the top 6 m of a chimney and either to replace it with acid-resisting brickwork or provide a stainless steel flashing plate over the brickwork for the same distance. When bricks are laid against the inner face of a concrete windshield, it is advisable to treat this face with a silicone solution so that rain, saturating the thinner sections of the windshield, does not penetrate the brickwork. Such moisture normally evaporates in the gas stream, but should the chimney be brought rapidly back on load, severe shrinkage can occur.

TABLE 6.4

Recommended lining heights for various thicknesses of Moler flue linings

Thickness of lining (mm)	Nature of construction	Maximum recommended height (m)†	
115	With air gap		13
115	Without air gap and in contact with structural shell		23
115	Bonded to structural brickwork		30
115	Used as permenent shuttering		45
150	With air gap		18
150	Without air gap and in contact with structural shell		30
150	Supported on corbels to depth of 40 mm	Lift height	6
		Total height	45
150	Used as permanent shuttering		60
230	With air gap		41
230	Without air gap and in contact with structural shell		45
230	Supported on corbels to depth of 75 mm	Lift height	7·5
		Total height	60
230	Used as permanent shuttering		75
230	Shuttering on corbels to depth of 115 mm	Lift height	7·5
		Total height unlimited	

† The maximum recommended heights are optimum values and are based on the assumption that due consideration has been given to the structural stability of the lining and that the diameter–thickness ratio is suitable.

Castable linings

Refractory materials have been used for many years as linings for single skin steel chimneys, and such linings generally consisted of ordinary Portland or high alumina cement and sand. In recent years the development of lightweight aggregates such as crushed moler brick, or Lytag and Aglite, the pulverized fuel ash derivatives, has led to the more popular use of castable refractory linings with low thermal conductivity and some structural strength when used with high alumina cement. A range of precast chimneys has now been developed incorporating such linings cast integral with an outer reinforced concrete shell. These are produced in easily handled lengths and capable of speedy erection. Various flue shapes can be cast to meet architectural requirements.

Linings of this type have been used at temperatures in excess of 500°C, but the choice of aggregate and aggregate–cement ratio needs careful consideration. There are also limitations of handling and transport for flues above approximately 3 m. Larger diameter chimneys are best lined with a refractory brick or with a lining 'gunned' on *in situ*. The latter method is also preferable when the flues are of a non-circular shape involving large radii of curvature. For example, for elliptical flues the stability of the brickwork is suspect.

Glass-reinforced plastics

The development of glass-reinforced plastic consisting of a resin, generally polyester, set to a hard solid and reinforced with strong glass or carbon fibres has led to the development of a range of chimneys. These are generally of limited height if free standing, but of pleasing architectural appearance since the basic material is capable of being moulded into multi-flue chimneys of variable shapes or as cladding over mild steel chimneys. Glass-reinforced plastic has long been used in the petrochemical field for storage vessels and can withstand temperatures up to 250°C. Its resistance to the majority of corrosive elements found in the chimneys serving industrial boilers is good, but temperature has a considerable effect on the chemical resistance of the resin employed.

The design of glass-reinforced plastic chimneys is specialized, since most of the expertise involved in the formulation of a suitable resin–fibre combination lies with the specialist moulding firms. Fillers such as wood flour or china clay are sometimes added to improve the physical properties; the curing or hardening is generally carried out by the use of a catalyst and an accelerator.[12,13] The formulation is then specific to the chimney concerned, after due consideration of the physical and chemical properties that it is required to have in service.

This section on the choice of materials has been confined mainly to a description of the materials, both lagging and lining, which are commercially available. It would be ideal to be able to say that under a specific combination of flue gas temperature and analysis a specific combination of lining material plus lagging should be employed. However, the uses for which chimneys are employed are varied. When one considers the industrial as well as the more domestic field then one is considering the processes employed in producing cement, bricks, oil and petrochemicals, gas, electricity, iron and steel among many and in destroying domestic waste and industrial debris among others. The operating conditions are varied hence the chimneys are varied. The universal chimney does not exist but a consideration of the basic requirements discussed in this chapter will, it is hoped, bring about a better understanding of the function of a chimney and the part which it plays in the control of industrial pollution.

References

1. Clean Air Acts 1956 and 1968, HMSO, London.
2. Ministry of Housing and Local Government, *Chimney Heights*, 1956, Clean Air Act memorandum, 2nd edn, HMSO, London, 1967.
3. The Clean Air (Height of Chimney) (Prescribed Form) Regulations 1969, HMSO, London.
4. *Steam—its generation and use* (37th edn.), Babcock and Wilcox, New York, 1963.
5. *Notes on fan engineering*, Davidson and Co. Ltd, Belfast.
6. Kern, 'Process Heat Transfer', McGraw-Hill, 1950.
7. *Specification for the Design and Construction of Reinforced Concrete Chimneys* (ACI 307—69), American Concrete Institute, 1969.
8. *Effects of thermal stresses on chimney-lining brickwork*, Laboratory Note No. RD/L/N/91/63, Central Electricity Research Laboratories, 1963.
9. Timoshenko and Goodier, *Theory of elasticity*, McGraw-Hill, 1951.
10. *Flue gas corrosion in boiler plants*, Flue Gas Corrosion Research Committee, Danish Academy of Technical Sciences, Danish Technical Press, 1960.
11. *Design of steel chimneys*, BS 4076, British Standards Institution, 1966.
12. *Crystic Monograph No. 1*, Chemical Resistance Crystic Research Centre, Scott Bader Co. Ltd, Wellingborough, Northants.
13. *Crystic Monograph No. 2*, Polyester Handbook Crystic Research Centre, Scott Bader Co. Ltd, Wellingborough, Northants.

Acknowledgement

The publishers wish to thank Shell-Mex and B.P. Ltd, for their assistance in providing certain information in this chapter.

7 Principles of reducing pollutants in gaseous discharges

J. BETTELHEIM, B. H. M. BILLINGE and W. S. KYTE

The removal of pollutants from gaseous discharges is often complicated by the fact that the contaminants are generally present in small concentrations with a large excess of inert diluent gases. This means that processes must be found that either remove the pollutant at these low concentrations or concentrate it in another phase. Direct conversion by combustion is sometimes possible or the pollutant may be concentrated in a liquid (absorption) or on a solid (adsorption).

The selection and design of the removal unit depend not only on the chemical equilibria and kinetics of the chemical reactions involved but also on the physical transport properties of the various species involved. These transport phenomena are common to all the processes and include the effects of both heat and mass transfer. Economic factors will also influence the selection of the type of process used and may place limitations on the design of particular units. Thus in the treatment of large volumes of gas, pumping costs may be high and the minimization of pressure losses may be of paramount importance.

In this chapter the basic principles underlying the design of efficient removal units are developed.

Absorption processes

One of the most frequently used techniques for controlling the composition of industrial waste gases before discharge to the atmosphere is absorption of gases in liquids. In general, waste gases are mixtures of gaseous components, some of which are soluble in a selected liquid phase, a major part being practically insoluble carrier gas. The direct contact of gas with liquid enables mass transfer to take place between the two phases in directions governed basically by the concentration gradients of individual components. In most cases absorption of one component is simultaneously accompanied by desorption of another one (e.g., absorption of SO_2 in an aqueous solution accompanied by desorption of water). Deliberate desorption can be achieved by a change of physical conditions (rise of temperature, decrease of pressure), stripping of the liquid with inert gas (or steam) or by chemical decomposition of the sorbent. Absorption followed by desorption provides one of the basic means of establishing a cyclic operation with reuse of the sorbent. Such a scheme acts as a device for separation and concentration of the selected gas.

Equilibrium and mass balance considerations

The design of a new unit or assessment of an existing plant is always based on a knowledge of the total gas flow and its inlet composition. The conditions of the outlet treated gas could be fixed by environmental requirements, economic factors or by a combination of both. The main objectives of the assessment of any particular case are:

1. to choose the most suitable liquid phase considering the required selectivity, saturation and stability;
2. to estimate the limiting concentrations (efficiencies) and the mass balance of the entire operation;
3. to select an appropriate apparatus for liquid–gas contact;
4. to estimate the main design parameters such as surface and volume of contact, diameter to length ratio and required pressure drop;
5. to estimate the capital and operating costs.

The selection of a suitable liquid solvent and the determination of the limits of absorption efficiency are based on solubility data. In most cases no formula is available for the calculation of solubility and only tabulated or graphical data can be used. The most frequently used expressions for concentrations in the gaseous and liquid phases of binary systems are summarized in Table 7.1.

From measurements of equilibrium conditions with low concentrations, it has been found that the partial pressure of gas varies almost linearly with its concentration in the liquid

$$p_A = H x_A, \tag{7.1}$$

where H is Henry's coefficient. The value of H depends on the definition of concentration units, but for isothermal conditions, H should be independent of composition and the total pressure of the system. In practice the linearity applies only within a limited range of concentrations and for a total pressure normally not exceeding 5 atm.

TABLE 7.1

Gas and liquid concentration units used in absorption

	Gas	Liquid
Partial pressure	p_A	—
Mass fraction	—	U_A
Mole concentration [mole volume^{-1}]	$C_A = \dfrac{p_A}{RT}$	$C_A = U_A \dfrac{\rho}{M_A}$
Mole fraction	$y_A = \dfrac{p_A}{p_A + p_1}$	$x_A = \dfrac{U_A M_B}{U_A M_B + (1 - U_A) M_B}$
Mole ratio	$Y_A = \dfrac{p_A}{p_1} = \dfrac{y_A}{1 - y_A}$	$X_A = \dfrac{U_A}{1 - U_A} \dfrac{M_B}{M_A} = \dfrac{x_A}{1 - x_A}$

ρ = Average density; $M_{B, A}$ = molecular mass; p_1 = partial pressure of inert gases.

If, however, mole ratios are used instead of partial pressures and mole fractions, Eq. (7.1) becomes

$$\frac{Y_A}{1 + Y_A} = \frac{H}{P} \frac{X_A}{1 + X_A}, \tag{7.2}$$

in which the relationship between Y_A and X_A is not linear and the equilibrium line is a curve. However, if X_A and Y_A are sufficiently small, then at constant total pressure the equilibrium can be approximated by a straight line.

For a countercurrent absorber (Fig. 7.1) the overall mass balance is expressed by

$$G(Y_1 - Y_2) = L(X_1 - X_2), \tag{7.3}$$

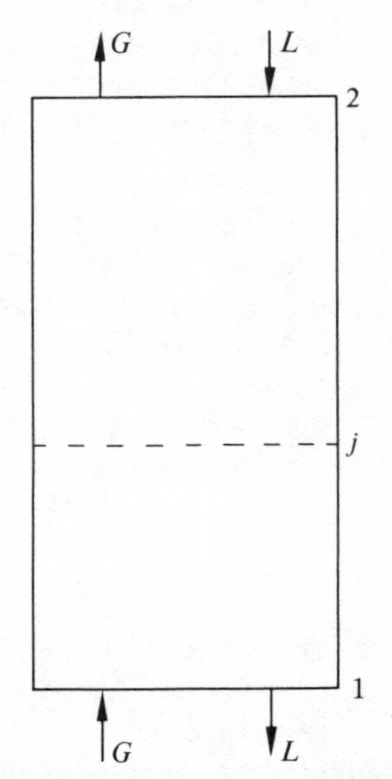

Fig. 7.1. Countercurrent absorber mass balance.

and for a balance between the bottom and any cross-section (j) of the absorber

$$G(Y_1 - Y_j) = L(X_1 - X_j), \tag{7.4}$$

$$Y_j = Y_1 - \frac{L}{G}(X_1 - X_j). \tag{7.5}$$

To define this function it is necessary to select, besides the given data on Y_1 and G, another three values for the variables X_1, X_2, L, Y_2. In many practical cases this procedure is simplified by

1. restrictions on the composition of the liquid feed supplied from another source independent of the absorption, i.e., $X_2 =$ constant,
2. required minimal efficiency of absorption, and
3. range of reliable liquid rates for a selected absorber.

To find the optimum operating conditions, it is necessary to balance the overall costs, as described for instance in Perry.[43] Generally the operating conditions cannot be changed arbitrarily, since at any point in the absorber the concentration of an active component in the gas phase must be higher than the relevant equilibrium concentration in the liquid. Figure 7.2 illustrates the case where the mass balance is restricted by the given values of Y_1, Y_2 and X_2. Equation (7.5) is represented by several straight operating lines, the slope of which varies with L/G. The minimum consumption of absorbent is

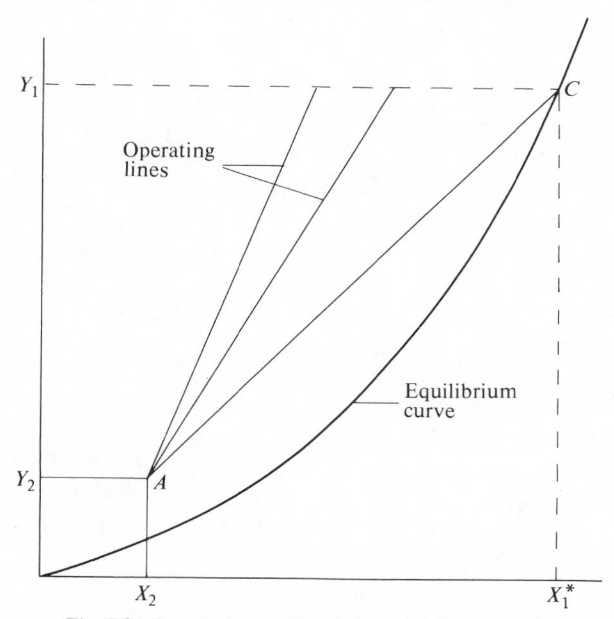

Fig. 7.2. Countercurrent steady state mass transfer.

determined by the slope $(L/G)_{min}$ of the line AC intersecting the equilibrium curve at level $Y = Y_1$, thus

$$\left(\frac{L}{G}\right)_{min} = \frac{Y_1 - Y_2}{X_1^* - X_2}, \tag{7.6}$$

where X_1^* is the mole ratio in the liquid phase in equilibrium with Y_1. Whilst packed absorbers are of the countercurrent type, cocurrent or crossflow units are also widely used. Fluidized beds or the individual stages of plate towers are examples of crossflow units. Cocurrent flow prevails in Venturi-type scrubbers and sometimes is used even with packed units for hydraulic reasons, e.g., shorter ducts, separation of droplets.

Conditions in a cocurrent absorber are illustrated in Fig. 7.3. The mass balance

$$G(Y_2 - Y_1) = L(X_1 - X_2) \tag{7.7}$$

and operating lines with a changing gas to liquid ratio are illustrated in Fig. 7.4. The minimum liquid rate in this case is defined by

$$\left(\frac{L}{G}\right)_{min} = \frac{Y_2 - Y_1}{X_1^* - X_2}, \tag{7.8}$$

and for any particular case is always greater than that given by Eq. (7.6).

In some instances a part of the outlet liquid is mixed with the fresh feed and recirculated to the absorber. This technique is frequently employed for treating waste gases when the estimated liquid rates lie below the value for reliable coverage of the contact surface. In other cases recirculation is necessary to maintain isothermal conditions or to

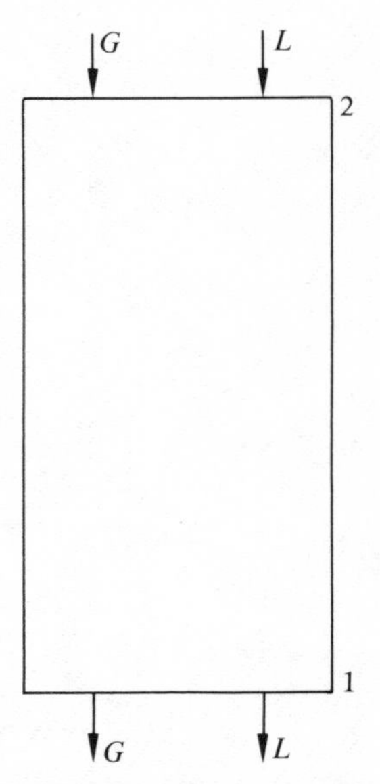

Fig. 7.3. Cocurrent absorber mass balance.

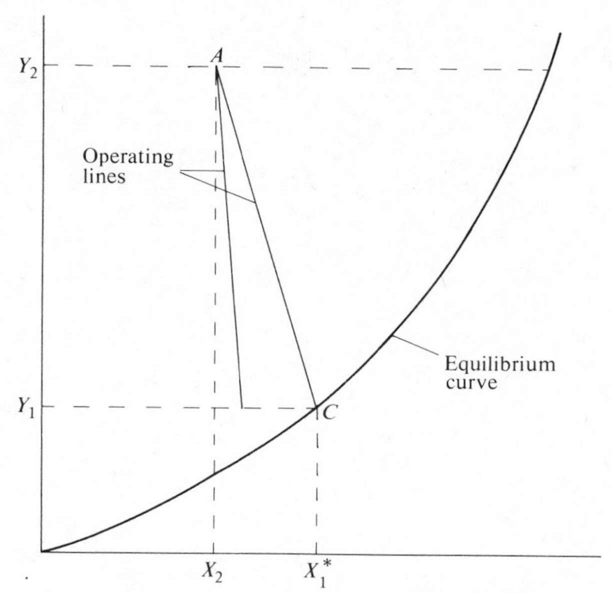

Fig. 7.4. Cocurrent steady state mass transfer.

prevent deposits forming. Any recirculation reduces the rate of mass transfer owing to the decrease in the driving force between the phases, and must therefore be counterbalanced by the introduction of additional absorption stages.

A block diagram of a two-stage absorption unit with separate recirculation is illustrated in Figs. 7.5 and 7.6. The mass balance is expressed by the following equations:

Overall mass balance	$L(X_1 - X_5) = G(Y_1 - Y_3)$,	(7.9)
First-stage mass balance	$nL(X_3 - X_4) = G(Y_2 - Y_3)$,	(7.10)
Second-stage mass balance	$nL(X_1 - X_2) = G(Y_1 - Y_2)$.	(7.11)

The total efficiency of an absorption unit with absorbent recirculation increases with the number of separate stages. There is an optimum sub-division for an absorption unit and it can be shown that, for cases where Henry's law applies, this division is into stages with an equal number of transfer units. The total number of stages used is determined by economic considerations.

For the treatment of waste gases it is particularly important to include in any overall mass balance an assessment of liquid entrainment. In many cases the concentration of an active component in the absorbent can be significantly higher than in the gas phase. Provided that the composition of droplets leaving the absorption unit equals that of liquid feed the minimal efficiency of a separator for a countercurrent unit (Fig. 7.1) is determined by

$$\left(\frac{L_{en}}{G}\right)_{max} = \frac{Y_1 - Y_2}{X_2}, \tag{7.12}$$

where L_{en} is the entrainment flow (mole of solvent time^{-1}).

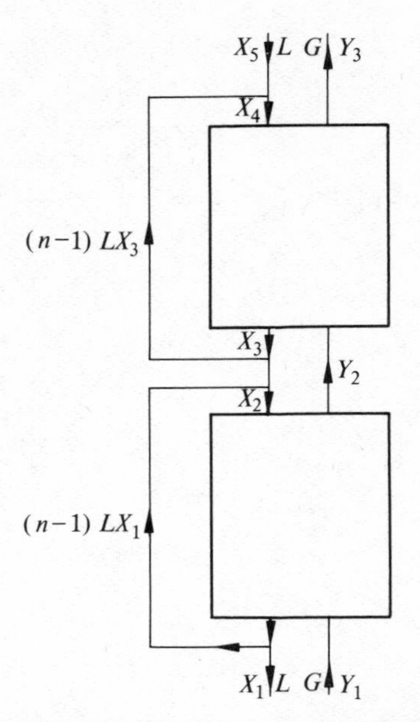

Fig. 7.5. Countercurrent absorber mass balance (with recycle).

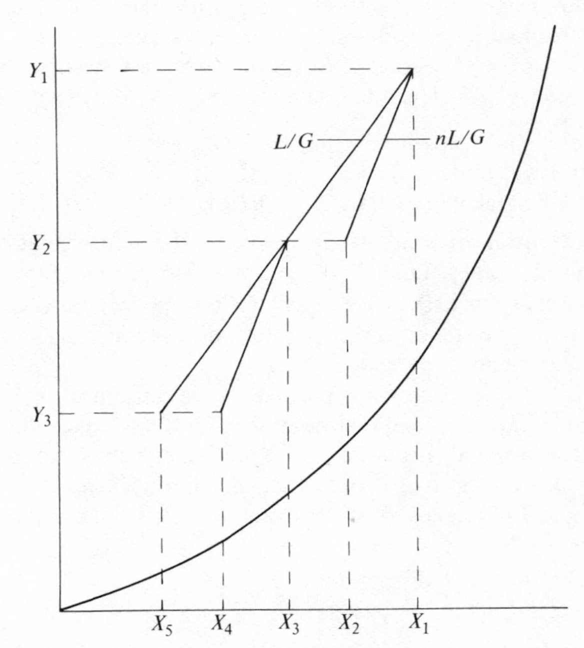

Fig. 7.6. Countercurrent steady state mass transfer (with recycle).

Details of the design of spray separators and eliminators are given by Stearman and Williamson.[54a]

Rate of mass transfer

Mass transfer between two fluids is carried out by eddy diffusion in the bulk of each phase and by molecular diffusion close to the interphase boundary. It is assumed that complete equilibrium is established instantaneously at the boundary, i.e., the concentrations in the minute layers of gas and liquid in direct contact are thermodynamically equal. From this boundary the active species are transported deeper into each phase by molecular diffusion followed at some depth by mixing of isolated elements caused by eddy currents in the bulk phase. The depth of molecular diffusion is generally controlled by the hydrodynamic conditions prevailing in each phase, i.e., in practice by the design of the contacting apparatus and the relevant flow rates. Eddy diffusion is invariably orders of magnitude faster than molecular diffusion. The overall rate of mass transfer is therefore controlled by mass fluxes within the molecular diffusion layers.

Two mathematical models have been developed from several basic assumptions. In the simpler theory it is assumed that molecular diffusion in both phases is predominantly in thin layers with laminar flow conditions in which the concentration gradient does not vary with time. These steady-state assumptions in both laminar layers were adopted by Lewis and Whitman[36] in a two-film theory. It was obvious that in many cases the idea of laminar steady flow cannot truly represent the nature of the process (e.g., the discontinuous character of liquid flow in packed absorbers) and hence the conditions of unsteady molecular diffusion have been incorporated into more advanced models. It is assumed that an element of liquid stays in contact with gas at an interphase boundary for a limited time, after which it is transported by eddy currents into the bulk of the phase and replaced by another element initially with a uniform concentration distribution. During the contact period, active species diffuse into an element developing a concentration gradient which changes both with time and position.

Film theory. The value of the mass transfer coefficient for both laminar films can be derived from Fick's law of diffusion. For steady molecular diffusion in one direction through a layer of thickness δ ($D =$ constant) it follows that

$$\frac{d^2 C}{dZ^2} = 0, \tag{7.13}$$

where the local mass flux N is determined by

$$N = -D \, dC/dZ. \tag{7.14}$$

Integrating Eq. (7.13) twice we obtain

$$\frac{dC}{dZ} = a_1 \tag{7.15}$$

and

$$C = a_1 Z + a_2. \tag{7.16}$$

Values of the constants a_1 and a_2 are determined by the boundary conditions (Fig. 7.7). At the interphase boundary the concentration in the liquid is in equilibrium with the gas phase, and at the other side of the layer it is equal to the concentration in the bulk phase. That is,

$$Z = 0 \quad C = C_i$$
$$Z = \delta \quad C = C_o.$$

After substitution the local mass flux is

$$N = \frac{D}{\delta}(C_i - C_o) \tag{7.17}$$

and the mass transfer coefficient, k, is given by $k = D/\delta$.

Penetration theory. The penetration theory model considers the unsteady mass flux into a layer of stagnant liquid generally described by

$$\frac{\partial C}{\partial t} = D\frac{\partial^2 C}{\partial Z^2}. \tag{7.18}$$

Initial and boundary conditions for this case are expressed by

$$t = 0 \quad Z > 0 \quad C = C_o,$$
$$t > 0 \quad Z = 0 \quad C = C_i;$$
$$t > 0 \quad Z \to \infty \quad C = C_o.$$

The solution of Eq. (7.18) for short times is

$$C = C_o + \frac{C_i - C_o}{\sqrt{\pi D t}} \int_Z^{\infty} \exp\left(-\frac{Z^2}{4Dt}\right) dZ. \tag{7.19}$$

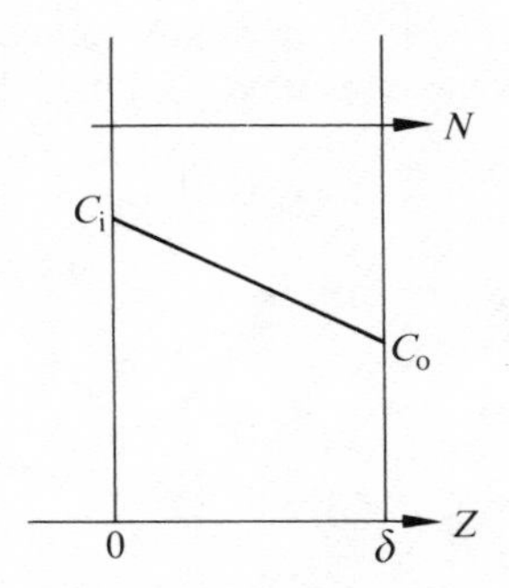

Fig. 7.7 Film theory.

The local mass flux through the interphase boundary is then defined by eqs. (7.14) and (7.19):

$$N = -D\left(\frac{dC}{dZ}\right)_{Z=0} = (C_i - C_o)\sqrt{\frac{D}{\pi\tau}} \tag{7.20}$$

where τ is the contact time of the given liquid element with the gas phase.

The mass transfer coefficient through the liquid boundary is now given by

$$k = \sqrt{\frac{D}{\pi\tau}}. \tag{7.21}$$

To estimate the average rate of absorption, it is necessary to define the distribution function of contact times at the interphase boundary. Different solutions are available in the literature by Higbie,[24] Danckwerts[12] and Davidson.[13]

In both basic models the local mass transfer rate is expressed as a product of the concentration difference between the interphase boundary and the bulk phase and a coefficient related to the diffusivity of the transported species. As it is not practicable to measure concentrations at the boundary, the absorption rate must be expressed as a function of conditions in both bulk phases.

From Eq. (7.17) the mass transfer rate in the gas phase equals

$$N = k_G(p_o - p_i), \tag{7.22}$$

and in the liquid phase

$$N = k_L(C_i - C_o). \tag{7.23}$$

On substitution from Eq. (7.1) and rearranging, this becomes

$$N = \frac{p_o - HC_o}{(1/k_G) + (H/k_L)} = (p_o - p^*)K_G, \tag{7.24}$$

and an overall mass transfer coefficient can be defined by

$$K_G = \frac{1}{(1/k_G) + (H/k_L)}. \tag{7.25}$$

Alternatively Eq. (7.24) can be expressed in the concentration units of the liquid phase.

In the case of highly soluble gases, the value of H (Eq. (7.25)) is small, and consequently

$$K_G \rightarrow k_G, \tag{7.26}$$

and for slightly soluble gases H is large, and

$$K_G \rightarrow k_L/H. \tag{7.27}$$

Basic design considerations

The amount of mass transferred through an interphase boundary must be equal to mass changes in each of the phases. In the general case, a mass balance over an infinitesimal element in an absorber with a defined contact area, dS, gives

$$G_t p = (G_t + dG_t)(p + dp) + dN, \tag{7.28}$$

where

$$dN = K_G(p - p^*)\,dS \tag{7.29}$$

and G_t is the total gas flow rate. If the second-order product $dG_t\,dp$ is neglected, this becomes

$$dN = -d(G_t p), \tag{7.30}$$

and thus

$$dS = -\frac{d(G_t p)}{K_G(p - p^*)}. \tag{7.31}$$

The capacity of packed absorption towers can be determined by integrating Eq. (7.31) over the concentration limits of the whole unit. Generally this calls for a knowledge of the variation of the mass transfer coefficient and total mass flow along the apparatus. The assessment is significantly simplified if it is assumed that G_t is constant:

$$dS = -G_t \frac{dp}{K_G(p - p^*)}. \tag{7.32}$$

The contact surface in packed absorbers is usually expressed as a product of the total volume V and the specific surface area per unit volume a, thus

$$dS = dV\,a. \tag{7.33}$$

Expressing partial pressures in units of mole ratios (Table 7.1) and substituting $G_t = G$, Eq. (7.32) becomes

$$-G \frac{(1 + Y)(1 + Y^*)}{(Y - Y^*)K_G}\,dY = Pa\,dV, \tag{7.34}$$

and over the whole unit (Fig. 7.1)

$$G \int_{Y_2}^{Y_1} \frac{(1 + Y)(1 + Y^*)}{(Y - Y^*)K_G}\,dY = PaV. \tag{7.35}$$

As was shown, the equilibrium surface is generally curved and hence the integral in Eq. (7.35) must be evaluated by graphical or numerical means. The solution is further simplified by assuming that

$$Y \ll 1 \quad \text{and} \quad Y^* \ll 1,$$

which are valid assumptions in practically all cases of waste gas treatment. Then Eq. (7.35) reduces to

$$G \int_{Y_2}^{Y_1} \frac{\mathrm{d}Y}{(Y - Y^*)K_G} = PaV \tag{7.36}$$

which still has no general analytical solution.

In many cases, however, even K_G can be considered constant or substituted by an average value within the given concentration limits. This is particularly applicable to cases where mass transfer is controlled by gas film resistance (e.g., fast irreversible reaction in liquid or only small changes of concentration in the liquid along the contact surface). Equation (7.36) now becomes

$$G \int_{Y_2}^{Y_1} \frac{\mathrm{d}Y}{Y - Y^*} = PaVK_G. \tag{7.37}$$

Many practical systems (e.g., absorption of SO_2 in milk of lime or sodium sulphite solution) are of the simplest type with negligible equilibrium partial pressure (i.e., $H \approx 0$). For these the analytical solution of Eq. (7.37) is

$$G \ln \frac{Y_1}{Y_2} = PaVK_G. \tag{7.38}$$

Similarly analytical solutions can be defined for cases where $Y^* = HX$ is approximately obeyed within the operating conditions. As the operating line and equilibrium lines (Fig. 7.2) are straight lines, the driving force in any position in the absorber can be expressed as

$$Y - Y^* = Y_2 - Y_2^* + (Y - Y_2)\frac{(Y_1 - Y_1^*) - (Y_2 - Y_2^*)}{Y_1 - Y_2}, \tag{7.39}$$

and Eq. (7.37) becomes

$$G \frac{Y_1 - Y_2}{(Y_1 - Y_1^*) - (Y_2 - Y_2^*)} \ln \frac{Y_1 - Y_1^*}{Y_2 - Y_2^*} = PaVK_G. \tag{7.40}$$

This result is frequently interpreted in the form of a logarithmic mean driving force

$$\Delta Y_m = \frac{(Y_1 - Y_1^*) - (Y_2 - Y_2^*)}{\ln((Y_1 - Y_1^*)/(Y_2 - Y_2^*))}, \tag{7.41}$$

and the total volume of absorber required for an outlet gas concentration of Y_2 is

$$V = \frac{G(Y_1 - Y_2)}{PaK_G} \Delta Y_m. \tag{7.42}$$

Another way of estimating the required capacity of a packed absorber is based on an assessment of overall transfer units N_{OG}, introduced by Chilton and Colburn.[9c] For the simplest case (K_G = constant, $Y \ll 1$), substituting the total packed volume V by the product of column height Z and gross cross-sectional area A, Eq. (7.37) can be rearranged to

$$N_{OG} = \int_{Y_2}^{Y_1} \frac{dY}{Y - Y^*} = \frac{PaK_GA}{G} Z. \tag{7.43}$$

The value of the integral varies with the height of the column, and it is obvious that for some concentration limits Y_2', Y_1' its value becomes unity. That is,

$$\int_{Y_2}^{Y_1} \frac{dY}{Y - Y^*} = 1. \tag{7.44}$$

The height of a column required for this change of concentration in the gas phase is called the height of a transfer unit H_{OG}:

$$H_{OG} = \frac{G}{PaAK_G}, \tag{7.45}$$

and thus $Z = N_{OG}H_{OG}$.

The primary assessment of the capacity of a plate tower is based on the mass balance over one theoretical plate. A theoretical plate is defined as a plate where the average composition of the gas leaving the plate is in complete equilibrium with the average composition of liquid leaving the plate. To estimate the number of theoretical plates required for a given separation effect, it is further assumed that there is no accumulation of mass within the absorber, and that the flow of inert carriers in both phases is constant.

The situation is illustrated in Fig. 7.8, where the indices refer to the plate the stream is leaving. The mass balance for the first plate is

$$G(Y_2 - Y_1) = L(X_1 - X_0), \tag{7.46}$$

and between any plate n and the top of the tower

$$Y_{n+1} - Y_1 = \frac{L}{G}(X_n - X_0), \tag{7.47}$$

which is again the equation of the operating line shown in Fig. 7.9. Contrary to the operating lines for packed towers (Eq. (7.5)), the operating line (Eq. (7.47)) for plate towers represents the actual conditions of gas–liquid composition in the space *between* successive plates. In a general case, the number of plates is estimated by starting from the top of the tower, where liquid is in equilibrium with the gas concentration Y_1. The value of Y_2 is determined from Eq. (7.47) or by the graphical method illustrated. For special cases where both operating and equilibrium lines are straight, general analytical solutions are available (Treybal[63]).

The application of the outlined principles of absorber capacity design for any particular case requires experimental data on mass transfer rates. The usual types of ab-

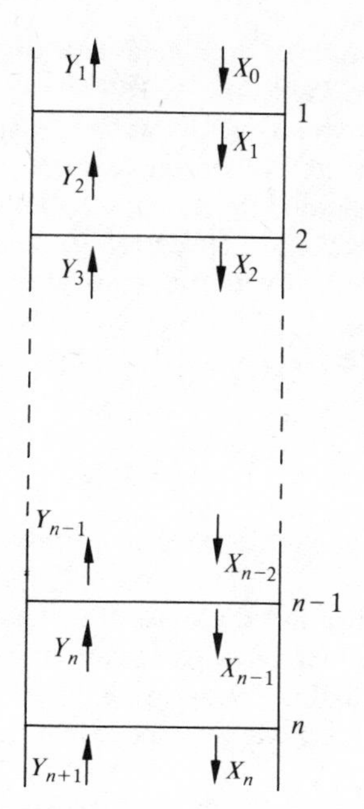

Fig. 7.8. Multistage countercurrent absorber mass balance.

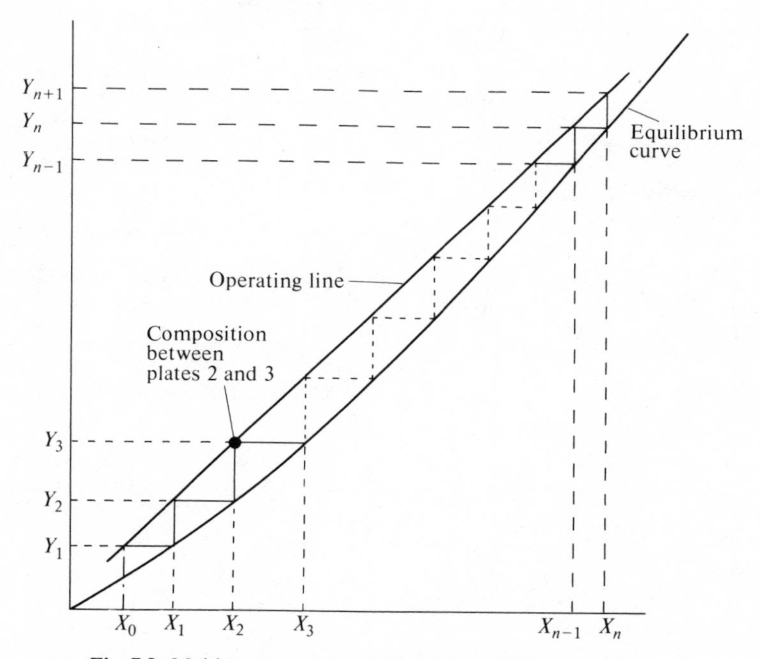

Fig. 7.9. Multistage countercurrent steady state mass transfer.

sorption equipment with comments on availability of transfer rate data and flow arrangements are summarized in Table 7.2.

For plate towers, data for two differently defined plate efficiencies are used. The simpler is an overall efficiency, related to the whole absorption unit, which is expressed as a ratio of the number of theoretical plates to the number of actual plates sufficient for attaining the separation required. The application to design is straightforward, but most published data are only concerned with the removal of hydrocarbon gases by oil washing (Perry[43]).

Efficiency related to a single plate was defined by Murphree[40a] as

$$E_M = \frac{y_n - y_{n-1}}{y_n - y^*_{n-1}}, \tag{7.48}$$

where y^*_{n-1} is the equilibrium gas mole fraction corresponding to the liquid composition x_{n-1}.

The relationship between E_M and the mass transfer coefficient can be illustrated in the following way. Mass transfer on a plate can be considered as mass transfer in a column of height h (effective liquid depth on plate) with a specific contact area related in some way to the plate's hydrodynamic conditions. Assuming that the liquid is ideally

TABLE 7.2

Characteristics of absorption units

Type	Typical flow arrangement	Superficial gas velocity (atmosp. press.) (m/s⁻¹)	Remarks
Packed towers Berl saddles, Raschig rings, wood grid, etc.	Countercurrent	0·5–1·5	Data available for assessment of most systems, primary importance of uniform liquid distribution, limited L/G range
Plate towers Bubble caps, perforated plates, impingement plates	Overall countercurrent Plate crosscurrent	0·3–1·5	Assessment comparatively easy as scale-up is limited, wide L/G range
Hydraulic dispersion Spray, cyclone, venturi washer	Cross or cocurrent	1·5–12	Data rather limited, difficult scale-up, wide L–G range
Mechanical dispersion Agitated tanks, Feld's washer	Crosscurrent	0·3–1	Data rather limited, size limited by moving mechanical parts
Fluidized bed principle Turbulent bed of hollow spheres, glass marbles	Cross and countercurrent	2·5–7	Recent development

mixed and x_{n-1} is constant at all points, the mass transfer rate in an infinitesimal layer dh is described in a similar way to Eqs. (7.30) and (7.37), that is,

$$K_G C_1 (y - y^*_{n-1}) dh = -G\,dy, \tag{7.49}$$

where C_1 is a constant. Hence

$$\ln \frac{y_{n-1} - y^*_{n-1}}{y_n - y^*_{n-1}} = -\frac{K_G}{G} C_1 h,$$

and after substituting into Eq. (7.48) this gives

$$E_M = 1 - \exp\left(-\frac{K_G C_1 h}{G}\right). \tag{7.50}$$

Murphree's efficiency varies widely from 0·01 to 0·85 for different gas–liquid systems. Comprehensive general information is available in the *Bubble Tray Design Manual* and in Perry.[43] For cases where no direct data are available, a first approximation can be obtained using a formula proposed by Walter and Sherwood:[71]

$$E_M = 1 - \exp(-m),$$

where

$$m = \frac{h}{(2{\cdot}50 + 0{\cdot}370/HP)\,\mu^{0.68}\,w^{0.33}},$$

and μ = viscosity of liquid (cP),
w = slot width (in),
h = effective liquid depth (in),
H = Henry's coefficient (1^b moles ft^{-3} atm^{-1}),
P = total pressure (atm).

Another empirical correlation has been presented by O'Connell.[42]

For packed absorbers, many data are directly available in monographs and in the literature for the more common liquid–gas systems (Hobler, Kohl and Riesenfeld, Norman). Where information is not available, a first approximation can be obtained based on liquid–gas equilibria only using Eq. (7.25). The mass transfer rate dependence on hydraulic conditions in conventional packed towers has been examined for gas-film resistance (NH_3–H_2O) and liquid film resistance (O_2, H_2–H_2O) separately (Perry).[43]

For a detailed design of absorbers, many data are directly available for the industrially important systems (Hobler, Morris and Jackson, Nonhebel, Norman, Riesenfeld and Kohl, Danckwerts). Technological and equipment aspects are also covered in a series of chemical process and pollution control reviews published by Noyes Data Corporation and in a handbook by Calvert.[5a] Specific problems of construction materials were considered by Sakol and Schwarz.[50a]

In cases where no experimental information on mass transfer rates is available a first approximation may be obtained using Eq. (7.25). Individual values of k_G and k_L for relevant hydraulic conditions can be found in the literature (Laurent and Charpentier[33a]).

Adsorption processes

Adsorption is a process whereby gases, vapours or liquids are concentrated on a solid surface as the result of surface or chemical forces. These solids, termed adsorbents, may thus be used to separate gases or vapours from gas streams. The extent of adsorption varies with the nature and character of the adsorbate–adsorbent system. The adsorbent may be used only once and then discarded, or it may be used and then reactivated (a process known as regeneration). Molecules adsorbed on to a surface are held by forces on that surface. These forces can be physical or chemical in nature thus leading to the terms *physical adsorption* and *chemisorption.* The principal properties of the two types of adsorption are summarized in Table 7.3.

TABLE 7.3

Comparison of physical adsorption and chemisorption

	Physical adsorption	Chemisorption
Forces involved	Van der Waals' forces	Chemical valence forces
Strength of binding	As cohesion in a liquid	Chemical bonds
Rate at which equilibrium is attained	Very rapid	Not necessarily rapid, activation energies may be large
ΔH	Exothermic	Exothermic
Reversibility	Usually complete and rapid	May be slow or irreversible
Specificity	Small	Great

As adsorption is a surface phenomenon, only those solids which contain large amounts of internal surfaces are likely to be useful, since the external surface area of even very finely divided solids is relatively small. Typically commercial adsorbents have surface areas in the region of 50–500 $m^2\ g^{-1}$ with an average pore radius ~10–1000 A. Other properties of the ideal adsorbent are

1. It must be readily available at reasonable cost to reduce loss make-up costs.
2. The adsorbent should be readily regenerable.
3. It must be chemically inert to the bulk component of the gas stream. It should also have adequate physical strength and hardness so that attrition losses are minimized.
4. The adsorbent must have a high capacity for the substance to be adsorbed, otherwise the size of the adsorption unit will be uneconomically large. A high selectivity is a useful asset if several components are present in the fluid stream to be treated.

The most important adsorbents in industrial use today and their uses are shown in Table 7.4. From this table it can be seen that adsorption can effect many separations.

TABLE 7.4

Modern uses of adsorbents

Adsorbent	Important industrial uses
Fuller's earth	Decolourizing of petroleum fractions, edible oils
Bauxite	Percolation of petroleum fractions, dehydration of gases
Bone-char	Sugar refining
Activated carbon	Decolourization of organic and inorganic substances, solvent recovery, elimination of industrial odours and noxious gases, purification of industrial gases, gas masks, purification of hydrocarbons, hypersorption
Activated alumina	Dehydration of gases and liquids
Silica gel	Dehydration and purification of gases, gas masks
Molecular sieves	Selective adsorption of polar molecules such as H_2O, CO_2, NH_3, C_2H_2, H_2S and SO_2

Several very good reviews of the field of adsorption have been given by Carter[6,7,8] which include several example design calculations. Bowen[3] has also presented a detailed analysis of adsorption processes.

Theoretical aspects

As in many chemical engineering operations, both equilibrium and kinetic data are required to characterize the design of an adsorption unit. As most adsorption operations are dynamic, equilibrium data for adsorbents are primarily of use for determining the absolute capacity of a particular adsorbent for a certain adsorbate. These data can then be used for a preliminary evaluation of the proposed system.

Adsorption equilibrium data are often readily available from the manufacturers or in the literature. When this is not the case adsorption equilibria can be easily measured. The equilibrium data can often be expressed in the form of one of the many theoretical isotherms which have been postulated. These isotherms relate the amounts of adsorbate on the adsorbent to the concentration of the adsorbate in the fluid phase at a given temperature.

The physical equilibria of the adsorption of gases on various adsorbents have been extensively studied, and numerous standard works are available on the subject. Notable in the field are those due to Brunauer,[4] Gregg and Sing[22] and Young and Crowell.[75] Experience has shown that the equilibrium distribution of adsorbate molecules between the surface of the adsorbent and the gas phase is dependent upon pressure, temperature, nature and area of the adsorbent, and the type of adsorbate.

The most well-known isotherms are shown in Table 7.5, which indicates the dependence of the heat of adsorption on surface coverage which each isotherm implies. Most experimental systems can be fitted to the Langmuir and Freundlich isotherms, though at low concentrations of adsorbate Henry's law is often obeyed and a linear isotherm is

TABLE 7.5

Isotherm models

Name	Equation	Heat of adsorption	Reference
Langmuir	$\frac{v}{v_m} = \frac{bp}{1 + bp}$	Constant	32, 33
Freundlich	$v = ap^{1/n}\ (n > 1)$	Experimental	20, 31, 76
Temkin†	$\frac{v}{v_m} = \frac{1}{a_t} \ln p$	Linear	53, 56
Brunauer, Emmett, Teller, BET	$\frac{p}{v(p_o - p)} = \frac{1}{v_m c} + \frac{(c-1)p}{v_m c p_o}$	Constant	5, 16
Henry's law	$v = Kp$	Constant	—

† This isotherm is generally more useful for chemisorption.

obtained. When the adsorbed layer is more than one monolayer thick the Freundlich and Langmuir isotherms break down. However in this region adsorption is adequately represented by means of the BET equation.

Usually the adsorbent manufacturer supplies, with an adsorbent, adsorption isotherms over a range of temperatures. This information may then be used for a preliminary process design; but it is usually advisable to make separate measurements under exact process conditions. If only a limited number of data are available the Clausius–Clapeyron equation may be used to predict the isotherm at other temperatures. Thus the equilibrium amount adsorbed at T_2 may be estimated from the isotherm at T_1, as follows

$$\left(\frac{v_{T_2}}{v_{T_1}}\right)_P = \exp\left(-\Delta H(s) \Big/ R\left(\frac{1}{T_1} - \frac{1}{T_2}\right)\right), \tag{7.51}$$

where $\Delta H(s)$ is the heat of adsorption (which may vary with surface coverage) but which, for physical adsorption, is usually close to the heat of liquefaction. Isotherms may also be extrapolated to other temperatures by making use of the Polanyi adsorption potential.[45,75]

Kinetics of adsorption

The overall kinetics of the adsorption system used are affected by both the type of gas–solid contacting method and the transport processes of the adsorbate from the bulk gas to the internal adsorbate surfaces. The following transport processes can affect the overall rate of the adsorption process and should be considered.

1. Transfer of adsorbate from the bulk gas stream to the external surface of the adsorbent (external mass transfer, film diffusion or mass transfer, interparticle diffusion). This effect is analogous to the gas-film resistance in gas absorption.

2. Transfer of material from the external surface of the particle to the internal surface of the porous particle by some diffusive mechanism (intraparticle diffusion).
3. The kinetics of surface adsorption, whereby the adsorbate is transferred from the gaseous phase to the adsorbed phase. The rate of adsorption can range from the very fast (as in most cases of physical adsorption) to the very slow (as may occur in some cases of chemical adsorption, where considerable molecular rearrangement may be needed on the adsorbent surface). This adsorption process is usually accompanied by the evolution of heat.
4. Transfer of the generated heat from inside the particle to the external surface of the particle by a process of solid and gaseous conduction (intraparticle conduction).
5. Transfer of heat from the external surface of the particle to the bulk gas stream by transport across the external gas film (external heat transfer).
6. Transfer of heat from the bulk gas stream to the vessel wall.

If all the heat transfer rates become very large then the bed is acting as an isothermal adsorber (this condition pertains in many laboratory experiments). However if the heat transfer rate from the bulk gas stream to the adsorber wall is zero, then the bed acts as an adiabatic adsorber (as in many large industrial adsorbers). In general, all these steps have markedly different rates and the overall rate of adsorption is most affected by the rates of the slowest steps. These steps are usually referred to as the rate-controlling steps. It should be noted that steps (1), (2), (4), (5) and (6) are consecutive but that steps (2), (3) and (4) occur in parallel.

External film mass and heat transfer. The flux of adsorbate across the external gas film surrounding the adsorbent particle can be expressed in terms of a mass transfer coefficient. Thus the flux through the particle surface can be equated to that across the gas film:

$$k_c(C-q)_{\mathrm{sur}} = -D_e\left(\frac{\partial q}{\partial r}\right)_{\mathrm{sur}}, \tag{7.52}$$

where k_c is the external mass transfer coefficient and D_e is the effective diffusivity within the particle.

The mass transfer coefficient can be readily determined from correlations such as that of de Acetis and Thodos[11] (for fixed beds) and Riccetti and Thodos[47] (for fixed and fluidized beds). In an analogous manner, the transfer of heat through the external gas film can be expressed in terms of a heat transfer coefficient h_f.

Intraparticle mass transfer. The rate of diffusion within a porous catalyst can be expressed by means of Fick's law. Thus for a spherical particle (to which most adsorbent particles approximate)

$$\frac{\partial q}{\partial t} = \frac{1}{r^2}\frac{\partial}{\partial r}\left(D_e r^2 \frac{\partial q}{\partial r}\right). \tag{7.53}$$

The type of diffusion that takes place within the pores of the particles depends on the relative magnitude of the diameter of the pores and the mean free path of the gas under the conditions that exist in the pores. The mean free path of a gas molecule (which is inversely proportional to pressure) is about 10^{-5} cm at one atmosphere. In pores of appreciably smaller diameter than this, or at low pressures, molecules will tend to collide with the pore walls rather than with other molecules and the flow is termed Knudsen diffusion. Knudsen[27,28] showed that under these conditions the diffusion coefficient is given by

$$D_{\mathrm{k}} = \frac{4r_{\mathrm{p}}}{3}\sqrt{\left(\frac{2RT}{\pi M}\right)} = \tfrac{2}{3} r_p \bar{u}. \tag{7.54}$$

In larger pores or at high pressures, the majority of collisions are between molecules, and the diffusive mechanism is the same as in the bulk gas. Usually values of the bulk diffusion coefficient may be obtained from the literature[15,51] or they may be calculated from a theoretical equation based on modern kinetic theory and the Lennard-Jones[25] expression for intermolecular forces. The transition region between Knudsen flow and bulk flow (10^{-4}–10^{-6} cm at 1 atm) can be represented by the exponential law presented by Wheeler:[74]

$$D = D_{\mathrm{b}}(1 - \exp(-D_k/D_{\mathrm{b}})). \tag{7.55}$$

Under Knudsen flow conditions, flow will be independent of pressure gradients, but under all other conditions, superimposition of forced flow on bulk flow can result from a pressure differential. If the normal Poiseuille[44] equation is rewritten in terms of a concentration gradient then a forced flow diffusivity may be defined as

$$D_{\mathrm{f}} = \frac{r^2 C_{\mathrm{T}} RT}{8\eta}, \tag{7.56}$$

where C_{T} is the total gas concentration.

In most adsorption processes, where pressure drops are small, forced flow can generally be neglected. These calculated diffusion coefficients are invariably much larger than those which have been measured experimentally for porous materials. This is because all the diffusion coefficients have been derived on the basis of free pore volume and straight uniform pores. In order to correct true diffusivities to effective diffusivities the following equation is often used:

$$D_{\mathrm{e}} = \frac{D\epsilon_{\mathrm{p}}}{\tau}, \tag{7.57}$$

where ϵ_{p} is the void fraction in the porous mass which corrects for the free volume available for diffusion and τ is the tortuosity factor, which allows for both tortuosity and varying pore cross section. In pellets it has been found to vary from about 0·85 to 10. Most of these measurements have been carried out by the method pioneered by Weisz,[72] in which two faces of a porous pellet are exposed to different gases and the steady state counter-diffusional fluxes are measured. Several theoretical models have been developed for bidisperse porous media, notably those by Wakao and Smith[69,70] and Foster and Butt.[19]

Surface flow can also play an important part in the flow of gases through porous media. In general, surface diffusivities will depend on the concentration of the sorbed phase, being greater at higher surface averages. The effects of surface and pore diffusion are additive and under equilibrium conditions pore and surface diffusivities can be combined analytically.

The role of diffusion in porous particles has been well summarized by Satterfield[51] and by Wakao *et al.*,[68] Hewitt[23a] and Youngquist.[75a] Diffusion in small crystallites can affect the overall rate and is characterized by the fact that the rate of diffusion is independent of macro-particle size. The overall diffusion may still be described in terms of an effective diffusion coefficient based on solid phase adsorbate concentrations.

Surface adsorption kinetics. In general, for physical adsorption the actual rate of adsorption is extremely rapid, as no chemical bonds are formed to influence the overall kinetics. In chemisorption the rate of adsorption is often comparatively slow and can appreciably affect the overall rate. The Elovich equation[55] is widely applicable and is one of the most convenient ways of describing chemisorption kinetics. In its differential form, the equation is written as

$$\frac{\mathrm{d}v}{\mathrm{d}t} = a\exp(-\alpha v), \tag{7.58}$$

where a and α are constants for each system at a given temperature.

Intraparticle heat transfer. Heat transfer within a pellet can be treated in the same way as mass transfer by defining an effective conductivity. Values of the effective conductivity have been measured by Sehr.[52] However the thermal diffusivity in porous particles is very much greater than the mass diffusivity, and in general in adsorption processes the pellets may be considered to be at a uniform temperature.

Overall rate of adsorption in single particles. An evaluation of the relative effects of interparticle and intraparticle heat and mass transfer can be gained from the values of the Nusselt numbers for heat and mass transfer ($k_c d/D_e$ for mass transfer and $h_f d/K$ for heat transfer). The Nusselt number defines the ratio of the resistance in the porous solid to that in the external gas film. Generally for mass transfer the Nusselt number is much larger than unity, showing that most of the resistance is in the solid, whilst the converse is true for heat transfer. The equations for heat and mass transfer within the pellet may be solved to give the rate of adsorption of adsorbate by a single particle. Under isothermal conditions and with the isotherm linear, it has been shown (Vermeulen[66]) that the diffusional resistances may be combined to give one overall resistance in a manner similar to the derivation of an overall mass transfer coefficient for absorption.

Analytical solutions of the simple linear single particle diffusion problem for various boundary conditions have been given by Crank.[10] Theoretical and experimental studies of adsorption in single particles where the isotherm is non-linear have been carried out by Timofeev *et al.*[60,61,62], Garg and Ruthven[21] and Loughlin and Ruthven.[34] Billinge *et al.*[2] have studied the case where diffusion and a slow chemisorption step determined the overall rate. Rimmer[48] has used Vermeulen's[65] quadratic driving force approximation to study the effects of simultaneous heat and mass transfer in porous adsorbents.

Process design of sorption units

The different techniques used for contacting fluid and solid are shown in Table 7.6.

TABLE 7.6

Classification of adsorption processes

Technique used	Fluid	Method of operation	Example
Stagewise contact	Liquids	Batch and continuous	Contact filtration
Continuous contact			
1. Steady state			
(a) Moving bed	Gases and liquids	Continuous	Hypersorption
(b) Fluidized bed			Landmark for CS_2
(c) Falling cloud			Reinluft process for SO_2
2. Unsteady state			
Fixed bed		Semi-batch	Solvent recovery Gas drying

1. Stagewise contact. In industrial practice, stagewise operations are limited to the treatment of dilute liquid solutions and thus will not be considered here.

2. Continuous contact. In these operations the fluid and adsorbent are in continuous contact without periodic separation of the phases. The process may be carried out in strictly continuous steady state conditions, characterized by movement of the solid. Alternatively owing to the rigidity of solid absorbent particles it is often advantageous to operate in a semi-continuous fashion, characterized by a moving fluid and a stationary bed (fixed bed). This results in unsteady-state conditions, with the result that the theoretical analysis of such units is complex.

(*a*) *Moving beds.* Beds in which solids move in plug flow relative to the containing vessel are designed to operate in the steady state, and thus the theoretical analysis of their process design is simpler than that of fixed beds. Under these conditions moving beds may be considered in terms of transfer units and the mathematical analysis then proceeds in a manner similar to that used for absorbers. A mass balance across an increment of the bed may be written:

$$u\epsilon_B A \, \mathrm{d}C = -k_c A(1-\epsilon_B)a_p(C-C_i)\, \mathrm{d}z, \tag{7.59}$$

which may be integrated over the whole bed to give

$$\frac{k_c a_p z_o}{mu} = \int_{C_o}^{C} \frac{dC}{C - C_i}, \tag{7.60}$$

where C_o and C are the fluid sorbate concentrations into and out of the moving bed. The integral represents the number of transfer units, but in general the interface concentration C_i is not known, though it is possible to express the rate of transfer in terms of an overall driving force $(C - C^*)$, where C^* is the fluid concentration in equilibrium with the mean adsorbed phase concentration. Then

$$z_o = \frac{mu}{Ka_p} \int_{C_o}^{C} \frac{dC}{C - C^*} \tag{7.61}$$

and the integral may be evaluated graphically as in the design of absorption units (K is an overall mass transfer coefficient).

(*b*) *Fluidized beds*. The complex fluid and solid hydraulic behaviour in a fluidized bed is outside the scope of this chapter and reference should be made to a standard text such as those by Davidson and Harrison,[14] Vanecek *et al.*,[64] Kunii and Levenspiel.[29] A rigorous treatment of continuous adsorption is quite complicated. If it may be assumed that the bed solids are in equilibrium with the leaving gas, then a simple mass balance for the bed may be written:

$$Au(C_o - C^*) = F(C_s - C_{se}). \tag{7.62}$$

This mass balance may be solved, together with the equilibrium relationship, to give the composition of the exit gas and exit solid. For the special case of a linear isotherm, Kunii and Levenspiel[29] show that the efficiency of gas removal and solid uptake are related by

$$\eta_s = 1 - \eta_g, \tag{7.63}$$

showing that a 99 per cent removal efficiency is possible only on acceptance of a 1 per cent efficiency in the use of the adsorption capacity of the solid. Thus it is highly advantageous to use multistaging, and under these conditions calculations can be carried out exactly as in the case of plate-tower absorption. For non-isothermal operating schemes, the design procedure has been outlined by Ermenc.[17]

When equilibrium cannot be assumed, a design method has been outlined by Kunii and Levenspiel[29] using a bubbling bed model. The average composition in the gas phase can be calculated from the transfer rates predicted by the bubbling bed model and from an assumed gas kinetic model. The behaviour of the solids can then be predicted from an appropriate model such as the shrinking core model[57] or the linear driving force model.[66] A material balance links the changes in the solid and gaseous phases. These equations do not in general have an analytical solution, but a solution can readily be found by trial and error, graphical or numerical methods.

Davidson and Harrison[14] give a solution assuming a first-order reaction but including the effects of bubbling in the fluidized bed for both fully mixed solid and for plug flow. These solutions are really for first-order catalytic reactions, but may be applied to the adsorption problem provided that the single particle adsorption rate can be expressed in a pseudo first-order rate equation. The fluidized bed has been used with success in the Landmark process.[46]

(*c*) *Falling cloud*. In the falling cloud process, where a cloud of particles is allowed to fall through an upstream of gas, the particles are essentially independent, and it may be considered as a dispersed-phase moving bed.

(*d*) *Fixed beds*. An understanding of the processes going on in a fixed bed is fundamental to the design of any sorption equipment. The problem has been receiving increasing attention, but the solutions developed fail to describe the behaviour of large beds accurately. However, fixed beds are the most widely used type of adsorption equipment because of their simplicity of operation. A very good summary of theoretical methods for fixed beds has been given by Vermeulen.[67] In order to specify conditions in a fixed bed the equations of continuity of heat and mass in the bed must be solved together with the particle rate equations. Taking a mass balance across an element of bed, the following equation of continuity may be devised:

$$u\left(\frac{\partial c}{\partial z}\right)_t + \left(\frac{\partial c}{\partial t}\right)_z = -\frac{1}{m}\left(\frac{\partial C_s}{\partial t}\right)_z , \tag{7.64}$$

and a similar equation may be derived for heat transfer, though this will include an extra term to account for heat losses to the wall. Bowen[3] has shown that under many conditions the bed may be assumed to be isothermal, and a mean temperature may be defined by

$$T \simeq T_o + \frac{\lambda C_o}{\rho_M c_p} . \tag{7.65}$$

This temperature may then be used to calculate the concentration history within the bed.

Solutions of the problem of non-linear adsorption in fixed beds have been described by Meyer,[39] Weber and Chakravorti,[71a] Carter,[9] McGreavy *et al.*,[40] Nussey,[41] Kyte,[30] and Antonson and Dranoff[1], in which the relevant differential equations were solved by numerical means. However, all these models are too specialized and complex for use in general design work. More tractable models have been presented by Garg and Ruthven[21a,b,c,d] and Kyte.[30a]

When resistance to the transfer of sorbate is negligible, then concentrations in the solid and fluid are related by the adsorption isotherm. Under these conditions the continuity equation may be rewritten in terms of the total volume that has entered the bed, and from this it may be shown that

$$\left(\frac{\partial z}{\partial t}\right)_c = \frac{u}{1 + (1/m)f'(C)} \tag{7.66}$$

where $f'(C)$ is the slope of the adsorption isotherm at C_s.

The fundamental importance of this equation is that it relates the way in which the shape of the sorption wave changes as it moves along the bed, and this is also valid for the more common non-equilibrium case. Thus, if the isotherm is concave to the fluid concentration axis (favourable isotherm), then the sorption zone tends to become sharper. If the isotherm is convex (unfavourable), then the sorption zone becomes wider as it moves through the bed. For a linear isotherm the sorption zone is unchanged. The development of the sorption zone under various conditions is shown in Fig. 7.10. This principle carries over into non-equilibrium adsorption, where the resistances to mass and heat transfer tend to disperse and widen the adsorption zone. Under certain conditions, for a favourable isotherm, these dispersive forces and the sharpening effect of the isotherm may give rise to a stable condition in which a zone of constant length is formed. This tendency has been used by several workers to predict the overall rate of adsorption in a fixed bed.

The transfer unit concept may also be used for fixed beds operating with a constant sorption zone moving at a constant speed through the bed. Under these conditions, if the bed is given an imaginary velocity sufficient to bring the sorption zone to rest, then the bed may be treated by the methods described for moving beds. The method has been fully described (with an example) by Treybal.[63]

It has been shown that for a favourable isotherm the sorption wave may take up a constant pattern form. Then, if the length of bed required to set up this constant pattern is small compared with the total length of the bed, great simplifications may be made in the mathematical treatment of the problem. Advantage of this simplification has been taken by Hall *et al.*[23] to obtain a solution for the case where the absorption isotherm is of the Langmuir type. Because of the constant pattern assumption, the whole problem reduces to the solution of one parabolic partial differential equation. This equation was solved numerically and solutions were presented for both pore and surface diffusion in tabular and graphical form. An attractive feature of this approach is that the solution

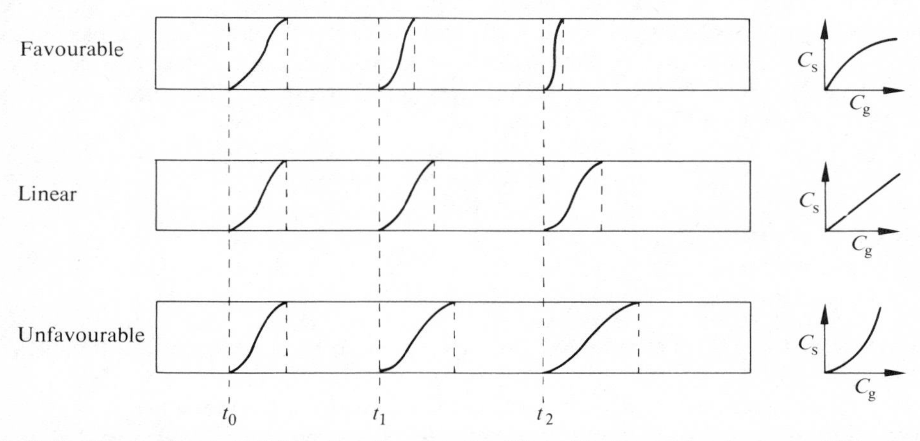

Fig. 7.10. The development of the adsorption zone for different isotherm types (under equilibrium conditions).

can be expressed in terms of two simple parameters: the equilibrium parameter (R) and the number of transfer units (N_{pore} or N_s). From the definitions of these parameters

$$R = \frac{1}{1 + bc_o},$$

where b is the constant in the Langmuir isotherm, and

$$N_{\text{pore}} = \frac{15D_p(1-\epsilon_b)z_o}{R_o^2 u}, \qquad N_s = \frac{15D_s K_o \rho_b z_o}{R_o^2 u}.$$

It can be seen that R is solely a function of the isotherm and inlet concentration, and N a function of bed parameters and the intraparticle diffusion coefficient. Thus the breakthrough curve (i.e., the outlet concentration of the bed as a fraction of time) is characterized by these two parameters only, and the prediction of the breakthrough simplifies to the determination of the values of R and N. Thus this method is of particular interest to the design engineer as many adsorbate/adsorbent systems obey the Langmuir isotherm and constant pattern conditions are often obtained. This treatment has recently been extended by Fleck *et al.*[18] to cover mixed resistances.

At low concentrations of adsorbate in the gas phase the adsorption isotherm is often linear and under isothermal conditions an analytical solution to the fixed bed problem is possible. Rosen[49,50] obtained a solution for the case in which external and/or intraparticle diffusion control the overall rate. This solution has been extended by Masamune and Smith[37] and Massaldi and Gottifredi[38] to include the effect of a finite surface absorption rate. Masamune and Smith found the solution in terms of an infinite integral of the form

$$\frac{C}{C_o} = \tfrac{1}{2} + \frac{2}{\pi}\int_0^\infty e^{\alpha_1} \sin \alpha_2 \frac{d\lambda}{\lambda}, \tag{7.67}$$

where α_1, α_2 depend on the rate-controlling mechanism and are detailed in Table 7.7. The infinite integral must be evaluated by numerical integration (e.g., Simpson's rule). The integrand converges to zero ($\alpha_1 < 0$) and thus the integration can be carried out over a finite range and the breakthrough curve determined. For the case in which external and intraparticle diffusion control the overall rate, Rosen[49,50] presented the following approximation which is valid for long beds.

$$\frac{C}{C_o} = \tfrac{1}{2}\left[1 + \text{erf}\left\{\frac{(3\tau/2\lambda' - 1)}{2/(1 + 5\chi/5\lambda)}\right\}\right], \tag{7.68}$$

where $\lambda' = 3D_e K_o z_o/mu R_o^2$,
$\tau = 2D_e(t - z/u)/R_o^2$,
$\chi = D_e K_o/R_o k_c$.

In general, these three approaches are sufficient to cover the design of most industrial adsorbers.

Multicomponent absorption in fixed beds is much more complex than single-component adsorption. This problem has been considered by Cater and Husain,[9a] Cooney and Strusi.[9d]

TABLE 7.7

Isothermal adsorption in fixed beds with a linear isotherm, Masamune and Smith[37]

General form of solution $C/C_o = \frac{1}{2} + \frac{2}{\pi}\int_0^\infty e^{\alpha_1}\sin\alpha_2 \frac{d\lambda}{\lambda}$

External diffusion Intraparticle diffusion Surface reaction	Intraparticle diffusion Surface adsorption	External diffusion Intraparticle diffusion	External diffusion Surface adsorption	External diffusion	Surface adsorption	Intraparticle diffusion
1	2	3	4	5	6	7
$\alpha_1 = -\frac{3\lambda_o}{h^2}\Omega_1$	$\alpha_1 - \frac{3\lambda_o}{h^2}\Omega_1$	$\alpha_1 = -\frac{3\lambda_o}{h^2}\Omega_1$	$\alpha_1 = -\lambda_o k_o \Omega_1$	$\alpha_1 = -\lambda_{of}\Omega_1$	$\alpha_1 = \lambda_o \Omega_1$	$\alpha_1 = -\frac{3\lambda_o}{h^2}\Omega_1$
$\alpha_2 = \theta_o\lambda_o\lambda^2 - \frac{3\lambda_o}{h^2}\Omega_2$	$\alpha_2 = \theta_o\lambda_o\lambda^2 - \frac{3\lambda_o}{h^2}\Omega_2$	$\alpha_2 = 2\theta_o\frac{\lambda_o}{h^2}\lambda^2 - \frac{3\lambda_o}{h^2}\Omega_2$	$\alpha_2 = \theta_o\lambda_o k_o\lambda^2 - \lambda_o k_o\Omega_2$	$\alpha_2 = \theta_o\lambda_{of}\lambda^2 - \lambda_{of}\Omega_2$	$\alpha_2 = \theta_o\lambda_o\lambda^2 - \lambda_o\Omega_2$	$\alpha_2 = 2\theta_o\frac{\lambda_o}{h^3}\lambda^2 - \frac{3\lambda_o}{h^2}\Omega_2$
$\Omega_1 = \left[\frac{\psi_1 + k(\psi_1^2 + \psi_2^2)}{(1 + k\psi_1)^2 + (k\psi_2)^2}\right]$	$\Omega_1 = \psi_1$	As 1	$\Omega_1 = \psi_1 = \phi_1$	$\Omega_1 = \psi_1 = \phi_1$	$\Omega_1 = \psi_1 = \phi_1$	$\Omega_1 = \psi_1$
$\Omega_2 = \left[\frac{\psi_2}{(1 + k\psi_1)^2 + (k\psi_2)^2}\right]$	$\Omega_2 = \psi_2$	As 1	$\Omega_2 = \psi_2 = \phi_2$	$\Omega_2 = \psi_2 = \phi_2$	$\Omega_2 = \psi_2 = \phi_2$	$\Omega_2 = \psi_2$
$\psi_1 = \left[\frac{\phi_1\sin 2\phi_1 + \phi_2\sinh 2\phi_2}{\cosh 2\phi_2 - \cos 2\phi_1}\right] - 1$	As 1	As 1	—	—	—	As 1
$\psi_2 = \frac{\phi_2\sin 2\phi_1 - \phi_1\sinh 2\phi_2}{\cosh 2\phi_2 - \cos 2\phi_1}$	As 1	As 1	—	—	—	As 1
$\phi_1 = -h\lambda\left[\frac{1}{2(\lambda^4 + 1)[\sqrt{(\lambda^4 + 1)} + \lambda^2]}\right]^{1/2}$	As 1	$\phi_1 = -\lambda$	$\phi_1 = \frac{\lambda^4}{1 + \lambda^4}$	$\phi_1 = \frac{\lambda^4}{1 + \lambda^4}$	$\phi_1 = \frac{\lambda^4}{1 + \lambda^4}$	$\phi_1 = -\lambda$
$\phi_2 = h\lambda\left[\frac{1}{2(\lambda^4 + 1)[\sqrt{(\lambda^4 + 1)} - \lambda^2]}\right]^{1/2}$	As 1	$\phi_2 = +\lambda$	$\phi_2 = \frac{\lambda^2}{1 + \lambda^4}$	$\phi_2 = \frac{\lambda^2}{1 + \lambda^4}$	$\phi_2 = \frac{\lambda^2}{1 + \lambda^4}$	$\phi_2 = +\lambda$

Regeneration of adsorbents

The regeneration of adsorbents is one of the most important operations in adsorption. The kinetics of regeneration have been considered by Carter[9b] and Zwiebel *et al.*[77] As the free energy of the system decreases during adsorption, energy must be supplied to the saturated adsorbent to drive off the adsorbate. The major part of the running cost of any adsorption process lies in this regeneration step. The optimum regeneration conditions must be found to give the greatest adsorbent efficiency and life with a minimum energy input during regeneration. There are several basic regeneration–adsorption cycles which are now examined in detail.

Thermal-swing cycle. This is the most commonly used cycle and is broadly defined as any cycle which employs different temperature levels for adsorption and desorption ($T_{des} > T_{ads}$). The heating for regeneration can be either direct or indirect. If indirect heating is used then the heating coils can also be used as cooling coils after regeneration and during adsorption. The major feature of this cycle is that high adsorption capacities may be obtained.

Pressure-swing cycle. Adsorption processes can be operated at constant temperature by utilizing a lower pressure for regeneration than for adsorption. This cycle can be relatively simple since heating and cooling steps are eliminated, and the cycle can be changed rapidly, thus reducing the overall adsorber dimensions.

Purge-gas stripping cycle. Purging an adsorbent with an inert gas reduces the partial pressure of the adsorbate in the vapour phase, and this causes desorption. The purge gas may be condensable, which greatly reduces the pumping requirements and simplifies the separation of the purge-gas and adsorbate after stripping.

Displacement cycles. Displacement cycles, in which a more strongly adsorbed adsorbate is used to displace the original adsorbate, are sometimes used when the original adsorbate is easily decomposed.

Combination cycles. The use of process steam is one of the most common ways of regeneration, since steam acts to create a thermal-swing, purge-gas and displacement cycle. Steam is also readily condensable and available in large quantities.

Usually fixed-bed adsorption units use the two bed system where one bed is being regenerated whilst the other bed is adsorbing, though if the regeneration cycle is long a three-bed system may be required. Bowen[3] has shown that for the satisfactory operation of a two-bed system the working capacity of the sorber must lie within the limits

$$x_b \geqslant x \geqslant x_l$$

where x_b is the breakthrough capacity and x_l is given by

$$x_l = \frac{[1 + x(S_s/S_c) + (W_c/W)(S_w/S_c)]MS_c(T_h - T_o)}{3C_m C_p(T_h - T_o)/4C_o - \lambda}. \tag{7.69}$$

If this criterion is not obeyed, then a three-bed system must be used.

Combustion processes

The fact that virtually all organic material can be burned to carbon dioxide and water vapour, both relatively harmless chemically, means that oxidation or combustion is a very satisfactory effluent removal process for organic materials. Certain inorganic materials can also be successfully burned. Combustion may be carried out either directly or indirectly or by catalytic methods.

Combustion

Many industries rely on combustion to decompose, by flame or high temperatures, gaseous effluents that are toxic or have foul odours. If the waste gas contains sufficient combustible material, it may be mixed with the requisite quantity of air and ignited. Precautions must be taken to prevent flame blow-back or extinction and to ensure good mixing of air and fuel. Therefore the design of the air injection and mixing systems is of importance and is discussed in detail by Knop and Teske.[26] If the heat output is large it may be used for steam raising; otherwise, direct stack burning is used. If high molecular weight organic compounds are being oxidized, then smoke emission may be a problem. This may be overcome by water or steam injection, which probably act either by suppressing polymerization or by the production of hydrogen and carbon monoxide which are then burnt.

When the waste gas is not readily combustible, fuel must be added to effect the combustion. Sulphur compounds (such as hydrogen sulphide and mercaptans) are generally oxidized to sulphur dioxide when emitted in quantities too small to be removed by other methods. Combustion temperatures of 500–1000°C are required for complete incineration, and hence the cost of fuel required may be high, particularly where large quantities of cold diluent gases also have to be heated. Direct or indirect incineration is also useful when high dust loads are present, or where catalyst poisons obviate the use of catalysts.

Catalytic combustion

Catalytic combustion is often used where particulate matter or catalyst poisoning are not a problem. The catalysts used are usually transition metals deposited on an alumina carrier. Some typical applications are the removal of trace synthetic nitrogen- and sulphur-containing compounds, phenol and formaldehyde from paper-treatment plants and lacquer residues from the manufacture of electrical insulation. Large efforts are at present being applied to remove carbon monoxide and hydrocarbons from automobile exhaust gases using platinum based catalysts. Nitrogen oxides can also be removed by a catalytic reaction with methane in the following reaction:

$$CH_4 + 2NO_2 \longrightarrow 2H_2O + CO_2 + N_2.$$

The basis of the design of catalytic converters is considered in detail in the following section.

Selection of method of combustion

In general for low contaminant concentrations catalytic combustion is to be preferred because of the lower combustion temperature required. However, a suitable catalyst must be available which must not be poisoned by any constituent of the waste gas. A large dust burden in the waste gas also rules out the use of catalytic combustion.

A mixture containing an inflammable gas is not necessarily directly combustible. The requirements for flame propagation are that a portion of the mixture must be heated above the ignition temperature and then sufficient heat must be generated to ignite a further portion of the mixture. After the initial ignition, a continuous self-propagating flame is established only if the combustible material–air ratio is within certain limits. These limits are themselves dependent on such factors as calorific values, specific heats and relative volumes of all gases present. The limits of inflammability for many gas–air mixtures under certain conditions are given in standard texts (e.g., Perry[43]) and it is apparent that many common gases such as methane (lower limit 5·3 per cent) and carbon monoxide (12·5 per cent) have a high lower limit. In many cases additional fuel has to be added to the mixture to sustain the combustion.

Catalytic reactions

As discussed earlier, the use of a catalyst increases the rate at which a chemical reaction reaches equilibrium and thus enables a lower temperature to be used. This factor is of great importance as it means that in many cases no extra heat input will be required and thus the removal process may become much more economically viable. In normal use a particle or pellet of solid catalyst is contacted with the reactant gas, using similar methods to those used for adsorption. It is often desirable to use porous catalysts in order to provide a large surface area for the reaction to occur. However, as the reactants need to diffuse through the porous solid, this internal surface may not be as effective as if all the catalyst surface were exposed on the external surface, since the chemical potential decreases as the reactant gas diffuses through the pellet.

Kinetics of catalytic reactions

Even a relatively simple treatment of the interaction between the mass and heat transfer and chemical reaction occurring in a catalyst bed is beyond the scope of this book, and hence only the barest outlines will be considered. For further treatment the reader is advised to consult standard texts such as Satterfield,[51] Thomas and Thomas[59] and Smith.[54] The mathematical theory is covered thoroughly by Aris.[1a]

The effect of diffusion upon the rate of reaction can be best seen from a consideration of Fig. 7.11. At low temperatures, the rate of intrinsic (i.e., surface) chemical reaction is so low that diffusional effects are insignificant. As the temperature rises, the intrinsic rate of chemical reaction increases much more rapidly than the diffusional flux, and thus the overall rate of reaction does not rise so rapidly due to diffusional

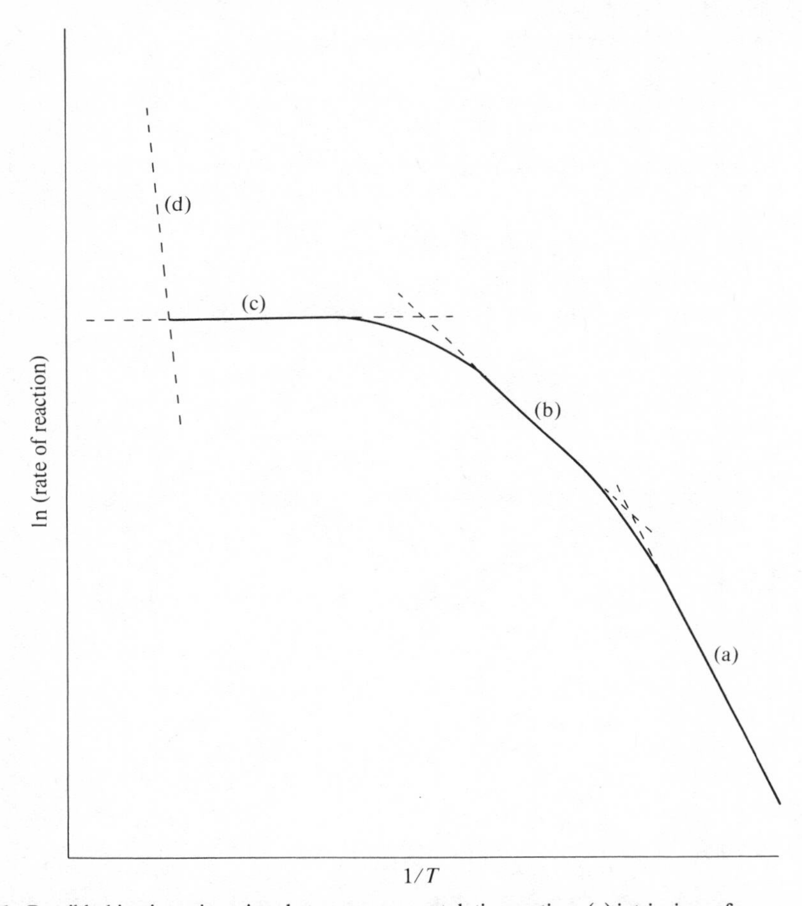

Fig. 7.11. Possible kinetic regimes in a heterogeneous catalytic reaction: (a) intrinsic surface reaction controlling (slope $= -E_{het}/R$); (b) pore diffusion predominant (slope $= -E_{het}/2R$); (c) external mass transfer (slope $\sim -(1\text{–}2)/R$); (d) Possible homogeneous reaction (slope $= -E_{homo}/R$).

limitations. It has been shown[51] that under these conditions the apparent activation energy of the reaction is about half that of the intrinsic reaction. As the temperature is raised further, the reaction beings to concentrate on the external surface of the particle and external mass-transfer limits the rate. Finally at very high temperatures the homogeneous reaction may become more significant than the heterogeneous reaction.

Thiele,[57] in a classical paper, recognized the effect that diffusion would have on the overall rate. These results may be presented in terms of an effectiveness factor (η) which is defined as the ratio of the actual reaction rate to that which would be obtained if all the surface of the catalyst were exposed to the reactant under the conditions pertaining at the surface of the particle. A similar factor (ζ) may be defined for the external gas film. Hence η and ζ may be regarded as correction factors to obtain the overall rate from the intrinsic chemical rate. In many cases η and ζ may be theoretically derived and thus used in design calculations.

It may readily be shown that[58] for a first-order reaction

$$\zeta = \frac{1}{1 + (k_v/k_c)/(R_o/3)}. \tag{7.70}$$

The effectiveness factor (η) as a function of the Thiele modulus for a spherical particle

$$\phi_s = R_o \sqrt{(k_v c_o^{m-1}/D_e)}$$

is shown in Fig. 7.12.

Under non-isothermal conditions two more parameters (γ, β) need to be considered, where

$$\gamma = \frac{E}{RT_o} \quad \text{and} \quad \beta = C_o \frac{(-\Delta H)\, D_e}{KT_o}.$$

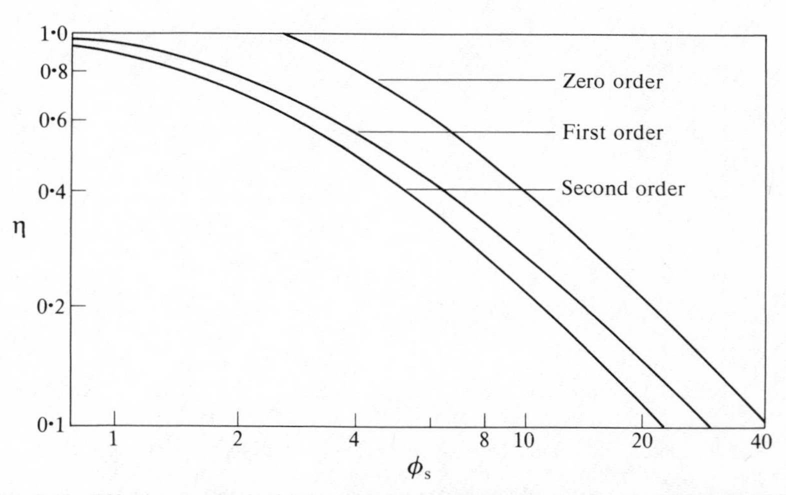

Fig. 7.12. Effectiveness factor (η) for zero, first and second order reactions (Satterfield[51]).

Weisz and Hicks[73] have presented solutions for η as a function of ϕ_s for various values of γ and β. It should be noted that under non-isothermal conditions the temperature in the centre of the catalyst particle may greatly exceed the surface temperature since ΔH can be high and thus may be much greater than unity under certain conditions. Figure 7.13 shows the effectiveness factor as a function of ϕ_s and β when $\gamma = 20$.

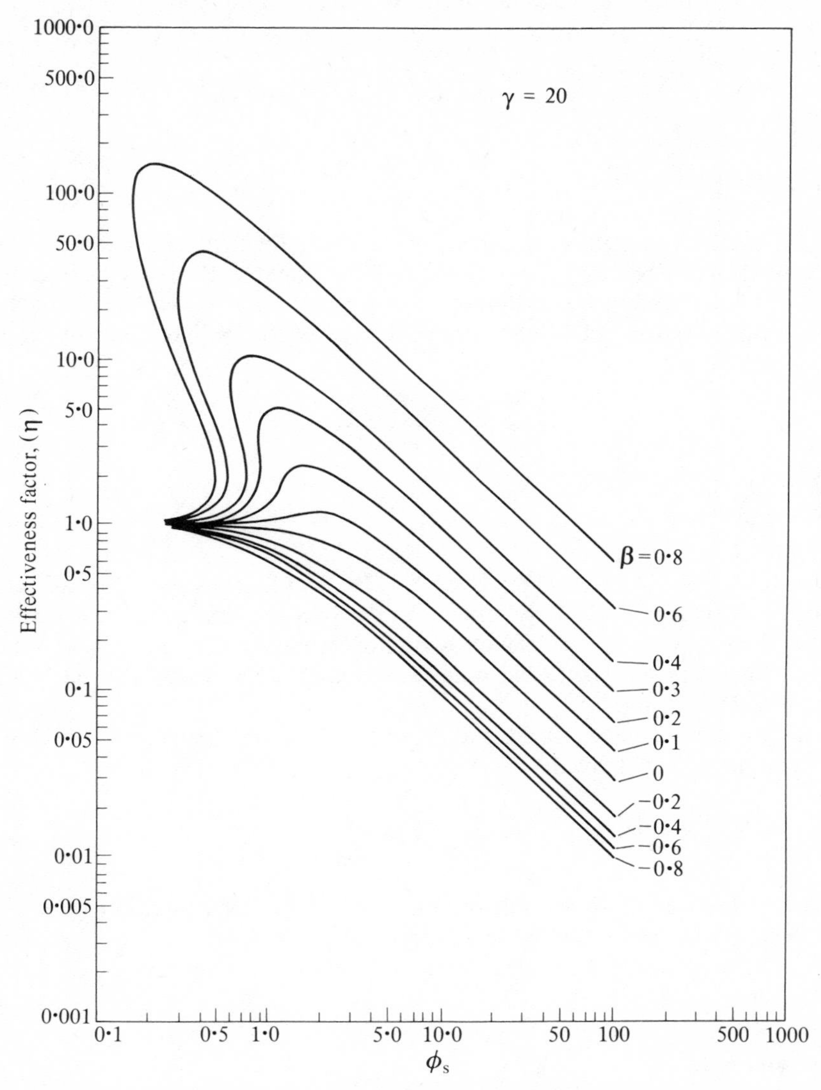

Fig. 7.13. Effectiveness factor (η) as a function of Thiele modulus (ϕ_s) for first-order reaction in a sphere (Weisz and Hicks[73]).

The maximum rise in temperature in a catalyst particle can readily be estimated from the following equation:[51]

$$T - T_o = \frac{D_e(-\Delta H)}{K} C_o, \tag{7.71}$$

by assuming that the concentration falls to zero at the centre of the particle.

Criteria for insignificant diffusional effects have been given by Thomas and Thomas[59] for the isothermal case and Weisz and Hicks[73] for the non-isothermal case

$$\phi_s < 0{\cdot}3 \qquad \text{isothermal}$$

$$\phi_s \exp\left(\frac{\gamma\beta}{1+\beta}\right) < 1 \qquad \text{non-isothermal.}$$

Design of catalytic reactors

In general, catalytic reactions are carried out in either fluidized or fixed beds. Detailed methods of fluidized bed design are given by Davidson and Harrison[14] and Kunii and Levenspiel.[29] If it is assumed that perfect mixing exists in the particulate phase, then the proportion of reactant which is left unconverted can be shown to be given by (Davidson and Harrison[14])

$$\frac{C}{C_o} = \beta e^{-x} + \frac{(1-\beta e^{-x})^2}{k' + 1 - \beta e^{-x}}, \tag{7.72}$$

where $\beta = 1 - u_o/u$,
$k' = \eta\zeta k_v h_o/u$,
$x = Q + k_b S$.

The design of fixed bed reactors, once the chemical kinetics and diffusional effects have been established, is relatively straightforward for the isothermal case. Even when the bed is non-isothermal, the relevant partial differential equations may be solved by standard numerical methods. Detailed design methods are given by Thomas,[58] Thomas and Thomas,[59] and Smith.[54]

By taking a mass balance in an isothermal tubular fixed-bed reactor, the following design equation may be derived:

$$z = \frac{G}{\rho_b} \int_{C_o}^{C} \frac{dC}{\eta\zeta\mathscr{R}}, \tag{7.73}$$

where $\mathscr{R}$ is the reaction rate per unit volume of catalyst. This equation gives the length of the reactor for the required conversion and may be integrated analytically, graphically or numerically depending on the form of the rate equation.

Catalyst poisoning

A poison is a species present in the feed stream or formed by reaction which is adsorbed on the catalyst and inactivates it. Particularly strong poisons are molecules with lone pairs of electrons capable of forming covalent bonds with solid surfaces. Examples are ammonia, phosphine, arsine, carbon monoxide, sulphur dioxide and hydrogen sulphide. The formation of carbonaceous deposits in the catalyst (coking) may also cause a marked deterioration in the catalyst activity.

Catalyst poisoning can markedly affect the overall rate of catalytic reactions, particularly where the reaction is strongly diffusion influenced. Theoretical and experimental investigations of catalyst poisoning have been well summarized by Satterfield.[51] Very often catalysts can be treated to reactivate them, thus coke and carbonaceous matter can be removed by burning off in oxygen or air.

Nomenclature

a	Constant in Elovich equation
a	Constant in Freundlich isotherm
a_p	Specific surface area/unit volume
a_t	Constant in Temkin isotherm
A	Cross-sectional area
b	Constant in Langmuir isotherm
b	Stoichiometric coefficient
c	Velocity of light
c	Constant in BET equation
c_p	Specific heat of fluid
C	Gas phase concentration
C_i	Interfacial gas concentration
C_o	Inlet gas concentration
C_s	Adsorbed phase concentration
C_{se}	Exit adsorbed phase concentration
C^*	Fluid concentration in equilibrium with adsorbed phase
d	Particle diameter
D	Diffusion coefficient
D_b	Bulk diffusion coefficient
D_e	Effective intraparticle diffusion coefficient
D_f	Forced flow diffusion coefficient
D_k	Knudsen diffusion coefficient
D_s	Surface diffusion coefficient
E	Activation energy
F	Solid flowrate
G	Gas flowrate/unit area
h	Dimensionless parameter $R\sqrt{(\lambda\rho_p/D_e)}$
h_f	Heat transfer coefficient
h_o	Bed height at incipient fluidization
H	Henry's Law coefficient

H_{OG}	Height of mass transfer unit
k	Dimensionless parameter $D_e/k_c R$
k_b	Mass transfer coefficient between a bubble and its surface
k_c	Mass transfer coefficient
k_G	Mass transfer coefficient for gas
k_L	Mass transfer coefficient for liquid
k_o	$\lambda_f/(\lambda + \lambda_f)$
k_r	Rate constant based on unit volume of solid
k_v	First-order rate constant
K	Overall mass transfer coefficient
K_G	Overall mass transfer coefficient
K_o	Equilibrium value of isotherm constant $(C_{s\,max}/C_o)$
L	Liquid flowrate per unit area
m	Intergranular void ratio $\epsilon_B/(1-\epsilon_B)$
M	Molecular weight
n	Freundlich constant
N_{OG}	Number of overall mass transfer units
N_{pore}	NTU for pore diffusion
N_s	NTU for surface diffusion
p	Partial pressure
p_o	Saturation vapour pressure
P	Total pressure
q	Pore gas concentration
Q	Rate of exchange between a bubble and the particulate phase
r	Radial parameter
r_p	Pore radius
$\mathscr{R}$	Reaction rate per unit volume of catalyst
R	Gas constant
R	Equilibrium parameter
R_o	Particle radius
S	Surface area of rising bubble
S_c	Specific heat of absorbent
S_s	Specific heat of adsorbate
S_w	Specific heat of vessel
t	Time
T	Temperature
T_h	Bed temperature during regeneration
T_o	Inlet or ambient gas temperature
u	Superficial gas velocity
u_o	Superficial gas velocity at incipient fluidization
$\bar{u}$	Mean molecular velocity
U_A	Mass fraction
v	Volume of adsorbate adsorbed
V	Volume

v_m Volume of adsorbate adsorbed for monolayer coverage
W Mass of solid
W_c Mass of container
x Fractional bed capacity
x Mole fraction in liquid phase
X Mole ratio in liquid phase
X Mean solid fraction conversion or adsorption
X^* Mole ratio in liquid phase in equilibrium with gaseous phase
y Mole fraction in gas phase
Y Mole ratio in gas phase
z Bed length parameter
z_o Total bed length
Z Length parameter

Greek symbols

α Constant in Elovich equation
β $C_o(-\Delta H)\, D_e / \kappa T_o$
γ E/RT_o
ΔH Heat of adsorption
ϵ_B Bed porosity
ϵ_p Particle porosity
η Effectiveness factor
η_g Gas removal efficiency
η_s Solid uptake efficiency
η Fluid viscosity
θ_o $\left(t - \dfrac{z\epsilon_B}{u}\right) \Big/ \left(\dfrac{\rho_B K_o z_o}{u}\right)$
κ Effective particle thermal conductivity
λ Molar latent heat
λ_f $3k_c/R\rho_p$
λ_o $\rho_B z_o \lambda/u$
λ_{of} $\rho_B z_o \lambda_f/u$
λ Forward adsorption rate constant
ρ_B Bed density
ρ_M Molar density
ρ_p Particle density
τ Tortuosity factor
τ Contact time
ϕ_s Thiele modulus for spherical particles
ζ External mass transfer efficiency factor

Further reading

The following books, most of which are not directly referred to in this chapter, are recommended for further reading.

Aris, R., 1975, *The mathematical theory of diffusion and reaction in permeable catalysts*, Vols. I and II, Clarendon Press (Oxford).

Astarita, G., 1967, *Mass transfer with chemical reactions*, Elsevier (London).

Bubble tray design manual, prediction of fractionation efficiency, 1958, AICHE (New York).

Control techniques for nitrogen oxide emission from stationary sources, 1970, Nat. Air Pollution Control Administration, No. AP-67 (USA).

Coulson, J. M. and J. F. Richardson, 1962, *Chemical engineering*, Vol. 2, Pergamon Press.

Coulson, J. M. and J. F. Richardson, 1971, *Chemical engineering*, Vol. 3, Pergamon Press.

Crank, J., 1975, *The mathematics of diffusion*, OUP (2nd edn).

Danckwerts, P. V., 1970, *Gas–liquid reactions*, McGraw-Hill (London).

Davidson, J. F. and D. Harrison, 1963, *Fluidised particles*, CUP.

Francis, W., 1965, *Fuels and fuel technology*, Vol. 2, Pergamon Press.

Gosline, C. A., 1952, *Air pollution abatement manual*, Manufacturing Chemists' Association, Washington, DC.

Gregg, S. J. and K. S. W. Sing, 1967, *Adsorption, surface area and porosity*, Academic Press.

Hobler, T., 1966, *Mass transfer and absorbers*, Pergamon Press.

Kjaer, J., 1958, *Measurement and calculation of temperature and conversion in fixed-bed catalytic reactors*, Jul. Gjellerups Forlag (Copenhagen).

Koch, R. and E. Kuciel, 1969, *Mass transfer in absorption columns*, Deut. Verlag Grundstoffind (Leipzig).

Kohl, A. L. and F. C. Riesenfeld, 1960, *Gas purification*, McGraw-Hill.

Kunii, D. and O. Levenspiel, 1969, *Fluidization engineering*, Wiley (New York).

Levenspiel, O., 1962, *Chemical reaction engineering*, Wiley (New York).

Morris, G. A. and J. Jackson, 1953, *Absorption towers*, Butterworths.

Nonhebel, G., 1972, *Gas purification processes*, Newnes-Butterworths.

Norman, W. S., 1961, *Absorption, distillation and cooling towers*, Longmans.

Perry, J. H., 1963, *Chemical engineers handbook*, McGraw-Hill (4th ed).

Ramm, W. M., 1968, *Absorption of Gases*, Israel Program for Scientific Translations (Jerusalem).

Reid, C. R. and T. K. Sherwood, 1966, *The properties of gases and liquids*, McGraw-Hill.

Riesenfeld, F. C. and A. L. Kohl, 1974, *Gas purification*, Houston Gulf. Pub. Co. (Houston).

Satterfield, C. N., 1970, *Mass transfer in heterogeneous catalysis*, MIT Press.

Sherwood, T. K. and R. L. Pigford, 1952, *Absorption and extraction*, McGraw-Hill.

Smith, J. M., 1956, *Chemical engineering kinetics*, McGraw-Hill.
Stabnikov, V. N., *Calculation and design of contact units of fractionating and absorption apparatus*, Tekhnika (Kiev).
Strauss, W., 1966, *Industrial gas cleaning*, Pergamon Press.
Strauss, W., 1971, *Air pollution control*, Wiley-Interscience.
Thomas, J. M. and W. J. Thomas, 1967, *Introduction to the principles of heterogeneous catalysis*, Academic Press.
Treybal, R. E., 1955, *Mass transfer operations*, McGraw-Hill.
Vulis, V. A., 1961, *Thermal regimes of combustion*, McGraw-Hill.
Young, D. M. and A. D. Crowell, 1962, *Physical adsorption of gases*, Butterworths.

References

1. Antonson, C. R. and J. S. Dranoff, *Chem. Eng. Prog. Ser.* (96), **65**, 20 (1969).
1a. Aris, R., *The mathematical theory of diffusion and reaction in permeable catalysts*, Vols. I and II, Clarendon Press, Oxford (1975).
2. Billinge, B. H. M. *et al.*, *Phil. Trans. Roy. Soc.*, A265, 309 (1969).
3. Bowen, J. H., in J. M. Coulson and J. F. Richardson, *Chemical engineering*, Vol. III, Pergamon Press (1971).
4. Brunauer, S., *The adsorption of gases and vapours*, OUP (1945).
5. Brunauer, S., P. H. Emmett and E. Teller, *J. Am. Chem. Soc.*, **60**, 309 (1938).
5a. Calvert, S., *Wet scrubber system study, scrubber handbook*, Environmental Protection Agency, PB 213016, 213017 (1972).
6. Carter, J. W., *Chem. and Proc. Engng.*, **47**, 37 (1960).
7. Carter, J. W., *Chem. and Proc. Engng.*, **47**, 70 (1960).
8. Carter, J. W., *Brit. Chem. Engng.*, **13**, 2, 229 (1968).
9. Carter, J. W., *Trans. Inst. Chem. Eng.*, **44**, T253 (1966).
9a. Carter, J. W. and H. Husain, *Chem. Eng. Sci.*, **29**, 267 (1974).
9b. Carter, J. W., *A.I.Ch.E.J.*, **21**, 2, 380 (1975).
9c. Chilton, T. H. and A. P. Colburn, *Ind. Eng. Chem.*, **27**, 255, 904 (1935).
9d. Cooney, D. O. and F. P. Strusi, *Ind. Eng. Chem.* (*Fund.*), **11**, 123 (1972).
10. Crank, J., *Mathematics of diffusion*, OUP (1957).
11. de Acetis, J. and G. Thodos, *Ind. Eng. Chem.*, **52**, 1003 (1960).
12. Danckwerts, P. V., *Ind. Eng. Chem.*, **43**, 1460 (1951).
13. Davidson, J. F., *Trans. Inst. Chem. Engrs.*, **37**, 131 (1959).
14. Davidson, J. F. and D. Harrison, *Fluidised particles*, CUP (1963).
15. Eerkens, J. W. and L. M. Grossman, Technical Report HE-150-150, Institute of Engineering Research, University of California (1957).
16. Emmett, P. H. and T. W. de Witt, *Ind. Eng. Chem.*, **13**, 28 (1941).
17. Ermenc, E. D., *Chem. Engng.*, 87 (29 May 1961).
18. Fleck, R. D., D. J. Kirwan and K. R. Hall, *Ind. Eng. Chem.* (*Fund.*), **12**, 1, 95 (1973).

19. Foster, R. N. and J. B. Butt, *A.I.Ch.E.J.*, **12**, 183 (1966).
20. Freundlich, H., *Colloid and capillary chemistry*, Duttons, New York (1922).
21. Garg, D. R. and D. M. Ruthven, *Chem. Eng. Sci.*, **27**, 417 (1972).
21a. Garg, D. R. and D. M. Ruthven, *Chem. Eng. Sci.*, **28**, 791 (1973).
21b. Garg, D. R. and D. M. Ruthven, *Chem. Eng. Sci.*, **28**, 799 (1973).
21c. Garg, D. R. and D. M. Ruthven, *Am. Chem. Soc., Adv. in Chem. Ser.*, **121**, 345 (1973).
21d. Garg, D. R. and D. M. Ruthven, *A.I.Ch.E.J.*, **21**, 1, 202 (1975).
22. Gregg, S. J. and K. S. W. Sing, *Adsorption, surface area and porosity*, Academic Press (1967).
23. Hall, K. R., L. C. Eagleton, A. Acrivos and T. Vermeulen, *Ind. Eng. Chem. (Fund)*, **5**, 2, 212 (1966).
23a. Hewitt, G. F. in P. L. Walker, *Chemistry and physics of carbon*, Vol. I, Marcel Dekker, New York (1965).
24. Higbie, R., *Trans. Am. Inst. Chem. Engrs.*, **31**, 365 (1935).
25. Hirschfelder, J. O., C. F. Curtiss and R. B. Bird, *Molecular theory of gases and liquids*, Wiley, New York (1954).
26. Knop, W. and W. Teske, *Technik der Luftreinhaltung*, Krausskopf-Verlag, Mainz (1965).
27. Knudsen, M., *Annin. Physik*, **28**, 75 (1909).
28. Knudsen, M., *Annin. Physik*, **35**, 389 (1911).
29. Kunii, D. and O. Levenspiel, *Fluidization engineering*, Wiley, New York (1969).
30a. Kyte, W. S., *Chem. Eng. Sci.*, **28**, 1853 (1973).
31. Laidler, K. J., *Catalysis*, **1**, 75 (1954).
32. Langmuir, I., *J. Am. Chem. Soc.*, **40**, 136 (1918).
33. Langmuir, I., *J. Am. Chem. Soc.*, **40**, 1361 (1918).
33a. Laurent, A. and J. C. Charpentier, *Chem. Eng. J.* (Lausanne), **8**, 85 (1974).
34. Loughlin, K. F. and D. M. Ruthven, *Chem. Eng. Sci.*, **27**, 1401 (1972).
35. Levenspiel, O., *Chemical reaction engineering*, Wiley, New York (1962).
36. Lewis, W. K. and W. G. Whitman, *Ind. Eng. Chem.*, **16**, 1215 (1924).
37. Masamune, S. and J. M. Smith, *Ind. Eng. Chem.* (*Fund*), **3**, 179 (1964).
38. Massaldi, H. I. and J. C. Gottifredi, *Chem. Eng. Sci.*, **27**, 1951 (1972).
39. Meyer, O. A., Ph.D. Thesis, University of New York (1966).
40. McGreavy, C., C. Nussey and D. L. Creswell, *Inst. Chem. Eng. Symp. Ser.*, **23**, 111 (1967).
40a. Murphree, E. V., *Ind. Eng. Chem.*, **17**, 747 (1925).
41. Nussey, C., Ph.D. Thesis, University of Leeds (1968).
42. O'Connell, H. E., *Trans. Am. Inst. Chem. Engrs.*, **42**, 741 (1946).
43. Perry, J. H., *Chemical engineers handbook*, McGraw-Hill (4th edn, 1963).
44. Poiseuille, J., *Inst. de France Acd. Sci. Mem.*, **9**, 433 (1846).
45. Polanyi, M., *Verh. Deut. Phys. Ges.*, **16**, 1012 (1914).
46. Rawson, H. M., *Brit. Chem. Eng.*, **8**, 180 (1963).
47. Riccetti, R. E. and G. Thodos, *A.I.Ch.E.J.*, **7**, 442 (1961).
48. Rimmer, P. G., Ph.D.Thesis, University of Wales (1970).

49. Rosen, J. B., *J. Chem. Phys.*, **20**, 387 (1952).
50. Rosen, J. B., *Ind. Eng. Chem.*, **46**, 8, 1596 (1954).
50a. Sakol, S. L. and R. A. Schwarz, *Chem. Eng. Prog.*, **70**, 8, 63 (1974).
51. Satterfield, C. N., *Mass transfer in heterogeneous catalysis*, MIT Press (1970).
52. Sehr, R. A., *Chem. Eng. Sci.*, **9**, 145 (1958).
53. Slygin, A. and A. Frumkin, *Acta. phys-chim, URSS*, **3**, 791 (1935).
54. Smith, J. M., *Chemical Engineering Kinetics*, McGraw-Hill (1956).
54a. Stearman, F. and G. J. Williamson, in J. Nonhebel, *Gas purification processes*, Newnes-Butterworth (1972).
55. Tamman, G. and W. Koster, *Z. anorg. allg. Chem.*, **123**, 196 (1922).
56. Temkin, M. I. and V. Pyzhev, *Acta. phys.-chim, URSS*, **12,** 327 (1940).
57. Thiele, E. W., *Ind. Eng. Chem.*, **31**, 7, 916 (1939).
58. Thomas, W. J., in J. M. Coulson, and J. F. Richardson, *Chemical engineering*, Vol. III, Pergamon Press (1971).
59. Thomas, J. M. and W. J. Thomas, *Introduction to the principles of heterogeneous catalysis*, Academic Press (1967).
59a. Thomas, W. J. and J. L. Lombardi, *Trans. Inst. Chem. Engrs.*, **49**, 240 (1971).
60. Timofeev, D. P., and I. T. Erashko, *Izv. Akad. Nauk SSSR Otd. Khim.*, **7**, 1192 (1961).
61. Timofeev, D. P. and O. N. Kabanova, *Izv. Akad. Nauk SSSR Otd. Khim.*, **10**, 1761 (1964).
62. Timofeev, D. P., N. A. Tverdokhleb and I. T. Erashko, *Russian J. Phys. Chem.*, **43**, 1, 137 (1969).
63. Treybal, R. E., *Mass transfer operations*, McGraw-Hill (1955).
64. Vanecek, V., M. Markvart and R. Drbohlav, *Fluidized bed drying*, Leonard Hill, London (1966).
65. Vermeulen, T., *Ind. Eng. Chem.*, **45**, 1664 (1953).
66. Vermeulen, T., *Adv. in Chem. Engng.*, **2**, 147 (1958).
67. Vermeulen, T., in J. H. Perry, *Chemical engineers handbook*, McGraw-Hill (4th edn., 1963).
68. Wakao, N., S. Otani and J. M. Smith, *A.I.Ch.E.J.*, **11**, 3, 435 (1965).
69. Wakao, N. and J. M. Smith, *Chem. Eng. Sci.*, **17**, 825 (1962).
70. Wakao, N. and J. M. Smith, *Ind. Eng. Chem. (Fund.)*, **3**, 2, 123 (1964).
71. Walter, J. F. and T. K. Sherwood, *Ind. Eng. Chem.*, **33**, 493 (1941).
71a. Weber, T. W. and R. K. Chakravorti, *A.I.Ch.E.J.*, **20**, 2, 228 (1974).
72. Weisz, P. B., *Z. Phys. Chem.* (Frankfurt), **11**, 1 (1957).
73. Weisz, P. B. and J. S. Hicks, *Chem. Eng. Sci.*, **17**, 265 (1962).
74. Wheeler, A., *Adv. Catalysis*, **3**, 266 (1951).
75. Young, D. M. and A. D. Crowell, *Physical adsorption of gases*, Butterworths (1962).
75a. Youngquist, G. R., *Ind. Eng. Chem.*, **62**, 8, 52 (1970).
76. Zeldowitch, J., *Acta. phys-chim, URSS*, **1**, 961 (1935).
77. Zwiebel, I., R. L. Gariepy, and J. J. Schnitzer, J. J. *A.I.Ch.E.J.*, **18**, 6, 1139 (1972).

8 Removal of grit and dust

K. DARBY

This chapter deals with the various forms of gas cleaning equipment available for removal of suspended material from gases. In addition to descriptions of the apparatus, some background theory is included to give an understanding of the principles of operation, the applications, and limitations of the different devices. In general it is found that, while theory can be useful in helping to understand the characteristics of the different systems, it is not possible to calculate the efficiency from first principles. Commercial designs are based almost entirely on practical experience, and for this reason users are strongly recommended to consult one of the specialist manufacturers when the need for gas cleaning equipment arises.

Information needed when selecting gas cleaning systems

The correct choice of gas cleaning system, from the points of view of both cost and performance, is possible only if sufficient information is available. This includes the process on which the gas cleaning plant is to be used, and physical and chemical properties of the gases and the dust suspended in them. In the case of new process plant where it is not possible to carry out analysis of gas or dust, identification of the process together with a description of the process cycle and analysis of raw materials enables the gas cleaning plant engineer to draw on experience to design the plant needed. The fullest possible analysis of the raw materials is needed, as in many processes where high temperatures are involved the dust and fume carried over to the gas cleaning plant can be determined by the presence of components of the raw material which are volatilized at the temperature of the operation and subsequently re-condense. The result is that solids suspended in the gas can consist of a high percentage of very fine particles, the presence of which has a profound effect on choice of gas cleaning system. The dust in such cases may have a chemical composition completely different from that of the raw materials used. For example, on a sinter plant producing iron sinter from ore, the ore contained less than 1 per cent of sodium and potassium compounds, but since this completely volatilized on the sinter machine, the dust in the gases leaving the sinter plant contained more than 50 per cent sodium and potassium salts in the form of sub-micron fume. This fume, in practice, completely dictated the gas cleaning plant requirements.

Other information needed on the process is whether it is cyclic or continuous in nature. Continuous processes are those such as cement kilns and power station boilers, where the nature of the gas and dust is reasonably constant over a long period. Cyclic processes are those where batches of material are subject to a variety of operations, during which the dust-laden gas can vary widely in temperature, quantity and composition. The effect of each stage must be taken into account; frequently one part of the cycle gives rise to the most difficult gas cleaning plant. Apart from information on the process and raw materials, the following information on dust and gas should be given if available. The significance of this information will become apparent when the principles of the various gas cleaning systems are described.

Gases to be cleaned

1. Temperature;
2. composition (the gas from most industrial processes consists mainly of oxygen, nitrogen, the oxides of carbon and water vapour).

It is usually of particular importance, from the point of view of gas cleaning, to know the moisture content of the gases. This can constitute a high percentage of the volume in some processes, particularly when the temperature of the gas exceeds 100°C. It is also extremely important to know whether there are acid components present such as hydrogen chloride, oxides of sulphur and oxides of nitrogen. Even if these constituents represent only small fractions of 1 per cent of the total volume, they dictate the temperature of operation of the gas cleaning process, particularly in the presence of moisture, and in many cases also decide the materials of construction both of the gas cleaning device and the container in which it operates.

Material suspended in gas

1. Whether solid or liquid;
2. if liquid, whether viscous or free flowing at the temperature of the gases;
3. if solid, whether a free flowing powder or tending to form agglomerates. Also, whether dust has any abrasive properties;
4. size and shape of suspended particles;

5. chemical composition;
6. electrical resistivity (electrostatic precipitators only).

One of the more significant factors affecting all types of gas cleaning devices is the size of the dust particles. It is also the factor over which most confusion arises due to the failure to understand the meaning of the way in which the results of particle sizing analyses are expressed, and the limitations imposed by a particular method of measurement of particle size distribution. It is proposed, therefore, to give some brief notes here of the methods used and the significance of the gradings produced.

While many forms of measuring apparatus are used these can be divided into two basic groups. Particle size is usually expressed in microns (μm, 10^{-6} m) which is convenient for the range of sizes involved, and this size is obtained by either of the two methods below.

1. Direct measurement using sieves for coarser particles. For finer particles the microscope or electron-microscope with a calibrated graticule; samples of the dust are mounted on microscope slides for counting and measurement.
2. Size calculated from measurements of the velocity of the particle in either gas or liquid when acted on by known forces. These are known as sedimentation methods.

The suspended particles in a gas are classified for convenience into four groups.

1. Liquid droplets or mists; due to surface tension effects these are all nearly spherical in form and vary greatly in size. A mist consists mostly of droplets less than 1 μm diameter.
2. Fume, solid particles usually spherical in form and less than 1 μm in diameter resulting from the volatilization and re-condensation of materials which, as a result of surface tension, form spheres. These usually arise from metallurgical processes.
3. Dust, particles in the size range 1–100 μm in the form of solid spheres, cenospheres (hollow spheres), irregular shapes including flakes, fibres, and spongy particles containing gas bubbles.
4. Grit, 100–1000 μm. These are mostly irregular particles and include solid homogeneous particles, spongy particles containing gas bubbles, flakes and fibres.

The method of particle sizing analysis used depends to some extent on size of particle, for example,

Sieving—restricted to sizes 50 μm and larger owing to difficulty in producing accurately smaller meshes.
Optical microscope—low limit not much below 1 μm.
Electron microscope—all particle sizes, particularly useful for fine fume 0·01–1 μm diameter.

Sedimentation methods—this includes sedimentation under gravity limited to above 1 μm. The range of this method can be extended by increasing the forces acting on the suspended particles by spinning the column of fluid.

Particle sizing theory and apparatus is a complex subject and can only be discussed briefly here. A useful review of the subject for those interested in pursuing the matter further has been produced by the Institute of Physics.[1]

When the particles are in solid spherical form, the different methods of particle sizing are generally in reasonable agreement, but with any other form of particle there can be very serious divergencies. This is readily appreciated when comparing the two systems. The sieve and microscope methods measure actual physical dimensions as graded by the sieves or as seen on the microscope slide; in the case of flakes this method of measurement tends to select the larger dimensions, while the sedimentation method gives the size expressed as the diameter of the solid sphere of the same material with the same free falling velocity.

When particles are not solid spheres, in particular when the dust particles are of a spongy nature or in the form of flakes, measurements may not give overall average dimensions. With spongy material a microscope shows the particles as being larger than by the sedimentation system, since gas bubbles reduce the free falling velocity and hence the diameter of the equivalent sphere as calculated for the sedimentation method. In the case of flake formations, again the microscope measures the largest dimensions; the flakes can be several millimetres in length and only a few microns thick. Such flakes have a very low free falling velocity and the sedimentation method represents them as comparatively small spheres. Thus it is very important, when carrying out dust gradings, to recognize that these differences can exist; for these reasons the method of dust grading used should always be recorded with the results obtained.

Most dust plant manufacturers and designers prefer to carry out dust grading measurements using their own systems, so that the results obtained may be more directly related to their previous experience. The question arises as to which method should be used, since neither fully defines the form of the particles.

Comparing the two methods from the point of view of what it is hoped to achieve with the gas cleaning apparatus which is most commonly used to reduce atmospheric pollution, there is much to be said for the sedimentation system which defines the dust particles in terms of free falling velocity. With regard to atmospheric pollution control, it is generally true that particles with lower free falling velocity travel farther from the stack, so the dilution is greater before the dust reaches ground level, and the potential nuisance value is correspondingly reduced.

In this chapter there are graphs showing the efficiency of various devices for removing particles of different sizes; in all cases the size given is the diameter of the equivalent sphere, that is the diameter of the sphere composed of the same material with the same free falling velocity as the dust particle. All determinations are by sedimentation methods, except for sub-micron fume, for which the electron microscope was used.

Gas cleaning methods

There are many types of industrial gas cleaning devices in service and it is possible here to give only a survey of the more important types. These can be grouped under six main headings, according to the principle on which the separation process depends:

1. settling chambers
2. inertial separators, including cyclones
3. electrostatic precipitators
4. filtration through a porous medium
5. wet washers and scrubbers
6. sonic agglomerators.

Settling chamber

The settling chamber is the simplest form of gas cleaning device, and depends on the fact that solid or liquid particles suspended in a gas fall through the gas under the action of gravity at a speed which is dependent on the size of the particles, their density and shape, and the density and viscosity of the gas. The particles rapidly reach a constant velocity referred to as the *free falling* or *terminal velocity*, which can be calculated for smaller particles from the formula derived by Stokes, which states that

$$W = \frac{gd^2(\sigma - \rho)}{18\eta}, \tag{8.1}$$

where W = free falling velocity,
g = acceleration due to gravity,
d = diameter of particle,
σ = density of particle,
ρ = density of gas,
η = viscosity of gas.

Since the density of gas is low compared with that of the dust usually met in gas cleaning applications, the effect of this in calculations from the formula is small and can usually be ignored. Figure 8.1 shows typical calculated values of terminal velocity using Stokes' law, together with actual measured values, from which it will be seen that Stokes' law is accurate for particles below 100 μm. It will also be seen that for coarse grit, of the order of 1000 μm or more, the free falling velocity is of the order of m s^{-1}, in contrast to that for 10 μm particles, where this velocity reduces to cm s^{-1}, and for very small particles becoming insignificant due to Brownian movement.

The graphs shown in Fig. 8.1 are based on spherical particles with a specific gravity of 1. For higher density materials, the free falling velocity increases proportionately, and the velocities shown can be multiplied by the new specific gravity in order to get the corresponding velocity. Generally, liquids have specific gravity of 1 or less, minerals

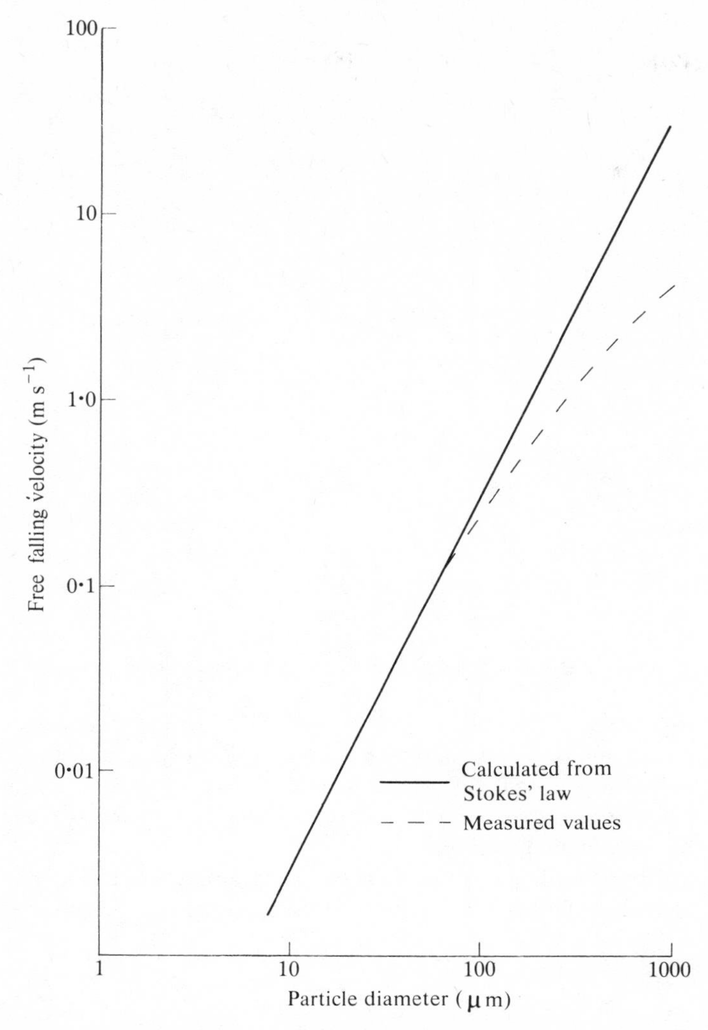

Fig. 8.1. Free falling velocity of spherical particles of specific gravity 1 in free air at 20°C, 760 mmHg.

such as silica, lime, ash approximately 2, while metallic oxides have specific gravities of 4 or more. Consideration of Fig. 8.1 shows that any device relying on the natural settling effect of particles under the force of gravity can only be effective on relatively coarse material, e.g., larger than 100 μm.

According to Stokes' law, the free falling velocity also decreases with increasing gas viscosity; Fig. 8.2 shows the variation of the viscosity of various common gases with temperature, from which it will be seen that the viscosity of the gas increases with temperature. The free falling velocity of the dust particles therefore reduces: for example, due to change in the viscosity, the free falling velocity decreases to one half if

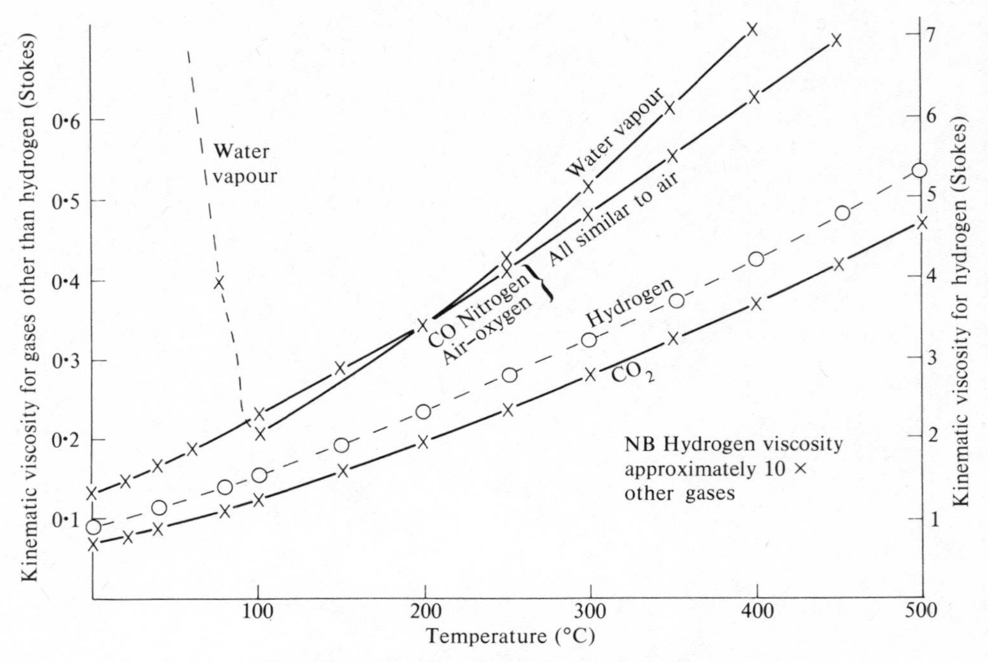

Fig. 8.2. Variation in gas viscosity with temperature.

the temperature of the gas increases from 0 to 300°C. The graph also shows that the viscosity of most common industrial gases is of a similar order, with the exception of hydrogen, which is nearly ten times greater, and carbon dioxide which is somewhat lower. For most industrial processes, therefore, only the figure for air is needed, corrected to the operating temperature.

A practical form of settling chamber is shown diagramatically in Fig. 8.3. The gas is uniformly expanded to fill the full height H of the gas chamber, which has a length L.

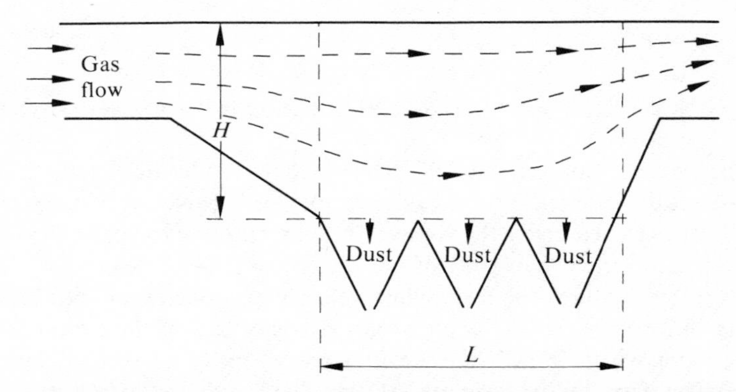

Fig. 8.3. Settling chamber. Gas velocity = V; effective height of settling chamber = H; effective length of settling chamber = L; time of residence of gases in chamber = L/V.

Below the settling chamber are the hoppers into which the dust falls when it leaves the gas stream; there should be no flow in these hoppers, and therefore, once dust falls below the level of the hoppers it is effectively retained in the settling chamber.

The diagram shows that the residence time of the gas in the settling chamber is L/V seconds; in this time, for complete collection of dust particles of any particular size, the retention time in the chamber must be equal to the time taken for a particle of that size to fall from the top of the chamber to the hopper level, that is, H/V. If the retention time is less than this value, then the proportion of dust collected is related to that contained in that volume of gas corresponding to the maximum height from which all dust would have time to reach the hopper level.

This simple calculation assumes perfectly uniform gas velocity throughout the chamber, the absence of turbulence and a uniform dust distribution. None of these conditions is completely achieved in practice and this has the effect of reducing the efficiency below the theoretical value. For practical purposes the settling chamber is limited to the collection of dust particles in excess of 100 μm, or particles with a free falling velocity in excess of 1 m s^{-1}. For smaller particles, the retention time and hence the size of the settling chamber becomes excessively large.

Inertial collectors and cyclones

There are many devices of different form falling into this category. All depend, for their increased effectiveness compared with the settling chamber, on the fact that when a dust particle is spinning in a circular path it is subject to an outward force given by (8.2).

$$F = MV^2/R, \tag{8.2}$$

where F = force acting on particle,
M = mass of particle,
V = gas velocity,
R = radius of spinning circle.

The acceleration of the particle is V^2/R, and by suitable choice of values of V and R this acceleration can be made many times the value of the acceleration due to gravity, g, and the force acting increases accordingly. Hence, replacing this increased acceleration for g in Stokes' law, the velocity of the particle relative to the gas is correspondingly increased. For instance, in a small sampling cyclone of about 3 cm diameter, designed for collecting samples of dust from flue gases from pulverized fuel boiler, it was estimated that the value was almost 1500g. The effect on the velocity of the dust particles will be seen by reference to Fig. 8.1. A 10 μm particle would have its velocity relative to the gas increased to a value greater than that of a 1000 μm particle falling under the force of gravity alone.

The use of this principle extends the possibility of separating particles from the gas down to 10 μm and even smaller. Since gas viscosity, particle diameter and density affect the velocity of the dust particle relative to the gas in a similar way to the settling chamber, they can be expected to have similar effects on the separation characteristics of inertial collectors and cyclones.

Inertial collectors

The simplest practical device using this principle is the inertial collector. The increased force on the particle is achieved by a sudden change in direction of gas flow, the dust particle tending to carry on for a short time in the original direction of the gas flow, passing into a dead space where it is allowed to settle, thus achieving a separation efficiency superior to the settling chamber.

Two forms of inertial collector are shown in Fig. 8.4. Type (a) is a simple baffle and is a logical development of the settling chamber. The gas is caused to make a sudden change of direction in order to flow round a baffle into the outlet flue. As a result of the change in direction the dust tends to carry on in its original direction until its inertia is destroyed by the viscous drag of the gas. If by this time the dust has reached the stagnant zone below the flow path it is effectively collected; otherwise it passes into the outlet main. Obviously this device is only marginally more efficient than the settling chamber.

The type shown in Fig. 8.4(b) is a labyrinth type separator consisting of elements such as those shown in the right-hand upper corner of the diagram. These elements are arranged vertically spaced at intervals with between three and six layers of elements in

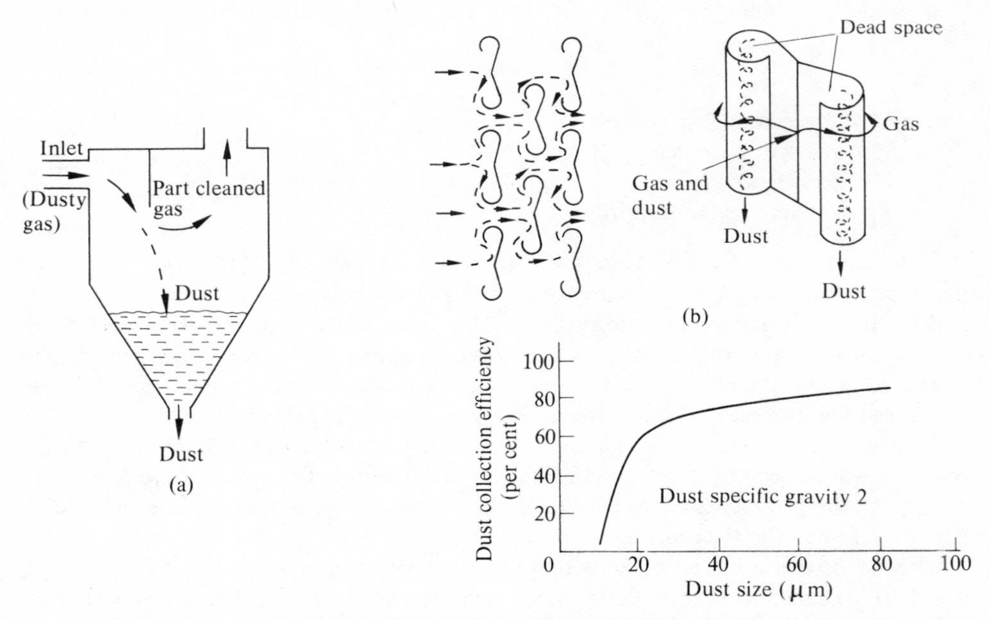

Fig. 8.4. Inertial separators; (a) simple inertial separator; (b) labyrinth separator.

depth. The gas impinges on the element and flows around it; the dust collects in the closed scroll and passes to the hopper below the gas collecting system. Such a device is reasonably efficient for dust in excess of 50 μm. The graph shows performance on ash from a pulverized fuel boiler, when the device was tested as a pre-collector to reduce the inlet dust concentration to an electrostatic precipitator. With present clean air legislation such devices have very limited application.

Compound collector

This type of collector is illustrated in Fig. 8.5. It uses both the inertial and cyclone effect, and consists of an outer cylindrical vessel to which the gas is introduced tangentially. The gas spins in this chamber, which concentrates the dust in the outer layer, and then escapes through louvres near the centre which ensure that the gas undergoes a sharp change of direction resulting in a further separation of any residual dust in the gases. This remains in the spinning column of gas, which continues to rotate in the chamber, with the result that any dust which does not escape through the louvres must eventually be caught by the collector. In the design of collector shown the outer

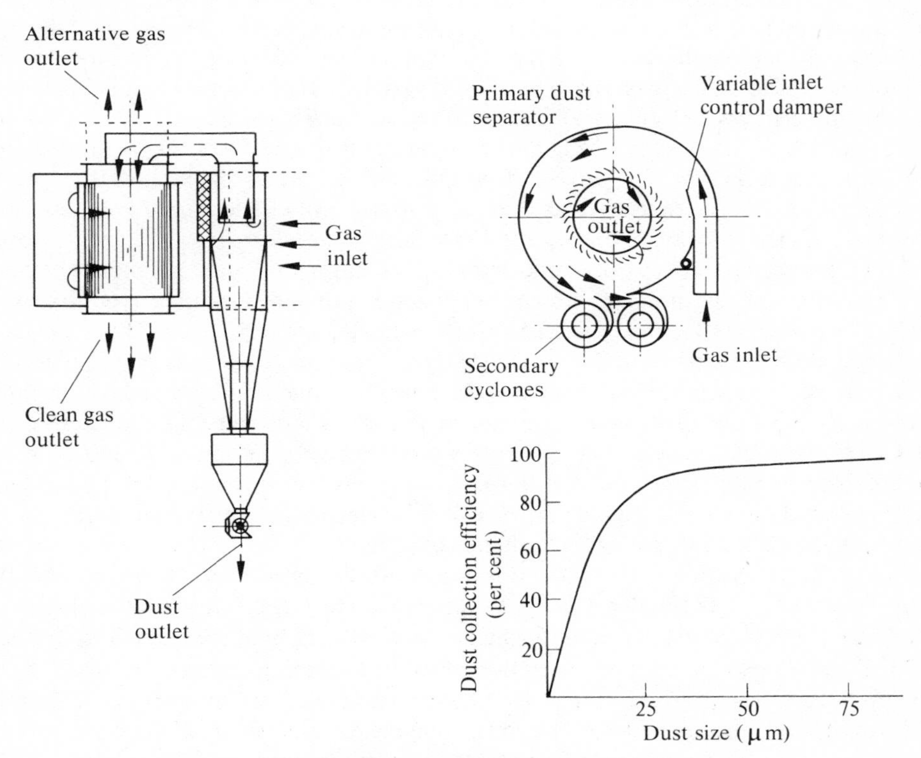

Fig. 8.5. Compound collector (Buell Ltd.).

layer of gas is skimmed off, and this gas, less than 10 per cent of the total volume, is treated in conventional high efficiency cyclones which will be described later.

Consideration of the performance of this type of collector shows its efficiency is determined by the separation efficiency of the first inertial stage of cleaning rather than the high efficiency cyclones. It is not suitable where high efficiency collection is required for particles much below 50 μm. The graph in Fig. 8.5 shows an efficiency of 90 per cent is possible.

Low pressure loss cellular collectors

Two samples of this type of collector are shown in Fig. 8.6. Both consist of a large number of separating cells arranged in parallel, each cell consisting of a cylinder roughly 30 cm in diameter with a fixed impeller at the inlet to the cell which imparts a spinning action to the gas. In both types, as a result of the spinning action, the greater part of the dust is concentrated in the outer skin of the gas.

In type (a) this outer skin, constituting about 10 per cent of the total gas volume, is extracted together with the dust contained in it and subjected to treatment in high efficiency secondary cyclones. The centre column of gas, which contains only a small amount of residual dust, passes right through to the clean gas duct. The gas from the high efficiency cyclone is fed back into the clean gas system following treatment. As in the compound collector the overall efficiency is determined by the separating efficiency of the primary cells rather than the efficiency of the secondary cyclones.

In type (b), no gas is extracted in the secondary system, the dust being separated only by the effects of inertia. The difference in efficiency of the two collectors is shown in Fig. 8.6; it will be seen that whereas type (a) is 90 per cent efficient for 25 μm particles, the same efficiency for type (b) corresponds to 50 μm particles.

The disadvantage of type (a) is the extra energy absorbed by the secondary cyclone system. Although the primary system has a lower pressure loss than type (b), extra energy is involved in the fan power needed to overcome the resistance of the secondary cyclones. Both systems have been widely used for cleaning the gases from pulverized fuel boilers. Type (a) suffers from the disadvantage that if the secondary cyclones become blocked, the dust carried over can cause serious erosion of the fan impellers. It was partly for this reason that type (b) was developed, the lower efficiency being acceptable since for pulverized fuel boiler applications the collector was followed by a further collection system, such as the bag filter or electrostatic precipitator, capable of removing the finer dust passing through the collector.

In Fig. 8.7 is shown a variation on this design; as the gases enter a large chamber the larger particles fall out under gravity. Further separation is effected by the change of direction of the gases as they enter the high efficiency section of the plant. This consists of small axial flow cyclones arranged in parallel; the spinning action is imparted to the gas by fixed vanes arranged around the perimeter of the cells. The spinning column of gas travels down the tube, then breaks away making a rapid change of direction into the discharge tube of the cyclone; the separated dust travels down into the hopper. Such a

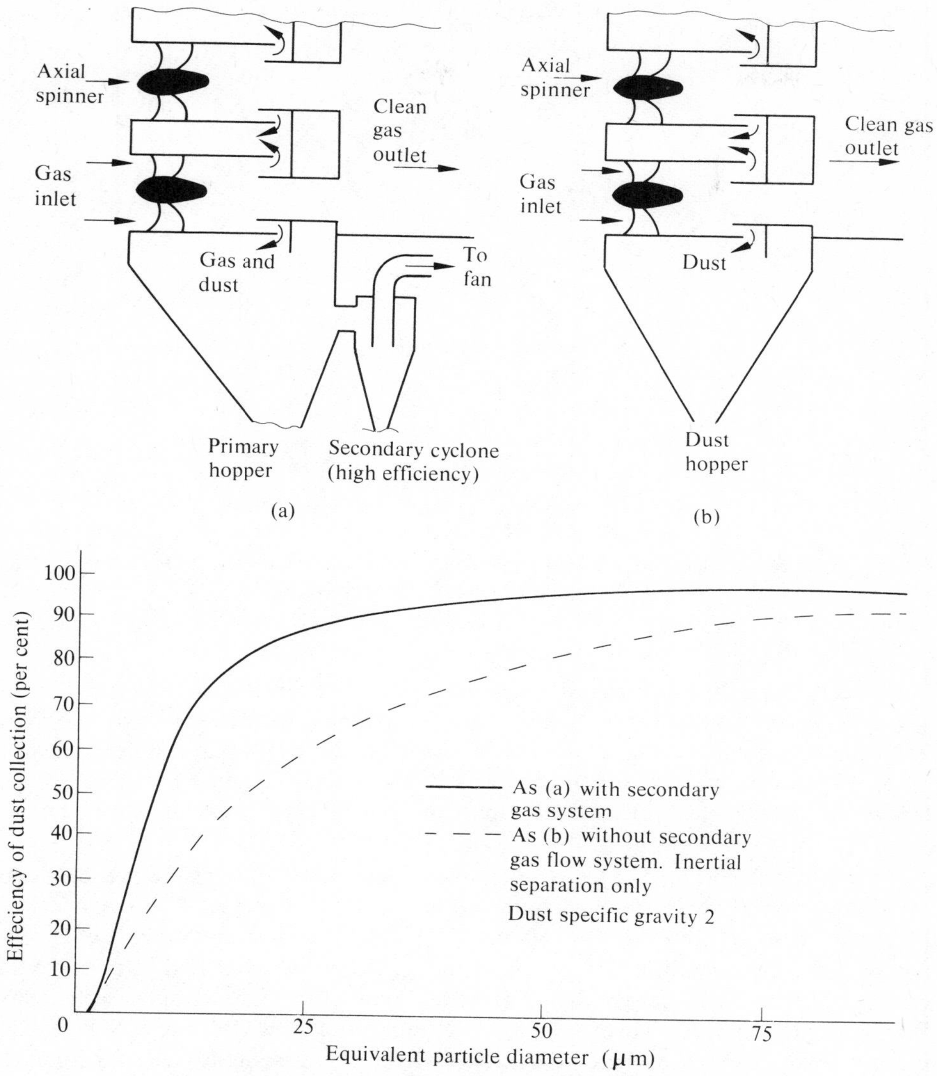

Fig. 8.6. Low pressure loss cellular collector (Davidson).

collector can have an efficiency of 95 per cent on 10 μm particles, but the pressure loss is considerably higher than for the types shown in Fig. 8.6.

Cyclones

The cyclone has generally the highest possible efficiency of the inertial devices. The simple cyclone, Fig. 8.8, consists of a vertical cylindrical vessel into which the gas is

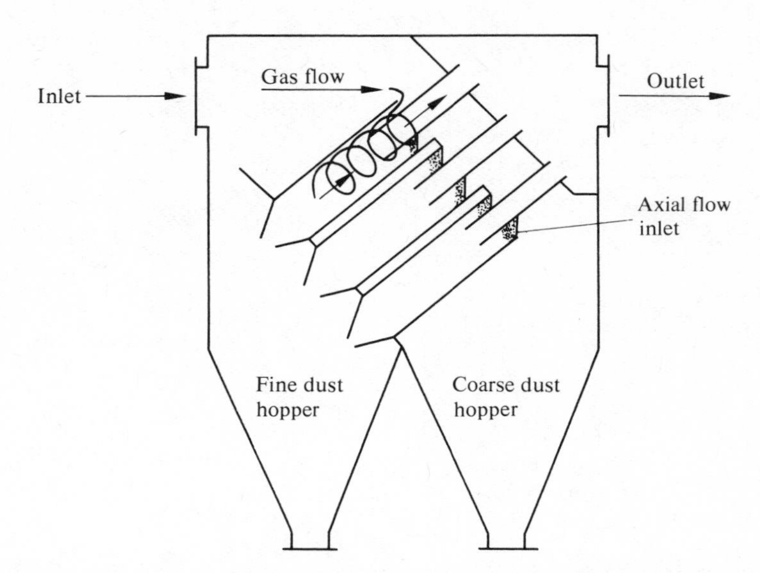

Fig. 8.7. Axial flow cyclones with inertial separation. Nest of small diameter, high-efficiency cyclones. Efficiency for 10 μm particles = 95 per cent.

introduced through a horizontal tangential inlet. The dust is concentrated by the spinning action into the gas layer in close proximity with the wall of the vessel. The spinning column of gas progresses along the vessel, finally breaking away to pass out through the exit duct on the axis of the vessel. The dust falls into a hopper below the moving gas column. Examples of high efficiency cyclones are shown in Fig. 8.9.

It is not possible to predict from purely theoretical considerations the particle sizing–efficiency curves for different types and designs of cyclones. For cyclones of a particular design, however, the effects of changes in cyclone size, gas throughput and the characteristics of the dust and gas can be calculated with reasonable accuracy, once basic design data have been established for certain selected dusts. Particle sizing analysis, which is one of the more important factors, is still far from being an exact science, and a considerable difference in grading efficiencies is liable to be obtained by different laboratories, even when using the same grading apparatus. It is advisable for this reason that, for determination of the efficiency of a mechanical or cyclone collector for a particular application, the particle sizing should be carried out by the manufacturer of the equipment. This reduces the requirements of the particle sizing analysis to that of maintaining relative standards of accuracy.

Many attempts have been made to calculate the effects on cyclone efficiency of changes of cyclone size, particle size and the physical properties of the transporting gas. Hawksley *et al.* (see Jackson[2]) derived an equation for the efficiency of a cyclone as follows.

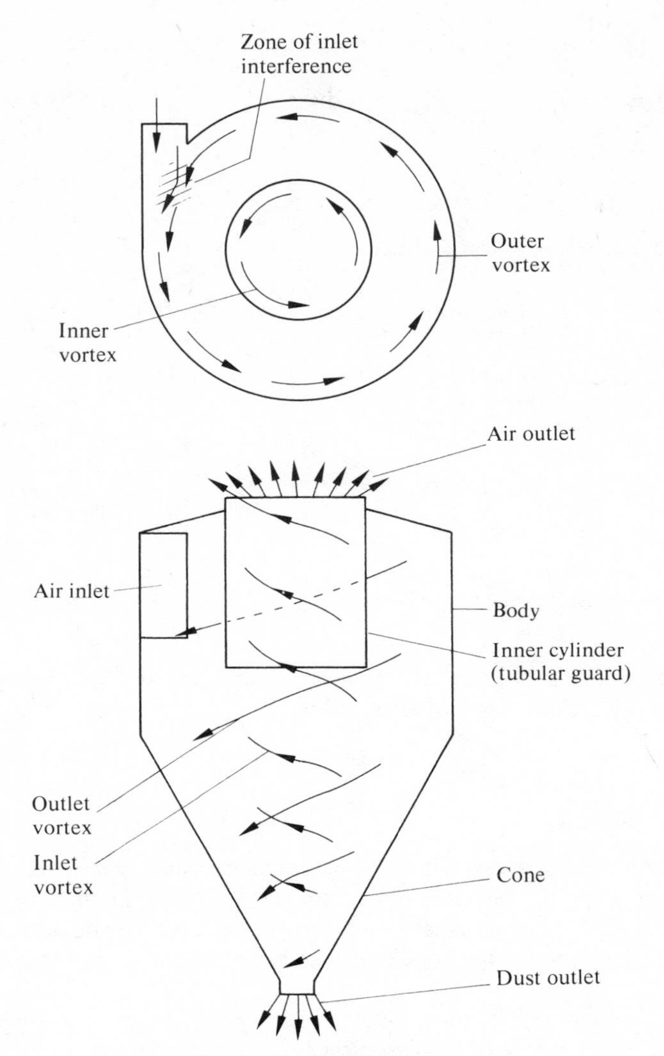

Fig. 8.8. Gas flow simple cyclone.

Efficiency of collection is a function of

$$\frac{(\sigma d^2 V_c)}{18\eta D}, \tag{8.3}$$

where d = particle diameter,
σ = density of particle,
V_c = gas velocity at cyclone inlet,
η = gas viscosity,
D = cyclone diameter.

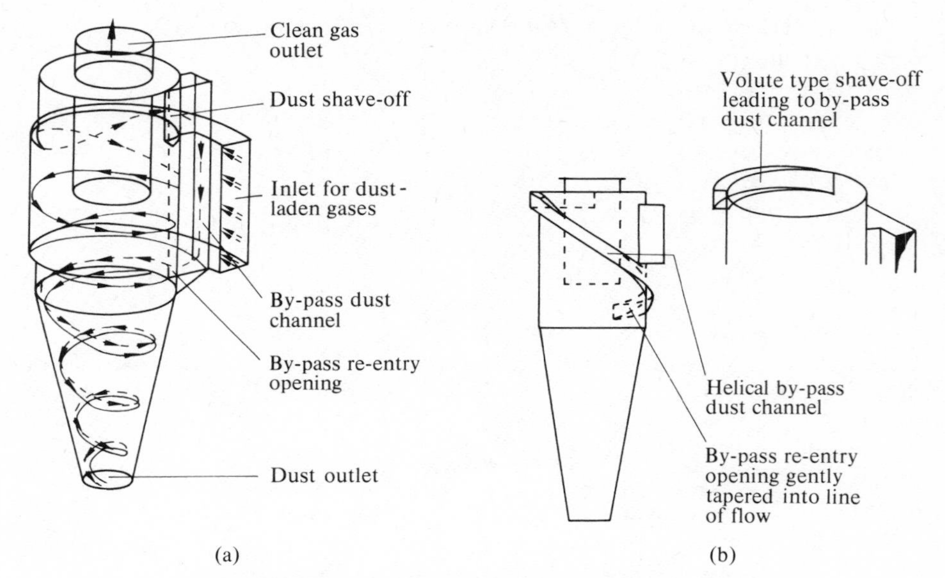

Fig. 8.9. High-efficiency cyclones (Buell); (a) original; (b) improved.

From this equation a number of important characteristics become apparent. The efficiency of the cyclone is clearly unchanged if

$$\frac{(\sigma d^2 V_c)}{18\eta D}$$

is constant.

The full relationship between efficiency and other operating characteristics cannot be calculated and must be determined experimentally for a particular cyclone configuration. Once this has been established, the performance of the cyclones of similar geometry for different operating conditions can be calculated. For example:

1. *Cyclone diameter*. Efficiency of dust collection will increase as cyclone diameter decreases, *or* diameter of dust particle for a given collection efficiency varies as cyclone diameter.
2. *Cyclone capacity*. Efficiency of dust collection increases directly with gas velocity at inlet which is a measure of cyclone capacity, *or* diameter of dust particle collected with a given efficiency decreases as inlet velocity and cyclone capacity increase.
3. *Gas temperature*. Since gas viscosity increases with increasing gas temperature (see Fig. 8.2), so the separating velocity of the particles decreases (Eq. (8.1)). This reduces the efficiency of dust separation. However, for a given mass of gas, volume increases with temperature; this increases gas velocity at inlet and tends to increase efficiency. The net effect is that if the *mass* of gas is kept constant, then

the diameter of particle collected with a given efficiency is proportionate to (gas temperature (°C) + 273)$^{0.15}$.

These effects are illustrated in Fig. 8.10; curve (1) shows a typical particle size–efficiency curve for a high efficiency cyclone of about 1 m diameter. Curve (2) shows the curve corrected according to the relationship given above for the same mass of gas at 300°C. The opposing effects of increasing gas viscosity and increasing gas velocity nearly cancel each other out, so there is little difference in efficiency. There is, however, a considerable increase in pressure loss, illustrating the point that for the same mass of gas it is more economical in terms of fan power to work at lower temperatures. Curve (3) illustrates the effect of scaling up the cyclone to twice the diameter of (1). Volume capacity increases fourfold with the same pressure loss as (a), but the efficiency–particle size curve shows an appreciably worse characteristic. It would be possible to give the same characteristic as (1) by doubling the volume again, giving eight times the capacity of (1), but the pressure loss would increase proportional to the square of the velocity, i.e., four times that of (1).

It is evident from these calculations that while the cyclone can operate at very high temperatures, limited only by materials of construction, a considerable economy in

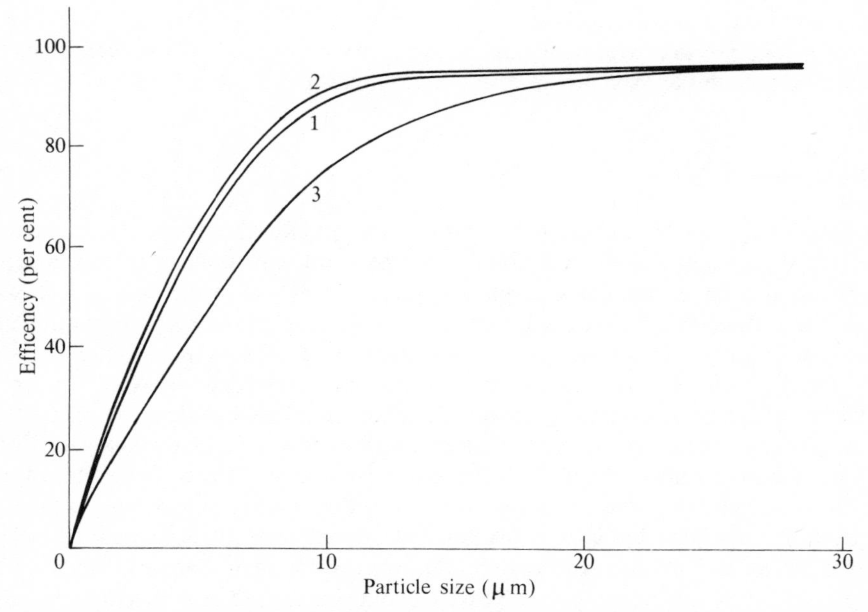

Fig. 8.10. Operating characteristics of high efficiency cyclones; (1) cyclone diameter = D, gas temperature = 20°C, inlet velocity = V; (2) cyclone diameter = D, gas temperature = 300°C, same mass flow as (1), i.e., inlet velocity and hence volume capacity increased by ratio 573/290 (absolute temperatures), pressure drop increased by same ratio; (3) cyclone diameter = $2D$, gas temperature = 20°C, inlet velocity V as (1), mass flow capacity increased by factor of 4 from (1), pressure loss as (1). To give same particle size, efficiency characteristic as (1) inlet velocity would need to be doubled, i.e., mass flow capacity would be 8 times that of (1), pressure loss proportional to V^2, i.e., 4 times that of (1).

terms of fan power can be effected by reducing the gas temperature. Since this may involve cost also, a careful consideration of the overall economics is necessary. Similarly, on the question of large and small cyclones, a comparison of the sum of operating and capital costs for one large cyclone, with higher operation cost, and that of a number of smaller cyclones in parallel, with lower operating costs, must be made. In the space permitted here, only a brief summary of cyclone theory is possible. For fuller information, Jackson[2] is recommended. Other relationships have been established, many of which are mentioned in a paper by Caplan,[3] who also derived equations for the prediction of the behaviour of cyclones. This stated that in addition to the factors already mentioned, efficiency of a cyclone would increase with increased cyclone body length, or length of cone, and increase with the increasing ratio of body diameter to gas outlet diameter. On the other hand, efficiency could be expected to decrease with increase in cyclone diameter, width or area of inlet duct to cyclone, and gas outlet diameter.

Other important factors influencing the efficiency of cyclones are as follows:

1. Roughness of cyclone body. By increasing turbulence of the spinning column of gas, the effect is to reduce both pressure loss and efficiency. This is important on some applications involving highly abrasive dusts or high temperature operation, where the cyclone is lined with a refractory material, the surface of which must be maintained in a smooth state.
2. It is reported that with heavy dust concentrations normal cyclone efficiency tends to be constant or even increase, and pressure drop is slightly reduced.

Commercial cyclones

The simple cyclone was found to suffer from a number of defects, including interference between the spinning column of gas and the incoming gas stream. It was also reputed to have undesirable types of eddy current within the body which would reduce the efficiency of dust collection. A considerable amount of work was carried out by various workers, resulting in improved forms of cyclone. Figure 8.9 shows schematically two designs resulting from the work carried out by van Tongeren to establish the optimum cyclone geometry. The gas inlets and outlets are proportioned and arranged to preserve a smooth uninterrupted flow of gas and to avoid undue turbulence and uncontrolled eddying effects which could interfere with the separation of dust from the gas. In addition, in type (a), a by-pass re-entry slot is built immediately before the point where the inlet gases join the cyclone chamber; this has the effect of giving a 'shave off' of dust. In type (b), this vertical by-pass channel is replaced by a volute type 'shave off', leading to a by-pass channel arranged in a helical form around the chamber of the cyclone, so that a proportion of the outer layer of gas, in which much of the dust is concentrated, is sheared off, by-passed through the connecting duct, and re-introduced into the main flow of circulating gases below the inlet where the beneficial downward flow current can take effect.

The grading efficiency curves for such cyclones vary considerably with their shape, e.g., with the proportions of the height of the gas inlet duct to the height of the cyclone chamber and the taper of the cone below it. In Fig. 8.11 is shown diagrammatically a commercial form of a high efficiency cyclone, and the performance curves for two typical cyclones, each with a body of 1 m diameter. The most efficient has a pressure loss of 14 cm water gauge and an efficiency of 90 per cent for particles of 7 μm,

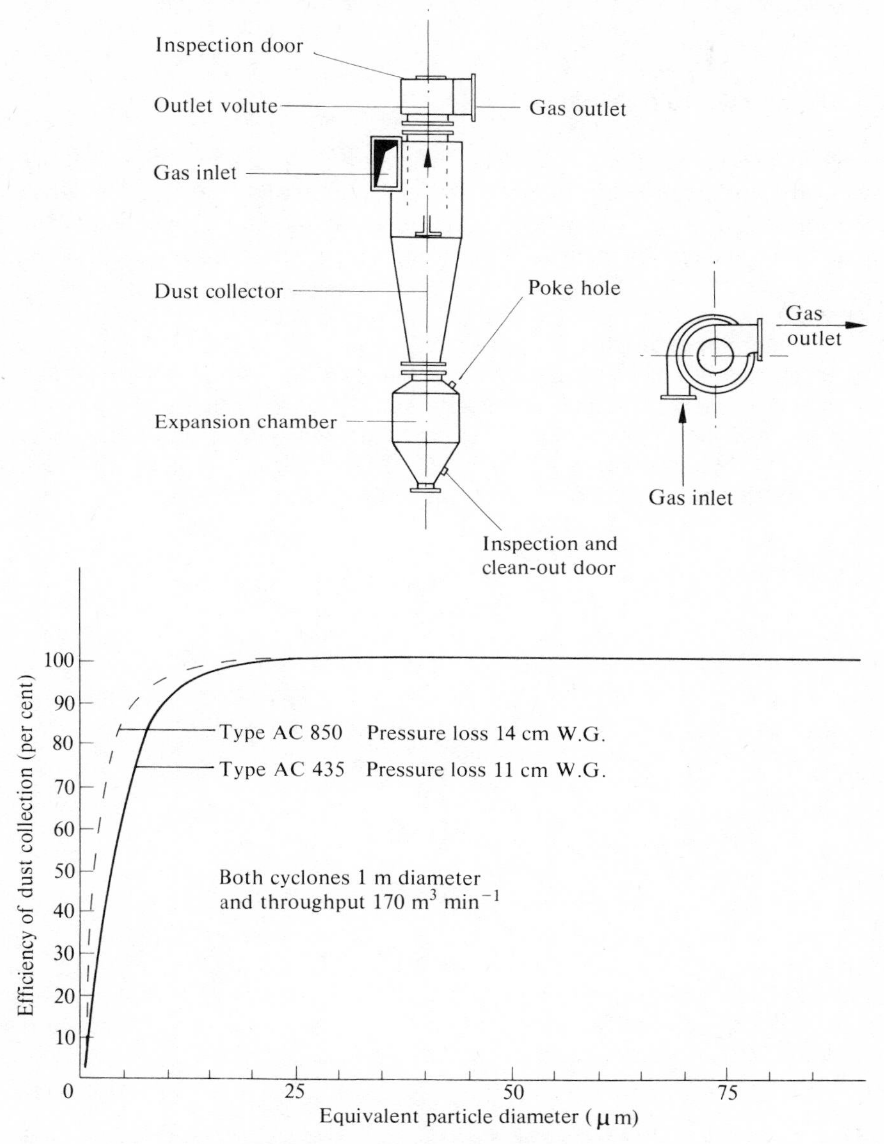

Fig. 8.11. High efficiency cyclones Type AC.

compared with the lower efficiency type which has a pressure drop of 11 cm W.G. and an efficiency of 90 per cent for 10 μm particles. The difference between these two cyclones is in the relative dimensions of the various parts, inlet, body, cone, dust hopper, etc. The reason for having two such cyclones is purely one of economics, the lower efficiency one being cheaper to construct and consuming less fan power because it has a lower pressure loss.

Practical application of the cyclone

The cyclone is probably the simplest high efficiency gas cleaning device available, and consequently also the lowest in capital cost. Temperature of operation is limited only by materials of construction. The most common material is mild steel, with a working limit of about 450°C. Operation at higher temperatures reduces to the question of economics of the use of refractory linings or materials more expensive than mild steel, together with consideration of the effect of temperature on pressure loss, and hence operating costs. The importance of this is illustrated in Fig. 8.12, which shows the electrical energy consumed as fan power to overcome the resistance of the gas cleaning device. The positions of the vertical lines in the left-hand corner at the base of Fig. 8.12 show the drop in water gauge pressure in (a) electrostatic precipitators, (b) cyclones and (c) bag filters, and the positions on the base line of the vertical lines in the middle of the diagram show the range of pressure loss in water gauge with high-energy scrubbers.

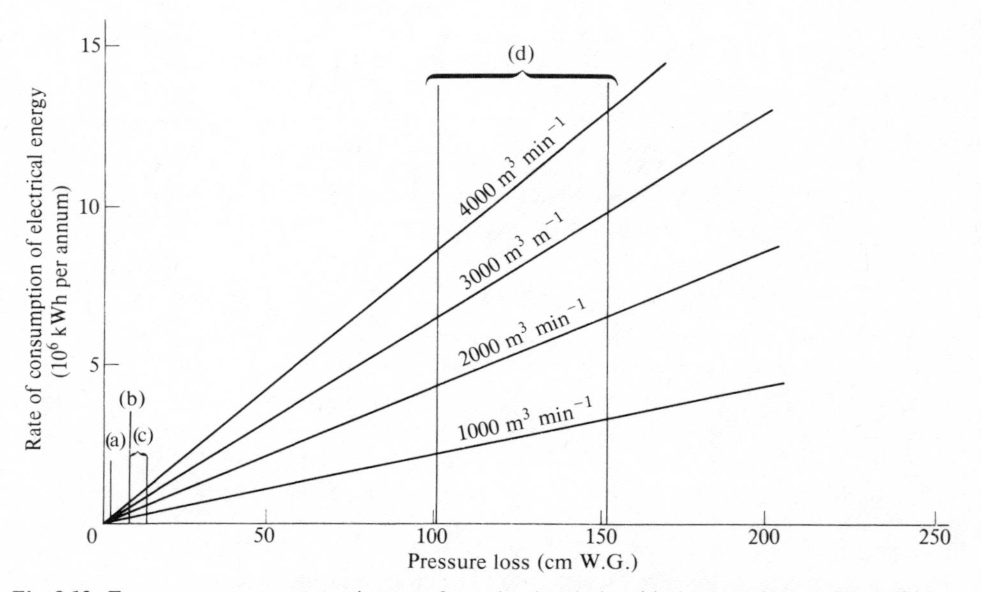

Fig. 8.12. Fan power to overcome resistance of gas cleaning device: (a) electrostatic precipitator; (b) cyclone; (c) bag filter; (d) high-energy scrubber.

Most materials caught in cyclones vary in specific gravity from 2 to 5, and this includes minerals and ores. While cyclones such as those built for the sampling of dust laden gases have efficiencies in excess of 99 per cent for particles of 1 μm, these cyclones are only a few centimetres in diameter with very small chambers. For cleaning industrial gases on a larger scale, cyclones are usually limited to dust predominantly in excess of 10 μm diameter if high efficiencies are required. The reason for this is simply that small cyclones have a very limited capacity, and when connecting large numbers of such cyclones in parallel to handle the gas volumes met in industrial processes, severe problems arise in ensuring equal loading of the cyclones and freedom from blocking, with the result that it is unusual to find commercial cyclones less than 30 cm in diameter. While the problem of equal loading is not as acute as the straight-through cellular-type collector (Fig. 8.7), the problems of blockage still remain, and this again imposes limitations on the minimum diameter of cell used.

Summarizing the important factors of commercial high efficiency cyclones: the balance of operating and capital costs, together with practical problems of blockage and gas distribution, dictate diameters of not less than 30 cm, an average figure being 1 m. Capacity is usually calculated to limit pressure loss to 15 cm W.G. or less. With these limitations, cyclones are not normally used when the dust contains significant proportions less than 10 μm. While operating temperature is virtually unrestricted, the cost of fan power and special materials of construction impose indirect limitations. Subject to the limitation on particle size, liquids can be collected if free flowing, and any type of solid material provided it does not adhere to the walls of the cyclone and promote blockage. Owing to the high velocity of the dust in the proximity of the walls of the cyclones, coarse abrasive dusts can cause acute abrasion problems.

Electrostatic precipitator

The electrostatic precipitator makes use of the fact that electrically charged particles subjected to an electrical field are attracted towards, and deposited on, the electrodes creating the field. The separation of the suspended particles from the gas requires three fundamental steps:

1. inducing an electrical charge on the suspended particles;
2. deposition of the charged particles under the influence of the electric field on the collecting electrodes;
3. transfer of the collected material from the electrodes to a storage hopper from which it may be withdrawn continually or at intervals.

Charging of the particles is achieved by passing the dust-laden gases through an electrode system designed to favour the production of corona discharge when energized by a suitable high voltage supply, which is usually unidirectional. Charging of the particles to nearly the maximum possible level takes a very short time, less than one-tenth of a second, compared with the time of several seconds during which the gas is subjected to the electric field. Once the suspended particles are charged, they pass through the electrode system and experience forces due to the electrical field. They therefore move towards and are eventually deposited on the electrodes of opposite charge to their own, which is generally that of the corona discharge.

In the case of liquid droplets, coalescence occurs, and the liquid film drains away from the electrodes. Where the material deposited is dry dust, a layer is allowed to build up which is periodically dislodged by hammering or vibrating the electrodes. From the electrodes it falls into storage hoppers below the moving gas stream, from which it is removed for disposal. Forces due to the electrical charge and molecular attraction ensure that the dislodged dust remains in agglomerated masses with high free falling velocity, sufficient to ensure that they reach the hoppers. If this did not occur the dust would be re-entrained in the gases and not caught.

There are basically different electrode systems used to achieve the process of particle charge and deposition. These are referred to as the *two-stage* and *single-stage* designs. The two-stage design shown schematically in Fig. 8.13(a) consists of a charging stage in which corona discharge is used to charge the particles. Following this there is a non-corona section where the electrical field is used to deposit the charged particles on the electrodes of opposite charge, usually connected to earth. The corona electrodes are usually fine wires, the diameter of which determines the corona current while the other electrodes are in the form of flat, non-discharging plates. This system has the merit that current in any quantity is only consumed in the corona stage, and as a

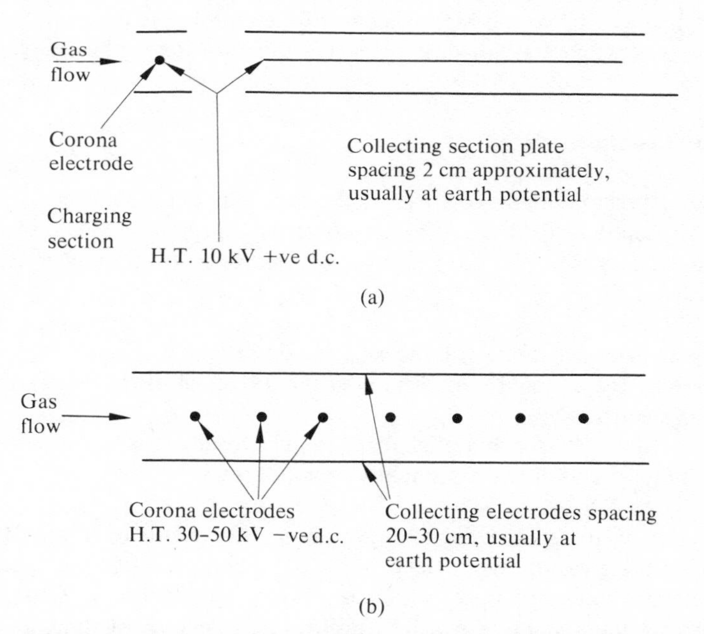

Fig. 8.13. Two-stage and single-stage precipitator. Arrangement of electrode systems: (a) two-stage; (b) single-stage.

result it has a relatively lower power consumption. In practice, application is limited to low dust concentrations, usually less than 100 mg m^{-3}. It finds particular use for air-conditioning systems and for cleaning air for clean rooms. Electrode spacing for this type of unit is only about 2 cm, which has the advantage of reducing the high tension supply needed to a level of about 10 kV positive d.c. Positive polarity is used where the gas being cleaned is air which is subsequently being used for ventilation, as positive corona, although giving somewhat lower efficiencies of gas cleaning than the negative corona, also produces very much less ozone. This is highly toxic in nature, and it is essential that the generation is strictly controlled in air-conditioning systems.

The second type of precipitator, the single-stage design, is the one most commonly used for large-scale industrial applications, and is capable of handling very high dust concentrations, up to several hundred mg m^{-3}, under a wide variety of operating conditions. In the single-stage design shown schematically in Fig. 8.13(b), the corona electrodes also act as the field electrodes, and deposition of the charged dust occurs on the earthed electrodes opposite to the corona electrode system. Figure 8.13(b) shows the plate form of precipitator which is the type most commonly used today in industry. It consists of vertical parallel plates with discharge electrodes hanging vertically midway between the plates. The plates are spaced much wider than in the two-stage system, usually 20–30 cm, and the voltage applied is in the region of 30–50 kV negative polarity d.c. Since the gases from this type of precipitator are discharged to atmosphere, the presence of traces of ozone is not very important, since this is diluted when discharged from the stack, and the ozone decomposes in a relatively short time. Hence the use of negative polarity corona discharge is acceptable. This gives a much higher working voltage, and hence higher (about 50 per cent more) electric field strength between the electrodes than for positive polarity. This has an important influence on the dust removal efficiency of the precipitator and on the economics of large plant.

In addition to the plate design, there is also the tubular design, which consists of vertical tubes acting as the receiving electrodes, with the corona electrodes arranged on the centres of the tubes. The tubes are either hexagonal or circular in form. The hexagonal form has the advantage of approximating closely enough to the circular form for the electrostatic field. It enables both sides of the tube to be used as receiving electrodes, thus resulting in an economy in the use of metal. The tubular design was the earliest used, but has been superseded by the plate type, except for certain smaller applications such as the de-tarring of coke oven gas and removal of acid mists on sulphuric acid plant. The plate design lends itself much more readily to the construction of multi-stage high efficiency precipitators capable of handling very high gas volumes.

Electrostatic precipitation process

The electrostatic precipitation process has been applied industrially for well over half a century. During this period many attempts have been made to obtain a satisfactory relationship between the motion of the charged particles in the electric field, calculated from purely theoretical equations, and the actual collection efficiency of the precipitator. A résumé of the work carried out, together with much fuller information than is possible here, is given by Rose and Wood,[4] and by H. J. White.[5] Although it is not intended to go deeply into the theory of precipitation, some reference to the basic theoretical equations is included to give a better understanding of the process. These equations are given in Table 8.1; Eqs. (1) and (2) give the forces acting on the charged

TABLE 8.1

Electrostatic precipitation basic formulae

Force acting on a charged particle in an electric field

$$F = qE_\rho, \tag{1}$$

where F = force
q = charge on particle
E_ρ = field strength

if q is limiting charge, then

$$q = \rho E_c a^2.$$

$$F = \rho E_c E_\rho a^2, \tag{2}$$

where a = particle radius
E_c = corona field strength
ρ varies with properties of particle and is 3 for conductors and varies from 1·5 to 3·0 for particles of dielectric constant for average insulating materials

Motion of a particle in a resisting medium—Stokes' Law

$$F = 6\pi d\mu\omega \tag{3}$$

where μ = gas viscosity
ω = velocity of particle referred to in electrostatic precipitation as migration velocity.

Hence

$$\omega = \frac{\rho E_c E_\rho a}{6\pi\mu}. \tag{4}$$

In a single stage precipitator $E_c = E_\rho$ very approximately

Hence

$$\omega = \frac{\rho E^2 a}{6\pi\mu} \tag{5}$$

Efficiency of dust collection of an electrostatic precipitator

$$\epsilon = 1 - e^{-\lambda}, \tag{6}$$

where ϵ = collection efficiency
e = exponential constant
λ is dependent on various factors related to precipitator design and dust physical properties.

Deutsch law

$$\epsilon = 1 - e^{-AS/Q}, \tag{7}$$

where A = area of receiving electrodes
Q = gas flow rate
S = effective migration velocity

particle, while Eq. (3) is Stokes' law, which gives the motion of a particle when a force is imposed on it, the motion being opposed by the viscous drag of the gas.

From these equations is derived Eq. (4), which gives the resultant velocity of the charged particle through the gas, this velocity being usually referred to as the migration velocity. Equation (5) indicates that the following effects might be expected:

1. Migration velocity is directly related to (electric field strength).^{2} This indicates that operating at the maximum electrode voltage without excessive flash over is important. This prediction is in line with practice and illustrations will be given later in the text.
2. Migration velocity decreases directly as the diameter of the particle decreases. This implies that fine fume particles could not be collected efficiently by an electrostatic precipitator. This is not found to be true in practice; the precipitator collects sub-micron fume almost as readily as the larger particles. An explanation will be given later in the text.
3. Migration velocity decreases as gas viscosity increases. This implies that better gas cleaning efficiencies are likely to be obtained if the process can be operated with the gas at a relatively low temperature. This is due to the fact that gas viscosity decreases with decreasing temperature (Fig. 8.2). Evidence indicates that practice is in line with theory on this point.
4. Migration velocity is related to a factor dependent on the electrical properties of the suspended materials. The true effect of this variable has not been investigated, but it is unlikely to be of great importance, since most materials caught in an electrostatic precipitator have fairly high electrical resistivities and behave in a similar fashion.

In practice, it is not possible to produce mathematical equations to enable the efficiency of an electrostatic precipitator to be calculated from first principles. Some of the reasons for this are as follows:

1. Difficulty in measuring particle size accurately as mentioned earlier, particularly when non-spherical.
2. Calculation of electric field strength for the single stage type presents great problems, and no satisfactory solution has been produced to date.
3. Probably most important is that the gas flow in a commercial precipitator is not laminar. Precipitators operate at relatively high values of Reynold's number and the flow is fully turbulent. In addition, further disturbance is created by the electrode system, both as a result of its physical disturbance of flow and the effect of electric wind caused by the corona discharge.

Statements (1) and (2) mean that the force acting on the suspended particle and the subsequent motion cannot be calculated accurately. Statement (3) means that the trajectory of the charged particle is not a straight line, contrary to the prediction of Table 8.1, Eq. (5). In practice the migration velocity due to the electric field is small compared with that due to the motion of the gas. The dust particle, as a result, travels a very tortuous path, and deposition on the collecting electrode depends on whether it approaches the electrode closely enough to be retained. The dust particle may, in fact, approach the electrode closely several times before being caught, as the gases progress through the precipitator. Taking this effect into account, it is not surprising that the effect of particle size on collection efficiency is much less pronounced than would be expected from Table 8.1, Eq. (5).

Since it is not possible to calculate the dust removal efficiency from purely theoretical considerations, the designer must rely on accumulated experience of industrial processes and data obtained from precipitators operating on those processes. It is useful to be able to derive from such data the design parameters of plant operating on a similar process but with different requirements; for example, a different efficiency of dust collection. It has been established that the efficiency of an electrostatic precipitator obeys an exponential law (Table 8.1, Eq. (6)). A particular form of this equation much used in the electrostatic precipitation industry is that due to Deutsch (Eq. (7)) which gives efficiency in terms of the following:

1. Effective migration velocity, not to be confused with migration velocity used in Eqs. (1–5). Effective migration velocity is calculated from actual plant performance data using the Deutsch law.
2. The ratio of total area of collecting electrodes, which is a direct measure of the size of the precipitator, to the gas flow rate.

By increasing the size of the precipitator, that is by increasing the area of collecting electrodes relative to the gas volume to be treated, the efficiency can be made to approach 100 per cent, but in practice can never reach it. The significance of this will be discussed later. This formula, while useful, has limitations which must be clearly understood, otherwise serious errors can be introduced into calculations. It can, in fact, be applied only to precipitators with identical electrode configurations, and even then other variables such as gas velocity through the electrode system, particle sizing and many design features of the plant must also be taken into account. The magnitude of the problem is illustrated by Table 8.2, which lists 27 variables known to affect the gas cleaning efficiency of the precipitator and hence the effective migration velocity calculated from the Deutsch formula. Apart from properties of dust and gas which are related to the process to which the precipitator is attached, purely practical considerations such as the correct operation of the electrode cleaning system, and the effective use of the whole electric system by ensuring that the gas flow is uniform over the whole cross section of the electrode system, can result in large variations in calculated effective migration velocity obtained from plant tests. It is therefore unsatisfactory to use any test results for design purposes without the fullest knowledge of the design of the precipitator and the conditions under which the test took place. These limitations are discussed in some detail in a paper by Darby.[6]

Consider now the effect of some of the more important variables under various headings given in Table 8.2.

TABLE 8.2
Factors influencing precipitator performance

(a) *The gas*
1. Density
2. Composition
3. Temperature
4. Viscosity
5. Velocity in field and distribution of velocity in field

(b) *Dust and fume particles*
6. Size
7. Composition
8. Shape
9. Concentration
10. Electrical resistivity
11. Specific gravity
12. Surface properties

(c) *Receiving and discharge electrodes*
13. Profile of receiving electrodes
14. Profile of discharge electrodes
15. Spacing of receiving electrodes
16. Spacing of discharge electrodes
17. Deposits of dust on receiving electrodes
18. Deposits of dust on discharge electrodes
19. Alignment of electrode system
20. Accuracy of manufacture of electrode system
21. Re-entrainment of dust caused by mechanism used to dislodge accumulations on receiving electrodes
22. Mechanism used to dislodge accumulations on discharge electrodes

(d) *Electrical*
23. Field strength
24. Corona current
25. Size of precipitator sub-section energized by each H.T. rectifier unit
26. Polarity of d.c.
27. Design of H.T. rectifier

1. *Properties of the gases.* Assuming that the gas is uniformly distributed in the electrode system, the physical properties of the gases can affect efficiency in two ways.

Gas composition affects breakdown voltage[7] which, in turn, affects electric field strength and precipitator efficiency (Table 8.1, Eq. (5)). Gas viscosity also affects particle mobility and hence precipitator efficiency. Thus, while the breakdown voltage of the gas is little affected by temperatures up to 450°C, the influence of gas viscosity, which increases considerably with temperature, indicates that the lower temperature condition would produce greater mobility of the dust particles, with the possibility of higher plant efficiency and higher effective migration velocities. Since for the same mass flow, gas volume would also be lower at the lower temperature, the combined effect considerably reduces the size of precipitator needed.

2. *Dust and fume particles*. The ability of an electrostatic precipitator to collect the finest fume particles, in contradiction to the prediction of the simple theory, is one of the reasons for its wide industrial application since otherwise it would have little advantage over the cyclone. Effective migration velocities calculated from the actual plant tests indicate that the migration velocity does decrease with decreasing particle size, but the variations are only in the order of 2–1 over the range 5 to 0·01 microns. This is in contrast to Table 8.1, Eq. 5, which implies a ratio of 500 : 1 might be expected.

The other property of the dust of great significance is electrical resistivity. While dust of any resistivity can be deposited on the collecting electrode, it is the effect of the deposited layer on the subsequent behaviour of the process which is important. If electrical resistivity is very low, as with carbon black, for instance, the dust particles lose their electrical charge immediately, and on acquiring the same charge as the earth plate are repelled back into the gas stream. In the case of liquid droplets, which are often of low resistivity, since these wet the electrode the problem of re-entrainment does not arise. Where the resistivity is higher, the dust particles are held in contact with the receiving electrode until dislodged by the shaking mechanism. As resistivity increases, however, the efficiency of the precipitator can be reduced due to the effect of the highly resistive layer of dust on the electrodes on the corona discharge.

Dusts that are poor conductors electrically, which includes nearly all industrial fume, have dust temperature–resistance characteristics like those shown in Fig. 8.14. This is for cement, but most dusts show a similar relationship, differing in absolute resistivity and in the temperature at which the peak occurs. In the absolutely dry state, with no water vapour in the gas, resistivity decreases with increasing temperature. This condition rarely exists in practice, due to the presence of moisture resulting from the fuel or from the original raw material. Moisture has the effect of modifying the curve such that a peak value is shown for temperatures in the region of 150–200°C, the resistivity corresponding to the peak varying with the water contained in the gases. The reason for this is that at temperatures below that corresponding to the peak, the electrical resistivity of the particle is determined by an adsorbed surface film of water vapour. At higher temperatures than the peak, the particle itself becomes increasingly conductive and the presence of water in the gas correspondingly less important.

Apart from water vapour, other gases such as sulphur trioxide and ammonia, present only to the extent of a few parts per million, can exert strong influence on the electrical resistivity. Where resistivity is relatively low, that is less than 10^{10} ohm cm, the layer deposited on the receiving electrode has negligible effect on precipitator efficiency. On the other hand, it has been shown[8] that when resistivity is in the region of 10^{12}–10^{13} ohm cm, a reduction in resistivity from 10^{12}–10^{13} ohm cm resulted in effective migration velocity being doubled. The significance of this is that (Table 8.1, Eq. (7)) the size of the precipitator to give the same efficiency with the same gas volume would be halved.

It might be expected that the precipitator efficiency could be improved to an equal degree by working at a different temperature on the curve from that giving peak resistivity, in order to reduce resistivity to a more acceptable level. In practice, for

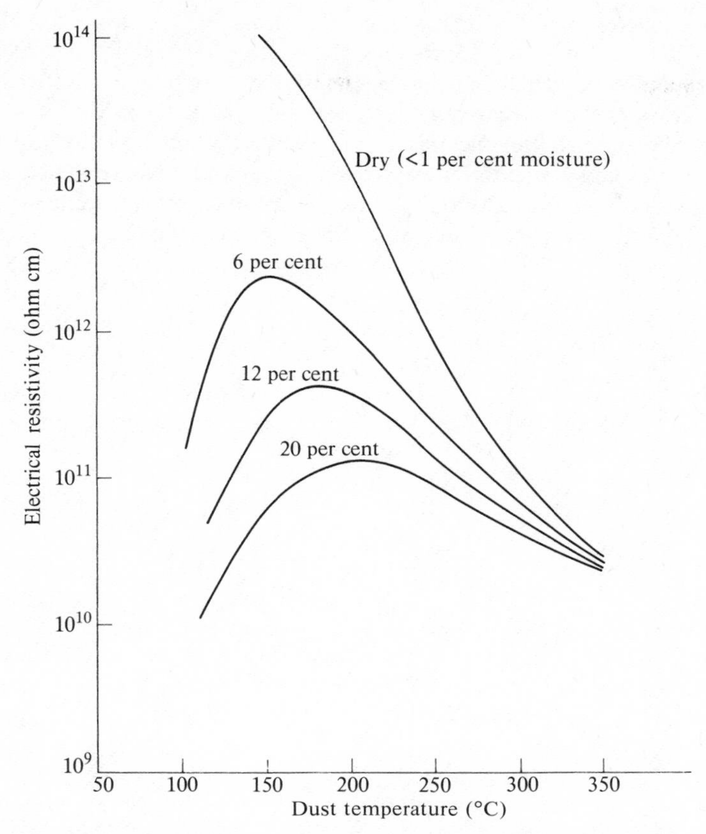

Fig. 8.14. Effect of moisture content of entraining gas and temperature on cement dust resistivity.

corresponding values of resistivity, at the high temperature side of the curve the effective migration velocities are generally lower than those obtained on the lower temperature side, where the better agglomerating properties of the dust help to reduce re-entrainment and the lower viscosity of the gas increases particle mobility and hence collection efficiency.

Electrical operating conditions

As indicated by the theoretical equations, the gas cleaning efficiency for both single and two-stage precipitators is directly influenced by the electrical field strength, which is proportional to the voltage applied between the corona and collection electrodes. For this reason, on industrial gas cleaning plant in which the presence of traces of ozone is not important negative corona is used, as the voltage which can be applied without arc-over is roughly 50 per cent greater than for positive corona.

A practical illustration of the importance of maintaining the electrode voltage as high as possible is shown in Fig. 8.15(a). This shows total current applied plotted against relative dust emission, which is a measure of the precipitator inefficiency. As the current increases the dust emission decreases only as long as the corona electrode voltage continues to rise. Beyond this point, despite the increase in total current, dust emission increases and precipitation *inefficiency* increases. It will be seen also that as the dust emission increases from 20 to 45 units, the corona electrode voltage falls from 48 to 45 kV. This clearly shows the importance of operating at maximum discharge

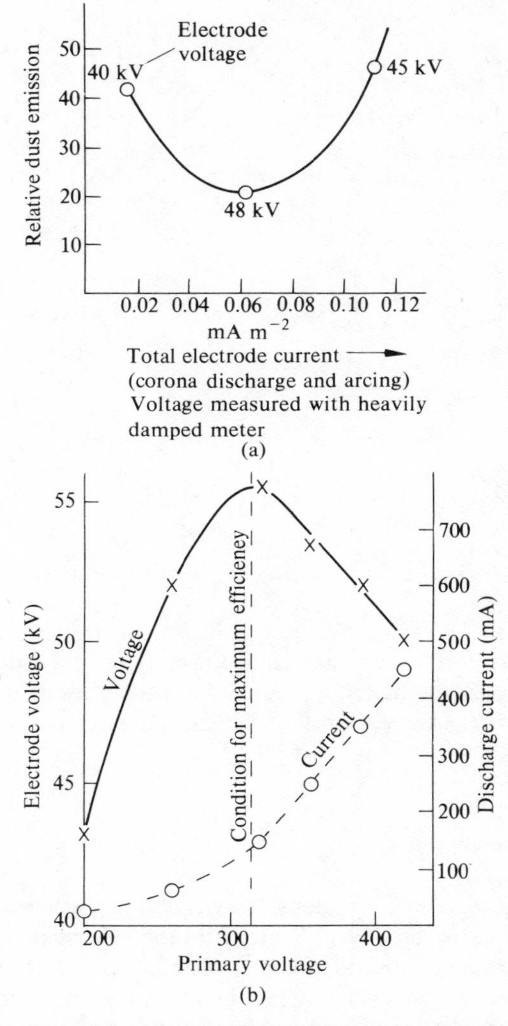

Fig. 8.15. Electrical characteristics of an electrostatic precipitator: (a) variation of dust emission with applied voltage and current; (b) selenium rectifier and transductor control.

electrode voltage at all times. Since this voltage varies with process conditions and does not correspond to a fixed current value, an automatic system to control the high tension supply to the precipitator should ideally operate to seek the maximum corona electrode voltage at all times. In order to understand the characteristic curve in Fig. 8.15(b) it should be noted that the current axis refers to *total* current, this being the sum of corona current and that due to arc discharge. The reason for this is simply that in large commercial precipitators arc-over takes place at voltages well below the best operating condition. Each arc represents a loss of electrical field for a brief instant, the voltage during this period falling to 50 per cent or less, or even to zero. The extent to which arcing can be tolerated without loss of dust collection efficiency is determined by the magnitude and frequency of the arcs.

Figure 8.16 illustrates this effect diagrammatically. As the corona voltage, measured instantaneously, is increased, the frequency of arc-over is also increased. At first, despite the loss of field when arc-over takes place, the *mean* corona voltage also increases, but eventually losses due to the arc-over start to reduce this mean voltage. At this point efficiency of dust collection starts reducing, and experience has shown this is the best operating point for the precipitator. In some automatic control systems the control points are set by selecting a pre-set value for the rate of flashover, which may or may not correspond to the peak mean corona voltage. It is good policy to fit high-voltage meters to measure directly the voltage at the discharge electrode, as this can be used to check the operation of the automatic control system.

For the reasons given, the high-voltage rectifier feeding the precipitator must be designed to give minimum loss of time of electric field during any arc-over. In practice it

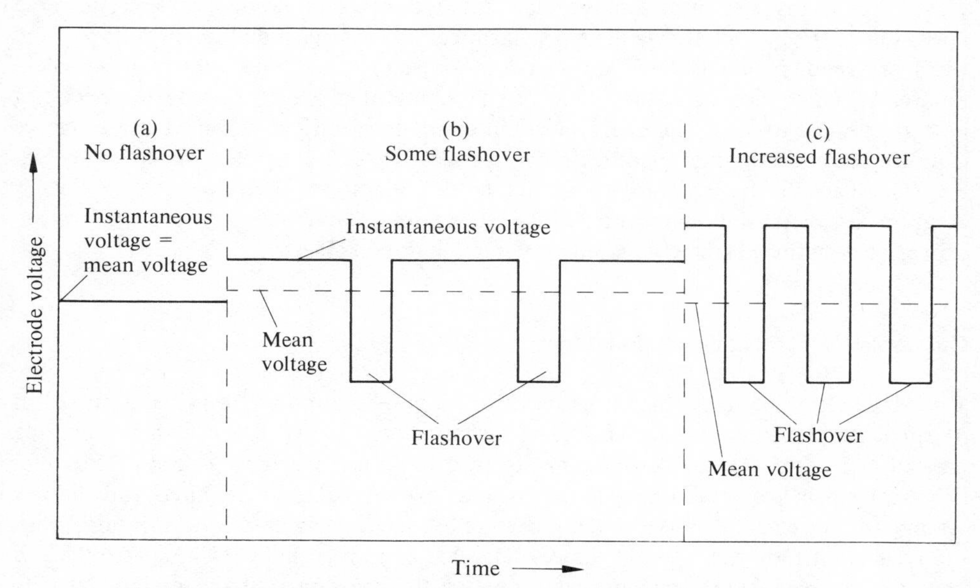

Fig. 8.16. Effect of flashover in precipitator on mean discharge electrode voltage.

is easy to give rapid arc extinction; the greater problem is in reducing the time needed to restore the voltage to full value. This is related to the design and type of high-voltage rectifier unit. Variations in the recovery time of the voltage can vary from a few cycles up to a considerable fraction of a second. Unless the recovery time is rapid, less arc-over can be tolerated and the precipitator must be operated below the optimum level.

Electrode system

The design of the collecting electrode is dictated by a number of requirements. Although for a uniform field, for instance, a simple flat sheet is adequate, the sheet must be flat within quite close tolerances, as any deviation from flatness reduces the design gap between electrodes and hence reduces the arc-over voltage. In full-scale precipitators, electrode spacing is in the range 10–15 cm. Some measure of the problem is given when it is realized that for large gas volumes, in excess of 30 000 m^3 min^{-1}, receiving electrodes can be 13 m high and 5 m long. Another requirement is to create a dead space in front of the collecting surface to reduce the re-entrainment of dust from the deposited layer by the gas flow. Two designs of collecting electrode which satisfy these requirements are shown in Fig. 8.17(c) and (d). The latter consists of vertical rolled channels with stiffened edges, a type of construction used by most manufacturers, although there are many different profiles which meet the basic requirements.

The corona electrode must be either in the form of relatively fine wires (2–3 mm diameter), or have sharp points to give the necessary conditions for corona discharge. When fine wires are used, these are often held in tension by weights at the lower end. This system has the disadvantage that the weights can swing, causing unstable operation of the precipitator with heavy flashover and serious reduction in voltage. Fine wires are readily damaged by arc-over and subject to a serious rate of mechanical failure. When a wire fails, the section of precipitator in which this occurs becomes inoperative, due to the broken wire short circuiting the electrical system. The alternative is to use larger sections of metal with sharp edges. Typical examples are shown in Fig. 8.17(e). Alternatively, barbed wire is often used. The semi-rigid electrodes shown in the diagram are usually mounted on tubular supporting frames. Such a system is more expensive than the weighted wire but very much more reliable.

Commercial electrostatic precipitators

Two-stage electrostatic precipitators used for air conditioning are designed in the form of small standard modules based on the format shown in Fig. 8.13(a), with a filtering area of 0·5–1 m^2 and a filtering capacity of a few cubic metres per second. Modules are arranged in series and parallel, according to the efficiency required and the gas volume to be treated. These filters are used only for very low dust concentrations, with the result that they can operate for considerable periods, usually 24 h, before there is any need to clean the electrode systems. For this type of unit the method usually

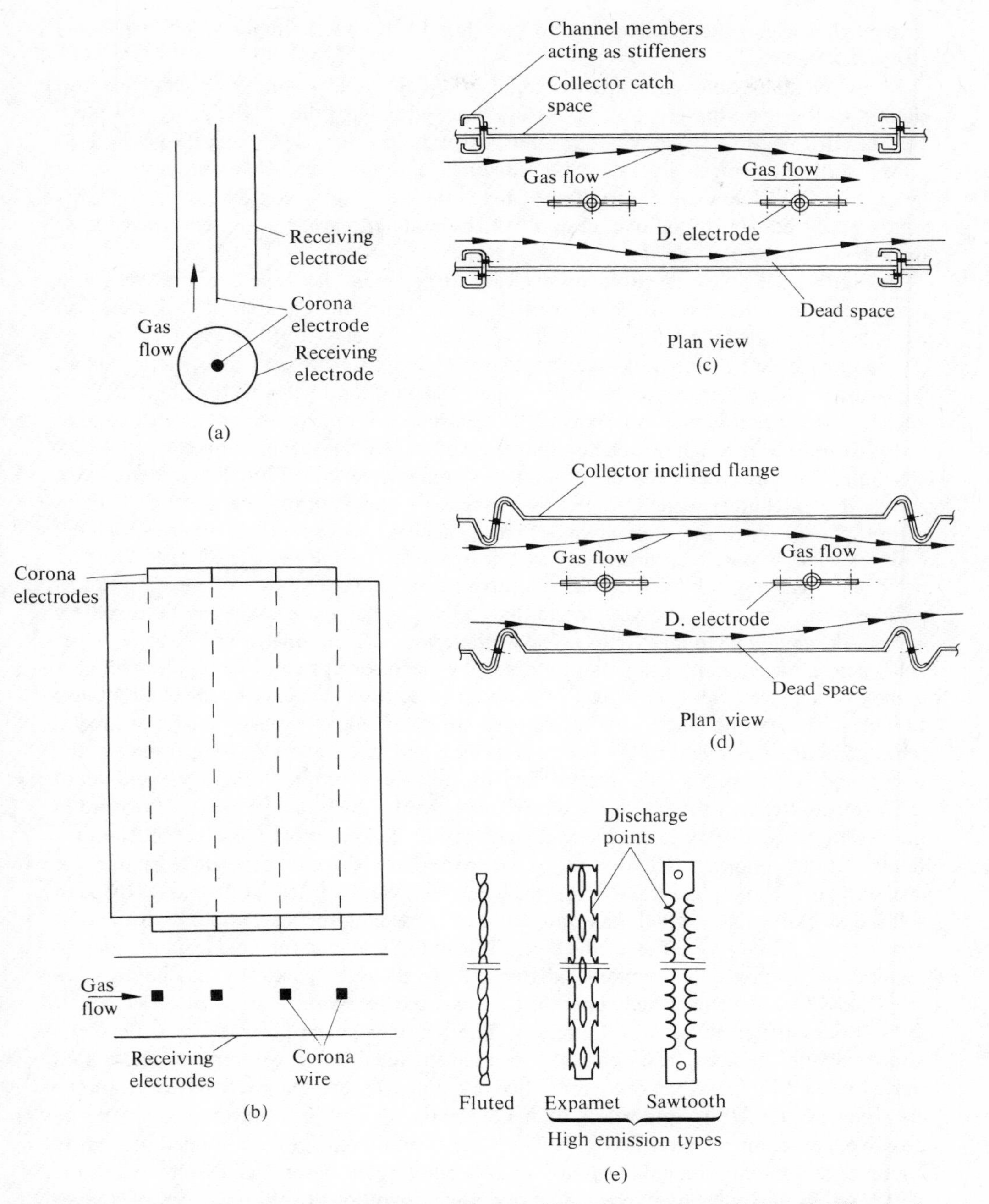

Fig. 8.17. Single-stage precipitators: (a) single-stage tube; (b) single-stage plate; (c) channel stiffened collecting electrode; (d) inclined flange; (e) types of discharge electrode.

employed to clean the electrodes is to switch off the power supply and to wash the electrodes clean.

While the single-stage precipitator can also be built in very small sizes and spacings similar to the two-stage type, it is usually associated with the cleaning of very large volumes of gas, often millions of cubic metres per hour, such as cleaning the gases from power stations, cement works, metal refineries and other industrial processes on a large scale. The following description of the plate precipitator contains many remarks which apply equally to tubular design, which as stated earlier has been replaced for most duties by the plate design.

In Fig. 8.18 is shown a cut-away drawing of a typical dry type precipitator, i.e., a precipitator in which the dust is deposited in a dry state and transferred to the hoppers by rapping or vibrating the electrodes. At the inlet end of the casing is the gas distribution system, consisting of triangulated splitters in this particular arrangement, which turns the gas from the vertical to the horizontal and helps to distribute the gas velocity evenly across the full area of the electrodes. From the triangular distributors the gas is passed through orifice distributors, which have a final smoothing effect on the gas flow. The design of such distribution systems is determined by flow models using standard wind tunnel techniques, to ensure that the gas entering the electrical field is within 25 per cent of the average value. This is generally accepted as being the practical limit of what is possible, and is a vast improvement on the older systems, including perforated plates which were assumed to give good control of flow. When wind tunnel tests were first carried out it was found that the perforated plate could give reverse flow in considerable percentages of the field with serious effects on the efficiency of dust collection. Downstream of the distributors is the electrode system. The type shown here is the catch space (Fig. 8.17(c)) with the corona electrode elements attached to tubular supports. The corona electrodes in this case are mounted on either side of stout metal tubes going the full height of the electrode system and attached to locating frames at the upper and lower ends. This system has the advantage that there is the minimum mechanical stress on the discharge electrode element, which is an important feature in the design, since discharge electrode failures are by far the most common precipitator faults. In the diagram the discharge electrodes are shown suspended from pillar insulators in a room placed above the gas chamber. The suspension tube passes through a fused silica insulator which has only the duty of sealing the gap, and makes the best possible use of the properties of the silica. For many applications, this insulator design requires no heating and functions satisfactorily in the dirty gases. In the chamber are shown the mechanisms which operate to clean the receiving and corona electrodes. These are in the form of drop hammers. The hammer heads protrude into the space above the gas chamber, in which is the mechanism for lifting the hammers. These pass through seals in the floor, and are lifted by cams at pre-set intervals to clean the electrode system. It is important to maintain the discharge electrodes as clean as possible, and comparatively little dust is collected on them. They are rapped at a fairly high rate and re-entrainment is small. The collecting electrodes, which catch most of the dust, are not very sensitive under most operating conditions to the existence of a layer of dust on the surface. On the other hand, due to the quantity of dust involved, excessive

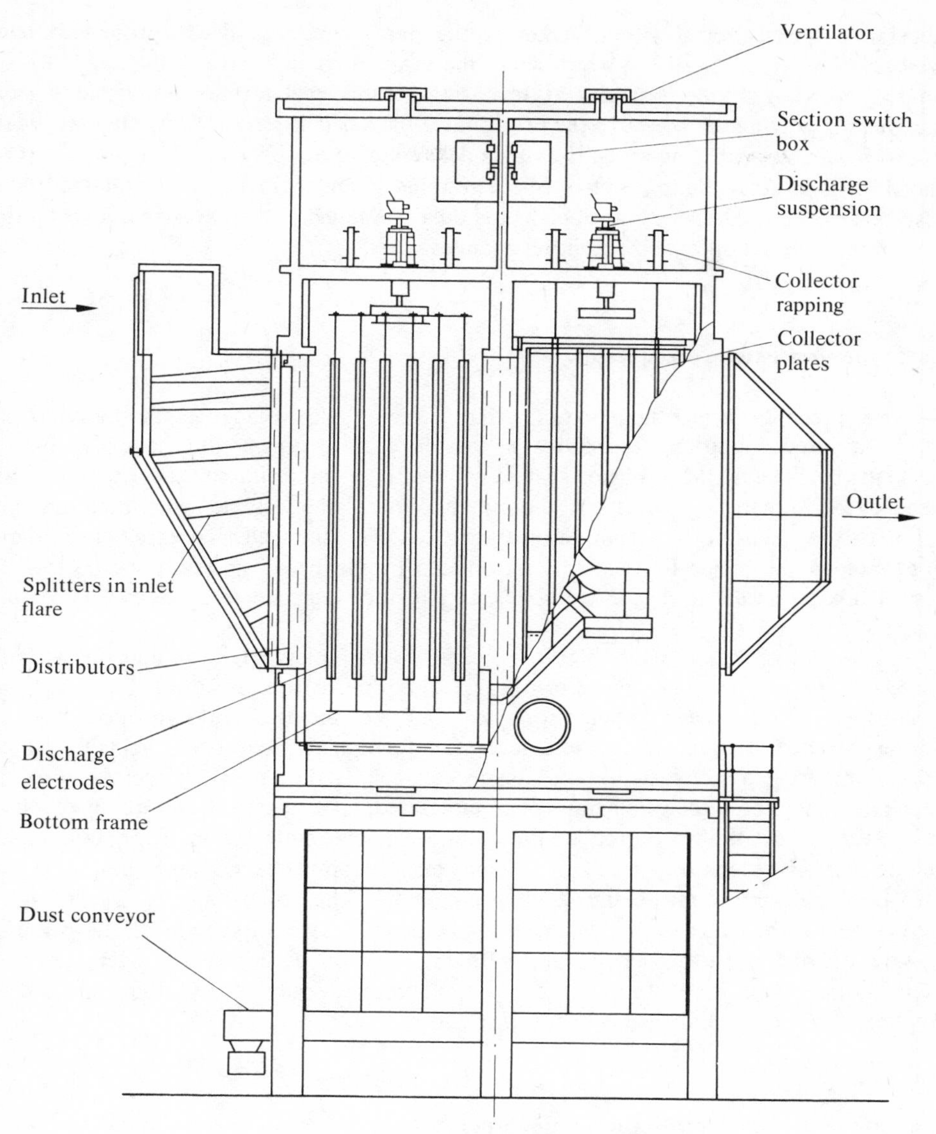

Fig. 8.18. Typical dry process electrofilter.

rapping can reduce precipitator efficiency by re-entraining dust. The frequency and intensity, therefore, of the collecting electrode rapping is carefully set to maintain the plant in a reasonably clean state without reducing the efficiency unduly by re-entrainment. The arrangement of the rapping mechanism in the room above the gas chamber has the advantage that the moving parts of the mechanism are in clean air and accessible for servicing without entering the gas chamber and are protected from weather. The presence of the warm air in the rapping room has the advantage also of limiting corrosion in the roof area of the casing. At this level are shown the high tension rectifiers supplying the power to the precipitator banks. The design shown has two fields in series with separate silicon rectifier units delivering up to 60 kV d.c. By putting two or more fields in series better control of the rapping re-entrainment loss is obtained, and also better electrical operation, since as the dust contained in the gases decreases so the optimum electrical operating condition changes.

Wet type electrostatic precipitator

The wet precipitator has the advantage that it treats the dust-laden gases in a relatively cold saturated condition. Resistivity is, therefore, no problem and since the dust is removed by flushing the electrode system with water or any suitable fluid, rapping is not needed and re-entrainment is not a problem. Where sticky or adhesive materials are involved the electrodes are continuously irrigated. This naturally increases the demand for water which, since the water is introduced by means of sprays, must be low by industrial standards in suspended solids and further increases the complexity of the water system.

The wet precipitator, while highly efficient on the finest of fume and on difficult materials, can be subject to corrosion problems due to the presence of mixtures of corrosive acid gases such as sulphur oxides, oxides of nitrogen and hydrogen chloride. In the presence of liquid water, these acid gases form dilute mixtures of acids for which special materials of construction may be needed, e.g., lead or stainless steel. Since in the interests of economy the water must be re-circulated, an expensive water treatment plant is needed to cool the water, reduce the amount of solid materials in suspension which might otherwise block the pipes and flushing sprays, control the dissolved materials at a suitable level and control the acidity–alkalinity. This often means discarding a percentage of the water to prevent build up. If the waste water has to be discharged into sewers or rivers, further chemical treatment may be necessary to meet pollution regulations. Wet precipitators are used, therefore, only on certain specialized applications for which the extra cost is justified.

Special features of electrostatic precipitators

The electrostatic precipitator is probably the most versatile of all filtering systems. The following are some of the important features:

1. All sizes of particle down to below 0·01 μm can be removed from the gases.
2. There is a very wide range of operating temperatures but usually restricted to a maximum of 450°C to enable mild steel to be used in the construction.

3. Can be designed for almost any corrosive condition by a suitable choice of materials. For example, electrodes have been constructed from all common metal alloys and even more sophisticated materials such as titanium and silver. Collecting electrodes have been constructed in addition in mild steel, from metal alloys of all types and non-metals such as rubber, PVC and wood.
4. Any type of suspended material can be removed by a precipitator. If the dust is in a dry free-flowing form, then the dry precipitator is used; when it is liquid in the form of a mist, the wet precipitator can produce a satisfactory solution to the problem.
5. Any efficiency required can be obtained, but reference to Fig. 8.19 and consideration of Table 8.1, Eq. (7) makes clear the need to design for an efficiency no higher than strictly required. It will be seen from Fig. 8.19 that to increase efficiency from 90 to 99 per cent requires the size of the precipitator to be doubled. Similarly, to increase from 99 to 99·99 per cent, the plant size is again required to be doubled, although the additional dust being caught by the increase in size, which would roughly double the capital cost of the plant, would be less than 1 per cent.

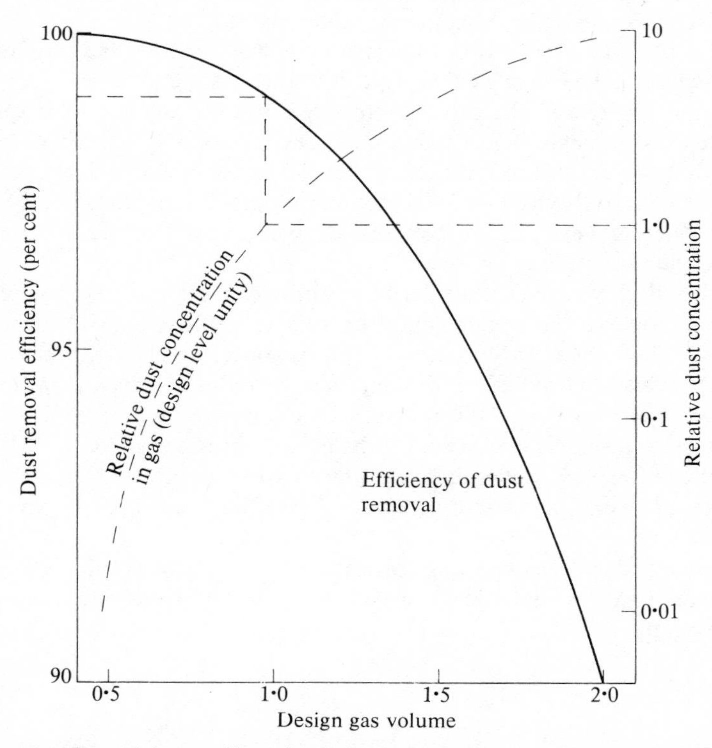

Fig. 8.19. Electrostatic precipitator: effect of variation in gas volume on efficiency and dust concentration in gas.

Filtration

Filtration is the oldest method used for the removal of suspended materials in gases, on the principle of passing the dust-laden gases through porous filter media in which the dust is trapped. The filter media can broadly be considered under the following headings:

1. gravel or sand aggregate bed with or without on-line cleaning;
2. porous paper and fibrous mats, usually 'fixed filters';
3. woven and felted fabric filters, usually with 'on-line cleaning'.

All of the filter media mentioned collect the dust by a combination of different effects, which are briefly as follows:

1. Particles are separated due to the sieving effect of the filter media; any filter medium has a particular pore size and all particles larger than this are separated.
2. Particles are separated by the inertial effect. Their passage through the filter medium is usually a very tortuous one and during the many changes of direction the dust particles are deposited on the material of the filter.
3. There is often an electrostatic charge on the dust particles and the finer particles in particular may be deposited as a result of this charge.
4. Very fine particles in the sub-micron range, which have a similar weight to the molecules of the carrier gas follow the stream lines of the gas flow and are not separated by inertial effects. Such small particles are subject to Brownian movement, and as a result of this motion are finally brought into close proximity to the filter material, where they are deposited and held by a combination of electrostatic and molecular forces.
5. Particles of size smaller than the holes through the filter are retained with high efficiency due to the accumulation of a layer of dust on and within the filter material. Once the layer is established, filtration takes place through the layer of deposited dust, but initially efficiency may be reduced until the layer is formed. This layer also increases the pressure loss compared with the clean filter. It is necessary to take this into account when the plant is designed, as this has a most important influence on the filter rate which can be used. The exposed area of filter is arranged so that the pressure drop across the filter bed is 10–15 cm W.G.

Filtration methods of gas cleaning form one of the largest families of gas cleaning devices, and are used for very widely varying conditions and appliances. These will now be reviewed briefly.

Practical forms of filtration equipment

Aggregate beds

Aggregate bed filters consist of uniform size particles of mineral materials such as sand or gravel and have been used for a variety of applications, although they are not currently in wide use for industrial gas cleaning. The efficiency of gas cleaning increases with decreasing size of aggregate and increases with depth of bed. One recent application uses beds of fine sand 2 m deep to filter fine particles of radio-active material from exhaust gases. Such a filter was found to be over 99 per cent efficient even on sub-micron particles.

Various arrangements of aggregate bed filters are used; where dust concentrations are low the bed is fixed and the filter medium is taken away periodically for cleaning. In other cases the filter is in sections, each of which can be isolated in turn from the gas flow. During the shut-down period, the bed is subject to vibration and/or back blowing to remove the dust which has been caught. The dislodged dust falls into hoppers below the bed. The aggregate filter has the advantage of simplicity and the ability to work at elevated temperatures. It can sometimes usefully replace the bag filter, made of woven fabric, which has a much lower working temperature.

When this type of filter is used on gases with high moisture contents, care must be taken that the aggregate is above dewpoint when the dust-laden gases are introduced, otherwise the bed becomes blocked by wet dust which normal cleaning methods do not dislodge. This is an important precaution for the aggregate bed filter owing to the large weight of the filtering medium requiring large quantities of heat to raise its temperature. The aggregate bed filter, therefore, is best used on processes where moisture content of the gases is low, but gas temperatures are high or highly variable. A typical application is for cleaning the gases arising from cooling by air of the clinker from cement kilns.

Attempts have been made to use fluidized beds of aggregate as the collecting filter, but these have been generally found to be relatively low in efficiency (of the order of 80 per cent) even on fairly coarse dust, due to the re-entrainment of dust caused by the motion of the bed.

Paper filters

Paper filters, as their name implies, use porous paper as the filtering medium. Paper has relatively poor mechanical properties, consequently the methods used for woven fabric filters to remove the collected dust cannot be used. It is normal, therefore, to use this type of filter on the lower temperature applications where dust concentrations are low (less than 5 mg m^{-3}), and where particularly clean gas is required, as in clean rooms, rather than for industrial pollution control requirements.

The filter remains in service until the accumulation of dust has caused the pressure drop to increase to the point at which either insufficient gas is passing or the pressure drop is unacceptably high, when the filter element is replaced. This is an acceptable practice for low concentrations at relatively low gas temperatures, where the volume of gas to be cleaned is not large, that is, in the range of a few cubic metres per minute. Paper filters can remove with very high efficiency particles down to 1 μm and below, according to the grade of the paper. With the correct choice of paper they can be used for the removal of virtually all suspended material, including bacteria, so that the air leaving the filter is essentially free from solid contaminants. The filtering rate of paper filters is of the order of 150 cm min^{-1}, the resistance being 2–3 cm W.G., rising to roughly double this before the filter is replaced. When filtering city air, for example, such a paper filter can have a life of several thousand hours. One such filter is shown in Fig. 8.20(a); the paper is in pleated form, the object being to expose the maximum filter area to the gases within the volume of the filter body. This is a form commonly used for cleaning air for internal combustion engines.

Fibrous mats

The fibrous mat filter consists of fibres which can be natural (cotton or wool), synthetic (nylon, etc.) or glass, in fact virtually any material capable of being produced in fibre form, including metal fibres such as stainless steel. The fibres are usually mounted in the form of mats between supporting sheets of gauze or perforated metal, as shown in Fig. 8.20(b). The characteristics of such filters vary according to the depth of the filter, the type of material from which the filter is constructed, and the density of packing of the filter. Filter rates can range from 20–50 m min^{-1} with a pressure loss of only a fraction of a centimetre of water gauge. Such filters will obviously only remove the coarser material, but when the packing density is such that the pressure loss rises to the range of 15–30 cm W.G., high collection efficiencies are possible for very fine materials, including the very fine sub-micron acid mist originating from sulphuric acid plants.

The advantage of this type of filter is that a vast number of different kinds of material are available to suit different conditions of operation, such as temperature and corrosion. As all filters of this type fall in the category of fixed filters, i.e., there is no on-line cleaning system, they can only be used for relatively low concentrations of dust and when the resistance rises to an unacceptable level must be replaced or cleaned.

Fabric filters

The fabric filter is the most widely used type in this group, being capable of treating large gas volumes with the highest dust concentrations met in any process on a continuous basis, restricted only by the efficiency of the on-line cleaning system and the properties of the fabric. The filter medium is in the form of woven fabric or felted materials manufactured from natural or man-made fibres. With materials commonly in use, the range of temperature is restricted (see Table 8.3); even glass fibre is limited to 270°C. With the inherently lower filter rate of woven glass compared with other

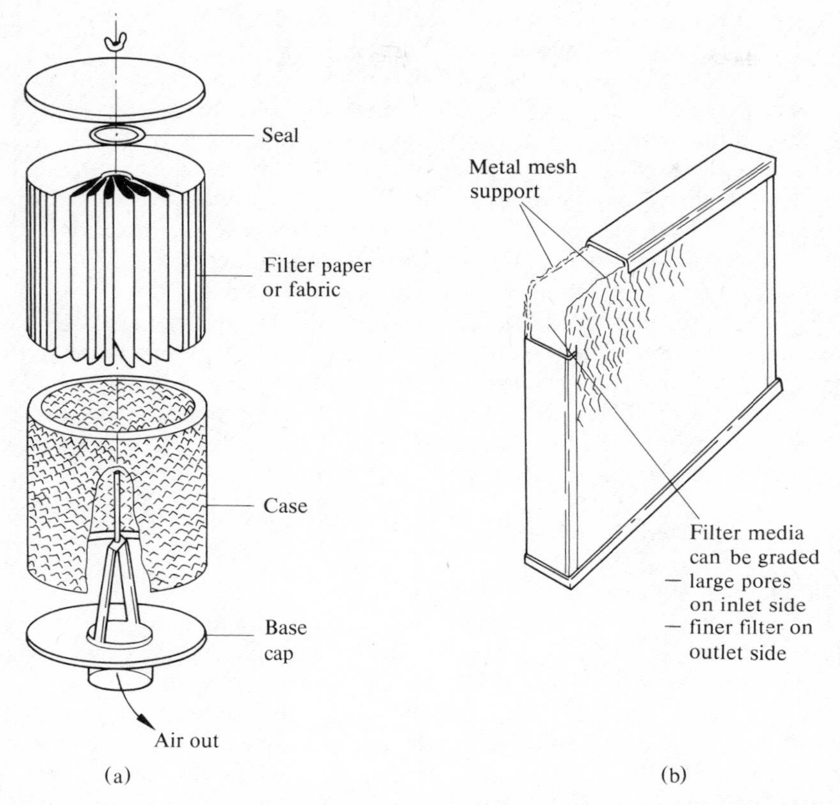

Fig. 8.20. Filters for low dust concentrations: (a) pleated fabric or paper type; (b) fibre mat sandwich type.

TABLE 8.3

Bag filters—properties of various fabrics

Material	Maximum operating temperature (°C)	Resistance	
		Acids	Alkalies
Cotton	90	Low	High
Wool	90	High	Moderate
Nylon	110	Low	High
Orlon	120	High	Moderate
Terylene	130	High	High
Nomex	200	Moderate	High
Teflon	230	High	High
Glass fibre (Silicone coated)	270	Moderate	Moderate
Stainless steel	+400		

fabrics, it is often considered equally acceptable from the point of view of capital cost to cool the gas to a level suitable for other fibres by allowing air to leak in.

Currently new materials are being developed which can be expected to extend the temperature range considerably. They include metal fibres such as stainless steel, high temperature synthetic fibres and alloys. The possibilities are indicated by Fig. 8.21, which shows the results of tests[9] on some of these. In addition, stainless steel is reputed to have an operating temperature in excess of 400°C. At present these new materials are expensive, and can only be economically justified for special applications. It can be expected, however, that with time and increasing demand, cost will decrease and the range of application of the fabric filter will increase considerably.

With all types of filter materials, the range of working conditions is determined by the mechanical properties of the materials at the operational temperature and on the incidence of chemical attack by corrosive elements in the gas and dust, which can result in rapid failure of the filters. As shown in Table 8.3, the choice of the correct fibre material is likely to infuence filter life significantly. Chemical attack is likely to become an even more important factor with the new high temperature materials.

Important characteristics of the fabric filter are:

1. Efficiency is always high for felted materials.
2. Efficiency is not greatly affected by dust concentration or by filter rate.
3. The filter rate, that is the volume per unit area of cloth, is usually designed to give a pressure loss of 6–12 cm W.G., depending on whether the cloth is in a clean state or has a layer of dust deposited on it. This is normally considered the economic range of pressure loss taking into account running cost in terms of fan power (see Fig. 8.12) to overcome the resistance of the filter and capital cost of the bag filter.
4. Pressure loss increases as (filter rate).^{2} This characteristic largely determines the economic filter rate and hence the filtering area needed.

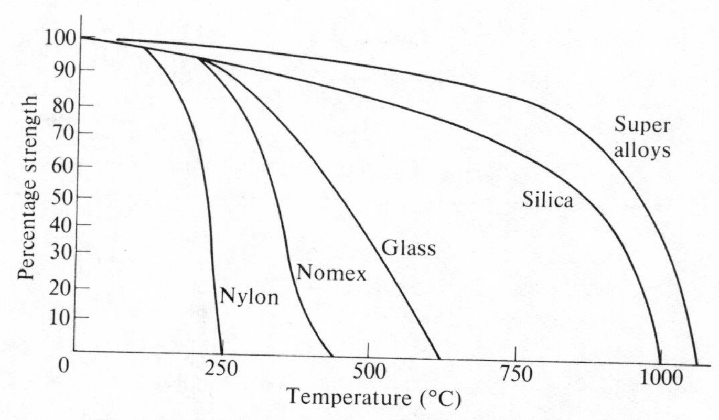

Fig. 8.21. Variation of strength of fibres with temperature.

Woven cloth, due to the method of manufacture, has relatively large gaps where the threads cross. These are often large compared with the size of dust particles to be caught. In practice this results in a relatively low efficiency when the bags are new, but improves as the holes become blocked as dust builds up on the fabric. Eventually filtration efficiency is determined by the thickness of the dust, so that particles well below 1 μm can be caught, despite the apparent open weave of the cloth.

Felted cloth consists of fibres laid in random fashion so that the problem of regular holes does not arise as with the woven cloth. Felted cloth also has a pile on the surface which has the effect of increasing the effective filtering area. Hence felted materials have a higher filter rate than woven cloths and are gaining favour as a result. Owing to the raised pile on the felt, effective cleaning is more difficult than for woven cloths. A development of the felted cloth is the so-called 'needle felt'. Synthetic fibres do not have natural felting properties, and it is necessary to use a special process involving the needle loom with many thousands of barbed needles which pass through the mat of fibres and produce a type of felt. Needle felts are usually reinforced with a layer of woven fabric to improve mechanical strength. It is claimed[10] that needle fibres combine the high filter rate advantage of felts with the easier cleaning properties of woven cloths.

Typical filter rates are

woven cloth	1–2 m min^{-1}
felted cloth	2–5 m min^{-1}.

These rates are determined by the pressure loss with the filter cloth in its operating condition, that is with a layer of dust deposited on the cloth, and are based on pressure losses of 8–16 cm W.G. This pressure loss is a compromise between capital cost and operating cost, the latter consisting of cost of fan power (Fig. 8.12) and of replacement filters. Since pressure loss increases as the square of the filter rate, the permitted rate for any set of operating conditions is fairly narrow.

Natural and synthetic fibres in felted or woven form in the dry state are good electrical insulators. This can pose serious problems when filtering explosive mixtures of gas as high static voltages can build up on the bags due to the passage of the gas and deposition of charged particles. Unless the electrostatic charge can leak away, the voltage increases until arcing occurs through the gas, and this can cause explosion or fire. This danger has now been overcome for both felted and woven materials by introducing a small percentage of metal, or other conducting fibre, into the cloth, rendering it anti-static.

Successful application of the fabric filter depends on two major considerations:

1. Ability to remove accumulated dust from the filter, otherwise the pressure drop will increase until either the filter fails due to pressure on the material, or the flow rate through the filter cloth falls to an unacceptable level.
2. A reasonably long life, preferably several years for the filter material.

Ability to remove the layer of dust depends on the type of dust being caught and the mechanism being used to dislodge the dust. In general, fabric filters must be confined to processes where at the operating temperature the dust is dry and free-flowing. In addition it is inadvisable to use fabric filters on cyclic processes where the temperature of the gas falls at intervals below the dewpoint. Condensation takes place during these periods, causing any residual dust on the fibres to become caked. If this state of affairs is allowed to progress the filter material becomes blinded and the bags need to be removed for laundering.

The life of the filter is determined by a combination of chemical attack by dust and gas and mechanical stress imposed by the cleaning mechanism. The latter involves obtaining by practical experience the correct balance between the degree of cleaning of the filter and long life, since the more vigorous the cleaning action the shorter the life of the filter.

Considering the commercial form of fabric filter, in most forms the filter material is in the form of vertical cylindrical sleeves 10–25 cm diameter, and a height up to 6 m. The main object of any commercial filter design is to build the maximum filter area within a given casing size, since the casing constitutes a high proportion of the total capital cost of the filter. In addition an on-load cleaning system is fitted and must satisfy the two basic requirements, effective removal of the dust from the filter combined with the minimum stress on the filter material.

In Fig. 8.22 are shown three methods currently used for cleaning the filter material. Gas is blown through the filter material in the opposite direction to the normal direction of the gas when being filtered. This removes the dust layer partly by deforming the filter bag and by blowing dust off the surface. In method (a), a ring continually travels up and down the bag which thus remains in a state of near constant pressure drop. In method (b), high pressure pulses of air are blown into the bag, the pulses are controlled by a valve, and this is reputed to give more effective cleaning than a continuous flow of air due to the repeated flexing of the filter bag. In method (c), a rotating arm periodically passes a nozzle over the bag outlet, reversing the flow of gas. This is a low pressure continuous gas flow. In all cases the removal of dust is effected partly by collapsing the bag and partly by blowing dust from the surface. In some designs, methods (b) and (c) are used simultaneously. When the gas has a fairly high moisture content, to avoid chilling of the filter which might induce caking of the dust on the filter, the reverse cleaning gas supply is heated or consists of clean gas from the bag filter outlet.

Summarizing, the important characteristics of bag filters are as follows:

1. Efficiency is always very high; for felted material in excess of 99 per cent, even for 0·01 μm particles.
2. Efficiency is independent of filter rate, but owing to the increase in pressure loss with increase in filter rate, this is set at a compromise level, usually 6–15 cm W.G.
3. Efficiency is not affected by dust concentration, either high or low concentrations can be handled successfully.
4. Fabric filters are suitable only for free-flowing dust. Even so, some types of dust which are free flowing but which have sharp edges should be avoided, as these can cause premature failure of the filter materials by cutting fibres during the reverse blowing cycles.

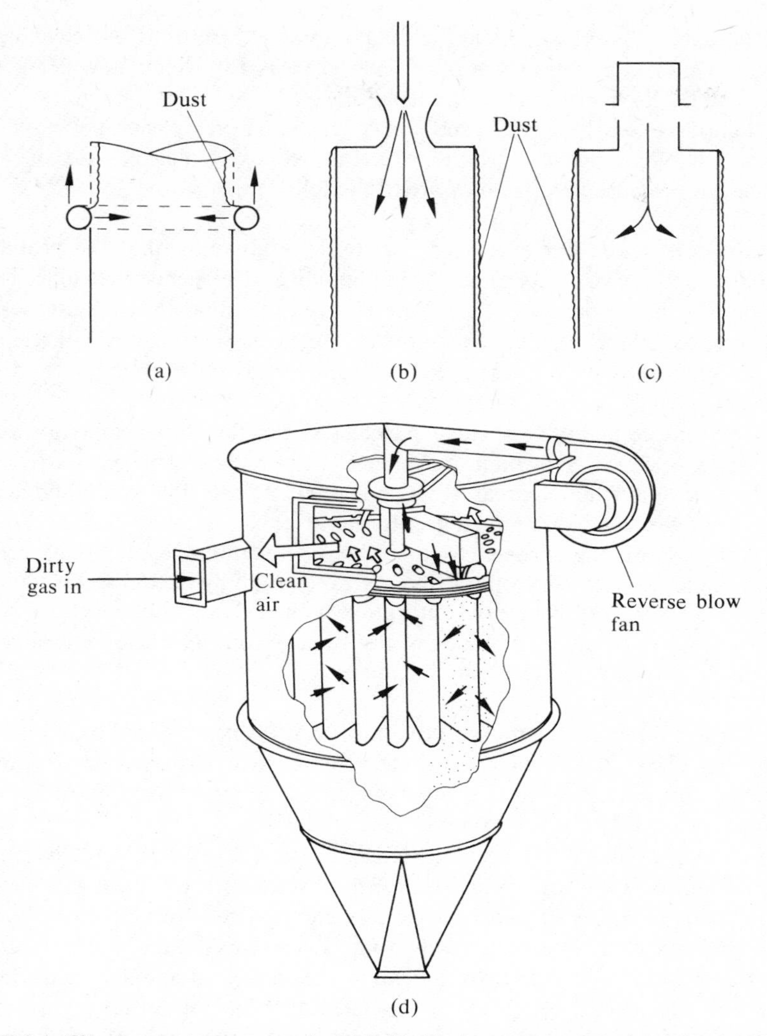

Fig. 8.22. Fabric filter for continuous operation: (a), (b), (c), reverse flow methods of cleaning filter fabric; (d) sectional drawing of typical reverse flow fabric filter.

5. It is important to ensure that the maximum working temperature for the fabric is not exceeded. Even comparatively small excesses can result in significant reduction in the life of the filter fabric.

Wet washers and scrubbers

Wet washers and scrubbers are devices using mixed phases of gas and liquid, the object of the scrubber or washing device being to transfer suspended particulate matter in the gas to the scrubbing liquid which can be readily removed from the gas cleaning device, leaving the clean gas to pass onwards to the process for which it is being used, or alternatively to be discharged to the atmosphere.

The action of wet washers and scrubbers is dependent on collision betweeen the dust particles and liquid droplets in suspension in the gas. Collision and the resultant coalescence can be caused by one or more of the following:

1. *Inertial effects and the effects of gravity.* It is apparent that the probability of collision between droplets of scrubbing liquids and suspended particles in the gas increases as the velocities of these two relative to each other also increases.
2. *Electrostatic effect.* Many dust particles carry an electrostatic charge, and this causes attraction between the liquid droplet and the particle, since the liquid droplet is usually at earth potential.
3. *Diffusion phenomena.* Very small particles in the fine fume range are subject to what is known as Brownian movement. This means that the particles are in random motion, which increases the probability of collision with liquid droplets of large diameter, which are not subject to this effect.
4. In many scrubbing processes the gases are initially at a relatively high temperature, consequently the effect of the scrubbing action is to cool the gas, saturate it, then cool below the saturation temperature. The net effect is that the dust particles behave as condensation nuclei on which condensing vapour selectively deposits, so increasing the mass.

Of the four ways in which the mass of the dust particles is increased by the addition of liquid, the first effect predominates for the coarser dust particles, electrostatic effect and Brownian movement influence most strongly the finer fume particles, and the condensation factor is likely to affect particles of all sizes.

A vast amount of work has been carried out over the years to establish the theory related to the performance of the various types of scrubbers, of which there are a great number of different designs. Much of this is summarized by Calvert.[11] Lapple and Kamack[12] showed that when the contacting device was in the form of an energy dissipation system involving a pressure drop in the gas, the dominant factor in contact efficiency was turbulence. Further work by Semrau[13] indicated that regardless of the source of energy, the contacting efficiency and the efficiency of removal of dust particles from the gas was related to the *total energy absorbed*, either in the form of pressure drop in the gas system or energy used to atomize the scrubbing water. This concept is subject only to the proviso that in the region where the pressure loss in the gas is incurred, that is the point of maximum turbulence, there is an adequate supply of scrubbing liquid to provide droplets in sufficient number for collision to occur with the dust particles. The significance of this is that efficiency of dust removal in the scrubber has little relation to the design and geometry of the scrubbers, provided adequate liquid for scrubbing is supplied in the contacting zone of the scrubber.

If this concept is accepted, and there appears to be a substantial amount of evidence to support it, particularly for the higher energy scrubbers, then little difference can be expected from variations in the form of the scrubber. This means that the skill of the designer is most usefully directed at practical features concerned with giving continuous cleaning with the minimum of maintenance and replacement costs. For example:

1. By designing the scrubber in the correct materials to minimize erosion, which is one of the most serious causes of failure of scrubbers, and is at a maximum where the turbulence is greatest. It is related to particle size and hardness, and to degree of turbulence of gas. Although with large particles turbulence, and hence contacting energy needed, is low, a matter of a few cm W.G., rapid wear occurs particularly with abrasive dust. In the case of fine fume, high pressure differentials are involved, often greater than 100 cm W.G. The situation can be serious in this case because very few industrial processes give rise to fine fume only; there is almost always a small percentage of dust in the coarse range. Contacting energy is determined in this case by the finest particles, but the small percentage of coarse particles determines the erosion potential. In either of the cases mentioned, experience has shown that incorrect design can result in a life of only hours or days.
2. When atomizing sprays are used to introduce the water and to break it down into suitable droplet sizes, the pressure drop at the spray increases with the fineness of droplet produced. This introduces the problem of wearing of the sprays, which is due to suspended contaminants in the water. In all industrial scrubber systems it is necessary to re-use the water. The treatment of water to remove suspended materials becomes increasingly expensive with the degree of cleanliness required and the cost of the water treatment plant constitutes a high percentage of the total cost of the scrubber system. It is preferable, therefore, for the water to be recirculated with the minimum of cleaning following its passage through the scrubber; this requires careful attention to spray design and method of introducing scrubbing liquid.
3. Many industrial gas processes give rise to gas volumes which can vary considerably. The scrubber is often required to maintain a constant low level of dust in the scrubbed gas, regardless of variations of flow rate, and in some cases also with variations in particle sizing. This requires the scrubber to be designed, in the case of varying volumes, to maintain a constant energy and hence constant gas cleaning level. If the particle sizing varies it may be necessary to operate at different levels of pressure differential. It is desirable that the scrubber design can accommodate these requirements.
4. Following any scrubber it is necessary to remove the entrained water droplets entirely. They contain the solid material which has been removed by the scrubber. No scrubber can be more efficient than its droplet separator. When the object is to prevent atmospheric pollution, failure to remove the entrained droplets can result in artificial rain in the area around the plant, which, regardless of dust, can present a considerable problem.

There are many varieties of gas cleaning device depending on liquids for removal of the suspended material, and it is proposed now to review the basically different types and comment briefly on their characteristics.

Gravity spray tower

The spray chamber consists of a large vessel through which the gas flows usually in the vertical plane at a relatively low velocity. The scrubbing liquid is introduced at the upper end of the tower and is usually contra-flow to the gas. A typical arrangement is shown in Fig. 8.23, where the liquid is introduced by a system of sprays near the roof of the tower. Above the sprays is a simple inertial droplet eliminator to prevent carry over of water droplets into the outlet flue.

It is to be expected that the cleaning efficiency of this simple device depends on the amount of water and the droplet size. In Fig. 8.23 is a graph which shows the target efficiency, that is the collision efficiency of dust particle and water droplet, for different sizes of water droplet and dust particle. This graph was derived by Stairmand.[14] It illustrates the fundamental limitations of the gravity spray tower, namely that this device is only really suited to the removal of dust particles greater than 10 μm diameter. There is little point is using very fine spray droplets even if they could be produced economically, as the important factor is to maintain a high relative velocity between the water droplets and the dust to increase the probability of collision and coalescence. With these limitations in mind, it is clear that the gravity spray tower, where the relative velocity of dust particle and water droplet is effectively equal to the free falling velocity of the water droplets, has a limited target efficiency. Reference to Fig. 8.1 indicates the variation of free falling velocity with particle diameter; this can be used to estimate the relative velocity of various sizes of particles and liquid droplets. The advantage of this type of gas scrubber is its very low energy consumption, usually of the order of 2–4 cm W.G. It is suitable only for coarser material; the pressure differential is related to design and can be varied only by varying gas flow through it.

Self-induced spray scrubber

The self-induced spray scrubber, as its name implies, has no spray system as such but uses the velocity and energy of the gas to induce the spray for the scrubbing action. A typical scrubber of this type is shown in Fig. 8.24(a). The dirty gas enters the scrubber through a venturi shaped throat which is connected by a narrow slot to the water contained in the sump below the scrubber. The effect is to induce a flow of water from the slot into the gas, where it is atomized and carried round the circular chamber of the scrubber, which acts as a contacting region for the droplets and dusty gas and as a cyclonic separator. The clean gas passes through a water separator of the inertial type towards the centre of the chamber, which has vanes causing the gas to change direction by roughly 180°, thus ensuring that no water is carried with the clean gas leaving the scrubber along the axis. With this type of scrubber. the same water recirculates and

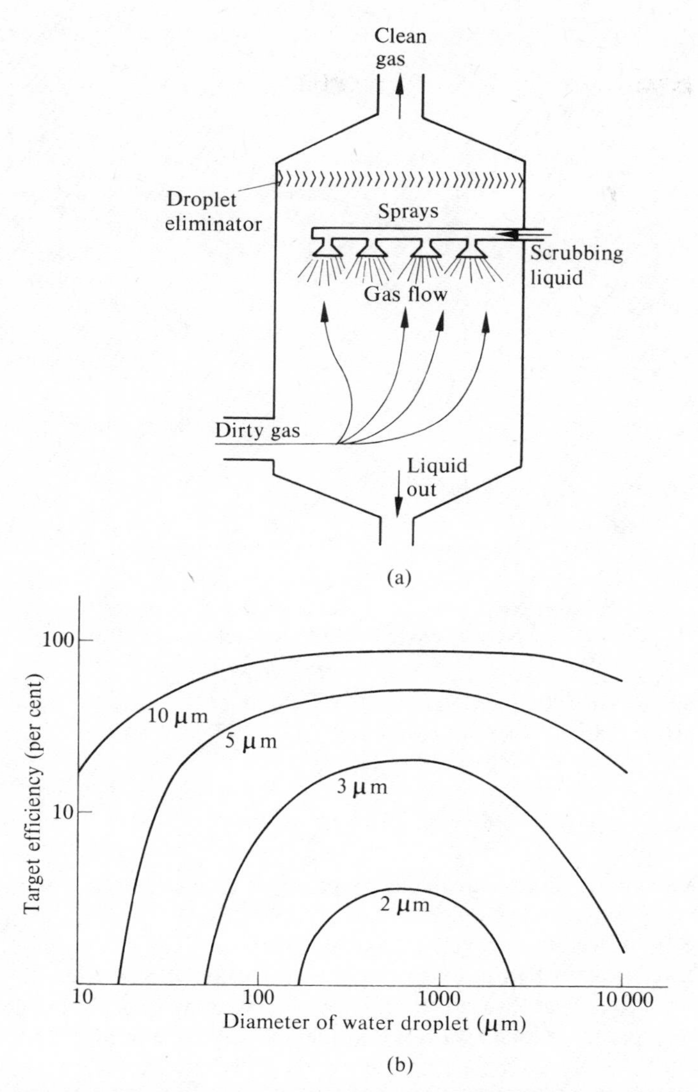

Fig. 8.23. Gravity spray tower: (a) gravity spray tower; (b) relationship between droplet size and collection efficiency for particles 2–10 μm diameter of dust of specific gravity 2 (Stairmand).[14]

reliance is placed on natural sedimentation to separate the dust particles in the water, the sludge being removed from the base of the chamber. Provision must be made for the level of water to be closely regulated within the scrubber if the self-induced spraying action is to continue uniformly. The particle sizing–efficiency relationship and pressure loss of such a scrubber usually fall between that of the impingement scrubber and

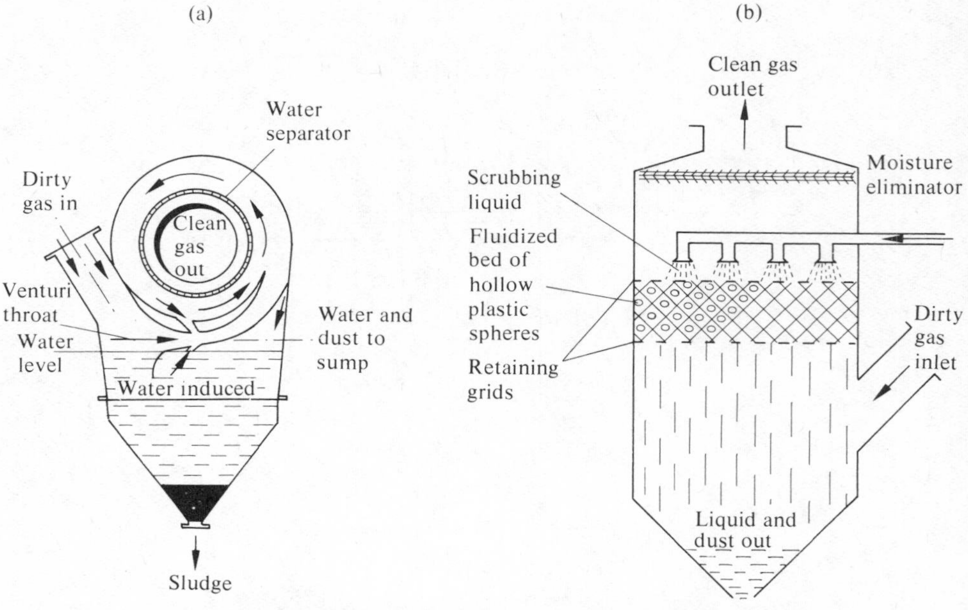

Fig. 8.24. (a) Self-induced spray collector; (b) fluidized bed scrubber.

gravity spray tower shown in Fig. 8.23. It is limited to particles of above 1 μm, if efficiency better than 90 per cent is required.

Impingement scrubbers

The impingement scrubber uses the same principle as the inertial collectors. One type, shown in Fig. 8.24(b), is the fluidized bed scrubber. In this design the dirty gas enters the base of the tower and travels upwards through a falling curtain of water, which removes the coarse dust in a similar way to the gravitational tower. At roughly the centre height of the tower, contained between two retaining grids, is a bed of low density hollow plastic spheres, which as a result of the uprising column of gas are maintained in a state of fluid motion. Over the upper retaining grid the scrubbing liquid is sprayed, so that the spheres are continually wetted by a falling spray of liquid. Since the spheres, although fluidized, are in close proximity, the gas path is of a tortuous nature and there are many changes of direction. The net effect is that as a result of the changes of direction the dust particles are separated from the gas and impinge on the wetted spheres. The advantage claimed for this type of fluidized bed dryer is that due to the motion of the spheres the system is self-cleaning. Other collecting devices in this group consist of stationery baffling systems of packed timbers, or plastic mouldings such as Raschid rings, which rely only on the tortuous nature of the gas passes for collection of suspended particles.

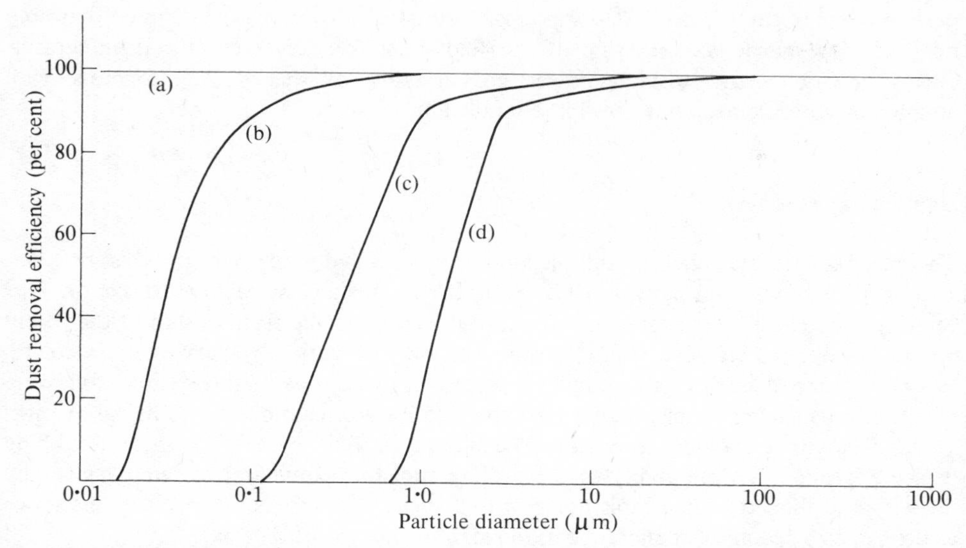

Fig. 8.25. Typical scrubber particle size–efficiency relations: (a) high energy scrubber, 150 cm W.G. pressure loss; (b) medium–high energy scrubber, 40 cm W.G. pressure loss; (c) impingement scrubber, 15 cm W.G. pressure loss; (d) spray tower, scrubber, 5 cm W.G. pressure loss.

There are other variations of the impingement scrubber. Pressure losses according to design can range from 20 to 80 cm W.G., and it is claimed that the higher pressure loss with certain designs gives efficiencies of dust removal by weight in excess of 99 per cent, even on dust down to and less than 1 μm. A typical particle sizing–efficiency curve for this type of scrubber is shown in Fig. 8.25, from which it will be seen that an impingement scrubber with a pressure loss of 15 cm W.G. would have a reasonably high efficiency (more than 90 per cent) for particles down to 1 μm.

Disintegrator scrubbers

Disintegrator scrubbers were used extensively some years ago to clean blast furnace gas. They consist of cylindrical rings of rods, alternate rows being either static or rotating at a high velocity about the axis of the device. Water is injected as a jet and broken up into fine particles by the collision with the moving and static rods. The gas passes through the system and is subject to bombardment by the rapidly moving droplets of water, turbulence in addition being promoted by the moving rods. The energy consumed by the disintegrator is almost entirely that of rotating the movable elements. Such a device can have a slight energy recovery effect and assist the fan driving the gas through the device so that there is no pressure loss.

Disintegrators have been used successfully to clean blast furnace gas, giving dust removal down to 10–20 mg m^{-3}. Due to need to avoid build-up in the disintegrator and to avoid abrasion which would result from the violent turbulence with larger dust particles, it is normal to precede distintegrators by high efficiency cyclones so that only relatively low burdens, consisting of the finer dust fraction, enter the disintegrator. Owing to the relatively high power consumption of this device, together with maintenance problems, it has now largely fallen into disuse.

High-energy scrubbers

The best known scrubber in this section is undoubtedly the venturi scrubber, one arrangement of which is shown in Fig. 8.26. It contains a classical venturi section. Gas entering the venturi accelerates into the throat. Owing to the fixed design of the throat pressure drop, gas cleaning characteristics can only be varied by varying gas velocity. Water is pumped in through atomizing sprays in the throat of the venturi, following which the suspended water and gas pass through a water eliminator, in this case shown in the form of a cyclone separator. The ultimate efficiency of the whole scrubbing device depends upon the ability of the cyclone moisture eliminator to remove virtually the whole of the suspended moisture. For high-energy scrubbers this problem increases as droplet size becomes smaller, due to the high turbulence of the gas.

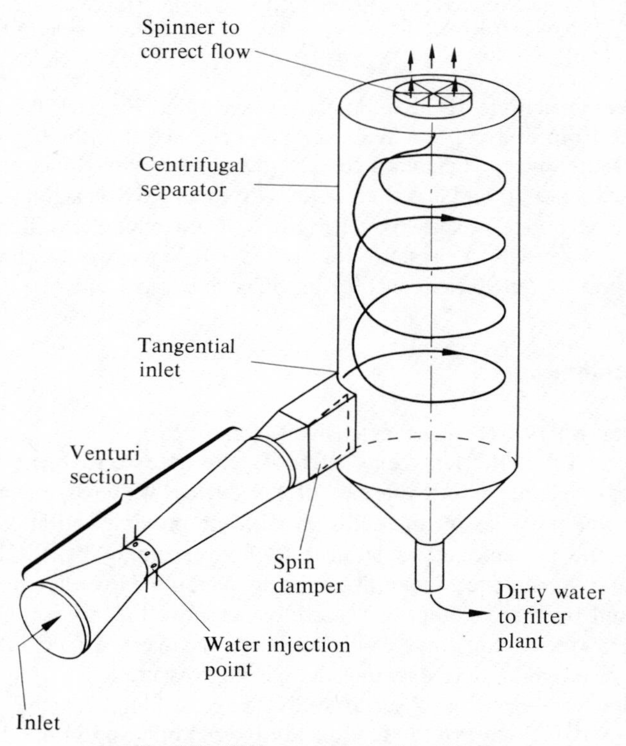

Fig. 8.26. High energy venturi scrubber.

The pressure loss, and hence the velocity in the throat of the venturi, will be determined by the particle size and the efficiency of removal required. This is illustrated by Table 8.4, showing the relationship between the throat velocity, the diameter of the water droplets which are induced by the turbulent energy of the throat, and the target efficiency, that is the efficiency of removal expected for 1 μm particles. The droplet sizes in this table were calculated by Stairmand using relationships derived by Nikiyama and Tanasawa.[15] The table indicates that energy requirements increase with decreasing particle size and with efficiency of removal required. The effect is illustrated in Fig. 8.25, curves (a) and (b). Curve (a) shows a high-energy scrubber with a pressure loss of more than 140 cm W.G.; under these conditions the scrubber was maintaining an efficiency of 99 per cent down to 0·02 μm diameter. The characteristics of the scrubber beyond this point were not determined due to the difficulties of carrying out particle sizing analysis for sub-micron fume, which involves special techniques using the

TABLE 8.4

Scrubbers—target efficiency for 1 μm dust particles

Gas velocity ($m\ s^{-1}$)	Induced droplet diameter (μm)†	Target efficiency (per cent)
15	320	42
30	160	75
60	80	95
90	55	99
120	40	99

† Calculated according to Refs. 14 and 15.

electron microscope. It is clear, however, that provided enough energy is available the high-energy scrubber is capable of approaching 100 per cent efficiency for particles down to the order of 0·01 μm. While these remarks are made in connection with the venturi scrubber they also apply to the various other types of scrubber to be described.

The disadvantage of the venturi scrubber is that for a particular size of venturi throat, efficiency cannot be varied, except by varying gas velocity through the throat. In many practical applications of scrubbers it is desirable to maintain the same pressure differential, that is the same cleaning energy for varying gas volumes, and numerous variations have been developed. Apart from the need to vary pressure drop for a constant volume, or to vary volume and maintain constant pressure drop, it is also desirable to be able to utilize only semi-clean water; the significance of this will be appreciated when practical aspects of design of the scrubber system are discussed. In Fig. 8.27 are shown some scrubber designs currently used to give the flexibility required to achieve this objective.

Figure 8.27(a) shows a slot venturi scrubber in which the dust laden gases are introduced through a circular pipe clear of the converging venturi walls. Scrubbing liquid flows over a weir on the converging wall of the venturi, forming a film of liquid

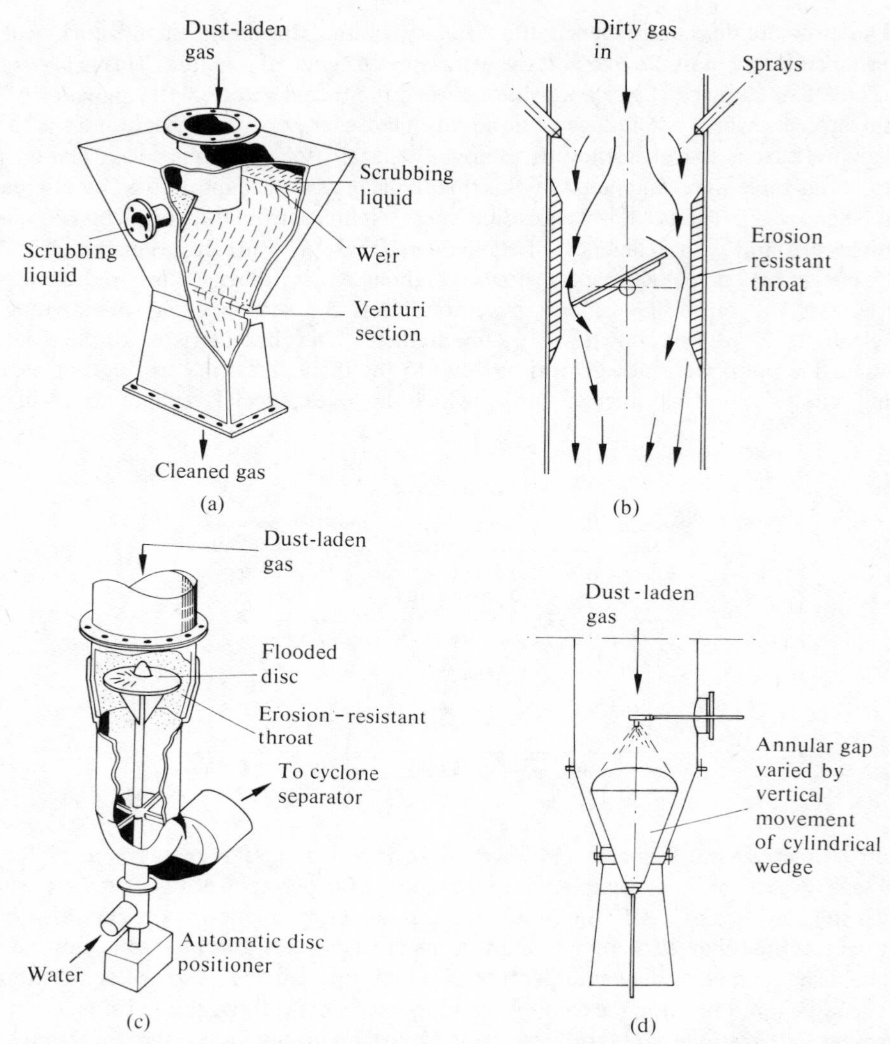

Fig. 8.27. Typical gas scrubber arrangements: (a) slot venturi scrubber; (b) flooded disc scrubber; (c) variable orifice scrubber; (d) annular clearance washer.

which is broken up and atomized by the turbulence in the venturi throat. The advantage claimed for this system is that it permits the use of liquids containing a high percentage of solid material, approaching a slurry condition. This scrubber, as illustrated, has a fixed throat, but versions exist in which the profile of the throat can also be varied to allow for variations in pressure loss or gas flow.

In Fig. 8.27(b) is shown the flooded disc scrubber. A horizontal disc mounted on a vertically moving shaft moves axially in the gas main through which the dust-laden gases flow. The disc is somewhat smaller in diameter than the gas main, which in the region of the disc is of conical section so that by varying the position of the disc in the vertical plane the gap between the disc and the walls can be varied. Thus the pressure loss across the scrubber, or the volume which can be handled for the same pressure loss, can be maintained constant as desired. The scrubbing liquid is introduced through the shaft of the disc and flows over the upper surface of the disc, atomization being brought about by the turbulence between the disc and the walls. Such a scrubber can also use liquids containing a high proportion of solid material.

Figure 8.27(c) shows the variable orifice scrubber, which is probably the simplest device of all those described. It consists essentially of a turn damper mounted in the flue carrying the dust-laden gases. This damper, when normal to the direction of flow, is smaller than the diameter of the main, so that an annular gap is maintained. The size of this annular gap is varied according to the pressure drop or flow characteristics required for the scrubber. Variation of pressure loss with varying flow is effected by rotating the damper blade, thus opening the gap between the edge of the disc and the wall of the gas main. Water is introduced through sprays mounted in a ring around the scrubber roughly one diameter upstream of the turn damper. These sprays are of a through flow type of simple construction and can handle liquids containing a fairly high percentage of solids, though probably not up to the standard claimed for scrubbers (Fig. 8.27(a)).

Figure 8.27(d) shows another variable annular throat design in which the gas flows into a throat of roughly venturi form. On the axis of the throat is a device in the form of a circular wedge with sides parallel to those of the throat. This device is mounted on a rod so that it can be moved in the vertical plane to vary the gap around the venturi throat. This gives the ability to accommodate varying flow conditions.

All the devices just described require a moisture entrainment separation device which varies in complexity with the energy of scrubbing. According to Table 8.4, the droplet size produced diminishes with increasing scrubbing energy and pressure drop across the scrubber, and hence the characteristic of the moisture eliminator must change to eliminate the smaller droplets with an efficiency of virtually 100 per cent. This is important, as high energy scrubbers are only used when high efficiency gas cleaning on fine fume is necessary.

Practical aspects of scrubber application

On the basis of the principle mentioned, that scrubber efficiency is determined by the total energy consumed by the scrubber, whether in the form of energy to atomize water or pressure loss across the scrubber, consideration must be given to the best way in which the water should be introduced. Water is an expensive commodity and is becoming increasingly so with increasing severity of pollution control legislation. It is now rare to find a scrubber in which water is passed through only once. In practice the water passes from the scrubber to a cooling and filtering system and is then recirculated. This total system to permit re-use of the water constitutes the greater part of the capital cost of the scrubber.

The water leaving the scrubber is passed over cooling towers then through sedimentation type filters. These remove the solid material which settles under gravity; the treated water is recirculated to the scrubber. It could be argued that for recirculated water there is little point in cooling, since whether the water is hot or cold does not materially affect the scrubber efficiency. Reference to Fig. 8.28 shows that if the water is permitted to recirculate and the temperature rises above 60°C, the amount of water carried forward with the gas in vapour form increases sharply, since the gas leaving the scrubber is saturated, hence the amount of make-up water required in the system can increase considerably. With the design of scrubber shown in Fig. 8.27(a), need for efficient sludge removal from the water may appear less important. However, while this is so for relatively low-energy scrubbers, in high-energy scrubbers solid particles originating from either the gas or the water, if this is not filtered, can cause a high rate of erosion of the scrubber throat. Various materials of construction are used, ranging from alloy steels to silicon carbide, a refractory material which approaches the hardness of the diamond. Since this is comparatively expensive it is used only when strictly necessary, but can be expected to have a life, even under arduous conditions, of years. Silicon carbide is a difficult material to fabricate, but fortunately in scrubber types shown in Fig. 8.27(b) and (c) this is necessary only for the containing throat in the region of maximum turbulence.

Water quantities required by a scrubber depend on the operating conditions, in particular on the temperature of the incoming gas, and usually range from 1 to 3 l m^{-3} of gas treated. Mention has been made of the risk of erosion of the throat of the scrubber. When the scrubber has an adequate supply of water, apart from providing the

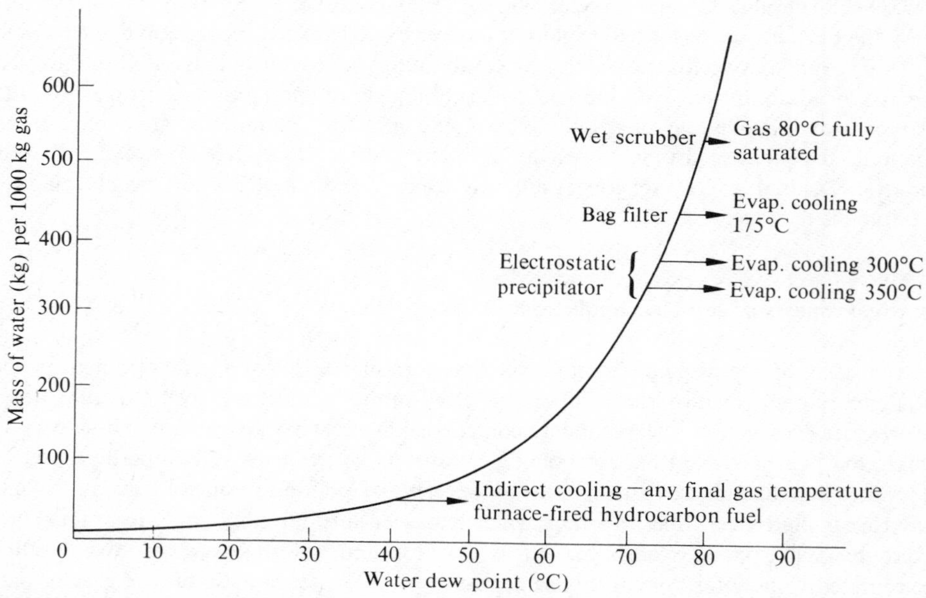

Fig. 8.28. Water content of stack gases after cooling from 1000°C to temperature to suit cleaning device.

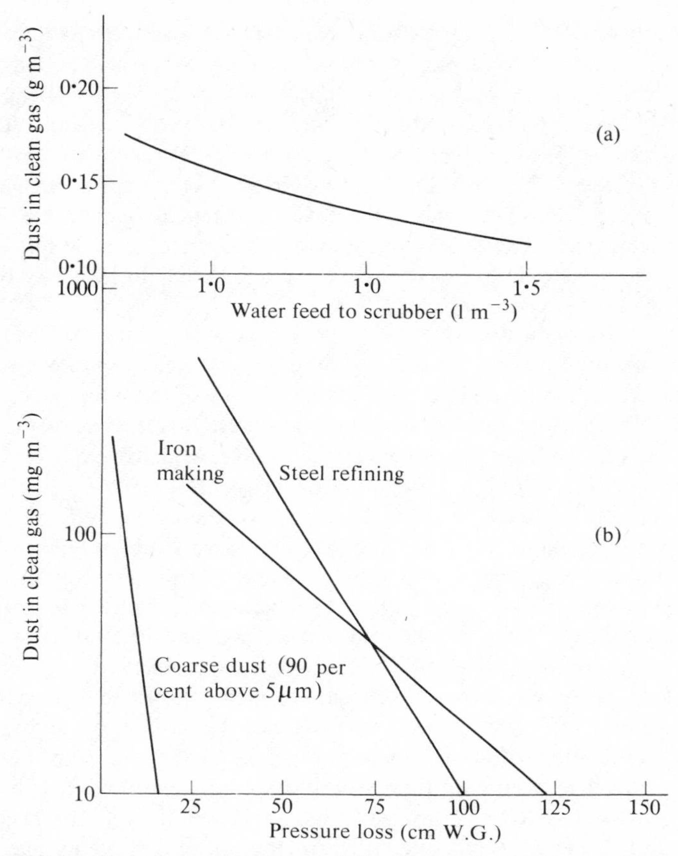

Fig. 8.29. Typical operating characteristics of venturi and orifice scrubbers.

contacting medium for the dust, it also acts in a secondary role of lubricating the throat of the scrubber and reducing erosion. It is important not to allow gas to pass through the scrubber with too little, or no water, as the life of the scrubber will be shortened considerably apart from reducing dust removal efficiency. Figure 8.29 shows the effect of varying water quantity for a scrubber of the variable orifice type. It will be seen that the final emission for the same pressure drop across the scrubber improves considerably as the water quantity increases. The choice of water quantities, therefore, becomes a matter of economics, balancing the cost of pumping the extra water against that of obtaining the increase in pressure loss necessary to maintain gas cleaning efficiency.

The lower graph in Fig. 8.29 shows actual results obtained with scrubbers on three industrial processes. The coarse dust is power station fly ash. For pollution control purposes a dust concentration of 0·1 g m^{-3} is normally considered adequate with this type of dust. A low-energy scrubber can produce this standard of gas cleanliness with a pressure drop of not much more than 6 cm W.G.

Similar gas cleanliness for dust from iron making and steel refining, consisting mostly of sub-micron size fume particles, require 35 cm W.G. and 50 cm W.G. respectively. In iron making, where the gas has value as a fuel, a dust content of 0·01 g m^{-3} is normally required, reducing to 0·001 g m^{-3} if the gas is used as fuel for a gas turbine. The scrubber pressure loss for these conditions is very high, and if developed by fan power represents high running cost in terms of electrical power (Fig. 8.12). In practice, when such scrubbers are used, the blast furnaces are operated at high pressure, up to two atmospheres. The gas is released at the furnace top at this pressure and provides more than enough energy to operate the scrubber to give any degree of gas cleanliness likely to be needed.

High-energy scrubbers used on blast furnaces also combine the function of controlling the gas pressure in the clean gas main, where the required pressure is usually less than 50 cm W.G. Under these conditions high levels of noise are generated, and this is currently a problem. During laboratory tests to investigate this problem, it was shown that the true venturi and designs such as Fig. 8.27(a) and (d) required higher throat velocities than Fig. 8.27(b) and (c) to give the same pressure drop due to the higher degree of turbulence in the latter type. The result was a very significant difference of noise level, types shown in Fig. 8.27(b) and (c) being considerably quieter.

As with batteries of cyclones in parallel to handle large gas volumes, it has been argued that small venturis arranged in parallel give more effective cleaning than single large gas cleaning units. However, the total energy concept does not support this view. Furthermore, the scrubbers in parallel are subject to problems of maldistribution of flow, so that the gas velocity can vary considerably in venturis in different parts of the unit, with consequent effects on cleaning efficiencies. There is also a serious problem of dust build-up and blinding of scrubbers in parallel, in a similar way to that experienced with cyclones, and it is becoming increasingly popular, therefore, for the single large unit to replace the multiplicity of smaller units. Scrubber units are currently in service which can handle several thousand cubic metres of gas per minute. Mechanical considerations rather than factors affecting performance dictate that for very large gas volumes such units are arranged in relatively small numbers in parallel; this situation, as with cyclones, presents no problems.

Summarizing, the important characteristics of scrubbers and wet washers are as follows:

1. There is no limit on temperature, but high temperatures increase water consumption, since gas is discharged at higher temperatures in a saturated condition. This leads to steam plume formation which can create secondary pollution problems.
2. Capital cost of the actual scrubber is moderate but, dependent on cost of water treatment plant, the overall capital cost can be high.
3. Pressure drop to give the required gas cleanliness increases with increasing cleanliness (Fig. 8.29), and with decreasing particle size.
4. For the same pressure drop, scrubber efficiency increases with water quantity (Fig. 8.29).

5. Although efficiencies are quoted for scrubbers, it is found that the dust concentration in the gas following the scrubber varies comparatively little for large variations in inlet concentration.
6. Variable scrubbers such as those illustrated in Fig. 8.27(b), (c) and (d) can maintain the same pressure drop, and hence gas cleaning, for variations of volume in excess of 5 : 1.

Sonic and ultrasonic agglomeration

If fine dust particles can be made to agglomerate, i.e., cling together, because of various natural forces including molecular attraction, or if liquid droplets can be made to coalesce into single large droplets, then the process of removing them from a gas stream is much simplified, as simple settling chambers or inertial or cyclone collectors can then be used.

Agglomeration takes place in the electrostatic precipitator which has been used in this way as the basis of a pre-collector for carbon black. Those agglomerated materials which are of high conductivity cannot be collected in the precipitator, and are readily removed by cyclones in a following stage; the original sub-micron particles could not be collected in this way. In recent years a number of attempts have been made to design gas cleaning systems specifically to promote the formation of agglomerates, by means of relatively low-frequency waves in the sonic and ultrasonic ranges. In essence the equipment consists of the generator of the sonic or ultrasonic energy, in a specially designed chamber in which the agglomeration takes place, followed by a cyclone or similar collector to remove the agglomerated material. Agglomerators in the sonic and ultrasonic group may be listed as:

1. The high-frequency oscillator using a moving coil transducer similar to a loudspeaker in a radio set. A suitable driving circuit can produce considerable amounts of energy at any desired frequency.
2. The high-frequency siren which operates on a similar principle to sirens used for warning purposes consisting of alternate rotary and stationary perforated drums.
3. The Hartland jet which uses vibrations produced by directing a jet of air at very high velocities at a specially shaped target located at the focal point of a parabolic reflector.

Using various arrangements of these sonic generators there has been reported considerable success with the agglomeration of sulphuric acid mists and similar liquids in the aerosol range. Very little information is available generally on the application of gas cleaning systems of this type. Design appears to present difficulties and the effect is unpredictable. There seem to be few advantages compared with the more conventional types of gas cleaning system. The power consumption is appreciable and as much attention is being given to the effect of both audible and ultrasonic oscillations on the nervous system of human beings, the use of these generators might create difficulties. Gas cleaning systems based on this method appear to have no great future.

Selection of equipment

The selection of the most suitable form of equipment for a particular gas cleaning problem involves consideration of capital cost, the actual purchase price of the plant and the cost of operation. The last covers all items related to the cost of running the plant including the following:

1. operators' wages;
2. electrical power—fans, electric motors, pumps, HT rectifiers (electrostatic precipitators);
3. water—either from town supply which is expensive, or from rivers, or sewage works effluent when some processing may be necessary;
4. chemicals—dosing of water to combat corrosion and other undesirable chemical reactions both when water is originally used and to permit recirculation and discharge back into sewers or rivers;
5. overhaul and maintenance—both labour and components.

In practice the capital cost is written off over a period of 10 or 20 years according to the anticipated life of the equipment. To this fraction of the capital cost is added the annual running cost. The sum of these two costs is taken to represent the actual cost per annum, and is used as a basis of comparison of cost of different ways of producing the same gas cleanliness. Capital cost estimates can be obtained from suppliers of equipment. It is possible to predict operating costs under headings (1)–(4) fairly accurately. The cost of overhaul and maintenance is more difficult to predict, particularly for new applications because of lack of past experience. The difficulty of assessing this cost is increased by its great dependence on correct choice of gas cleaning system. For example:

Cyclones. Abrasive dusts can reduce life to hours or days, whereas with correct applications life is years.

Electrostatic precipitators. Corrosion is the biggest single factor. If the correct materials of construction are used, no major replacements may be necessary in the life of the equipment (often more than 20 years).

Fabric filters. Life of the filter fabric can be considerably reduced by abrasive dusts, chemical attack or excess temperature. Cost of filter fabrics constitutes a considerable proportion of the total purchase price.

Scrubbers. Erosion due to dust and chemical attack due to use of wrong materials in the construction of the scrubber throat are the main considerations.

Gas cleaning plant is installed to recover valuable material from the gases (metallurgical processes), to permit the use of the gas where the dust would be undesirable (gasification of coal, coke oven gas) and to reduce the level of dust in the gas when discharged to the atmosphere to a level sufficient to avoid pollution problems and to conform to antipollution legislation.

Whatever the reason for gas cleaning plant being installed, the minimum standard of cleaning required must be established, as usually capital and operating costs increase as efficiency required increases. This was demonstrated by Stairmand[16] who produced the graph shown in Fig. 8.30, in which efficiency of gas cleaning is plotted against a figure based on the sum of capital and operating costs. The various designs and type of gas cleaning equipment are represented on the graph at efficiencies at which they can be economically used. The graph shows that cost increases with increasing efficiency. Accurate assessment of costs such as those plotted are difficult to obtain in practice, and some of the scatter of the points could be due to this problem. However, it appears that there is a large variation in cost for devices of similar order of efficiency or type but of different design. For example, types 7, 8 and 9, which are different forms of scrubber, show that even if the type of gas cleaning system has been laid down, there is still considerable economy to be achieved in selecting the best design. The slope of the line in the graph shows that as efficiency increases through 90–99–99·9–99·99 per cent, cost increases in the ratio 1:2:3:4; hence the importance of specifying the *minimum* efficiency needed.

While cost increases generally in this way it is important to realize that the efficiency required determines to some extent the type of equipment. For example, fabric filters

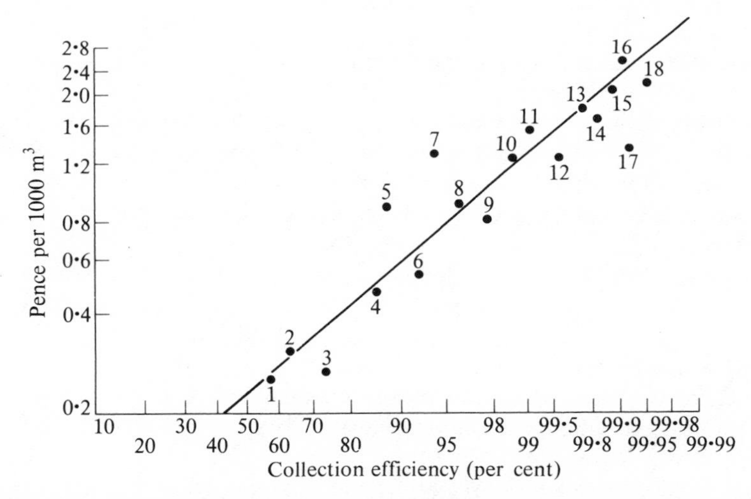

Fig. 8.30. Cost of gas cleaning equipment treating fine industrial dust. Annual capital cost taken as 10 per cent installed cost. Operating cost on basis 8000 hours per year (Stairmand)[16]. (1) Inertial collector; (2) medium efficiency cyclone; (3) low resistance cellular cyclones; (4) high efficiency cyclone; (5) impingement scrubber (Doyle type); (6) self-induced de-duster; (7) void spray tower; (8) fluidized bed scrubber; (9) irrigated target scrubber (Peabody type); (10) electrostatic precipitator; (11) irrigated electrostatic precipitator; (12) flooded disc scrubber, low energy; (13) flooded disc scrubber, medium energy; (14) venturi scrubber, medium energy; (15) high efficiency electrostatic precipitator; (16) venturi scrubber, high energy; (17) shaker type fabric filter; (18) reverse jet fabric filter.

only become competitive when high efficiencies are required, i.e., over 99 per cent, at which level cost is of similar order to that of electrostatic precipitators and scrubbers.

As efficiency required decreases, the cost of scrubbers and electrostatic precipitator decreases in cost roughly as indicated by Fig. 8.30, but for different reasons. With electrostatic precipitators, size of plant and hence capital cost decreases (Fig. 8.19). With scrubbers the reduction is in fan power due to the decrease in pressure drop (Figs. 8.12 and 8.29). In contrast, the fabric filter cannot be readily varied in efficiency, and cost cannot be reduced significantly.

Once the degree of gas cleanliness has been decided, the choice of gas cleaning device can proceed. For relatively coarse dusts, provided abrasion is not a problem, some form of inertial or cyclonic collector is the logical choice. As the dust becomes finer, the choice rests between the dry electrostatic precipitator and wet washers or scrubbers. At the highest efficiencies a further choice, the fabric filter, becomes economic.

The high efficiency systems for fine dust and fume divided into two groups: dry systems, including the dry electrostatic precipitator and fabric filter, and wet systems, including wet washers, scrubbers and wet precipitators. Apart from economic considerations, the dry systems have advantages when used to prevent atmospheric pollution. When the gas is discharged hot, better dispersion of the stack gases takes place due to the increased buoyancy, in contrast to the wet systems which cool and saturate the gases. Cool saturated gases, in addition to lack of buoyancy, have a tendency to steam plume formation which can create a secondary pollution problem.

Temperature of operation and cooling of gases

The normal temperature ranges for the different types of equipment have already been mentioned. It is sometimes necessary, when temperature is excessive, to reduce it to the level best suited to the equipment from the viewpoint of efficiency and of capital and operating costs. Costs must include any equipment installed to provide controlled cooling of the gas.

When cooling is necessary to suit the gas cleaning plant, it can be achieved by three methods:

1. mixing with atmospheric air;
2. introducing atomized water, which causes cooling by evaporation;
3. indirect heat extraction, such as waste heat boilers.

The method employed can seriously affect the final gas volume to be treated by the cleaning device, as shown in Fig. 8.31. For a final temperature of 200°C, as required for instance to suit a fabric filter, the final gas volume presented to the gas cleaning plant varies from 30 per cent of the original volume when cooled by indirect heat extraction, to 200 per cent of the original volume when cooled by air inleakage. This implies very nearly a 7 to 1 difference in the capacity of the plant required to handle the gas. While the dust concentration is diluted in the case of the air dilution method, the capital cost

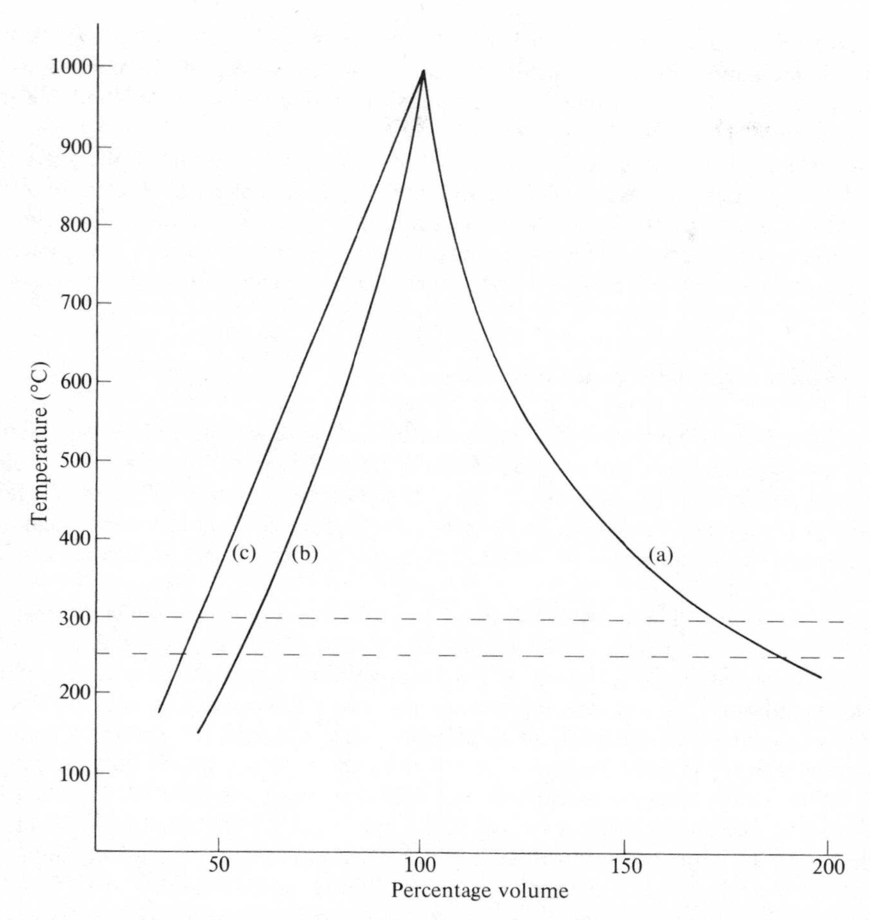

Fig. 8.31. Variation of gas volume cooling from 1000°C: (a) cooling by air dilution; (b) evaporative cooling tower; (c) indirect heat exchange, e.g., waste heat boiler.

of equipment when using air inleakage is several times that obtained when using indirect heat extraction. In many processes, in particular smaller applications and cyclic processes, indirect heat extraction is not considered possible or economical, and cooling of the gas is obtained by spraying finely atomized water, which evaporates completely, leaving the gas at a suitable temperature for the gas cleaning device.

For a temperature of 200°C using this method, the gas volume is 50 per cent of that at 1000°C, compared with 30 per cent by indirect heat extraction. While clearly not as favourable, this constitutes the only practicable alternative method of indirect heat extraction. This method of cooling involves the use of specialized cooling towers with automatic spray control systems, which have been described in detail elsewhere.[17] They are used, for instance, on municipal incinerator plants, where the gas temperature is reduced from around 1000°C to a constant 300°C at the inlet to the precipitator. As the final gas temperature is reduced, the size and hence cost of the cooling towers increases rapidly. This must be balanced against the reduction in cost of the gas cleaner due to the smaller volume.

With wet washers and scrubbers and wet electrostatic precipitators, the gas can be cooled directly after processing or following a heat recovery system. Heat recovery has the effect of reducing the water demand, since the higher the initial temperature, the higher the gas temperature leaving the gas cleaning device, and hence the greater the quantities of water carried away in vapour form, since the gas is saturated.

Condensation and vapour plume formation

Plume formation occurs when the gases fall below the dewpoint and vapour starts to condense. Normally this is the water dewpoint of the gases and occurs below 100°C, though in exceptional cases, when there is a significant quantity of sulphur trioxide present in the gases, a sulphuric acid mist may be formed, and this can condense at temperatures well in excess of 100°C, giving a visible plume of sub-micron acid droplets.

The main concern is the water dewpoint condition, however, and while the plume of water vapour is in itself fairly innocuous, it can, according to the amount of condensation and under certain climatic conditions, create secondary pollution problems, although the gas cleaning process preceding the stack may have removed all but trace quantities of the particulate emission. The problem of secondary pollution arises owing to the lack of buoyancy of the relatively cold saturated plume, which as a result tends to descend fairly close to the stack, so that the pollution is in a rather concentrated localized form. Any traces of corrosive gases such as sulphur dioxide and sulphur trioxide, hydrogen fluoride, etc., together with the traces of dust remaining, can considerably aggravate this secondary pollution. The extent of the problem depends on the type of process involved.

The problem is particularly acute when wet washers and scrubbers are applied to the cleaning of gases at high temperature, as for instance from hot blast cupolas, or arc furnaces. From an arc furnace, the gases are discharged at a temperature of about 80°C (Fig. 8.28) and contain roughly one-third by weight of water vapour. Due to the characteristics of saturated air, a fall of gas temperature of only 10°C results in a condensation of 50 per cent of this water vapour. When the gas temperature is well above the water dewpoint, a considerable amount of dilution with cold air can take place before a visible plume appears. Under these conditions, while the plume may attract comment it rarely gives rise to problems. When wet cleaning devices are used on ambient temperature gases, the question of visible plume formation is not so serious, although the problem of dispersion of the plume, particularly if it contains objectionable odours or dust, can still create a problem.

Although most legislation is aimed at the elimination of particulate emission, there is growing interest in the elimination of certain gaseous contaminants. The most common of these are the oxides of sulphur which arise from the burning of fossil fuels, hydrogen fluoride from certain processes such as the primary smelting of aluminium, and hydrogen chloride which can arise from the secondary smelting process of aluminium scrap and the burning of chlorinated plastics (PVC).

Gaseous contaminants have been removed by fabric filters by coating them with a replaceable layer of a chemical reagent in powder form, designed to react with the contaminant, and therefore remove it. The most commonly accepted approach is a scrubbing method. By chemical dosing of the water used in the scrubber, this can be used to remove both the solid and the gaseous contaminants to a high degree. It is likely, therefore, that the wet washing systems will increase in favour if legislation requiring control of gaseous pollutants is strengthened. In this case the question of the plume dispersion must be taken seriously. Under such conditions, consideration must be given to the use of a more complicated system, whereby the gases are part cooled by indirect heat extraction before passing into the wet cleaning device. This would have the effect of reducing the amount of vapour carried by the gas and hence the intensity of the plume. It might be that the heat so removed could be used for other purposes. On the other hand, if the heat extracted were used, for example, to heat air mixed with the gas leaving the wet cleaning device, the saturation temperature of the gases would be further reduced and the temperature of the gases would be raised. The result would be that the gases would leave the stack well above dewpoint. This system would add considerably to the cost of the gas cleaning system and is unlikely to be generally used until more stringent pollution legislation requires it.

References

1. *Physics of particle sizing analysis,* Institute of Physics Conference, 1954.
2. Jackson, R., *Mechanical equipment for removing grit and dust from gases,* B.C.U.R.A., 1963. Now obtainable from Inst. of Fuel, London.
3. Caplan, K. J., 'All about cyclones', *Air Engineering,* September, 1964.
4. Rose, H. E. and A. J. Wood, *An introduction to electrostatic precipitation in theory and practice,* Constable, London, 1966.
5. White, H. J., *Industrial electrostatic precipitation*, Pergamon Press, Oxford, 1963.
6. Darby, K., 'The use of electrostatic forces in the separation of suspended materials in gases and liquids', *Symposium, Electrochemical Engineering,* Newcastle University, April, 1971.
7. Darby, K. and K. R. Parker, 'The electrostatic precipitator and its range of application', *Symposium, Filtration Society,* September, 1971.
8. Whitehead, C. and K. Darby, 'High efficiency precipitator performance on modern power stations firing fuel oil and low sulphur coal', *International Clean Air Congress,* Washington, 1970.

9. Idris Jones, J., *Proc. Filtration Society,* March–April and April–June, 1972.
10. Dietrich, H., 'Dust filtration in needle felts', *Symposium, Filtration Society,* September, 1971.
11. Calvert, S., *'Source control by liquid scrubbing',* Air Pollution, Academic Press, 1968.
12. Lapple, C. E. and H. J. Kamack, *Chem. Eng. Progr.,* **51**, 110, 1955.
13. Semrau, K. T., *J. Air Pollution Control Ass.,* **10**, 200, 1960.
14. Stairmand, C. J., Personal communication.
15. Nikiyama, S. and Y. Tanasawa, *Trans, Mech. Eng. Japan,* **5**, 68, 1939.
16. Stairmand, C. J., 'Selection of gas cleaning equipment: a study of basic concepts', *Filtration Society Conference,* April, 1969.
17. Whitehead, C. and K. Darby, 'Cleaning of gases from the incineration of waste materials', *Institute of Fuel Conference,* November, 1969.

9 Removal of sulphur oxides from flue gases of power stations and industrial boilers

J. BETTELHEIM and B. H. M. BILLINGE

Among the gaseous air pollutants the sulphur oxides have historically attracted major attention because of their common occurrence and known harmful effects at high concentrations. More recently it was realized that sulphur dioxide is, apart from carbon monoxide, the only major man-produced pollutant which exceeds emission from natural sources. Various estimates indicate a global annual sulphur dioxide emission of about 10^8 tonnes, out of which not more than 2 per cent originates from natural sources (Robinson and Robbins,[71] Rasmussen *et al.*).[67,68] The situation with hydrogen sulphide, the only other important sulphurous gas, is quite the opposite. Natural processes produce hydrogen sulphide at a rate equivalent to about $1{\cdot}5 \times 10^8$ tonnes of sulphur dioxide. Anthropogenic sources contribute only insignificantly. If the present rate of industrial development continues, then the natural emission of all sulphurous compounds might be exceeded by industrial sources at the end of this century.

The close link between sulphur dioxide emission and industrial societies is illustrated by the fact that over 90 per cent of the total is emitted in the northern hemisphere, and at least 70 per cent of this originates from fossil fuel combustion. As sulphur is one of the basic biogenic elements, sulphur dioxide is also a natural constituent of the atmosphere. The lowest background concentrations, detected over the oceans and Antarctica, amount to 1–5 $\mu g\ m^{-3}$ of sulphur dioxide at STP. It is readily absorbed by soil and at low concentration can be directly used by vegetation, particularly in sulphur-deficient areas (Ross,[72] Kamprath[40]).

Extensive research on health effects and vegetation damage indicates that generally deleterious effects are not noticeable below a 400–500 $\mu g\ m^{-3}$ limit for a 24-hour period (Jones *et al.*,[38] Tabershaw-Cooper Associates).[82] Many of the legislative measures which have been adopted aim at ensuring air quality standards in which this range is a maximum. The question of long distance transport and possibility of accumulation of the oxidation products in rainwater or snow are questions which still remain unresolved (Grennard and Ross,[26] Norwegian Ministry of the Environment[58]). The drive to recover sulphur from combustion as an economic source of a valuable raw material lost its urgency owing to the development of new abundant sources of this element in the late 'sixties. Even with the fluctuation of World prices it is very unlikely that extraction of sulphur from combustion processes could become truly competitive. At present the cost of sulphur produced by air pollution abatement methods could be up to an order of magnitude more expensive.

The oldest and most convenient method for preventing environmental damage and nuisance is dispersion of waste gases by chimneys of sufficient height. This height is determined by the parameters of the source (e.g., emission rate) and the local geography in order to avoid excessive ground level concentrations. In cases where resulting stacks would be impractically high, an intermittent control system might be considered. This takes advantage of high level dispersion during favourable meteorological conditions and introduces combustion of low sulphur fuel or partial reduction of load in adverse dispersion conditions. In order to be effective. such a system must include a sulphur dioxide monitoring network, meteorological measurements and a model relating all parameters to ambient air quality standards (Paulus and Roffman,[59] Montgomery *et al.*[53]).

Another more difficult approach is to reduce emission generated per unit of energy produced directly at the source. Apart from the use of naturally occurring low sulphur fuels, which are in short supply, the available techniques can be considered in three categories: extraction of sulphur from fuels, sulphur reduction within the combustor and finally treatment of the flue gases. The concentration of sulphur in fossil fuels is an order of magnitude higher than that in flue gas. There is a prima facie case for cleaning fuel, but the practical difficulties encountered have forced greater attention to be paid to flue gas cleaning. The only large-scale processes which have been operated on a long-term basis are the flue gas washing plants on Battersea and Bankside Power Stations in London with up to 30 years of operational experience. In spite of the substantial new capacities commissioned in the US and Japan since the late 'sixties, still only a small fraction of the sulphur emission is controlled by these processes. This contrasts with the large investment on research and development over the last 15 years, emphasizing the technical difficulties encountered and the controversial character of these measures.

Reduction of sulphur in coal

It has been pointed out that the sulphur content of fuels is an order of magnitude higher than in flue gases. This advantage for early separation of sulphur is partly vitiated by problems of treating non-homogenous, chemically complex solids and liquids. This is well illustrated by the case of coal cleaning. Sulphur in coals appears in both organic and inorganic forms. Organic sulphur, present in the form of cystin, thiols, sulphides and some cyclic compounds, can be intrinsically and almost uniformly bound with most of the carbonaceous material and can be removed only by chemical processing. Relevant industrial experience has been gained in producing liquid fuels from coal under high pressure and temperature. In contrast to the liquid fuel production the sulphur removal technique is aimed at minimizing hydrogen consumption, accepting liquefaction as an unnecessary effect. For example, catalytic hydrogenation of coal suspended in tar at 100–250 atm at 450°C can achieve 75 per cent desulphurization with the consumption of 20 kg hydrogen per tonne of feed coal. Only 5 per cent of this hydrogen is transformed to hydrogen sulphide (US Bureau of Mines[89]). This type of processing is technically sound, but it is unlikely that it could be economically acceptable for the purpose of lowering sulphur content alone.

Practically all inorganic sulphur exists in coals as iron disulphide, the occurrence of sulphates being generally low. Two forms of disulphide are fairly common in most coals. Pyrite (metallic yellow, cubic, density 5·2) is both the more stable and more common in the majority of coals. Marcasite (rhombic, density 4·8) oxidizes more easily and may promote the disintegration of coal exposed to air. Pyrites occur in a wide range of sizes, from massive lumps suitable for hand picking; to finely divided veins and small particles only microns in diameter. Detailed studies have been made on coal briquets by optical methods (McCartney[49]) correlating the size distribution with reduction in pyritic sulphide by a float and sink method. As all pyrites separation methods require mechanical disengagement of the sulphide particles from the combustible matter, the float and sink method at density 1·6 offers a better measure of practically achievable efficiency than chemical methods. In most coals of moderate sulphur content (1–3 per cent), pyrites form 40–60 per cent of the total sulphur but their relative fraction increases with high total sulphur (Wandless,[92] Walker and Hartner[91]).

The feasibility of pyrites removal on a large scale was demonstrated during the Second World War, when about 20 000 tonnes of pyrites were produced annually in Britain (Davies[18]). As the objective was to supply acid manufacturers with high quality pyrites (over 40 per cent sulphur) at prices comparable with pre-war imports, the processing was limited to relatively few profitable sites. All these plants used a combination of conventional cleaning equipment, e.g., jigs and concentrating tables. Use of these techniques, combined with froth flotation, is still dominating most of the recent feasibility studies (Lemke *et al.*,[43] Zimmerman[97]). A study of 90 American coals (Bituminous Coal Research[8]) indicated that up to 50 per cent reduction of total sulphur might be possible for 45 coals crushed to less than 10 mm size. A detailed examination of the sulphur removal potential of German coals (Mandel *et al*[47]) concluded that 25 per cent reduction was achievable even with low sulphur (1·3 per cent) coals. The total operating cost at 1974 prices was about 80p per tonne of coal. About half of the cost is attributed to loss of combustible matter (2·5 per cent). The greatest disadvantage of these techniques is the use of wet washing, which would be difficult to incorporate into the coal handling scheme on a power station site. Separation of dry, ground solids avoids cumbersome filtration, cuts energy consumption and reduces installation costs.

The efficiency of the dry processes is more dependent on the size distribution of pyritic particles than in the case of wet treatment. In air the adhesion forces between particles below 100 μm increase rapidly and make separation of clean fractions very difficult. In principle the density difference offers a sufficient basis for effective separation. Denser pyritic particles of the same diameter as coal have a falling velocity about three times higher and are separable by a variety of methods (Ross[73]). The major difficulty is production of a reasonably monodisperse system, because sieving large quantities of coal in the 500 μm ranges is not practicable.

Magnetic and electrostatic separation could considerably improve chances of large-scale dry separation. The important physical properties are summarized in Table 9.1 (Monostory *et al.*[52]).

TABLE 9.1

	Magnetic vol. susceptibility $\times 10^6$	Conductivity (ohm^{-1} cm^{-1})	Dielectric constant
Coal	−0·6	10^{-13}	4–7
Pyrite	+20 + 60	10^{-3}	5·2
Silica	−20 + 12	10^{-12}	5·1
Pyrrhotite	+250 000		

The difference in magnetic properties is not sufficient for direct separation. Chemical conversion of small quantities of disulphide to pyrrhotite ($4FeS.Fe_3S_4$) results in a large increase of susceptibility. This change can be achieved by dielectric, steam or hot air heating (Ergun[22]). Trindade *et al.*[87] demonstrated the separation of untreated pyrites from a water slurry passing through a high gradient magnetic field. Promising results were reported by Abel[1] who combined centrifugal and electrostatic separation. Even these schemes depend on the suppression of excessive amounts of dust (less than 10 per cent) and benefit from treatment of a narrow size distribution of the fuel. None of the advanced techniques has been developed to a scale which would allow a realistic economic assessment. The major drawback of pyrites removal is the limited achievable efficiency, realistically expected to be in the range of 20–30 per cent of the total sulphur. This obviously cannot compete with the performance of flue gas desulphurization or production of low sulphur fuel oils. Consequently only fairly restricted effort has so far been applied to this field. The prospects for pyrites separation processes would be much more favourable if a moderate reduction of the total emission was the required action. In such a case coal treatment processes might become economically attractive.

Fuel oils

The sulphur content of crude oils varies widely, but high sulphur sources are prevalent. Only about 10 per cent of world supplies contain less than 1 per cent of sulphur, and consequently considerable effort has been applied in the area of oil desulphurization.

Physical separation processes, typically distillation and extraction as used in all petrochemical refineries, give several choices for reducing sulphur in fuel, depending on the initial content and maximum allowable sulphur levels required. Figure 9.1 shows a simplified process flow sheet, with schematic distribution of yields and sulphur contact (Hobson,[29] Jimeson,[35] Horne,[30] Wolfe,[94] Wada[90]).

Crude oil is initially subjected to atmospheric distillation producing a light fraction usually containing less than 8 per cent of the total sulphur (90 per cent of this is in gas oil). The residue, with a boiling point above 350°C, can be directly used as heavy fuel oil or further processed, e.g. by distillation under vacuum. The distillate, vacuum gas oil, contains 30–40 per cent of the original sulphur. The sulphur content of the remaining vacuum residue is further reduced by asphalt extraction. This refining does produce a reduced sulphur vacuum gas oil with sulphur content dependent upon the level in the initial crude oil. Overall economic advantage might arise from concentrating desulphurization techniques on flue gas from the combustion of the high sulphur vacuum residue only. However, conventional heavy fuel oils represent various blends of atmospheric and vacuum residues together with residues from secondary processes, e.g., cracking or reforming. To date, this selective blending satisfactorily meets composition requirements in Europe.

Further reduction of fuel oil sulphur content has been achieved by hydro-desulphurization, a method developed from experience with desulphurization of distillate oils. The technique, now well established on a large commercial scale, treats

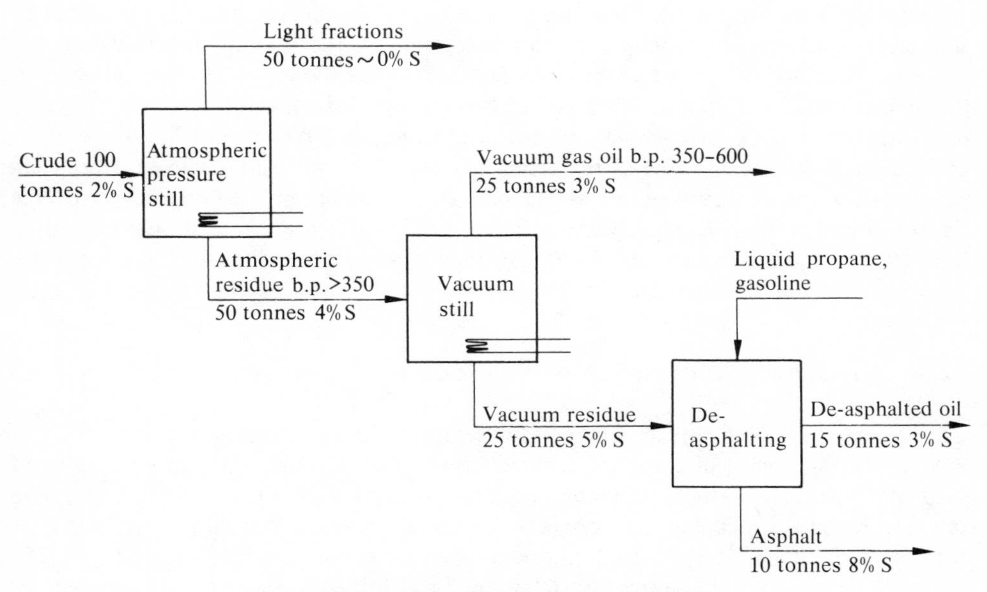

Fig. 9.1. Distribution of sulphur in oil.

fuel oils with hydrogen under high pressure (40–150 atm) and elevated temperatures (300–500°C) which results in partial cracking and consequent release of hydrogen sulphide. The final product of the sulphide is elemental sulphur.

Hydrodesulphurization is a major method of sulphur reduction in fuel oils. It is complicated by the presence of organo-metallic compounds which cause poisoning of the cobalt–molybdenum based catalyst. Asphaltenes, containing sulphur in a molecular structure fairly resistant to hydrogenation, also cause problems. They are the primary reason for high consumption of hydrogen (80–300 Nm^3 $tonne^{-1}$ of feed). The energy requirements of hydrodesulphurization inclusive of hydrogen production represent 3·5–8 per cent of the feed fuel. Two main techniques have been developed, basically distinguished by the type of feedstock. Direct desulphurization uses atmospheric residue and typically achieves 70–80 per cent sulphur removal. The indirect route treats vacuum gas oil which is easier to desulphurize because of low concentration of metals and asphaltenes. The indirect desulphurization product contains only about 0·3 per cent sulphur, and the final fuel is then prepared by blending the low sulphur fraction with other residues. The overall sulphur extraction is correspondingly limited.

TABLE 9.2

Fuel oil HDS capacities, 1975 (10^3 tonnes per day)

	Japan	Caribbean	US	Other
Direct	95	3	5	12
Indirect	146	141		10

Major progress in this field has been made by oil companies. Extensive research is still under way, mostly concerned with extension of the operating periods of the catalyst, new design of the reactor vessels and improvement of energy utilization (Ginneken *et al.*[24]). Table 9.2 summarizes known installed capacities in 1975, totalling over 400 000 tonnes of fuel oil per day, equivalent to 80–100 000 MW of power generation. A substantial part of the product is used for industrial purposes. About 60 per cent of world capacity is situated in Japan, clearly underlining the importance of the environmental issues. The probable total operating cost for direct desulphurization (mid-1975) is in the range of £6–10 per tonne, making no allowance for possible revenue from sulphur sales.

Sulphur reduction in the combustion process

The high temperatures prevalent in modern industrial furnaces make direct extraction of sulphur during combustion impractical. At temperatures above 1200°C even calcium sulphate becomes thermally unstable, limiting the theoretical removal efficiency. The only technique applicable to conventional combustors is the limestone injection technique, in which finely ground limestone particles are suspended in the flow of flue gases. The temperature of the gases falls rapidly with their passage through the boiler.

However, no further development of this process has taken place because of slagging and the inefficient use of limestone. Because the reactions take place in the flue gas downstream from the furnace this technique will be discussed briefly in a later section on the treatment of flue gases.

To exploit the conditions accompanying combustion two techniques, distinguished by the extent of oxidation of the fuel, are being developed. With complete combustion, favourable conditions for sulphur extraction can be achieved only in a fluidized bed, which allows full burn out at a uniform temperature in the range of 700–1000°C. Combustion with substoichiometric amounts of oxygen (and steam injection), lead to fuel gasification producing lean gas suitable for conventional boilers or, preferably, as fuel for gas turbines. Gasification can be performed in a variety of reactors, including fluidized beds.

Fluidized combustion has a long history of industrial application, but boilers with fluidized combustion at atmospheric and higher pressures are only in the early stages of commercial application. Fluidized bed coal gasifiers (the Winkler system) were built in Germany in the late 'twenties to provide lean gas for driving huge process compressors (equivalent of 130 MW). Fluidized bed roasting of pyrites has become acceptable practice, and a number of fluidized bed furnaces of the 'Ignifluid' type (Godel[25]) have been commissioned since 1950. In fluidized bed boilers a complete combustion of fuel occurs in a levitated bed of ash, sand or limestone with significant (over 50 per cent) extraction of total heat by steam tubes submerged in the fluidized bed.

All combustion air passes through the support grit as no secondary air injection is required. Typically fluidization velocities of 1·5–2 m s^{-1} correspond to combustion rates of 200–300 kg m^{-2} of coal, similar to that for conventional stokers. Substantially higher specific output can be achieved with pressurized fluid bed boilers.

In spite of the enhanced heat transfer within the bed, resulting in considerable saving of the steam tubing surface, the capital cost of an atmospheric fluidized bed boiler will probably not be lower than that of a conventional boiler. The major advantage prompting the commercial exploitation is the use of low grade fuels and above all, potential control of sulphur and partially even nitrogen oxides emission. The possibility of sulphur extraction during adverse weather conditions is attractive. From this aspect the fluidized boiler offers a unique solution of the sulphur emission problem with major financial advantages over other techniques. Limestone injected into the bed reacts with sulphur dioxide and forms calcium (magnesium) sulphate.

$$CaO + SO_2 + \tfrac{1}{2}O_2 \rightleftarrows CaSO_4.$$

A sulphur removal efficiency of 90 per cent should be achievable up to 1100°C, but the experimentally established optimum temperature is around 850°C. Moss[56] suggested that above this temperature the process is controlled by sulphur dioxide diffusion through the sulphated shell, causing a fall in rate with increasing temperature. For similar reasons only partial sulphation is achievable in one step. Removal of 70–90 per cent of sulphur requires a calcium to sulphur molar ratio of 2–2·5, resulting in substantial limestone consumption. It should be noted that the fluidized bed system is twice as efficient as the dry limestone technique, which has a similar chemistry.

If the limestone is ground to the same size as coal, then a practically inseparable mixture of ash and lime is produced. An alternative is to fluidize comparatively coarse particles of lime forming a desulphurizing layer in which either oil or pulverized coal is burnt. The fine ash can be elutriated and the partially sulphated lime regenerated:

$$2CaSO_4 + C \rightarrow 2CaO + 2SO_2 + CO_2.$$

This regeneration, which requires temperatures in excess of 1000°C, would substantially suppress the consumption of limestone (Ca/S 0·14 instead of 2·5). The feasibility of this step has yet to be demonstrated on a satisfactory scale. Several commercial prototypes of the atmospheric boiler (Pope, Evans and Robbins; Babcock-Wilcox; Foster/Wheeler) equivalent to up to 30 MW(e) capacity have recently been commissioned and commercial exploitation is most likely before 1980. The pressurized version has been also studied intensively, particularly in connection with combined cycles and gas turbines. The utilization of limestone for sulphur removal will be worse than at atmospheric pressure because of the increased limestone decomposition temperature in higher carbon dioxide partial pressures. The pressurized systems (up to 10 atm) have not yet reached the stage of full-scale demonstration (Roberts *et al.*[70]).

Combustion with substoichiometric amounts of oxygen is carried out in the Esso Chemically Active Fluid Bed Gasifier. Sulphur-containing oil is fed into a fluidized bed of lump lime and partially burned with 20 per cent of stoichiometric air at 900°C. Under the reducing conditions in the bed, sulphur is removed as calcium sulphide which is regenerated by oxidation yielding gas with 8–9 per cent sulphur dioxide. The process calls for a comparatively high degree of recirculation as the calcium material withdrawn from the fluid bed gasifier contains only about 5 per cent calcium sulphide (Moss,[54] Moss *et al.*[55]).

In coal gasification, partial oxidation is accompanied to a greater or lesser extent by hydrogen formation from steam. The primary products contain a mixture of carbon dioxide, carbon monoxide, hydrogen and hydrogen sulphide. Most industrial activity has been historically concentrated on the production of high calorific gas using oxygen and promoting formation of methane. In order to remove sulphurous compounds, the raw gas has to be chemically treated before combustion. The only commercially available pressurized gasifier (20 atm) is the Lurgi fixed-bed system, classically using oxygen. For the production of low calorific gas for power generation, extensive tests have been completed with air blowing. After extraction of sensible heat from the fuel gas, hydrogen sulphide is removed, e.g., by absorption in a solution of potassium carbonate, and the resulting sulphur-free gas is used for steam raising and/or gas turbine drive. A large testing unit for 70 MW(e) generator has been constructed at the Powerton Station of the Commonwealth Edison Co. The Consolidated Coal Company's[15] 'carbon dioxide acceptor process' attempts to avoid the hydrogen sulphide absorption step. The gasifier, operating at 15 atm, uses a combination of air and steam, and carbon dioxide and hydrogen sulphide are extracted by reaction with calcium oxide within the generator.

An entirely new design, suitable exclusively for power generation, has been pioneered by the Applied Technology Corporation.[2] Coal is dissolved by injection into a molten iron bed and is simultaneously oxidized by injected air. Under overall reducing conditions a gas rich in carbon monoxide is produced and sulphur is transferred to a slag of coal ash and added limestone.

The presently available commercial experience indicates that adaptation of conventional gasifiers (such as Lurgi) is likely to dominate industrial applications for the near future. The estimated cost compares unfavourably with flue gas treatment and might add more than 50 per cent to capital cost of generating capacity.

Flue gas treatment processes

Apart from combustion of low sulphur fuels, flue gas desulphurization (FGD) is the only presently commercially available technique for limiting sulphur oxides emissions. Since the introduction of Clean Air Regulations in the US and Japan in the late 'sixties many desulphurization plants have been built in both countries. The capacity installed in the utilities in US at the end of 1975 was not less than 8000 MW, and it is expected that by 1980 it will be 30–40 000 MW (Brocke,[11] US Environmental Protection Agency[21]). The situation in Japan has not been monitored so precisely but in 1975 operating plants represented waste gas treatment corresponding to 12 000–14 000 MW. A major part of this capacity is probably on industrial rather than power generating plants. No commercial plants outside Britain have been operated in Europe. Small units have been tested in many countries, and at least 50 different processes have been tried on a pilot plant scale. The most advanced systems will be discussed in some detail. Further data are available in the literature (Slack[79]).

Apart from limestone injection, dry methods received little interest until the late 'fifties. The much quoted disadvantages of wet methods, namely localized pollution increases because of lost plume buoyancy and excessive water requirements, directed research towards dry methods, particularly of the regenerative types which avoided the problem of waste disposal. The technical problems encountered have now caused a shift back in favour of wet methods, although a few dry processes are being operated on a commercial scale. The re-heat of washed gases is a common design feature of all modern processes. After efficient elimination of carry-over droplets, the usual re-heat is 20–40°C above the scrubbing temperature (Kellog[41]).

As in many other cases the original economic assessment of all FGD processes has proved to be grossly over-optimistic (Templeton[83]). Capital costs in the range of £5–15 kW h^{-1} were frequently quoted before the new units were actually commissioned. Detailed studies were carried out for comparison of the technically feasible processes (McGlamery *et al.*[51]) and FGD was compared with oil and coal treatments (Mandel *et al.*[47]). Because of inflation, rapidly rising costs of energy and a shortage of reliable data on most of the processes, the situation is rather confused, and no attempt has been made to evaluate the financial merits of individual techniques.

The limestone washing processes may be taken as a rough indicator because of their relative simplicity, low cost of the raw material and because data on the actual costs incurred during the last few years are available. The capital cost of FGD in the late 1975 prices (£1= $2·1) varies between £35–60 kW h^{-1}, inclusive of waste disposal. The total operating costs, related to coal with 2 per cent sulphur, are probably in the range of £200–300 per tonne of sulphur dioxide removed. It is evident that FGD is a very expensive technique, adding 20–30 per cent to the installation costs of modern power stations.

Limestone–lime washing

The only process which has been operated on a large scale over a long period is the aqueous scrubbing method first installed at the Battersea power station (London) in 1931 (Rees[69]); with the exception of the war years it has been operating continuously. A plant working on the same principle was installed at Bankside Power Station, London, in 1951 and a second unit on the same site was commissioned in 1962. The removal efficiency of sulphur dioxide has been over 90 per cent. The Bankside flue gas washing plant operates in two parallel sections, each corresponding to 150 MW capacity. The washing is carried out in cast iron, timber-lined towers. Each tower is divided into three passes, two packed with cedar wood scrubber banks and the intermediate pass diverting the partially cleaned gas from the top of the first pass to the base of the second washing section. On leaving the spray eliminators the flue gas is at a temperature of 20–30°C. Alkalinity for sulphur dioxide neutralizaiton is partially supplied in the natural alkalinity of the River Thames and 50 per cent is fed in as chalk delivered in a form of waste sludge from water softening plants. The water leaving the towers is acid and contains mostly calcium sulphite. The effluent is therefore oxidized in four aeration chambers, where some iron sulphate is added as catalyst. The outlet from the aeration chambers is no longer deficient in oxygen but is still acid. Before discharging into the river it is mixed with 10 times its own bulk of used cooling water.

The vast quantities of water needed in this once-through process led to trials for development of a recirculation scheme (Pearson *et al.*[60]). The ICI Howden Process, as it is known, tried first at Billingham and later at Fulham Power Station, London, revealed the necessity of close control of crystallization of calcium salts which could block the system by solid deposits. Good removal efficiency was reached both with lime and chalk. Limestone processes have again received considerable attention more recently, because of their relative simplicity, cost and the availability of limestone.

Over 80 per cent of all FGD capacity installed utilizes the fixation of sulphur dioxide by a calcium salt. It is significant that most of the new plants have suffered lengthy commissioning problems, mostly due to uncontrollable deposition of solids. At the end of 1975 only some 25 per cent of the US plants were operating with satisfactory reliability.

Washing processes using calcium salts can be classified from two aspects, the nature of the primary neutralizing agent and the degree of recirculation of the absorption liquor. A comprehensive document on this technology prepared by the Tennessee Valley Authority[85] was later expanded by a series of theoretical and experimental papers, the most important of which is the study by Lowell.[44]

Either lime (CaO) or finely ground limestone ($CaCO_3$) can be used for fixing dissolved sulphur oxides. The marked difference in the reactivity of these materials is not reflected in the overall performance of desulphurization plant. The basic chemistry occurring in the reaction slurries is very similar for both materials as lime, after hydration, is quickly transformed into carbonate because of the high concentration of carbon dioxide in the flue gas. This also explains why only minor (approximately 10 per cent) differences in absorption efficiencies are achievable. A major advantage of lime usage is much lower consumption as compared with limestone factor (1·5–2), because of precipitation of active, finely dispersed carbonate within the absorption tower. In the earlier stages of process development, attempts were made to produce reactive lime by injection of limestone into the furnace followed by wet scrubbing removal, together with sulphur dioxide removal, but this was unsuccessful.

The chemistry of sulphur oxide absorption is complex and not yet fully understood. The important difference of this system as compared with the magnesium or sodium processes is the very low solubility of calcium sulphite (less than 1/100 of the magnesium sulphite). It is probable that the controlling step is the reaction of sulphur dioxide with calcium bicarbonate. The other important distinction between these technologies is in the degree of water recirculation or water feed rate. Attempts to minimize water consumption are limited by water evaporation into the flue gas and losses accompanying treatment of products.

In all processes using recirculation of absorbent the temperature in the washers equilibrates between 50–58 °C, and the minimum water evaporation accounts for consumption of 0·1–0·15 m^3 MW h^{-1}. The once through, non-recycling processes, such as at Bankside or in sea water washing (Bromley,[13] Bockman *et al.*[9]) result in a low gas discharge temperature (determined by the water feed temperature) thus avoiding any significant water loss by evaporation. Discharge of the products of desulphurization accounts for the more important part of the water balance. For once-through processes the typical consumption is of the order of 10–15 m^3 MW h^{-1}.

The recirculating processes fall within the range defined as 'open' or 'closed' loop operations. Practically all modern installations fall into these categories. High concentrations of sulphur in the circulating absorbent are achieved by building up the calcium salts content, to an ultimate value of 10–15 per cent of solids in the slurry. The concentration and composition of the absorbent is critical to the prevention of uncontrolled crystallization.

Open loop processes are systems in which removed sulphur is discharged as the operating (10–15 per cent solid) slurry and the water of this discharge is not recovered. The previously mentioned Fulham process was a typical example, where concentration of the slurry in the absorption circuit was maintained by discharging part of the slurry into barges and replacing the volume by fresh water and lime. Before 1939 barges were ultimately emptied into the sea. It is, however, equally possible to deposit the solids in settling ponds (as in the case of fly ash) and discharge the overflow to watercourses. For coal with 2 per cent sulphur content (typical for British conditions) the water consumption amounts to 0·2–0·3 m^3 MW h^{-1}. The major objection to this reliable process is the danger of water pollution. The drain water contains 2–4 g l^{-1} of mainly calcium salts accompanied by a variety of metals and trace elements originating from the fuel used. It should be emphasized that for many cases the open loop might be acceptable and each case should be judged on its merits.

In theory, closed loop systems completely eliminate process water penetration, other than by evaporation, into the environment. After settling the solid products in a pond, the overflow is returned into the absorption circuit. Because of the nature of the sludge, particularly the gelling tendencies of calcium sulphite crystals, even the 'dewatered' waste contains substantial quantities of water. The water retained amounts to 0·025–0·075 m^3 MW h^{-1}. Considerable supersaturation of the circulating liquor with calcium salts can occur, promoting the possibility of uncontrolled crystallization and deposition of solids, particularly on mist eliminators and the internal structure of absorbers. Under steady state conditions the 25–75 l of water retained in the disposal pond must also contain all reactive materials produced by combustion of approximately 350 kg of coal or 220 kg of oil. Thus it is particularly important to select corrosion resistant construction materials (Tice[86]). Disposal of the reaction sludge is still a major problem. The mixture of sulphites, sulphates and unreacted limestone is thixotropic and calls for large packing volumes. Starting with 10–15 per cent slurry, settling alone produces only 30–50 per cent solid content with starting settling rates about 5 cm h^{-1}.

Vacuum filtration and centrifuging results in about 60 and 75 per cent solids respectively. The dewatered sludge shrinks but returns to its original state when rewatered (Rossoff and Rossi[74]). Under present conditions 1·5–2 m^3 are needed for the disposal of 1 tonne of the solid product (i.e., 40 MW h). It is expected that substantial compaction will be possible by aeration, oxidation or chemical treatment of the sludge. Evans[23] estimated that in 1980 over 70 million tons of this material will be generated annually in the US alone.

Limestone scrubbing is also used on a large scale for sulphur dioxide emission control in the USSR. After preliminary trials of the cyclic method an industrial unit, comprising 21 scrubbers each of 6 m diameter, was installed for desulphurization of 3×10^6 m^3 h^{-1} of waste gas from an agglomeration plant. Attempts to use the waste sludge for production of building materials were unsuccessful (Brodski *et al.*[12]).

Magnesium regeneration methods

The magnesium regenerative method represents one of the most advanced processes. It has been developed to an industrial scale in the US by Chemical Construction Corporation (Chemico) in cooperation with Basic Chemicals Inc., and in the USSR by two research institutes, Niiogaz and Niiuif.

All variants of the magnesium regenerative process are based on the low solubility of magnesium sulphite in water. Magnesium sulphite produced in hydrated form in the absorption cycle is separated in solid form and calcined to produce sulphur dioxide and magnesium oxide for recycling. Magnesium sulphite is always accompanied by varying amounts of magnesium sulphate due to the direct reaction with sulphur trioxide and secondary oxidation of sulphite. There are four distinct operations: absorption of sulphur dioxide from flue gases, separation of insoluble components, drying of a filter cake and finally calcination.

The main chemical reactions can be expressed by the following equations, neglecting the different hydrates possible:

$$MgO + SO_2 = MgSO_3$$
$$MgSO_3 + SO_2 + H_2O = Mg(HSO_3)_2$$
$$Mg(HSO_3)_2 + MgO = 2MgSO_3 + H_2O$$
$$MgO + SO_3 = MgSO_4$$
$$MgSO_3 + \tfrac{1}{2}O_2 = MgSO_4$$
$$MgSO_3 = MgO + SO_2$$
$$MgSO_4 + \tfrac{1}{2}C = MgO + SO_2 + \tfrac{1}{2}CO_2.$$

There is a distinct similarity between magnesium regenerative methods for sulphur dioxide removal from waste gases and the magnesium process used in paper pulp production, known commercially as the 'Magnefite process' (Darmstadt, Tomlinson,[17] Hull *et al.*,[31] Rydolm[75]). About 1 million tonnes per annum of pulp are produced in the US by the Magnefite process, and operating experience with absorption and calcination in the magnesium oxide cycle is therefore available. There are, however, several differences:

1. 'Cooking liquor' in the Magnefite process requires magnesium bisulphite solution with excess sulphur dioxide and a pH of about 3·5, which is achieved by absorption of fairly concentrated sulphur dioxide gases. Long-term loss of magnesium oxide reactivity, which is possible in the Chemico process, is not observed in the Magnefite process.
2. No slurry of magnesium salts is produced and consequently no filtration is required in the Magnefite method.
3. Intimate contact between magnesium sulphate and carbonaceous material is ensured in the Magnefite process, since the waste liquor solution contains lignins, sugars and added material before combustion. This could result in better reduction conditions than in the Chemico technique.

Chemico have developed two variants of the process (Chemico,[14] Dennis,[19] Maurin and Jonakin[48]).

1. The magnesium oxide slurry process. This is designed for ash-free gases, such as tail gases from sulphuric acid production or oil-burning furnaces, and is schematically illustrated in Fig. 9.2. Sulphur dioxide is washed out in a Venturi scrubber and precipitated crystals of magnesium sulphite and sulphate are separated from the absorption slurry. The filter-cake is dried and after addition of carbonaceous material finally decomposed in a calciner.
2. The clear liquor process. This is designed for gases containing substantial amounts of particulate matter. The scrubbing is carried out by a solution of magnesium salts; suspended fly-ash is separated from the liquid phase and the

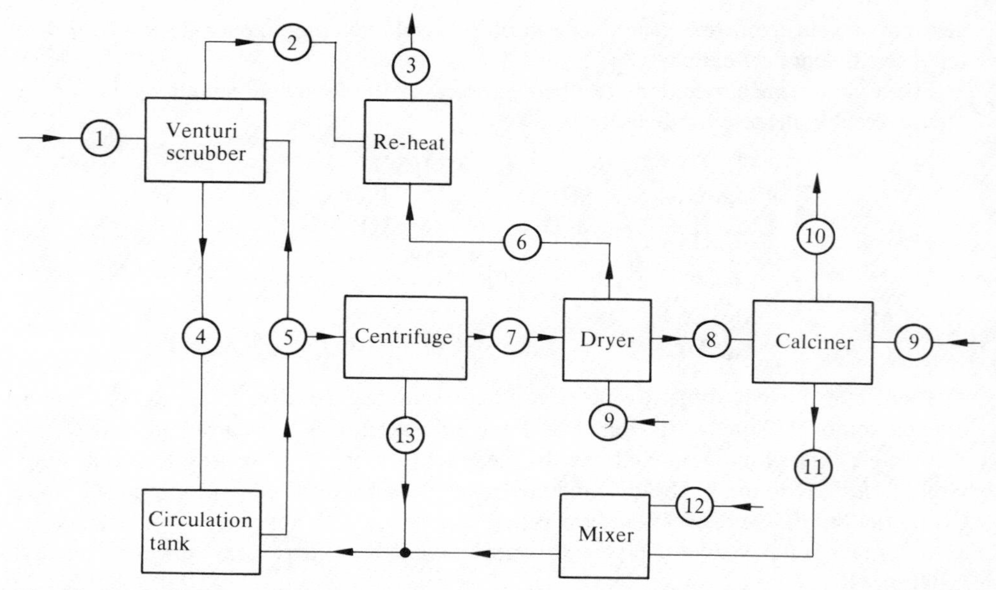

Fig. 9.2. Block diagram: Chemico slurry process. (1) SO_2 gas; (2) cold desulphurized gas; (3) reheated desulphurized gas; (4) outlet slurry; (5) built-up slurry; (6) hot wet gas; (7) filter cake; (8) dry Mg material; (9) fuel; (10) rich SO_2 gas; (11) MgO; (12) H_2O; (13) mother liquor.

absorbed sulphur dioxide is then precipitated by addition of magnesium oxide. The calcination section is identical with the slurry process.

The main achievements of the Chemico processes are:

1. Considerable improvement in the plume rise as a result of re-heating desulphurized flue gases with the outlet from the dryer at 200°C, made possible by separating the drying and calcination operations.
2. Chemical processing on power generation sites might be reduced by the use of 'a central sulphur dioxide recovery plant concept'. Dried magnesium sulphite and sulphate mixture is transported to a central site for calcination and sulphuric acid production, and the recovered magnesium oxide is returned to individual generation sites for adsorption. This concept would require
   (a) the transport of about 1000 tonnes per day of magnesium oxide, sulphite and sulphate mixture for a 1000 MW plant,
   (b) facilities for cooling of magnesium sulphite and sulphate after the dryer, and
   (c) slightly higher heat consumption in calcination.
3. Electrostatic precipitators may not be required if a sufficiently high ash removal efficiency is achieved in the Venturi scrubbing system.

Since 1972 Chemico have extensively tested the process on 155 MW oil fired Mystic Station No. 6 of Boston Edison Co. The magnesium oxide recovery and acid production were carried out at a converted sulphuric acid plant of Esso Chemical (Shah and Quigley[76]). After three years of intermittent operation the development programme was completed and the facility shut down. Another retrofitted Chemico plant was built at the 100 MW Dickerson 3 Station of the Potomac Electric and Power Co.

A distinctive feature of the Russian process is the direct use of slurry scrubbing even for gases containing ash. A combination of hydrocyclones and vacuum belt filters is used for classification and separation of the solid phase from the absorption slurry. After each removal the clarified magnesium salts are returned to the production cycle (Pinaev and Mosharov[64]). The absorption and separation processes have been tested since 1960 on a pilot scale using flue gases from a lignite burning plant. Following poor performance of a rotary calcining furnace Niiuif used a fluidized bed system in which fuel gas was burnt directly. Combined drying and calcination in a single unit was attempted but the tests have not been scaled up to larger units.

In spite of the fact that magnesium sulphite hexahydrate should be metastable at temperatures above 40°C, a mixture of hexa and trihydrate was precipitated under most operating conditions of flue gas scrubbing (i.e., well above 50°C). Hexahydrate contains more than 50 per cent of water by weight, which must be evaporated before calcination. Precipitation of trihydrate therefore appears preferable from the heat balance point of view. In the USSR such a process, called 'the high temperature method' was developed (Pinaev and Mosharov[64]) but no report about its application on large pilot plant tests is available. The main drawback is apparently the separation step. Filtration of magnesium sulphite trihydrate is reportedly difficult and the advantages gained by depression of the water of crystallization may be more than counterbalanced by mother liquor retained in the filter cake. The reason for this is obviously different crystal size as the hexahydrate is much larger.

Pinaev[62] used the dynamic method with solutions of similar concentration to actual operating conditions, and measured equilibrium sulphur dioxide pressure for the range of temperatures form 30 to 60°C, pH 4·02–5·88. The minimum equilibrium pressure of sulphur dioxide (P_{SO_2}) observed was 0·03 mm Hg, maximum 0·295 mm Hg.

For solutions saturated with magnesium sulphite the following expression was presented:

$$\log P_{SO_2} = b \log C_{SO_2} - a,$$

where P_{SO_2} is the equilibrium pressure of sulphur dioxide (mm Hg),
C_{SO_2} is the total sulphur dioxide concentration (g l^{-1}) (i.e. magnesium bisulphite and sulphite).

Values of constants a and b for different temperatures and compositions of the solution are given by Pinaev.[62]

Higher partial pressures were reported by earlier investigators (Kuzminikh[42]). Pinaev and Pitelina[63] also published data for a technologically interesting range of conditions for a system unsaturated with magnesium sulphite. They argue that in practice this is a more realistic case as the rate of reaction between dissolved sulphur dioxide and magnesium sulphite is much higher than the rate of solution of solid sulphite crystals, existing always in excess in the slurry. For both saturated and unsaturated solutions, it was concluded that over 90 per cent absorption efficiency requires molar ratios of sulphur dioxide to magnesium in solution of less than 1·6.

Grillo process

In the Grillo process sulphur dioxide is removed from flue gases by washing with an aqueous suspension of magnesium and manganese oxides, followed by thermal regeneration of the spent reagent.

It was stated (Lowicki,[45] Lowicki and Husman[46]) that the oxides used have the general formula xMgO.MnO_2 in which x varies between 3 and 6. A compound Mg_6MnO_8 has been detected by X-ray studies. A sorbent of this composition enables high efficiency absorption even at contact times less than one second (Zentgraf[96]) with thermal regeneration at temperatures preventing magnesium oxide sintering and consequent loss of activity.

Assuming that, in a water slurry, these compounds react as individual oxides, the main reaction of magnesium salts in the absorption circuit would be identical as in the magnesium methods, accompanied by reactions of manganese dioxide

$$\begin{aligned} MnO_2 + SO_2 &= MnSO_4 \\ 2MnO_2 + SO_2 &+ Mn_2O_3 + SO_3 \\ Mn_2O_3 + 3SO_2 &= Mn_2(SO_3)_3 \\ Mn_2(SO_3)_3 &= MnSO_3 + MnS_2O_6. \end{aligned}$$

However, in the solid phase these basic compounds would be found as hydrates, their form depending on the reaction temperature. If the absorption takes place at about 60°C they would probably be magnesium sulphite and sulphate hexahydrates, and manganese sulphate monohydrate. Both sulphates are fairly soluble in this range of temperature (35–40 per cent by weight) and could be separated in solid form only after saturation of the liquid has been reached. The absorption slurry at pH 6·5–7 is reported to remove 90–95 per cent of the sulphur dioxide. Lowicki[45] reports that the sulphur content of the loaded mass for this pH range is 17–21 per cent. Of the sulphur present, 50 per cent should be in the form of sulphite ($MgSO_3$) and the rest mainly as sulphate.

Figure 9.3 shows that hot flue gas (stream 1) passes through the absorber concurrently with a spray of aqueous suspension, which is recirculated (4, 5), from the circulation tank. Desulphurized gases (2) enter highly efficient droplet eliminators, and after mixing with effluents from the dryer (9), and possible re-heat, are discharged to the atmosphere (3).

A major departure from conventional design is the total omission of a separation of solid phase from the circulating suspension. A fraction of the absorbent saturated with sulphur dioxide (7), equivalent to the sulphur removed, is injected into a spray dryer using hot gases from a separate oil-fired furnace (10). The spray dryer produces a powder material (8) which could be stored and transported to a roasting plant. The equipment up to the dryer must be adjacent to the power station. Regeneration is carried out in a fluidized bed using oil both as a source of energy and a reducing agent (14).

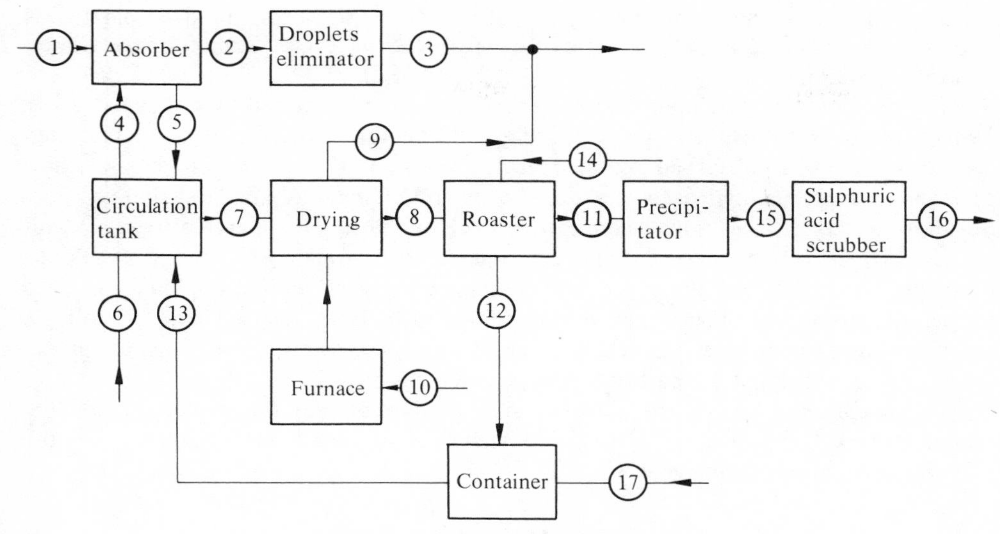

Fig. 9.3. Block diagram: Grillo process.

Sulphur dioxide gas (11) is drawn off for final treatment to give liquid sulphur dioxide, sulphuric acid or elemental sulphur (15, 16). Fresh absorption mass from the roaster (12) is conveyed to a container, where it is added to the make up (17) and fresh water (6) and recirculated (13) (Husmann and Hänig[32]).

Initial laboratory research started in 1964, and was followed by pilot plant tests by Union Rheinische Braunkohlen Kraftstoff AG, Wesseling, between 1967 and 1969. The gas handling capacity of the pilot plant was 27 000 $m^3 h^{-1}$.

A comprehensive cost and design study of the Chemico and Grillo concepts was carried out by McGlamery *et al.*[50] The report presents available process design data and critically outlines the main features of several variants. An economic comparison of an on-site and central recovery plant indicates the competitiveness of this type of technology with lime–limestone washing.

The Wellman–Lord process

This removes sulphur dioxide by washing flue gases with an aqueous solution of sodium sulphite. This treatment of waste gases containing sulphur dioxide is quite common practice in the chemical industry. The solutions produced are used as intermediate products or for direct production of sodium pyrosulphite ($Na_2S_2O_5$). The process was tested on a pilot scale at the Gannon Power Station of Tanysa Electric Co., Florida and a 25 MW prototype was subsequently installed at Crane power station of the Baltimore Gas and Electric Co. (Maurin and Jonakin,[48] Dennis[19]). Commercial units were installed at the Paulsboro works of Olin Mathison Chemical Corp. and at the Chiba

Works of the Japan Synthetic Rubber Co. (Tsushima[88]). In 1973 a plant was commissioned on a 220 MW boiler of the Nishi-Nagoya Power Station of the Chubu Electric Power Co.

The gas containing sulphur dioxide (Fig. 9.4) is at first washed by circulated water which removes fly ash and partially takes out sulphur trioxide. In the second step sulphur dioxide is absorbed by a saturated solution of sodium sulphite, which is converted to sodium hydrogen sulphite. The absorption liquid is piped to a double effect evaporator–crystallizer, where concentrated sulphur dioxide is liberated together with steam. About 8–10 tonnes of steam are required for stripping off 1 tonne of sulphur dioxide. The outlet from the evaporator contains crystals of sodium sulphite, which are separated from the mother liquor. After addition of make up, caustic soda crystals are dissolved in the condensate from the cooler and returned to absorption. To maintain the desired level of sulphates a part of the recycle steam is purged.

The equilibrium partial pressures of sulphur dioxide and steam, determining the sulphur dioxide removal efficiency and steam consumption, are expressed by the following equations (Johnstone *et al.*[36]):

$$P_{SO_2} = M \frac{(2C - S)^2}{C - S},$$

where

$$\log M = 3{\cdot}519 - 1987/T$$

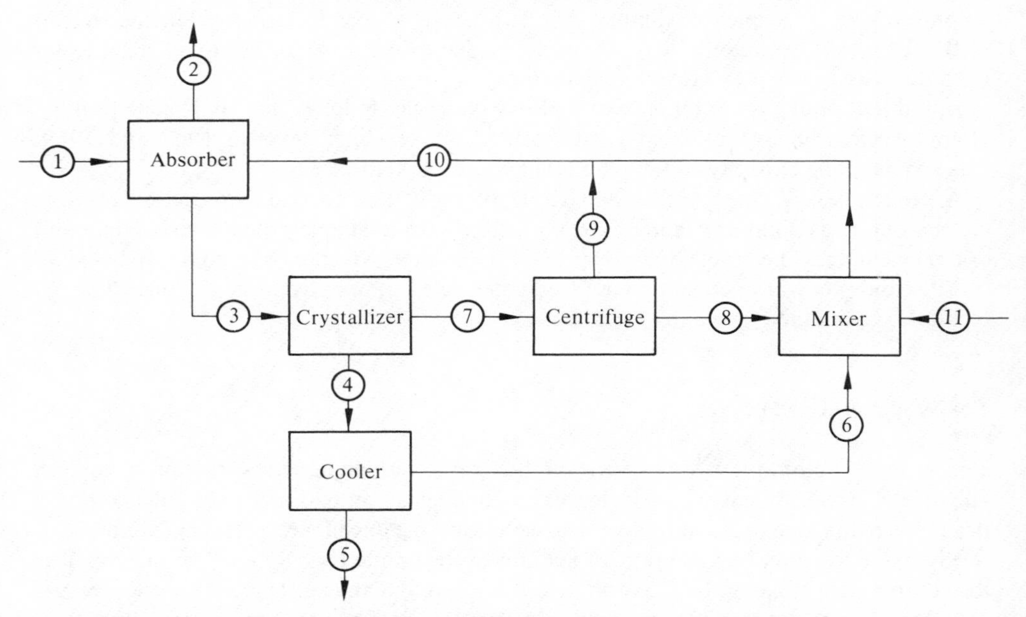

Fig. 9.4. Block diagram: Wellman–Lord process. (1) SO_2 gas; (2) desulphurized gas; (3) saturated absorbent; (4) wet SO_2 rich gas; (5) dry SO_2 gas; (6) condensate; (7) slurry of Na_2SO_3 crystals; (8) filter cake; (9) mother liquor; (10) recirculated absorbent; (11) Na_2CO_3.

and

$$P_{H_2O} = P_W \frac{100}{100 + C + S},$$

where P_{SO_2} = equilibrium partial pressure of sulphur dioxide (mm Hg),
P_{H_2O} = equilibrium partial pressure of steam (mm Hg),
P_W = equilibrium pressure of steam at given temperature (mm Hg),
C = moles of sodium per 100 moles of water in the solution,
S = moles of sulphur dioxide per 100 moles of water in the solution,
T = temperature in K.

The complicated liquid–solid phase equilibria in the system Na_2SO_4–Na_2SO_3–$NaHSO_3$–H_2O were measured by Jager *et al.*[33,34]

The dry limestone technique

This represents the simplest on-site treatment for limiting sulphur dioxide emission from combustion processes. It requires the provision of cheap limestone–lime material which is finely ground and injected into the furnace at suitable temperatures to permit reaction with sulphur dioxide. The reacted solid is collected in conventional particulate arresting equipment and disposed of together with the flue fly ash.

The complex mass and heat transfer process is characterized by the following equations:

$$CaCO_3 + SO_2 = CaSO_3 + CO_2$$
$$CaCO_3 = CaO + CO_2$$
$$CaO + SO_2 = CaSO_3$$
$$CaCO_3 + SO_2 + \tfrac{1}{2}O_2 = CaSO_4 + CO_2$$
$$CaO + SO_2 + \tfrac{1}{2}O_2 = CaSO_4.$$

The limits of the thermal stability of the relevant compounds for typical flue gas conditions are summarized in Table 9.3.

TABLE 9.3

Maximum temperatures for compound stability in flue gases†

Compound temperature (K)	$M(OH)_2$	MCO_3	MSO_3	MSO_4
Ca	680	1045	729	1480
Mg	373	660	522	1130

† 2000 ppm sulphur dioxide; 15 per cent carbon dioxide; 1% oxygen (volume basis)

The system has been studied extensively throughout the world on the laboratory scale and on industrial trials (Slack and Falkenberry,[78] Williams and McLaren,[93] Zentgraf,[95] Hartman and Coughlin,[27] Borgwardt,[10] Pigford and Slinger,[61] Coutant *et al.*,[16] Bellot *et al.*[3]).

It was found that only 20–65 per cent removal of sulphur dioxide can be achieved using a feed of limestone up to 2·5 times stoichiometric. Adding of limestone to coal burning furnaces results in formation of silicates and the consequent decrease in melting temperature of slags. Thus great care has to be taken not to interfere with the boiler operation (Plumley *et al.*[66]). In general, smaller particle sizes, less than 100 μm, lead to better removal of sulphur dioxide, but this advantage is off-set by the cost of grinding.

In spite of great economic advantages predicted for this process in the early stages of development (Tennessee Valley Authority[84]) it has never been adopted on a continuous basis on a large scale. Apart from technical difficulties, such as uniform distribution of injected limestone in a suitable temperature zone, several financial constraints have been indicated for typical modern coal burning stations (Bettelheim and Collins[4]).

The cat–ox process

This is also referred to in the literature as the Penelec or Portland or Monsanto process; it is based on direct oxidation of sulphur dioxide to sulphur trioxide on a vanadium-based catalyst. The sulphur trioxide produced is absorbed in circulating sulphuric acid and the desulphurized gases leave the system at temperatures of 100–120°C. The main differences between the conventional contact process for sulphuric acid production and cat–ox are shown in Table 9.4. The development of the cat–ox process by Monsanto started in 1961 (Stites *et al.*[81]) with a capacity of about 700 $m^3 h^{-1}$ of flue gas at the Seward Generating Station of Pennsylvania Electric Co. (Dennis[19]). In seven years it reached the pilot stage with a plant at Portland station treating 6 per cent of flue gas of 250 MW plant burning bituminous coal (Slack,[77] Stites[80]). At present a 125 MW plant is operated on the coal-fired Wood river Plant.

Two different layouts are being offered for the cat–ox process.

TABLE 9.4

Main design differences between the cat–ox and the contact processes in the conversion of sulphur dioxide to sulphuric acid

	Contact process	Cat–ox
Sulphur dioxide concentration (per cent)	7–11	0·2
Final step in purification	Wet washing	Electrostatic precipitator
Raw sulphur dioxide gas: water/sulphur dioxide molar ratio	1	30
Operating temperature of absorption	$<$ 50°C	$>$100°C
Converter	Multi-stage	One-stage

*1. **Reheat cat–ox.*** This is envisaged for existing power plants with particulate matter precipitators working at the end of the flue gas line, e.g., in the range 150–200°C. As the 'ignition temperature' (i.e., temperature at which the catalyst becomes effective) of a vanadium catalyst is much higher (above 400°C), the reheat cat/ox system provides the necessary steps for the reheating of comparatively cold flue gas (Fig. 9.5). The flue gas, cleaned of fly ash in the precipitator (A), is drawn through a Ljungstrom regenerative re-heater (B) and mixed with the required quantity of hot flue gas from an oil- or gas-fired furnace (C). As the sulphur dioxide content is very low, the converter (D) would operate under almost isothermal conditions. The heat of the sulphur trioxide-containing gas is partially recovered in the Ljungstrom heat exchanger and the flue gas is then passed into an absorption tower (E) operating between 105 and 120°C where the sulphur trioxide is absorbed with high efficiency together with a corresponding amount of water vapour. The sulphuric acid mist leaving the absorber is removed in a glass-fibre-packed eliminator (F) and the clean gas is then passed into the atmosphere.

*2. **Integral cat–ox system.*** This has been designed for new installations. To avoid the complicated re-heat system, mechanical and electrostatic precipitators have been

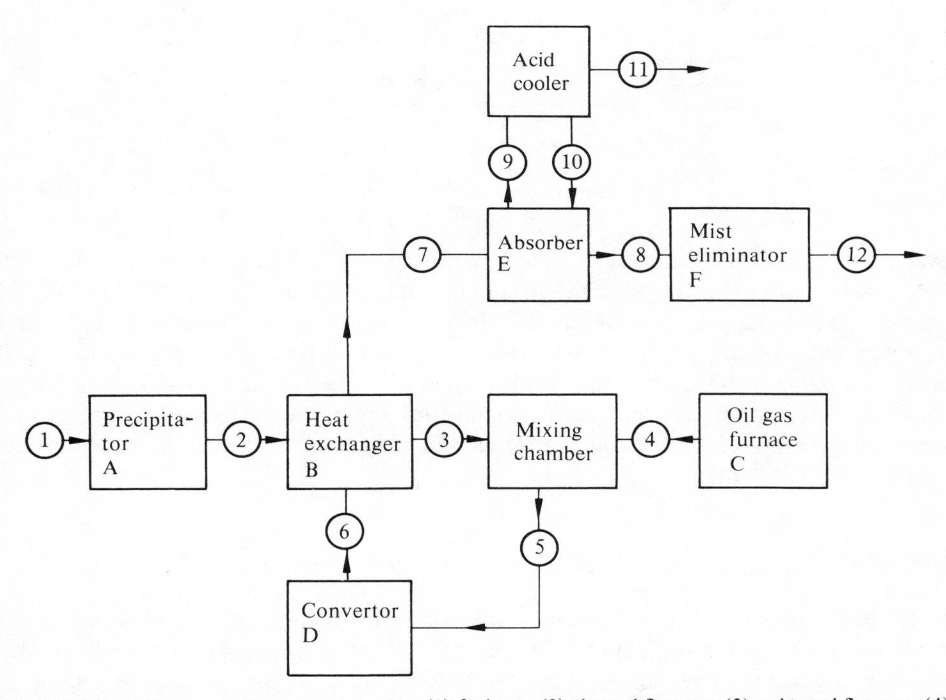

Fig. 9.5. Block diagram: reheat cat-ox system. (1) fuel gas; (2) cleaned flue gas; (3) re-heated flue gas; (4) hot combustion gas; (5) hot flue gas (>400°C); (6) hot gas containing sulphur trioxide; (7) cooled sulphur trioxide gas; (8) gas outlet from absorption 120°C; (9/10) recirculation of absorbing acid; (11) production acid; (12) cleaned gas to atmosphere.

designed to operate at about 500°C. The heat of the converted flue gas is recovered in an economizer followed by an air re-heater with tail end processing as above.

The theoretical limits of sulphur dioxide oxidation can be estimated for the reaction

$$SO_2 + \tfrac{1}{2}O_2 \rightleftarrows SO_3$$

from equilibrium conditions.

Table 9.5 shows the theoretical limit for sulphur dioxide conversion within the temperature range 400–500°C, for 1, 2 and 4 per cent oxygen in the flue gas. Hence, the lower operating temperatures, at constant oxygen concentration, favour higher sulphur trioxide yields and are also preferable for fly ash precipitation. However, the rate of conversion decreases markedly with increase in temperature above about 400°C. By careful choice of catalyst, the optimum minimum operating temperature is achieved in the range 420–480°C. The change of flue gas temperature during the

TABLE 9.5

Equilibrium conversion of sulphur dioxide in flue gas

Temperature (°C)	400	450	500
Percentage of oxygen		Percentage conversion	
1	97·7	93·2	83·3
2	98·4	95·1	87
4	98·8	96·4	90·9

isothermal oxidation of sulphur dioxide is negligible under the conditions specified with only 1–4 per cent oxygen in the flue gas and the converter can consequently be designed as a simple one-stage isothermal reactor. However, it should be emphasized that a drop in temperature of the converter (e.g., due to substantial decrease in the station load) below the value at which the catalyst becomes effective would stop the oxidation of sulphur dioxide. In spite of the 99·7 per cent efficiency of the electrostatic precipitators, the accumulation of particulate matter in the converter calls for screening of the catalyst every 3 months.

Dry methods

In the late ’fifties the widely publicized disadvantages of wet methods, particularly for application on very large generating stations, led to an upsurge of interest in dry removal processes. The developments followed two main paths; chemisorption of sulphur dioxide by single or mixed metal oxides followed by thermal regeneration or sulphate reduction; and chemisorption on active carbon which was then regenerated by thermal or washing methods.

Metal oxide methods

The alkalized alumina process typifies these methods, and illustrates the problems which stopped their development. The sorbent, a porous form of sodium aluminate, was exposed to flue gases over the temperature range 130–330°C in a shallow fluidized bed or a 'falling cloud' reactor (Bienstock *et al.*,[5] Newell[57]):

$$2NaAlO_2 + SO_2 + \tfrac{1}{2}O_2 \rightarrow Na_2SO_4 + Al_2O_3.$$

The spent material was transferred to a moving bed regenerator, where it was treated in a reducing gas at 650–700°C:

$$Na_2SO_4 + Al_2O_3 + 4H_2 \rightarrow 2NaAlO_2 + H_2S + 3H_2O.$$

The product hydrogen sulphide was oxidized to elemental sulphur via the Claus process:

$$2H_2S + SO_2 \rightarrow 3S + 2H_2O.$$

The high cost of the oxide sorbent made its chemical and physical integrity a primary economic factor. Cycling tests showed that the capacity fell to a steady level, but X-ray studies revealed a loss of sodium, probably by evaporation of sodium hydroxide, an intermediate in the regeneration step (Billinge *et al.*[7]). This would have eventually caused a reduction in capacity. The rigorous temperature and chemical cycling to which the sorbent was subjected led to its deterioration as a result of crystallite growth and loss of surface area. These changes not only affected the diffusion-controlled sorption kinetics but caused physical weakening, resulting in uneconomically high attrition rates.

The Shell FGD process (Ploeg *et al.*[65]) overcomes some of the major disadvantages by using an inert base (alumina) to support the reactant copper oxide and evolving a sorption regeneration cycle which can be carried out without change in temperature. Third, the particulate adsorption bed is fixed, reducing attrition loss to a minimum. As the reactant is impregnated onto an inert base, the overall capacity of the sorbent is low, so that frequent regeneration is required. The sorbent is held between wire gauze plates which are mounted parallel to the direction of flue gas flow. Sulphur dioxide (and oxygen) react with the copper oxide impregnant at a temperature of ~400°C to form copper sulphate:

$$SO_2 + \tfrac{1}{2}O_2 + CuO \rightarrow CuSO_4.$$

This is regenerated, at the same temperature, in a hydrogen-rich gas:

$$CuSO_4 + 2H_2 \rightarrow Cu + SO_2 + 2H_2O.$$

The resultant sulphur dioxide stream is converted to elemental sulphur via the Claus process. This process is now being operated successfully on a commercial scale (115 000 $Nm^3\ m^{-1}$) in Japan. The development has been mainly aimed for refinery and oil-burning station use, but trials are in progress on coal-fired plant.

Carbon processes

Carbons have been used for sorption from liquids and gases for many years. Their usual mode of action is physical absorption on a large internal surface area, as in the Landmark process for recovering carbon disulphide from rayon plants. They were, therefore, an obvious choice for sulphur dioxide sorption. One of the first larger developments using carbon—the Reinluft process—was due to Johswich.[37]

A cascading or slow-moving bed flowed countercurrent to the flue gas flow. The flue gas temperature was held at about 100°C. Carbon emerging from the system held from 1 to 5 per cent sulphur dioxide by weight. The carbon particles, manufactured from a lignite coke, were fed into a regenerator where the sulphur dioxide was stripped off by heating the solid to 400°C in inert gas or cleaned flue gas. The high temperature of the sorption and regeneration stages and the presence of carbon dioxide in the regenerant gases showed that the process was one of chemisorption. The increasing surface area of the carbon with cycling, claimed as one of the advantages of the process, was also indicative of chemisorption. Laboratory examination of the water–oxygen–carbon–sulphur dioxide system showed that sulphuric acid forms on the surface of the carbon. Thermal regeneration causes oxidation of the carbon surface, with a consequent increase in surface area and sorptive capacity (Billinge[6]):

$$2(C—H_2SO_4) > (C) + CO_2 + 2H_2O + 2SO_2.$$

Whilst this increase is initially to the advantage of these processes it leads directly to their major disadvantages: the increase in surface area makes the carbon more sensitive to oxidation, with a consequent fire risk within the bed. The loss of carbon reduces the physical strength of the material, resulting in a high loss by attrition.

These problems caused development of the process to be discontinued, but a number of similar processes which attempt to avoid the drawbacks have progressed further. In the Hitachi Process (Hishinuma *et al.*[28]) the spent carbon is regenerated by water washing, thus eliminating the loss of carbon by oxidation. The product, dilute sulphuric acid, is not an easy one to dispose of. The Japanese concentrate it and use it in the manufacture of gypsum, for which there is a market in that country. Even though product processing is expensive, a unit treating flue gas from 150 MW of plant is being successfully operated at the Kashima plant of Tokyo Electric. The Bergbau-Forschung Process (Jüntgen *et al.*[39]) offers alternative regeneration systems. The spent carbon from the moving bed adsorber may be regenerated by washing as above or by a thermal method where the absorbent is contacted with hot sand in an inert atmosphere. This method allows a compact regenerated high temperature and minimizes fire risk. A high sulphur dioxide content gas results which may be used to produce liquid sulphur dioxide, sulphuric acid or elemental sulphur via the Claus process. As carbon dioxide is also a product, it is obvious that chemical loss of the absorbent occurs. A 20 MW pilot plant is operating in Florida, US.

Other types of FGD process utilize carbon as a catalyst in wet washing processes. In both the Lurgi (Ellwood[20]) and Hitachi wet processes (Hishinuma *et al.*[28]) the flue gases contact wetted carbon beds or carbon–water slurries. The chemistry apparently follows the same route as dry adsorption/wet regeneration processes in that the carbon surface catalyses the oxidation of dissolved sulphur dioxide to sulphuric acid.

The large quantities of dilute sulphuric acid resulting from these processes create considerable disposal problems except in certain locations.

References

1. Abel, W. T., *Removal of pyrite from coal by dry separation methods*, PB 221 627, 1973.
2. Applied Technology Corporation, *SO_2 free two stage coal combustion process*, PB 211 888, 1972.
3. Bellot, J. A., *et al.*, *Review of the dry limestone injection process*, PB 211 431, 1972.
4. Bettelheim, J. and A. C. Collins, *Chem. and Ind.* 122–4, February 1975.
5. Bienstock, D., *et al., Trans. Amer. Soc. Mech. Eng.* 350, July 1964.
6. Billinge, B. H. M., *Second Ind. Carbon and Graphite Conf.*, 289, London, 1965.
7. Billinge, B. H. M., *et al., Phil. Trans. Roy. Soc.*, **A265**, 309–18, 1969.
8. Bituminous Coal Research, *An evaluation of coal cleaning processes and techniques*, PB 192 484, 1970.
9. Bockman, O. K., *et al.*, 'Removal of sulphur dioxide from flue gases by absorption in sea water', *Eng. Int. Scand. Congr. Chem. Eng.*, **AII**, 1–11, 1974.
10. Borgwardt, R. H., *Environ. Sci. Technol.*, **4**, 59–63, 1970.
11. Brocke, W., *Brennstoff-warme-Kraft*, No. 3, 102–8, 1974.
12. Brodski, Y. U., *et al.*, 'Desulphurization of flue gases', *ECE Seminar*, Geneva, 16–20 November 1970.
13. Bromley, L. A., *Int. J. of Sulf. Chem.* **B7**, 77–84, 1972.
14. Chemical Construction Corporation, *Magnesium base processes for SO_2 recovery and fly ash removal from stack flue gases.*
15. Consolidated Coal Co., *Development of the CO_2 acceptor process*, PB 210 840, 1971.
16. Coutant, R. W., *et al.*, *Investigation of the reactivity of limestone and dolomite*, PB 204 385, 1976.
17. Darmstadt and Tomlinson. *Tappi* **43**(8), 647, 1960.
18. Davies, D. T., *J. Inst. Fuel*, **21**, 301–11, 1948.
19. Dennis, C. S., *Combustion*, **42**(4), 12–21, 1970.
20. Ellwood, P., *Chem. Eng.*, 62–4, 16 June 1969.
21. EPA-US Environmental Protection Agency, *Flue gas desulphurization—installations and operations*, 1974.
22. Ergun, S. and E. H. Bean, *Magnetic separation of pyrite from coals*, US Bureau of Mines, Report Inv. 7181, 1968.
23. Evans, R. J., *Potential solid waste generation and disposal from desulfurization processes*, US Bureau of Mines Report iC 8633, 1974.

24. Ginneken, A. J. J., *et al.*, 'The Shell process for hydrodesulphurization of residual oils', *National Petroleum Refiners Association Annual Meeting*, March; San Antonia, Texas, 1975.
25. Godel, A. A. *Rev. Général de Thermique*, **5**(52) 349–59, 1966.
26. Grennard, A. and F. F. Ross, *Combustion*, 4–9, January 1974.
27. Hartman, M. and R. W. Coughlin, *Ind. Eng. Chem. Proc. Des. Devel.*, **13**(3), 1974.
28. Hishinuma, Y., *et al., Paper on the 67th Annual Meeting of the APCA*, Denver, Colorado, 1974.
29. Hobson, G.D. and W. Pohl, *Modern petroleum technology.* Applied Science, 192–5, 1973.
30. Horne, Th. de Menten. 'Les fuels à basse teneur en soufre', *International Symposium on Modern Power Stations*, Liege, 1970.
31. Hull, W. Q., *et al., Ind. Eng. Chem.*, **43**, 2425–35, 1951.
32. Husmann, K. and G. Hänig, *Brennstoff-Wärme-Kraft*, **23**, 85–91, 1971.
33. Jager, L., *et al., Chem. Prum.*, **9**(7), 361–63, 1959.
34. Jager, L., *et al., Chem. Prum.* **10**(10), 518–20, 1960.
35. Jimeson, R. M. and L. W. Richardou, 'Census of oil desulfurization to achieve environmental goals, *A.I.Ch.E.*, No. 19-C, Meeting Vancouver, B.C., September 1973.
36. Johnstone, H. R., *et al., Ind. Eng. Chem.*, **30**(1), 101–9, 1938.
37. Johswich, F., *Combustion*, 18–25 October 1965.
38. Jones, H. C., *et al.*, Paper 74, *67th Annual Meeting of the Air Pollution Control Association*, Denver, Colorado, June 1974.
39. Jüntgen, H., *et al., E.C.E. Seminar on the Desulphurization of Fuels and Combustion Gases*, 16–20 November 1970.
40. Kamprath, E. J., *Combustion*, 16–17, October 1972.
41. Kellog Co., *Applicability of SO_2 control processes to power plants,* PB 213 421, 1972.
42. Kuzminikh, I. N. and M. B. Babushkina, *3 Annual Chem. USSR*, **30**(3), 495–98, 1957.
43. Lemke, K., *et al., Gluckauf*, **105**, 141–6, 1969.
44. Lowell, P. S. *A theoretical description of the limestone washing process*, PB 193 029, 1970.
45. Lowicki, N., 'The desulphurization of gases by Grillo-AGS-Process 1970', *ECE Seminar on the desulphurization of fuels and combustion gases*, Geneva, 1970.
46. Lowicki, N., and K. Husmann, Paper presented in Essen, *Haus der Technik*, 11 November 1971.
47. Mandel, H., *et al., VGB Kraftwerkstechnik* **55**(6), 355–9, June 1975.
48. Maurin, P. G. and J. Jonakin, *Chem. Eng. Deskbook*, 173–80, 27 April 1970.
49. McCartney, J. T. and H. J. O'Donnell, *Sabri Ergurn*, US Bureau of Mines Report Inv. 7231, 1969.
50. McGlamery, G. G., *et al., Conceptual design and cost study, Magnesia scrubbing*, EPA Report R-2-73-244, May 1973.

51. McGlamery, G. G., *et al., Detailed cost estimating for advanced effluent desulphurization processes*, EPA-600/275006, 1975.
52. Monostory, F. P., *et al., Aufbereitungs-Technik*, 207–20, 599–614, 1970.
53. Montgomery, T. L., *et al., J. Metals*, 1–7, June 1973.
54. Moss, G., 'The desulphurizing of fuel oil in fluidized beds of lime particles', *Economic Commission for Europe Seminar on the desulphurization of fuels and combustion gases*, Geneva, November 1970.
55. Moss, G., *et al.*, 'The desulphurising fluidized bed gasifier', *International Conference on the Control of Gaseous Sulphur Compound Emission*, Salford, 1973.
56. Moss, G., 'The mechanism of sulphur desorption in fluidised beds of lime', *Fluidised Combustion Conference*, London, 1975.
57. Newell, J. E., *Chem. Eng. Prog.* **65**(8), 62, 1969.
58. Norwegian Ministry of the Environment, *Acid precipitation and its effects in Norway*, 1974.
59. Paulus, J. J. and A. Roffman, 'Supplementary control systems', *37th Annual Meeting American Power Conference*, Chicago, April 1975.
60. Pearson, J. L., *et al., J. Inst. Fuel*, **8**, 119, 1935.
61. Pigford, R. L. and G. Slinger, *Ind. Eng. Chem. Proc. Des. Develop.* **12**(1), 85–91, 1973.
62. Pinaev, V. A., *Zh. Prikl. Khim.* **36**(10), 2116–21, 1963.
63. Pinaev, V. A. and N. P. Pitelina, *Zh. Prikl. Khim.* **38**(8), 1652–8, 1965.
64. Pinaev, V. A. and V. Ya. Mosharov, *Zh. Vses. Khim, Obschest*, **14**(4), 394–9, 1969.
65. Ploeg, J. E. G., *et al., Petr. Inst.* **14**(7), 50–8, 1974.
66. Plumley, A. L., *et al., Combustion*, 16 July 1968.
67. Rasmussen, K. H., *et al., Sources and natural removal processes for some atmospheric pollutants*, PB 237 168, 1974.
68. Rasmussen, K. H., *et al., Water, Air and Soil Pollution*, **4**(1), 33–64, March 1975.
69. Rees, R. L., *J. Inst. Fuel*, **25**, 350–6, 1953.
70. Roberts, A. G., *et al.*, 'Fluidized combustion of coal and oil under pressure', *Fluidized Combustion Conference*, London, 1975.
71. Robinson, E. and R. C. Robbins, *Sources, abundance and fate of gaseous atmospheric pollutants*, Stanford Research Institute Report, 1968.
72. Ross, F. F., *Combustion*, 6–11, August 1971.
73. Ross, F. F., *Combustion*, 16–20, July 1973.
74. Rossoff, J. and R. C. Rossi, *Disposal of by-products from non-regenerable flue gas desulphurization systems*, PB 237 114, 1974.
75. Rydolm, S. A., *Pulping processes*, Interscience, New York, 1965.
76. Shah, I. S. and C. P. Quigley, Paper at 70th National Meeting A.I.Ch.E., September, 1971.
77. Slack, A. V., *Chem. Eng. Dec.*, 188–96, 1967.
78. Slack, A. V., and H. L. Falkenberry, *J. Engng. Pwr.*, January 1970.

79. Slack, A. V., *Sulfur dioxide removal from waste gases*, Noyes Data Corporation, 1971.
80. Stites, J. G., *National Eng.*, 8–9, May 1970.
81. Stites, J. G., *et al., Chem. Eng. Progr.* **65**, 74–9, 1969.
82. Tabershaw-Cooper Associates, *A critical evaluation of current research regarding health criteria for sulphur oxides* (for US Federal Energy Administration), April 1975.
83. Templeton, R. E., *Power Engineering*, **78**(7), 42–5, 1974.
84. Tennessee Valley Authority, *Sulphur oxide removal from power stack gas-sorption by limestone of lime: dry process*, PB 178 972, 1968.
85. Tennessee Valley Authority, *Use of limestone in wet scrubbing process*, PB 183 908, 1969.
86. Tice, E. A. *Materials Performance*, **13**(4), 26–33, 1974.
87. Trindade, S. C. *et al., Fuel*, **53**, 178–81, 1974.
88. Tsushima, Y., *Chem. Econ. and Engng. Rev.* **3**(11), 1971.
89. US Bureau of Mines, *Low sulphur fuel oil from coal*, PB 203 889, 1971.
90. Wada, A., 'Une usine de desulfuration directe des huiles lourdes', *Chemie et Industrie–Genie Chimique*, **101**, 25–30, 1969.
91. Walker, F. E. and F. E. Hartner, *Forms of sulphur in US coals*, U.S. Bureau of Mines, Ing. Circ. 8301, 1966.
92. Wandless, A. M. *J. Inst. Fuel*, **32**, 258–66, 1959.
93. Williams, D. F. and J. McLaren, *J. Inst. Fuel*, **42**, 303–8, 1969.
94. Wolfe, H. C. 'Beurteilung der Moeglichkeiten zur Entschwefelung von Schwerem Heizol', V.G.B.—Conference *Kraftwerk und Umwelt*, 1973.
95. Zentgraf, K. M., *Z. Ver. Dt. Ing.* **109**(35), 1686, 1967.
96. Zentgraf, K. M., *Combustion*, 6–11, November 1969.
97. Zimmerman, E., *Coal Preparation*, 105–12, May/June 1967.

10 Reduction of emissions of grit, dust and acid smuts from power stations and industrial boilers

A. A. BARRETT

This chapter is based on a study by the design staff of the British Central Electricity Generating Board of the factors affecting the efficiency of methods and equipment in use for the removal of grit, dust and acid smuts from the flue gases of power stations and industrial boilers.

The burning of any fuel can cause a nuisance by the emission of smoke, dust, grit, agglomerates or gaseous pollutants unless careful thought is given to the design of the plant as a whole. The problem varies with the type of fuel being fired. The main purpose of this chapter is to give general guidance on the steps that need to be taken in the design and assessment of an installation and to give a practical guide to the important features of dust collectors to be taken into account to avoid unacceptable chimney emissions.

There are only two general types of collectors used on generating stations, mechanical collectors and electrostatic precipitators. Bag filters are not generally used for cleaning combustion gas because, so far, there has been no economic solution to the problem of installing bags that will withstand the damage caused by attack by acid, the effect of high temperature, the damage caused by the shaking mechanism necessary to remove the collected dust and difficulties with condensation during start up which cause blinding of the surface. They are used with success in removing fuel dust from the displaced air from dust silos where the temperature is not much above ambient. Gas washers or scrubbers for the removal of grit and dust have been investigated, but have not been generally adopted for large boilers because of the very large quantities of water that would have to be used, cleaned and recirculated.

Gas firing

Gas firing does not give any problems of smoke and dust, but when very large quantities of certain types of gas are burned, it is necessary to disperse the waste gases at a high level.

Oil firing

Oil firing can give rise to problems caused by the emission of smoke, fine dust or acid smuts. The flue gas solids burden is a combination of the ash contained in the fuel and the unburnt carbon; the carbon content usually accounts for about 80 per cent of the total dust quantity.

Dust burden

Careful design of the boiler, with particular attention to the size of the combustion chamber, the design of the burners to give complete atomization of the fuel over the whole operating range, and satisfactory control and distribution of the air supply ensure that the total dust emission from a large boiler does not exceed 0·11 g m^{-3} (0·05 grain ft^{-3}) under normal operating conditions. This is generally accepted as being satisfactory.

Smoke

This is only a problem on a well-designed boiler if the amount of air is insufficient for satisfactory combustion or if the distribution of air is such that certain burners have insufficient air. The overall control of the air quantity is by total flow measurement with fine adjustment derived from flue-gas meters, preferably backed up by carbon monoxide meters.

Acid smuts

If they are emitted from the chimney acid smuts are likely to cause complaints in the immediate locality of the power station. It is for this reason that a great deal of work has been done to find the best way of eliminating this problem. Current practice is to keep excess oxygen down to 0·5 per cent by volume at the furnace exit; to keep all surfaces in contact with the gases at a temperature above the acid dewpoint; to ensure that the burners are capable at all times of complete combustion of the fuel and that the burner dampers are gastight; to design the ducting with a sufficiently high gas velocity and suitable splitters or vanes on each bend to ensure that the dust is not deposited in the flues or chimney. During shut-down periods, dampers are closed to retain as much heat as possible so that on start-up the temperature of the whole system is raised above the acid dewpoint as quickly as possible. This rapid heating is assisted by pre-heating the main air heaters by recirculation through the steam air heaters, and by pre-heating the furnace purge air and by-passing the air heaters on the gas side until the metal temperature is above the acid dewpoint.

With all large boiler installations, the products of combustion are widely dispersed in the atmosphere by discharge from high chimneys.

Coal firing

Coal firing causes the biggest problem because the ash content of the fuel can vary between 10 and 20 per cent and causes high dust and grit burdens. When burnt on a chain grate or in a stoker fired boiler, much of the ash is collected under the grate, and the dust carried forward by the gas stream tends to be coarse; sufficient can be removed by a mechanical collector in some instances to give a satisfactory standard of gas cleaning. When pulverized coal is burnt, some 80 per cent of the ash is carried forward by the gas stream. This necessitates the use of an electrostatic precipitator to give the necessary degree of gas cleaning.

Mechanical collectors

Stoker fired boilers

On older power stations in the UK, where relatively small boilers generating only 10–170 tonnes/h of steam and equipped with travelling grates, spreader stokers or retort stokers, some form of mechanical grit collector was usually employed. Such power stations represented only a small proportion of the generating capacity of the British Central Electricity Generating Board. Mechanical collectors are used on small industrial boilers. In some cases, dust is removed by a scroll type collector on the induced-draught fans. These collect some 40–50 per cent of the grit and dust presented to them, but they are not very effective on the finer fractions. Other small boilers fitted with centrifugal or reverse flow types of collectors, usually with a secondary circuit to collect the dust separated out of the main gas stream by the primary system, have an efficiency of grit and dust removal of 50–80 per cent.

Oil-fired boilers

On two large oil-fired installations where it has been anticipated that the dust burden would be unsatisfactorily high, low efficiency mechanical collectors were installed. They consisted of a primary circuit which separated the dust from the gas stream by means of centrifugal force, but had no secondary collecting system. These collectors had an efficiency of 50 per cent. On later oil burning boilers, improved designs of the burners and of the furnaces, with improvements in the methods of controlling the combustion process, have made it possible to obtain the necessary low dust burdens without using a mechanical collector.

Pulverized coal-fired boilers

Most central power station boilers firing pulverized coal are equipped with electrostatic precipitators to collect the fly ash, since between 70 and 80 per cent of the ash content of the fuel is carried over in the gas stream and would otherwise be discharged to the atmosphere. At one time it was the practice to install mechanical collectors prior to the electrostatic precipitators with the idea of reducing the dust burden so that smaller precipitators could be used. However, this was not so attractive as had been thought and the designs reverted to using larger precipitators without mechanical collectors. This decision was taken because the mechanical collectors were prone to blockage and the capital cost of the installation was thus reduced. In addition, the drop in pressure of 4 cm WG. across the mechanical collector was eliminated, resulting in a saving in the cost of the power consumed by the induced-draught fans. It also avoided the maintenance required on the mechanical collectors, which was largely due to erosion of the impellers on the fans and the cones on the cyclones of the secondary system. A further reason for the change was that the early promise of being able to install smaller electrostatic precipitators was not realized, mainly because performance was adversely affected by the tendency of the mechanical collector to remove the larger dust particles and thus present only the finer dust to the precipitator. On some stations, where mechanical collectors have been installed, it has been found that by removing the spinners in the primary system of the mechanical collectors and blocking off the secondary circuit, the improvement in collecting efficiency of the electrostatic precipitators is sufficient to compensate for not removing the larger particles in a mechanical collector.

On one power station where the electrostatic precipitators were not giving the required collecting efficiency, an experiment was made with the installation of mechanical collectors placed after the electrostatic precipitators; this assisted under ideal conditions but was not generally successful, owing mainly to the finer dust choking the mechanical collector.

Early installations, particularly where banks of small cyclones were used, suffered from dust choking the collector, causing loss of efficiency and in some cases a rise in the resistance to the gas flow. It is important to ensure that the total gas volume is equally divided between the banks of the collector installation, and to make sure that the distribution of the gases across the face of the collector is such that a uniform gas velocity is obtained. Bad distribution results in loss of collecting efficiency and the possibility of dust falling out of the gas stream and blocking parts of the installation in the low velocity areas. The hoppers for collecting the dust have an important role in the satisfactory operation of a mechanical collector. If they are allowed to become over-filled they cause the dust to choke the apertures through which the dust has been separated from the gas stream, and the collecting efficiency is reduced. The recommendations given later in this chapter for the fitting of hopper level indicators and heating elements to the electrostatic precipitator hoppers is equally applicable to mechanical collector hoppers.

It is important that air is not allowed to leak in through the dust hoppers, since this causes loss of collecting efficiency. It is necessary also to pay particular attention to the valve arrangement on the hopper outlet so that having discharged all collected dust there is no ingress of air when the valve is closed. It is normal to use balanced flap valves which automatically open when a pre-determined head of dust is collected. There are available rubberized self-sealing valves which have no moving metal parts to wear or corrode.

Electrostatic precipitators

On modern power stations burning pulverized fuel, plate type electrostatic precipitators are used. On a number of older, smaller stations where the required collecting efficiency was only 96–98·5 per cent, the tubular type of electrostatic precipitator was used. The principle of collection was the same as in the plate type, but in the tubular type the gas passed vertically up cylindrical or hexagonal tubes, through the centre of which were suspended the discharge electrodes. Tubular precipitators are now seldom used except on very small commercial installations.

Collecting efficiency and guarantees

On central power stations in the UK where large units burning solid fuel have rated capacities of 500 MW or more, the collecting efficiency of the gas cleaning equipment is specified as a minimum of 99·3 per cent removal of grit and dust. This efficiency has to be obtained whilst the plant is operating on automatic control, and has to be obtained soon after commissioning and again twelve months later, with reasonably continuous operation. This requirement ensures that the installation is reliable and will continue to give the required performance. During the year's guarantee operation between the two tests, no cleaning of the inside of the precipitators and no repair work is permitted if the guarantee is to be fulfilled. If a major repair or a large number of small repairs become necessary, whether they are due to faulty design, material or manufacture, consideration is given to extending the twelve months' guarantee period from the completion of such repairs.

Arrangement of plant

With small industrial installations a precipitator usually consists of one casing, through which all the gas flows. The larger the gas volume to be treated, the larger the casing required, and since there is a limit due to structural reasons on the size of the casing, more than one is needed to cope with large gas volumes. On modern large power stations it is usual to have three casings in parallel, each handling a third of the total gas. Each of these casings is split into four sections in series, giving a total of twelve sections for the boiler unit, each having its independent rapping gear on both the discharge and collecting electrodes and each having its own high tension equipment (Fig. 10.1).

The number of sections in series is dictated to some extent by the efficiency required, but it should never be below two sections, even for a small precipitator, because any dust that escapes the first section has then a chance of being collected by the second. The more sections there are, moreover, the less is the reduction in efficiency if a fault develops in one of them; even on a two-stage precipitator, the failure of either zone would only reduce the collecting efficiency by about 20 per cent.

The gas cleaning equipment is installed between the outlet from the air heater, where the gases have been cooled to the region of 120°C, and the induced-draught fans, so that the casings are under suction. There is interconnecting ducting, prior to and

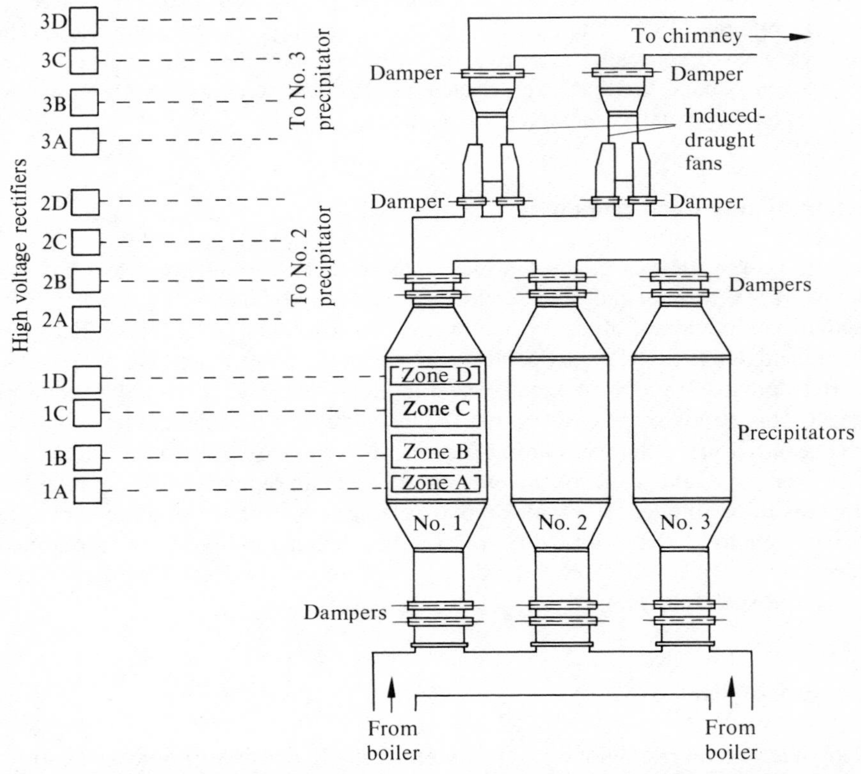

Fig. 10.1. Schematic diagram: typical electrostatic precipitator arrangement for a 500 MW boiler.

subsequent to the electrostatic precipitators, and on each of the legs it is usual to install double isolating dampers both sides of each precipitator casing. If a fault develops in one part of the electrostatic precipitator, the boiler load can be reduced and the dampers on the particular precipitator flow can be closed to facilitate maintenance, whilst the products of combustion on the boiler are treated by the remaining flows.

It has been suggested that the precipitator could be installed before the air heaters to prevent the latter fouling. This would not be satisfactory because the precipitator would have to be larger to handle the extra volume of the gas due to the higher temperature, and the changed characteristics of the gas and dust would be detrimental to the collecting efficiency. In any case, air heaters can be designed so that they do not become fouled by the dust and grit.

Casings

The casings on modern plant are usually in steel which has to be carefully lagged and weather-proofed to guard against acid attack, but reinforced concrete casings are also used. The choice is governed by economic considerations, since both are technically acceptable. There are obvious advantages in using steel casings if the speed of erection on site is important, or if there is limited space that can be made available for erection purposes. Steel casings also have an advantage if the total weight has to be kept to a minimum owing to limitations on the loads that can be carried by supporting structures or foundations.

Dust hoppers

The dust hoppers for collecting the dislodged dust should be of sufficient capacity to house the dust produced in 24 h, but it is desirable that the dust is removed every 8 h or less. This procedure reduces the risk of the dust becoming hard packed in the hopper and difficult to remove; it also allows some time to rectify a fault on the ashing equipment before the boiler needs to be taken off load. The heating of the bottom part of the hoppers helps to keep the dust dry and free flowing during the emptying process; means of aerating the dust is an advantage. It is essential that the whole of the hopper is lagged to prevent condensation and the consequent damping of the dust which makes removal difficult.

Each hopper should be equipped with a high level and a low level dust indicator; the former to prevent overfilling of the hoppers which will immediately short circuit the electrode system; and the latter to indicate when the hoppers have been properly emptied. These indicators should be reliable and of sufficiently robust design to ensure that they will not be damaged during routine overhauls of the plant.

Collecting electrodes

Many types of collecting electrodes, for which differing claims have been made about their advantages, have been designed and used by various contractors. For the most part it is considered that for the collection of pulverized fuel ash with a carbon content generally below 1 per cent, a flat plate is just as efficient as any complicated configuration. The only exception is in the collection of a high percentage of carbonaceous and coked material which tends to roll along the plates in the direction of the gas stream. This can be prevented by ridges or traps.

One of the important parameters of an electrostatic precipitator is the plate area available for the collection of the dust. This area is the projected area, and not the developed area, if a rolled section is used. If the area is less than that reasonably required for the gas volume to be treated and the efficiency to be obtained, then no amount of refinement in the way of sophisticated control or enhanced rapping systems will overcome such a disadvantage. The height of the plates for power stations in the UK is usually limited to about 9 m to ensure that the mechanical difficulties of assembly, in particular the alignment of the electrodes, does not become too difficult.

The farther the dust has to drop across the gas stream before being collected in the hoppers, the more likelihood there is of its being re-entrained in the stream. Further experience on the height of plate would dictate whether higher plates could be used and if so, how much the performance would be impaired. Unless there were severe restrictions in the ground area available for the installation and unless economic advantages were to be obtained it is unlikely that the current figure would be much exceeded.

All plates, in view of their height, have to have some method of making them reasonably rigid so that they hang vertically and are not displaced by the gas flow. At the same time they must remain suitably flexible to transmit satisfactorily the rapping impulses which periodically remove the collected dust. The plates must also be robust enough to withstand the force imparted to them by the rapping gear.

The spacing of the collecting electrodes is usually dictated by the standard design of a particular manufacturer, but is generally between 20 and 30 cm. As long as the associated electrical equipment, such as the high tension sets, the insulators and the necessary electrical clearances, are designed to suit a particular spacing, variations in this dimension do not appear to be critical.

Discharge electrodes

On large power stations the precipitators are equipped with various types or combination of types of discharge electrodes. These include round wires, square wires, concave-sided wires, barbed wires or barbed steel strips. Research and a study of test data have shown that the type of electrode is not particularly important, provided that the rest of the equipment is designed to suit. If barbs are employed, then the operating voltage needed to give a satisfactory corona discharge is less than if a round wire is used, and the high tension sets would be designed accordingly. Because of the alternatives in operating voltage, attention would have to be given to the electrical clearances and to the size of the electrical insulators. The weakest link in most precipitators is the reliability of the discharge electrode system, and although it is difficult to generalize, investigations have shown that failures come mostly under the headings of faulty manufacture, corrosion, uneven tensioning, and fatigue. Faulty manufacture can occur either on the factory floor, or as a result of design faults when, for instance, sharp edges cut into the wire, and sudden changes of section create highly stressed points. In most installations mild steel discharge electrodes are satisfactory; but there have been instances of corrosion when a boiler has been firing high sulphur coal or large quantities of oil for low loads and coal ignition and the uncleaned precipitator has been shut down for a protracted period. The dust can then be acidic and cause corrosion as the casing cools. If this is a regular feature of operation, it is desirable to fit stainless steel wire electrodes.

When a number of electrodes are tensioned in a frame, it is most important that the tension of each wire is checked prior to the commissioning of the precipitator. The over-stressed wires might fail, but they can cause slackness in adjacent wires which will oscillate and suffer severe electrical burning with consequent failure. Fatigue failure is usually created by the rapping system. Since mechanical rapping has supplanted high frequency electric vibrators, little trouble has been experienced. It is, nevertheless, important that the cross-sectional area of the wires is sufficient to withstand the rapping forces.

Rapping methods

Having obtained satisfactory discharge and collecting electrodes and having established that the plate area is sufficient for the duties required, the next vital factor in the electrostatic precipitator is the rapping of both surfaces to remove the collected dust. A thin dust layer on the discharge electrodes suppresses the corona; it is necessary to keep clean both the discharge electrodes and the collecting electrodes. If this is not done all other features of the design are negated.

It is essential that the distribution of rapping effort is as uniform as possible because if the dust is removed from some parts leaving deposits in others, the operation of the plant suffers just as much as if the whole area had the poorer rapping. A further factor, particularly in relation to the discharge electrodes, is that high impact forces, local to the point of impact, cause fatigue failure. Usually the rapping is by mechanical means since it has been found that electric vibrators or high frequency hammers have brought about considerable metal fatigue and have often caused the dust to compact rather than to be removed. It is essential that any steelwork for transmitting the rapping be kept to a minimum so that the maximum force is transmitted direct to the plates and electrodes, and no unnecessary losses are incurred. Much investigation has gone into establishing a satisfactory method of measuring the rapping force on both the discharge electrodes and the collecting electrodes and into the measuring of the thickness of dust that is left on the surfaces after the rapping cycle has been completed. It is now generally accepted that the effectiveness of rapping can be measured by the maximum acceleration or g value at a particular position on the electrode. The work also indicates that acceleration in the plane of the electrode producing a shear in the dust layer is the most important component. It is possible that the wave form, the induced frequency, or the time for the vibrations to die away are further factors, but they seem to be insignificant when compared with the effect of the maximum g. As stated previously, it is important that the distribution of rapping effect is as uniform as possible, and it is therefore important that the measuring of the acceleration during the rapping cycle should be checked at several points and not only near the point of impact. The overall effectiveness must be considered on the basis of the lowest reading and not on the basis of the maximum or average figure.

Rapping of collecting electrodes. Only dust in the hoppers can be considered as collected. It is therefore necessary to have suitable apparatus to dislodge the deposited dust from the collecting electrodes. Generally, this is achieved by striking the plates with hammers or by physically moving the plates, sometimes against a spring and allowing them to return against a stop. Each successive zone of a precipitator collects less dust. Consequently the rate of build-up in the thickness of the dust layer slows down in each successive zone, and the rapping cycle needs to be progressively less frequent. The design must therefore incorporate some means of adjusting the frequency of rapping in each stage, so that an optimum layer of dust is collected before it is removed. The first stage may need rapping every five minutes whilst the last stage may only need rapping every one or two hours.

It is impossible to make hard and fast rules on what is a satisfactory dust layer for a particular installation. Ideally the collecting electrodes when rapped should be completely cleaned, but in practice this is impossible and some thickness of dust is always left behind. If this layer is highly resistive, constituting an electrical barrier to the flow of current, then the permissible thickness of dust is very thin—probably 1 mm would be too much to maintain a satisfactory performance. If the dust is more conductive, then a thicker layer—even 10 mm—would be acceptable. This characteristic of the dust is generally known as the *resistivity* and on coal-fired power stations it is largely influenced by the sulphur content of the coal. A sulphur content in excess of 1·5 per cent is usually sufficient to give a low resistivity, but below this the operation of the precipitator is impaired, necessitating the installation of a larger plant for a given set of conditions. The carbon content of the dust, which with modern boilers is generally below 1 per cent, does not affect the resistivity of the dust. However, if the carbon content is high, it is usually in the form of coked particles which are not so readily collected by an electrostatic precipitator and thus the efficiency is lower.

As it is difficult to obtain any reliable figures by observation of the thickness of the dust layer, it is necessary to use a simple gauge, having a flat plate which can be held against the surface of the dust whilst a probe is inserted through the plate to penetrate the dust layer and thus obtain a reliable measurement. The dust build-up is unimportant if the required performance is obtained, but the thickness of dust should be checked if an improvement is necessary. Investigations on numerous precipitators have shown vast differences in g values in the direction of the rapping blow imparted. On some older plants values as low as $0{\cdot}7\ g$ have been found. This is an extremely low figure and is perhaps unusual, but since it is considered that a minimum of $30\ g$ is desirable, this suggests that much unsatisfactory operation is accounted for in this way. At one power station a series of checks was carried out and although there was a great deal of scatter in the results obtained, it clearly demonstrated that on this particular installation, a value of $30\ g$ would probably result in the thickness of the dust layer being below $1{\cdot}3$ mm, and this was not detrimental to the collecting efficiency required. Table 10.1 gives some results that have been obtained.

TABLE 10.1

'g' value in the direction of rapping blow			
At point of impact	1·5 m away	Minimum found	Plate height (m)
55	28	15	7·6
140	32	31	7·9
490	not recorded	11	8·2
300	210	60	8·5
330	280	65	8·5

It is evident from Table 10.1 that providing a rapping gear is satisfactorily designed, neither the height of the plate nor the intensity of the blow at the point of impact is the dominant factor in affecting the minimum values. The figures clearly indicate how quickly the forces are absorbed within the plate and underline the necessity of not relying on one point of rapping. The high figures obtained at the point of impact are much greater than those measured only $1{\cdot}5$ m away, and even these are considerably higher than the minimum values obtained. It is essential that if a programme of work to measure acceleration of collecting electrodes is to be undertaken, the accelerometers must be firmly bolted to the plates and they and the associated recording equipment must be identical during each test. The g value should be measured in three mutually perpendicular directions if possible, and at a minimum of ten points over the surface of the plate being tested. Before the work proceeds too far, a random check on other plates should be made to ensure that the chosen plate is reasonably representative.

A study of rapping effects has shown that with high plates two positions of rapping gives an improved distribution and the minimum value is more satisfactory. This is because when the effect of one hammer is diminishing, the effect of the second is increasing and vice versa. It is not difficult to obtain a minimum value of $100\ g$ anywhere on the surface of the plate. If the minimum rapping effect is unsatisfactory even with the maximum blow recommended by the manufacturer, then extensive modifications have to be contemplated. It may be necessary to reinforce the plates to take a greater impact without suffering mechanical damage, or even to duplicate the rapping gear to obtain better distribution.

Rapping of discharge electrodes. Relatively little dust is collected on the discharge electrodes, but rapping gear, in the form of hammers striking the steel framework, is essential to prevent an accumulation of dust which would ultimately suppress the electrical discharge and stop the dust-collecting process. The wires should preferably be kept completely clean or, at the worst, the residual dust layer should be kept below 0·5 mm. No adjustment in the frequency of rapping is necessary. It is generally found that the wires are kept sufficiently clean by rapping every two or four minutes. The distribution of rapping effect can be measured by bolting an accelerometer to the supporting steelwork. It is desirable to check the *g* value on the actual discharge wires because it is at this point that the dust collects; but this is complicated by the flexibility of the wire necessitating the use of very light accelerometers because their own mass affects the reading. In most instances it is satisfactory to take readings on the steelwork and correlate these with the build-up of dust on the electrodes.

Control

With the increase in size of installations it has been the practice to install remote control so that the start up and close down of an electrostatic precipitator can be made from a central control rather than from controls local to the precipitator installation. The controls consist of the necessary switches to start up and shut down each of the high tension sets and the rapping motors. The control desk usually incorporates a high tension voltmeter, a milliammeter and a meter to read the primary current supplied to the sets. Where space is limited, it is possible to have one set of instruments and a selector switch to take readings from each individual set. Instrumentation lights are necessary to indicate what plant is energized. In addition there must be some form of automatic control of the high tension sets so that under all operating conditions, the optimum power is supplied to the precipitator. This automatic control must be such that the corona discharge is kept to a point just below continuous flash-over, a level which varies according to the condition of the gas and of the dust being collected.

Sectionalization

It has been realized for many years that there is a minimum size of high tension set for a particular section, if sufficient power is to be impressed on the discharge electrodes. At the same time, it is important that the size of the sections is not too large. Much research has been done at a power station where there was the facility of connecting differing numbers of sections to high tension sets of different sizes. By careful analysis of the results obtained, useful parameters have been established.

These parameters can be used in the design to reduce the total size of a precipitator if the number of sections with their associated high tension sets is increased. Since the cost of additional sectionalization is high, there is obviously a point at which the reduction in cost because the installation is smaller is offset by the increased cost of sectionalization. On large installations it is generally economical to restrict the size of section to about 3700 m^2 of collecting electrode area. A slight improvement in performance can be obtained by halving each zone electrically and connecting both to a single high tension set via independent surge resistors. This practice ensures that flash-overs in one half of the zone do not greatly affect the other half; but it is essential that the resistors used are rated for the full output of the set and that they are reliable.

Gas distribution

It is necessary to ensure that the correct distribution of the gases is obtained; not only must the right proportion of the total gas volume be fed to each casing, but the flow through each casing must be uniform. A low velocity in one particular part of the cross-sectional area would give a higher efficiency locally, but this would not compensate for the reduction in efficiency in the consequent high velocity areas. When a precipitator is designed, it is assumed that gas distribution will be satisfactory, but if it is not, then the required efficiency will not be obtained. If the precipitator is reasonably small, it is quite possible to check the distribution after the plant is erected and then to fit in necessary baffles or splitters to correct the gas flow. When the precipitators are large and consist of several casings, it is more economical to build a model varying between one-eighth and one-sixteenth scale to check gas distribution not only to split the gas into each casing but also to check the uniformity of the distribution over the cross-sectional area of each casing. Normally there is an expansion piece connecting the duct to the casing of the precipitator and it can be shown that it is necessary to fit splitters or vanes at the beginning of the taper to obtain a coarse control of the gas flow. It is then possible to obtain the final degree of correction required by installing one or two perforated screens at the larger end of the taper adjacent to the precipitator. The perforations in such baffles can be varied to correct the observed abnormalities in flow and will only be effective if the gas approaching them is reasonably distributed.

The time and effort spent on work with a model is justifiable for a large plant, as the cost of adjusting baffles, splitters or vanes on the full-scale plant would be very high. Provided close attention is given to the accuracy of the model, it is seldom necessary to do any extensive alterations on the installed plant when the distribution is checked. It is desirable to ensure that the gas distribution is such that individual readings do not exceed ± 25 per cent of the average gas velocity.

High tension equipment

The discharge electrodes have impressed upon them a negative d.c. voltage of 40–50 kV and this is supplied from a number of high tension sets. Each section of the precipitator must have its own high tension set, but if the plate area within a section exceeds 3700 m^2 it is preferable to have more sets. Having established the number of zones and thus the number of high tension sets, the capacities must be checked to ensure that they are capable of delivering the required power; it has been found that on power station installations the plate area per installed kVA should not exceed about 90 m^2. On older installations, mechanical type rectifiers were used, but these needed costly housing, requiring a large ground area. In addition it was necessary to have forced ventilation to cool the chamber and to remove unwanted gases. On later installations, static rectifiers, which can be placed in the open, take up less space, require less maintenance, and give a marginally improved performance, have been used. The first static rectifiers had banks of selenium, but with the advent of larger sets silicon is usually used.

Satisfactory operation

Care should be taken to ensure that the maintenance recommended by the manufacturer is carried out, if consistent satisfactory performance is to be attained. With a modern installation very little maintenance is required but it is essential to lubricate all the moving parts, most of which are associated with the rapping gear, as recommended by the manufacturer.

The most common failing is that hoppers are allowed to become overfilled, which not only causes an immediate short circuit of the discharge electrodes, but may do damage by causing burning of the wires, necessitating remedial work. Periodically all insulators should be cleaned and checked for cracks to ensure that the power being supplied by the high tension sets is being impressed on the electrodes and not being dissipated across the insulators to earth. If low electrical readings are experienced on any particular zone, particularly if it is associated with instability, every effort should be made to find the cause. In most cases it will be found that an electrode has broken or has become misaligned, but if this is not the trouble, then the electrical circuits and equipment of the transformer/rectifier and the associated control gear should be checked. Particular attention should be given to the automatic control system. During the annual overhaul it is advisable thoroughly to wash the plant, particularly if deposits of dust are adhering to the plates or electrodes. An additional incentive is that it facilitates examination of all the internal parts for wear or faulty operation before putting in hand the necessary maintenance work.

Information required by the designer

There are many factors which affect the design of the precipitators when applied to a particular process. The more important points are discussed in the following paragraphs.

Collecting efficiency. It is important to ensure that the requested collecting efficiency is no higher than necessary, because the efficiency has a large influence on the size of the plant and thus the cost. It is perhaps more useful to consider the amount of particulate matter which is not going to be collected, rather than that it is desired to collect. It is easy to think that there is little difference between 98 or 99 per cent collecting efficiency, but in fact there is a considerable difference because the percentage slip is halved, and to obtain this decrease in emission means that the precipitator would have to be some 15 per cent larger. To avoid any misunderstanding when the contractural acceptance tests are carried out, the collecting efficiency should include any necessary tolerances due to inaccuracies in the testing procedure.

Gas volume. The cost of a precipitator is proportional to the gas volume to be handled. Before proceeding with the design considerable thought should be given to the gas volume. A precipitator will not give its guaranteed collecting efficiency if the design volume is exceeded; therefore the volume must be that which covers all operating conditions. Conversely, an over generous sizing is a waste of capital resources. If the volume is to be measured, it is as well to make sure that there are no leaks in the flues which would allow the entry of excess air to increase the total volume. It should also be borne in mind that if the gas is to be conducted to the precipitator by long runs of flues, the temperature, and thus the volume, will be reduced to something less than that leaving the boiler.

Gas temperature. Gas temperature as well as influencing the actual volume of gas may, in itself, have a direct effect on the precipitation process, and the normal operating temperature should always be given. The maximum gas temperature must be given to ensure that the casing is designed with a satisfactory margin of safety.

Fineness of dust. The precipitator collects coarse dust more readily than fine dust. It is therefore necessary to specify the grading of the dust as a basis for design. If the dust grading is to be checked on site on an existing process, then it is essential that the sample is taken by inserting a probe into the duct and sampling in accordance with BS 3405 and not by removing a sample of dust from convenient hoppers or disposal points which would not give a representative sample.

The dust burden. It is important to assess the dust burden from the boiler because it dictates the collecting efficiency required to give an acceptable chimney emission. Variations in dust burden are not thought to influence the performance of a precipitator over the range usually experienced, but if the burden falls below about 2 g m^{-3} of gas, then some fall off in the percentage removal of grit and dust is to be expected.

Sulphur content of coal. It is now accepted that the sulphur content of the coal used has an effect on the efficiency of precipitators at power stations, and it is usual to specify the sulphur content for the purposes of the efficiency guarantee. A correction curve is supplied by the manufacturer showing the variation of efficiency with the sulphur content of the coal.

Carbon content of dust. The maximum carbon content under normal operating conditions should always be given, but the carbon content is unimportant unless it exceeds about 5 per cent.

Space available. The ground space occupied by the electrostatic precipitator is dependent upon the design factors, but to some extent the shape of the occupied rectangle can be varied should this be necessary. Any restriction in length can be compensated by additional width and vice versa, and even the ground area can to some extent be reduced by increasing the height of the plate. It is, therefore, essential that the designer is given as much information as possible on the space available, so that he can design the most economical plant. He should also be given as much information as possible on the site layout, so that he can allow for satisfactory runs for the gas flues from the boiler.

Basis of design

When it is necessary to compare relative merits of precipitators that have been offered to suit a particular requirement, it is difficult to compare the offers from the basic dimensions. One plant might be longer and this would be a good point, but on the other hand, it might be narrow or have less height, and this could be a bad point in design. Certainly the smallest plant, probably at the lowest price, may not be the best. On the other hand, there is no point in paying for larger equipment than is required.

The basis of all precipitator design is what is known as the *effective migration velocity* (EMV) and this takes into account the gas volume, the size of the precipitator and the efficiency required. In simple terms, this means the velocity at which the dust proceeds towards the collecting electrodes measured perpendicular to the plate surface. The calculated velocity includes the effect of re-entrainment and thus is lower than if a particle of dust proceeded to the plate and stayed there. The velocity is usually expressed in cm s^{-1}.

The most commonly used equation for calculating the effective migration velocity is due to Deutsch (1922) and can be expressed:

$$\text{Collecting efficiency} = 1 - 1/e^k,$$

where $k = \dfrac{AW}{Q}$

A = area of receiving electrodes (m^2)
W = effective migration velocity m s^{-1}
Q = gas volume at operating temperature m^3 s^{-1}.

If, from previous experience or by trials, it is known what EMV can be achieved for cleaning the gases from the firing of a particular coal, the plate area to suit the gas volume and the efficiency required can be calculated. Since this calculation is vital, the calculation of the plate area for a typical precipitator is given below by way of example, using Naperian logarithms where necessary.

Efficiency 99%, gas volume $= 47{\cdot}17\ \mathrm{m^3\,s^{-1}}$
Effective migration velocity $= 0{\cdot}0935\ \mathrm{m\,s^{-1}}$

$$\text{Efficiency} = 1 - 1/e^k$$

$$1/e^k = 1 - 0{\cdot}99$$

$$e^k = \frac{1}{0{\cdot}01} = 100$$

$$k \log 2{\cdot}718 = \log 100$$

$$k = \frac{2{\cdot}00}{0{\cdot}4343} = 4{\cdot}6$$

$$k = \frac{AW}{Q}, \quad A = \frac{kQ}{W}$$

$$A = \frac{4{\cdot}6 \times 47{\cdot}17}{0{\cdot}0935}$$

$$\text{Plate area} = 2320\ \mathrm{m^2}.$$

Obviously the same calculation can be used for checking a designed EMV for a given precipitator. This calculation is applicable to any type of precipitator irrespective of the industrial processes involved, and is therefore an important item to consider when precipitators for a particular application are being evaluated. If experienced manufacturers design a particular plant on the basis of an EMV of 9 cm $\mathrm{s^{-1}}$ it is most unlikely that a plant based on a figure of 12 cm $\mathrm{s^{-1}}$, which will be smaller and thus cheaper, will prove to be satisfactory in service. If such a plant were installed, it is more than likely that the higher EMV would not be obtained, and thus the collecting efficiency would be less than that required. Having installed a plant that is too small in the first place, it is in most cases difficult and always very expensive to make the plant any larger. For the most part the EMV is established from the experience gained on precipitators operating on similar or preferably identical coals.

If no previous experience is available, it may be necessary for a pilot plant to be installed on the site. A portion of the gases to be cleaned can be drawn off from the main flue and passed through the pilot plant, on which tests can be carried out to establish a satisfactory basis for design. It must be pointed out, however, that the calculated EMV from a pilot plant is always very much higher than that which would be obtained on a full-scale plant. This factor is known by the designer and he must make due allowance. This is not a marginal difference, as will be seen by the fact that when a pilot plant handles pulverized fuel ash a migration velocity of the order of 20 cm $\mathrm{s^{-1}}$ will be obtained, but on full-scale plant it is possible that only 8 cm $\mathrm{s^{-1}}$ will be obtained.

Factors influencing effective migration velocity

The factors influencing EMV must be carefully considered for any particular installation, otherwise a basis that has proved satisfactory for one plant may, due to slight changes in design or conditions, be unsatisfactory for another. The following are some of the more important items that must be considered.

Specific surface area. The specific surface area, usually designated F, is the area of collecting electrode for a given volume of gas per second, measured at the operating temperature. So far as collecting efficiency is concerned, the larger this figure the greater is the collecting efficiency, but the actual EMV is slightly less and account must be taken of this factor in the design. The reason for this reduction in EMV is that the law of diminishing returns applies. If a higher efficiency is being strived for, this can only be achieved by trying further to clean gas that has very little entrained dust in it, and this dust is much finer, which makes it more difficult to precipitate.

Size of the precipitator. With larger gas volumes it is obviously necessary to have large precipitators to give the required EMV, but this size factor in itself, has an effect on the design because, as previously stated, with pilot plants at one end of the scale and large installations at the other, there is a considerable change in performance. The reason for this is not clearly defined, but it is possibly due to smaller plants having better gas distribution and less difficulty in checking every item during erection, thus ensuring that every section works perfectly.

Gas velocity. The gas velocity through the treatment zones is important only if it is too high and scours off the collected dust, particularly during the rapping cycle. For boiler application the gas velocity is frequently restricted to a maximum of $2{\cdot}15$ m s^{-1} under any conditions. Since the most important item is the receiving electrode area, a precipitator with a high gas velocity and thus a small cross-sectional area must be longer to give the required collecting surface. The ratio of length to width may also be affected by the ground area available.

Sulphur content of coal. One of the more important factors is the sulphur content of the coal used, because this influences the amount of sulphur trioxide in the gases. The lower the sulphur content, the lower is the migration velocity. In some cases, where the coal being burnt has a low sulphur content and the precipitators are too small to give the required performance, the difficulty might be overcome by injecting a carefully controlled amount of ammonium sulphate or sulphur trioxide into the gas stream. This remedy is only worth trying if the precipitator is mechanically and electrically in good order.

Ash content of coal. The ash content of the coal influences the quantity of dust to be collected but, in addition, it has an effect on the migration velocity.

Dust grading. The fineness of the dust has a considerable effect on the performance of a precipitator, and whereas a very high migration velocity is obtained with dust above 20 μm, it is low for the smaller fractions. The sizing or grading of the dust is usually expressed in terms of the percentage of the total in various size ranges. This is satisfactory for most purposes but it is useful to have a single figure for each dust when comparisons have to be made. For this purpose, the calculation shown in and below Table 10.2 can be used to convert the grading to an equivalent surface area per unit weight of dust, assuming that the dust density is 2 g cm^{-3}. This calculation has been used in the analysis of test results given later in this chapter.

TABLE 10.2

1	2	3	4
Micron size	Percentage in range	Factor	2 × 3
−5	18	0·750	13·50
5 to 10	13	0·400	5·20
10 to 20	22	0·220	4·84
+20	47	0·095	4·46
		Total =	28·00

$$\text{Specific surface area } g = \frac{\text{total of (4)}}{100}$$

$$= 0{\cdot}28 \text{ m}^2 \text{ g}^{-1}$$

Carbon content of dust. Within the range of carbon content usually experienced on modern boiler installations (below 5 per cent) there is little effect on performance, but if the percentage is high it reduces the effective migration velocity.

Electrical equipment. The design of the electrical equipment, such as the form of control, the size of the sets and the size of the zones, affects the performance, and the designer has to make sure that the balance between the various electrical requirements is satisfactorily evaluated.

Analysis of test results

Object. As can be seen from the foregoing, there are several factors that influence the effective migration velocity and form the basis for design of a precipitator for collecting pulverized fuel ash. Although it is easy to identify the factors, it is difficult to assess precisely the effect of any particular factor on the overall result. It was with this problem in mind that the British Central Electricity Generating Board in 1966 embarked on a detailed study of the efficiency test results obtained for electrostatic precipitators handling pulverized fuel ash. The object was to produce a series of correction curves, one for each variable, which would give a calculated EMV for any combination of conditions for a particular application. Care was taken to select only the variables that would be known to a designer and would be of particular assistance to him. For instance, it would have been easy to refer to such items as the resistivity of the dust or the conductivity of the gas, but since the effect of these on the precipitation process would not be known, and unlikely to be known for a new installation, the excercise would have been pointless.

Data studied. Initially the work was based on 74 sets of test data from 19 different power stations. An analysis gave useful information and encouraged further work on 109 sets of data from 23 different power stations. The variety and quantity of the data ensured the inclusion of considerable variations in the dust to be collected, the volumes and compositions of the gas, the collecting efficiencies, designs and sizes of precipitators by various manufacturers, and other basic factors. This number of test results and the varieties enumerated ensured that the analyses were representative.

Method. The analyses were made on a computer using linear multiple regression to obtain formulae which, when applied to the known data for each of the 109 tests, would give figures as near as possible to the EMVs obtained during the tests.

The computer analyses were based on the formula

$$\text{EMV} = k - x_1V_1 - x_2V_2 - x_3V_3 - \cdots,$$

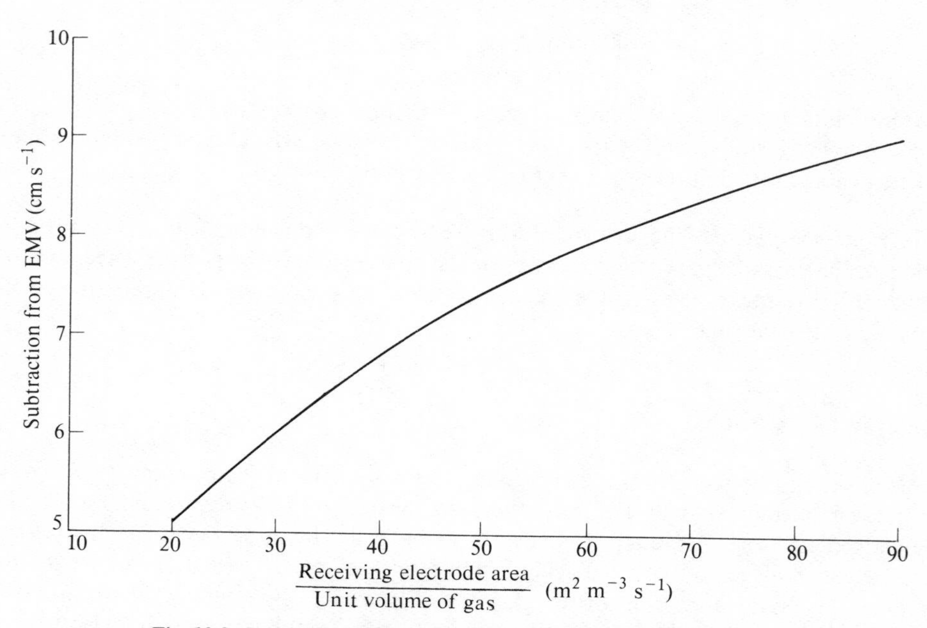

Fig. 10.2. Receiving electrode area per unit volume of gas per second.

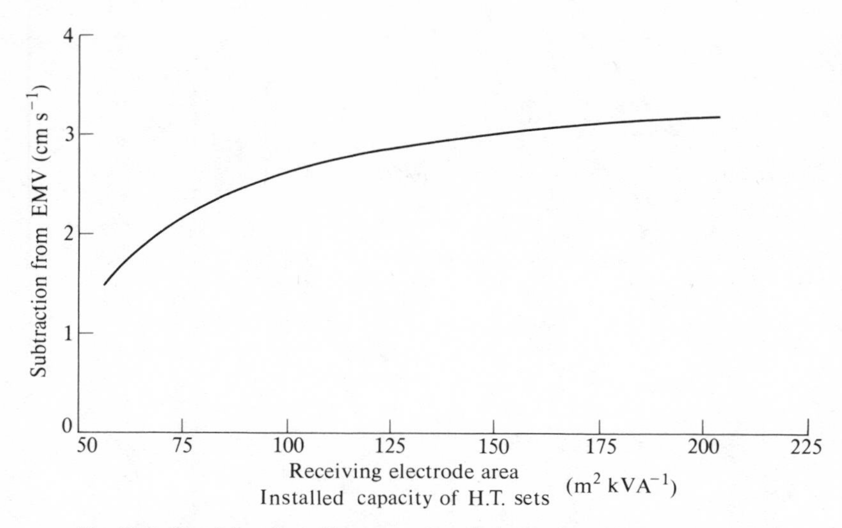

Fig. 10.3. Receiving electrode area per installed capacity of high tension supply.

where each variable, V_1, V_2, etc., represents the known variable such as sulphur in coal, ash in coal, receiving electrode area, etc., and k, x_1, x_2, x_3 are the constants that the computer arrived at to give the best solution. The program allowed any number of variables to be used in any combination, but unless a variable improved the overall agreement, it was not retained in the solution.

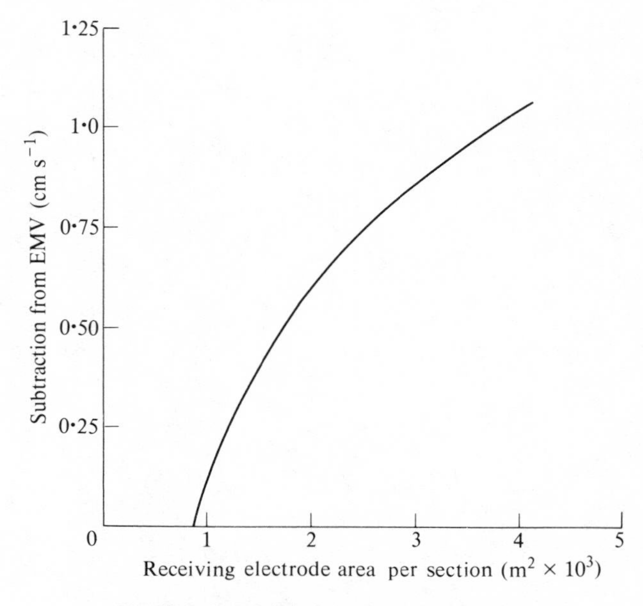

Fig. 10.4. Receiving electrode area per section.

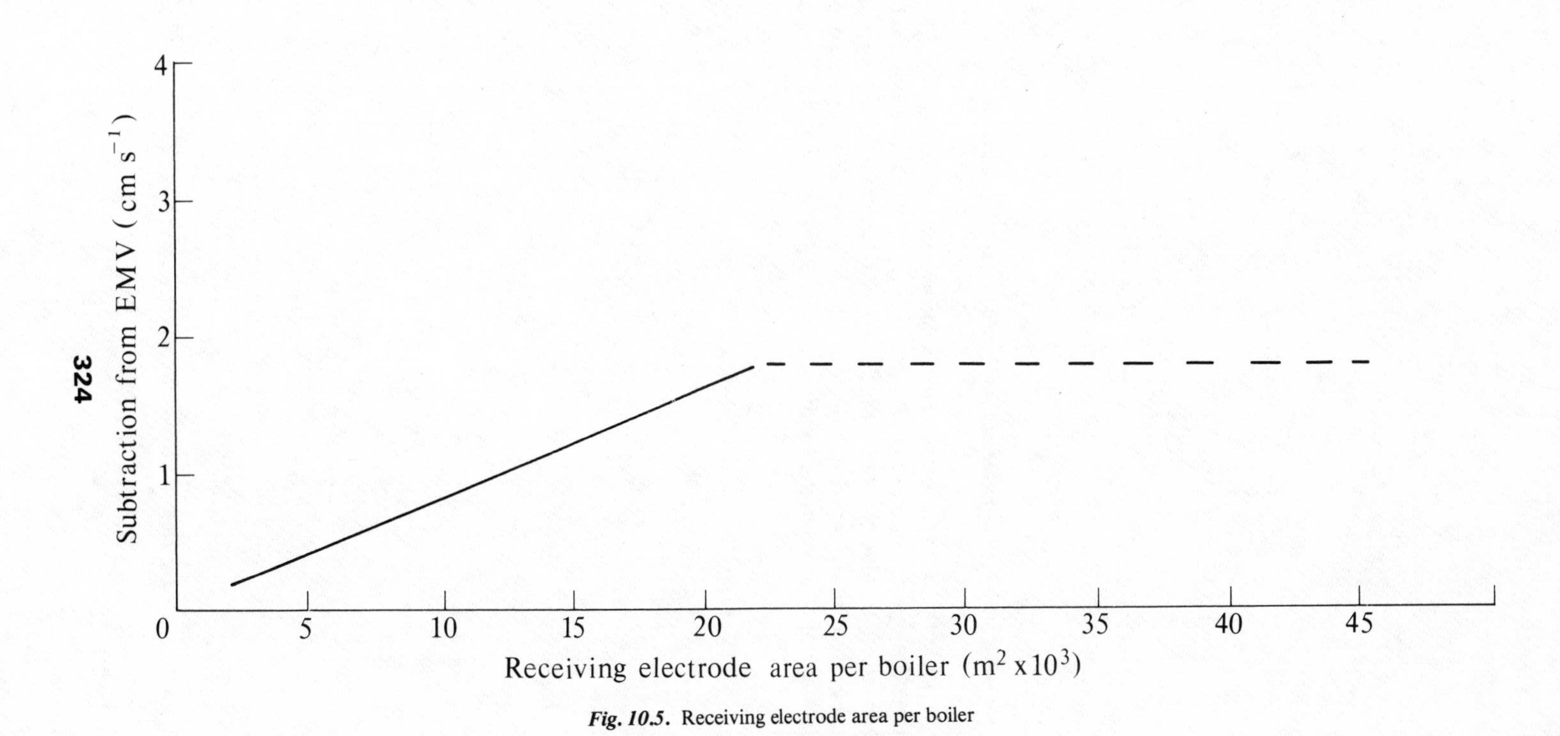

Fig. 10.5. Receiving electrode area per boiler

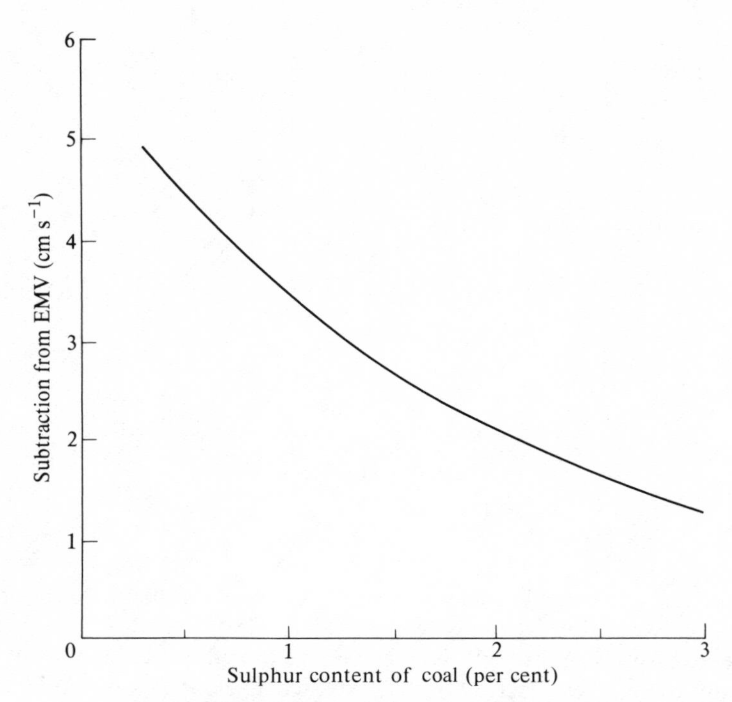

Fig. 10.6. Sulphur content of coal per cent by weight.

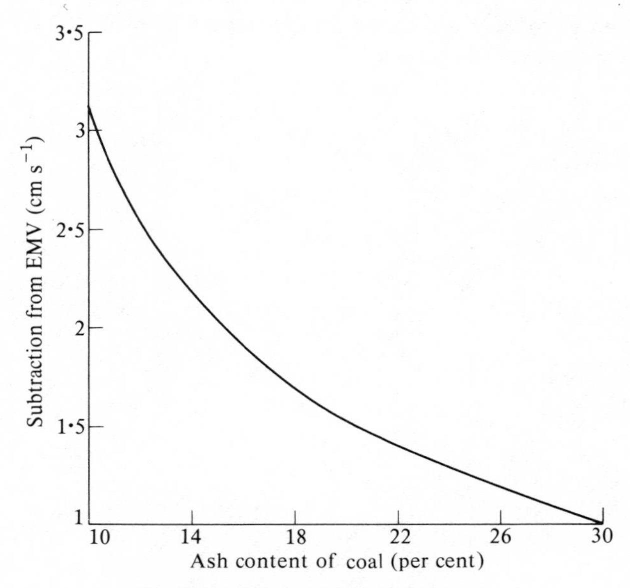

Fig. 10.7. Ash content of coal per cent.

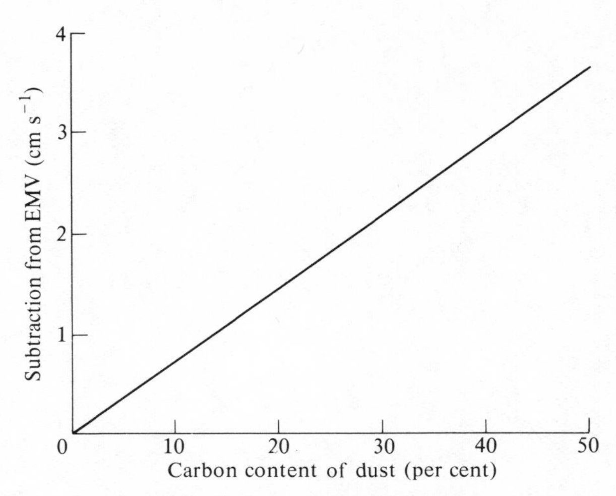

Fig. 10.8. Carbon content of dust per cent.

Having arrived at a suitable solution, a study was made of the difference (known as the residual) between the calculated result and the test result for each set of data. Plotting these residuals against any factor that was thought to be significant showed if there was any trend. If the factor was not significant, the points would be completely random either side of the x-axis. Because the computer has the facility of giving graphical output for the relationships, any number can be studied in a short time. Many factors in addition to those given in the solutions were investigated and rejected when they did not add to the overall agreement.

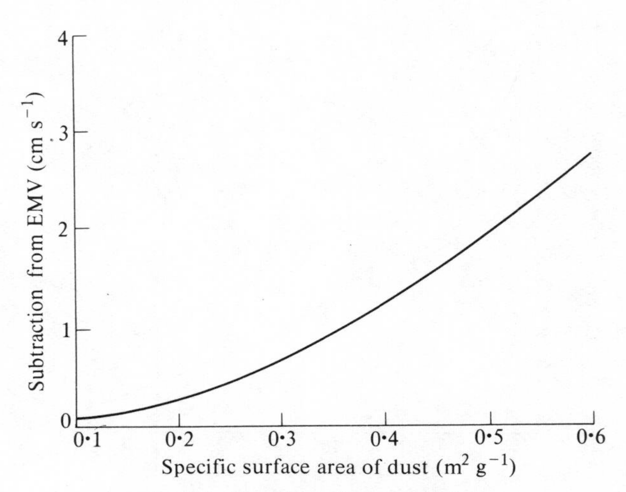

Fig. 10.9. Specific surface area of dust.

Solution. The solution given by the computer consisted of the constant k from which the corrections from each variable have to be subtracted. For simplicity, this formula has been converted into a series of graphs Figs. 10.2–8 and can be used as follows: the constant is 27·2 and adjustments have to be made by subtracting the appropriate figure from each graph. It is essential that a correction from each of these curves is used and that they are only used within the limits shown.

Comments. Figure 10.10 is a plot of the calculated effective migration velocity against the tests results. As can be seen, the agreement between theory and practice is remarkably good, bearing in mind that complete agreement is not possible because the results are obtained from site, and not laboratory tests.

It will be noted that Fig. 10·5 only gives a suggested value above 22 000 m^2 because there are not sufficient data on these very large plants to be quite sure of the trend beyond this figure. The dotted line does agree with the information that is available.

All the curves are self-explanatory and apart from Fig. 10·7 give relationships that would be expected. No immediate explanation for the results shown by Fig. 10·7 can be seen, but any solution arrived at could be improved by the inclusion of this factor.

The method can be used either to check the design parameters of a precipitator before it is installed or to check the performance of an existing precipitator. In the latter case, if an improved collecting efficiency is desired, a great deal of time and money can

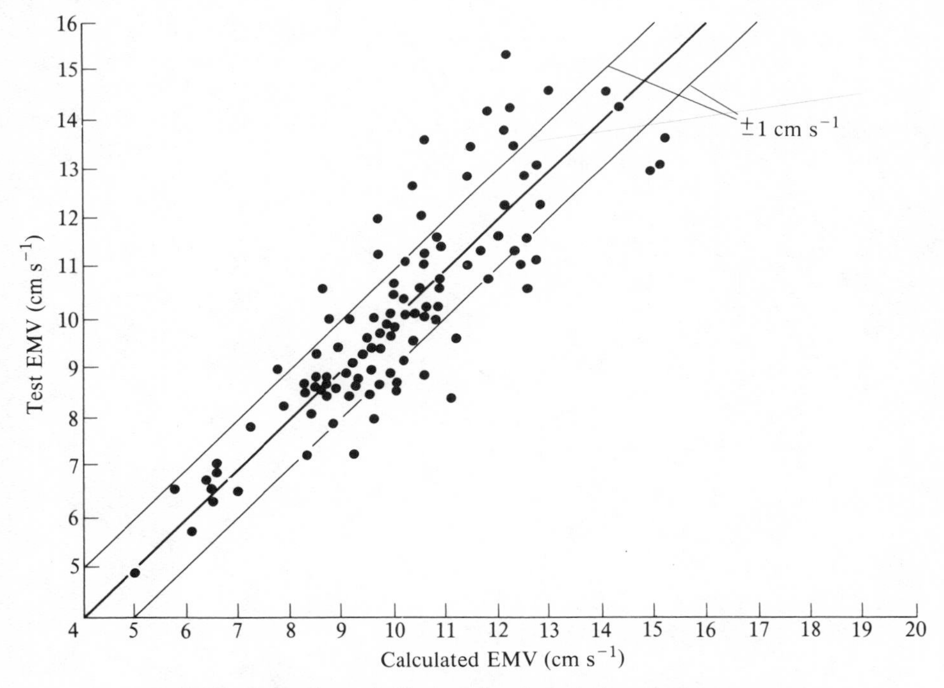

Fig. 10.10. Comparison of effective migration velocity calculated per test.

be saved by using the suggested method to see if it is likely that improvement can be made without increasing the size of the precipitator.

When investigations are being carried out on the effect of a particular variable, the method can be used to correct the test results in respect of other variables, thus bringing all the tests in a series to a common basis. A true appraisal of the particular investigation can then be made.

11 Nuclear power industry and miscellaneous uses of radioactive material

PAMELA M. BRYANT

All production and use of radioactive material ultimately leads to some form of radioactive waste, which may conveniently be classed as airborne, solid, or liquid waste. Much radioactive waste is ordinary material, e.g., paper, metal, air or water, contaminated with radioactive nuclides. Thus, for example, liquid and gaseous waste streams from a chemical process involving radioactive material or discarded equipment may need to be considered as radioactive waste. The radionuclides may be those which occur naturally or those artificially produced in various nuclear processes, such as the fission of uranium by neutrons and activation of stable elements. Only a small fraction of the radioactivity produced in the nuclear power industry is discharged as waste to the environment, but the industry is described here in some detail to explain the control of all the radionuclides produced, not only those discharged. Wastes of high radio-toxicity are not discharged to the environment; they are stored in conditions which allow for the possibility of future treatment and ultimate disposal. Less detailed consideration is given here to other sources of radioactive waste, e.g., hospitals, research laboratories, the luminizing industry, which are many in number but give rise to wastes in relatively small quantities.

The nuclear power industry[1]

Uranium ore milling

Uranium ore usually contains 0·1–1 per cent uranium, in the form of U_3O_8. Natural uranium is composed of the two isotopes uranium-238 and uranium-235 and uranium-234 in the proportions 99·274 per cent and 0·720 per cent and 0·006 per cent respectively. Uranium-238 is the parent of a long chain of radioactive daughters, all of which are present in the ore. The daughter products include uranium-234, thorium-230, radium-226, radon-222 (a gas), and lead-210.

The uranium is extracted from the ore at uranium mills by a series of processes including crushing and washing, leaching, ion-exchange and solvent extraction. Most of the uranium is leached from the ore; the uranium is purified and concentrated in subsequent chemical treatment. The daughter products remain with the bulk of the feed material, which is present as suspended solids in the liquid waste resulting from the extraction processes. This liquid waste is discharged to tailings piles from which the water drains into tailings ponds. Practices at present-day mills cause little contamination of the air; however, old tailings piles can cause airborne contamination as a result of erosion or of leakage of radon.

Fuel fabrication

Fuel elements are fabricated from the uranium concentrate. Uranium-238 is not fissionable by neutrons of thermal energy and most modern reactors designed to operate with thermal neutrons need a fuel which is enriched in uranium-235. The uranium, in the form of the hexafluoride, is normally enriched by a multi-stage gaseous diffusion process, after which it is converted to the chemical form required for the fuel elements. There is very little discharge of uranium to atmosphere from these processes; the uranium in gaseous waste streams is usually present as solid particles which can be removed by conventional air-cleaning methods. The radiotoxicity of natural and slightly enriched uranium is very low and their chemical toxicity is of greater significance.

The fuel in most present-day reactors is uranium dioxide, though in some, such as Magnox reactors operated in the UK by the Electricity Generating Boards, it is the metal itself; the use of carbide is also being studied. The fuel, often in the form of pellets, or rods of the metal, is encased in a cladding material, for example stainless steel or a zirconium alloy, to retain fission products formed in the fuel during irradiation. Some reactors under development use a mixture of uranium-238 and plutonium-239 as fuel. Because of the high radiotoxicity of plutonium, plants making fuel elements for these reactors have to be designed to provide thorough containment at all stages.

Power production[2–5]

Nuclear power stations utilize the energy released by neutron-induced fission of fissile nuclei into two or, occasionally, three parts (ternary fission). The fission process is accompanied by the emission of additional neutrons which maintain the chain reaction. The products of fission of uranium-235 and plutonium-239 have mass numbers mainly in the ranges 80–110 and 125–155; in addition, hydrogen-3 (tritium) is produced in ternary fission. Most fission products are parents of radioactive decay chains and there are over 200 radionuclides present in reactor fuel after a short time on power.

The core of a thermal reactor has four main components:

1. fuel elements enclosed in a cladding material;
2. a coolant (liquid or gas) circulating through the core to remove the heat generated by fission;
3. a moderator to slow neutrons to the thermal energies required for fission of uranium-235;
4. a reflector to reduce loss of neutrons from the core.

As well as fission, radionuclides are produced by neutron activation of materials such as coolant and the metals used in the reactor's construction. Examples of such activation products are tritium formed from deuterium (hydrogen-2), lithium or boron in cooling water; argon-41, carbon-14 and nitrogen-16 from constituents of air or carbon dioxide coolants; and zinc-65 and zirconium-95 from constituents of metal components. Radionuclides are also produced by activation of stable fission products in fuel, such as casesium-134 from caesium-133. In the fuel itself, the eight nuclides from plutonium-238 to plutonium-245 are part of the complex heavy element chain built up by neutron interactions from the uranium-235 and uranium-238 originally present.

Fast breeder reactors, which are currently under development, do not have a moderator, because the energy is generated by fission of plutonium-239 by fast neutrons; these reactors generally use a fuel containing about 15 per cent plutonium-239 in uranium-238. The fuel is surrounded by a blanket of uranium-238, which captures escaping neutrons and can produce more plutonium-239 than is used in the fission process. Fast reactors have a larger heat output per unit core size, and are normally cooled by liquid sodium metal or sodium–potassium alloy.

In thermal and fast reactors, neutron absorbers are inserted into the core, or are present in the primary coolant, to control the power generation rate by regulating the neutron flux in the core. The primary coolant is used to drive conventional electric generators, either directly or by producing steam in heat-exchangers.

Primary coolants in reactors become contaminated with small quantities of fission products leaking from the fuel elements, even if the cladding remains essentially intact. If the cladding fails, there will be a relatively larger loss of activity to the coolant. Such increases in the activity of the coolant are monitored and defective fuel elements replaced.

The primary coolant in Magnox reactors and advanced gas-cooled reactors (AGRs) in the UK is carbon dioxide gas under pressure. This is cleaned by means of bypass filters, usually cyclones, ceramic candles or sintered iron, to prevent build-up of activity, and recycled. Primary coolant gas is discharged during deliberate depressurization; this takes place in the main reactor system about once a year and in ancillary systems several times a week. The high efficiency blow-down filters used in UK reactors are ceramic candles, sintered bronze or sintered stainless steel. High temperature gas-cooled reactors (HTRs) under development use much more expensive helium as the primary coolant; the cost of replacing this provides an incentive to minimize its discharge by means of appropriate circuit design and mode of operation.

The main activity discharged to atmosphere from UK reactors cooled by carbon dioxide is argon-41. The most important source of this activity is air used to cool the steel pressure vessel exterior and primary concrete biological shielding surrounding the core in Magnox reactors of early design. The air is discharged after filtration by oil-impregnated glass-fibre to remove particulates. The amount of argon-41 activity discharged depends upon reactor design, and can be up to about 100 000 Ci per year.* In Magnox reactors of later design and AGRs, the core, pipework and boilers are enclosed in a pre-stressed concrete pressure vessel which is water cooled. A less important source of argon-41 is the argon-40 present as an impurity in the carbon dioxide coolant. Some of this argon-41 is discharged during depressurization and by any natural leakage from the pressure circuit. The radiological significance of argon-41 discharges and noble gas discharges from water-cooled reactors (described in the paragraphs which follow) is discussed in the section on estimated doses to members of the public.

Boiling water reactors (BWRs) use light (ordinary) water as the primary coolant, which boils in the reactor core and is used directly to drive the turbines before being condensed and recycled. Any non-condensable gases in the stream are discharged to atmosphere via filters, and sometines via delay tanks to reduce activity content by radioactive decay. The activity discharged, which varies from a few thousand to several hundred thousand curies per year, depending on reactor design and delay time, consists mainly of short-lived noble gas fission products, their particulate radioactive daughters being removed by filtration.

Pressurized water reactors (PWRs) have light water coolant circulating through the core under sufficient pressure to prevent boiling; steam for the turbines is generated in a secondary system. Non-condensable gases from the primary circuit, released from the coolant during depressurization, are usually compressed and stored for up to 120 days before being discharged to atmosphere. After delay and filtration, the discharge consists mainly of a few thousand curies of xenon-133 per year. In pressurized heavy-water reactors (PHWRs) tritium is produced in the heavy water coolant as a result of neutron interaction with deuterium. Some of this tritium is discharged to atmosphere but the amounts are of low radiological significance.

The life of fuel elements is limited by failures of the cladding material caused by thermal or radiation damage, by depletion of the fissile material, and by build-up of fission products which may absorb neutrons. When withdrawn from the reactor, fuel elements are very highly radioactive and are 'cooled' (i.e., stored to allow radioactive decay) for periods up to six months before being transported to the fuel reprocessing plant.

* The special unit of activity is the curie (Ci); one curie is numerically equal to $3{\cdot}7 \times 10^{10}$ disintegrations per second.

Fuel reprocessing

At reprocessing plants such as the Windscale plant of British Nuclear Fuels Limited (BNFL) in the UK, plutonium produced in the reactor and the unused uranium remaining in the fuel are separated from the fission products for re-use. The cladding is removed from the fuel elements either mechanically, or chemically during acid dissolution of the fuel. All the noble gas fission products remaining in the fuel are released during dissolution to the dissolver cell ventilation air, along with 5–10 per cent of the iodine and very small proportions of other fission products, mainly caesium and strontium. This ventilation air stream and those from subsequent extraction processes are filtered before discharge to remove particulate matter, iodine and other volatile fission products present. The filtration of iodine from airborne wastes presents a special problem because of the variety of physical and chemical forms in which it may be present. It can be removed by a composite filter consisting of copper mesh to remove elemental iodine vapour, a millipore filter to remove iodine adsorbed on particulate matter, and charcoal-loaded filter paper and a charcoal plug to remove gaseous compounds of iodine. In practice the amount of iodine present is usually small and a glass-fibre filter is often found to be adequate, though the composite filter is used for monitoring. Other constituents are removed from the streams by conventional methods, for example, electrostatic precipitators, glass-fibre filters, and scrubbers, as appropriate.

The chief noble gas fission product remaining in fuel which has been cooled for about six months is the long-lived krypton-85 (half-life of about 10 years). Xenon-133 (half-life about $5\frac{1}{2}$ days) is a significant constituent only when the fuel is cooled for much shorter periods. At present all the krypton-85 in fuel is released to cell ventilation air during dissolution, and is discharged to atmosphere. Conventional filtration methods are obviously ineffective for the removal of noble gases from airborne wastes; methods have, therefore, been developed to remove krypton-85 from the cell during dissolution and to store it; possibly absorbed on activated carbon, for a long period of time to allow the radioactivity to decay.[6] The current production rate of krypton-85 does not justify the use of special methods to reduce the quantity of activity discharged.[6,7] The radiological significance of krypton-85 discharged is discussed in the section on estimated doses to members of the public.

Nuclear research laboratories

Nuclear research and development laboratories associated with the nuclear power industry are operated in the UK by the United Kingdom Atomic Energy Authority (UKAEA), BNFL, the Electricity Generating Boards and other organizations. Studies at these laboratories range from tracer experiments involving very small quantities of radionuclides to development work on new types of reactor involving comparatively large quantities of radionuclides, either within reactors on site or in specially equipped handling areas. All exhaust air from buildings where there are substantial amounts of radioactivity, for example, caves in which irradiated fuel elements are examined and exhaust gases from specially designed incinerators for combustible solid waste are filtered before discharge to atmosphere.

Production of radioactive materials for use outside the nuclear power industry

Much of the UK production of radioactive materials in the form required by users described in the next section is carried out by the Radiochemical Centre Limited (TRC) at Amersham. The radionuclides required as raw materials are usually obtained by activation of stable nuclides or by extraction from mixed fission products. These radionuclides are then prepared by chemical or physical processes to produce the form required. They may be as encapsulated radiographic sources, metal foils for thickness gauges or a range of plaques, grains, needles and tubes for clinical application. In addition, hundreds of radiochemicals and pharmaceuticals incorporating specified radionuclides are prepared for use, usually in solution or as solids.

These processes give rise to varying quantities of airborne wastes. Much of the work is carried out in glove boxes or remote handling bays with filtration of exhaust air and sometimes of room exhaust air, as required.

Miscellaneous uses of radioactive materials[8]

There are many applications of radionuclides in hospitals, general research laboratories, and the luminizing and other industries. The amounts of activity used in these applications are commonly several orders of magnitude less than those encountered in the nuclear power industry. These applications are described briefly in the paragraphs which follow, but the amounts of radioactive waste involved are too small to require consideration in the section on estimated doses to members of the public.

Hospitals[9] and general research laboratories

Sources of external radiation, for example, cobalt-60 and caesium-137, are used in hospitals in the treatment of disease, particularly the many forms of cancer. Such sources are returned to the supplier when they are no longer required or become ineffective through decay; no waste arises from these at hospitals.

Many radionuclides are used for treatment and diagnosis of disease. For example, iodine isotopes are used in treatment and diagnosis of thyroid disorders, for kidney function studies and for liver and lung scanning. Carbon-11 is used in heart, lung, blood and bone diagnostic techniques; phosphorus-32 is used in the treatment of bone metastases and as a cell label. Technetium-99m is widely used for scanning the brain, liver, bone, kidney and spleen. The procedures involved are usually simple. Discharges to atmosphere from these generally low-activity sources of short half-life are of no radiological significance.

General research laboratories use various radionuclides in tracer and other experimental work. As with hospitals, small quantities of activity or short half-lives are used whenever practicable and the amount of activity released to atmosphere is seldom of significance. It is not usually necessary to filter the exhaust air from the fume cupboards and glove boxes used in this kind of work.

Luminizing industry

Radium was extensively used in luminizing but has now generally been replaced by tritium. Glove boxes and fume hoods equipped with good ventilation are used during processing. The processes lead to discharge of only small amounts of airborne waste which are of no radiological significance.

General industry

Radionuclides are extensively used in industry for tracer work and control of manufacturing processes. Sources of external radiation are used in quality control as components of thickness and density gauges to test welds and detect faults in components. Such sources are returned to the supplier when no longer required and do not give rise to airborne wastes. Other radionuclides are used in tracer work and are added to components under test, for example, in assessing leaks or rates of flow in liquid or gaseous systems; or they may be produced by neutron activation of appropriate components, for example, in assessing lubricant efficiencies and wear of gears and bearings. The short half-lives of most nuclides involved in tracer work normally ensure that the discharges of airborne wastes are trivial.

Evaluation of radiation exposure of man

Effect of ionizing radiations on man[10,11]

Much of the knowledge about the effects of ionizing radiations on man comes from studies of people exposed to radiation in connection with medical practice, the survivors of the atomic bombs in Japan, and experiments on animals. All these studies are of the effects of high doses and high dose rates. They show that, at doses of a few hundred rads* delivered in a short time, there are acute somatic effects, principally to the blood-forming organs, which can prove fatal. Doses at this level given over a longer period induce delayed somatic effects, including leukaemia and other cancers, skin damage, cataracts and impaired fertility. There is some evidence that, at lower doses than this, irradiation of the foetus can increase the risk of abnormalities and cancer in the child.

* The rad is the unit of absorbed dose, equal to an absorbed energy of 10^{-2} J kg^{-1} of tissue. The millirad (mrad) is one thousandth part of this.

Irradiation of the gonads, leading to mutations in the genes or chromosomes of the germ cells, can result in hereditary defects influencing any part of the range of inherited characteristics. All the above somatic and genetic effects are also present in populations where the only radiation source is the natural background (about 100 mrad per year).

Restriction is placed on radiation exposure of members of the public from all controlled sources of activity; exposure resulting from natural background and medical procedures is excluded from this restriction. For somatic exposure of individuals, some of whom may be children, the restriction takes the form of dose limits recommended by the International Commission on Radiological Protection (ICRP); these are endorsed by the UK Medical Research Council. The annual dose limit for the whole body is set at 0·5 rad; for certain organs it is somewhat higher. Dose limits are set low enough to ensure that there is no possibility of acute somatic effects. The risk to the individual of delayed somatic effects associated with irradiation up to the annual dose limit is small compared with the ordinary risks of life.

Genetic changes affect society as a whole and the effects on populations are dependent upon the average dose which members of those populations receive between conception and mean age of childbearing. This is the basis for the requirement that the average genetic dose to the whole population of the UK from waste disposal must be limited to 1 rad per 30 years.[12]

Pathways to man involving initial dispersion in the atmosphere

The several possible pathways by which activity discharged as airborne waste can lead to radiation exposure of man are shown in simplified form in Fig. 11.1. Dispersion of the activity in air near the discharge point may be followed by deposition on surfaces of radionuclides which are present as vapours, attached to dust particles or in liquid droplets. The noble gases are not deposited to any appreciable extent and spread ultimately throughout the whole of the troposphere if they are of long radioactive half-life. Gamma-emitting nuclides in air or on surfaces cause external radiation exposure of the whole body, while beta-emitting nuclides cause exposure of skin and superficial tissues. If tritiated water, i.e., water containing the beta emitter tritium is deposited on skin, some is absorbed into the body resulting in internal radiation exposure of the whole body. Inhalation of activity in air causes radiation exposure of the lungs and of other parts of the body to which the activity may subsequently be transported.

Deposition of airborne activity on food crops and soil in dry weather or by washout by rain leads to internal radiation exposure of man following ingestion of contaminated foodstuffs. Examples of these pathways involve initial deposition on leafy vegetables, such as lettuce, and on pasture grazed by dairy cows. The latter pathway may be important for iodine, strontium and caesium, because cows graze large areas of pasture each day and these elements are readily transmitted from cows' diet to milk. The two nuclides strontium-90 and caesium-137, which have long radioactive half-lives (about 30 years), pass from contamined soil to pasture grass in addition to contaminating the grass by direct deposition. Iodine-131 has a short half-life (8 days) and its brief

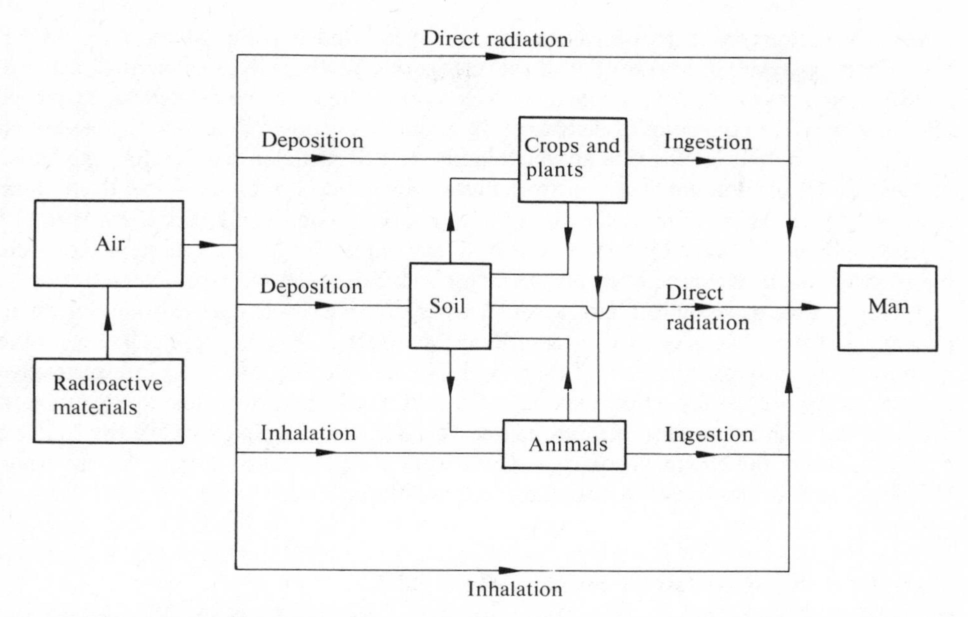

Fig. 11.1. Simplified pathways between radioactive materials released to atmosphere and man. (Courtesy Dr F. D. Sowby (ICRP) and Pergamon Press. ICRP; Principles of environmental monitoring related to the handling of radioactive materials. A report by Committee 4 of the International Commission on Radiological Protection, ICRP Publication 7, Pergamon Press, Oxford, 1965.)

presence in soil does not contribute to iodine-131 in grass. Other animal products, for example meat and eggs, may also be contaminated, but it is usually found that for any one nuclide in a mixture discharged at a given site, one particular pathway is of much greater importance than others. If this is the case, it is possible to base monitoring programmes and control of the discharge on this single pathway. The pathway involving iodine-131 and cows' milk is often used for this purpose.

Limits, called derived working limits (DWLs), can be calculated giving the rates of discharge of activity which, if released steadily, would give annual doses not exceeding the ICRP recommended dose limits. The ICRP recommendations are given in terms of annual doses so the DWLs are derived in terms of annual discharges, although they may be expressed over a shorter period of time, for example, curies per week, for operational reasons. It is often sufficient for routine control purposes to compare measured discharge rates with the relevant DWLs for discharge; if, however, the discharge is a significant fraction of the relevant DWL, monitoring of environmental materials may well be required in addition.

As an example, consideration is given here to the derivation of the DWL for discharge of iodine-131 for the pathway via pasture and cows' milk.[13] Idealized dispersion formulae are available giving the downwind concentration in terms of the average wind speed and the horizontal and vertical spread of the cloud of discharged material. From meteorological observations, categories classifying types of weather have been defined in terms of values of wind speed and stability of the atmosphere. Using the observed frequency of these categories in the UK, the annual average concentration resulting from a constant discharge rate has been calculated for distances up to 100 km from the point of discharge. In addition, average values have been derived from published data for the rate at which iodine-131 deposits on pasture for a given air concentration, the amount taken up from pasture by the cow, the fraction of this uptake appearing in each litre of milk, the amount of milk that people drink, and the fraction of ingested iodine-131 taken up by the thyroid. These values for the various steps along the pathway enable a discharge rate corresponding to the dose limit to be calculated.

When setting acceptable limits to discharges consideration must also be given to ecological systems. It is generally found that the effect of a given dose is larger on the more developed living species and that the limits derived for man are more restrictive than would be needed for other species. In the context of waste disposal, an exception to man being the limiting case applies at one fuel fabrication plant, where the limiting consideration is the chemical toxicity of uranium affecting grazing animals, this route being more important than those to man via inhalation or food.

Examples of estimated doses to members of the public

To assess the radiological significance of discharges of short-lived noble gases from reactors, it is appropriate to estimate annual doses to the usually small critical group living within about 6 km of the site. Annual doses decrease quite rapidly with distance as a result of dispersion in the atmosphere and radioactive decay. Annual gonad doses to this group are estimated to be, typically, about 10^{-5} rad for PWRs, about 10^{-3} rad for GCRs, and to vary from about 10^{-4} to 10^{-2} rad for BWRs according to reactor design and the delay time before discharge.[14,15] The highest of these doses is only 2 per cent of the ICRP recommended dose limit for individual members of the public and it applies only to BWRs of early design.

Annual doses arising from discharge of the long-lived krypton-85 from fuel reprocessing plants have been calculated for the UK power programme.[16] The annual gonad dose to the small critical group living within 6 km of the reprocessing plant is about 4×10^{-5} rad, which is less than 0·01 per cent of the dose limit for individual members of the public. The entire UK population receives, on average, a much smaller dose due to the continuing dispersion of UK krypton-85, and receives, in addition, a dose due to recirculating krypton-85 from both the UK and world power programmes; the total annual gonad dose is estimated to have been about 10^{-6} rad in 1972. This may be compared with the annual dose of 3×10^{-2} rad corresponding to the 1 rad per 30 years genetic dose limit set by UK waste disposal policy.

Summary and conclusions

The review in this chapter of current practices in control of pollution by radioactive waste includes descriptions of the various parts of the nuclear power industry and of miscellaneous uses of radioactive materials. The amounts of activity involved in the latter category, which includes use in hospitals, general research laboratories and the luminizing and other industries, are commonly several orders of magnitude less than those used in the nuclear power industry. In practice discharges of radioactive waste to atmosphere from these sites are extremely small, often approaching zero, and are of no radiological significance.

Discharges of radioactive wastes to atmosphere from the nuclear power industry may be of radiological significance and the review includes a description of pathways by which activity so discharged can lead to radiation exposure of man. Control of amounts discharged is based on annual dose limits recommended by the ICRP for radiation exposure of the general public; the members of the public who provide the limiting case are often those who live near the site in question or who consume exceptionally large quantities of foodstuffs which become contaminated by specific radionuclides readily transmitted along a particular pathway. Examples are given of estimated annual doses to local populations near different types of nuclear power stations. Annual doses to both local and country-wide populations from the noble gas krypton-85 discharged from the Windscale reprocessing plant are also given. None of these estimated annual doses exceeds 2 per cent of the relevant dose limit.

The conclusion to be drawn from this review is that the conservative approach adopted in radioactive waste management, involving strict legislative and operational control over discharges; and where appropriate; environmental monitoring, has resulted in levels of radiation doses to members of the public which are generally far below recommended limits. There is no evidence to date that radiation due to waste disposal has caused any injury whatsoever to any member of the public. It is certain that this inherently safe approach will continue to be applied to the disposal of radioactive wastes.

References

1. Glasstone, S. and A. Sesonske, *Nuclear reactor engineering,* Van Nostrand, New York, 1963.
2. IAEA, 'Management of radioactive wastes at nuclear power plants', *Safety Series 28,* International Atomic Energy Agency, Vienna, 1968.
3. Mawson, C. A., *Management of radioactive wastes,* Van Nostrand, New York, 1965.
4. IAEA, 'Techniques for controlling air pollution from the operation of nuclear facilities', *Safety Series 17,* International Atomic Energy Agency, Vienna, 1966.
5. IAEA, 'Basic factors for the treatment and disposal of radioactive wastes', *Safety Series 24,* International Atomic Energy Agency, Vienna, 1967.

6. Dunster, H. J. and B. F. Warner, *The disposal of noble gas fission products from the reprocessing of nuclear fuel,* United Kingdom Atomic Energy Authority Report AHSB(RP)R 101, HMSO, 1970.
7. NEA, *Radioactive waste management practices in Western Europe*, The European Nuclear Energy Agency, Paris, 1970.
8. IAEA, 'The management of radioactive wastes produced by radioisotope users', *Safety Series 12 and 19*, International Atomic Energy Agency, Vienna, 1965 and 1966.
9. Wagner, H. N., Jr. (ed.), *Nuclear medicine,* H.P. Publishing Company, New York, 1975.
10. ICRP, 'The evaluation of risks from radiation', A report by Committee 1 of the International Commission on Radiological Protection, *ICRP Publication 8*, Pergamon Press, Oxford, 1966.
11. ICRP, 'The recommendations of the International Commission on Radiological Protection', *ICRP Publication 9*, Pergamon Press, Oxford, 1966.
12. 'The control of radioactive wastes', *Cmnd 884*, HMSO, London, 1959.
13. Bryant, Pamela M., 'Derivation of working limits for continuous release rates of iodine-131 to atmosphere in a milk producing area', *Health Physics,* **10**, 249–57, 1964.
14. Commission of European Communities, 'Radioactive effluents from nuclear power stations in the community: discharge data, radiological aspects', *CEC Doc. V/1973/74e*, 1974.
15. Bryant, Pamela M. and J. A. Jones, 'Estimation of radiation exposure associated with inert gas radionuclides discharged to the environment by the nuclear power industry', p. 91, *Proc. IAEA/NEA/WHO Symp. Environmental Behaviour of Radionuclides Released in the Nuclear Power Industry*, International Atomic Energy Agency, Vienna, 1973.
16. Bryant, Pamela M. and J. A. Jones, 'The future implications of some long-lived fission product nuclides discharged to the environment in fuel reprocessing wastes, p. 131, *Proc. NEA/IAEA Symp. Management of Radioactive Wastes from Fuel Reprocessing,* Paris, 1973.

12 Coke and gas industries

Part 1 Coke ovens

G. E. HALL

World production of coke-oven coke from the carbonization of coal, mainly for use in metallurgical industries, particularly iron and steel works, in 1972 was 340 million tonnes, of which 85·7 million tonnes was produced in Western Europe, 79·8 million tonnes in the USSR, 58·8 million tonnes in the USA and 36·2 million tonnes in Japan.[1] In the UK in 1972/3 the carbonization of 22·8 million tonnes of coal in coke ovens gave 15·0 million tonnes of coke.[2]

All the methods of production of coke and gas from coal involve problems of mitigation of air pollution to a greater or lesser extent. The most troublesome process, partly for technical reasons and partly because it is operated on a considerable scale in several countries, is that of the carbonization of coal in coke ovens to provide coke suitable for the metallurgical industries. In his annual report for 1971 the Chief of the Alkali and Clean Air Inspectorate for England and Wales said 'Coke ovens continue to be uneasy neighbours ...'.[3]

Coking plants range in size from those carbonizing a few hundred tonnes of coal per day at the small works to works treating 5000 tonnes or more per day at certain large integrated iron and steel works. It might be expected that the incidence of air pollution would be proportional to the scale of operation, since a high proportion of the pollution arises from the ovens and is related to the quantity of coal processed. The larger works, however, are usually of more modern construction and benefit from the inclusion of equipment designed to mitigate pollution, so that though a large modern plant may give rise to more pollution than a small older plant, the incidence of pollution per tonne of coal carbonized may be appreciably less.

To appreciate the problems and difficulties of eliminating or very greatly reducing the discharges of air pollutants in the operation of the classical coke-oven process, brief reference must be made to the nature of the equipment and method of working. Carbonization in coke ovens involves the destructive distillation of coal in closed chambers, the gaseous products being conducted from the chambers through a pipe system to a gas treatment plant to remove condensible tar and ammonia liquor and leaving in the chamber the solid coke.

A single coke oven is a rectangular chamber, typically some 450 mm wide, 4–7 m in height and 12–15 m long. Each vertical end of the chamber is closed with a door occupying almost the full cross-sectional area and there are three, four or five charge holes and a gas offtake in the roof of the chamber. The construction is of silica refractory shapes, and flue-ways are built into the wall separating each pair of ovens; in these ovens temperatures between 1100 and 1400°C are maintained. A battery of ovens so built with heating walls between each pair of adjacent chambers may include as many as 70 or 80 chambers, each of which is charged with coal periodically by machines travelling on the top of the battery. Other machines are provided to remove the end doors of each chamber and to transfer the coke to a quenching station and storage wharf. It is these operations, carried out in a pre-arranged sequence, which cause most concern in relation to air pollution. Figure 12.1 is a photograph of a typical battery of coke ovens.

In conventional practice there is the following sequence of operations. An oven stands empty with the doors on, the charge-hole lids off and the gas offtake disconnected from the gas-collecting system. A charging car or lorry carrying a volume of coal sufficient to charge the oven and having a set of hoppers to correspond with the charge-holes is brought above the oven. Sleeves connecting each hopper with a charge-hole are lowered and the coal, previously crushed usually to give about 80 per cent below 25 mm in size, is allowed to fall into the chamber. At the commencement of this operation the gas offtake is connected to the collecting system and a steam jet is used to assist in drawing gas from the oven into the system. When a proportion of the coal has been gravity charged into the oven, a levelling bar, located on the pushing machine, is introduced into the top of the chamber through a small door provided for the purpose and is moved backwards and forwards through the space at the top of the oven as the last part of the charge of coal descends; the object of this operation is to maintain a passage above the coal charge to facilitate the removal of gas, the evolution of which begins immediately. The pusher machine then moves to discharge coke from an oven in which the coal has been fully carbonized and the charging car travels to a coal service bunker to collect a fresh charge of crushed coal for that oven.

Conventional high temperature carbonization of coal in coke ovens is thus essentially a batch process, although by having ovens arranged and operated in batteries the flow of products from the overall plant is virtually continuous in practice. The carbonizing cycle in each oven usually occupies about 16–18 hours. When carbonization is completed and the coke is ready to be discharged, the gas offtake is disconnected from the gas collecting main and the doors are removed from each end of the oven. A pusher or ram is lined up at one end of the oven and a guide frame at the other end to direct the hot coke into a coke car which stands ready at a lower level (Fig. 12.2). On a signal from the coke car operator the ram is advanced into the oven, thus pushing the coke

Fig. 12.1. Modern battery of coke ovens. (Acknowledgements to Simon Engineering Ltd., Stockport, England, and the British Steel Corporation.)

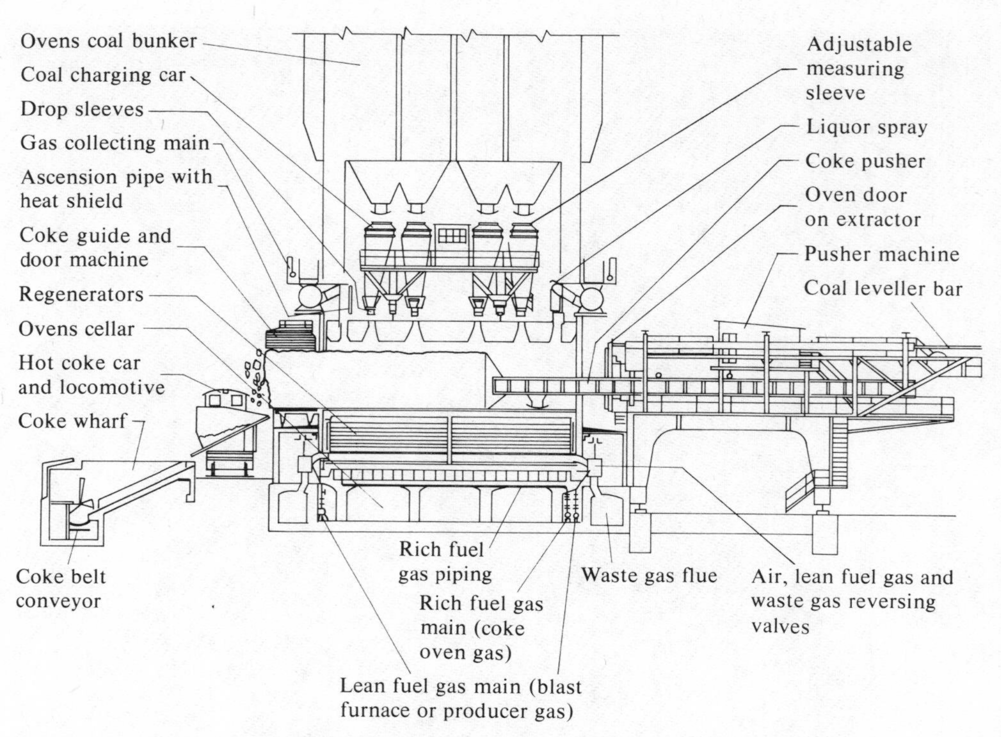

Fig. 12.2. Cross-section of a typical battery of coke ovens.

through the guide into the coke car. The car then begins to travel in the direction of the quenching station, thus distributing the red-hot coke evenly along the floor of the car, which having received all the coke travels to a quenching station or tower. At this station 20–30 m^3 of water is sprayed over the coke in some 90 s, thereby extinguishing it and permitting subsequent handling. The oven doors are cleaned and replaced on the oven, which is then ready to receive a fresh charge of coal.

12.1.1. Emissions from charging coke ovens

During the charging operation, airborne emissions of smoke and fume can occur around the base of the charging car hoppers, and following the discharge of coal from the hoppers, from the hoppers themselves. Prompt replacement of the charge-hole lids is an obvious requirement if emissions are to be minimized and clearly the time taken for the whole operation has an important bearing on the overall volume of emissions to the atmosphere.[4] The design of modern charging equipment takes these points into account and technical improvements during the past few years have made significant progress in achieving a much greater degree of smokeless charging than that obtained in the past.

Most emission control systems are devices such as steam jets or liquor sprays, or both, located at the main junction of the ascension pipe and gas collecting main at one or both ends of the oven (Fig. 12.3). The purpose of these devices is to assist in the aspiration of the oven on charging the coal so as to maintain balanced pressure or slight suction at all points of emission. These steam jets and liquor sprays are insufficient to provide complete control of emissions but they are an essential part of most charging systems.

The approach towards smokeless charging of coal into the ovens has resulted in the development of two basic systems, namely the method of sequential charging and the washer-car system. Sequential charging is based on the concept that the aspiration of gases at the oven ascension pipe controls the pressure balance at only one opening in the oven and emissions are prevented by restricting the extent of openings and ensuring that the openings are under positive suction. Desirable features of this system are the provision of adequate suction at both ends of the free space in the oven above the coal charge and careful individual control of the lifting and replacing of charge-hole lids and of the discharge of coal from the hoppers on the larry car.

There are several variants of this system, which was developed primarily by the operators of coking plant. At the plant in the UK where this system originated in 1961 the hoppers are discharged in a pre-determined sequence, 1–4–2–3, with the hoppers numbered consecutively from the pusher side of the ovens (Fig. 12.4). The capacity of

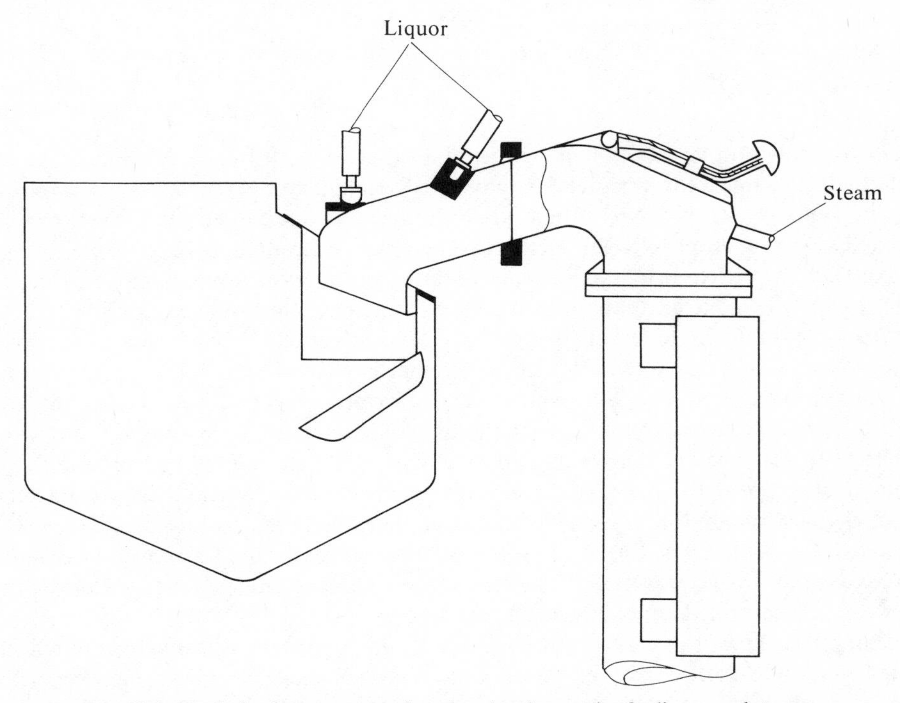

Fig. 12.3. Typical coke oven ascension pipe showing entries for liquor and steam.

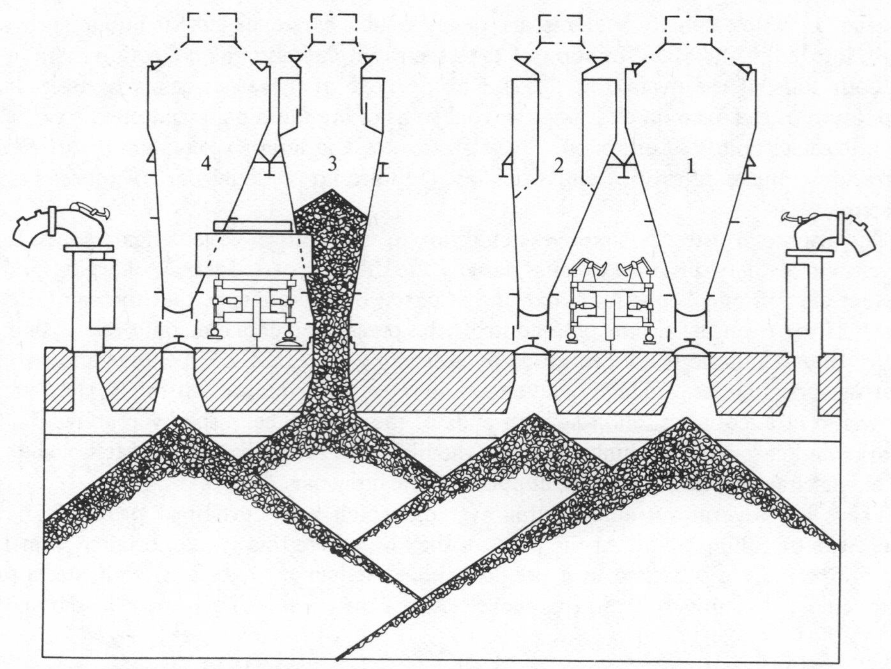

Fig. 12.4. Diagram to explain charging of coal to a coke oven through hoppers 1, 4, 2 and 3 in sequence. (Acknowledgements to M. R. Meades and G. E. C. Randall.)

each of the two outer hoppers is greater than that of each of the two inner hoppers. Reliable mechanical or magnetic handling of each charge-hole lid is an important requirement, as each lid is lifted as a hopper is discharged and then replaced immediately that hopper is emptied.[5] In the absence of dual collecting mains other means are adopted to provide adequate removal of the gases evolved during charging, such as the use of a portable breeches pipe to connect the oven being charged to an adjacent oven. This pipe, in the form of an inverted U-bend, may be carried independently on a mono-rail track or on the charging car.

Washer car systems,[6,7] which have been developed by the builders of coke-oven plant, in general require more complex equipment. They are available in a number of different forms. They are all designed basically to collect the emissions from the charge-holes at the top of the oven, ignite and/or scrub the emissions and then exhaust the cleaned gas or steam through one or more chimneys mounted on the charging car (Fig. 12.5). This is achieved by arranging the equipment so that some form of telescopic tube descends from the hopper outlet into the charge-hole aperture and exhausts the evolved gas from the shrouded annular space created.

Charging cars of the washer type require a supply system or tank for the washing water and a vessel to receive the fouled water that has been passed through the system. For each charging of coke the water supply tank has to be re-filled and the tank

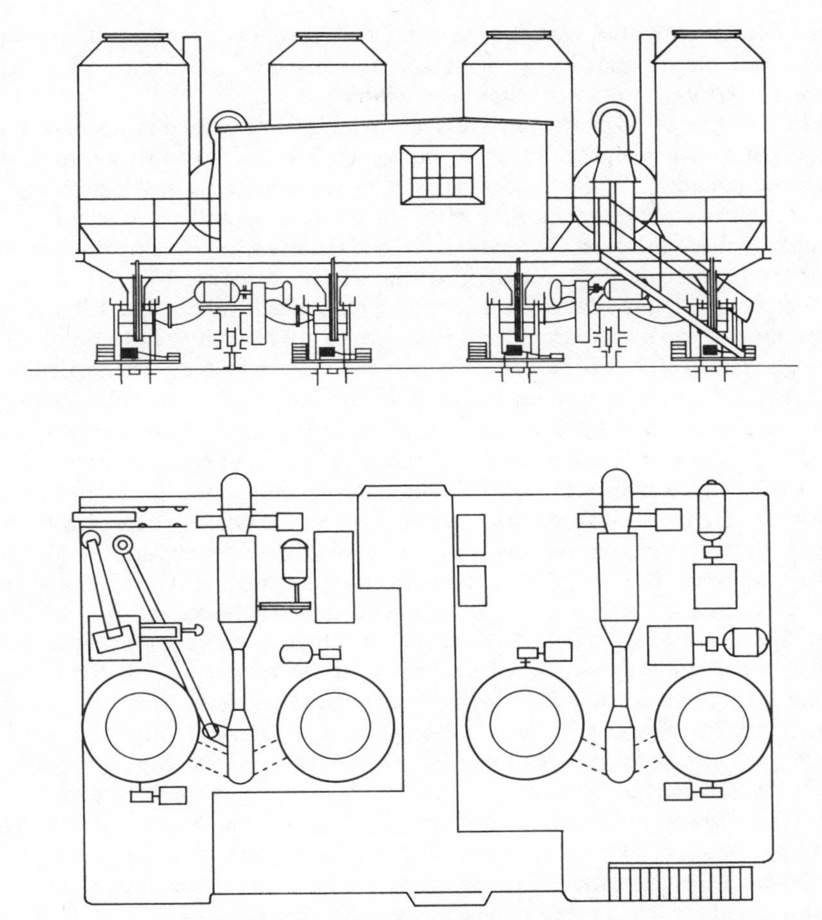

Fig. 12.5. Washer car for charging coal to coke ovens. (Acknowledgements to Hartung, Kuhn Maschinenfabrick GMBH, Dusseldorf.)

receiving the fouled water has to be emptied. Automatic handling equipment for charge-hole lids is a normal feature of the system using washer cars for coke charging. The development of washer charging cars has involved very careful consideration of the dimensions of all the components, particularly in the gas collecting zone around the hopper mouth. Early cars of this type frequently failed to sustain combustion of the collected gas and together with inadequate washers gave rise to coloured plumes from the discharge pipes. Attention to design detail in the matching of dimensions of equipment with the volumes of gas to be treated, combined with relatively powerful fans and washers of generous size, have resulted in the reasonable certainty of smokeless charging, the discharge to atmosphere being in the form of plumes of water vapour, which soon disperse. The weight and complexity of a washer charging car are considerable as compared with the systems using a simple charging car or sequential charging and the advisability of such a heavy machine moving on the oven tops, supported only by the brickwork, has been questioned.

The new technique of charging coke ovens with coal preheated in a suitable form of drier/preheater to a temperature up to about 250°C deserves mention in relation to atmospheric pollution.[8] If this process is used the residence time of the charge of coal in the oven is greatly reduced, since the early stage of drying and initial heating will have taken place outside the oven. The substantial reduction of carbonizing time so obtained is sufficient to increase the throughput of the battery of ovens by as much as 40 per cent. The economic importance of this development needs no emphasis. Its fairly wide adoption may be expected, particularly as improved coke quality can be obtained.

An important feature of the system is that the fluidity of the preheated coal is such that levelling of the charge to maintain a free gas space at the top of the charge in the oven is not required. The handling of preheated coal, however, presents special problems, and three different methods of charging are being adopted, each designed to cope with the high initial rate of gas evolution that occurs when the coal meets the hot brickwork of the oven. All are enclosed systems and the coal is charged with the oven connected to the gas collecting system so as to avoid the tendency for the leakage of gas to the atmosphere. The first and perhaps most direct method of charging employs a conventional charging car; though washer type cars have been used[9,10] for charging preheated coal, a method is being advocated whereby gas in excess of that taken by the normal gas collecting system will be taken from the remote end of the oven into a 'charging main', whence it will be washed to remove pollutants. The second method, known as pipeline charging,[11] delivers the coal to each oven through fixed pipes with valves at each oven and provided with steam jets at intervals along the pipe runs to maintain the transport of the coal. The third method[12] uses an enclosed conveyor of the Redler type, arranged longitudinally over the top of the battery and capable of discharging coal to each oven as required through a divided chute connecting with two charge-holes in the oven roof. The second and third methods require a longer time to charge an oven than when using a conventional charging car.

Whatever systems or equipment are used to charge coke ovens, several factors influence the effectiveness of the methods adopted to reduce to the minimum discharges causing air pollution. These include the particle size and the moisture and volatile matter contents of the coal and the bulk density of the charge. In addition, the physical design of the oven, the type and number of charge-holes, the speed of movement of machines and, above all, the effectiveness of the operators have an important bearing on the overall level of pollution from the charging of coke ovens.

12.1.2. Emissions from discharging coke ovens

The pollution arising from the discharge of coke from the ovens is a major problem and difficult to overcome in practice. It may be tackled in various ways. The methods used include total or coke-side enclosure, enclosed discharge combined with an enclosed quenching system or a hood connected to a duct leading to an independent scrubber-extractor arrangement. Fogging spray systems have been tested as a means of mitigating pollution from discharging coke from ovens but are not regarded as a serious contender for a permanent solution. The problem is how to handle in a pollution-free way some twenty tons of red-hot coke falling about 9 m from a coke guide into a coke car 15 m long and travelling through a distance equal to its length during the period of discharge of the coke into the car.

Total enclosure of the oven battery was used at a few plants in the UK during the Second World War. Though there was no extraction of fumes and only dispersion by natural ventilation, it is understood that the working conditions, though not good, were reasonably acceptable in the particular circumstances. More recently, suggestions have been made for complete battery enclosure with arrangements to collect and clean all the coke-oven emissions. There have been studies to determine whether such a system is feasible, for example by the Koppers Company in the USA. The merits of such a system, which may be successful in collecting and treating the emissions, have been doubted because of the technical problems involved in addition to the high capital costs. The structure for a large battery of coke ovens would be vented with a fan driven by a motor with power requirements in the range of 7·5–11 MW. Lighting would be required within the structure and the possibility of tar condensation and high rates of corrosion would have to be taken into account. In the event of a power failure the enclosure may be unsafe in the absence of lighting and air circulation and the possibility of the formation of explosive mixtures would have to be considered.

Coke-side enclosures have been or are being adopted at a number of plants in West Germany and the US with a reasonable degree of success (Fig. 12.6). One such has been built at the St Louis plant of the Great Lakes Carbon Corporation in the USA to cover the coke side of a battery of 40 ovens. The enclosure has a profile, which allows smoke and the smaller particles of grit evolved when coke is discharged from an oven to rise into the upper part and to spread along the ridge through which the scoops of the main suction duct protrude. The emissions, with the exception of the larger solid particles, which fall down in the enclosure and onto the coke car track, are drawn through the duct on the top of the ridge by large fans and are passed to the quenching tower, which is fitted with continuously operating sprays. Since the commissioning of this system, however, some authorities in the US have questioned the soundness of the concepts because men are obliged to work inside the enclosure in an objectionable atmosphere.

One of the alternative systems now being favoured is that of enclosed discharge coupled with enclosed coke quenching. In this system the coke is enclosed during the whole period from its discharge from the oven until the completion of quenching. The controlled quenching ensures that the moisture content of the quenched coke is reasonably uniform and it eases the elimination of grit from the quenching plume. In an installation (Fig. 12.7) at the Weirton plant of the National Steel Corporation in the UK the coke is pushed into a fully enclosed transfer car through an enclosed guide. A telescopic quadrant hood forms a seal at the oven face and at a corresponding mating surface on the transfer car. The car is in the form of a hopper lined with refractory

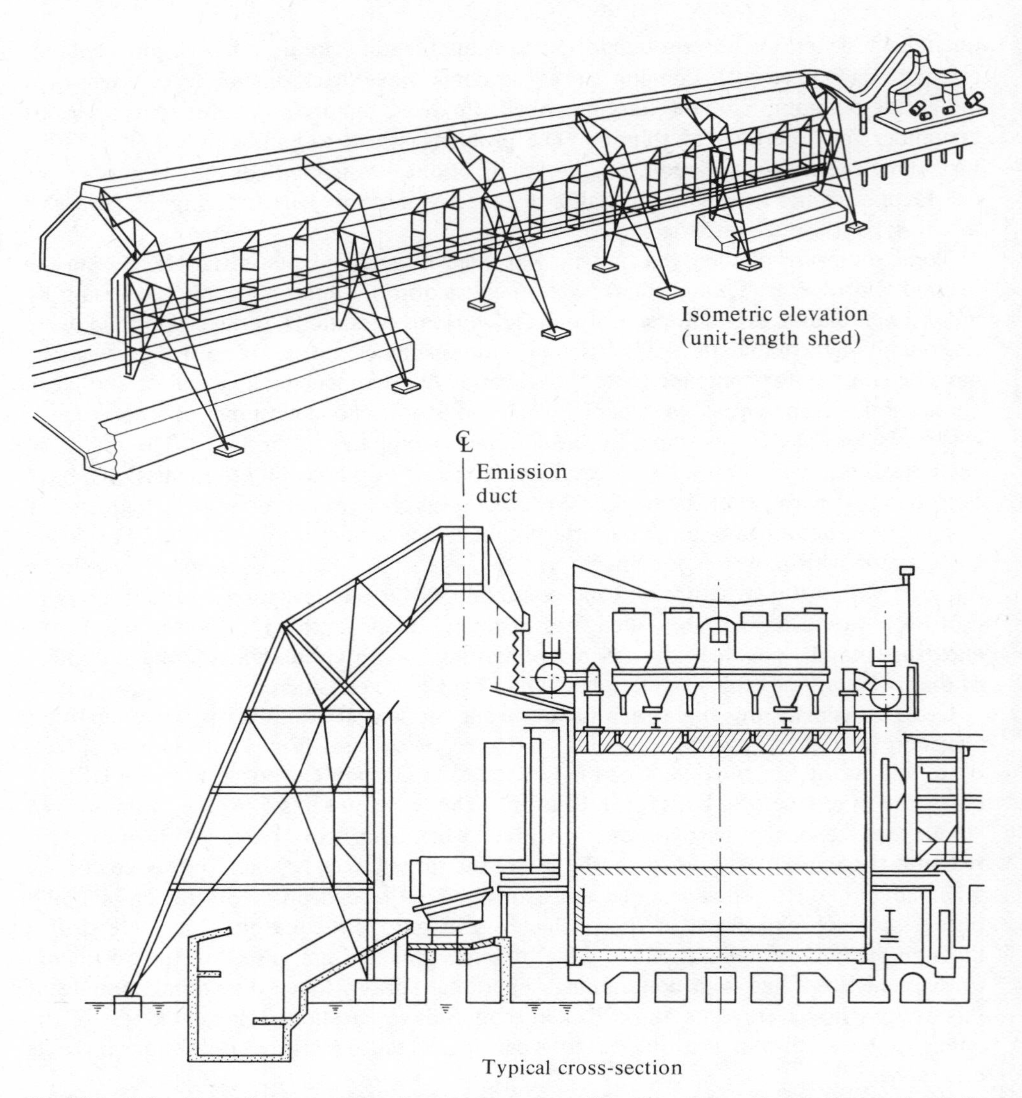

Fig. 12.6. Coke-side enclosure system at the coke-oven plant of the Great Lakes Carbon Corporation at St Louis, US.

material. As a guide is withdrawn the inlet of the car is closed by a stainless steel curtain. Fumes arising from the coke are drawn into a cleaning system as the coke is being pushed into the transfer car and conveyed to the quenching station. When the car reaches the quenching station the coke is discharged through two openings, equipped with hinged doors at the bottom of the car, into one of four track hoppers above the quencher. Fumes from the hoppers are exhausted to a gas cleaning unit. The hot coke is

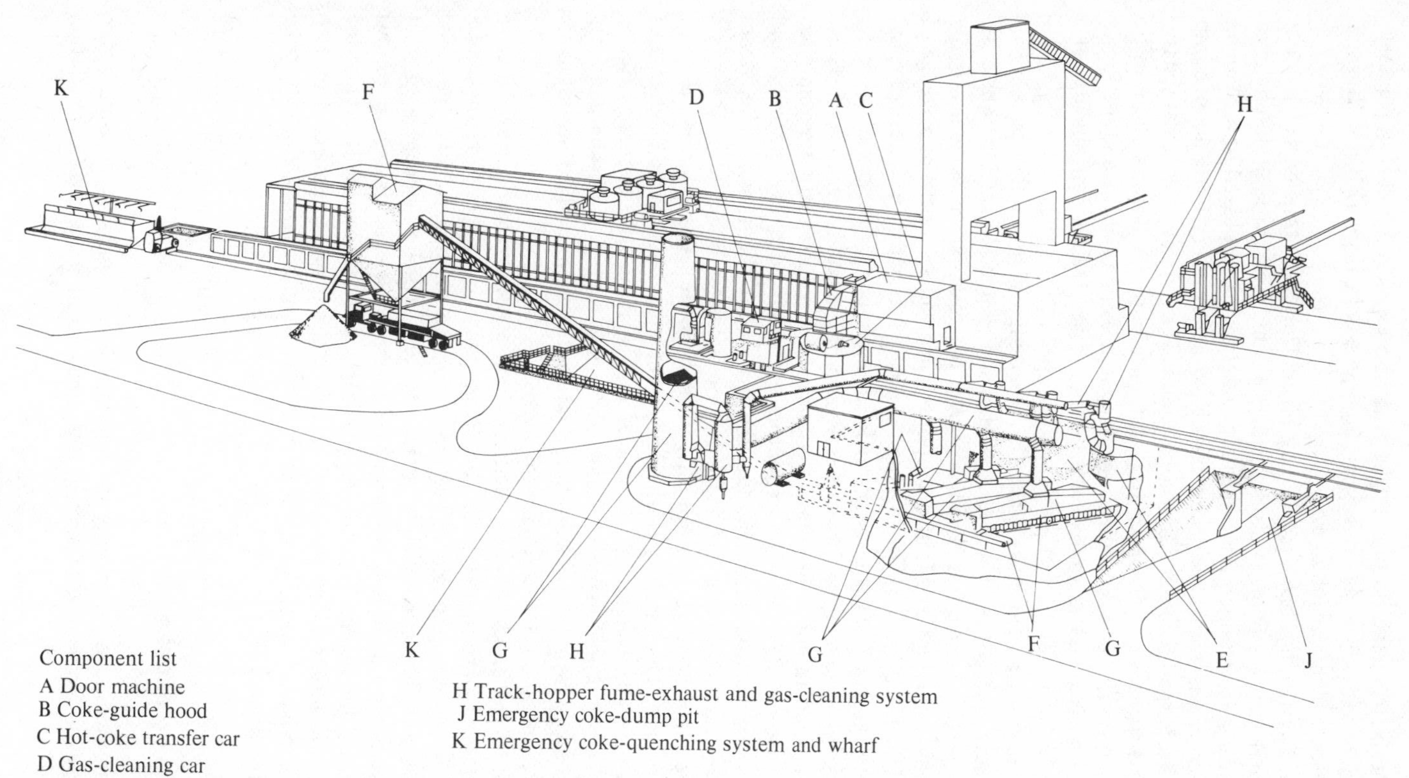

Fig. 12.7. Enclosed discharge and quenching system for coke at the Weirton plant of the British Steel Corporation. (A) Door machine; (B) coke guide hood; (C) hot coke transfer; (D) gas cleaning car; (E) track receiving hoppers; (F) coke handling; (G) spray-water and steam exhaust system; (H) track-hopper fume exhaust and gas cleaning system; (J) emergency coke quenching system and wharf. (Acknowledgements to Koppers Co., US and the US Environment Protection Agency who sponsor the system.)

extracted continuously from the bottom of the track hoppers by hooded, refractory-lined vibrating feeders onto vibrating conveyors. Here, the coke is quenched with water sprays and the stream is withdrawn by fans.

A similar enclosed system has been installed by Firma Carl Still at the Osterfeld plant in Germany. The coke in this plant is discharged from the oven through an enclosed guide into a hopper car from which the fumes are withdrawn and passed to a cleaning system. The hopper car is then closed and travels to a refractory-lined hopper into which the coke is discharged through a telescopic chute. The hopper is closed at the top by sliding gates and the coke is delivered from the base onto a vibrating chute to be quenched by water sprays. There is further quenching in a second hopper and the fumes from quenching are passed to the atmosphere through a stack fitted with baffles and sprays.

Hoods and scrubbers are also included in the Gneisenau plant at Bergbau-AG of Dortmund, Germany that was installed by Hartung Kuhn Maschinenfabrik of Dusseldorf. A mobile hood extending completely over the coke car and guide is rigidly connected to the guide machine and is also supported on a rail above the upper edge of the coke wharf. The fumes arising when coke is pushed into the car are extracted through five ducts in the roof of the hood, three in the main part of the hood where the coke falls into the car and two in the forward extension of the hood over the car. These five ducts are connected to eight automatically-controlled venturi extractors supplied with superheated steam and water. The extractors are mounted above the hood and provide the necessary suction and separate the grit from the fumes. The steam generator for the extractors can be carried either on a tender attached to the guide machine or on the main hood. Tanks for fuel oil and boiler feed water can also be carried in either position. The capacity of the equipment is matched to the discharge of coke from six ovens. The weight of the whole unit is about 130 tonnes. Coke is quenched in the conventional manner. Steam from the venturi is discharged at a position about 6 m away from the coke discharge side of the ovens and is thus not likely to interfere with the operations on the top of the ovens.

Another variant of the hood system is that in which a hood extends over the coke guide and over at least part of the coke car. From the top of the hood there is a connection into a fixed duct mounted at oven-top level. An example of this type is the Mitsubishi–Amagasaki–Shinwa smokeless pushing system, which has been installed at nearly all coking plants in Japan. A hood supported on the guide extends over all or part of the coke car during the pushing operation. The free space between the hood and the car is limited to avoid intake of excess air. The top of the hood is connected by two hydraulic cylinders to a corresponding opening in a suction main running along the coke discharge side of the battery.

12.1.3. Dry cooling of coke

The most common method of quenching the hot coke is a simple system using water. This method has two major demerits, namely the emission of a large volume of water vapour[13] carrying with it fine particles of coke and the wastage of a considerable quantity of sensible heat. The alternative approach, that of dry cooling,[14] has been used from time to time over a period of many years at coking plants in several countries. The aim basically is to improve the economy of operation by recovering the sensible heat of the hot coke discharged from the ovens, which amounts to about one half of the energy input to the ovens from the underfiring gas. The only example of dry cooling at a coking plant in the UK is the Ford Motor Company's installation at Dagenham, where a dry-cooling unit of Sulzer design came into operation in 1934. It is claimed at the Dagenham works that dry cooling is preferable to wet quenching, as it provides coke of less variability in size and a somewhat smaller quantity is required for a given load in the blast furnace for steel production.

Dry cooling, however, has failed to secure much wider adoption for several reasons: first, the high capital cost; second, with dry cooling as hitherto practised at most of the plants, the steam raised from the hot coke tends to be variable in quantity and quality; third, the units available often offended even against the lower environmental standards of earlier years; and fourth, maintenance was held to be a severe problem. However, in a dry-cooling unit of Sulzer design, which has operated satisfactorily since 1925 at the Homecourt plant in France, a turbo-alternator uses most of the steam from dry-cooling and a supplementary gas-fired boiler keeps the alternator at maximum output whatever the rate of production of steam from dry cooling. Improved designs and

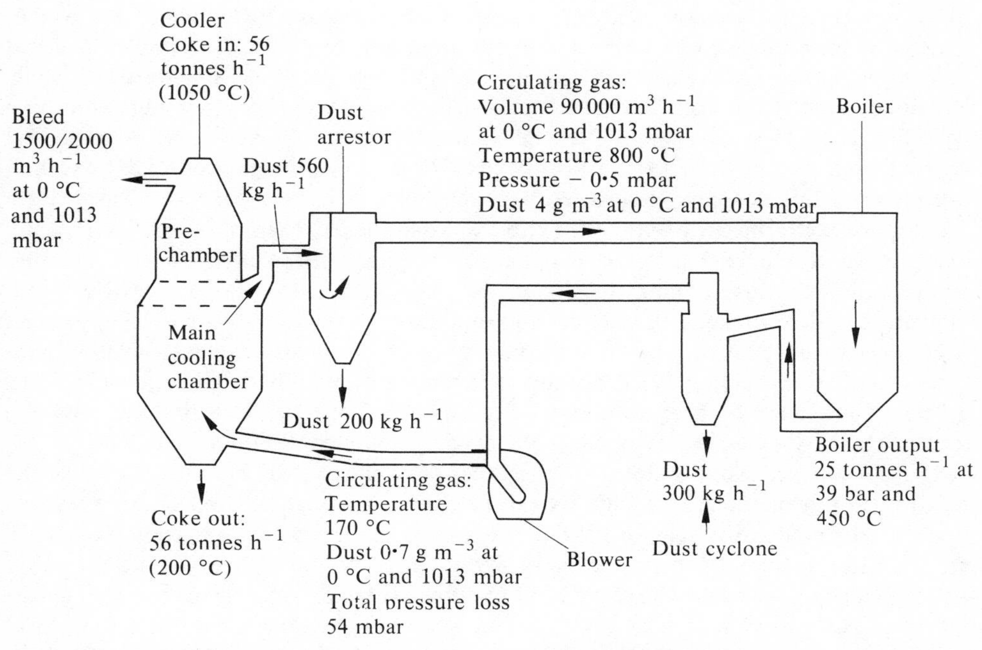

Fig. 12.8. The Giprokoks systems of dry cooling of coke. (Acknowledgements to Giprokoks, USSR, and Woodall-Duckham, Ltd., England.)

methods of operation have now enabled the other operational disadvantages to be overcome.

Recent developments in dry cooling have been introduced by Giprokoks, the USSR Institute for the Design of Coke-oven Plant. Units of Giprokoks design are in operation at a number of plants in the USSR (Fig. 12.8). It is now State policy to use dry cooling at all new plants. The advantages claimed for the Giprokoks system are that the rate of steam production is substantially stable, does not fluctuate from one charge of hot coke to the next and is not immediately affected by stoppages in the supply of hot coke to the chambers. It is also claimed that the dry-cooled coke is of more uniform size and quality than coke quenched with water and that the process is now economically attractive. According to some figures that have been issued, a typical plant cools 56 tonnes of coke from 1050°C to 200°C, collects 200 kg h^{-1} of coarse dust and 300 kg h^{-1} of fine dust and raises 25 tonnes h^{-1} of steam at 39 bar and 450°C. The design does not by itself offer any control of the emissions during oven discharge or transport of the hot coke from the oven to the dry-cooling unit, but it avoids the production of the plume that is characteristic of wet quenching.

12.1.4. Sulphur dioxide in waste gas from heating coke ovens

Coke ovens are normally heated by the use of a proportion, in the region of 40 per cent, of the gas made in the ovens. The most common alternative is blast furnance gas, which is often used in those cases where the ovens form part of a large iron and steel plant complex. In rare cases producer gas made from coke may be employed. Sulphur compounds are minor constituents of each of these gases and their combustion in the heating system of a coke-oven battery gives rise to sulphur dioxide in the waste gases emitted from the chimney stacks of the ovens. Fortunately, draught considerations dictate the use of high chimney stacks. With modern plant of large capacity the waste gases are dispersed at a height of 90–100 m above ground level. These chimney heights are comparable with those that would be required if sulphur dioxide dispersal were the only consideration in determining height on consideration of exit gas velocity and temperature, the proximity of tall buildings and the configuation of surrounding terrain. It is unlikely that a coking plant of a capacity of 5000 tonnes per day would be built other than on reasonably level ground and remote from other buildings of heights greater than 50 m. To provide draught for a plant of this size at least four chimney stacks, spaced over some 500 m and of the height mentioned, would be required.

Dispersal considerations alone[15] suggest that one chimney 100 m in height should be adequate for the amount of sulphur dioxide discharged through the four or more stacks. Even so, the amount of sulphur dioxide discharged in the waste gases from a plant of the capacity mentioned may reach 15–20 tonnes per day when the firing is by unpurified coke-oven gas. The removal of hydrogen sulphide from the underfiring gas is the best (but costly) means of making a substantial reduction in this emission. At a number of works, when it has been necessary for other reasons to purify gas sent out from the plant, purification equipment for treating the total make of gas has been installed so that firing with purified gas could be practised. It should be noted, however, that most purification systems remove only the hydrogen sulphide and leave in the gas some organic sulphur compounds which ultimately appear as sulphur dioxide in the waste gas discharged through the chimneys.

12.1.5. Other emissions from coking plant

The foregoing has dealt with the important points from which polluting emissions may arise from the ovens of a well-conducted coking plant. In practice other emissions may arise.[16] Doors, ascension pipe caps and charge-hole lids may all suffer from leakage. Small amounts of coal spilled on the tops of the ovens may become heated and cause smoke. Dark smoke, denser than Ringelmann No. 2, may on occasion issue from the chimney stacks, particularly if there are leakages through holes in the refractory walls of any of the ovens that need repair. All these faults have been termed lack of good housekeeping and should be avoided by day-to-day attention. Door seals will leak if they and their corresponding frames are inadequately cleaned after each discharge of coke and not given regular maintenance. Charge-hole lids require to be kept clean and sealed with grout, rather than with fine coal spillage, as is sometimes the practice. As already mentioned there is smoke from the chimneys if leaks are allowed to develop in the chamber walls, so that when ovens are gassing freely following charging, gas passes to the flue system where it is incompletely burnt before entering the chimney. Coal and coke handling plants are now normally enclosed by dust-collecting units, and only when such materials are tipped to stock or reclaimed in the open is there the risk of air pollution by dust. In such cases the manner of tipping and reclaiming and the weather conditions at the time have an important bearing on the amount of pollution caused by dust.

It is bad practice to site stockyards upwind from a housing estate. Studies have been made on the effects on the immediate environment of such operations and have shown that tipping and reclaiming are the chief sources of the generation of dust. It is only when there are strong winds that dust is blown from a quiescent tip. Tipping from a height, such as from a ropeway, causes a greater production of dust than when the tipping is from a point close to the surface of the tip, as from an adjustable boom tipper. Reclaiming by means of an inclined harrow with minimum disturbance of the tip is a less dusty procedure than the use of a mechanical shovel which must discharge each shovel load from a height sufficient to clear whatever vehicle is used.

Part 2 Coal carbonization for production of solid smokeless fuel, gas and by-products

ALBERT PARKER

In the UK there has been a demand over centuries by householders for open firegrates burning solid fuel. For a long period the fuel was bituminous coal. During the Second World War and for a few years afterwards supplies of coal and coke were rationed, so householders began to use more gas and electricity for heating, and many permanently reduced their demand for coal for open fires. The use of bituminous coal for domestic heating was one of the main causes of great pollution of the air near ground level by smoke and sulphur oxides, especially at times of natural fog as described in chapter 1.

Many attempts were made during the period 1920–40 to produce reactive cokes as smokeless fuels suitable for the open domestic fire. The subject was investigated from 1920 to 1938 by the UK Government Fuel Research Station. Tests were made by the Research Station on sixteen processes that had reached the stage of large pilot plants. All the methods were dependent on carbonizing coal at a temperature between 500° and 700°C as compared with about 1000°C on carbonizing coal in coke ovens and in retorts at gas works. Only two of the systems examined were developed to the stage of commercial operation and they now produce more than 4 million tonnes of smokeless fuel per annum. Based on their own researches since about 1950, the National Coal Board developed another system which has now been in full-scale operation for a few years.

The three methods in operation differ considerably from one another. In the first to reach the commercial stage, the coal is carbonized in tubular iron retorts heated externally by the gas produced. In the second, the coal is in a large chamber and is heated by direct contact with the products of combustion of the gas made. In both cases the product reactive coke is screened to give sizes suitable for the open fire and for closed stoves. In the third method, the coal is carbonized by fluidization with hot gas from combustion of the coal gas made, and the relatively small particles of coke produced are pressed to form briquettes.

During the coal carbonization all three systems emit some smoky fume, but during the last few years there have been improvements in the design and operation of the plants that have greatly reduced the emission of fume. When the third process mentioned was first used, dense fumes were emitted in briquetting the carbonized product, with unpleasant conditions for the men operating the briquetting machines. The conditions were improved by drawing the fumes away through hoods by suction fans to enter ducts for discharge through high chimneys. Even so, there have been complaints on occasions of the deposition of tarry droplets from the gases discharged. Combustion of such gases and vapours in a specially designed chamber before they enter the chimney has so far proved to be the best means of dealing with the problem.

There are also systems of making solid smokeless fuel in which certain types of coal, for example anthracite duff, are briquetted with pitch at a suitable temperature and then carbonized. In such plants the same precautions must be taken as with other carbonization plants to avoid objectionable emissions of smoke, fume and odour during the processes of charging the briquettes into the carbonizing chambers and discharging and cooling the carbonized briquetfes. Though well-made briquettes are robust enough to withstand serious breakage on handling they can give rise to dust when handled in the open, as the fine dust on their surfaces may be released and carried considerable distances in windy weather. Rough vegetation around an open stockyard can help to keep such dust within the works perimeter.

As with other systems of carbonization of coal, the waste gases from the heating phases contain appreciable amounts of sulphur dioxide and should be discharged at sufficient velocity from a chimney of appropriate height. Each unit of the several systems of low-temperature carbonization plant uses much less coal per day than is carbonized in a large battery of coke ovens, and does not usually require chimneys of a height greater than about 60 m to provide adequate dispersion of the waste gases in the atmospheres.

12.2.1. Gas for town supply

The carbonization of coal at gas works to provide gas for street lighting and for factories and houses was begun in the UK during the first 20 years of the nineteenth century. Gas companies were established to provide gas for several towns. At first gradually and then more rapidly the quantity of gas made for supply to towns was increased. In 1951[17] the volume made from coal in Western Europe was 23 Gm^3, including 14 Gm^3 in the UK, and the total for the world was 33 Gm^3. The quantity reached a maximum of 44 Gm^3 for the world in 1968 and then declined to about 24 Gm^3 in 1972.[1]

In the nineteenth century, town gas was made by the carbonization of coal mainly in horizontal retorts with intermittent charging of coal to and discharge of coke from each retort. Inclined and vertical retorts were also developed and used. Each retort of the several intermittent types held less than one tonne of coal and groups of retorts were heated by producer gas made from coke.[18] All these intermittent systems gave rise to the emission of fumes during the charging and discharging process and to pollution from the quenching of hot coke with water.

Early in the twentieth century there were developed and built systems of continuous vertical reports with coal in a hopper at the top that could be closed as required to supply coal continuously to each retort.[18,19] The coke produced was continuously extracted by mechanism from the bottom of each retort to enter a cooling chamber, which was opened periodically at its base to discharge the cooled coke into wagons. A little water was then sprayed on the coke to avoid the risk of the coke igniting. There was no undue pollution by fumes or from coke quenching with continuous vertical retorts. A large city would have several gas works. For example, in 1923 one of several gas works in a city in the Midlands of the UK had six ranges of continuous vertical retorts with an overall capacity for carbonizing 1350 tonnes of coal per day.

All the systems of intermittent and continuous retorts were heated by producer gas made from coke. The waste gases discharged from the combustion of the producer gas contained sulphur dioxide derived from the sulphur in the coke, and the weight of coke used in making the producer gas was about 12–14 per cent of the weight of coal carbonized. The waste gases from heating a range of retorts carbonizing 1000 tonnes of coal per day would contain a quantity of sulphur dioxide in the region of 2·5 tonnes per day; the waste gases were discharged to the atmosphere through the chimneys of the range of retorts.

Many gas works also had plant for the production of blue water gas made by heating coke by combustion of some of the coke with air and then passing steam through the bed of hot coke. The waste gas from the heating phase contained nitrogen, carbon monoxide, carbon dioxide, sulphur dioxide and coke grit and dust. This waste gas, usually untreated, was discharged into the atmosphere, which was thereby polluted. The water gas made during the steaming process contained in percentages by volume about 4 carbon dioxide, 42 carbon monoxide, 49 hydrogen, 0·5 methane and 4·5 nitrogen, and had a calorific value of about 2·7 Mcal m^{-3}, which is little more than one half of the calorific value of coal gas made in retorts. In some instances the blue water gas was enriched by a petroleum fuel oil[18,19] that was cracked to form gas in a combination of carburettor and superheater. The blow gas formed by combustion of part of the coke in the blue water gas generator was mixed with air to burn the combustible gases it contained and used to heat the carburettor and the superheater and the gaseous products were discharged to the atmosphere. The blue water gas was then passed through the heated carburetter into which the oil was sprayed and then through the superheater. The object of carburetting the blue water gas was to increase its calorific value to about that of the coal gas made in retorts. The advantage of having plant for the production of blue or carburetted water gas was that it could be brought into operation from cold within a few hours to meet sudden increases in demand for gas, whereas it required at least two weeks to bring spare coal gas retorts into operation. Usually the water gas was mixed with a larger volume of the gas made from coal, and the mixture was purified to remove most of the sulphur compounds it contained before it was distributed.

Since about 1930, several processes have been developed and operated on a large scale for the production of town gas from various grades of petroleum oil, ranging from heavy fuel oil to light distillate and liquefied petroleum.[20] Most of the processes involve the use of a catalyst. Those using the heavier oils are cyclic in character, involving a heating phase with the waste heating gases used to raise steam in a boiler before discharge to the atmosphere and a gas-making phase in which steam and oil react to provide gas of the required calorific value. Heavy fuel oils may contain as much as 3 per cent of sulphur by weight. In such case, the waste gases discharged into the atmosphere from the combustion of oil during the heating phase of the cyclic process contain an appreciable amount of sulphur dioxide. Fortunately, the cyclic and continuous processes which use light oil containing less than 0·05 per cent of sulphur by weight have been preferred for various practical reasons. The quantity of sulphur dioxide discharged into the atmosphere during the heating phase of the cyclic processes using such light oil is insignificant, and there is no discharge of sulphur from the continuous processes.

12.2.2. By-products from coal carbonization

In all methods of carbonization of coal the combustible gases evolved at high temperature carry vapours of tar and water and compounds that are soluble in liquid tar and water. The tar and most of the water are removed by cooling the gas to condense them to liquid. Closed cooling systems are used. It is now forbidden in the UK to build plant incorporating spray type coolers the water from which is cooled over open frames from which toxic emissions can arise. The condensed tar and water are separated from one another, and the tar is treated at a distillery by various methods to recover phenols and other products of value. The condensed water contains compounds of ammonia, which are often recovered as ammonium sulphate. In competition with ammonia made by synthetic processes, it is far from profitable to recover ammonia from gas liquor, but there are not always facilities for disposal of the gas liquor without causing serious water pollution as the liquor also contains phenolic compounds and thiocyanates.

The basic plant in a modern tar distillery for the recovery of phenol, cresol and other compounds of value consists mainly of closed units. Tar products generally have strong odours, and it is not easy to avoid the odours being added to the atmosphere of the distillery. Some vapour necessarily escapes when tanks are loaded with products for despatch, and small leaks quickly produce an observable odour. Good housekeeping is the only effective way of keeping such odour nuisance to the minimum. Fortunately, with care and attention to detail, most tar works have avoided causing an odour nuisance outside the area of the works.

The residue of tar distillation is liquid pitch, which solidifies on cooling. The solid pitch may be soft, medium or hard according to the final temperature to which the tar is distilled. In the liquid state pitch evolves fume. In the solid state, especially with hard pitch, an unpleasant dust may be produced during handling and transport, and this dust can cause cancer. Special care is therefore taken when handling pitch to safeguard workpeople against this danger.

There is no real problem of air pollution from the treatment or disposal of ammonia liquor, but as already mentioned there are questions of treatment and disposal to avoid undue water pollution.

12.2.3 Natural gas

In those parts of the world in which good resources of natural gas have been found this gas has replaced gas made from coal and oil for town supply. World production of natural gas rose from 228 Gm^3 in 1951[17] to 1213 Gm^3 in 1972.[1] Examples of the effects of good availability of natural gas on the production of town gas from coal and oil are shown by the changes over the years in the US and the UK. The volume of natural gas used in the USA in 1949[21] was 153 Gm^3 (equivalent in heating value to 340 Gm^3 of town gas made from coal) and the volume of coal gas made at gas works was 11 Gm^3. In 1972[1] the quantity of natural gas was 655 Gm^3 and the amount of gas made at gas works was only about 0·2 Gm^3. In the UK in 1949 the gas made at gas works was 13·1 Gm^3 and the amount in 1963 was 14·5 Gm^3. The import of natural gas was begun in 1963 and the use of natural gas from the North Sea was started in 1964. In 1972 the volume of natural gas used in the UK[2] was 59 Gm^3 and the amount of town gas made from coal and oil was only 3·6 Gm^3.

The use of natural gas in place of gas made from coal and coke has virtually eliminated the discharge into the atmosphere of sulphur dioxide in waste gases from gas works. Natural gas from many sources contains little or no sulphur, though the gas from some sources contains a considerable amount which must be removed before the gas is distributed for general supply. For example, the natural gas discovered in 1949 at Lacq near Pau in the south-west of France[20,22] contained the unusually high proportion of 15·3 per cent of hydrogen sulphide by volume. The sulphide was removed from the gas mainly by the solvent action at ordinary temperature of monoethanolamine, from which it was recovered in the regeneration of the solvent by heating. The hydrogen sulphide gas so obtained was then partially oxidized in a Claus kiln to produce elemental sulphur. More than 95 per cent of the sulphur would be recovered in this way for sale, and the gaseous effluent from a kiln would contain some sulphur dioxide. With natural gas containing a smaller but appreciable amount of sulphur compounds the sulphur recovered by solvent would be oxidized to sulphur dioxide for use in the production of sulphuric acid.

References

1. *World energy supplies, 1969–1972*, Statistical Papers Series J No. 17, United Nations, 1974.
2. *United Kingdom energy statistics*, Department of Trade and Industry, HMSO, London, 1973.
3. *108th Annual Report on Alkali, etc. Works for 1971*, Dept. of the Environment, HMSO, London, 1973.
4. Sellars, J. H. and M. P. Hornsby-Smith, 'Studies of smoke emissions during charging coke ovens', *Yearb. Coke Oven Managers' Asstn.*, Mexborough, England, 1962.
5. Meades, M. R. and G. E. C. Randall, 'Smokeless charging', *Yearb. Coke Oven Managers' Asstn.*, Mexborough, England, 1962.
6. Murphy, A. D., 'Coke oven machinery and the control of air pollution', *Yearb. Coke Oven Managers' Asstn.*, Mexborough, England, 1972.
7. Barnes, T. L. *et al.*, *Summary report on control of coke-oven emissions,* American Iron and Steel Institute and Batelle Institute, Columbus, Ohio, 1973.
8. Graham, J. P. and V. J. Pater, 'Application of preheating to British coals', *Yearb. Coke Oven Managers' Asstn.*, Mexborough, England, 1972.
9. Webster, J. R. and V. J. Pater, 'Considerations on methods of charging preheated coals', *Developments in Ironmaking Practice,* Iron and Steel Institute, London, 1972.
10. Staniforth, W., 'Preheating at the Brookhouse coking plant', *Yearb. Coke Oven Managers' Asstn.*, Mexborough, England, 1974.
11. Evans, D. A. E. and D. G. Harting, 'The Coaltek system of preheating and pipeline charging coal to coke ovens, *39th Blast Furnace Conference*, British Steel Corporation, 1971.
12. Beck, K. G., 'Precarbon process: blast furnace coke from preheated coal', *Yearb. Coke Oven Managers' Asstn.*, Mexborough, England, 1974.
13. Jackson, R. and E. R. Waple, 'Elimination of dust and drizzle from quenching towers', *Yearb. Coke Oven Managers' Asstn.*, Mexborough, England, 1961.
14. Graham, J. P. and J. E. Barker, 'Recent developments in dry-cooling coke', *Yearb. Coke Oven Managers' Asstn.*, Mexborough, England, 1974.
15. *Chimney Heights* (2nd edn.), Clean Air Act Memorandum, HMSO, London, 1967.
16. 'Practical suggestions for reduction of emission of smoke, dust and grit at coke ovens', *Special Publications No. 5* (3rd edn.), British Carbonization Research Association, Chesterfield, England, 1974.
17. *World energy supplies, 1951–54,* Statistical Papers Series J No. 2, United Nations, 1957.
18. Davidson, W. B., *Gas manufacture,* Longmans, London, New York and Toronto, 1923.
19. Ward, E. R., *Gasmaking,* Technical Services Branch, British Petroleum Co., London, 1965.

20. *Gas making and natural gas,* B.P. Trading Ltd., London, 1972.
21. *World energy supplies, 1929–1950*, Statistical Papers Series J No. 1, United Nations, 1952.
22. Delsol, R. and A. Renauldon, *Transactions of the Institution of Gas Engineers, 1958–59,* London, 1959.

13 Petroleum refineries

M. T. WESTAWAY and G. J. BROCKIS

World oil refining capacity over the 10 years to the end of 1974 increased at an average annual rate of about 7 per cent. At the end of 1974, it was a massive 3409 million tonnes per annum, more than half of which was located in Western Europe and the US (Table 13.1).[1]

TABLE 13.1

Refining capacity of main world groups

	Annual refining capacity at end 1974		Refinery crude oil throughput, 1974
	(tonnes × 10^6)	(percentage share)	(tonnes × 10^6)
Western Europe	973	28·5	701
US	745	21·9	610
Rest of world	1691	49·6	1400
	3409	100·00	2711

The development of the oil refining industry in size, location, concentration, process type and product pattern and quality are amongst factors which need to be considered when assessing the environmental effects of any particular refinery. Within Western Europe, for example, less than half the countries account for over 90 per cent of refining capacity, some of which is heavily concentrated in areas which have a high level of other industrial activity and high population density; similarly, in the US, over 70 per cent of refining capacity is located in seven States (Table 13.2).[2]

TABLE 13.2

Refining capacity of various countries

	Number of refineries	Capacity† (BPCD × 1000)	Percentage share of total capacity	Cumulative (per cent)
Western Europe				
Italy	34	3953	20·3	20·3
W. Germany	33	2987	15·3	35·6
France	24	3342	17·1	52·7
UK	20	2783	14·3	67·0
Netherlands	7	1841	9·4	76·4
Spain	9	1165	6·0	82·4
Belgium	8	867	4·4	86·8
US				
Texas	40	3733	26·3	26·3
California	34	1809	12·7	39·0
Louisiana	18	1667	11·7	50·7
Illinois	11	1151	8·1	58·8
Pennsylvania	11	690	4·9	63·7
New Jersey	5	619	4·4	68·1
Ohio	7	572	4·0	72·1

† 1 tonne of crude oil equivalent to approximately 7·33 barrels. BPCD = barrels per calendar day.

Variations which may occur in the product pattern are illustrated by the widely divergent requirements of Western Europe and the US (Table 13.3).[1]

TABLE 13.3

Percentage on crude, by weight

	W. Europe	US
Gasolines	21	48
Middle distillates	37	30
Fuel oils	42	22
	100	100

The product pattern from any particular refinery may, of course, be considerably different from the above general picture given for these areas, and is largely influenced by market demand. The required product pattern and quality influence the relative size, number and complexity of the processing facilities installed. For example, the greater demand for gasoline in the US, as indicated by the above figures, results in greater use of catalytic cracking in the US than in Europe. The installed facilities, in turn, together with other factors such as type of crude oil processed, refinery location and age and the degree of control required by national or local authorities, have an important bearing on the extent and kind of pollution control measures applied and on the capital and operating costs involved.

Although the level of control required for individual pollutants from specific refineries varies according to local or national circumstances, the same pollutants are, with few exceptions, common to all refineries in kind, if not in degree. Those of most interest from the air pollution point of view are sulphur oxides, hydrocarbons, particulates, malodorous emissions, carbon monoxide and nitrogen oxides; and they may arise from storage areas, individual processes and operations, leakage and the use of fuel for production of the refinery's own heat and power.

It is worthwhile to consider briefly at this stage the general magnitude of refinery emissions and, particularly as none of the pollutants mentioned above is exclusive to the oil refining industry, their relative contribution to overall pollutant emissions. For various reasons, some of which have already been outlined above, there is considerable variation in the degree of control exercised by individual refineries or groups of refineries. The following example, referring to refineries in Los Angeles, indicates to what extent reductions in emission levels have been made in a heavily populated area which is notorious for its adverse meteorological circumstances and consequent pollution effects, and which therefore demand special attention. The estimated daily Los Angeles refineries' emissions in tonnes per 100 000 barrels crude oil capacity,[3] excluding refinery fuel emissions, were as shown in Table 13.4. The improvements attainable with increasing application of strict control measures are immediately apparent. The emissions, however, taken in isolation, still appear large. Even so, refineries' contribution is still only relatively small in relation to the overall emissions' burden, as is indicated by Table 13.5, taken from Elkin and Constable.[3] These authors

TABLE 13.4

Pollutants from Los Angeles refineries (Tonnes per 100 000 barrels crude oil capacity.)

	Without controls	After rigorous controls 1958	1966	1971
Hydrocarbons	100	14	7	7
Nitrogen oxides	16	—	—	8
Sulphur dioxide	145	20	5	7
Carbon monoxide	220	—	13	7
Particulates	14	<2	—	1

TABLE 13.5

Extent of refineries' contribution of pollutants

	All US sources (10^6 tonnes per annum)	Estimated US refinery emissions (10^6 tonnes per annum)	
Carbon monoxide	65	5	(7·7%)
Sulphur oxides	23	1·6	(7·0%)
Hydrocarbons	15	1·8	(12·0%)
Particulates	12	0·13	(1·1%)
Nitrogen oxides	16	0·33	(2·1%)

emphasize that, although the estimates given were made on incomplete data, they still provide a useful approximation. The nitrogen oxides data have been calculated from data for 1967 given by the US Department of Health, Education and Welfare.[4]

It will be recognized that the contribution from a refinery to total emissions may sometimes be a larger factor than indicated here within a particular community; but it should also be appreciated that the per cent contribution *at ground level*, that is, the contribution to immission values, is generally much lower than indicated in the Table 13.5 when the high level dispersion factors of refineries are taken into account. Additionally it should be noted that the majority of refinery hydrocarbon emissions are essentially photochemically inert and non-toxic. Only a small percentage (about 10 per cent of the total hydrocarbon evaporation loss from the Los Angeles refineries,[5] for example) can be considered 'reactive' in the sense of their possible involvement in the production of photochemical smog. Most such reactive hydrocarbons derive from other sources, in particular evaporation from, and incomplete combustion of, fuel in vehicles.

The importance of immission as well as emission values has been mentioned. Indeed, many authorities use calculation methods based on what they consider an acceptable contribution of a refinery or other installation to immission values to determine a minimum height for emission rather than, or in addition to, directly limiting the total emission. That is, legislation is in some cases based only on calculated immission values. The calculation methods which may be used are dealt with in chapters 5 and 6. In addition to the theoretical approach to control, it is increasingly common practice to institute direct measurements of immission values to assess the effects of the control measures adopted, sometimes in the interest of good public relations but often to conform to local regulation requirements. In the latter case, the responsibility for defining the methods to be used and for executing them may rest either with the authority or the refiner.

Monitoring is particularly valuable in the case of a grass roots refinery or where an existing refinery is to be expanded. If measurements are made of general pollution levels in the area for a year or more before construction or expansion is effected, a more valid assessment of a refinery's contribution to subsequent immission levels may be possible.

The measuring schemes which may be adopted are essentially of two types. In the first, simultaneous measurements at a number of fixed points are related to meteorological data collected at the same time. Alternatively, measurements are made with a single instrument or group of instruments in a random fashion, both in time and position, according to a predetermined statistical programme, and no account is taken of meteorological data since it is assumed that the process of randomization covers a range of weather conditions. The latter approach, obligatory in some countries, gives long-term average pollution data for the whole area and an indication of peak values. The fixed points approach, however, produces a larger amount of information (at greater cost), and because of this is generally to be preferred. It may, for example, be possible to determine whether pollution is coming from a refinery or not and, by comparing upwind and downwind measurements in steady weather conditions, to obtain some indication of the contribution of a particular refinery.[6]

Refining operations

Before discussing refinery emissions in some detail, a very brief description of refining and some of the processes used is given here. It is hoped that this will provide the reader with a suitable background to the subsequent sections on emissions.

Crude oil is predominantly a complex mixture of hydrocarbons, but may contain minor proportions of compounds containing sulphur, oxygen and nitrogen as well as trace metals such as vanadium and nickel. No two crude oils are alike, but usually the percentages by weight of the main materials are as follows: carbon 85–90, hydrogen 10–14, sulphur 0·1–7, oxygen 0·06–0·4 and nitrogen 0·01–0·9. The function of a refinery is to produce from this mixture a very wide range of products which meet market needs for automotive and aviation fuels, fuels for power and heat generation, lubricants, bitumens, chemical feedstocks and other special products according to an overall oil company plan.

To the layman, a refinery appears to be a maze of pipelines, valves, pumps, towers and vessels, but the entire operation can be simply broken down into four major steps: separation, conversion, treatment and blending. Expanding this a little before dealing with these steps in more detail, crude oil is first separated into selected fractions, according to boiling range, in proportions determined by the crude oil used. These proportions may not meet the quantitative demand for products in their boiling range and the composition within a particular boiling range may need to be substantially altered to produce products of the right quality. Consequently some of the less valuable streams are converted or up-graded by splitting, rearranging, or polymerizing component molecules. At some stage refinery streams may be treated to remove contaminants, such as sulphur. In the final stages of product preparation, refinery streams may be blended together, often with additives, to meet specification requirements.

Each of these four stages will now be examined in more detail to give the reader some idea of the various processes used.

Separation

Crude oil distillation. Primary separation of crude oil is accomplished by simple distillation in which the preheated crude is passed to fractionation towers for separation, according to boiling range, into 'straight-run' fractions such as refinery gas, gasoline, kerosine, diesel oils and residual oils. Heavy or residual fractions of the crude oil which do not vaporize at this stage may be further fractionated by distillation in the presence of steam and under vacuum. The main uses of vacuum units are to make feedstock for catalytic crackers (see later), to produce lubricating oil and to make residues for bitumen manufacture.

Conversion

As mentioned earlier, yields and quality from distillation are usually quite inadequate to meet product demands and consequently a range of conversion processes is available.

Cracking. In cracking processes, large molecules are broken down into smaller molecules under the influence of heat and pressure and usually in the presence of a catalyst. The prime function of cracking is to increase refinery gasoline output. Catalytic cracking, for which the feedstock is usually in the gas oil or vacuum distillate range, essentially yields a mixture of gaseous hydrocarbons and liquid products boiling through the gasoline to gas oil range, which are subsequently separated by distillation. Hydrocracking is an alternative process using a hydrogen atmosphere and different catalyst yielding a more paraffinic or saturated hydrocarbon product from a similar feedstock. The less popular thermal cracking processes, which do not use a catalyst, are used for converting gasoline feedstocks to ethylene and similar products and for treating heavy residue feedstocks, the trace metal content of which may poison catalysts. In the latter case products include gaseous hydrocarbons, gasolines, gas oil, residual fuel oil, and sometimes coke.

Catalytic reforming. Reforming is a method of upgrading feedstocks in the gasoline range, by rearranging component molecules in the presence of a catalyst. At the same time large quantities of hydrogen are produced, which may subsequently be used in processes such as desulphurization.

Polymerization, alkylation, isomerization. Polymerization and alkylation are catalytic processes, used to produce gasoline by the bonding together of primarily olefinic gaseous hydrocarbons produced in cracking operations. Isomerization, a molecular rearranging process, converts low grade light gasoline to high grade gasoline blending components.

Treatment

To improve product quality, it is often necessary to remove minor concentrations of constituents which may be of crude oil origin or produced during processing. Catalytic and other chemical treatments are available. Removal of sulphur is the most important treatment process in a refinery and this is effected most commonly by hydrogen treatment in the presence of a catalyst. Hydrogen sulphide (H_2S) is produced and the sulphur is subsequently recovered from it in elemental form.

As far as chemical treatments are concerned, the following categories may be mentioned with some specific examples.

1. Acid treatment—removal of aromatic compounds from special lubricant fractions, using sulphuric acid;

2. Solvent extraction—removal of asphaltenes from residues using liquid propane and the removal of aromatics from streams in the kerosine-to-lubricating oil range, using polar solvents;
3. Sweetening—conversion of sulphur compounds to less obnoxious materials in streams boiling below about 250°C.

Finally, physical methods such as filtration, absorption and air blowing may also be used to remove undesirable components.

Blending

The refining industry produces several thousand different products each year to specifications which include limitations on vapour pressure, viscosity, specific gravity, sulphur content and octane number among many others. The refinery produces relatively few separate product streams and consequently the market demand is met by blending these 'base stocks' to give the required products. Each refinery is unique in design and may employ just a few or very many of the types of processes already described. A typical simplified layout of a complex refinery designed to give a wide range of products is shown in Fig. 13.1.

Refinery emissions and control

It is hoped that the foregoing has given the reader some idea of what a refinery does. Emissions to the atmosphere occur from many of these operations and, as has been mentioned earlier, those of most interest are sulphur dioxide, particulates, hydrocarbons, odours, carbon monoxide and nitrogen oxides. The remainder of this chapter is devoted to each of these in turn.

Sulphur dioxide (SO_2)

Of all air pollutants, sulphur dioxide is the one which gains most attention worldwide and the responsibility of the refinery here is two-fold. It must ensure that it meets regulations imposed on it to minimize emissions from refinery processes and combustion sources, and it must prepare fuels and other products to specifications imposed by authorities which safeguard against pollution arising from their subsequent use. Thus in many countries in Europe, specifications limit sulphur contents of gas oils down to as low as 0·3 percent by weight, and there is a growing requirement to reduce the sulphur contents of the heavy fuel oils particularly for use in dense urban and industrial areas.

Some of the naturally occurring sulphur compounds in crude oils are malodorous and if they were not removed or converted to innocuous compounds during processing they would cause odour problems. By their very nature these compounds are found in

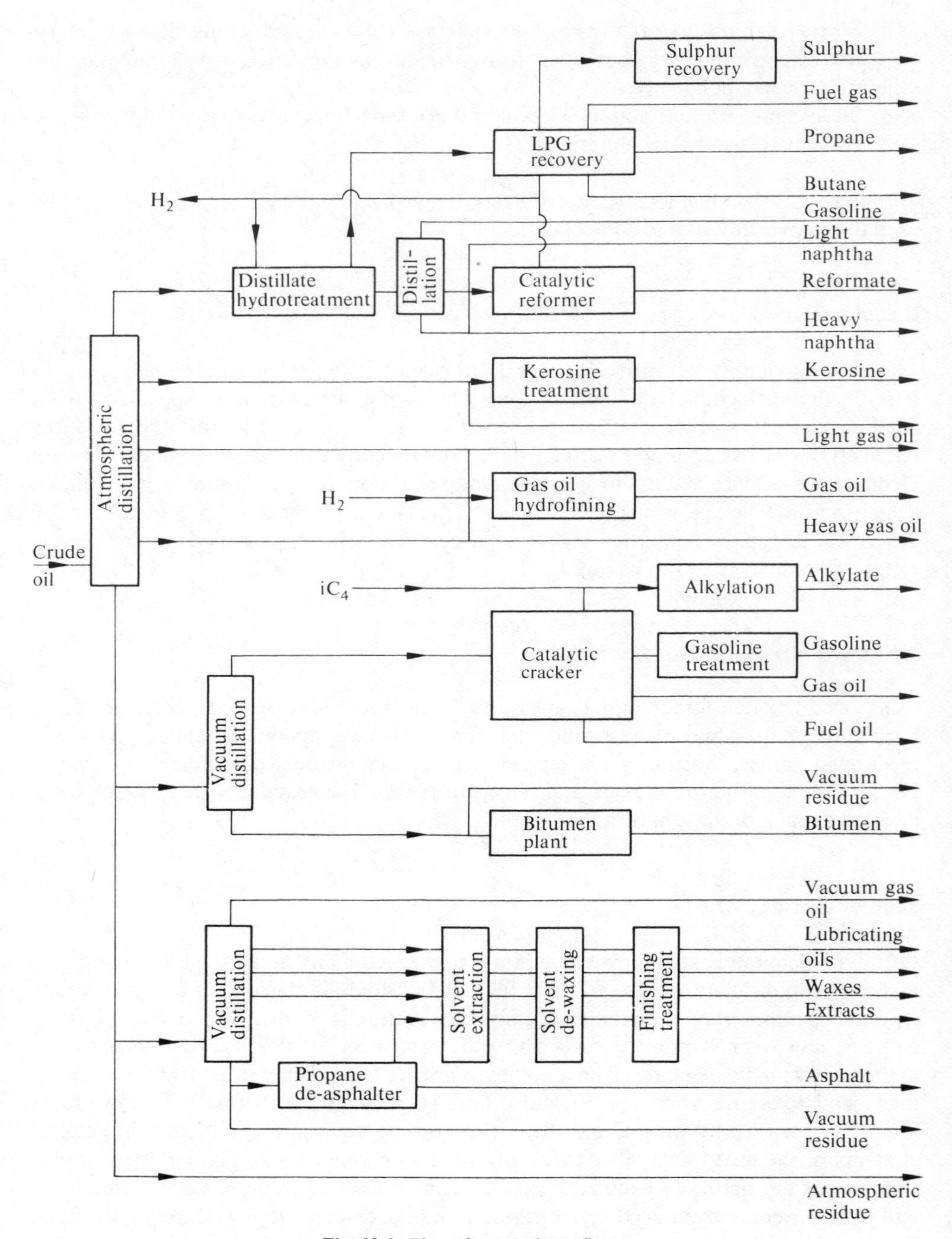

Fig. 13.1. Plan of a complex refinery.

the lighter fractions from crude oils such as gasolines. In the heavier fractions, gas oils for example, the compounds are not malodorous, but sulphur is removed to minimize emission during subsequent use. However, it should be stressed that the removal of sulphur compounds from products is not practised for pollution abatement reasons only; it is also necessary to remove sulphur to improve product quality (for example improving gasoline anti-knock properties) and to reduce corrosion in refinery plant.

Sulphur in crude oil occurs as the particularly malodorous hydrogen sulphide and mercaptans, and as thiophenes, sulphides and polysulphides, although the distribution and concentration of these different types differ for each crude oil. The sulphur content varies according to the origin of the crude as exemplified in Table 13.6.

TABLE 13.6

Sulphur content of various crude oils

Crude oil	Sulphur (per cent by weight)
Kuwait	2·5
Iranian (heavy)	1·58
Abu Dhabi (Umm Shaif)	1·38
Nigerian (Forcados)	0·21

During the processing of crude oil, further hydrogen sulphide and mercaptans can be produced by the breakdown of larger sulphur-containing molecules but, in general, sulphur compounds remain distributed amongst the products in proportions which increase with increasing product boiling point and in a way which depends on the spectrum of compound types in the original crude. This is demonstrated in Table 13.7 for Iranian and South Iraq crude oil products.

TABLE 13.7

Sulphur content in various refinery products (per cent by weight)

	South Iraq	Iranian Heavy
Crude oil	1·95	1·58
Gasoline	0·013	0·14
Kerosine	0·057	0·26
Diesel oil	1·11	1·01
Heavy fuel oil	3·40	2·58

It should also be appreciated that, although some of the sulphur contents do not appear high, sulphur can be a small part of a large molecule and hence the content of sulphur-containing compounds can be very large, perhaps half the total crude oil. When sulphur removal is necessary it is normal practice, therefore, to use processes which break the molecules down and remove the sulphur in the form of hydrogen sulphide rather than removing the sulphur compounds themselves.

Treatment of refinery process streams. Treatment of refinery streams boiling below about 250°C eliminates the odour problem by converting mercaptans to disulphides. In other words, the sulphur compounds are not removed but merely changed chemically. Typical of the older 'sweetening' methods is the copper chloride process. Oil and air are brought into contact with a copper chloride catalyst deposited on a porous material, Fuller's earth. The mixture is then separated under gravity and the catalyst re-cycled. Products such as jet fuels may be treated in this way.

It is more common to find the Merox 'sweetening' process used for feedstocks low in hydrogen sulphide.[7] Essentially the mercaptan forms a complex with the catalyst. The complex then decomposes to disulphide in the presence of oxygen (air) and an alkaline environment, usually caustic soda. Liquid–liquid and fixed bed processes are available. In the latter, 'sour' oil with dissolved air is passed through a bed of activated charcoal impregnated with catalyst and saturated with aqueous caustic soda. The charcoal absorbs the mercaptans and brings them into closer contact with the catalyst. For this reason the fixed bed process is suitable for a wider range of feedstocks than is the liquid–liquid process. The disulphides formed pass from the bed in the oil streams. Feedstocks with high hydrogen sulphide contents are usually scrubbed with aqueous caustic soda before Merox treatment.

An effective method used more in older refineries for removing the simpler sulphur compounds, hydrogen sulphide and the more volatile mercaptans, is washing with aqueous caustic soda. Solubility considerations limit the use of the process to gas and to liquid streams such as gasoline and kerosine boiling below about 250°C. 'Solutizers' are added to the aqueous caustic soda to improve mercaptan solubility. Caustic soda which has been used for mercaptans extraction only may be regenerated by steam stripping or, preferably, by oxidation.

As mentioned earlier, it is not practical to extract the more complex sulphur compounds found in streams boiling above about 250°C. These compounds have to be broken down to get at the sulphur; catalytic processes have been developed which do this and convert the sulphur to easily removed hydrogen sulphide. In addition to extending the range of refining desulphurization capability, these processes are also preferred in modern refineries to caustic soda washing of the more volatile gaseous and liquid streams on both technical and economic grounds. Consequently, catalytic desulphurization is now widely used for all distillates up to a boiling point of about 350°C and sometimes for distillates up to 550°C.

Common methods of catalytically desulphurizing the overhead products and gas oil products from crude oil distillation are known as *hydrotreating* and *hydrofining* respectively.[7] The processes are basically similar. Taking hydrofining as an example, the gas oil feedstock is passed with hydrogen over a pelletted alumina catalyst containing cobalt and molybdenum oxide at a temperature of 300–450°C and 40–70 bar (600–1000 lbf in^{-2}). The sulphur in the feedstock is converted to hydrogen sulphide, which is then removed by fractionation for subsequent treatment in the

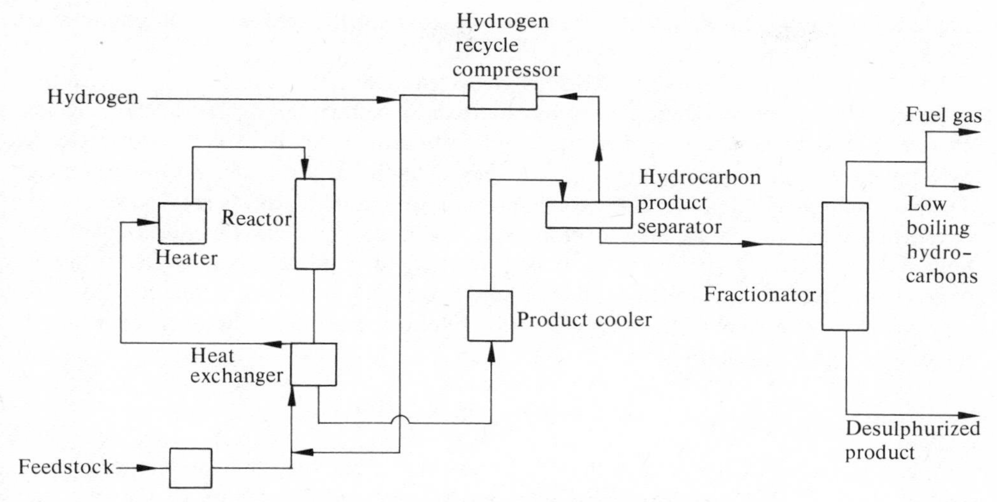

Fig. 13.2. Flow diagram for desulphurization process.

sulphur recovery units of the refinery (see next section). The rest of the feedstock is virtually unaffected and only about 2–3 per cent of low-boiling hydrocarbon products are formed, which are also stripped in the fractionation stage. Fundamental to the process is an ample supply of hydrogen; it is therefore usually most economic to construct units in which this is available from catalytic reformers. A diagram of a typical process is shown in Fig. 13.2.

Further increases in the proportion, the complexity and the stability of sulphur compounds in refinery streams boiling above 350°C, such as the residues from crude oil distillation, make catalytic desulphurization even more difficult and costly. To achieve a reasonable level of sulphur removal the feedstock has to be treated under more severe conditions of pressure and temperature. In order to keep pressures down, high purity hydrogen is required and producing plant must be installed. A further problem is that of catalyst poisoning by trace metals and asphaltenes in the feedstock, resulting in a loss of catalyst activity. However, because of demands on the refining industry for low sulphur residual fuel oils, a few processes are in operation, mostly in Japan.[8]

Sulphur recovery. The quantity of hydrogen sulphide in gas streams from desulphurization units from which the sulphur is to be recovered depends on the sulphur content of the feedstock to the units and the degree of desulphurization required. Thus, the production of hydrogen sulphide from the hydrotreating of distillates is lower than from gas oil hydrofining. In the latter case, reducing the sulphur content of gas oil from, say, 1·0 to 0·1 per cent by weight in a 1300 tonnes per day unit would produce about 11·7 tonnes of hydrogen sulphide daily. It is common practice to recover the sulphur in elemental form rather than to burn the hydrogen sulphide and release the sulphur dioxide which would be formed to the atmosphere through a tall chimney or stack.

In the first stage of recovery, the sour gas streams at about 35–50°C are scrubbed with an aqueous amine solution, such as monoethanolamine and diethanolamine, which absorbs hydrogen sulphide. The sorbent, rich in hydrogen sulphide, is then regenerated by heat and/or stream stripping and is returned to the absorber. In the second stage stripped hydrogen sulphide at a concentration of about 90 per cent by volume is converted to sulphur in a Claus reactor. Here, the hydrogen sulphide is first burnt in a limited supply of air so that a third is converted to sulphur dioxide. The resultant mixture of hydrogen sulphide and sulphur dioxide is then treated catalytically to produce elemental sulphur (normally sold for sulphuric acid manufacture) according to the following relationship:

$$2H_2S + SO_2 \rightleftharpoons 3S + 2H_2O.$$

The catalyst normally used in the Claus reaction is bauxite, and the temperature is held above 210°C. Conversion efficiency is dependent on the quality of the sour gas stream and control of the air supply to the combustion stage. Trace impurities such as carbon dioxide and water vapour can upset the reaction equilibrium, but an 85 per cent conversion can be expected from a single catalytic stage, rising to about 94 per cent conversion from two stages. Unconverted hydrogen sulphide in the tail gas from the Claus unit is incinerated and converted to sulphur dioxide before dispersal to the atmosphere via a tall stack. For situations where further clean-up of tail gas may be needed, several processes are available,[10] including the Beavon and the Institut Françaíse du Petrole (IFP) processes.

In the first stage of the Beavon process[11] the Claus tail gas stream containing hydrogen sulphide and sulphur dioxide is hydrogenated over a cobalt–molybdenum catalyst to hydrogen sulphide. Impurities such as carbonyl sulfide and carbonyl disulfide react with water vapour to give hydrogen sulphide so that all the sulphur is eventually converted to hydrogen sulphide. The gas is then cooled and the hydrogen sulphide extracted in the second stage, where the hydrogen sulphide reacts with sodium carbonate solution to give sodium hydrosulphide as follows:

$$H_2S + Na_2CO_3 \rightleftharpoons NaHS + NaHCO_3.$$

The sodium hydrosulphide is then oxidized to sulphur by sodium vanadate also in solution:

$$NaHS + NaHCO_3 + 2NaVO_3 \rightleftharpoons S + Na_2V_2O_5 + Na_2CO_3 + H_2O.$$

Sulphur appears as a froth and is skimmed off. Subsequently the vanadate is oxidized to the pentavalent state by blowing with air in the presence of an oxidation catalyst.

In the IFP process,[12] the Claus tail gas at 130–140°C is contacted countercurrently in a single stage with a liquid stream containing dissolved catalyst. A liquid phase Claus reaction continues until the hydrogen sulphide concentration in the gas phase is reduced to 2000–3000 ppm. This residual concentration is incinerated. In an emergency, the sour gas streams may be burnt in the fuel gas system or may be released to an elevated flare system (see later) where the hydrogen sulphide is burnt to sulphur dioxide. For a lengthy emergency, the amount of flaring would, if necessary, be reduced by cutting back on desulphurization operations.

Production of heat and power. In a refinery, heat and power are generated by combustion of oil and gas. Processes require high pressure steam and heated feedstocks and a wide range of types of heaters and boilers are in use. A simple refinery of 5 million tonnes of crude oil throughout per annum can require for heating about 200 000 tonnes of heavy fuel oil per annum or the equivalent from gas or a mixture of oil and gas. This fuel, indigenous to the refinery, represents in this simple case about 4 per cent of the total throughput of crude oil, or an average of the equivalent of about 24 tonnes h^{-1} of heavy oil. For a very complex refinery, the requirement could be much higher. It is general policy to burn the least valuable products as refinery fuel. The amount of gas available is variable, and if it exceeds sales at any given time, it is burnt because it is expensive to store. Importantly, local regulations governing emission and immission levels must also be taken into account.

The amount of sulphur dioxide released from combustion sources depends on the sulphur content of the fuel. This can vary from about 0·01 per cent by weight of sulphur for gas to 3–4 per cent by weight for heavy fuel oil. Typical emission from a 5 million tonnes per annum refinery might be of the order of 15 tonnes per day. About 3 per cent of the sulphur oxides released is in the form of sulphur trioxide. The favoured means of control of sulphur oxides is by the use of tall stacks designed to give adequate dispersion of these and other gases, so that effects at ground level are negligible. The height of stack selected depends on local environment. Reflecting the tendency to construct fewer stacks and because of increasingly stringent controls, stack heights have been increased in refineries over recent years as in many other industries. It is becoming common in the highly integrated new refineries to provide only one or two stacks to serve all the process furnaces so that the dispersal of emissions is optimized. Figure 13.3 is a view of one such refinery in Holland.

In some areas total emissions to the atmosphere are controlled. In the US, for example, while no precise Federal emission requirements for refineries have been published at the time of writing, other industries, such as power generation, are faced with controlling sulphur dioxide content of flue gases to a maximum level of about 1000 ppm. Such levels equate to fuel sulphur contents of less than 1 per cent by weight, and the oil industry markets lower sulphur heavy fuels to meet this need. In theory, an alternative would be to use high sulphur fuels and desulphurize the flue gases. In practice, however, no acceptable high efficiency flue gas desulphurization process is yet available[13] as an economic proposition.

Other sources of sulphur dioxide. Along with other gases, sulphur dioxide is produced during the regeneration of catalysts used in catalytic cracking and desulphurization processes. The reasons for, and some details of, regeneration are given in the section on

Fig. 13.3. Single stack serving the process area at BP's Rotterdam refinery.

particulates, but some comment is made here on sulphur dioxide released during such operations.

The amount of sulphur dioxide released during continuous regeneration in catalytic cracking depends on the sulphur content of the feedstock, catalyst circulation rates and other operating conditions. A survey of six refineries in Los Angeles[14] showed that, from the six fluid catalytic cracking units handling between them 20 500 tonnes of feedstock per day, total daily emissions were about 42 tonnes. Concentrations in the flue gases ranged from about 300 to 2200 ppm. A more recent survey[15] of 122 units in the US showed concentrations ranging up to 3300 ppm. As with other gaseous emissions, the regenerator gases are disposed of via tall stacks to minimize immission levels.

In fixed bed processes such as desulphurization and catalytic reforming, catalyst regeneration is carried out periodically rather than continuously. Although the emission rate of sulphur dioxide during the regeneration step in desulphurization is similar to that obtaining for continuous regeneration processes, it occurs for only about a day at a time—and at a frequency of perhaps once per year. Thus the total sulphur dioxide emission from this fixed bed process of catalyst regeneration is very much less than that from continuous regeneration. Emissions from reforming processes are even lower.

Particulates

A major potential source of airborne particulate matter from refineries is catalyst regeneration, and steps are therefore taken to ensure that emissions are minimized. Gases from heaters and boilers also contain particulates but to a much lesser degree.

During catalytic processes, increasing contamination of the catalyst surface occurs with time, due to deposition of breakdown products such as coke and sulphur. As a result, catalyst activity is impaired. To restore activity, the catalyst must go through a regeneration stage in which the contaminants are burnt off to a satisfactory degree. Catalyst attrition also occurs.

Basically, there are two types of refinery catalytic process—circulating or fixed bed. In the first, catalysts are continuously circulated between reactor and regenerator because of the rapid build-up of deposits. This is typical of modern catalytic cracking processing, which may be further classified according to the method needed to transfer catalyst.[7] In fluid catalytic cracking, the catalyst is moved suspended in a gas stream, contrasting with the moving bed process where catalyst is transferred mechanically or by an air lift arrangement. Fluid cracking is more commonly used, and a simplified flow plan is shown in Figure 13.4.

In a fluid cracking process handling 4000 tonnes of feedstock daily, a fine supported catalyst would be circulated at a rate in excess of 1000 tonnes h^{-1}. Regeneration is carried out in an air stream at about 660°C and 1·7 bar (25 lbf in^{-2}) giving typical gaseous emission rates up to about 4200 m^3 min^{-1}.

In the second type of process, deactivation of catalyst occurs much more slowly, and regeneration is carried out batchwise at infrequent intervals from three months

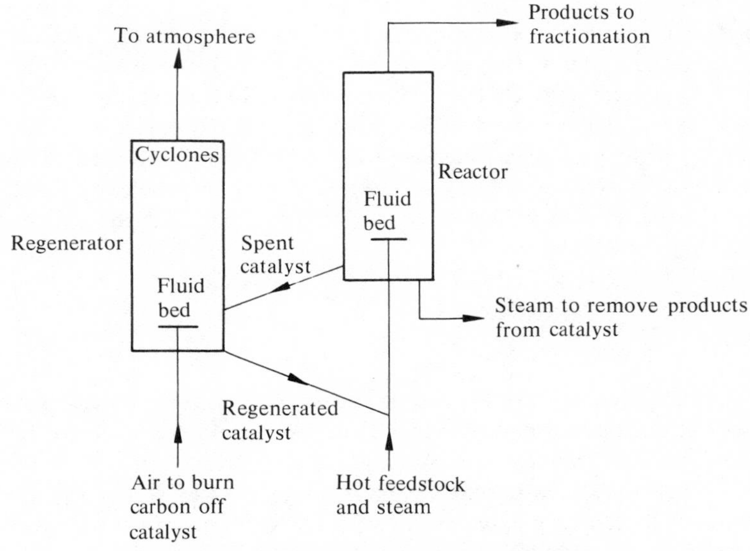

Fig. 13.4. Fluid catalytic cracking process.

upwards; examples are desulphurization and catalytic reforming. A hydrofiner processing 2000 tonnes of gas oil per day contains, say, 13 tonnes of catalyst. Regeneration takes up to 24 h, using a stream of air and steam at 370°C and atmosphere pressure. In current residue desulphurization processes,[7] however, regeneration is not practicable, because of excessive metal contamination, and the catalyst is discarded. In reforming processes where, for example, 1300 tonnes per day units contain some 25 tonnes catalyst, regeneration is carried out by purging with an inert gas–air stream at about 400°C and atmospheric pressure. For both desulphurization and reforming processes the flue gases from the regenerator are led to an adjacent stack and discharged or burnt in a furnace. No particulate control is practised, as quantities are insignificant in view of infrequency of regeneration.

With continuous catalyst regeneration plant, as in catalytic cracking, cyclones are fitted to control emission of particulates. These may be located within the regeneration vessel or externally. With such equipment, losses of catalyst dust have been confined to 45–450 kg h^{-1}, depending on size, age and design of unit.[16] Losses of catalyst are continually made up.

In addition to cyclones, electrostatic precipitators (chapter 8) may be fitted to improve efficiency of collection of fine dust particles. Where so fitted, the precipitators are arranged in parallel to deal with the large volumes of gas which have to be handled. Gases are cooled to 260°C in waste heat boiler (see later in carbon monoxide section) before they are passed to the electrostatic precipitator, prior to which ammonia injection may be practised to increase the conductivity of the gas stream. Using both cyclones and electrostatic precipitators for solids recovery, greater than 99·9 per cent removal can be obtained. Losses from large catalytic cracker units treated in this way have been reported to be as low as 22·5 kg h^{-1}.[15] After treatment, the flue gas is dispersed to the atmosphere through a refinery stack.

The emission of particulates from modern heaters and boilers can be kept down to acceptable limits. The rate of emission depends upon the types of fuel being burnt and furnace operating conditions. Of these, the air–fuel ratio is most important, and many refineries fit oxygen analysers and smoke detectors to their heaters to ensure adequate control. More recently, however, increased use of high-intensity burners[17] with more efficient combustion characteristics has reduced operating problems. Typical control required in Europe for particulates emissions from combustion sources is about 1–2 g kg^{-1} of liquid fuel.

Hydrocarbons

Within the refinery fence there are several sources of gaseous emissions which are predominantly hydrocarbon in character, although other materials such as oxygen- and sulphur-containing compounds are often associated with them. It is these minor constituents which may give rise to localized odour problems. The main sources of hydrocarbons are dealt with in some detail later, and a few comments are then made on other sources of hydrocarbons and odour.

Storage. The journey of oil through the refinery begins and ends in storage areas or tank farms where crude oil and products are stored (Fig. 13.5). Tanks are also needed to store intermediate products used for feedstocks to process units and for product blending purposes. In addition, the most highly volatile products are kept in closed spherical or cylindrical pressure tanks. A typical refinery processing 5 million tonnes per annum might well require to hold 300 000 tonnes of crude oil at any one time and have storage capacity of 400 000 tonnes for a range of intermediate and finished products. Cylindrical storage tanks for crude oil, with capacities of up to 100 000 tonnes, are common. These are some 85 m in diameter and 20 m tall. The tanks are designed to minimize losses of hydrocarbons which occur in storage either during filling or on standing.

Losses during filling occur by displacement of air containing hydrocarbon vapour. On standing, changes in temperature and pressure affect the liquid and vapour volume within a tank, and again air, containing hydrocarbons, can be displaced. This is known as breathing loss. Factors which affect losses are, therefore, vapour pressure, temperature and pressure changes, rates of throughput, the colour of the tanks (since this affects the amount of heat absorbed and consequently liquid temperature) and the condition and type of tank.

Fig. 13.5. A view of part of a storage area. Note the fixed roof tanks in the foreground with floating roof tanks and sealed spherical tanks in the background.

For products in the gas oil range and above, vapour losses due to either filling or breathing are small because vapour pressures are low. It is therefore quite acceptable to store these products in cylindrical tanks with fixed conical roofs. These are designed to accommodate only very small changes in pressure (about 5 mbar) before venting occurs. The actual venting pressure is determined by the structural characteristics of the tanks.

For liquids with higher vapour pressures, for example crude oil and gasoline, losses are minimized by virtually eliminating the vapour space above the liquid by employing roofs which float on the liquid surface. The roof therefore rises and falls during filling and emptying respectively. Seals are provided between the roof perimeter and the tank wall, and it is here that evaporation losses can still occur. A simple calculation shows that some 340 tonnes of gasoline per annum could be lost from a 10 000 tonnes fixed roof tank during normal service. It is possible to prevent 90–95 per cent of this loss by using a floating roof tank so that, quite apart from the obvious environmental benefit, it makes good economic sense to choose this design. Another important feature is that, by eliminating the vapour space, the fire hazard is greatly reduced.

Several designs of floating roof are available. The basic types are pontoon and double deck (Fig. 13.6). The type shown in Fig. 13.6(a) has an annular pontoon round the edge and a complete single deck in the centre. The upper pontoon surface is pitched down for rainwater drainage and the lower is pitched upward to the centre. In addition to buoyancy, the pontoons provide space insulation from the sun's heat and inhibit boiling in the annular area. The deck traps any vapour formed. The double deck floating roof (Fig. 13.6(b)) has two complete decks, the upper sloping downwards to the centre for rainwater drainage and disposal and the lower sloping upwards to trap vapours should they form. The air space between the two decks provides good insulation over

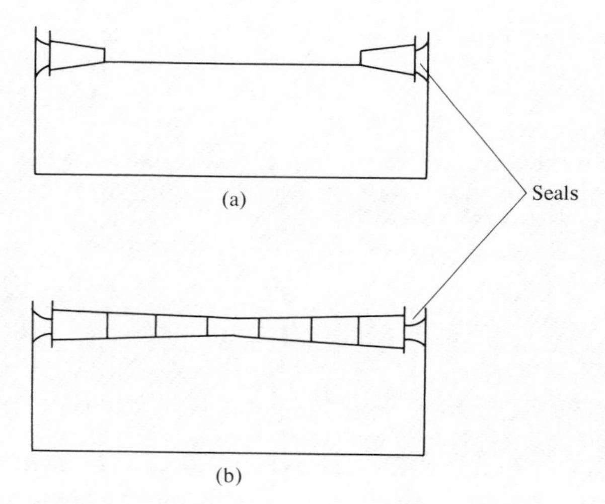

Fig. 13.6. Basic types of floating roof tank: (a) pontoon; (b) double deck.

the entire surface and this feature means that this design is more efficient and more applicable to storage of volatile products such as gasoline.

It can therefore be seen that to reduce losses it is very important that the peripheral space between the floating roof and tank is effectively sealed. The seal must be effective over long periods and remain so despite the distortion of the tank which can occur. A range of seals available[18] which will not be described here, includes metallic, resilient foam and liquid-sealed fabric seals. The metallic seal, for example, is effective over a rim space variation of 25 cm (and this can result from distortion due to uneven settling of the tank). Reflective paints are used to keep down tank skin temperature and reduce heat input, thereby reducing emissions significantly.

In addition to the above, other approaches to minimizing vapour loss from storage tanks have been developed. For example, internal floating covers, normally made of aluminium or plastic, may be installed inside fixed roof tanks. Resilient peripheral seals are provided between cover and wall. Properly designed internal covers provide more effective vapour conservation than standard open pontoon type floating roofs, due to added protection from wind and rain.

Variable vapour space tanks have also been developed, but are little used because they are expensive and effective control can usually be achieved with floating roof tanks. Here expanding vapours are stored temporarily in a gas holder and vented to atmosphere only if the holder capacity is exceeded. On cooling, the vapours are drawn back into the tank. Two designs of variable vapour space tank are used, the lifter roof and the flexible diaphragm. In the former the roof is telescopic and, not being rigidly fixed to the shell, it moves up and down as the vapour above the liquid expands or contracts. Again, sealing is very important and several seals are available.[19] Expansion capacity in the flexible diaphragm tank is provided through the movement of the diaphragm. There are two types of tank—the integral and the separate. In the former, the diaphragm unit is held in a shell located on the roof of a fixed roof tank. In the latter, the diaphragm unit is connected to one or more fixed roof tanks.

It is worth mentioning here that in Europe increasing attention is being given to the storage of crude oil and, to a lesser extent, products in underground rock or salt caverns[20] which may hold as much as 100 000 m^3 of liquid. The oil is stored wholly below the water table, the pressure of which is always greater than the pressure exerted by the oil and which, therefore, retains the oil. The vapour spaces for several caverns may be interconnected. As a result, vapour venting is minimized but, when this is necessary, caustic scrubbing facilities are available to remove malodorous compounds from crude oil vapours.

Loading facilities. Products leave a refinery either by pipeline or in bulk or in drums by road, rail or ship. During the filling of product tankers, hydrocarbons can be lost to the atmosphere in the same way as during storage, by displacement of air containing hydrocarbons by the incoming liquid.

The volume of vapour displaced is greatly influenced by the method of loading or filling, and losses can be reduced by using techniques which minimize evaporation and contact between expelled air and the incoming product. In practice, tanks can be filled through either overhead hatches or a bottom connection. Overhead loading can be further divided into splash and submerged filling. In the former, the delivery pipe exit is above the liquid during all or most of the loading but, in the latter, the pipe exit is near the tank bottom. Splash filling creates considerably more turbulence, greater evaporation and contact between liquid and air than does submerged filling. According to the literature[21] this can give rise to differences in gasoline vapour losses of a factor of three between the two techniques. Further reductions in vapour losses are obtained with bottom loading. Emissions from bottom and top loading can be controlled by isolating the tanks from the atmosphere, collecting all the vapours and piping them away to a recovery unit. Several companies have produced vapour recovery systems[22] for road tankers and rail cars in which the recovered air–vapour mix is compressed to condense out the hydrocarbons. Collection efficiencies are about 90 per cent.

Emergency safety measures for waste gas disposal. Large quantities of gas are produced in refineries. They are used to meet, at least in part, the demand for refinery fuel and as feedstock to process units. Production and consumption are balanced as far

Fig. 13.7. An elevated flare on emergency use with steam injection to minimize smoke.

as is practicable. Scheduled shut-downs and upsets in process plant caused by compressor, pump or power failures can, however, upset this balance and overload the recovery system. To meet this eventuality and safeguard personnel and equipment, as well as to reduce emissions to the atmosphere, refineries have waste gas disposal systems. Such a system can comprise a manifolded pressure-relieving or blow-down system and a blow-down recovery system and a system of flares for combustion of excess gas.

Equipment in a refinery operates over a wide range of pressures from less than atmospheric to greater than 70 bar (1000 lbf in^{-2}). To ensure the safe removal of gas or liquid in an emergency from the process area, both manual and automatic safety and relief valves are provided. Vented gas is then led to a blow-down drum, where any entrained liquid is collected, and from here the gas is led to a flare where it is ignited and burnt. For reasons of economy and safety, vessels and equipment venting to blow-down systems are usually segregated according to operating pressure.

Flares have been designed to operate either at ground level or at an elevated position, commonly about 80 m high. High-level flares (Fig. 13.7) are particularly suited for emergency release of gas and combustion of sulphurous gases when necessary. They also minimize radiation effects at ground level. Ground level flares may be used for the operational disposal of sulphur-free 'sweet' gas although, in some raised ground flares, some sour gas can be tolerated.

Incomplete combustion can give rise to smoke and unburnt hydrocarbons. In flare operation, this can be overcome by manually or automatically injecting steam into the flare combustion zone to provide turbulence and to entrain air. It is also believed that steam reacts with gaseous fuel to form oxygenated compounds which burn more readily. Several designs of flare are available, essentially differing in the manner and location of steam injection.[22]

Oily water systems

Considerable quantities of water are used in refineries for process and cooling purposes, although nowadays very significant reductions are being achieved using air cooling. The water effluent streams, together with rainwater run-off, contain oil and have to be treated before discharge to the sea, estuary or river. A refinery therefore has a complex drainage system leading the water streams to a treatment area. Treatment varies according to discharge conditions but, as a minimum, gravity separators are required. Basically these are open concrete or metal lined multi-channel tanks designed for a specific water throughput to give sufficient time for oil to separate out under gravity. The separated oil is then skimmed from the water surface.

Clearly in such an open system, which for a single channel is typically 6 m $\times$ 30 m, it is possible for gas and vapour evolution to the atmosphere to occur. In modern refineries the drainage systems leading the oily water to the separators are covered, thereby reducing emissions. A survey by Stichting CONCAWE[23] showed that open separators used in European refineries caused little or no odour problem. Losses of hydrocarbons from separators are very difficult to estimate, although equations have been published[24] but they are not normally considered significant. If, however, control is considered necessary, as in some areas which are particularly sensitive to hydrocarbons emission, covers may be fitted to the forebays or primary separator section, where the major part of the oil is separated. Fixed and floating covers have been used. It is not easy to seal fixed covers effectively, however, and explosion hazards may arise as well as vapour release. Floating covers, fitted with access points, with skimmers to recover separated oil and with supporting legs, overcome these problems but may interfere with oil separation. Gaps between roof and separator walls are sealed with fabric or rubber.

Significant reduction in gravity separator dimensions have been obtained using the tilted and corrugated parallel plate separators.[25] Liquid effluent is passed through a series of closely spaced parallel plates which provide coagulating surfaces for the oil. In this way gravity separation is speeded up, although problems can occur through clogging with oil. The smaller size means that these can be more readily covered and hoods are generally provided.

Other sources of hydrocarbons and odours

Catalyst regeneration. Catalytic processes and catalyst regeneration have been mentioned in the discussion on particulate emission. Of these, catalytic cracking and catalyst regeneration produce the greatest quantity of hydrocarbons, mainly in the form of methane and ethane.[22] Concentrations up to 9000 ppm have been reported in a recent survey,[15] but where carbon monoxide boilers (see Carbon monoxide emissions) are used, the emission is virtually eliminated.

Pumps, valves, compressors, flanges. Such equipment is used extensively in the petroleum industry. Due to normal wear and tear, pumps and compressors can leak vapour or liquid products between moving shafts and stationary casing, despite the widespread use of mechanical and packed seals. Pipeline and pressure release valves and flanges can also leak under the influence of heat, pressure, vibration and so on. These intermittent leakages are reduced by regular inspection and maintenance.

Vacuum jets. Some refinery processes are operated under vacuum which is produced by steam-driven vacuum jets in conjunction with a barometric condenser. Light hydrocarbons which do not condense and are discharged with exhaust steam can be routed to fire boxes on boilers and heaters.

Air-blowing of asphalt. Asphaltic material produced as residue from a vacuum distillation unit sometimes needs further treatment before it is suitable for paving and roofing. It is upgraded by passing a current of air through it, which produces further polymerization and condensation of component molecules. The effluent air contains hydrocarbons and odorous sulphur and nitrogen compounds amongst others. Control of these has been accomplished by scrubbing and incineration.[22]

Cooling towers. Before re-using water for refinery cooling purposes, heat absorbed during passage through heat exchangers must be removed. This is usually accomplished by allowing the water to cascade through a series of decks or slotted grids in a cooling tower. If the water has passed through a leaking heat exchanger, then it might contain dissolved or free oil. Subsequently this could be released to the atmosphere from the cooling tower. A survey[14] of over 90 cooling towers showed that more than 80 per cent of them did not emit hydrocarbons in detectable amounts. The low levels of hydrocarbon emission from the remainder varied from only 1·4 to 68 kg per day. Nowadays most new refineries use air cooling and minimal water recycling takes place.

Sour condensate and spent caustic solutions. Many processes in refineries use steam in distillation and as a diluent to reduce hydrocarbon partial pressure in catalytic or thermal cracking. As a result of subsequent condensation in the presence of process liquid and vapour, the waters are soured or fouled primarily by hydrogen sulphide, ammonia and phenols. Levels reported for 13 sour waters[26] varied fron 390 to 8250 ppm, from 135 to 6550 ppm and from 100 to 900 ppm by volume respectively. These condensates can cause odour problems or possibly upset biological treatment in the refinery effluent system and consequently a stripping pretreatment is often used. Where possible, however, sour water recycle to other refinery processes is considered.

In a typical countercurrent process, the stripping medium may be steam or flue gas, with design conditions in the range $6{\cdot}9$–345×10^{-2} bar (1–50 lbf in^{-2}) and from 38° to 130°C. Alternatively the sour water may be acidified to 'fix' the ammonia, as ammonium chloride for instance, prior to stripping hydrogen sulphide. The non-condensable products from stripping are incinerated or led to sulphur recovery. Another method of treating sulphidic sour condensates is by oxidation in an air stream at elevated temperature and pressure, when thiosulphates and sulphates are produced in solution and any excess mercaptans stripped and incinerated.[27]

Oxidation in air is also a satisfactory method of treating spent caustic soda solution used for the removal of hydrogen sulphide and mercaptans from product streams (see Sulphur dioxide emissions). Phenols, however, interfere with the oxidation process, and consequently this method cannot be used for treating phenolic spent caustic from catalytic cracker product scrubbing. This is best treated in a two-stage flue gas stripping column. The soda is first neutralized with flue gas containing carbon dioxide, and then stripped with flue gas. Phenols and cyanides are partially stripped and then incinerated with the other contaminants in the stream. When hydrogen treatment has been used for the removal of sulphur compounds this has significantly reduced the scale of caustic soda handling problems and associated atmospheric emissions.

Carbon monoxide (CO)

Combustion sources, boilers, heaters and flares in a refinery emit carbon monoxide to the atmosphere, although in quantities which are generally not considered to be significant.

More significant, however, are the emissions from catalytic cracking regenerators. Coke is burned off the catalyst in a limited supply of air to control the operating temperature and, as a result, considerable quantities of carbon monoxide are formed. Gas analyses have shown[14,15] that carbon monoxide concentrations of 10 per cent by volume are not uncommon when no control is practised. Concentrations depend on catalyst circulation rate amongst other operating parameters, and this is one reason why regenerator gases from the moving bed type of catalytic cracker contain less carbon monoxide than those from a fluidized bed system.

When control of carbon monoxide emission is necessary, waste heat boilers are used[22] in which the carbon monoxide is converted exothermically and almost completely to dioxide. Other combustibles in the regenerator gases such as hydrocarbons are also burnt. Subsidiary refinery fuel may also be used to maintain the carbon monoxide gas combustion temperature at about 1000°C. In the carbon monoxide boile , therefore, heats of combustion are utilized as well as the sensible heat of the gases. For the future, improved technology, including better catalysts, will make higher regenerator bed temperatures possible. This will allow complete combustion of carbon monoxide to dioxide and eliminate subsequent conversion treatment of carbon monoxide from the regeneration.

Oxides of nitrogen (NO_x)

Whenever fuel of any type is burnt, nitrogen oxides are formed in two ways: by the combination of nitrogen and oxygen in the input air, and by the oxidation of nitrogen present in combined form in the fuel.

To put refinery nitrogen oxide emission into perspective, it was shown in Table 13.6 that in 1967 only about 2 per cent of total nitrogen oxide emissions in the US was due to refinery operation, the major sources being heaters and boilers. A survey of 21 refinery stacks serving 14 heaters, 10 boilers and 1 fume burner gave average nitrogen oxide emissions of 0·0068 g m^{-3} of fuel oil and 3·7 g m^{-3} of gas. On this basis a simple 5 million tonnes per year refinery would emit of the order of 5 tonnes of nitrogen oxide daily, but emission of these quantities from tall refinery stacks from which larger quantities of sulphur dioxide are released have not, according to the authors' knowledge, caused any significant problems.

Major factors affecting nitrogen oxide formation are the proportion of excess air, the flame temperature and residence time in the combustion zone. Some control of emissions can therefore be achieved in existing heaters by controlling air supplies, but actual levels also depend on the burner and furnace design.[28]

References

1. *British Petroleum Company's statistical review of the World oil industry*, 1974.
2. *Oil and gas J*, **72,** 1 April 1974 and 30 December 1974.
3. Elkin, H. F. and R.A., Constable, Presentation at A.I.Ch.E. Air Technical Forum, Dallas, Texas, February, 1972.
4. *Control techniques for nitrogen oxide emissions from stationary sources.* US Department of Health, Education and Welfare publication AP-67, 1970.
5. *Air pollution data for Los Angeles county*, Los Angeles County Air Pollution Control District, Los Angeles, California 1967.
6. *Surveying air pollution around refineries*, Stichting CONCAWE (Conservation of Clean Air and Water in Western Europe), Report No. 14/72, December, 1972.
7. *Refining Process Handbook* **51**(9) published by Hydrocarbon Processing Magazine via Gulf Publishing Co. Houston, US, 1972.
8. *Japan Pet. Weekly*, 31 May 1971.
9. Asselin G. F. and D. H. Stormont, Treating light refinery products, *Oil and Gas J.* **63** (1), 90–93, 1965.
10. Barry, C. B., *International Hydrocarbon Processing*, **51** (4), 102, 1972.
11. Beavon, D. K. and R. P. Vaell, Paper presented to 37th Mid-year Meeting of API Division of Refining, New York, 9 May 1972.
12. Bartel, Y., *et al.*, Paper presented to 36th Mid-year Meeting of API Division of Refining, San Francisco, 12 May 1972.
13. Rosenberg, H. S., *et al.*, *Chem. Eng. Prog.*, **71** (5), 66, 1975.
14. *Emissions to the atmosphere from petroleum refineries in Los Angeles County*, Joint District, Federal and State Project Final Report No. 9, Los Angeles County Air Pollution Control District, Los Angeles, California, 1958.
15. *Catalytic cracking emission survey*, compiled by Crossley Surveys Inc. for API Committee of Air and Water Conservation, 1971.
16. *Control techniques for particulate air pollutants*, US Department of Health, Education and Welfare Publication AP.51, 1969.
17. Standley, C. A., 'Torroidal combustion for fired heaters', *Chem. and Process Eng.*, **49** (2), 106, 1968.
18. 'Evaporation loss from floating roof tanks', API Bulletin No. 2517, February, 1962.
19. 'Evaporation loss in the petroleum industry—causes and control', API Bulletin No. 2513, February 1959.
20. 'Rock Caverns for Oil Storage', Tanker and Bulk Carrier **19**, September 1972.
21. API Bulletin No. 2514, November 1959.
22. *Air Pollution Engineering Manual*, Air Pollution Control District, County of Los Angeles, published by US Department of Health, Education and Welfare, 1967.
23. *An investigation into the causes of refinery smells*, Stichting CONCAWE Document No. 3875, June 1969.
24. Litchfield, D. K., 'Controlling odors and vapours from API separators', *Oil and Gas J.*, **69** (44), 60–62, 1971.

25. Morrison, J., 'Tilted-plate separators for refinery waste water', *Oil and Gas J.* **68** (50), 86–88, 1970.
26. *Aqueous wastes from petroleum and petrochemical plants*, *Beychok*, M. R. Wiley, 1967.
27. Fassbender, H., Effluent pretreatment and biological effluent treatment in the oil industry, *Chem. and Industry*, 1539, 1967.
28. *Abatement of nitrogen oxides emissions from stationary sources, National Academy of Engineering publication*, Washington DC, 1972.

14 Iron and steel works

G. E. SPEIGHT

The manufacture of iron and steel from iron ore to finished steel in an integrated iron and steel works involves many processes, some of which have high potentiality for atmospheric pollution. This may arise from the following:

1. combustion of fuels which may lead to emissions of fly-ash, sulphur dioxide, black smoke, etc.;
2. physical degradation of raw materials, e.g., grit and dust particulates of coal, coke, iron ore, lime, etc.;
3. chemical reaction between process ingredients, e.g., the fine red iron oxide fume which is produced by injection of high-purity oxygen into iron melts.

Table 14.1 presents a broad classification of the main types and sources of air pollution which may arise in an integrated iron and steel works. Coke-making and associated processes which are also operated in integrated works are not included, since they are discussed in chapter 12. As the processes mentioned in Table 14.1 are not necessarily operated at every iron and steel works, the overall nature of the emissions from particular works may differ.

During the last two to three decades, control of air pollution in the industry has substantially improved, such that the traditional picture of the industry as a complex of tall, smoking chimneys no longer exists. Technological advances in fuel efficiency have been made and the pattern of fuel usage has markedly changed. Coal-firing has been substantially replaced by electrical, gas or oil heating and this had led to very evident reductions in the low and medium level emissions of black smoke. At the same time, other important technical changes have taken place, some of which have added to the

TABLE 14.1

Main pollutants and sources of air pollution at integrated iron and steel works

Process	Metallurgical fume	Black smoke	Dust	Grit	Sulphur dioxide etc.
Iron works					
1. Ore crushing, screening and drying			Dust	Grit	
2. Sintering and pellitizing			Dust	Grit	Sulphur dioxide
3. Blast furnace			Dust	Grit	
4. Blast furnace gas	Toxic metallic oxides, e.g., Zn		Dust	Grit	Carbon monoxide, cyanides
5. Direct reduction			Dust	Grit	
Iron and steel foundries					
Cupolas	Toxic metallic oxides	Smoke	Dust	Grit	Sulphur dioxide
Steel works					
Furnaces and converters	Red fume		Dust	Grit	Sulphur dioxide
Rolling mills					
1. Deseaming	Red fume				
2. Reheating and heat treatment		Smoke	Dust		Sulphur dioxide
Services					
1. Locomotives, cranes		Smoke			
2. Boilers		Smoke	Dust	Grit	Sulphur dioxide
3. Heating and power generation plant		Smoke	Dust	Grit	Sulphur dioxide

Note Coke oven and associated processes not included.

industry's pollution control problems. These include the increasing use of specially prepared burdens in place of raw materials for the blast furnace charge and the impact of high-purity oxygen in steelmaking. This latter change has led to much greater rates of steel production, to the development of new steelmaking processes and to the concentration of production in a smaller number of larger capacity furnaces.

Although legal aspects of air pollution control in the UK are discussed in chapter 3, it may be of interest to mention the dual supervisory system which applies to the iron and steel industry. Thus when the pollution arises mainly from fuel-burning operations and an acceptable solution is available, supervision is exercised by the local government authority. Processes which are considered to present special technological difficulties in the control of air pollution—mainly those where the emissions arise from chemical and physical changes in the process materials—are 'scheduled' and supervised by the Alkali and Clean Air Inspectorate. The Alkali and Clean Air Inspectorate requires that the 'best practicable means' be used to control atmospheric pollution; it sets 'presumptive' standards, which can be modified as pollution control technology and general conditions improve. The present 'presumptive' standards for chimney emissions for the scheduled processes of the iron and steel industry are summarized in Table 14.2.

Iron making

Present practice in UK integrated works is based on a series of operations involving reduction of iron ore in a blast furnace to produce iron containing 3–4 per cent carbon, followed by refining in conjunction with some steel scrap to yield the initial steel product. An alternative but as yet less exploited route involves direct reduction[1] of rich iron ore, size-graded lumps or pellets, to yield a highly metallized iron for subsequent steelmaking, either alone or in admixture with steel scrap, in an electric arc furnace. The number of such installations is rapidly increasing throughout the world, including the UK.

Iron ore preparation

Until about 40 years ago, preparation of iron ore before charging to the blast furnace was minimal. It usually involved separating and discarding the ore 'fines' from the 'run-of-mine' or dried ore to avoid choking the furnaces and thus reducing the rate of iron output. Modern practice requires more uniform size grading of the charge materials. Iron ore is normally crushed and graded to 50–10 mm, the undersized fraction being agglomerated by the sinter or pelletizing processes and again graded. The size-graded iron ore and sinter are then charged to the blast furnace together with the necessary graded coke and fluxes.

The preliminary operations of iron ore unloading, stocking, crushing, screening and general handling generate substantial emissions of relatively coarse dust and grit. At the unloading and stocking stages, windbreaks and liberal use of water sprays are used to control air pollution.[2] The crushing, screening and general handling operations, e.g., belt systems and transfer points, involve close fitting hoods or total enclosure fitted with dust extraction units as appropriate.

Sintering

Sintering is now the most important pre-treatment for the blast furnace burden. It not only facilitates blast furnace operation, but permits the utilization of fine materials (ore, coke and sinter) which would otherwise be discarded. Sintering also improves the burden materials chemically by elimination of volatiles including some sulphur, and by partial reduction of iron oxides, both of which improve the economics of blast furnace operation.

TABLE 14.2

Presumptive limits for grit, dust and fume in emissions from chimneys (1975)

Process	Presumptive limits: $mg\ m^{-3}$ (15°C and 760 mm)	Presumptive limits: General	Comments
Ore dryers			
Existing	458		Future possibility of more stringent standards for new, larger units.
Sinter plants			
1. Existing installations before 1970	458	Chimney height in accordance with sulphur dioxide loading	
2. New (or rebuilt) installations after 1970	115		
Blast furnaces			
Integrated works	458	Chimney height sufficient to minimize hazards from carbon monoxide with provision for prevention of liquid sprays	These standards are applicable to blast furnace gas discharged to the atmosphere
Hot blast cupolas[16]			
New installations	115	Waste gases burnt before discharge. Chimney height 21–43 m, depending on capacity between 7–100 tonnes h^{-1}	Existing plants not equipped for full fume arrestment shall install this by 31 December 1978 (one extra year in some agreed cases)
Steelmaking furnaces			
1. *Top-blown oxygen converters*	115	Near-invisibility of red fume	Secondary fume control required at new plants
2. *Open hearth furnaces*			
(a) Not using injected oxygen	458	Adequate chimney height for dispersal of pollutants	Cleaning units not yet mandatory; may be required in future
(b) Using injected oxygen	115	Near-invisibility or red fume	
3. *Electric arc furnaces*			
(a) Not using injected oxygen	458		Cleaning units not yet mandatory; may be required in future if standards are not achieved
(b) Fumeless refining	458	Near-invisibility of red fume during refining	If these standards are achieved cleaning units are not required
(c) Using oxygen for refining	115	Near-invisibility of red fume	Secondary fume controls required at large new plants
Scarfing	115	Near-invisibility of red fume	

Description of process. Sintering is normally carried out on a travelling grate or 'strand' 3–5 m wide by 50–100 m effective horizontal length. In effect, the 'strand' is an endless chain of cast iron pallets, placed broadside-on, moving continuously over a series of wind-boxes connected to an exhaust system. A mixture of iron ore and sinter fines (less than 6·5 mm) with 6 per cent coke breeze, moisture and sometimes a little limestone, usually prepared in a rotating drum, is charged to the entry end of the 'strand' to yield a bed depth of about 1 m. The coke breeze in the surface of the bed is ignited under the hood at the entry end and as the grate travels forward, the 'combustion front' gradually penetrates the bed, driving off volatiles and fusing the mixture into a clinker-like material until the combustion is complete at the end of the 'strand'. At this point, the pallets overturn and the hot sinter is discharged into an enclosed chamber where it is broken and screened. Final cooling on up-draught circular coolers is the normal present practice, but some plants are now adopting 'on-line' cooling, which entails extending the sinter strand in order to achieve cooling prior to discharge and breaking.

Nature of emissions. Two main sources of air pollution from a sinter plant are first, the waste gases from the combustion and sintering reaction on the strand, and second, the grit and dust arising at the discharge end. Further sources of dust emissions are final sinter cooling and screening, storage and miscellaneous transfer systems which require hoods, covers, or enclosure with extraction units as appropriate.

Selection of cleaning unit. High efficiency cleaning units are required for both strand and discharge-end waste gases. In the UK dry electrostatic precipitators are preferred for these applications.

The sinter grit in the strand waste gas is very coarse and highly abrasive and if not satisfactorily removed can be very destructive to the exhaust fan. In comparable sinter plants with similar burdens, exhaust fan lives of only 15 months were obtained when grit removal was unsatisfactory and up to 8 years with an efficient electrostatic precipitator.[3] Efficient gas cleaning thus yields economic benefits from lower plant maintenance costs and down-time, and by providing more recovered sinter grit for re-use. Problems of high resistivity dust have been encountered in Europe and the US, where high basicity sinter is produced. High basicity sinter, in which the $CaO–SiO_2$ ratio is high (e.g., 1·8–2·9), has high resistivity, and such dust is more difficult to collect

---

Notes For discharges of waste gases containing sulphur dioxide, the chimney heights must be in accordance with the recommendations given in *Memorandum on chimney heights*, (2nd edn.), HMSO, London 1967. Above this range the following relationships are used:

Rate of emission (tonnes SO_2 per day)	3·7	7·6	13·2	21·3	30·2	40·1
Basic chimney height (m)	30·5	45·2	60·3	75·3	90·4	102·1

To allow for interference from nearby buildings, the following correction is applied:

$$H = 0{\cdot}625A + 0{\cdot}935B,$$

where H = final chimney height, A = basic chimney height, B = building height.

Fig. 14.1. Iron ore sintering strand. Ignition hood at bottom right-hand corner. (*Courtesy* Lurgi (UK) Ltd.)

in electrostatic precipitators. In the UK, CaO/SiO_2 ratios have been generally lower, and fewer problems are encountered with electrostatic precipitators.

The water vapour and sulphur dioxide contents of the strand waste gases give rise to important effects. Whilst they help in 'conditioning' the gas for efficient dust removal in the electrostatic precipitator, they can lead to condensation and corrosion difficulties unless suitable precautions are taken. Thus the gas temperature must be maintained above 150°C, and this must be carefully watched during shut-down and starting-up periods, the precipitator being by-passed if possible. The interior of the precipitator shell may be lined with a protective coating such as an aluminous cement and the exterior is usually thermally insulated. The electrode plates also may sometimes be made of low-alloyed corrosion-resisting steel.

In some older UK practices based on local ores, the sulphur dioxide emission was substantial; values of 96 tonnes of sulphur dioxide from a daily sinter output of about 10 000 tonnes were reported.[4] The present UK practice for control of sulphur dioxide pollution is dispersal from a chimney of sufficient height as specified by the Alkali and Clean Air Inspectorate.[5] Thus chimney heights of up to 110 m have been installed in some UK sinter plants. The increasing use of rich imported iron ores of low sulphur contents, however, has led to marked reductions in sulphur dioxide emissions, and hence there may be less need for these very tall chimneys. In Japan and elsewhere, processes for the removal of sulphur gases from sinter plant waste gases are being developed and applied where necessary.[6]

Dry electrostatic precipitators have also been adopted for dust removal from the sinter discharge-end gases, which are low in water vapour and virtually free of sulphur dioxide. Consequently, in cold dry conditions, the gas 'conditioning' may be inadequate and precipitation efficiency impaired. In these circumstances it is advantageous to ensure adequate 'conditioning' by steam injection or water sprays. In some installations this need has been met by venting the hot moist dust-laden gases from the quenching of hot return fines into the discharge-end waste gases for joint electrostatic precipitation.[7] Since sulphur dioxide is virtually absent from these gases, a lower chimney height in this system is acceptable.

Pelletizing

Agglomeration of iron ore fines by pelletizing is becoming increasingly practised as an alternative to sintering, and plants with annual outputs of up to 4 million tonnes of pellets are now operating. The process is particularly useful for very fine iron ore concentrates which may be unsuitable for sintering. It can be operated in several different forms, such as the vertical shaft furnace, the straight grate and the grate kiln systems.[8,9] A brief outline of the latter system is as follows: A feed mixture is first prepared from crushed and ground iron ores with 1 per cent of binder (bentonite) and 8–10 per cent of moisture. This mixture is fed into inclined drums to produce green balls of about 9–16 mm diameter. The green balls are charged on to a continuous chain grate which first traverses a drying zone (370°C) and then a preheat zone (1300°C). The preheated balls are passed into a rotary kiln, for hardening at about 1300°C, then into an annular cooler, and are finally discharged at less than 210°C. The air pollution problems are similar to those of the sinter process, namely grit, dust and sulphur dioxide, and similar pollution control techniques are adopted

Note

Unless otherwise specified in the text, measurements of pollution in gases given in this chapter are in the equivalent of 1 m^{-3} at NTP (0°C and standard barometer of 760 mm mercury).

Blast furnaces

The blast furnace has several potential sources of atmospheric pollution, the two most important being the exit gas from the furnace top, for which a high efficiency cleaning system is required, and the furnace tapping arrangements.

Description of process. The essential features of the blast furnace are indicated in Fig.

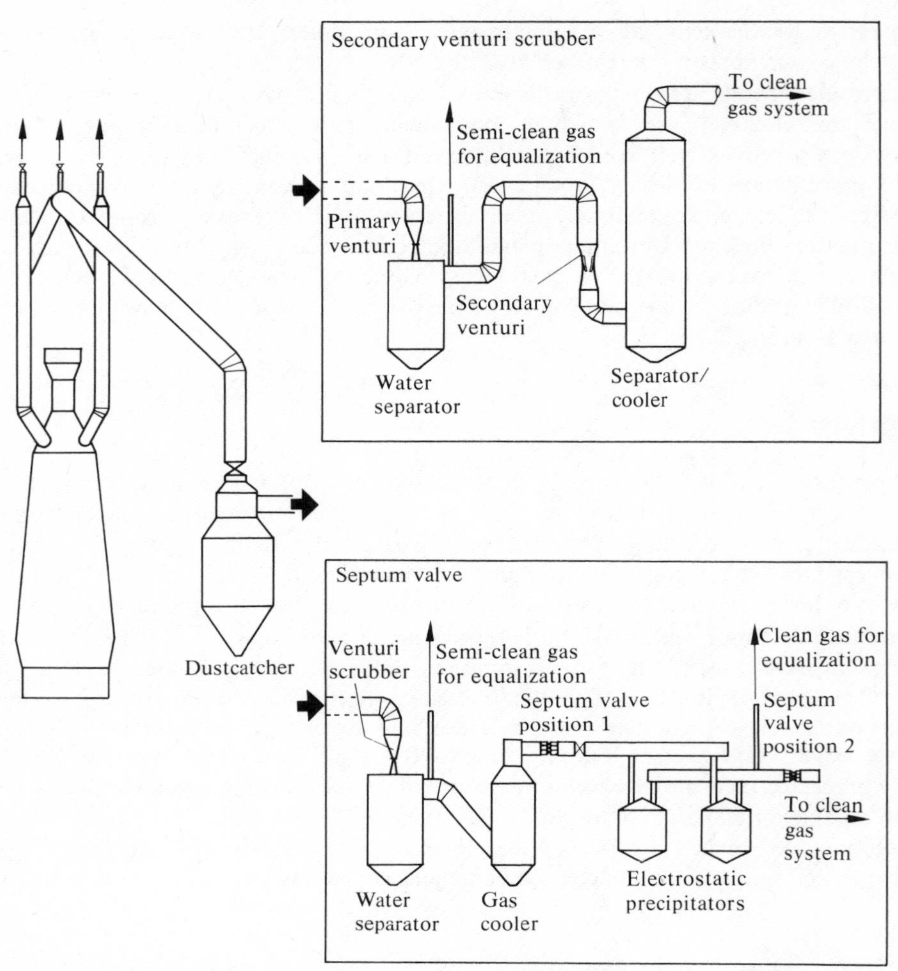

Fig. 14.2. Blast furnace and gas cleaning system. Alternative pressurization and gas cleaning systems for high top pressure operation. (*Courtesy* Davy Ashmore International Ltd.)

14.2. The furnace has a steel shell, which is shaped like two truncated cones joined at their bases and lined with refractory. Iron ore sinter and size-graded iron ore, coke and limestone as necessary are charged into the top of the furnace through a double bell or other system, designed to minimize leakage of dirty gas. Preheated air is blown through tuyères into the furnace, thereby producing carbon monoxide, which reacts with the heated charge to form molten high-carbon iron, slag and blast furnace exit gas. The molten iron and slag collect in the hearth and are periodically discharged from tap holes. Blast furnace gas passes through an off-take into a cleaning system. The process is continuous, and as the burden gradually passes down the shaft, charging is continued to maintain a substantially constant level of burden in the furnace.

Nature of furnace exit gas. The blast furnace top gas contains about 28 per cent of carbon monoxide with entrained particles of iron ore, sinter and coke normally amounting to about 15–30 g m^{-3} (NTP), depending on the nature of the burden and on operating practice. For instance, the use of high top pressure considerably reduces the particulate content of the exit gas. The particle size ranges[10] from very coarse grit (greater than 100 μm) down to sizes less than 10 μm. Exit gas temperature is 150–200°C. At most integrated works, the cleaned exit gas is used as a low-grade fuel for underfiring coke ovens, boilers and for the blast furnace stoves which preheat the air blast to the furnace.

Selection of cleaning system. In its original dirty condition, blast furnace gas is unsatisfactory for use as a fuel, and to avoid blockages or undue erosion of burners, a high standard of grit and dust removal is required. In the UK, a multi-stage cleaning system is usually employed, comprising a dust-catcher and a spray or water washing tower, which reduce the dust loading to about 0·5 g m^{-3} (NTP), followed by the final, third stage cleaning to about 0·01 g m^{-3} (NTP) in a wet electrostatic precipitator. In new or rebuilt blast furnaces which are equipped for high maximum gas pressure operation, greater use is made of high energy wet washers of the venturi type, either alone or in combination with a wet electrostatic precipitator. Figure 14.2 outlines alternative pressurization and gas cleaning systems for blast furnace operation with high maximum pressure.

Furnace emissions. Irregular furnace operation may occasionally give rise to direct emissions to the atmosphere. Thus, sudden falling or 'slipping' of the burden may produce violent pressure fluctuations in the furnace and gas cleaning system. Pressure relief valves in the gas off-takes release dirty gas to the atmosphere when these pressure fluctuations occur, thereby guarding against an explosion and consequent damage to the plant. The use of sinter promotes smoother furnace operation and consequently reduces 'slipping', and hence these irregular emissions of dirty gas. Similar improvements are also obtained by regular 'checking' or easing of the blast to produce a minor controlled 'slip'.

Cast house dust and fume control. Modern large multi-taphole blast furnaces cast almost continuously and considerable amounts of dust and fume are emitted at tapholes, runners, ladle spouts and slag notches. It has become necessary to install some pollution abatement measures.

In Japan, where most of this development has occurred,[11] the pollution control measures include hoods at tapholes and ladle spouts, and almost total enclosure hoods on the runners with exhaust systems leading to a dust collector, generally a filter bag unit. Since these hoods do not provide complete capture of the fumes, etc., they have been supplemented in most cases by general cast house extraction systems leading to filter bag collectors.

Similar systems are being applied to modern large blast furnaces elsewhere, but the layout of some existing cast houses may often preclude the installation of the comprehensive system outlined above.

Direct reduction of iron ore

Of the many processes[12] that have been devised for making pre-reduced iron, only five have so far reached commercial application, namely the Hyl, Midrex, SL/SN, Lurgi and the Krupp sponge iron processes. At the moment, two Midrex plants,[13] each designed to produce 400 000 tonnes of metallized pellets per year, are being installed in the UK.

The Midrex process is operated in a continuous shaft furnace with reformed natural gas, i.e., carbon monoxide and hydrogen, as reductant. The spent reducing gas is cooled, mixed with more natural gas and passed to a reformer to produce fresh reducing gas. The advantages of the process are its continuous nature, low energy consumption and low manning requirements. The product contains about 93 per cent iron and 1–2 per cent carbon. Air pollution is confined to dust generation at several points for which collection units are required.

Iron foundries

Cupolas

The cupola is the most important source of molten iron for making iron castings. As with other iron and steel making processes, it presents special air pollution problems.

Description of process. The cupola is in many ways similar to a small blast furnace. It is usually a simple vertical cylinder of plate steel, lined with refractory. Near the base there are inlets (tuyères) for combustion air and tapholes for metal and slag. Near the top is a door for charging the raw materials and above that a gas cleaning cap and extension cylinder for high level discharge of the waste gases.

A bed of coke is first charged on the bottom of the cupola to a height of 1–2 m. When this has been fired and is burning, alternate layers of metal (cast iron and steel scrap) and of coke and flux (usually limestone) are added to fill the cupola to the charging door. The combustion air through the tuyères burns the coke and melts the metallic charge. As the burden descends, more charge materials are added at the top. Molten iron containing 2·5–3·5 per cent carbon flows out through the tap hole at the bottom.

There are two main types of cupola, namely the 'cold-blast' and the 'hot-blast' cupola. The cold-blast cupola is the more common in the UK. This utilizes a non-heated air blast, and an open-topped shaft, preferably with direct combustion or after-burning arrangements for the cupola stack gas, which is then discharged through a grit and dust arrester cap at the top of the stack. The hot-blast cupola has a closed top with off-take for the gas, which after removal of grit and dust is then passed to a combustion/

recuperator system for preheating the incoming air blast to about 500°C. In the hot-blast cupola, a greater proportion of steel scrap can be melted, thus economizing on the more expensive iron, but at the same time accentuating the air pollution problems.

Cupolas are rated on output per hour, which is usually 2–10 tonnes h^{-1}; cold-blast cupolas of up to 50 tonnes h^{-1} output are, however, operating in the UK.

Nature of stack emissions. The products emitted from the stack include the following:

1. Smoke from oil, grease, paint, etc., arising from the steel scrap, wood, etc.
2. Grit and dust particles arising from charged materials and their impurities, i.e., sand, coke, rust, limestone. About 15 per cent of the solids emitted are smaller than 10 μm and over 50 per cent greater than 50 μm. Typical emissions for a normal sized cupola would be about 7–11 kg per tonne of metal produced.[14]
3. Metallurgical fume, which consists largely of metallic oxides, slags, ash, etc., and is greater from the hot-blast cupola than from a comparable cold-blast cupola.
4. Sulphur dioxide arising from the sulphur in the coke. The amount emitted, which is required for chimney height estimations,[15] depends on the amount of sulphur absorbed by the metal and the sulphur dioxide retained by the arrester. For a coke with 0·7 per cent sulphur, 50 per cent is absorbed by the metal and 30–50 per cent of the remaining sulphur is retained by a wet arrester, none by a dry arrester.

The stack gas, which contains 10–20 per cent of carbon monoxide is usually burned at the chargehole. Sometimes, after-burners (gas or oil) are necessary when the gas is lean, and when combustion cannot be maintained the height of the discharge stack has to be extended.

Control of stack emissions. Different means are required for control of the various pollutants in the cupola stack gas. Thus for a cold blast cupola, smoke can usually be reduced by combustion of the stack gas, or its effect mitigated by dispersal from a tall stack. Simple arresters, dry for the smaller cupolas and wet for cupolas above 5 tonnes/hour, usually provide satisfactory retention of the grit and dust particles. Metallurgical fume, which is often brown in appearance, causes the greatest difficulties; the simple arresters are inadequate and only high efficiency cleaners are really capable of removing these fine particles. At the moment, dispersal of this fume from a tall stack of 36·5 m minimum height is acceptable, although the installation of an appropriate cleaning unit is recommended. The sulphur dioxide emission is controlled to some extent by absorption in the wet arrester—water treatment may be required to reduce plant corrosion difficulties—and by subsequent dispersal from a tall stack.

Due to the greater amount of metallurgical fume, hot-blast cupolas require better gas cleaning equipment than cold-blast cupolas. Hot-blast cupolas are controlled by the Alkali and Clean Air Inspectorate, whose requirements have gradually become more stringent. Present requirements[16] are that all new hot-blast cupolas are to be fitted with combustion and high efficiency cleaning systems to achieve a nearly invisible discharge,

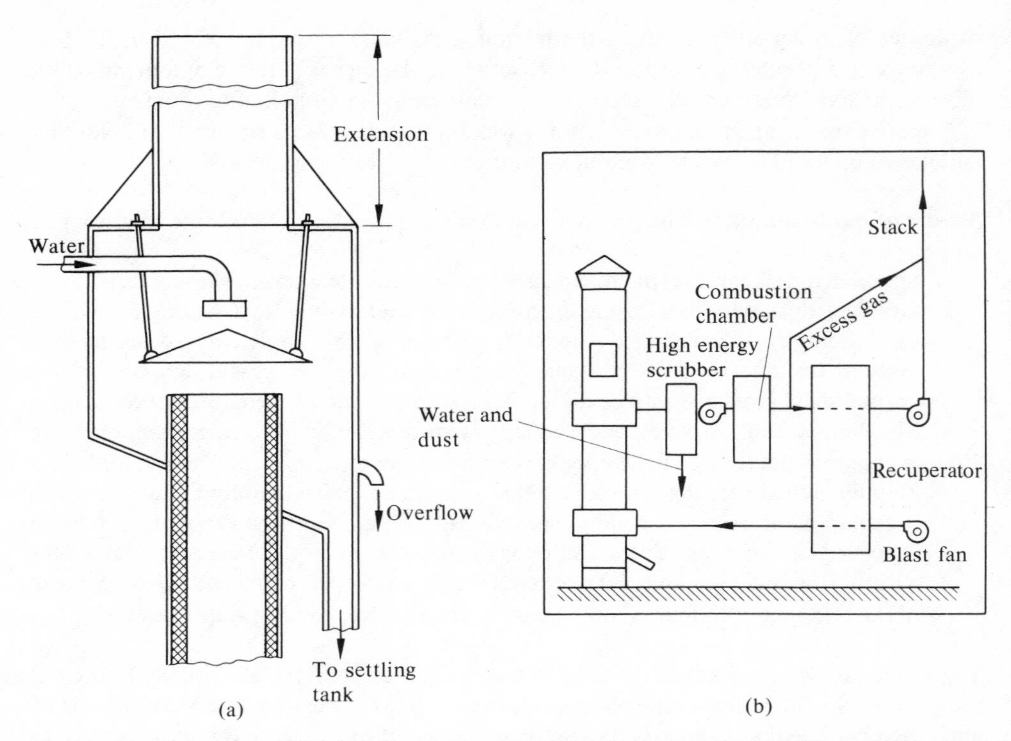

Fig. 14.3. Typical waste gas cleaning systems for cold- and hot-blast cupolas: (a) simple wet arrester for a cold-blast cupola; (b) gas cleaning recuperative system for a hot-blast cupola. (*Courtesy* British Cast Iron Research Association.)

i.e., 0·115 g m^{-3} (15°C and 1 bar) from a chimney of 21–43 m height depending on the cupola rating. Existing hot-blast cupolas must conform to this requirement within a time limit agreed with the Alkali Inspectorate, normally 31 December 1978. Figure 14.3 outlines an appropriate gas cleaning system.

Other foundry operations

Atmospheric emissions also arise from other foundry operations, e.g., smoke and odours mainly from organic additives to the core-making and mould-pouring processes, and dust and grit including metallic particles from grinding and shot-blasting of iron castings. In regard to control of smoke and odours, the present practice in the UK is to collect them in exhaust hoods for subsequent discharge at high levels. Shaw,[17] however, discusses the use of collection and discharge systems involving a combustion stage. Dust and grit emissions are usually entrained in an exhaust system for subsequent collection by fabric filter or wet washers.

Steelmaking

The production of steel from high carbon-iron or high carbon-iron/steel scrap mixtures is fundamentally the elimination of several per cent of impurities by oxidation. The oxygen required may be supplied as iron oxide, air, or as high-purity gaseous oxygen.

Although the older open hearth and bessemer steelmaking processes which were based on refining with iron oxide-air respectively had substantial capability for atmospheric pollution, the availability of high-purity oxygen at economic prices has not only changed the world pattern of steelmaking during the past 25 years but has added a new dimension to the pollution problem. Under the generic name of *basic oxygen steelmaking* (BOS), new processes involving top-blowing with oxygen have been developed of which the LD converter is the most popular. Moreover, the injection of high-purity oxygen into open hearth and electric arc furnaces has greatly increased their steel output rates. During the last few years, revival of interest in oxygen bottom-blowing through fuel-gas shielded tuyères has resulted in improved open hearth (SIP)[18] and converter (Q-BOP)[19] operations.

When oxygen is injected onto molten iron, refining is accelerated and high rates of steel output are obtained together with the fine red oxide fume which characterizes modern steelmaking and is now one of the industry's most difficult and costly air pollution problems. Waste gas cleaning to the required standard, near-invisibility of the exhaust plume, requires high efficiency cleaning to the order of 99·5 per cent fume removal. The high temperature of the waste gases, their flammable nature and the fineness of the fume, 0·05–1·0 μm, create very difficult operating conditions for the cleaning unit. Some of the problems and the pollution control measures used are discussed later.

Open-hearth steelmaking

For virtually a century, open hearth steelmaking has been the mainstay of UK bulk steel production. Although the rate of output of open-hearth steel was increased by use of high purity oxygen, these furnaces are now being replaced by converter or electric arc processes.

Description of process. Essentially, the open-hearth furnace is a long shallow refractory hearth, enclosed in refractory-lined roof and walls, with ports at each end for admission of fuel and preheated combustion air, and the emission of hot waste gases. The furnaces are fired from each end in turn, reversals taking place at intervals of 20–30 min. The original fuel was producer gas, but this was replaced by coke-oven gas, or by liquid fuels injected from a burner. The hot waste gases pass from the furnace to a checker brick heat exchanger (which pre-heats the incoming combustion air on the next reversal), and thence through a waste heat boiler to the stack. Open-hearth furnaces vary in capacity from 50 to 500 tonnes and in charge materials from 80 per cent hot molten iron with 20 per cent cold scrap at one extreme to 15 per cent cold pig iron with

85 per cent cold scrap at the other. The metallic charge is melted together with slag-making materials, such as lime, silica and other fluxes, and is then refined by iron oxide additions or oxygen injection. The resulting impurities either pass into the slag, or in the case of carbon, into the waste gas as carbon monoxide which is then burned to carbon dioxide. High-purity oxygen is used in open-hearth furnaces in two ways:

1. enrichment of the combustion air, and
2. injection via a steel lance into the molten bath.

The first use makes no substantial change to the nature and level of particulate emissions, but the second produces the copious fine red fume for which cleaning equipment is essential.

Nature of emissions. In normal practice without oxygen injection, the dust loading averages 0·2–0·6 g m^{-3} (NTP), with 50 per cent of the particles larger than 5 μm. With oxygen injection at the rate of about 40 m^3 $tonne^{-1}$ for about 4 h, the amount of particulates, i.e., dust and fume averages about 8 g m^{-3} (NTP), with peak loadings of up to 16 g m^{-3} (NTP) during the oxygen lancing period.[20] These particulates are mainly fine iron oxide fume in the size range 0·1–1 μm, with mean size 0·2–0·3 μm. The concentration of fume in the waste gas is influenced by several factors, including the amount of excess air which may leak into the furnace waste gas system. The composition of the waste gas varies according to the fuel used; from the gas cleaning aspect, water vapour, which may vary between 2 and 25 per cent, and sulphur dioxide, between 1 and 2 g m^{-3} (NTP), are important.

Control of atmospheric pollution. For open hearth practice without oxygen injection, the installation of waste gas cleaning equipment has not been mandatory. Particular furnaces using high charges of commercial steel scrap from which the emissions may contain excessive amounts of toxic metals, i.e., lead or zinc, would be expected to adopt control measures. Chimneys of 60–70 m height are required, however, to ensure adequate dispersion of the grit and sulphur dioxide emissions.

For furnaces charged with hot metal and with full oxygen injection practice, cleaning units are required; in the UK, dry electrostatic precipitators have been generally favoured. The presence of water vapour and sulphur dioxide, though useful for conditioning the gas, can introduce corrosion difficulties should condensation occur. To minimize these difficulties, a number of precautions should be introduced in the design and operation of the electrostatic precipitator, mainly related to achieving and maintaining a safe operating temperature (above 150°C)[21] and to avoiding air in-leakage to the waste gas system.[22]

Bottom-blown open-hearth furnaces (SIP). The *submerged injection process* developed recently in Canada by adopting the Q-BOP principle of bottom-blowing with oxygen through propane-shielded concentric pipe tuyères—actually two tuyères through the sides of the hearth—provides a further way of extending the life and improving the

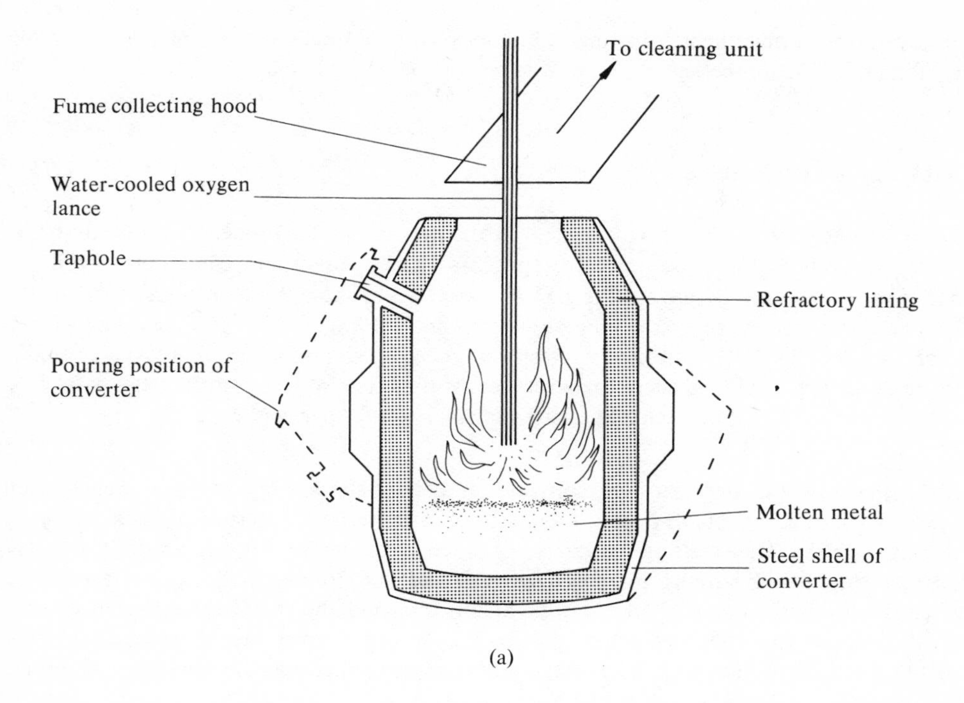

(a)

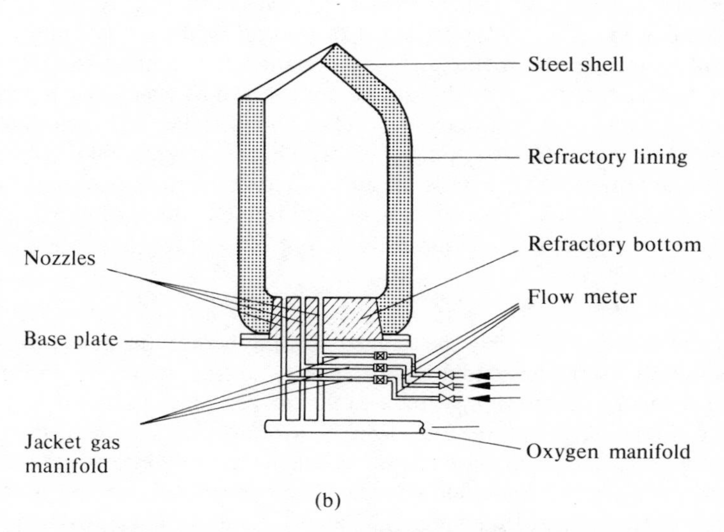

(b)

Fig. 14.4. Basic oxygen steelmaking: (a) top-blown LD converter; (b) bottom-blown Q-BOP converter.

performance of open hearth furnaces. Pollution control facilities are similar to those for BOS and Q-BOP processes.

Basic oxygen steelmaking

The first and now the most widely used of the new converter processes based on high-purity oxygen is the top-blown LD process developed in Austria about 1950. In addition to this and its variant the LD-AC or OLP process for high phosphoric irons, basic oxygen steelmaking now includes the recently developed bottom-blown processes such as the Q-BOP converter. Basic oxygen steelmaking is increasing rapidly throughout the world; it accounted for nearly half (i.e., 49 per cent) of the total UK steel output in 1974, and seems likely to approach 70–80 per cent by 1980.

Description of LD process. The converter is a pear-shaped open-topped steel vessel, lined with basic refractory and mounted on trunnions to permit vertical rotation through 360°. The melting capacity of such vessels, 30–40 tonnes in the early installations, has gradually increased to 220 and 300 tonnes in the most recent UK plants. Charging is through the mouth at the top, and the metal is tapped through a small hole in the shoulder when the vessel is turned into the horizontal. Typical installations comprise either two converters (one operating and the other out of service for relining) or three converters with two operating.

Steel scrap, 20–25 per cent, is first charged, followed by molten high-carbon iron and then by the slag-making additions lime and fluxes. The mouth of the vessel is brought underneath a fume hood, which forms the first component of the waste gas collection and cleaning system. A vertical lance projecting through the fume hood directs high pressure oxygen at the metal surface and starts the refining reactions. Impurities are oxidized and either pass into the slag or are evolved as waste gas, together with fine red iron oxide fume and entrained lime, slag and metallic particles. This refining stage normally occupies 16–20 min, after which the oxygen blowing ceases, finishing additions are made and the metal is tapped. The converter operating cycle, tap-to-tap, is 35–50 min and the oxygen usage about 3 m^3 $tonne^{-1}$ min^{-1} of blowing time. Depending on the size of converters, melting shop output rates of up to several hundred tonnes per hour can be achieved.

Nature of emissions. During the oxygen blowing period, the gas at the converter mouth arises mainly from the oxidation of carbon, and assuming no entrained air or other source of oxygen may approximate to 90 per cent carbon monoxide with 10 per cent carbon dioxide at about 1650°C, with a normal volume of less than twice that of the oxygen blown. The particulate emissions are mainly fine iron oxide fume with some grit or dust from charge materials or slag. The concentration may approach 100 g m^{-3} (NTP) of the unburned converter gas at NTP with about 85 per cent below 1·0 μm, the average being in the 0·05–0·20 μm range. There is some evidence[23] that the size and composition of the collected particulates are affected by the nature of the gas cleaning process, i.e., whether or not the converter gas is burned before fume collection.

Selection of cleaning equipment. The high temperature and combustibility of the initial waste gas, the extreme fineness of the fume and the intermittent nature of the process all contribute to the difficulties of this complex air pollution control problem, on which much research and development effort has been expended.

In the earliest fume collection systems, the initial CO-rich waste gas was burned, cooled in a heat recovery boiler system having auxiliary oil-firing for the intermittent non-blowing periods, and then after 'conditioning' was passed to the cleaning unit, which was either a high energy scrubber or electrostatic precipitator. Air for combustion was drawn in between the converter mouth and gas collecting hood, and in some installations secondary air was fed into the boiler section. This increased both the volume and the temperature of the waste gas, and consequently the amount of heat to be dissipated was greatly increased. Due to operational difficulties with these earlier installations, the boiler systems with auxiliary firing and wet washers were replaced in later installations by water-cooled heat exchangers and dry electrostatic precipitators. Other improvements have also been made as a result of operating experience.

In the latest LD converter installations in the UK there has been a swing away from dry electrostatic precipitators and fully burned waste gases to systems involving high energy wet scrubbing of a substantially unburned waste gas. The final cleaned gas, which may contain more than 70 per cent carbon monoxide, is either burned at a flare stack or is collected for subsequent use as a fuel gas. There are three main variants of this system, the OG system[24] developed in Japan, and the IRSID-CAFL[25] and Krupp[26] 'minimum gas' systems developed in Europe. These systems comprise a close-fitting hood with a movable skirt over the converter mouth, so that the extent of combustion of the converter gas can be governed by controlling the air in-leakage at the mouth of the converter. The means adopted to achieve this include control of gap width, pressure regulation to match the gas drawn off by the exhaust fan with that produced in the converter, and the provision of a curtain of burned waste gas or injected nitrogen. Compared with a full combustion system, the four-fold decrease in the volume of the waste gas leads to reductions in size and cost of the cleaning unit.

The first installation of this type in the UK was the OG system in 1969 on 300-tonne converters at the Port Talbot works.[27] The system comprises a water-cooled movable skirt above the converter mouth, leading to a lower hood and thence to the upper hood, which has two holes, one for the oxygen lance and the other for the additions chute. The radiation section follows and leads up to the spark box—all units up to this point are water-cooled. The waste gas, which at this point is at about 900°C, is evaporatively sprayed to cool to 20°C and then passed to the two parallel saturation venturi washers for rough cleaning. Excess water is removed in an elbow separator and the gas then passes to the second venturi which completes the cleaning and controls the pressure at the mouth of the converter. The clean gas first passes through mist separators to an induced-draught fan, which can be switched to lower speed during non-blowing times, and thence to the flare stack fitted with coke-oven gas pilot flames.

Fig. 14.5. Close-fitting hood and outer movable skirt over mouth of converter. (*Courtesy* BSC General Steels Division.)

The success of these minimum combustion systems with high-energy wet scrubbers has stimulated interest in similar gas collection systems with electrostatic precipitators, both wet and dry. Although at first sight there might appear to be an explosion risk, two dry installations are reported to have operated successfully in Germany.[28]

Secondary fume emissions

In addition to the oxygen-blowing fume which is substantially collected by the primary hood and gas cleaning system, fume and dust arise from other converter operations such as hot metal charging, scrap charging, converter teeming, bath sampling and

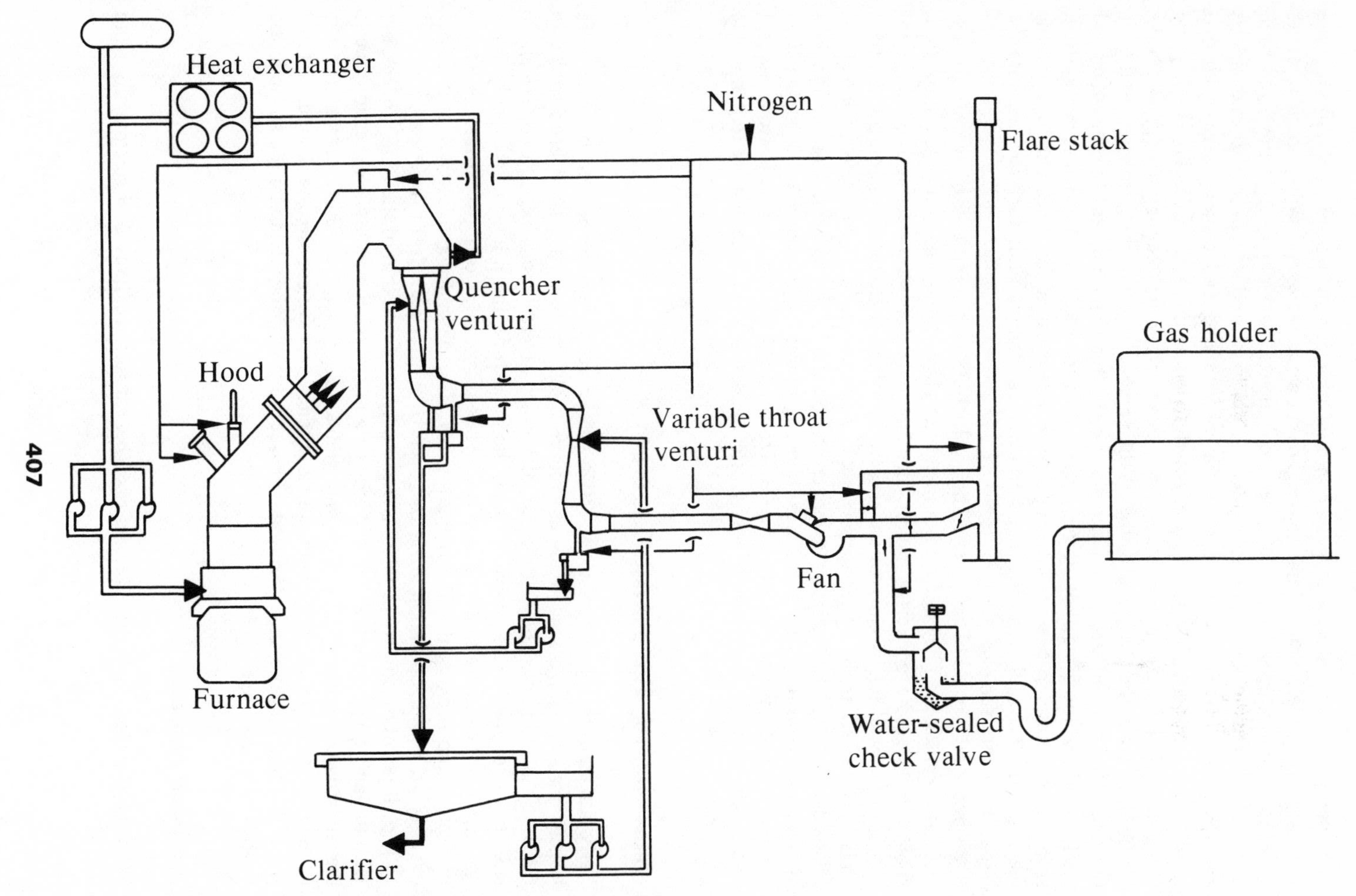

Fig. 14.6. O.G. waste gas cleaning and collection system. (*Courtesy* Davy Ashmore International Ltd.)

observations. These emissions, together with the irregular puffs of oxygen-blowing fume that escape capture by the primary hood, are known as *secondary fume*, which amounts to about 300–400 g per tonne of steel, i.e., about 2–2·5 per cent of the total fume generated.[29] The accepted control system is via local hoods, adjacent to and almost encircling the primary hood, one each for the charging and teeming areas with two smaller auxiliary side hoods, all connected to common ducting leading to a fume extraction plant, usually a bag filter plant. The charging and teeming hoods must be as large as possible compatible with the available space around the converter, and their optimum location and size for a particular plant often require detailed assessments and analysis of the hot fume flow pattern. Even with the most carefully designed system, some of the charging fume may escape the hood; to cater for this fraction, it is now usual to install over the converter and crane track a canopy hood which leads into the initial secondary fume system.

Several auxiliary operations associated with BOS steelmaking also give rise to dust and fume emissions. They include hot metal transfers, deslagging and desulphurization, skull cutting at the converter mouth, flux handling and steel casting, and the term *secondary ventilation* is now often used to embrace both the containment of these emissions as well as the converter secondary fume as above. For these emissions local hoods are used which feed into the converter secondary fume system. This combination system may consume very large quantities of power unless operated in a regulated manner by appropriate use of dampers. However, since the emissions from the hot metal pits may be substantial, a separate extraction and cleaning system may be preferred.

Botton-blown converter processes, e.g., Q-BOP[30]

Early attempts in the 'fifties to adapt Bessemer converters to oxygen blowing were largely unsuccessful, owing to tuyère failures. The problem has since been solved in Germany in the late 'sixties by the use of bottom tuyères consisting of two concentric steel tubes, oxygen being blown through the inner tube and a hydrocarbon-containing gas through the outer annulus. The endothermic dissociation of the hydrocarbon serves to protect and prolong the life of the tuyères. Commercial plants of over 200 tonnes capacity are now in operation in various parts of Europe and North America.

The air pollution control problem is essentially similar to that of the BOS process, but with the alleged advantages of more regular and smoother waste gas collection due to the submerged blowing. The OG and similar suppressed combustion systems are suitable for the primary fume with converter enclosures to capture the secondary fume from charging and tapping operations. Pearce[31] gives further detail on the air pollution control aspects.

Electric arc furnaces

With the decline of the cold-charged open-hearth furnace, its role of steelmaking from steel scrap charges has been taken over by the electric arc furnace using high purity oxygen injection. Thus in addition to the traditional two-slag process for special or alloy steels, electric arc furnaces now operate a single-slag, open-hearth-type process for the tonnage low carbon steels. Furnaces of up to 180 tonnes capacity are operating in the UK with output rates of over 60 tonnes h^{-1}, far exceeding those of the open-hearth furnaces they have replaced.

Description of process. The electric arc furnace consists of a shallow cylindrical shell lined with basic refractory to give a dish-like hearth. The roof comprises a large steel ring filled with refractory to form a shallow dome, through which three carbon electrodes are inserted. Steel scrap plus some iron is charged and melted by electric arcs between the tips of the electrodes. The smaller furnaces are usually charged through a side door, whereas the roofs of newer larger furnaces are raised and swung aside to permit top-charging from buckets with sectional leaf bottoms. The furnace can be tilted for tapping the refined steel.

The normal process for special steels involves two distinct stages: first, an oxidizing stage in which carbon and some impurities are removed; second, a reducing stage in which sulphur is removed; the metal is then deoxidized and alloy additions made. High-purity oxygen may be used to speed reactions in the oxidizing stage, thereby giving rise to the fine red fume characteristic of oxygen refining. For making tonnage low-carbon steels, however, only the oxidizing stage with oxygen injection is normally required.

Nature of emissions. The air pollution problem of electric arc furnaces, particularly the large, oxygen-lanced, fast-working ones, has two main components:

1. the primary red fume which is generated during oxygen lancing and extracted from within the furnace shell;

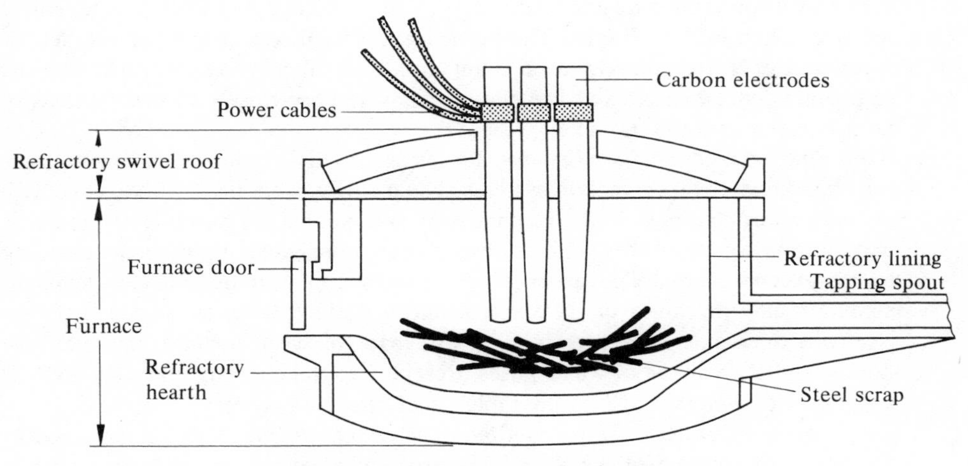

Fig. 14.7. Electric arc steelmaking furnace.

2. the various fumes discharged during the charging and tapping operations, which together with miscellaneous fumes escaping into the melting shop atmosphere are known collectively as secondary fume.

The primary component amounts to about 95–97 per cent of the total fume and dust (about 10 kg per tonne of steel) and the secondary to the remaining 3–5 per cent.[32] The fume composition can be variable; it may include slag, lime, fluxes, oxides of zinc and lead. The iron oxides content is usually lower than in other steel-making fumes. The size analysis depends largely on the iron oxide fume content, which tends to increase the proportion of sub-micron grades.

Selection of fume containment system. The earliest form of fume control was by extractor fans in the building roof, i.e., melting shop ventilation whereby dust and fumes were discharged to the external atmosphere. More sophisticated systems have since been developed.

1. *Building roof-mounted hoods*. Building extraction systems, often with canopy hoods attached to the building roof trusses, connected to a large exhaust duct and then to a bag filter plant are used at a number of plants in the US.[33] These systems aim to deal with both primary and secondary fumes.
2. *Furnace roof-mounted hoods*. Close-fitting hoods over the furnace roof successfully capture fume-laden gases but also draw in large amounts of extraneous air which dilutes and cools the gases sufficiently to permit the use of a fabric filter-bag cleaning system. As the costs of this system increase substantially with furnace size, hood collection is now used only on the smaller furnaces below about 10 tonnes, in which the use of injected oxygen is limited.
3. *Direct extraction from furnace*. This comprises an elbow-shaped off-take on the furnace roof through which the primary fume is extracted. As the extracted gases contain carbon monoxide and hydrogen, air for combustion is introduced, usually via a slice-gap in the off-take. The burned gas is then cooled and conditioned by water sprays before entering the cleaning unit. Careful pressure control is required to prevent fume emissions at furnace openings and conversely to avoid excessive air infiltration into the furnace.
4. *Semi-direct extraction from furnace*. This is essentially a compromise between the roof hood and direct extraction systems. The main extraction is by hoods over one or more short chimneys in the furnace roof, with subsidiary hoods over the doors and pouring spout; all these collection points are connected to a main duct leading to the cleaning unit. The absence of direct suction on the furnace tends to eliminate air infiltration to the furnace and to reduce fume emissions from the electrode openings. The gases burn in the space between furnace and vent-hole hoods, and the volume of waste gases presented to the cleaning unit may be 2–2½ times greater than that of a comparable direct extraction system.
5. *Direct extraction and building roof-mounted canopy hoods*. This has emerged as the preferred solution of the total fume problem of large, oxygen-blown electric

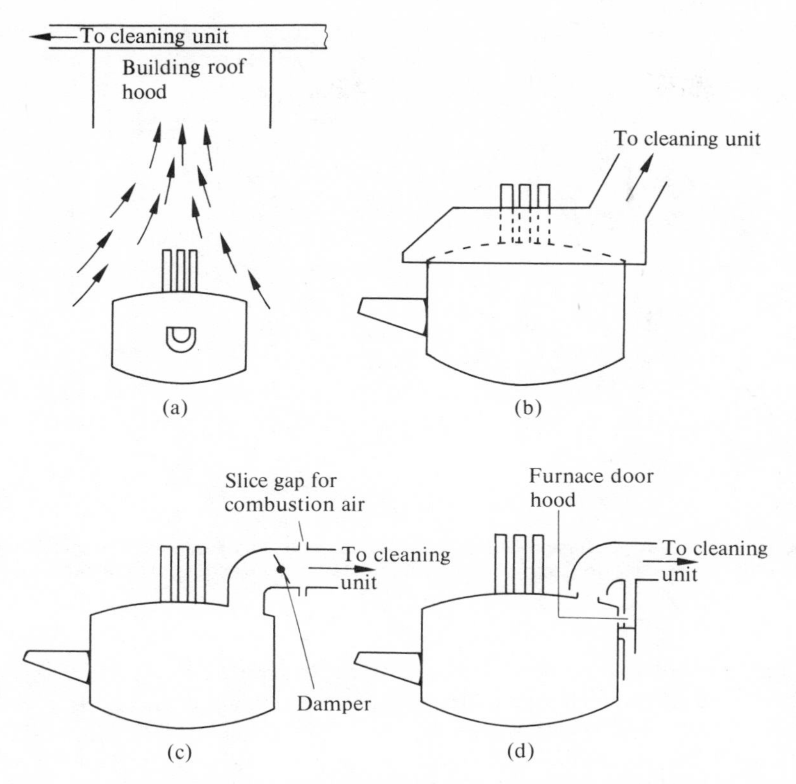

Fig. 14.8. Fume extraction systems for large electric arc furnaces: (a) building roof hoods; (b) furnace roof-mounted hoods; (c) direct extraction; (d) semi-direct extraction. (*Courtesy* BSC Special Steels Division.)

arc furnaces. With existing furnaces already fitted with direct extraction, the canopy hood system is usually a separate supplementary installation; with new plants, however, the two systems can be initially planned as a combined system leading to a common waste gas cleaning system, usually a bag filter unit.

Five BSC large electric arc furnace shops, already equipped with direct extraction systems,[34] have been or are being equipped with the supplementary building canopy hood system, and one new shop of two furnaces[35] with a combined furnace and building canopy hood system. The design of the building canopy hoods depends largely on building configuration and rate of operation. Usually two hoods, situated respectively over the furnace and casting bays, are prepared by isolating the appropriate sections of the building roof with vertical sheeting. To eliminate effects of cross draughts on the rising plume, the vertical sheeting should be extended as low as space permits, and care taken to ensure that all openings such as doors, windows, etc, in the building are kept closed as much as possible.

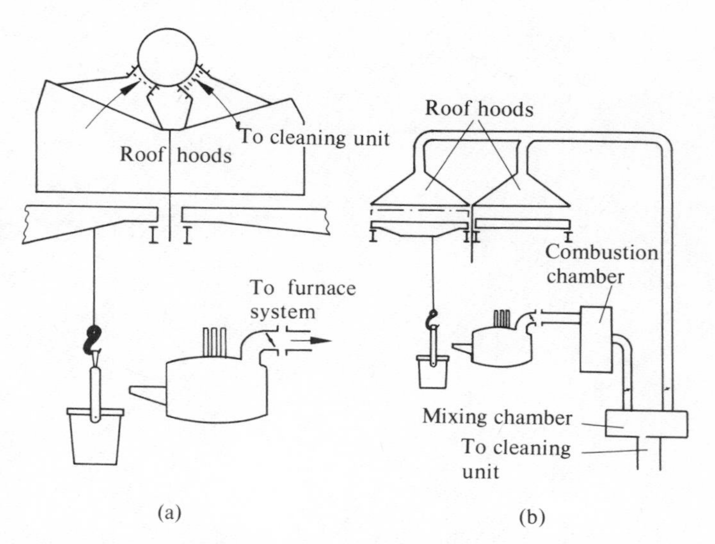

Fig. 14.9. Electric arc furnace fume collection: (a) separate building roof and direct furnace extraction system; (b) combined building roof and direct furnace extraction system. (*Courtesy* BSC Special Steels Division.)

6. *Direct extraction and movable hoods or air screens.* In view of the large volumes to be handled and consequent high costs of the building roof canopy system, other ideas for the control of secondary fume are being studied. Thus Marchand[36] discusses the use of movable hoods for both the charging and tapping operations and of a horizontal air screen for the tapping ladle.

Selection of gas cleaning system

1. Main furnace extraction systems. For similar furnace operating rates, direct-furnace extraction gives the lowest waste gas volume, but cooling, usually with water sprays, is required. Successively higher waste gas volumes arise from semi-direct extraction, furnace roof hoods and building roof hoods; with the two latter methods the dilution air often provides adequate cooling. For wet cleaning methods, the use of small spray towers ensures a fully saturated gas and hence reduces the tendency for the deposits to stick to the metal parts of the cleaning unit. For dry dust collection units, the gases are usually spray-cooled to just above the condensation temperature.

The selection of the cleaning system involves many factors, and it is possible here only to give broad indications. For oxygen refining practices, high efficiency cleaning units are required, such as wet or dry electrostatic precipitators, high efficiency wet washers and high temperature fabric filters. When fluorspar additions are used, wet methods are preferred in order to suppress emissions of fluorine compounds. In the absence of fluorspar additions, dry cleaning methods may be used. Direct extraction

Fig. 14.10. Electric arc furnace steelmaking. Large electric arc furnace equipped with combined building roof and direct furnace extraction system. View shows also building roof hood above the furnace. (*Courtesy* American Air Filter (Great Britain) Ltd.)

may be preferred with large furnaces using the one-slag process, but when infiltration of air into the furnace has to be avoided, semi-direct extraction should be used. With the smaller furnaces, many of which are used for making special steels, there is a tendency towards high-efficiency wet washers for furnaces in the 30–50 tonnes range and towards fabric filters for the smaller sizes.

2. Secondary fume. The most suitable cleaning unit for secondary fume systems seems to be the bag filter on account of its high efficiency and simplicity, as well as the dryness and low temperature of the gas stream.

Fumeless refining

Accelerated oxygen-refining methods have been developed, by which the formation of the copious red fume is substantially suppressed. One such method,[37] developed by the British Iron and Steel Research Association, is the injection of an oxygen-fuel (i.e., oil or carbon monoxide–hydrogen gas) mixture into the metal. This has been used successfully for carbon steels, but resulting metallurgical difficulties with high chromium alloy steels have limited its commercial application.

A similar method has been developed subsequently by Steel Casting Research and Technical Association.[38] This is based on the injection of iron oxides, or other appropriate metallic oxides, entrained in a high-pressure air stream into the metal. It has been applied successfully to both small and large electric arc and open hearth furnaces, but it is likely to find its main use on the smaller arc furnaces making alloy steels where very high refining rates are not imperative. Although red fume emissions may be substantially reduced, excessive grit and dust emission may necessitate the installation of an arrestment plant.

Re-use of iron oxide dusts

The amount of dust and fume collected from oxygen injection steelmaking processes is about 1–1½ per cent by weight of the metal charged. These dusts have high contents of iron oxides (75–85 per cent) with smaller amounts of slaggy matter and refractory oxides, and in some cases oxides of lead, zinc and other non-ferrous metals depending on the nature of the metallic charge to the furnace.

Re-use of these iron oxide dusts, particularly those from large arc or open-hearth furnaces which consume substantial amounts of miscellaneous steel scrap, is limited because of their relatively high contents of oxides of deleterious tramp elements such as zinc and lead. On the other hand, dumping of such wastes may present other problems such as pollution of adjacent land and waterways. Processes for the removal of these constituents are, therefore, attracting much research attention.[39,40]

In certain steelmaking plants where the amount and quality of steel scrap used were strictly controlled in order to limit the content of non-ferrous metals in the steel product, recycling of the iron oxide dusts was practised. Thus such dusts after being agglomerated have been incorporated in sinter plant charges[41] or were returned as feed ore to the open hearth process.[42] In another instance, the iron oxide was pumped as a slurry to the sinter plants.[43] The use of high-pressure-tube filters to transform iron oxide slurries for re-use in sinter plants has also been studied.[44]

Rolling mills

Scarfing or de-seaming

Intermediate products such as ingots, slabs, blooms and billets often contain surface defects, which are best removed by burning with an oxy-flame. This operation, which is known as scarfing or de-seaming, is intermittent; it produces considerable sub-micron red iron oxide fume entrained in large volumes of low temperature air. Both fume loadings and volumes of gas can vary widely, and volumes of 600–4000 $m^3\ min^{-1}$ and fume loadings of 0·25–2·5 $g\ m^{-3}$ (NTP) are normal for a de-seaming operation of 0·5–1·0 min. In the scarfing operation, scale and slag particles are flushed away by high-pressure water, and consequently large amounts of water vapour are entrained in the exhaust gases. As the waste gases are therefore substantially saturated, a wet method of fume arrestment is normally used. In the UK, wet electrostatic precipitators have been used mainly, but high energy scrubbers would also be satisfactory.

Reheating and heat-treatment furnaces

Intermediate steel products, ingots, slabs, etc., are normally reheated for further processing. The type of reheating furnace depends on the size and nature of the intermediate product and the subsequent processing. Coal-fired furnaces are now comparatively rare, and are no longer significant sources of air pollution. Reheating furnaces are usually fired by low sulphur gas—blast furnace gas, desulphurized coke-oven gas or natural gas—or by oil, and the resultant particulate and sulphur gas emissions are thus related to the nature of the fuel. Automatic combustion controls are often fitted so that conditions can be checked to ensure satisfactory combustion.

References

1. Scholey, R., *The present situation regarding pre-reduced iron and cokemaking technology,* International Iron and Steel Institute, Sixth Annual Conference, 8–12 October 1972.

2. Carter, G. C. and G. Punch, 'Gas cleaning in iron and steel industry, engineering aspects of pollution control in the metals industries', *The Metals Society Book No. 172,* p. 73, 1975.
3. Calvert, W. J., 'Dry electrostatic precipitators at Scunthorpe', BSC Midland Group Conference on Prevention of Atmospheric Pollution, Sheffield, 10 September 1969 (private communication).
4. Brooks, S. H. and W. J. Calvert, *External pollution from an iron and steelworks and measures towards its reduction*, ISI Special Report No. 61, p. 9, 1958.
5. Ireland, F. E., 'The determination of chimney heights in Britain', *106th Annual Report on Alkali, Etc. Works, 1969,* Appendix V, p. 56, HMSO, London, 1970.
6. 'Sintering Plant', *NKK News*, **15**(12), 3, 1975.
7. Punch, G., 'Elimination of fumes in iron and steel industry', *Steel International,* 9, July/August 1967.
8. English, A., 'Pelletizing at the Hanna Mining Co., *Iron and Steel Engineer,* 110–14, January 1968.
9. Battelle Memorial Institute, *Iron ores and ironmaking in the world,* Vol. 14, 31 July 1968.
10. Jennings, R. F., 'Blast-furnace gas cleaning. An analysis of performance', *J.I.S.I.*, 307, March, 1950.
11. The Japan Iron and Steel Federation, *Prevention of secondary pollution caused by dust produced in blast furnace operations in Japan*, Report to IISI Environment Committee Meeting, London, November 1974 (private communication).
12. Pitt, R. S., 'Direct reduction of iron ore', *Iron and Steel International,* 242–51, June and 347–53, August, 1973.
13. Willars, H. M. and R. C. Madden, 'The development of a new steelmaking process utilising highly metallised sponge iron, *Iron and Steel International*, 313–21, August 1975.
14. Shaw, F. M., 'Iron foundries and the requirements of clean air legislation, *Iron and Steel,* 183–84, June 1971.
15. *Cold blast cupolas, emissions from cold blast cupolas at iron foundries*, Appendix 1, HMSO, London, 1968.
16. Ireland, F. E., 'Notes on best practicable means for hot blast cupolas', *110th Annual Report on Alkali, Etc. Works, 1973,* Appendix VI, p. 73. HMSO, London, 1974.
17. Shaw, F. M., 'Iron foundries and the requirements of clean air legislation,' *Iron and Steel*, 186–7, June, 1971.
18. 'Bottom-blown open hearths?', *33 Magazine*, 30–31, August 1972.
19. 'Bottom-blown steel processes now number three; Q-BOP, LWS and SIP', *33 Magazine*, 35–8, September 1972.
20. *Problems of air and water pollution arising in the iron and steel industry,* UN Publication E70 11E6, p. 38, New York, 1970.
21. Watkins, E. R. and K. Darby, *The application of electrostatic precipitation to the control of fume in the steel industry,* ISI Special Report 83, p. 34, 1964.
22. Jackson, A., *Fume cleaning in Ajax furnaces,* ISI Special Report 83, p. 61, 1964.

23. Rowe, A. O., H. K. Jaworski and B. A. Bassett, 'Waste gas cleaning systems for large capacity basic oxygen furnace plant. *Iron and steel exposition,* Cleveland, Ohio, p. 20, June 1968.
24. Morita, S., *Operation and economy of the oxygen converter gas recovery process (OG process),* ISI Special Report No. 83, p. 109, 1964.
25. Namy, G., J. Dumont-Fillon and P. A. Young, *Gas recovery without combustion from oxygen converters; the IRSID–CaFL pressure regulation process,* ISI Special Report No. 83, p. 98, 1964.
26. Hoff, H. and J. Maatsch, *Converter waste gas cleaning by the 'minimum gas' method at Fried Krupp,* ISI Special Report No. 83, p. 104, 1964.
27. Boyce, H. P., *The OG gas cleaning system,* ISI Special Report, Operation of Large BOFs, p. 1, London, 1972.
28. Urban, G. and F. Fillies, *The Krupp system,* ISI Special Report, Operation of Large BOFs, pp. 20–1, London, 1972.
29. Pilkington, S., 'Collection of secondary fume in BOF steelmaking. Engineering aspects of pollution control in the metals industries', *The Metals Society Book No. 172,* p. 25, London, 1975.
30. Rowe, A. D. and S. C. Desai, Bottom-blown Q-BOP furnaces see a major revival for tonnage production (Private Communication from Davy Ashmore International Ltd.).
31. Pearce, J., 'Q-BOP steelmaking developments', *Iron and Steel Engineer,* 29–38, February 1975.
32. Flux, J. H., 'The control of fume from electric arc steelmaking', *Iron and Steel International,* 185, June 1974.
33. Venturinl, J. L., 'Historical review of the air pollution control installation at Bethlehem Steel Corporation's Los Angeles plant', *Air Pollution Control Association Annual Meeting,* St Paul, Minnesota, June 21–27, 1968.
34. Flux, J. H., 'Containment of melting shop roof emissions in electric arc furnace practice. Engineering aspects of pollution control in the metals industries', *The Metals Society Book No. 172,* p. 32, London, 1975.
35. Ref. 32, pp. 187–8.
36. Marchand, D., 'Dust and gas evaluation in arc furnace steelmaking and possible alternatives for reducing emission with suitable collection and cleaning systems. Engineering aspects of pollution control in the metals industries', *The Metals Society Book No. 172,* p. 47, London, 1975.
37. Spenceley, G. D. and D. I. T. Williams, 'Fumeless refining with oxy-fuel burners', *Steel Times,* 150–8, July 29, 1966.
38. 'Fumeless refining by oxide injection', *The British Steelmaker,* 14–21, December 1970.
39. Wetzel, R. and G. Mayer, *Processing of steelworks dust and slurry, operation of large BOFs,* ISI Special Report, p. 44, 1972.
40. Ref. 2, p. 81.
41. Boyce, H. P., *Dry electrostatic precipitators on open-hearth furnaces,* ISI Special Report 83, p. 52, 1964.

42. Jackson, A., *Fume cleaning in Ajax furnaces,* ISI Special Report 83, p. 62, 1964.
43. Wass, E. A., *Rotor-plant slurry reuse,* ISI Publication No. 128, pp. 120–3, London, 1970.
44. Pickin, G. A., 'Recovery and reuse of BOS fume in the sinter plant. Engineering aspects of pollution control in the metals industries', *The Metals Society Book No. 172*, p. 65, London, 1975.

15 Production of non-ferrous metals

A. K. BARBOUR, J. F. CASTLE and S. E. WOODS

Scope

This chapter identifies the main air pollution problems which accompany the smelting and refining of non-ferrous metals, and the technology available and under development to minimize these problems. Both primary processing from mineral concentrates and secondary processing from scrap and residual materials are covered, but the air-pollution problems directly associated with mining and concentrating operations are excluded, as are the medical aspects of industrial hygiene.

Organization

In terms of production volume (Table 15.1)[1] the major non-ferrous metals are aluminium, copper, zinc, lead and nickel. Aluminium occurs in its ores as the oxide and the winning of the metal from alumina involves, uniquely among the major metals, electrolytic reduction from a molten fluoride bath. The main chemical pollutants of concern in aluminium production are fluorides. The other metals occur as sulphides and the feed to the metal winning plant is typically a sulphide concentrate, the treatment of which involves oxidation to yield sulphur dioxide which is the main chemical pollutant to be controlled. The following approximate quantities of sulphur must be eliminated per tonne of metal produced from concentrates: lead 0·2–0·3 tonnes; zinc 0·5–0·6 tonne; copper 1·0–1·5 tonnes: nickel 3·0–4·0 tonnes. A sulphur-free source of nickel that is becoming increasingly important is the treatment of the oxide/silicate nickel laterite ores. Since all processing techniques involve the handling of large tonnages of solid feeds and, usually, large tonnages of molten metals and slags, the control of particulate emissions, both dusts and fumes, is an important problem common to all current techniques.

The proportions of the above metals produced by secondary processing (see Table 15.1) varies considerably, due to a combination of both economic and technological factors. Conservation of mineral resources will undoubtedly enforce recovery of an increased proportion of total metal production by secondary processing in the future.

TABLE 15.1

World non-ferrous metals production: approximate data

Metal	Production from mined ore (Millions of tonnes per year) (1975)	Percentage of Total Metal product obtained from secondary sources
Aluminium	12·7	22
Copper	7·3	38
Zinc	5·5	20
Lead	3·6	37
Nickel	0·7	20

Source: Metalgeselschaft, AC Metal Statistics, 1965–1975

The control of particulate emissions to the out-of-plant atmosphere is effected by the use, either alone or in combination, of conventional trapping equipment, such as venturi scrubbers, wet and/or dry electrostatic precipitators and bag-filter plants prior to exhausting the purified gases through stacks which are sufficiently tall to achieve acceptable ground-level concentrations by dispersion. Typical installations are exemplified in the appropriate sections of this chapter.

Adequate control of sulphur dioxide emissions poses more fundamental problems, particularly in the copper smelting industry, where much obsolescent reverberatory smelting plant remains operational. In the primary zinc and, to some extent, lead smelting industries, economic pressures have already compelled the replacement of the older processes by modern electrolytic and blast furnace processes which feature the production of a stream of sulphur dioxide of reasonably constant strength and sufficiently rich for ready conversion to sulphuric acid. As is detailed in the section on copper, it will be increasingly uneconomic for reverberatory smelting and batch converting of copper concentrates to meet current US sulphur dioxide emission regulations, and this will undoubtedly accelerate the replacement of such processes by flash smelting, electric smelting and the newer continuous processes described in that section.

Aluminium

Primary aluminium is produced entirely by the Hall–Héroult process, discovered in 1886.[2] This involves reduction by carbon anodes, of alumina (Al_2O_3) dissolved (at about 970°C) in a melt of sodium fluoride and aluminium fluoride approximately the composition of cryolite (Na_3AlF_6). The lining of the cells is also carbon and a pool of liquid aluminium contained in it forms the cathode. The cell reaction is

$$3C + 2Al_2O_3 = 4Al + 3CO_2.$$

Some CO_2 is reduced in the melt so that a gas containing carbon dioxide and carbon monoxide leaves the cell. The electrolytic cells (pots) are continuously operated. Three types of anode are used; horizontal spike Soderberg (HSS), vertical spike Soderberg (VSS) and prebaked (PB) (Fig. 15.1). Soderberg anodes are baked in the cell during operation and are completely consumed; prebaked anodes are periodically replaced after partial consumption. Most modern plants have been built with PB or VSS anodes. Plant sizes vary from 10 000 to 400 000 tonnes per year, a typical 50 000 tonne per year plant having about 220 pots of 100 kA each. Pots are connected in series to form a pot line; the plant may contain several pot lines.

Production of 1 tonne of aluminium consumes 1·95 tonnes alumina, 0·5 tonne carbon anodes, 35 kg fluorine in make-up salts, and up to 17 000 kWh of electric power (14 000–15 000 kWh on modern plants). Numerous attempts to replace the Hall process with alternative routes either with a lower demand for electrical energy input, or capable of treating lower grade raw materials, or involving less objectionable gas borne emissions, have been made, but none has yet been brought to large-scale operation. Electrolysis of aluminium chloride from a melt offers perhaps the best prospect of a replacement technology[3] (a semi-commercial plant is at present (1977) being commissioned, and would have the potential of reducing power consumption by as much as one-third and, by replacing fluorides by chlorides, of easing the problems of air pollution. However, the predominant problem at present concerns the design and operation of the molten fluoride electrolysis process, to minimize the dispersion of gas borne fluorides.[4]

Air pollution is associated with emission from the cells, in the issuing anode gases, of gaseous and particulate fluorine compounds. Fluorine compounds contributing to gas borne emission are also obtained in ancillary processes, notably the treatment of anode butts (in PB plants), the smelting of refuse from pot room and foundry to recover aluminium, and the burning of dusts and froth from the pots to recover cryolite. Pollutants other than fluorine compounds include tar aerosols and sulphur dioxide, and these are significant when considering the design of gas cleaning equipment. Air pollution effects may be reduced (and on modern plants avoided) by a combination of careful cell operation, gas cleaning, limiting the size of plant, favourable site location and use of tall stacks for dispersal. The various sources of emission in the overall process are indicated in the block diagram, Fig. 15.2, and their control is discussed in the following sections.

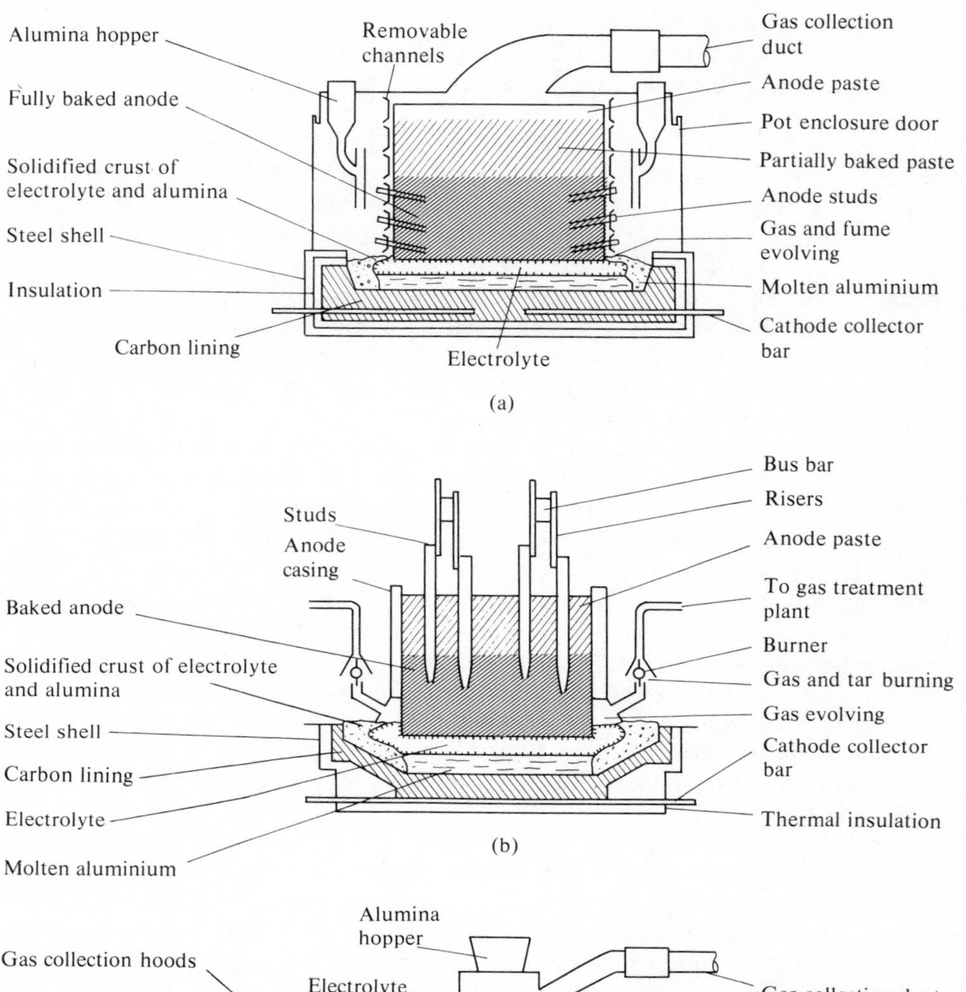

Fig. 15.1. Anodes used in Hall-Héroult process: (a) horizontal Soderberg; (b) vertical Soderberg; (c) prebaked.

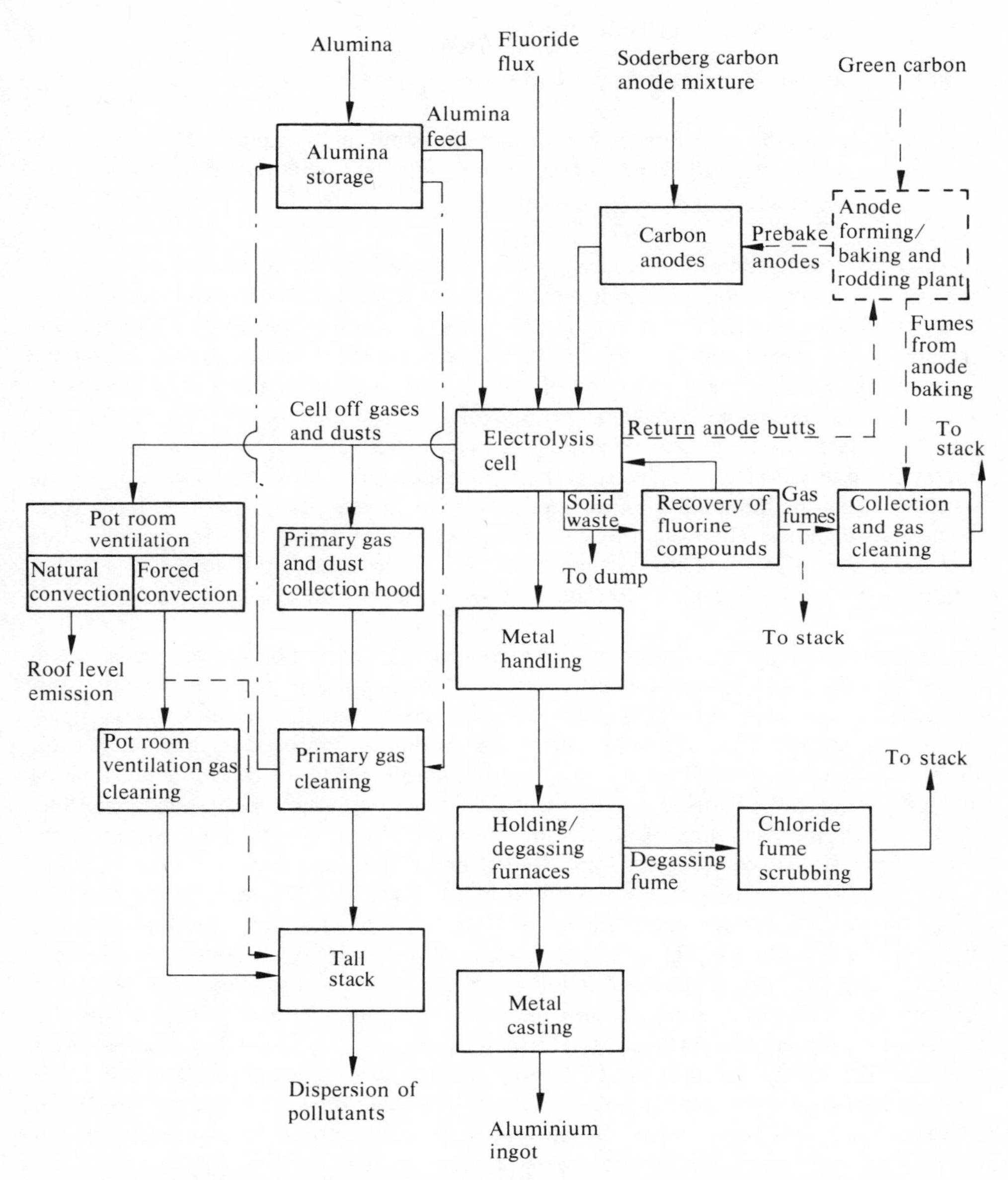

Fig. 15.2. Aluminium smelting: emission sources: --- alternative routes and processes; —·—·— route of alumina used for gas cleaning.

Pot gas collection and pot room ventilation

Fluoride emissions from the cell arise from

1. the small but significant vapour pressure of fluorides in the melt at working temperature;
2. passage into the cell of hydrogen compounds (water in the alumina feed, or hydrocarbons from the carbon anode) which are converted to hydrogen fluoride;
3. entrainment of liquid droplets and solid dust.

Associated with low power efficiency of the cell, a large amount of heat has to be dissipated into ventilation air. If this air is to be kept down to an acceptable temperature level, the volume requirement is very large—some 2 million m^3 per tonne of aluminium produced—compared with an evolution from the electrodes of about 2000 m^3 $tonne^{-1}$. There is thus a *primary* gas collection of cell gas diluted a limited number of times with air, and a *secondary* gas, much larger in volume, in which both the heat and the emissions escaping primary collection are removed. In many pyrometallurgical operations one can deal separately with the question of process gas stream and with ventilation, but in aluminium production the two are so interrelated that they have to be considered together. The design of cell influences the efficiency of primary collection and scrubbing. The emission escaping primary collection then fixes what sort of treatment is needed for dealing with the large secondary volume.

1. Gas emissions from cell, and primary collection.[5] The important factors in plant operation affecting fluoride emission are melt temperature and composition, the methods used in crust breaking and in feeding cells, and the amount of hydrogen entering the system. There is a substantial range of fluoride emission on different plants. Prebake pots may range from 12 to 33 kg of fluorine per tonne of aluminium: about one-half as hydrogen fluoride, one-half as solids. Soderberg electrodes range from 18 to 27 kg of fluorine per tonne of aluminium; here the effect of hydrogen predominates, hydrogen fluoride accounting for some 90 per cent of total emission.

A variety of types of primary collection hood is used. HSS pots require individual hoods over each cell, with large volumes of air to remove tar aerosols from the working atmosphere. Operation of the VSS system does not permit complete coverage by hoods; present designs have a gas collection skirt surrounding the anode, in which a concentrated cell gas is collected and burned with air in a burner. Prebake pots are sometimes designed without a separate primary gas collection (this type is termed open prebake-OP). On closed prebake (CP) pots the primary collection system can range from crude local collection at anodes to complete enclosure of the pot. The design of the hooding has to be considered in relation to the operational sequence in which opening of the hoods may be necessary during feeding of alumina, crust breaking and metal tapping. Average collection efficiency depends upon the proportion of hood-open time and upon a number of operating features. On modern plants, use of automatic alumina feeding, crust breaking and aluminium tapping can much reduce the

TABLE 15.2

Pot room gas collection and ventilation[6]

Type of pots	VSS	HSS	CP	OP
Pot exhaust rate (10^3 Nm^3 per tonne A1)	18	275	150	—
Temperature °C	120	50	100	
Collection efficiency on Pot exhaust (per cent)	75	85	90	
Pot room ventilation rate (10^6 Nm^3 per tonne A1)	(1·8) 2·2	2·2	1·2 (2·0)	(1·8) (2·2)

Notes Figures in brackets refer to conditions when the air is scrubbed. Some CP installations report collection efficiencies of >95%.

hood-open time. Typical efficiencies for various types of plant are given in Table 15.2. Individual plants can range considerably about the typical values cited.

2. Pot room ventilation. Gases which escape the primary collection systems, or those from open pots, are removed in the pot room ventilation air, which generally leaves the pot room at 15–30°C above ambient temperature. Typical ventilation rates are given in Table 15.2; note that the units for ventilation are millions of cubic metres per tonne of aluminium, those for pot exhaust rate are thousands of cubic metres per tonne. Ventilation requirements are higher in plants with open pots than in plants with efficient primary collection. Ventilation can be achieved by natural or forced convection. Forced convection using fans provides more stable ventilation with change in outside conditions and is essential when ventilation gas is to be scrubbed or discharged from a tall stack. If the gas is to be scrubbed, the volume should be kept to a minimum consistent with acceptable working conditions. The Threshold Limit Value (TLV) for fluoride[7] (measured as fluorine) adopted in the UK and in the US is 2·5 mg m^{-3}.

Primary and secondary gas cleaning

The volumes of gas to be cleaned are very large and require the largest gas cleaning equipment used in the non-ferrous industry. A 50 000 tonnes per year aluminium smelter based on VSS anodes would need $\simeq$100 000 m^3 h^{-1} of pot primary exhaust and 10 000 000 m^3 h^{-1} of secondary ventilation air. Modern CP plants of the same capacity operating at 90 per cent primary collection efficiency would need to clean 1 200 000 m^3 h^{-1} of primary exhaust. Pot room ventilation air is not at present cleaned on CP installations, though there are indications that in particularly fluoride sensitive areas with stringent requirements, air cleaning may be necessary. The decision whether secondary cleaning is necessary depends upon the relation between the total emission escaping after primary gas cleaning together with the emissions in pot room ventilation, and the permitted total mass emission from the plant. The size of the plant enters into the relationship, and as plant size increases it becomes more difficult for primary gas cleaning alone to match some of the more stringent standards. Thus if a plant with capacity of 150 000 tonnes per year of aluminium has to emit less than 700 kg per day of fluorine in fluoride, ventilation air scrubbing would have to be practised even with a CP primary collection efficiency of 95 per cent.

Gas cleaning can be carried out by wet or dry methods. Hydrogen fluoride is highly soluble in water and can be chemisorbed on alumina. In order to deal with the combination of solid and gaseous emissions normally encountered, the cleaning train may contain several types of plant in series. Wet gas cleaning in packed towers, spray towers, and various types of wet gas scrubber are used for hydrogen fluoride. Solids are removed with electrostatic precipitators, bag filters and wet scrubbers. The first element in the train is often a cyclone to remove coarse dust. Reliability of operation and back up capacity to run when units have to be taken out of line for maintenance are important in practice.

Dry collection methods are of increasing importance, since they permit collection of particulates and hydrogen fluoride (provided tar aerosols are not present), in a re-usable form and without setting in train the complex of water effluent problems associated with wet scrubbing. Use is made of the reaction of primary gas with alumina, whereby hydrogen fluoride is tenaciously absorbed. An early system using filter bags coated with high surface area alumina achieved limited acceptance. Present day systems are based on reaction in a fluidized bed (Fig. 15.3)[8] or in a cloud of suspended particles; the reacted gas passes to the bag filter. Metal grade sandy alumina can be used; its adsorptive capacity depends upon degree of calcination as indicated by specific surface area (around 50 $m^2 g^{-1}$) or alpha alumina content (20–40 per cent). Recovery of 97–99 per cent of the total fluoride in the gas stream treated can be achieved.

This type of system does not collect sulphur dioxide present in the gas, but this is advantageous, since recycling of sulphur to the cells would be undesirable. The fluoride adsorbed into alumina is fed back to the cells; its value helps to offset the operating costs, or can even yield a credit with certain electrode systems. A consequence of operating the system is that the alumina feed to cells is enriched in fluoride content so

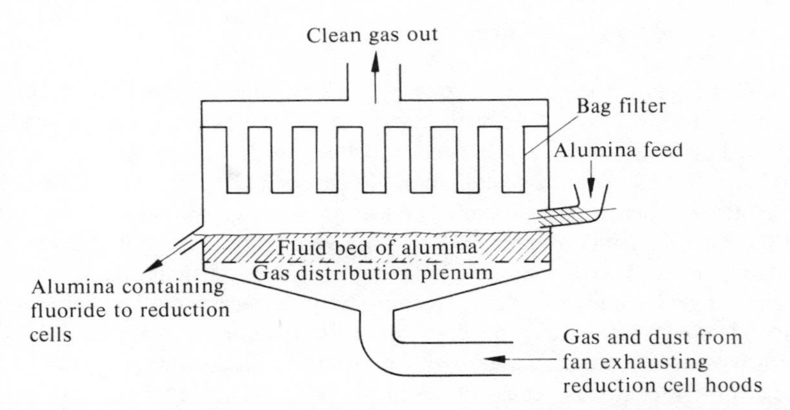

Fig. 15.3. Schematic layout of Alcoa 398 process reactor. (*Courtesy* Aluminum Company of America.)

that the plant systems for handling alumina feed must be made dust tight to reduce fluoride emission into the working atmosphere. This type of system is applied to PB or to VSS cells; if sulphur dioxide emission has to be controlled, the gas after dry hydrogen fluoride removal may pass to wet scrubbing. Some typical efficiencies on different types of gas cleaning equipment, as applied to different types of pot room, are shown in Table 15.3.

TABLE 15.3

Cleaning efficiencies on pot primary ventilation gas[6]

Type of pot	Category of gas cleaning equipment	Cleaning efficiency for fluoride (per cent)		
		HF	Solid F	Total F
VSS	1. Dry electrostatic precipitator + spray tower	99	99	99
	2. Venturi scrubber	98	92	97
	3. Alumina + bag filter	98	97	98
HSS†	1. Wet electrostatic precipitator + spray tower	98	98	94
	2. Floating ball scrubber	95	65	92
CP	1. Dry electrostatic precipitator + spray tower	96	(‡)	
	2. Floating ball scrubber	96	64	80
	3. Alumina + bag filter	97	96	97

† These data refer to old installations, modern equipment will probably operate at high efficiencies.

‡ There appears to be little experience of electrostatic precipitators with CP pots and claims for cleaning efficiency for solid fluoride range from 85 to 98 per cent.

Pot room ventilation (secondary) gas volumes are so large that efficient cleaning is prohibitively difficult. This applies particularly to the dust burden, which is notably harmful to vegetation. Actual scrubbing of secondary gas is confined to low back pressure wet scrubbers, usually located in the pot room roof. Typical efficiencies are given in Table 15.4. The capital cost of scrubbing plant can amount to 6–12 per cent of that of the smelter, and operating costs to 2–5 per cent of the aluminium production cost. The choice of the most economic scrubbing process to meet local environmental requirements is thus of major significance.

TABLE 15.4

Typical scrubbing efficiencies for pot room ventilation air[6]

Type of pots	Category of scrubber	Cleaning efficiency for fluoride (per cent)		
		HF	Solid F	Total F
VSS	1. Spray screen	93	45	88
	2. Packed tower	93	45	88
OP	Spray screen	95	45	70

With good primary collection and good gas cleaning, the emission, on prebake installations, can be as low as 1 kg total fluorine per tonne of aluminium produced. Vertical Soderberg installations treating both primary and secondary gas and using wet scrubbing report emissions as follows per tonne of aluminium metal produced: 1·0 kg total fluorine, 2·0 kg sulphur as sulphur dioxide, 5·0 kg dust and 2·0 kg tar.

Ancillary processes

Ventilation of the anode-making plant requires careful design and operation to maintain toxic pitch volatiles at an acceptable level. This is also an important problem in the pot rooms of Soderberg plant. External emmission from ancillary processes is much smaller than that from the pot room. On some plants the gases are passed to the reduction plant cleaning equipment. Anode baking plants recycling anode butts report emissions of 0·5 kg of fluorine per tonne of aluminium metal without gas cleaning; wet scrubbing reduces the figure to about 0·01 kg. Suitable wet scrubbing of gases from smelting skimmings, froth and dust yields emissions of 0·01–0·015 kg of fluorine per tonne of aluminium.

Chlorine is often used for degassing aluminium prior to casting.[9] Gas is injected through carbon lances. The resulting fumes contain aluminium chloride, which hydrolyses to alumina and hydrogen chloride. Venturi scrubbers are often used on the ventilation gas. Chloride emission is controlled by the scrubbing, but the corrosive gas results in high maintenance requirements.

Copper

Most of the world's copper is mined from large low grade disseminated sulphide deposits which occur with certain types of acid and intermediate igneous intrusive rocks. The main copper minerals are chalcopyrite and secondary chalcocite. Mineable ore grades vary from around 0·4 to + 1·0 per cent. A second major source is found in stratiform sulphide orebodies, grades varying from around 1 to greater than 3 per cent, the main copper minerals being chalcopyrite and bornite. By physical processes of crushing, grinding and flotation, a concentrate is obtained in which copper, iron and sulphur are present in similar concentration (approximately 20–40 per cent) and with usually significant amounts of precious metals (silver, gold).

The oxidation of the concentrate involves successively:[10]

1. production of a molten matte, a liquid composed of copper sulphide and iron sulphide, by loss of pyritic sulphur which is oxidized;
2. oxidation of iron sulphide preferentially with loss of iron oxide from the liquid sulphide to a coexisting molten silicate phase, to yield a sulphide approaching the composition copper sulphide.
3. oxidation of the sulphur in molten copper sulphide (white metal) to yield liquid copper.

Oxidation in the solid state in, for example, a fluidized bed roaster with controlled heat removal can yield, depending on the mineralization and the temperature range, a variety of products. A partially roasted product, essentially with pyritic sulphur removed, may be the feed to reverberatory or electric smelting; a partially sulphated product may be used as feed to hydrometallurgical treatment; a dead roasted product containing a substantial content of copper ferrite is not amenable to leaching, but can be smelted with carbonaceous fuel (e.g., Brixlegg electric furnace process).

The copper produced by the complete oxidation process is refined, usually by electrolysis in an aqueous bath. The unrefined metal anode yields refined copper cathode and an anode sludge concentrated in precious metals.

The pyro-processes by which over 90 per cent of the world's copper is currently obtained, and some modifications already developed, have as their aim the best simultaneous achievement of the following criteria:

1. Low loss of copper in reject slag. This requires exposing the slag to reducing conditions, preferably by contacting it with a molten sulphide relatively rich in iron sulphide.
2. Producing gas rich in sulphur dioxide (in the range of 5–14 per cent by volume) and reasonably constant in both concentration and volume rate.
3. Minimizing consumption of purchased fuel to reduce costs and to avoid unnecessary dilution of the off-gases for sulphuric acid manufacture.

Some processes produce a high copper level in slag from the primary matte smelting step and do not practise pyrometallurgical slag reduction as described in (1) above. Copper values are recovered from these slags by slow cooling, grinding and flotation to produce a copper concentrate for recycle.

The classical system, by which the major proportion of primary copper continues to be produced, involves carrying out the early steps of oxidation in a reverberatory furnace fired by solid, liquid or gaseous fuel and fed with copper concentrates and highly oxidized slag returned from the subsequent converting stage. The matte produced in reverberatory smelting is fed as liquid to the converters, which are refractory lined cylindrical vessels mounted with their axes horizontal. Air is blown through refractory tuyères placed along one side, and the vessel can be tilted for tapping to bring the tuyères above the matte level when not blowing. Diagrams of these units are shown in Figs. 15.4 and 15.5.

This system achieves objective (1), a low level of copper in reject slag, but finds difficulty in achieving (2) and (3). With regard to fuel economy, the reverberatory furnace is a large structure and the requirement of carbonaceous fuel is substantial (calorific value $1{\cdot}6 \times 10^6$ kcal per tonne of dry concentrate); on-site power generation from waste heat boilers is usually practised. With regard to sulphur dioxide utilization, converter off-gases can (by use of proper hooding) be produced at relatively rich sulphur dioxide tenor, but the volume and strength are subject to wide fluctuations due to the batch nature of the operation. Despite this, in some locations converter gases have proved suitable for conversion by feeding to a sulphuric acid

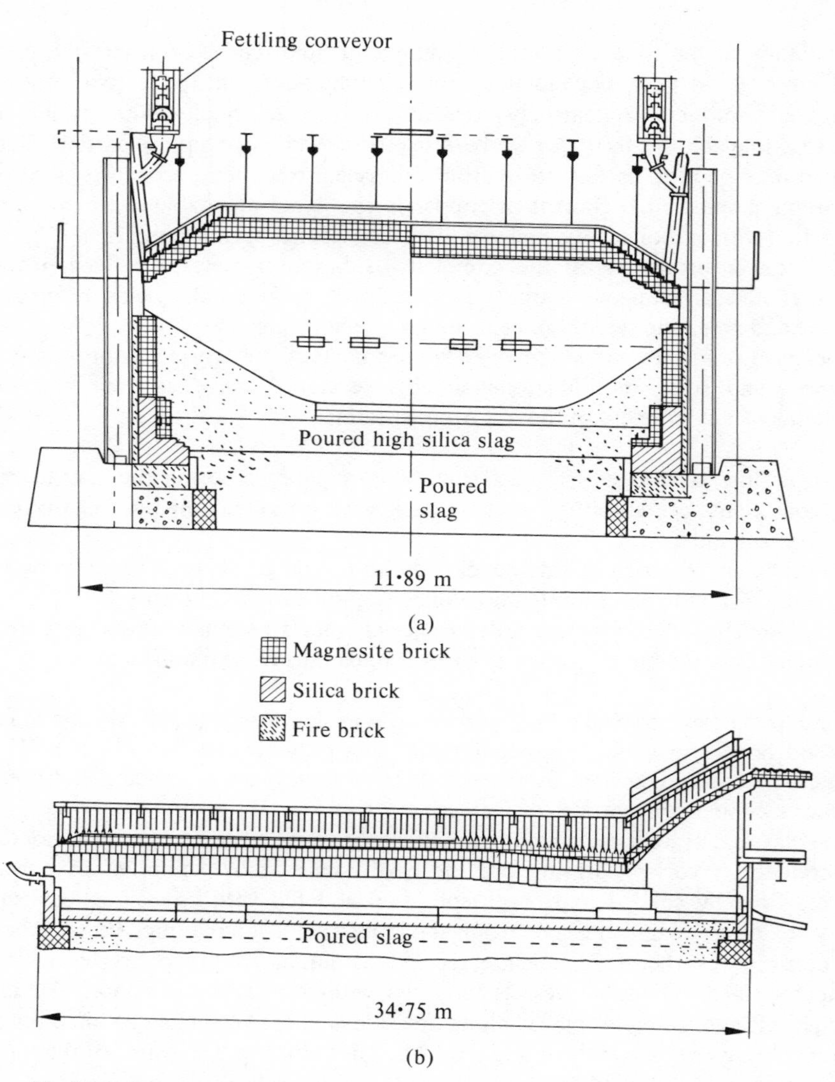

Fig. 15.4. Copper reverberatory furnace: (a) cross-section; (b) longitudinal section.

plant. Reverberatory gases are too dilute to sustain an acid plant, and only one smelter, as far as is known, attempts to collect these gases for acid production, and then in special circumstances. Thus, many reverberatory–converter smelting operations have been associated with dispersion to atmosphere of the whole sulphur dioxide output through tall stacks. Under most current and envisaged air pollution regulations, this type of technology will become increasingly uneconomic to operate. A typical reverberatory–converter flow diagram is depicted in Fig. 15.6.

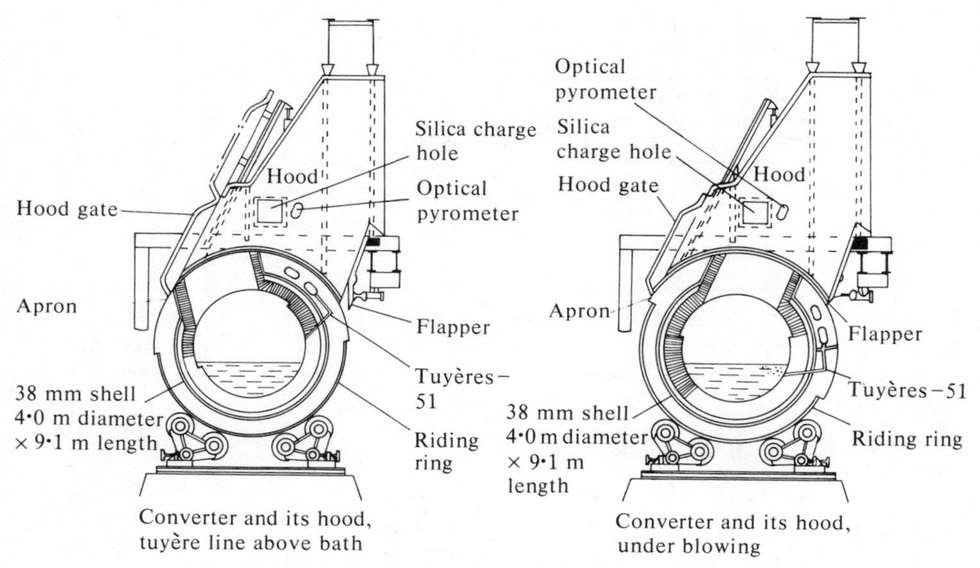

Fig. 15.5. Pierce Smith converter. (*Courtesy* Mitsubishi Metal Corporation.)

Modification of the pyrometallurgical route to copper has proceeded along the following lines:

1. Improved hooding and air volume control in converter operation, so that converter gas can now be fed to acid plants, particularly if combined with a high quality gas produced in the concentrate smelting stage. Methods of gas cooling other than by dilution (i.e., air-cooled high velocity flues or, where economic, waste heat boilers) are recent innovations at the converter stage to control gas volume and strength. The traditional Pierce-Smith type of converter is shown in Fig. 15.5[11] and newer Hoboken design in Fig. 15.7.[12] The latter relies on removing converter off-gases through a siphon flue which can be permanently connected to the main flue via a specially designed airtight swivel joint. The top-blown rotary converters (Inco)[13] as used in nickel production may find future application in copper smelting.
2. Minimizing the sulphur dioxide dispersed from the reverberatory operation by partially roasting sulphide concentrates in a separate fluid bed roaster, the roaster product being fed to the reverberatory furnace.
3. Replacing the reverberatory furnace by concentrate treatment in either flash smelting furnaces which yield strong sulphur dioxide gas (about 80 per cent Inco,[14] 10–18 per cent Outokumpo[15]) or electric smelting furnaces[16] (with dry green feed 4–6 per cent sulphur dioxide). Flash furnaces of the Inco and Outokumpu designs are shown in Figs. 15.8 and 15.9, and an electric furnace in Fig. 15.10.

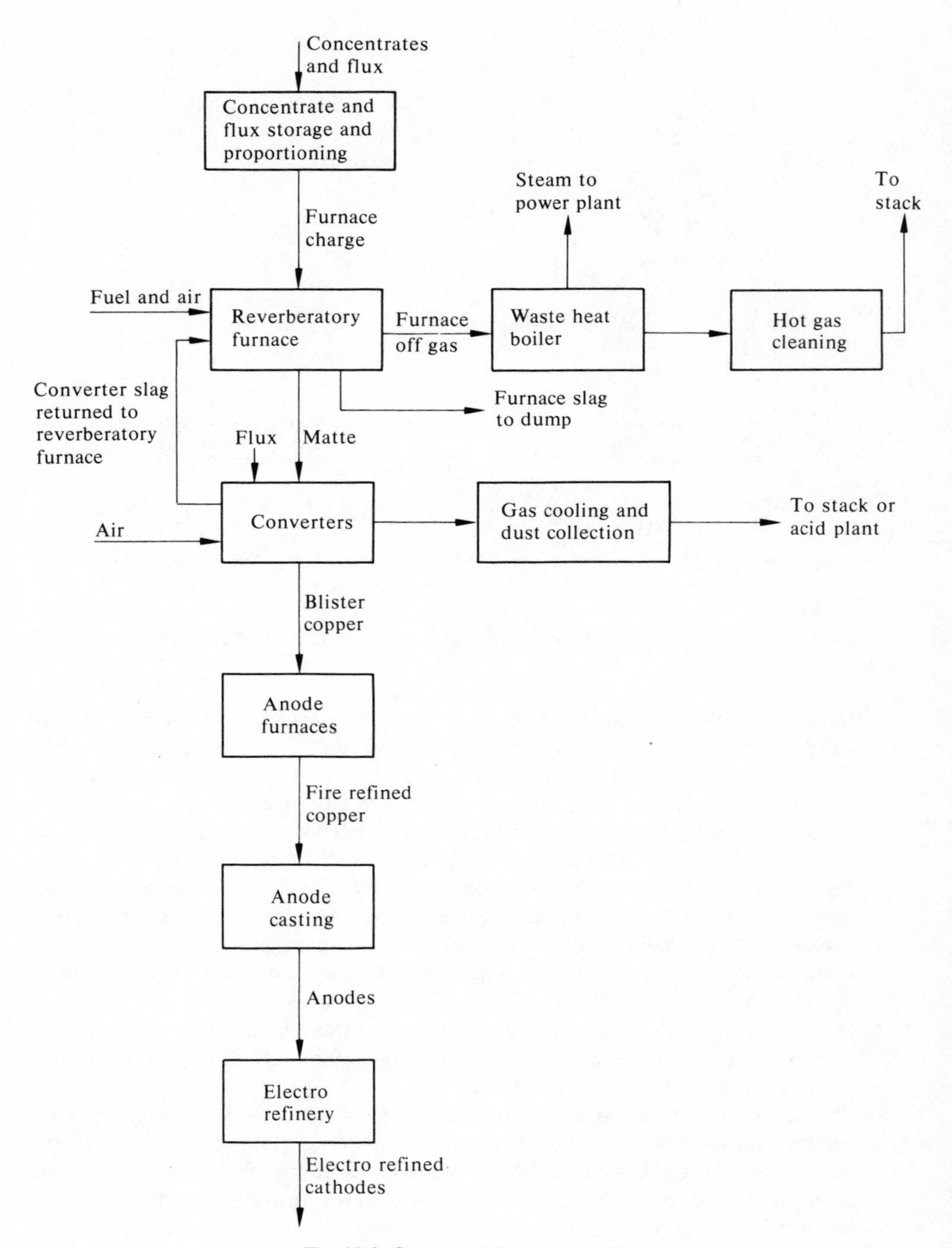

Fig. 15.6. Copper reverberatory smelting.

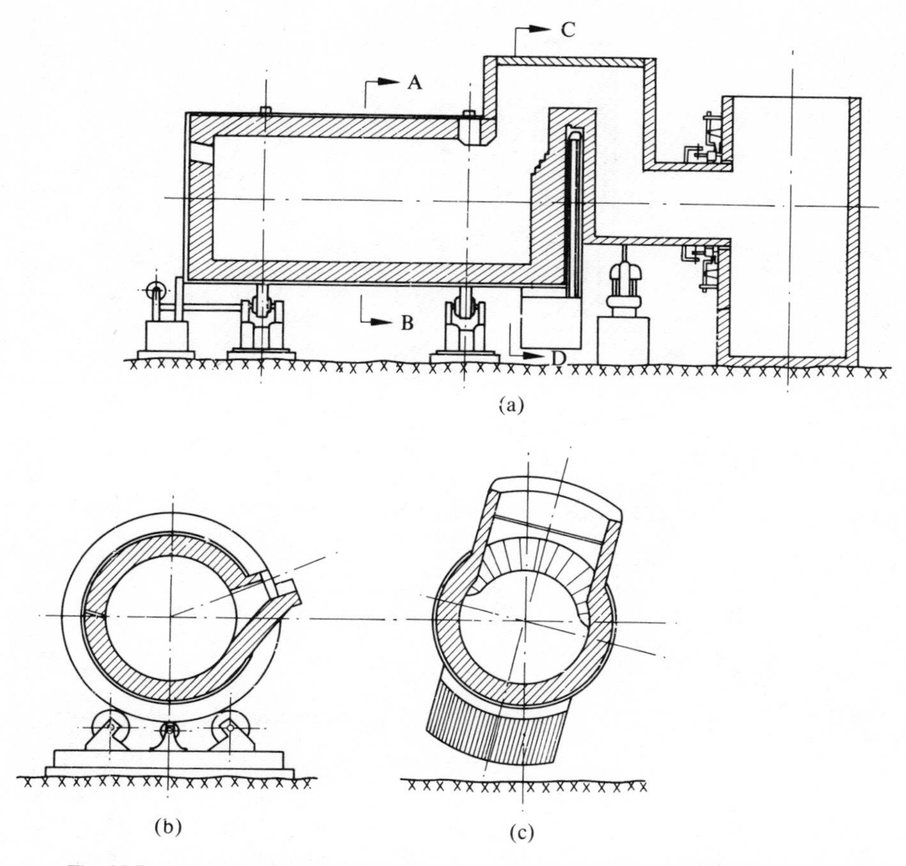

Fig. 15.7. Hoboken type siphon converter. (*Courtesy* Metallurgie Hoboken-Overpelt.)

4. Replacing the combined reverberatory furnace–converter installation by a continuous matte-making–converter operation which may be either single stage (Noranda[17] combined with slag cleaning by flotation) or multistage (Mitsubishi)[18]. The countercurrent flow system (Worcra)[19] had similar objectives but development has been discontinued. These processes are illustrated in outline in Figs. 15.11 and 15.12.

The replacement processes referred to in (3) and (4) are characterized by greater intensity of operation than the reverberatory furnace. The flash smelting furnace can produce high grade matte (thus reducing the work required by converter) by oxidation of much of the iron and sulphur in the feed, hence deriving fuel value from the concentrate. Outokumpu furnaces are being designed/constructed to produce directly white metal and, with certain types of concentrate, blister copper. Thus with the aid of air preheat or oxygen enrichment, a heat balance is achieved in which relatively little

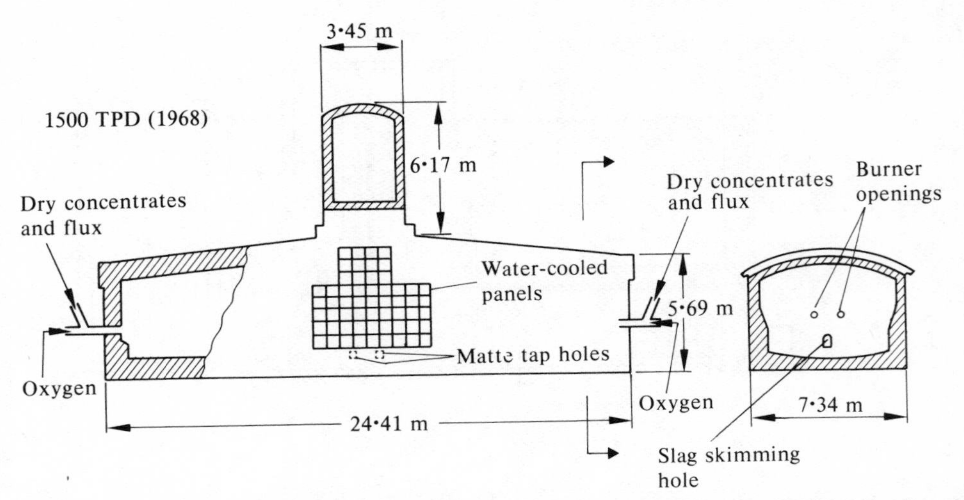

Fig. 15.8. Inco flash furnace (capacity 1500 tonnes per day). (*Courtesy* International Nickel Co. Canada Ltd.)

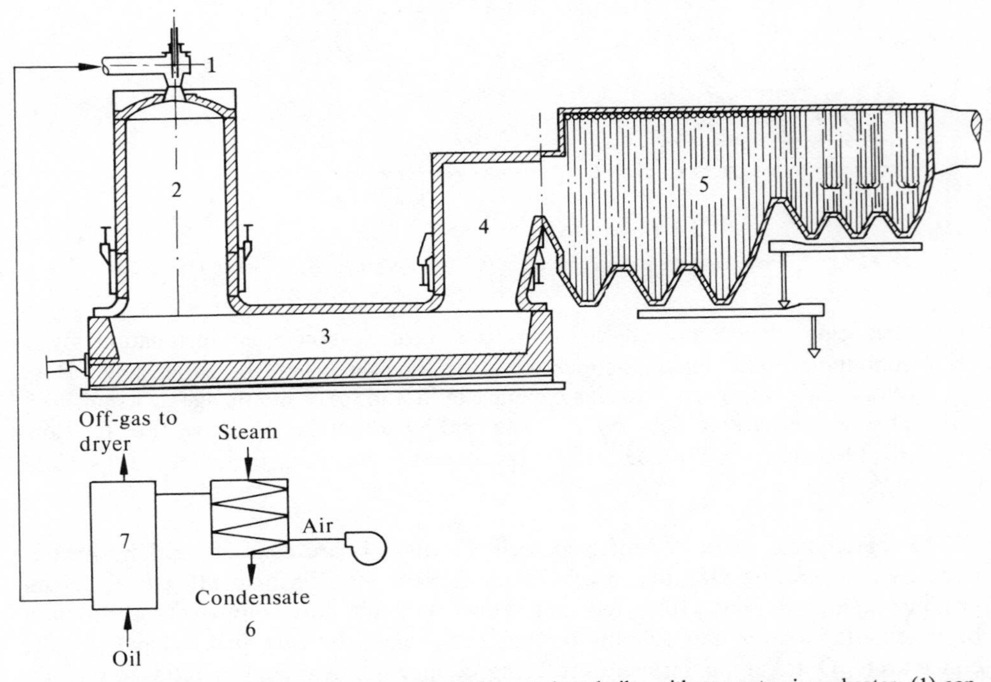

Fig. 15.9. Outokumpu flash smelting furnace and waste heat boiler with separate air preheater: (1) concentrate burner; (2) reaction shaft; (3) settler; (4) uptake; (5) waste heat boiler; (6) steam air pre-heater; (7) oil-fired air pre-heater. (*Courtesy* Outokumpo)

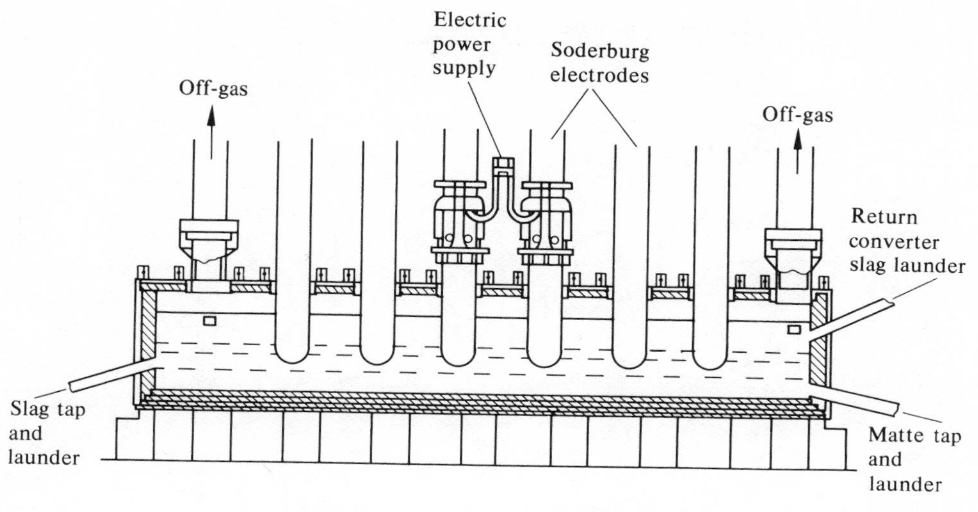

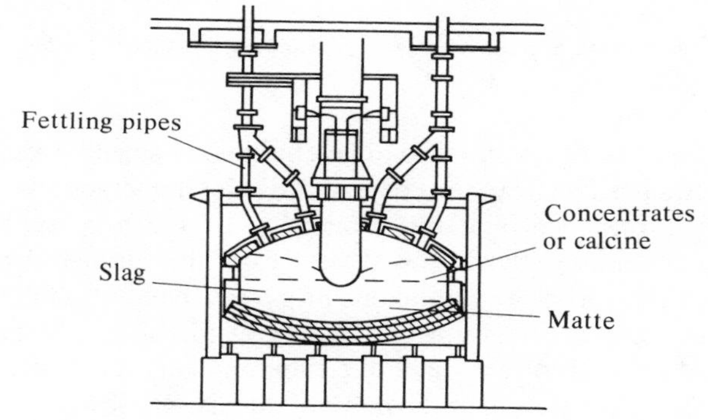

Fig. 15.10. Electric furnace.

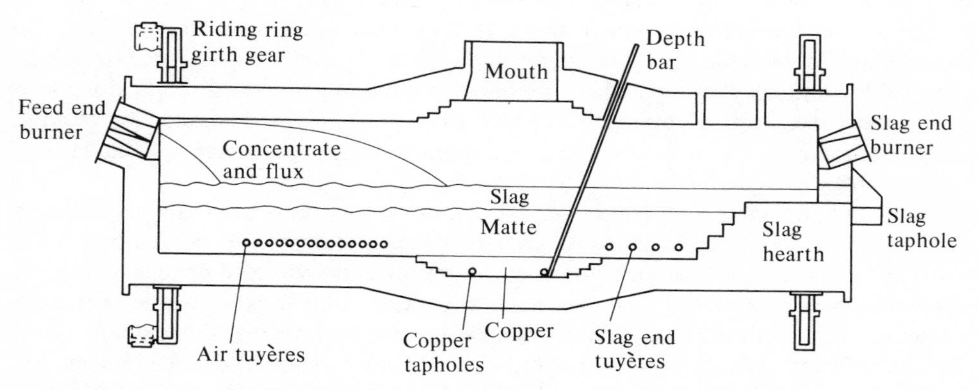

Fig. 15.11. Noranda pilot plant reactor—longitudinal cross-section. (*Courtesy* Noranda Mines Ltd., Canada.)

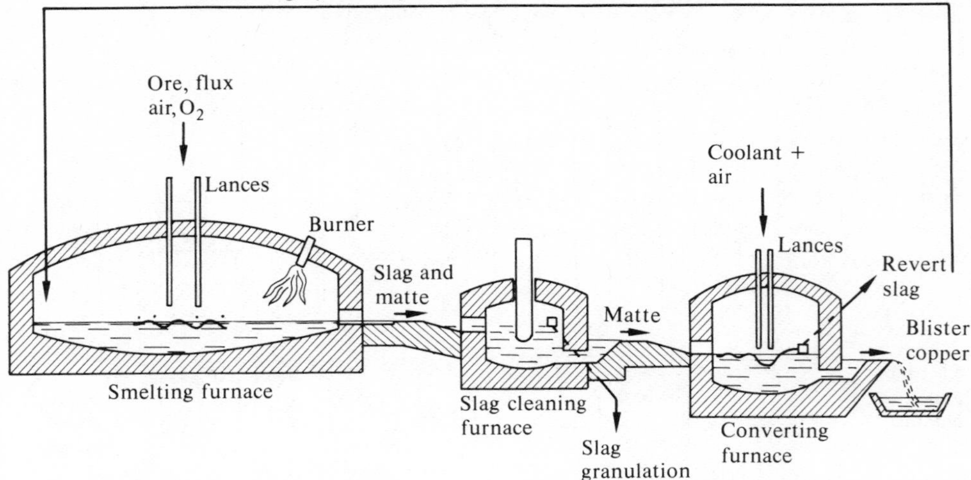

Fig. 15.12. Mitsubishi continuous process. (*Courtesy* Mitsubishi Metal Corporation.)

carbonaceous fuel has to be supplied. The continuous smelting processes also derive much of their required fuel from the concentrate oxidation reactions.

In the case of flash smelting the proportion of the total sulphur which can be eliminated as a continuous steady gas stream from the furnace depends on the matte grade chosen, whereas with the continuous processes the total sulphur is eliminated in a continuous steady gas stream from one or several furnaces. The gases from all these processes should be eminently suitable for sulphuric acid manufacture. With a double contact acid plant the final sulphur dioxide emission is very low.

If sulphuric acid is not marketable within an economic distance from the smelter it may be desirable to reduce the sulphur dioxide to elemental sulphur. This question is discussed later, but in this section dealing with copper extractive metallurgy, it should be noted that the same factors, i.e., percentage of sulphur dioxide and constancy of composition and delivery rate, which are important for sulphuric acid production, have even greater weight in relation to production of sulphur by gaseous reduction. A vital additional factor is the proportion of free oxygen in the off-gas, since this can lead to considerable wastage of reductant.

Many hydrometallurgical routes for copper extraction have been studied and this development effort is continuing. At least two such processes (Duval ferric chloride leach and Arbiter–Anaconda ammonia leach) are currently operated or scheduled to be operated on a commercial scale. In such processes sulphur is converted either to elemental form or to sulphate, and hence an important advantage is that sulphuric acid disposal is avoided. Such processes avoid air pollution by sulphur dioxide, but require careful evaluation with regard to liquid effluents and solids disposal. Energy requirements, quality of the metal produced, and efficiency of recovery of both copper and precious metals are other criteria on which data from full-scale operation are awaited with interest. On present evidence it is unlikely that hydrometallurgical routes will be generally competitive with the modern pyrometallurgical routes described above.

Significant quantities of copper are recovered from the dump leaching of oxidic materials followed by solvent extraction. In solvent extraction a specific organic chemical dissolved in a diluent, typically kerosene, extracts the desired metal, in this case copper, from an aqueous solution in which it may be associated with undesirable impurities such as iron, and yields a separate organic phase loaded with chemically combined copper. Reaction of the organic phase with an acid solution regenerates an aqueous solution of copper which is typically richer in copper content and acidity than the original solution, and is amenable to efficient recovery of metallic copper by electrolysis.

In overall effect, the solvent extraction system transfers copper from one aqueous solution to another using the following reactions:

$$\underset{\text{aqueous}}{CuSO_4} + \underset{\text{organic}}{2HR} \rightleftharpoons \underset{\text{organic}}{CuR_2} + \underset{\text{aqueous}}{H_2SO_4}.$$

The rection in simple terms can be described as a reversible one between the copper ion Cu^{++} and organic reagent HR to give an uncharged chelate CuR_2. The position of the equilibrium depends on the pH, and high H^+ concentration forces the reaction to the left. Thus high acid concentrations tend to strip the copper from the organic reagent. The organic reagents used for copper, the LIX, KELEX and other developing reagents, form series with a range of properties so that the appropriate reagent can be selected for a given flowsheet situation defined in terms of the copper concentration, acidity and impurity levels in the aqueous stream available for treatment, and of the corresponding levels required for the output copper solution. Solvent extraction plant for copper from acid leach solution has been engineered up to 300 tonnes of copper per day at Nchanga. In addition to solvent extraction from acid solutions there are applications in ammoniacal solution (Arbiter),[20] and in these processes secondary metallic copper can also be extracted.

Some processes combine both pyro- and hydrometallurgical stages. For example, in sulphate roasting in fluidized beds, a concentrate can yield basic copper sulphate and iron oxide:

$$2CuFeS_2 + 7O_2 = CuO.CuSO_4 + Fe_2O_3 + 3SO_2.$$

The gas stream from a roaster of this type can sustain an acid plant (although this is not in fact current practice in the few locations where sulphation roasting is carried out). Copper is leached from solid product in neutral solution with low solubility of iron. In the typical electrolytic circuit free acid is generated which has to be neutralized, preferably with a supply of oxidized ore.

Modern flash smelting–acid plants are capable of meeting the most stringent current air pollution regulations. A recent Japanese installation recovers over 96 per cent of input sulphur, divided as in Table 15.5.

TABLE 15.5

Recovery of sulphur

Recovery source	Percentage of input sulphur
In slag	2·7
In sulphuric acid	91·9 (single contact plant)
In aqueous effluent from gas cleaning	1·6
In tail gas scrubbing product	2·8†
In converter dust (EP)	0·1
Measured gas emissions	0·3
Unaccounted balance	0·6

† With a double contact acid plant the sulphuric acid production should be 94·5 per cent, the gaseous emission 0·5 per cent, and the bleed of dilute acid from gas cleaning 1·6 per cent.

Furthermore, by using the fuel value of the concentrates, the demand for purchased fuel is small. In environmental terms, such plants provide a difficult target for hydrometallurgical routes to match.

Zinc and lead

Primary zinc is produced by both electrometallurgical and pyrometallurgical processes. In 1976, the total production capacity of the western world was about 5·4 million tonnes, the installed processes being estimated in the proportions shown in Table 15.6. It is expected that in the future, increasing proportions of zinc will be produced by the electrolytic and blast furnace processes. The former has the advantages of not requiring

TABLE 15.6

Primary Zinc production capacity by processes in the western world in 1976

Method	Percentage
Electrolytic	74
Imperial smelting blast furnace (ISF)	12
Horizontal distillation	2
Vertical retort	5
Electrothermic	6

metallurgical coke, which is expensive or impossible to obtain locally in many areas, and of producing the higher grades of zinc without subsequent refining, but it requires high grade concentrates as feed, and cheap electricity. The ISF blast furnace process is more flexible in being able to smelt a wide range of feeds. It is a highly economical smelter of lead simultaneously with zinc, but its natural product is the lowest grade of zinc which requires subsequent processing to refine it to Zn I or Zn II standards.

Electrolytic process[21]

The electrolytic process consists of the following main stages:

1. Concentrate roasting to give zinc oxide calcine plus sulphur dioxide which is converted conventionally to sulphuric acid. This operation is carried out in fluid-bed or suspension roasters which are illustrated in Figs. 15.13 and 15.14. Such techniques have largely replaced the multiple hearth roaster (Fig. 15.15).

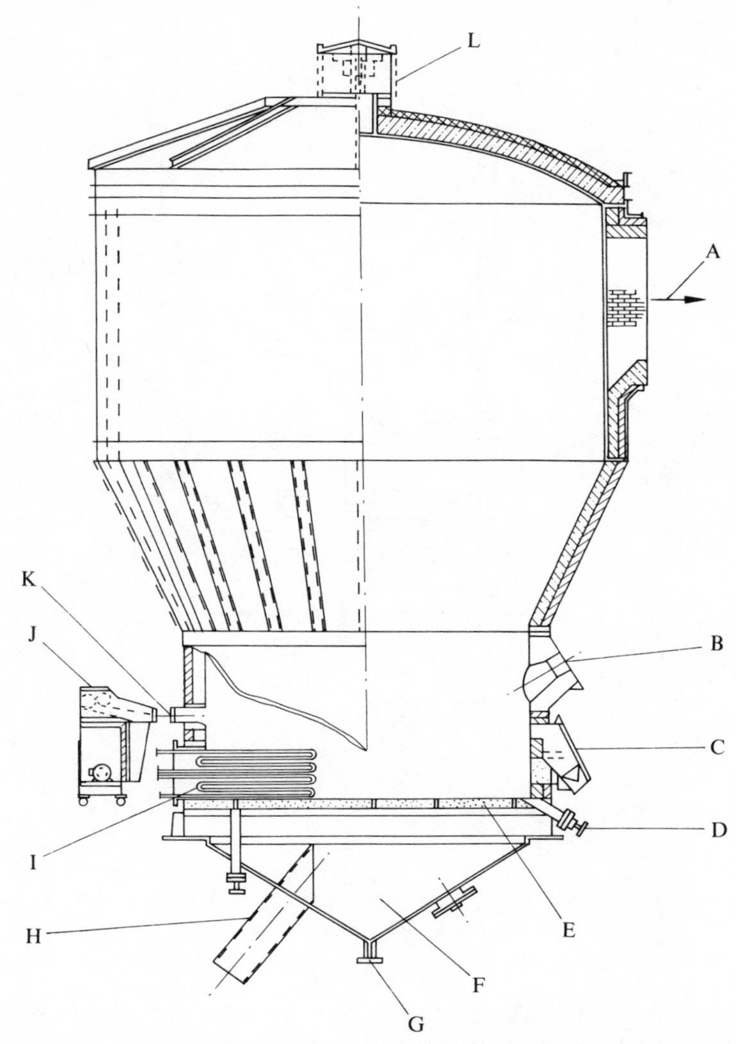

Fig. 15.13. Fluid bed roaster: (A) gas outlet; (B) oil burner nozzle; (C) bed overflow discharge; (D) underflow discharge; (e) bed grate; (F) wind box; (G) wind box discharge; (H) air inlet; (I) bed cooling coils; (J) slinger belt; (K) charging port; (L) safety valve.

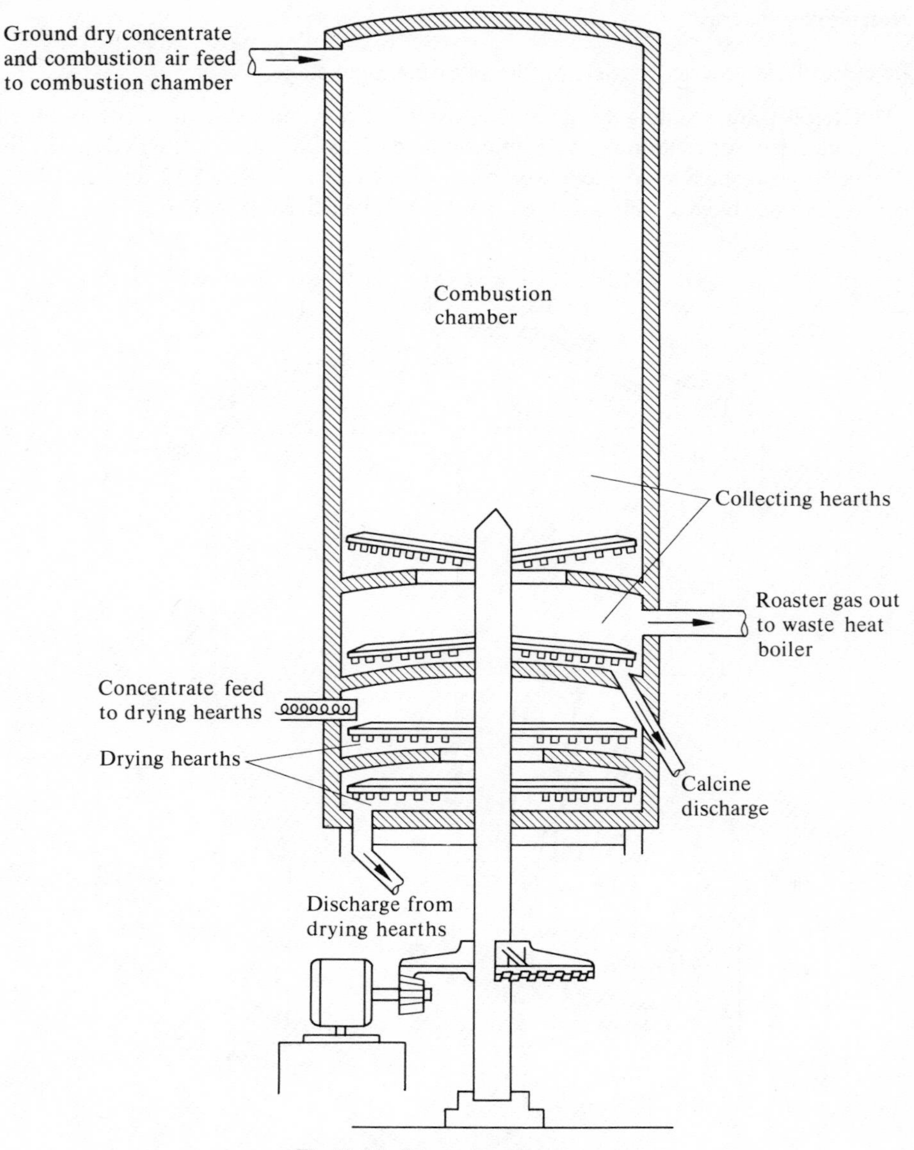

Fig. 15.14. Suspension roaster.

2. Calcine leaching with sulphuric acid to give an impure zinc sulphate solution.
3. Solution purification.
4. Electrolysis of the solution using aluminium cathodes to give zinc metal together with sulphuric acid for recirculation to the leaching stage.

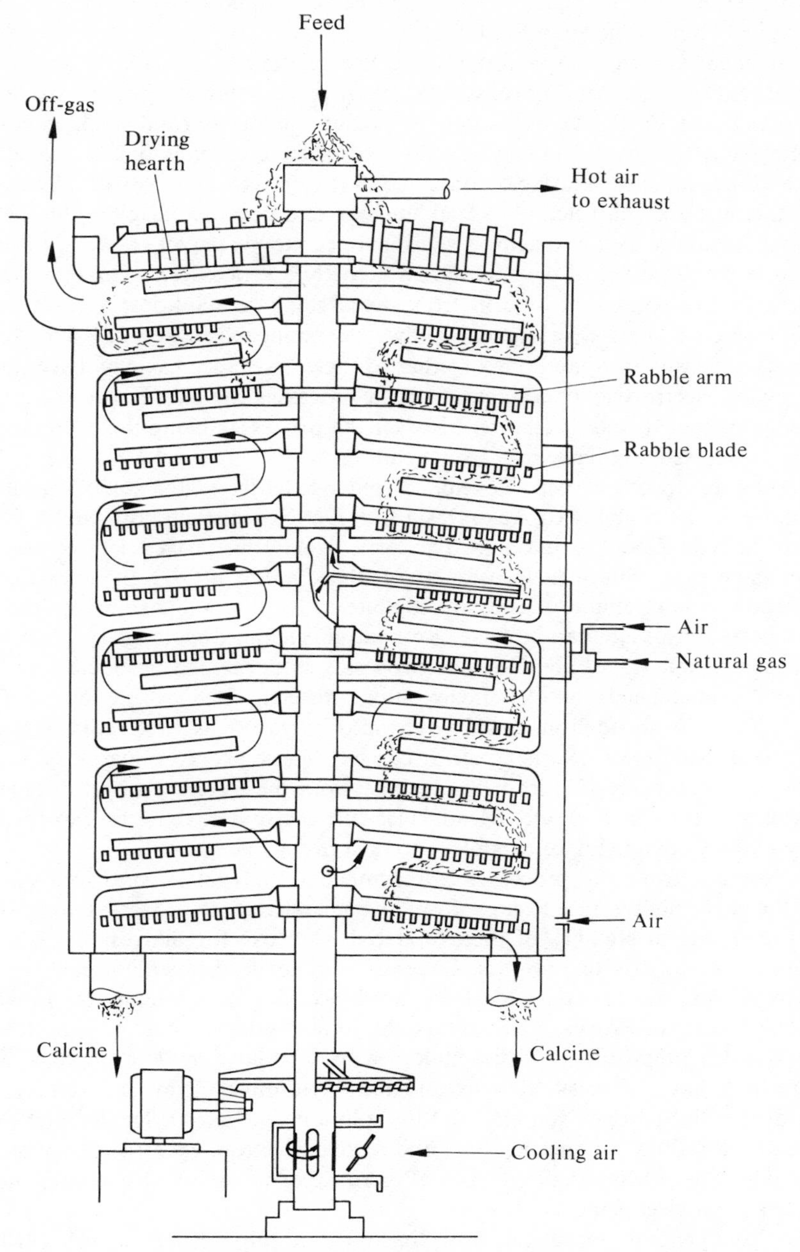

Fig. 15.15. Multiple hearth roasting furnace.

The cathodes are then stripped, melted, adjusted for composition and cast. A flowsheet of the typical circuit is given in Fig. 15.16.

Zinc concentrates of the type used in the electrolytic process conventionally contain 5–12 per cent iron, and during roasting most of this combines with zinc oxide to form zinc ferrite ($ZnO.Fe_2O_3$) which is largely insoluble under normal leaching conditions. Thus the ferrite removes from the circuit significant quantities of zinc, together with nearly all the lead, silver and gold contained in the original concentrates. Such residues from leaching were unattractive to lead smelters and became unsightly dumps at most electrolytic plants as well as placing a significant economic penalty on the process.

Recently development of the Jarosite and Goethite Processes has made it possible to remove high concentrations of iron from metal sulphate solutions in easily separated crystalline forms. This development allows economical recovery of zinc from raw materials containing iron, and one or other of these processes has now been adopted in most existing electrolytic zinc plants. In both processes the zinc ferrite is decomposed using sulphuric acid, the difference between the processes being the method used for removal of the resulting ferric iron from solution.

The Jarosite process was developed independently with some variations by Electrolytic Zinc Company of Australasia Ltd, Det Norske Zinkkompani A/S Norway and Asturiana de Zinc S.A. in Spain. The ferric iron in this process is precipitated from solution as an easily filterable Jarosite $2(NH_4, Na)Fe_3(SO_4)_2(OH)_6$ by adding ammonia, caustic soda or ammonium or sodium sulphate to the hot acidic solution. The Jarosite process flowsheet design can be based on neutral leaching first stage (Fig. 15.16) or on hot acid leach first stage. The Goethite process was developed by Société de la Vieille Montagne, and depends on the precipitation of ferric iron as goethite (FeOOH or $Fe_2O_3.H_2O$). The ferric iron dissolved from zinc ferrite is first reduced using zinc sulphide and then reoxidized using air or oxygen. These processes can yield lead, silver and gold as a relatively rich iron-free residue from the leaching stage. This residue is technically acceptable feed for a lead blast furnace, but it requires careful handling because its finely divided nature can lead to dust problems.

Iron residues from the processes are dumped usually in a controlled manner, as unlike the inert highly fused slags from pyrometallurgical processes little is known about the long-term stability of these materials. No use for the residues is known to date, although work is in progress. Recently another hydrometallurgical process for treatment of zinc plant residues has been developed by Dowa Mining Co. Japan, and is called the Hematite Process. This treats the total residue of zinc ferrite, precipitated iron and other impurities by dissolution with acid and sulphur dioxide in a low pressure autoclave. Excess sulphur dioxide is stripped from solution, copper is precipitated with hydrogen sulphide and the solution neutralized. The ferrous solution is then heated to 200°C in an autoclave and oxidized with oxygen to precipitate iron as ferric oxide. The precipitated hematite after roasting to remove trace sulphate can be sent as feed to a steel plant.

The calcine generally produced from roasting zinc sulphide concentrates in fluid-bed roasters is a fine dust. To protect the sulphuric acid catalyst from the adverse effects of such dust and from fume resulting from the partial volatilization of lead, cadmium and

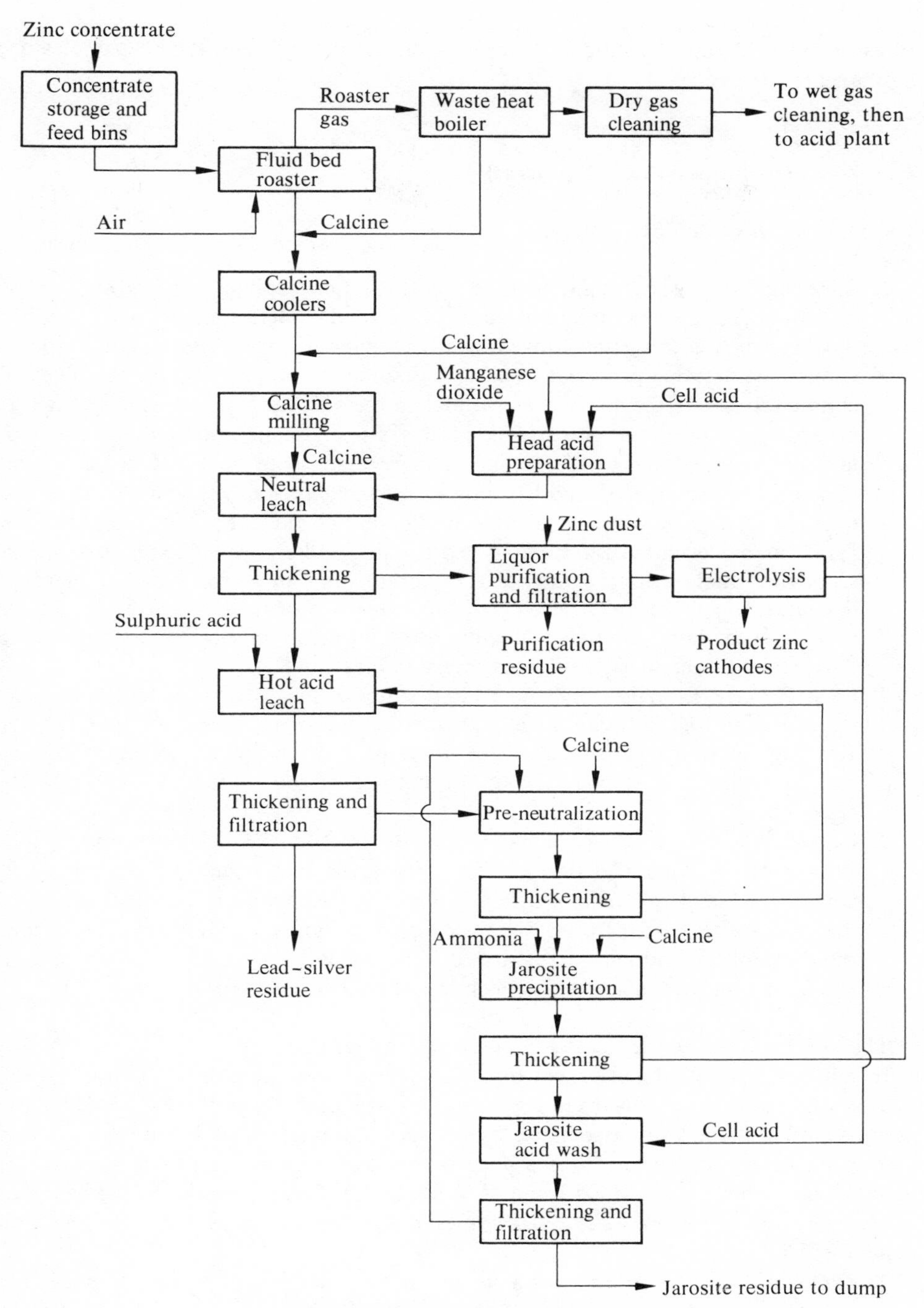

Fig. 15.16. Electrolytic zinc production.

other volatiles during roasting, a complex train of disengagement and gas cleaning equipment is necessary.

Zinc–lead blast furnace (the ISF process)[22]

The basic stages of the ISF comprise:

1. Sintering by the updraught method in which the sulphide concentrate feed is desulphurized and decadmiumized to give hard sinter, together with sulphur dioxide which is converted conventionally to sulphuric acid. The sintering process is an agglomerating process as well as a desulphurizing process, and the physical characteristics of the sinter are essential to the success of the smelting stage. If necessary, up to 25 per cent of oxide materials can be incorporated into the feed, and sinter acceptable to the furnace can be made with up to 25 per cent lead content. The presence of lead has, however, to date precluded the use of fluid-bed roasting instead of sintering, for the production of ISF feed. However, low lead feeds can be desulphurized by fluid-bed roasting and can be agglomerated (as can secondary and residual materials) by briquetting in a roll-press at elevated temperatures. When fully developed, this route will provide a useful process for treating secondary materials which are often ineligible for electrolytic processing. A typical up-draught sinter plant is shown in Fig. 15.18.
2. Blast furnace reduction, in which lump sinter is fed to a refractory-lined steel furnace shaft together with metallurgical coke and fluxes. Reduction by CO gives zinc vapour, which passes out of the upper part of the furnace to the condenser, and lead plus slag is collected at the bottom of the shaft and is periodically removed.
3. Condensation in which the zinc vapour issuing from the shaft is absorbed in a shower of hot, fine lead droplets for subsequent zinc separation by cooling in launders. This shock-chilling of zinc vapour is an important technical aspect of the process, because it freezes the reversion reaction by which the zinc undergoes rapid oxidation, and allows zinc metal to be condensed. All previous attempts to condense zinc from a zinc-oxidizing atmosphere had failed for this reason.

The ISF shaft and condenser lay-out is shown in Fig. 15.18.

In addition to the sulphuric acid plant, ancillary stages include equipment for the recovery of cadmium volatilized in the sintering stage, plant for refining the lead bullion formed and dealing with associated by-products (see later), and for cleaning and recovering heat from furnace gases and the optional refluxer plant for refining the liquated zinc from Zn IV to Zn I and II. A vacuum refining process (VDZ) has been operated to the semi-commercial scale to produce Zn II directly from the zinc-rich condenser lead.

The smelting of lead concentrates alone is typically carried out in blast furnaces fed by sinter produced on up-draught sinter machines.[23] The general process scheme is

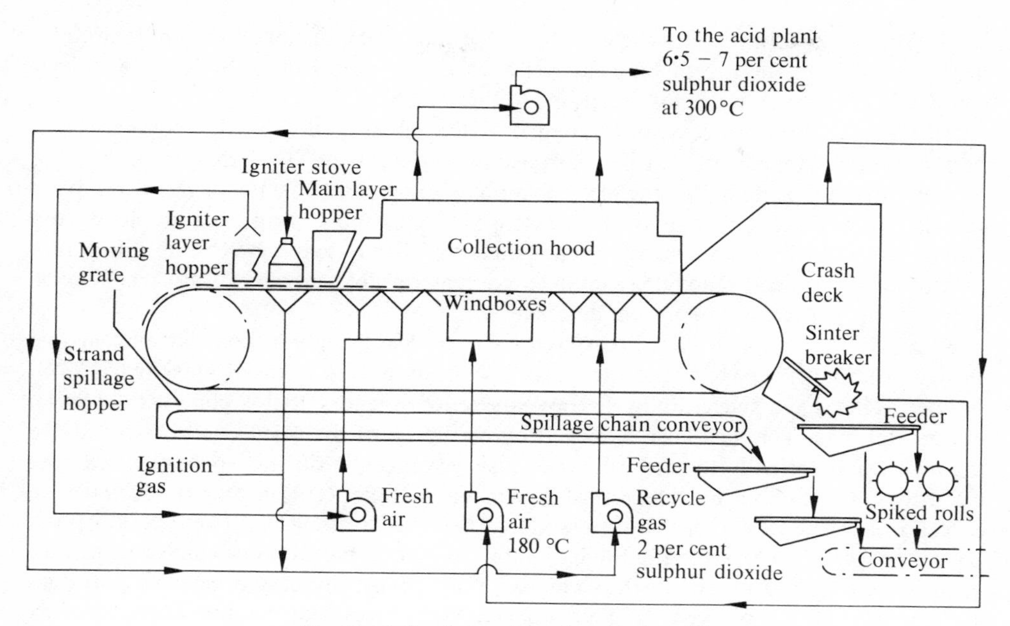

Fig. 15.17. Lead-zinc sinter plant (updraught).

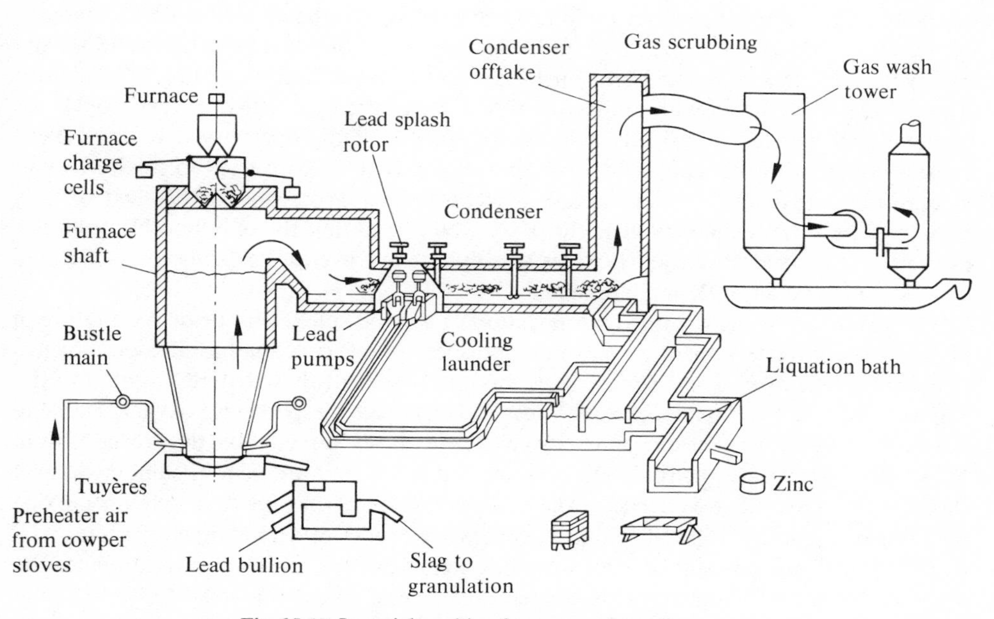

Fig. 15.18. Imperial smelting furnace—schematic.

similar to the zinc–lead blast furnace with the important distinction that no product metals issue from the top of the shaft, as does zinc vapour in the ISF. A typical lead smelter installation flowsheet is illustrated in Fig. 15.19.

All these smelting processes are applied to the metal oxides, so that the necessary first step with sulphide concentrates is roasting to yield oxide. The reaction achieved in copper smelting, oxidation of the molten sulphide to yield the metal and sulphur dioxide gas, is achievable only in a restricted range of conditions with lead sulphide, and impossible with zinc sulphide. With lead, smelting of this type is beset with difficulties (volatilization of lead sulphide, excessive content of PbO in reject slag) which have prevented its general use in present day plants.

Some of the problems of direct treatment of lead sulphide concentrates in one furnace without a separate roasting stage have been tackled at the Rönnskär works of Boliden. They have successfully directly smelted rich lead sulphide concentrates in an electric furnace. This has been achieved by oxidation of the dried concentrates above the melt using specially directed air jets causing vortices in the gas spaces between the electrodes. The sulphur dioxide gases (approximately 6 per cent) leaving the furnace are cooled in a waste heat boiler and pass via further gas cleaning to an acid plant. In the Kivcet process developed in the Soviet Union, oxygen combustion of lead concentrates in suspension yields a melt which is reduced in an electrically heated bath. The product gas contains 50–80 per cent sulphur dioxide. The process can be applied to complex ores containing zinc, lead and copper.

Roasting of zinc sulphide concentrates is successfully carried out in a number of ways, yielding a well desulphurized product and a roaster gas of adequate concentration for sulphuric acid manufacture. Fluidized bed roasting can be applied either to material of flotation concentrate sizing or to pelletized feed. Suspension roasting with the removal of sulphur completed on hearths is also practised. Sinter roasting of a mixture of sulphide concentrate with some five times its weight of crushed product yields an agglomerated product physically suitable for some smelting operations. With all these roasting operations the whole of the sulphur dioxide can be converted to sulphuric acid in conventional plant. The gas cleaning necessary for protection of catalyst beds is stringent, with particulates typically 0·0001 g Nm^{-3}, so that the only notable pollutant in the gas dispersed at stacks is oxides of sulphur; the level of gas emission is dictated by acid plant converter design (see later).

The gas theoretically obtained by oxidation of a metal sulphide to oxide is a mixture of air and the completely spent gas with composition 14 per cent sulphur dioxide. For use in acid manufacture, oxygen must be present in excess of that required to oxidize the sulphur dioxide to trioxide; about 8 per cent dioxide, 9 per cent oxygen is ideal. Fluidized bed or suspension roasting can yield gases directly of this composition. With sinter roasting a recirculation of some tail gases that are lean in sulphur dioxide through the roaster bed is needed. These gases are somewhat lower in sulphur dioxide than the ideal, though still well within the operating range for acid plant converters.

The typical zinc concentrate feed contains some lead but insufficient to influence the roasting operations. Lead sulphide concentrates or the mixed zinc–lead feed smelted in the zinc–lead blast furnace cannot be roasted in fluidized beds because of fusion and

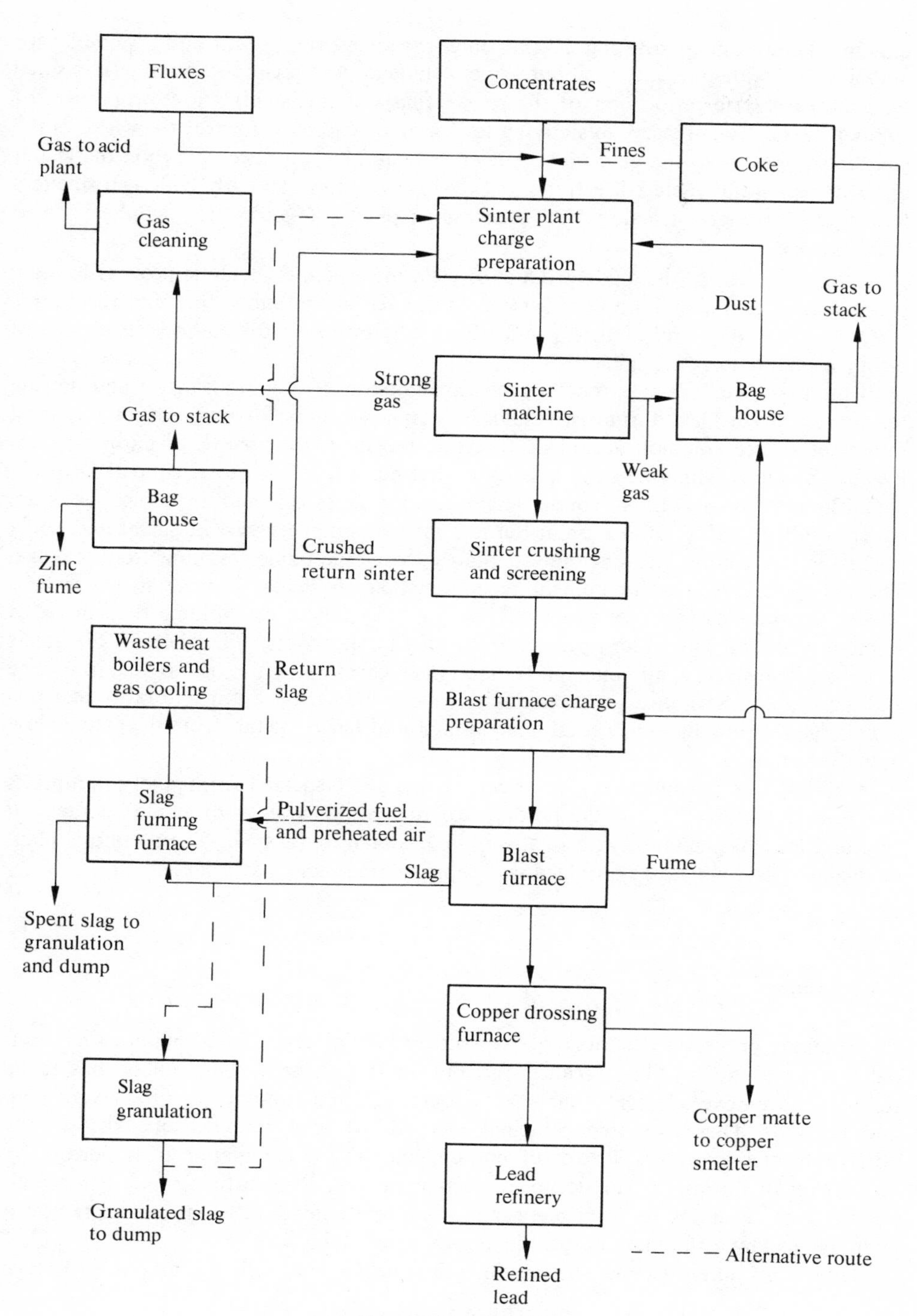

Fig. 15.19. Lead smelting.

agglomeration. Sinter roasting is the only roasting process currently suitable for sulphide feed with greater than about 12 per cent lead. As the proportion of lead in sinter feed increases, the proportion of the roaster gas obtained with a low proportion of sulphur dioxide is increased. In sintering for the zinc–lead blast furnace the whole of the roaster gas passes to the acid plant, but in the lead blast furnace operation the leaner gases are generally vented directly to stack. However, some recently built lead smelters utilize the stronger fractions of their sinter-plant off-gases for sulphuric acid manufacture.

In the treatment of the oxide product to yield metal, the electrolytic process for zinc production is associated with production of oxygen at the anode (the nett reaction is $ZnO \rightarrow Zn + \frac{1}{2}O_2$) and a small production of hydrogen at the cathode. Emission of airborne pollutants is negligible.

In the pyrometallurgical processes the consumption of fuel, both inside and outside retorts, or inside blast furnaces, results in a major volume of process gas requiring treatment before emission at stacks. In retort processes the volume of carbon oxides leaving the retort with zinc vapour is somewhat more than the molar equivalent of the zinc. The retort gases pass to condensers to deposit liquid zinc and the cooled gases are water scrubbed before utilization at burners for heating processes. In gas-heated retort processes the heating gas inevitably contains some zinc oxide resulting from leakage. The hot gases are used in preheating combustion air before passing to waste heat boilers. In the blast furnace process a substantially larger gas volume than in retort smelting passes through the furnace and through condensers. The whole of the gas is scrubbed intensively with water in Thyssen disintegrators before use as fuel in charge and air heaters. Scrubbed gas contains not more than 0·05 g Nm^{-3} total solids; this level is related to maintaining total mass emission of heavy metals from the plant below permitted limits.

In all the zinc pyrometallurgical processes the gas dispersed from stacks comprises two general streams. One is the products of fuel combustion, with a small content of zinc and lead particulates, with relatively little intake of false air. The second is plant ventilation gases which are provided with separate treatment, as discussed later.

Lead refining

The smelting processes described above recover lead as crude lead bullion either from the bottom of a lead blast furnace or from an ISF furnace. This bullion has to be refined to remove elements such as copper, arsenic, antimony, tin, silver, gold and bismuth. There are two principal methods of lead refining, one electro, the other pyro-metallurgical. Electro refining depends on the elimination of impurities by dissolution of an impure anode in an electrolytic cell. Pyro refining uses oxidation–precipitation reactions to form physically separable compounds that can be removed from the lead by differences in specific gravity or solubility.

There is no purely electro separation technique for lead refining, and all processes require a certain amount of pyro-metallurgical treatment in their flowsheets. The main elements in any flowsheet are

1. decopperizing,
2. removal of arsenic, antimony and tin if present,
3. desilverizing,
4. debismuthizing.

The pyrometallurgical route is the only one available for carrying out decopperizing. In the case of electrolytic lead refining. Stages (2)–(4) can be carried out in the electrolytic cell to produce a pure cathode and an anode slime containing arsenic, antimony, bismuth, silver and gold. The slime has to be processed subsequently by pyrometallurgical methods for separate recovery of the precious metals.

Pyrometallurgical refining removes the above elements (with reasonably selective separation) on a continuous or batch plant. The continuous plant because of capital and maintenance cost is only justified for large steady tonnage throughputs, i.e., around 100 000–150 000 tonnes per annum. However, particular steps in the batch refinery can be made continuous with advantage, e.g., reverberatory softening furnace (arsenic, antimony removal) and debismuthizing.

Decopperizing, a process common to both electro and pyro lead refining, is carried out in two stages. The first stage (removal of copper to 0·1–0·2 per cent) in most cases involves batch cooling and agitating the bullion to produce a dry dross containing the majority of the copper. This dross can also contain much of the arsenic and some of the antimony and tin present in the bullion. The second stage (removal of copper to 0·01–0·005 per cent) involves stirring the lead at its melting point with additions of elemental sulphur to produce a mixed copper sulphide–lead sulphide dross. Bullion from the ISF furnace and from a lead blast furnace differ in their sulphur content, the former being low in sulphur. The low sulphur content causes the arsenic to be precipitated with the copper dross, leaving a resultant low arsenic tenor.

Treatment of first stage copper dross from lead blast furnace bullion is usually by production of a lead sulphide–copper sulphide matte in a reverberatory or short rotary furnace, with attendant small emissions of sulphur dioxide requiring scrubbing. The matte is often disposed of to a copper smelter. A continuous copper drossing furnace has been developed by BHAS Port Pirie, in which a continuous stream of metal is cooled and the precipitated copper compounds are simultaneously smelted to matte in a heated part of the furnace. First stage copper dross from ISF bullion has in the past been difficult to process to matte, because of the large sulphur additions required. Recently a vacuum cleaner technique has been developed to remove the finely divided dross from the refining kettle and collect it in bag filters. This finely divided dross, containing a high proportion of oxide and elemental copper, has been treated by acid leaching or ammonia leaching–solvent extraction to produce cathode copper. Second stage copper dross can be produced as a dry (dusty) or wet (metallic) material which is usually returned to the first stage. This has not as yet been practised with treatment involving vacuum removal of first stage dross.

Electrolytic refining of decopperized bullion is carried out by treatment of anodes in a hydrofluorosilicic acid electrolyte. Other electrolytes have been used—sulphamic acid, hydrofluoroboric acid—but for reasons of lower cell voltage, electrolyte stability, use of high current density, and lower acid consumption, hydrofluorosilicic acid is generally chosen for industrial operations. Electro refining is the only process in current operation that can reduce bismuth levels in lead to below 10 g $tonne^{-1}$. It is selective against all more base metals, e.g., zinc, cadmium, nickel, cobalt, which pass into the electrolyte solution, and other elements, e.g., gold, silver, copper, bismuth, arsenic, antimony, selenium and tellurium, which pass into the slimes. Tin cannot be separated and passes to the cathode. Therefore if tin is too high in the feed it must be removed by oxidation in a pyrometallurgical step. The slime production is a small percentage of the metal throughput, usually 2–3 per cent, and thus precious metals are very concentrated. As there is no selective refining, all metals collect together, necessitating complicated oxidation–reduction pyrometallurgical processing stages for recovery of individual values. Some plants use electro refining to separate bismuth from lead, and selective pyro refining to recover other values. To retain the slimes on the anode and avoid slurry formation in the electrolyte, a reasonably rigid but porous structure must be formed. This is done by control of impurities in the anode with antimony maintained at 0·8–1·3 per cent. Wtih bismuth only present, much higher impurity levels are required, say 2·5 per cent or more. In electro refining about 30–40 per cent of the anode lead is recycled in spent anodes.

Pyrometallurgical refining carried out in batch kettles or continuous steps is selective. Continuous steps can be applied to arsenic, antimony, silver, gold and bismuth removal, and usually give better metallurgical grade control and richer by-product streams (slags, intermetallic crusts) than do batch processes. Bismuth in lead can only be reduced to approximately 100 g $tonne^{-1}$ with pyro refining using the Kroll Betterton process (addition of calcium and magnesium to produce a $CaMg_2Bi_2$ crust), reagent costs and operating difficulties making lower levels impracticable. This refining process is best done (even when other refining steps are batch) using the Hoboken continuous Kroll Betterton process, where rich crust is produced and better metallurgical control obtained. A recently published extension of the Hoboken work by B.H.A.S. achieves much lower bismuth levels. Arsenic, tin and antimony can be removed by continuous or batch oxidation with air lancing in a reverberatory furnace to produce a slag containing the metals; or in a dry dross from a batch kettle by stirring in $NaOH/NaNO_3$ (caustic soda sodium nitrate) mixtures. Another method is by use of molten NaOH in a Harris machine, which in association with hydrometallurgical treatment of the salt melt, provides a means of separating arsenic, antimony and tin without subsequent pyro treatment of oxide drosses or slags. The metals are oxidized in the order arsenic, tin, antimony, and some degree of selective removal can be obtained. Silver and gold are collected by addition of zinc, and removed in an intermetallic crust. The crust is liquated and zinc is distilled in a retort or under vacuum. The resulting rich silver–lead alloy is oxidized in cupels to remove lead and other impurities until silver and gold remain (dore bullion). These metals are separated by electrolysis in aqueous solution.

Although the pyro refining processes are reasonably selective in impurity removal, the drosses are generally much more dilute in precious metals than slimes from the electrolytic refinery. However, as can be seen from the preceding descriptions, both pyro and electro refining have to deal with similar plant hygiene and gas cleaning problems with regard to heavy metals. Most lead refineries in operation today are in the process of installing gas cleaning equipment to meet recent environmental regulations. On pyro or electro refineries there are two types of gas streams that need cleaning:

1. Ventilation gases from hoods over kettles, furnace tapholes, dross screens, etc., which involve large gas volumes required to maintain good inplant hygiene conditions (see later).
2. Process off gases (usually hot) that are produced by by-product treatment furnaces and contain some of the volatilized elements in the charge. These include gases from reverberatory furnaces, short rotary reduction furnaces, oxidizing cupels, zinc distillation retorts; treating slags, drosses, and alloys which contain the whole spectrum of elements removed in the refinery.

Ventilation gases covered by (1) can amount to a large volume, 200 000–250 000 $Nm^3\ h^{-1}$ for a 70 000 tonnes per annum pyro or electro refinery. The gases are usually cleaned using dry methods, i.e., bag filters.

Process gases (2) can vary in volume, depending on the type of refining and the by-product quantity and treatment flowsheet chosen. These gases are hot and their volume when presented for cleaning depends on the cooling methods adopted. Most cooling is by dilution, where possible using some of the hygiene ventilation air, water cooled and air cooled flues are also used. The furnaces are normally of the direct-fired type and so the off-gases contain products of combustion. Electric furnace reduction of electro refinery slimes is practised and affords a possible volume reduction.

The process gases are usually cleaned by dry methods, i.e., bag filters or electrostatic precipitators, as these avoid water treatment problems associated with heavy metal control. The major problem associated with dry cleaning is hygienic handling of the dusts produced for recycle, and the hygiene problems associated with equipment maintenance such as bag replacement. Flue dust is usually conditioned by wetting or, in more recent plants by passage through a flare. Wet cleaning methods such as venturis and wet electrostatic precipitators are used, particularly where the more volatile metal compounds are concerned.

One of the most difficult metals to trap is arsenic, as even at temperatures as low as 120°C, As_2O_3 would only exist as 98 per cent in the solid phase. Furthermore, this relationship may be changed if highly volatile chlorides or fluorides are present. Dry electrostatic precipitators are inefficient for the collection of arsenic, but test work with high pressure drop felt bag filters has proved more successful if temperatures can be suitably lowered without dew point problems.

The main metallic pollutants from a lead refinery are lead and arsenic but smelters treating zinc and lead concentrates should also monitor zinc and cadmium. Although emission regulations vary in different countries, both maximum concentrations and maximum allowable mass emissions are usually specified. The exit gas volume from the plant is considered to be the total of the ventilation and process gas volumes. For lead in the UK for large works with total gas emission above 140 000 SCFM (Class III works) the concentration limit in exit gas is 11·5 mg m^{-3} (volume measured at NTP). This consent limit does not apply to the main stack exit gas, but to individual flues from gas cleaning equipment before they enter the stack. The mass emission limit is 450 kg per week. Monitoring of heavy metal fall out around a non-ferrous smelter and correct interpretation of the results requires highly experienced personnel and is an expensive operation.

Apart from heavy metal emissions another pollutant from lead refining which requires control is fluoride. Fluoride emissions are particular to electro-refining, from decomposition of the hydrogen silicon fluoride (H_2SiF_6) electrolyte. The sources and distribution of fluoride gaseous emissions are given in Table 15.7.

TABLE 15.7

Emissions of fluoride

Source	Percentage of total emission
Recycle spent anode melting kettle	20
Cathode melting kettle	25
Heat treatment of electrolytic slimes	50
Unaccounted	5

Depending on the size of refinery and allowable emission, fluoride has to be removed from some or all of these gas streams. The total amount of emission and the above proportions depend very much on how much washing can be carried out in each stage prior to pyrometallurgical treatment. The amount of washing is very dependent on the plant water balance and can vary in different climatic conditions. The normal electrolytic make-up is 2·0–2·5 kg H_2SiF_6 per tonne of refined lead cathode. A large proportion of this is lost by adherence to electrodes and in slimes and is volatilized in subsequent heat treatment due to decomposition into hydrofluoric acid and silicon fluoride. With correct electrolyte temperature control, there is no problem of high fluoride levels in the cell room environment, and precautions for ventilation are not required.

Although there are impregnated dry bag filter methods of fluoride control, these are not practical in this application and wet scrubbing methods are necessary. The proportions of hydrofluoric acid and silicon fluoride which reach the scrubbers is uncertain, but the problems of scrubbing are similar to those encountered in the fertilizer industry where dilute H_2SiF_6 is used as scrubbing medium. The water effluent control problems associated with wet scrubbers depend very much on the plant water balance and how much fluoride can be recycled. Technology for the control of fluoride emissions from lead refineries is still under development. Work is being carried out on improved slime washing techniques. In some locations the slimes are roasted ($\simeq$300° C) in a kiln fitted with a wet scrubber prior to high temperature reduction in a short rotary or electric furnace. The latter approach avoids sending fluoride-containing materials to furnaces that use dry cleaning methods for dust collection.

Nickel[24]

There are two types of nickel ore, sulphides and oxide/silicate. Sulphides are currently the major feeds, but by far the most common type of nickel ore, accounting for about 75 per cent of the known nickel reserves and formed in large deposits, is the oxide or laterite ore.

Nickel sulphide processing

Sulphide ores are widespread geographically but only a few large deposits are known. The main nickel mineral is pentlandite (nickel iron sulphide, $(NiFe)_9S_8$, containing a theoretical maximum of 33 per cent nickel). Pyrrhotite and chalcopyrite are almost always found with nickel sulphides, but the amounts vary enormously from one location to another. Copper–nickel ratios vary from 2 : 1 to 1 : 50 and the nickel–cobalt ratios are higher in sulphide than in oxide ores. The sulphide ores often contain valuable quantities of precious metals. Grades of sulphide ores vary from approximately 1·6 per cent at Sudbury, Canada, to 2–3 per cent in Western Australia. The Australian ores contain a small percentage of copper. The ore grade is less important than the pyrrhotite–pentlandite ratio, as this affects the grade recovery relationship in flotation. In some cases where there is a high pyrrhotite–pentlandite ratio, a separate pyrrhotite concentrate is produced (0·8–1·5 per cent nickel) for separate treatment.

Production of nickel from sulphides causes generation of large quantities of sulphur-containing gases, as the preliminary stages of treatment are mainly pyrometallurgical and equipment similar to that used for copper pyrometallurgy is used. A block diagram indicating the various treatment routes is given in Fig. 15.20. The basic preliminary process is smelting to 'Bessemer' matte, usually done in three stages. Roasting in a fluid bed roaster or on a sinter plant reduces the sulphur content (30–40 per cent sulphur removed for nickel concentrates) of the concentrate to that required for matte production. The second stage is smelting the calcine or sinter to produce a matte and throwaway slag. Furnaces used for this stage are blast, reverberatory, or rectangular electric, the choice depending on economics. The electric furnace requires cheap power but gives the lowest nickel in slag. The third stage consists of blowing the matte, which contains about 25 per cent sulphur and a ratio of iron to nickel and copper of about 3 : 1, in a Pierce–Smith converter to remove iron to about 1 per cent and reduce the final sulphur to about 20 per cent. The converter slag, which is high in nickel, is returned to the matte smelting furnace. Thus the two stages of treatment, matte smelting and converting, permit production of a reject slag of low metal content and a matte of low iron content. Techniques for gaseous sulphur oxide capture are the same as for copper smelting.

As with copper production the matte-making furnaces can be replaced by flash smelting (Outokumpu) with the advantages of producing a strong sulphur dioxide gas stream for treatment. Ores with high pyrrhotite–pentlandite ratios need not usually be processed to supply a separate pyrrhotite concentrate if flash smelting is used. Pyrrhotite is used as a fuel in the flash smelter and the matte grade can be controlled by degree of oxidation supplied in the flash tower. Matte grades (as with copper smelting) are usually higher in the flash furnace than from the old matte smelting furnaces (nickel and iron to copper ratio approximately 2). Thus much more of the sulphur contained in the concentrate can be collected into a single concentrated continuous gas stream, and the slag produced and sulphur eliminated by the converters reduced. The flash furnace slag requires cleaning by reduction in an electric furnace before discarding to dump.

These treatments produce a 'Bessemer' matte low in iron, but containing appreciable quantities of copper. This is because of the intimate association of copper in the nickel mineral. It is not possible to produce a copper-free nickel concentrate without high nickel losses. Thus the next stage of nickel production after converting (unlike copper smelting) requires separation of nickel and copper from the sulphide matte. This is achieved in various ways at different smelters. The evolution of the many treatment methods has been influenced by final-product specification, local conditions, scale of operations, and copper–nickel ratios.

The methods of separation of copper–nickel matte include controlled slow cooling and flotation roasting and selective acid leaching, selective chloride leaching, and direct electrolysis of sulphide anodes. In the first three of these processes, some sulphur-containing gases are emitted during the process or subsequent treatment steps. Slow cooling of nickel–copper matte promotes crystal growth, and the copper and nickel sulphides can be separated by flotation. About 12 per cent metallics exist in the slowly cooled matte and act as a collector of precious metals. The copper concentrate produced in flotation is processed by conventional copper technology already described. The separated nickel sulphide concentrate, Ni_3S_2, after pelletizing is roasted to nickel oxide in fluid bed roasters and sulphur is removed down to 0·01–0·4 per cent. Nickel oxide is cleaned from the roaster gases in cyclones and electrostatic precipitators. Some nickel oxide is processed through chlorinators to remove copper, arsenic and lead, and then treated with hydrogen in fluid bed reducers to produce high grade nickel–nickel oxide sinter products. The gases from the chlorinator are scrubbed with lime to produce a sludge containing copper, lead and arsenic.

Roasting of copper–nickel matte is carried out on hearth roasters. The resultant dilute sulphur dioxide is treated by the dimethile anilene (DMA) absorption–regeneration process to produce a concentrated stream for liquid sulphur dioxide production. The calcine is selectively leached with sulphuric acid to remove copper and the leach residue is reduced in an electric furnace to produce anodes which are electro-refined to pure cathodes in diaphragm cells. Production of pure nickel cathodes by electro-refining anodes requires extensive electrolyte purification for removal of iron, copper, cobalt, arsenic, etc., before feeding to the cathode compartment of the diaphragm cell. This purification often involves gaseous chemical reagents such as chlorine and hydrogen sulphide. These are absorbed, however, and in the case of chlorine they are regenerated during electrolysis and recycled, so that there is no substantial problem with emission of gaseous pollutants.

Direct chloride leaching of crushed copper–nickel matte is a recent development. In this process the nickel is made soluble in hydrochloric acid solution with evolution of hydrogen and hydrogen sulphide gases from the leach reactor. These gases are scrubbed to remove traces of chloride and then incinerated in a waste heat boiler. The sulphur dioxide gas stream is passed to DMA absorption for liquid sulphur dioxide production. Copper sulphide is not made soluble and is separated for conventional roast–leach–electro treatment. Nickel chloride is crystallized from solution after it has been purified by solvent extraction of iron, cobalt and traces of copper. The $NiCl_2.4H_2O$ crystals are hydrolysed in a fluidized reactor to nickel oxide, which is subsequently reduced with hydrogen in a kiln. The hydrochloric acid gas produced in hydrolysis is absorbed and recycled. The hydrochloric acid absorber is fitted with a tail gas scrubber to reduce chloride emissions.

As previously mentioned, it is often necessary to make a separate nickel-containing pyrrhotite concentrate. This concentrate is typically 0·8 per cent nickel, 57 per cent iron, 37 per cent sulphur, and is increasingly difficult to treat because of the large amount of sulphur for disposal in relation to the nickel content. The pyrrhotite is roasted in fluidized bed roasters to remove sulphur. In one approach the dead roasted material is partially reduced in a kiln and the nickel is leached with ammonia–ammonium carbonate solution to leave an iron ore concentrate. Another treatment involves a selective sulphation roast of the pyrrhotite at a lower temperature to convert the nickel sulphide to sulphate for recovery by water leaching. Sulphur dioxide gases from these roasters are suitable, after cleaning, for conversion to acid.

A recent nickeliferous, pyrrhotite, fluidized roasting plant uses the Allied Chemical process (see later) to reduce the sulphur dioxide gas emissions to elemental sulphur using natural gas. The roasted pyrrhotite is subsequently reduced in a kiln to nickel–iron pellets for production of low alloy steel. This plant has been shut down, not, it is reported, because of problems with sulphur dioxide reduction, but because ring production in the kiln severely reduced design output.

A recent development in sulphide nickel technology is the Top Blown Rotary Converter (TBRC) (Fig. 15.20) based on the rotating Kaldo steel converter. Top blowing permits the use of oxygen on sulphide mattes so that high temperatures can be generated. The rotation of the vessel provides the necessary mixing and protects the refractories from hot spots. These features permit nickel matte to be blown to low sulphur contents, an operation that cannot be conducted in the Pierce–Smith converter. Low copper feeds can be blown to about 1 per cent sulphur in the TBRC, leading eventually to a 'fire-refined' nickel metal. Less pure feeds can be blown to 3–4 per cent sulphur and quenched in granular form to feed the nickel carbonyl refining process. The use of oxygen gives a relatively concentrated gas stream suitable for feed to an acid plant similar to the Pierce–Smith converter operation. In addition to the advantages of surface blowing and mechanically induced turbulence, the converter atmosphere may be adjusted to reducing or oxidizing conditions. This is achieved through the use of an auxiliary fuel burner. The operator can use the TBRC for melting, converting and

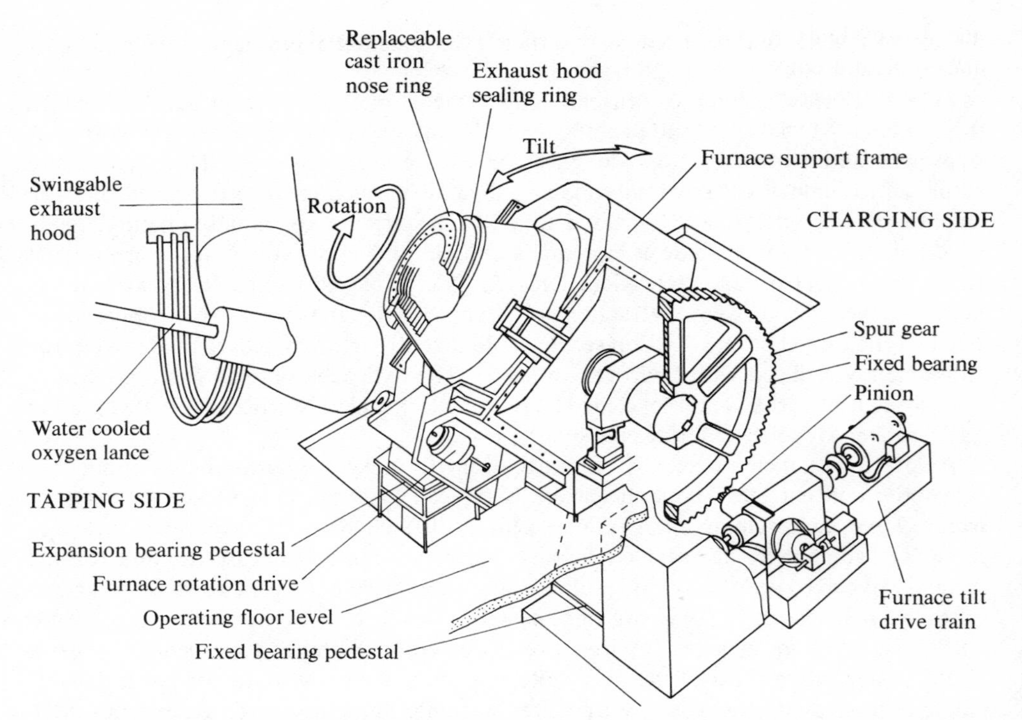

Fig. 15.20. Arrangement of top blown rotary converter, at Copper Cliff Works of Inco. (*Courtesy* International Nickel Co. Canada Ltd.)

reduction cycles as required by process parameters or process upsets. This flexibility is lacking in the Pierce–Smith converter.

Nickel forms a stable carbonyl, $Ni(CO)_4$. Carbonylation, i.e., reaction of nickel-containing materials with carbon monoxide gas, is the basis of some nickel-refining treatments. High pressure batch operation produces nickel carbonyl together with some iron and cobalt carbonyl. These can be separated into refined products by distillation. Copper, sulphur, cobalt, precious metals, etc., are left behind in the residue from carbonylation. Both the reagent carbon monoxide and the metal carbonyls are toxic and hazardous. They are handled in closed circuit using equipment specially designed to prevent the release of toxicants to atmosphere.

Oxide nickel processing

Nickel occurs in lateritic accumulations, produced by weathering, overlying basic rocks. Ores of the limonite type have high iron content, low magnesium and silica, and nickel up to 1·5 per cent. Ores of the garneritic type have low iron content, high magnesium and silica, and nickel in the range 1·6–3·5 per cent. Extraction processes face difficulties due to variability of composition, inapplicability of physical separation methods and high moisture content. Hydrometallurgical routes are preferred for limonite ores; the developed processes, pressure leaching with sulphuric acid, and ammonia leaching of a partly reduced material, do not introduce substantial air pollution problems and will not be discussed further. Since the high magnesium content of the garnerite ores would entail excessive acid consumption, they are better treated pyrometallurgically, either by

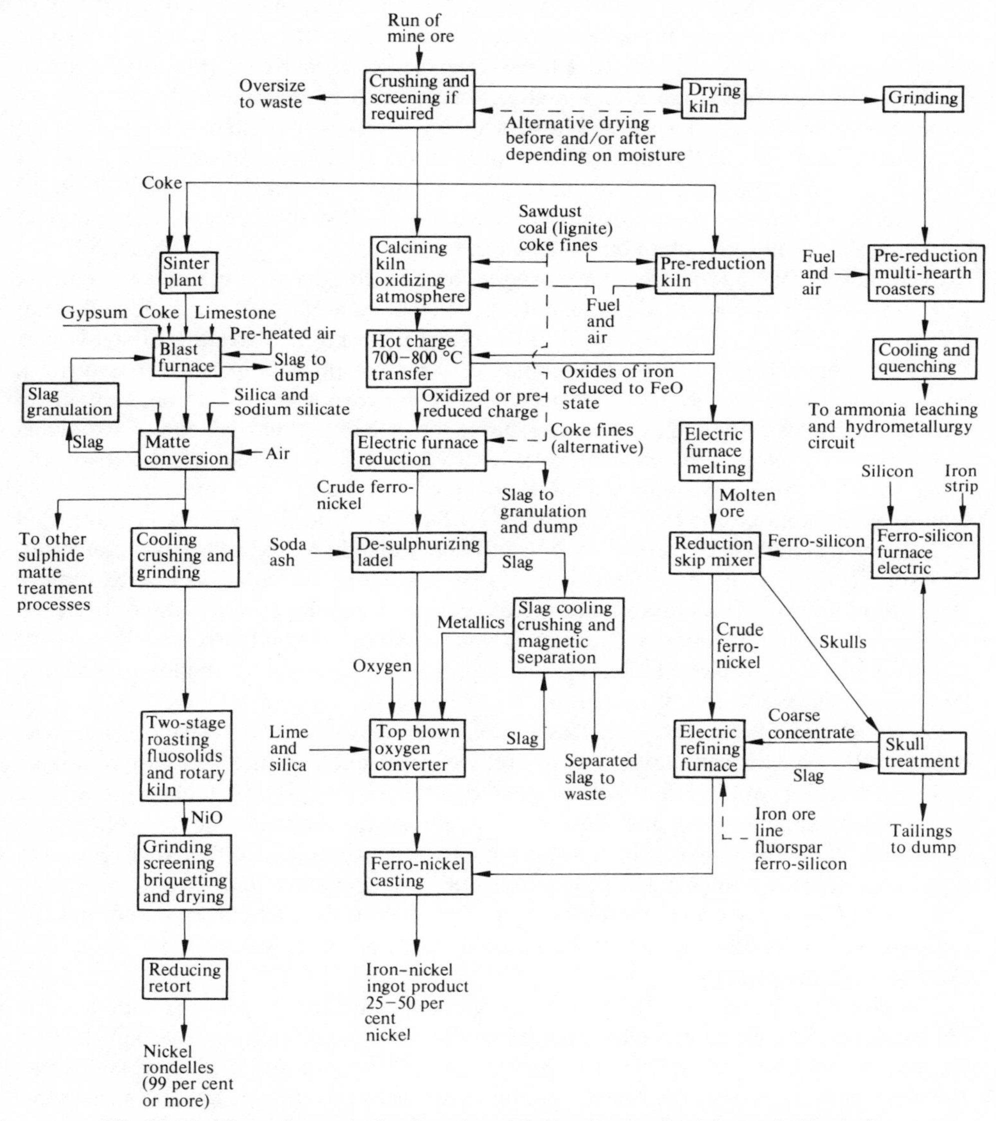

Fig. 15.21. Alternative methods of treating nickel laterite ores (mainly garnieritic type).

matte smelting or ferronickel smelting. The main pyrometallurgical process steps are indicated in Fig. 15.21.

In the matte smelting practice of Le Nickel in New Caledonia, gypsum is used as the source of sulphur to convert nickel to a sulphide matte. A similar operation in Japan uses nickel concentrate as sulphidizer. The ore is sintered, using coke fuel. The sinter, with added gypsum (concentrate) limestone, returned converter slag, and coke is smelted in a blast furnace, yielding a matte ranging 10–27 per cent nickel, and a reject slag containing about 0·3 per cent nickel and cobalt. The nickel–iron ratio in the ore must be within the correct range (SLN ore 2·8 per cent nickel, 13 per cent iron, dry basis) to produce a reasonable grade nickel matte, otherwise there is a high consumption of sulphidizing materials and coke and large sulphur dioxide volumes from converters for treatment. Treatment of the matte follows procedures already discussed in the previous section. The gas cleaning problems arising in producing the matte are firstly those of dust cleaning the sinter plant exit gas and cleaning of blast furnace gas, which is then burned in the combustion chamber of an air pre-heater.

Ferronickel smelting is based upon production of a liquid slag containing the gangue and part of the iron content, with most of the nickel and a part of the iron recovered in a ferronickel alloy. A typical procedure is to effect drying and heating of solids in a rotary kiln, and melting and partial reduction of the hot product in an electric furnace (in one case, in a blast furnace). The processes all use carbon for reduction, apart from that of Hanna Smelting Co., which is based on silicon as ferrosilicon. Ferronickel production is a large consumer of both hydrocarbon fuel and electrical energy, the latter mainly used in smelting a large tonnage of slag (a 1·3 per cent nickel ore can require approximately 55 000 kWh tonne^{-1} of nickel produced as crude ferronickel from oxide calcine). The ferronickel is usually upgraded in an oxygen blown converter. Desulphurization may be effected by ladle additions of soda ash, etc., before converting. Converter operation parallels oxygen steel-making practice, impurities such as silicon, carbon, phosphorus, chromium, being removed. A final ferronickel may have 25 per cent or 45–50 per cent nickel. At Larymna in Greece the ferroalloy is blown to 90 per cent nickel and cast as anodes for electrolysis.

Cleaning is required for off-gases from kilns, from electric furnaces and from converters. Hygiene ventilation is required in connection with ladle desulphurizing (sulphur dioxide) with handling of hot dusty calcines (700–800°C), in charging electric furnaces, in tapping metal and slag, and to remove tar emissions at the (Soderberg) electrodes. Kiln treatment calls for burning a large quantity of fuel (possibly with significant sulphur content) and the combustion products leave with steam and with a substantial (8–12 per cent) dust carry over. The sulphur dioxide content is not usually high enough to demand removal by scrubbing. Cleaning is generally effected with electrostatic precipitators.

The electric furnaces are designed to be gas tight and top pressure is kept close to atmospheric. The gas containing approximately perhaps 30 per cent carbon dioxide, 70 per cent carbon monoxide, is usually cleaned in a venturi scrubber and either flared at a stack or used in firing the kiln. Cleaning converter gases poses the same problem of fine iron oxide fume as in the steel industry. This is associated with control of dilution air supply to dissipate heat from combustion of carbon monoxide in the converter hood. High pressure venturis are typically used for cleaning.

Types of reactor used in non-ferrous pyrometallurgy

Most of the processes of pyrometallurgy involve reacting a metalliferous solid (possibly present in the reactor as a liquid) with a gas, either oxidizing (air, oxygen-enriched air, pure oxygen) or reductant (e.g., containing hydrogen or carbon monoxide). There is usually a product gas stream passing after treatment to atmosphere, and this is of particular importance in relation to eventual airborne pollution. A variety of reaction methods have been developed for carrying out these processes, and there is typically a choice available between alternative methods for any particular operation. Each type has strengths and weaknesses in relation to specific tasks; among the constraints will be suitability either for cleaning gaseous effluents to an acceptable level or for treatment in byproduct production processes. The following brief analysis presents the characteristics of the different reaction methods available (Table 15.8).

By considering the kinetics of the desired reaction over a range of temperatures, the melting points of the different species involved, and the place desirably occupied by the reaction in an overall flowsheet, one can arrive at the classification of the reaction in one of the categories:

solid feed—solid product
solid feed—liquid product
liquid feed—liquid product.

In fact the product is often a solid containing appreciable amounts of liquid, or a liquid with appreciable suspended solid. The validity of one type of reactor compared with another is typically concerned with the flow characteristics of the product, with

TABLE 15.8

US and UK standards

Type of emission	TLV
Copper fume	0·1 mg m^{-3}
Copper dusts and mists	1·0 mg m^{-3}
†Lead	0·15 mg m^{-3}
†Cadmium metal dust and soluble salts	0·2 mg m^{-3}
Cadmium oxide fume	0·1 mg m^{-3}
†Arsenic and compounds (as arsenic)	0·5 mg m^{-3}
Zinc oxide fume	5·0 mg m^{-3}
Sulphur dioxide	5·0 ppm
Carbon monoxide	50·0 ppm (to be reduced to 18 ppm)

† These values are under review in the US.

solids ranging from quasifluids to sticky agglomerates, and liquids of a range of viscosity.

Solid feed yielding solid product

In a number of reactors, reaction takes place at the top surface of a bed of material with gas flowing over it and some arrangement for stirring the bed to ensure that all solid particles are periodically exposed to reacting gas. Of this type are the following:

1. *Hearth roasters* (Fig. 15.15). An extended surface is obtained by forming beds of solid on a succession of vertically superimposed circular hearths, with mechanical raking. The reacting gas, usually air, is introduced at controlled rate at a number of levels. The intensity of reaction is relatively low and there is limited tolerance for sticky conditions in the solid.
2. *Rotary kilns*. In these, rotation of the kiln causes a slow renewal of the top surface at which gas reaction takes place. Heat is transferred to the solid both directly from gas and indirectly via heating of the refractory lining. In the Waelz kilns used for zinc oxide fuming, a reducing atmosphere is maintained within the solid charge at the same time as a highly oxidizing one above the bed. In large kilns (say 5 m in diameter, 95 m long) used for calcining or pre-reduction where the charge can be heated near softening point, excess fines in the charge can cause ring formation; and longitudinal segregation, thus producing batches of charge with variable composition for subsequent treatment.

A second type of reactor has reacting gas flowing through a bed of broken solids. The solid is typically agglomerates of material (pellets, briquettes, sinter lumps) characterized by a surface–weight ratio related to the sizing of the fine material and gas–solid exchange coefficients related to the size of agglomerate. Of this type are the following:

3. *Shaft furnaces*. Solid charge descends countercurrent to a stream of gas which may be introduced hot, or supply heat by a combustion process in the shaft. This is the traditional mode of reaction, more particularly in reactions with liquid products. The countercurrent nature of the process results in efficient use of fuel. The temperature history of the solid in this type of reactor is not under direct control, but derives from the simultaneous processes of heat and mass transfer between gas and solid.
4. *Fluidized beds* (Fig. 15.13). A wide range of fluidized bed systems exists. The solid may be of fine flotation concentrate sizing or agglomerated, e.g., by pelletizing; the bed may consist of the solid subjected to reaction or of an inert material such as sand in which the metalliferous solid reacts during a delayed passage through the bed. The product is collected both from an overflow of bed material and as suspended solids carried over in the gas stream, the proportion in

each fraction ranging widely depending upon the particular reaction carried out and the detailed design. The fluidized reactor can be contained in a welded gas-tight shell and gives good plant hygiene conditions. The reactor can treat a dry or slurried feed. The extensive solid mixing in a fluidized bed results in a reaction history with very closely controlled temperature range. The exchange processes are rapid and this permits, for example, oxidation of a sulphide to be maintained at high rate in a bed in which the sulphide is greatly diluted. The presence of very moderate amounts of molten product results in interparticle agglomeration, and upsets smooth operation, so that operation is restricted to temperatures below the melting point of species present in significant quantity.

5. *Sinter bed* (Fig. 15.17). By adjustment of the proportion of fuel in the bed of material on a moving grate one can establish, by preliminary heating of a shallow ignition layer, a regime in which continued flow of air through the bed maintains a hot reaction layer passing through the bed. This reaction system allows operation with maximum temperature above the melting point of some of the solids present; partial fusion, followed (after a period of the order of one minute) by cooling of the solid by flowing gas, results in a porous but strong structure of product suitable for treatment in shaft furnaces.

   This type of reaction arrangement is particularly important in relation to lead sulphide, when relatively low melting points prevent adoption of systems such as fluidized beds. The sintering process is associated with extensive solids handling since, in order to obtain a solid bed of the requisite fuel content, some 75–80 per cent of the solid product is crushed, mixed with new concentrates and returned to the sintering operation. A sinter machine is more difficult to keep gastight than a fluidized reactor. This combined with the large materials handling required for recycle, makes this type of operation more demanding on hygiene ventilation, and subject to dilution of process gases by air ingress. The heat generated in the roasting is retained in the solid, and is not practically recoverable for steam raising and power recovery.

A third family of processes provides a suspension of finely divided solids which ignites and reacts in its passage down an empty column.

6. *Suspension (flash) roasting* (Fig. 15.14), as applied for example to zinc concentrates, is of this type. The roasting is typically completed on hearths at the base of the suspension roasting column. This type of reactor requires ignition of the gas–solid mixture, and since the reaction time between gas and solid in the flowing stream is necessarily small, it is clear that the solids must be finely divided and reactive (dried). The completion of reaction on hearths results in some of the limitations of hearth roasters applying to this type also.

Solid feed yielding liquid product

Some of the systems already considered are applied also to situations in which solid feed is transformed to liquid products. Thus:

7. *Shaft furnaces* (Fig. 15.19) with liquid products provide the family of blast furnace processes. These are fed with agglomerates (see above) and can be of the open top type or closed top with charge fed through a double bell mechanism. The closed top type can be run under slight pressure to enable collection and treatment of off-gases without ingress of outside air. Furnaces can be round or rectangular in cross-section. Most non-ferrous furnaces are rectangular to give more chance of eliminating dead spaces (inaccessible to blast) for a given hearth area. Normally two liquids are tapped at the base of the furnace, a slag composed of unwanted gangue material together with silica or lime additions adjusted to provide a melt of the desired characteristics, and a liquid metal and/or sulphide melt. The fuel may be carbonaceous or provided by sulphides. Carbonaceous fuel in the charge is in the form of metallurgical coke, and in some cases can be supplemented by liquid or gaseous fuel injected with the blast through tuyères at the furnace bottom.
8. *Flash smelting* (Figs. 15.8 and 15.9) is a reaction carried out in the suspension mode but yielding liquid products. This type of process is applied to copper and nickel smelting. Flash smelting can be carried out either in a predominantly horizontal flowing gas stream using approximately 95 per cent oxygen in a modified reverberatory type furnace (Inco), or in a vertical reaction shaft down which the dried concentrates and preheated air (possibly with oxygen enrichment) are passed (Outokumpu).

   The Outokumpu process is in more general use and is licensed to many operators. The reaction gases and molten droplets of matte and slag pass down the reaction shaft and are separated into two phases in a settler. The degree of oxidation of the concentrates is controlled by the concentrate air ratio. Flash furnaces are under design to produce white metal (Cu_2S) directly from concentrates and in some cases (low iron concentrates) blister copper. The typical reactions taking place in the reaction shaft and settler are as follows:

Reaction shaft

$$6CuFeS_2 + 13O_2 \longrightarrow 3Cu_2S + 2Fe_3O_4 + 9SO_2$$
$$2CuS + O_2 \longrightarrow Cu_2S + SO_2$$
$$FeS_2 + O_2 \longrightarrow FeS + SO_2$$
$$3FeS + 5O_2 \longrightarrow Fe_3O_4 + 3SO_2$$

Settler

$$FeS + 3Fe_3O_4 \longrightarrow 10FeO + SO_2$$
$$2FeO + SiO_2 \longrightarrow Fe_2SiO_4.$$

In flash smelting the fuel value of the concentrates is fully used, and gas streams with strong sulphur dioxide and low free oxygen contents are produced that are suitable for acid making or direct reduction. Improved efficiencies have resulted from the use of shorter reaction shafts together with the use of higher speed oxygen concentrations permitted by better shaft temperature control.

9. *Reverberatory furnace smelting* (Fig. 15.4) may be considered the analogue, for a reaction with liquid products, of such processes as hearth roasting or kiln processes in which there is a large reaction area between flowing gas and an underlying solid. In reverberatory furnace smelting the solid is supplied at a number of feed points along a great part of the length of the furnace. This low intensity type of operation yields gases which are not suitable directly for either acid or sulphur production.
10. *Cyclone furnace.* In this type of reactor the feed of air, fuel and finely divided metalliferous material is directed at high velocity on a tangential (cyclonic) path. Intense reaction is achieved (due to high relative velocities of gas and solid) with rapid separation of liquid droplets out of the gas onto the walls, over which a liquid slag film drains into a collecting bath. Strong sulphur dioxide gas streams similar to flash smelting can be produced. Dust carryover from the furnace is less than for other types of flash reactor.
11. *Electric furnaces* (Fig. 15.10). Electric heating can be achieved by passing current through the solid charge or through liquid slag. In this type of heating the voltage of supply is adjusted to the resistance characteristics of the solid or liquid. Alternatively, heat can be supplied via an arc or a plasma. In all of these methods the heat requirements are supplied without the need for a large volume of gas sweeping through the furnace.

    Electric furnaces can be of the open or closed top variety. Most non-ferrous metal furnaces are closed top in order to reduce ventilation air requirements. However, the volume of gas leaving the furnace may be more than theoretically required for the reactions, because of the practical considerations of arranging for delivery of consumable electrodes, feeding the charge and mechanical access required. Electrodes can be 2000 mm in diameter, and present gas sealing problems. In some cases where furnace atmosphere is favourable, i.e., iron–nickel smelting, a good seal can be obtained between the electrode contact assembly and the roof. Some furnaces use gas recirculation to the electrode space to avoid in-leakage of air. On other furnaces, where sulphur dioxide atmospheres prevail, good sealing between electrode casing and roof is more difficult.

    There are two shapes of electric furnace, round (usually three electrodes) and rectangular (usually six electrodes in line). The non-ferrous industry uses both types, but for matte smelting the rectangular type is predominant. The round furnace has the advantages of improved structural strength, less cost to make air-tight, less refractory wear on walls, and possible capital cost reduction. Many electric furnaces tend to have hot spots, and the content of volatilized species in the gas is dictated by these high temperature regions rather than by average temperature.

12. *Converters: slag baths* (Figs. 15.5, 15.7 and 15.20). There are types of furnace in which metallurgical reaction is carried out between a gas flow and a liquid pool of slag and/or matte. These can be blown with gas either through submerged tuyères, submerged tip lances, or top blown with downward directed gas streams. Such baths can also be electrically heated, if the required reaction is between liquid slag and a second liquid or solid. While the reaction is between gas and liquid, it is frequently possible to operate with solid feed delivered either on to the surface or suspended in the inlet gas stream. Such reactions are typically very intense (Mitsubishi continuous copper smelting). If continuous operation is feasible it is preferred, but it has the limitation that in a continuous operation a reaction such as oxidation of a sulphide takes place with a liquid in which the sulphide may be present at low thermodynamic activity. For this reason, or because of either the development of accretions during the final stages of oxidation or control of end point, it is often necessary to operate in a batchwise manner. This results, if a sulphide is being oxidized, in a variable, and average low, sulphur dioxide content. Mixing of the liquid can be effected by rotation of the converter, or by the action of the gas jets. Rotation makes mixing independent of gas flow rate, i.e., oxidation rate.

Liquid feed yielding liquid product

Reactors with liquid feed are virtually all of the slag bath or converter type, just discussed. The typical hot liquid feed in metallurgical reaction is not readily adapted to feeding in subdivided form although the development of spray steelmaking indicates that if there were advantages in so doing one could feed liquid in the suspension mode.

Pyrometallurgy and process selection

Today, processes selected for a particular ore–concentrate treatment and location have to take into account the dramatic recent increases in energy, labour and capital costs. Labour and capital cost effects can be reduced by use of simpler, larger scale, more intensive operations. In the case of production of metals from disseminated sulphide ore bodies, the major expenditure of energy takes place in grinding the ore in the concentration process, and depends upon the grade of ore to be treated. However, the cost and efficient use of energy in the extraction process is becoming an important criterion in process selection and overall environmental impact. In these circumstances, a detailed analysis of the energy and environmental aspects of pyrometalurgical and other alternate steps is particularly necessary.

Treatment of bauxite to obtain aluminium and the pyro-processing of nickel laterite ores are unique in that they require large quantities of electrical energy in the reduction to metal stage. In the case of aluminium, the industry is experiencing difficulty at present in locating sources of cheap energy. There may be a movement of aluminium production away from the markets to developing countries with hydro-electric potential, and development of less energy intensive processes may be accelerated. But the change will be slow, and the only alternative process with potential at present is the Alcoa chloride process, which is also a molten salt process with gaseous reaction products. Laterite nickel treatment processes are largely determined by the chemical composition of the ore as already described. Pyrometallurgical processing is necessary for some types of ore, and in general pyrometallurgical processes give higher nickel recoveries. Ferronickel production from an electric furnace is a low capital cost route compared with hydrometallurgy; and although it requires a large amount of carbonaceous fuel as well as electricity, some operations take place near low grade lignite deposits which are used directly in the kiln or gasified for fuel. Practicable hydrometallurgical treatment routes of laterites are even more energy-intensive directly in fuel; and indirectly by use of high energy content reagents and high pressures.

Sulphide concentrates have a high fuel value in pyrometallurgical processes. The energy imported into a fully developed flash smelting operation is in the form of carbonaceous fuel. The concentrate fuel value is fully exploited in autogenous smelting of nickel and copper. The thermal content of fuel required to produce refined copper from an autogenous process might be $1{\cdot}4 \times 10^6$ kcal per tonne of concentrate ($\simeq$25 per cent copper) if sulphuric acid were produced and $2{\cdot}3 \times 10^6$ kcal/tonne^{-1} if sulphur dioxide were reduced to sulphur. These fuel quantities, if used to generate electrical energy (at 25 per cent efficiency), correspond to 360 and 590 kWh per tonne of concentrates. The pyrometallurgical extraction process and subsequent electro-refining demand electric power, but this is matched, with moderate deficit or surplus, by the power generated in waste-heat recovery on the smelter. The total energy requirement for copper hydrometallurgical processes is partly dependent upon energy input implicit in the use of reagent chemicals. A major common requirement is the energy for electro-winning, which from cupric solution is about 600 kWh per tonne of concentrates. Thus, even if the ideal hydrometallurgical route to yield elemental sulphur could be approximated, it would be difficult to compete on energy economy with the best available pyrometallurgical process for sulphur from copper concentrates.

Modern zinc production processes (electrolytic and blast furnace) both have pyro-roasting first steps, and hence require sulphur dioxide emission control. Energy requirements for the two processes are comparable assuming similar grades of zinc product and the additional equivalent lead smelting requirement for the electrolytic process.

Although process acceptability is closely related to economic and environmental requirements of a particular location, it can be said that purely hydrometallurgical processes, which avoid sulphur dioxide emissions, have the following characteristics compared with pyro-processing: lower metal recovery, more complex precious metal recovery, larger volumes of contaminated liquid waste, and less stable residues compared with slags. From the above, and with the added advantages, in the form of improved kinetics and phase equilibria, offered by high temperature processing, pyrometallurgical processing of the type described in this chapter will maintain its position for many years into the future, with a consequent need to treat hot off-gases for pollution control.

Sulphur dioxide problem

Non-ferrous smelting operations based on sulphidic feeds generate large quantities of sulphur dioxide diluted with air or fuel combustion products. The smelter has to fix the sulphur content in some form and limit the emissions to stringent limits. The problems are (1) the form in which the sulphur is to be fixed, and (2) economic methods of ensuring an acceptably low emission. These problems are also encountered in connection with power station gases, with many oil or gas refining plants, and in sulphuric acid plant practice, and have attracted very substantial research and development attention.

The only general sulphur dioxide disposal routes for the non-ferrous smelter, handling large annual tonnages of sulphur, are:

1. to convert the sulphur dioxide to sulphuric acid for sale;
2. to convert the sulphur dioxide to elemental sulphur;
3. to neutralize, probably with lime, to yield a stable solid product.

If transport of sulphuric acid to a market is not too expensive, this is the preferred product; new technology, now being commercialized, will make the second option a practicable possibility, with neutralization a last resort. Other possibilities, e.g., conversion into liquid sulphur dioxide, ammonium sulphate or alkali sulphite–bisulphite–sulphate compounds may exist in special local circumstances.

The gas stream produced in smelting is often rich enough in sulphur dioxide ($>3 \cdot 5$ per cent by volume) to sustain a contact sulphuric acid plant. For use in making sulphur by direct reduction, a proportion of sulphur dioxide greater than 10 per cent and of oxygen less than 1 per cent is desirable. Many smelting gases fall outside these limits, and a system of absorption–regeneration may be needed in order to produce acid or sulphur. In addition, in order to meet final emission standards, a treatment of tail gases, low in sulphur dioxide, may be needed.

Sulphuric acid manufacture

The sulphur dioxide-containing gas from a smelter contains impurities (arsenic, chlorides, fluorides, metal sulphide or oxide fumes, selenium, tellurium, mercury, etc.) which have to be removed to avoid rapid deactivation of the vanadium pentoxide converter catalyst. The high standard of cleaning the process gas is such that the impurities in the tail gas are well below the permitted environmental limits. Wet gas cleaning can be carried out by scrubbing with large volumes of water containing some sulphur dioxide in once through systems. Heavy metals have to be removed before the water is discharged. Alternatively, closed dilute acid scrubbing circuits may be used, from which contaminated acid is bled for treatment. Sulphur trioxide scrubbed out of the gas stream in these installations can amount to 1–3 per cent of total input sulphur. For gas borne emissions from the acid plant, the eventual problem is entirely that of the efficiency of the acid plant process in removing sulphur dioxide and trioxide from the gas. The former depends on the conversion efficiency of the plant and the latter can be dealt with by mist eliminators or absorbers.

Mercury, because of its volatility, finds its way into byproduct sulphuric acid even when the level of mercury in the sulphide concentrate is quite low. Low mercury in acid is required by the sulphuric acid users in food manufacture (0·5 ppm mercury limit in acid), animal feed and fertilizer production. Some mercury is precipitated in the form of particulates when roaster gas is cooled and cleaned by normal sulphuric acid plant gas cleaning methods, but unless special precautions are taken the bulk of the elemental mercury vapour and particulate compounds contaminate the product acid. Methods have been developed to remove mercury from the gas stream prior to conversion to acid and from the product acid itself. Outokumpu have developed a method for sulphating the mercury out of the gas stream by contacting with hot concentrated sulphuric acid; the mercury sulphate is separated from a saturated solution and further processed to elemental mercury. Boliden have also developed methods for removal of mercury from the gas stream, and from the product acid.

In a typical single-contact plant for making sulphuric acid, in which the gas stream passes through a succession of catalyst beds (usually four) and thence to absorption in sulphuric acid (single contact), the efficiency of converting sulphur dioxide to trioxide is about 97–98 per cent. A plant of this type requires gas strength greater than 3·5 per cent sulphur dioxide for autothermal operation. After absorption, the tail gas emitted from stacks contains 1000–3000 pp million sulphur dioxide compared with the maximum of 500 ppm permitted by current US regulations.

Passing the gas at an intermediate stage to an absorber to remove sulphur trioxide and returning it after heat exchange to further catalyst stages permits a higher efficiency of conversion of sulphur dioxide to trioxide. Plants of this type (double contact, double absorption, DCDA, Fig. 15.22) require higher sulphur dioxide gas strengths (>6 per cent) for autothermal operation. Conversion efficiencies up to 99·8 per cent (usually 99·5 per cent) are obtainable from the DCDA systems, and tail gas concentrations below 500 ppm are obtainable. The conversion–sulphur trioxide absorption section of a DCDA plant is higher in capital cost (15–20 per cent) than a single contact installation. However, a plant of this type is preferred, if feed gas strength permits, to a single contact operation followed by tail gas scrubbing.[25]

In sulphuric acid plants the conversion/absorption section operates optionally with a sulphur dioxide concentration of approximately 7 per cent. The capital cost of the gas-cleaning section depends on volume treated (i.e. sulphur dioxide concentration) but cleaned gas can be diluted to the 7 per cent level if necessary. The cost of the acid making section depends directly on the production of acid. It is worth noting that this is not the case with sulphur dioxide direct reduction plants (see later) where increases in gas strength reduce the cost of the entire reduction plant.

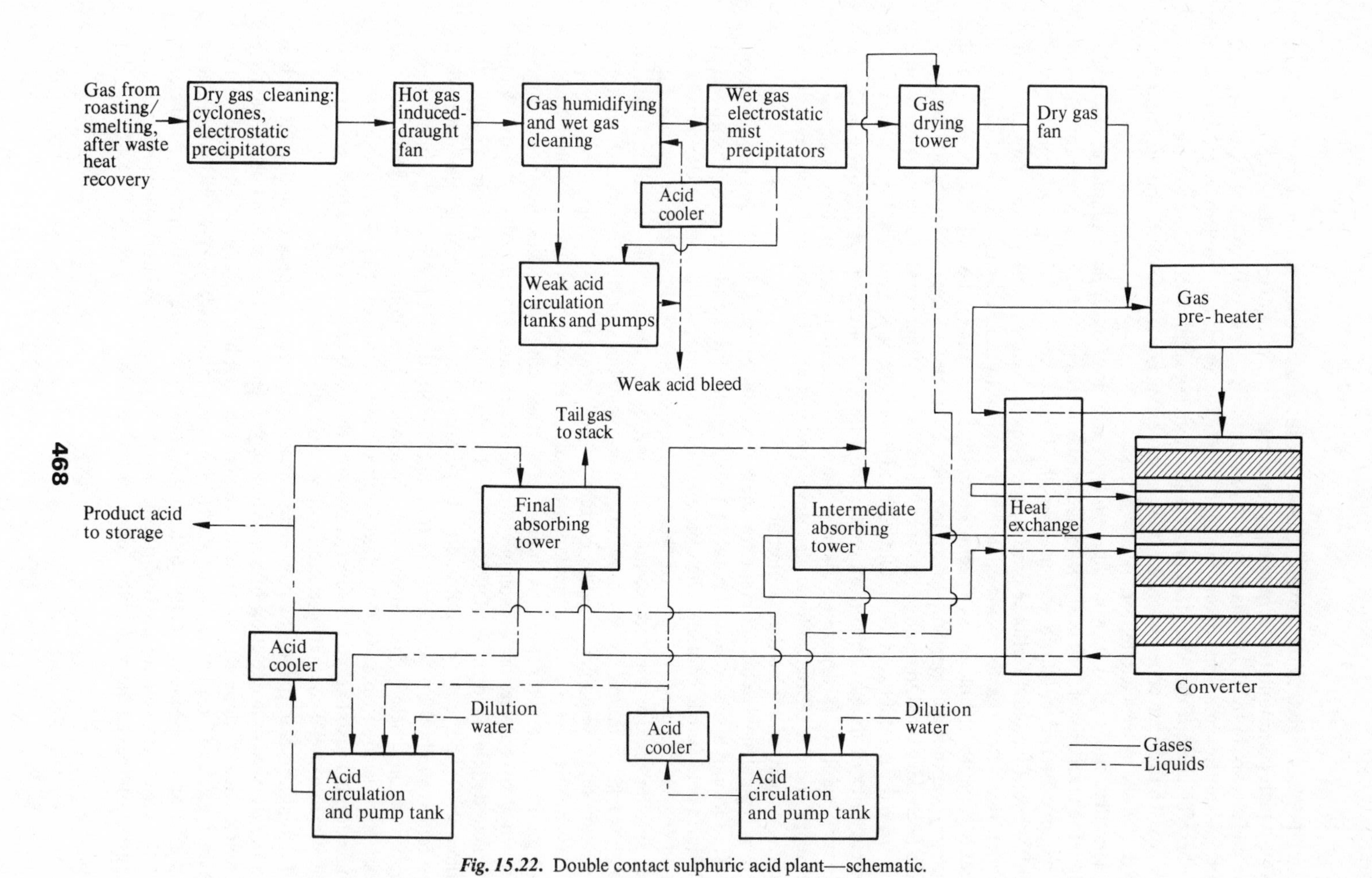

Fig. 15.22. Double contact sulphuric acid plant—schematic.

Production of elemental sulphur

Sulphur is the most attractive end product for plants remote from sulphuric acid markets. It can be stored easily with minimal environmental impact, and can be transported much more economically than the corresponding sulphuric acid (which will be three times the weight).

The sulphur production processes in the most advanced stage of development, already having been operated on a commercial scale, are based on direct reduction of a strong sulphur dioxide gas stream of constant volumetric flow and low in free oxygen content. Gas meeting these requirements can be obtained directly from flash smelting, fluosolids, and suspension roasting and from some continuous copper smelting processes. With some other processes, including reverberatory smelting, batch copper converting, sintering, electric furnaces for matte smelting, because of low sulphur dioxide concentration, high oxygen content or volumetric variation, the gas has to be subjected wholly or in part to absorption–regeneration treatment.

The size of sulphur dioxide reduction plant is related to the total gas volume treated, and is smallest for very concentrated gas. The gas produced in absorption/regeneration processes can approach 100 per cent sulphur dioxide, and allows the possibility of recycling the reduction plant tail gases after incineration back to the absorption stage, yielding a closed circuit. Strong gases are also produced in the processes in which oxygen or highly enriched air streams can be used; thus Outokumpu flash smelting can yield 18 per cent sulphur dioxide gas, Inco oxygen flash smelting about 80 per cent sulphur dioxide. If a strong sulphur dioxide–low oxygen gas stream of this type can be reduced directly without an absorption–regeneration step, this presents the cheapest route to synthetic sulphur. A serious deficiency of all established pyrometallurgical copper processes is their failure (because of batch converting) to yield total sulphur input in a form suitable for direct feed to a sulphur-producing process. This is a major target for some of the newer continuous copper processes, and elimination of an absorption–regeneration stage will be a major achievement of such processes. Development of continuous converting with or without a continuous matte feed, but supplying a continuous gas stream, is a worthwhile goal, and may be achievable with small modifications to existing processes.

Direct reduction processes produce a tail gas requiring further treatment (see later). However, the US Bureau of Mines sodium citrate process is unusual in that the use of this absorbent allows regeneration of the scrubber effluent directly to elemental sulphur.[26] Smelter gases containing 1–3 per cent sulphur dioxide and free oxygen are processed as follows:

1. drying and dedusting; cooling to 50°C;
2. absorption in buffered citric acid/sodium citrate at pH 3·8 (at pH values above 4, conversion to sulphate is accelerated);
3. liquid-phase Claus conversion to sulphur:

$$SO_2 + 2H_2S \rightarrow 3S + 2H_2O;$$

4. removal of sulphur from the citrate solution by thickening/filtration and melting.

Hydrogen sulphide is generated as required, usually from methane:

$$CH_4 + 4S + 2H_2O \xrightarrow[\text{alumina}]{700°C} CO_2 + 4H_2S.$$

Two pilot plants are currently being operated in the US to evaluate this technology. Like other absorption-based processes for treatment of dilute gases, it is unlikely to be competitive for sulphur with processes based on direct reduction of strong gas streams. However, it may well be attractive as an adjunct to a process where production of dilute sulphur dioxide containing oxygen is unavoidable.

The direct reduction processes do not necessarily provide a profitable treatment if one bases the process notionally on a zero cost supply of smelter gas. The costs of sulphur fixation must be achieved at the lowest net cost to the primary objective of copper production. Thus the overall costs have to be compared with alternative methods of sulphur dioxide disposal. Comparative studies have been made of a flash smelter–batch converter complex making acid and one using absorption–regeneration to collect batch converter gas and feed it plus flash furnace gas to a direct reduction plant. Incineration of sulphur plant tail gas was included in these studies but no tail gas treatment. The following conclusions were drawn:

1. Recovery of sulphur values to elemental sulphur depended on initial gas strength to the reduction plant, but was about 85 per cent, i.e., the tail gases contained 1–2 per cent sulphur dioxide, equivalent to a reverberatory furnace.
2. The location must supply a cheap source of suitable reductant, in the above cases natural gas at US \$0·5 per 1000 SCF (17·66 US\$ per 1000 m^3).
3. Assuming no revenue from by-products (acid or sulphur) the overall direct operating costs per tonne of cathode ($\simeq$100 000 tonnes per annum) for the sulphur production scheme were about 15–20 per cent higher than for acid production.
4. The capital cost of the copper–sulphur producing plants was 15–20 per cent higher than the copper–sulphuric acid plant.
5. The difference in operating costs would pay for long-distance acid transport and every effort should be made to find and develop markets over a large area.
6. Production and neutralization of acid is more expensive than sulphur production.
7. Provision of a steady gas feed to sulphur dioxide reduction is more important than for acid production. The reduction process is more intimately associated with the smelting operation, and in order to avoid production difficulties in the primary operation (copper production) a sophisticated control system (computer) is preferable for the reduction plant.

The production of sulphur is worth considering if acid markets cannot be found even over extended distances, and environmental control can be relaxed to lower the quantity of sulphur to be fixed. Incorporation of sulphur plant tail gas treatment (see later) will increase the costs and hence the breakeven point with acid transport costs. A copper smelting process that would provide a single continuous gas stream (development of continuous converting) would much reduce the cost of sulphur production.

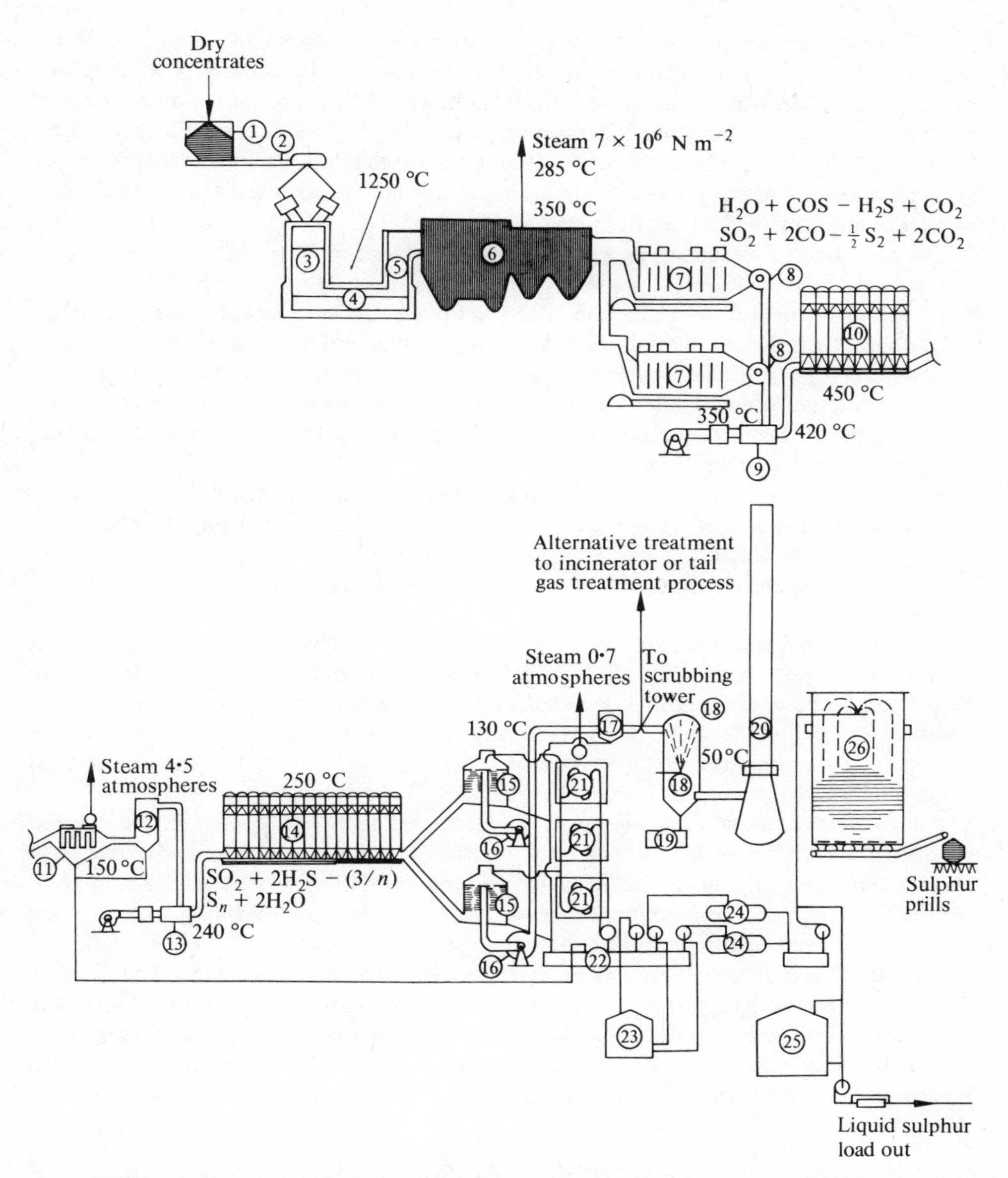

Fig. 15.23. Outokumpu process for direct high temperatures reduction of sulphur dioxide: (1) burner feed bin; (2) redler conveyer; (3) flash reaction tower; (4) flash furnace settler; (5) flash furnace uptake and SO_2 reduction chamber; (6) waste heat boiler; (7) electrostatic precipitator; (8) hot gas induced-draught fan; (9) direct fired gas re-heater; (10) hot catalyzer; (11) gas cooling boiler; (12) mist eliminator; (13) direct fired gas reheater; (14) cold catalyzer; (15) sulphur irrigated condensing towers; (16) cold gas fans; (17) mist eliminator; (18) water irrigated scrubbing tower; (19) settling tank; (20) stack; (21) liquid sulphur cooling boilers; (22) sulphur sump; (23) sulphur circulation tank; (24) arsenic removal autoclaves; (25) sulphur storage tank; (26) prilling tower. (*Courtesy* Outokumpu Oy, Finland.)

The two most developed direct reduction processes are those of Allied Chemical[27] and Outokumpu.[28] The Outokumpu process is based on technology developed for pyrite (FeS_2) processing at Kokkola (Fig. 15.13). The Allied Chemical process as used to treat gases from a fluidized bed roasting of pyrrhotite (FeS) is shown in Fig. 15.24. The Outokumpu process can use natural gas, naptha, oil or pulverized coal; the reductant for the Allied process is natural gas, but use of other gaseous reductants is under development.

The overall direct reduction process typically includes four stages:

1. a high temperature reduction step, catalytic (Allied) or non-catalytic (Outokumpu), reducing sulphur dioxide to sulphur but yielding also compounds such as carbon oxysulphide, hydrogen sulphide, carbon monoxide and hydrogen;
2. an intermediate catalytic reaction step at lower temperature in which carbon oxysulphide, hydrogen and carbon monoxide react further with sulphur dioxide to increase reduction efficiency;
3. a low-temperature catalytic reaction between hydrogen sulphide and sulphur dioxide (the Claus reaction) to minimize the residual content of sulphur gases;
4. a tail gas treatment, to bridge the gap between the best available Claus reaction system, and the environmental requirements for tail gas emission.

To be suitable for a non-catalytic first stage, the smelter gas temperature should be at least 1250°C (preferably 1300–1350°C). Copper flash smelter gas is suitable and can be reacted without gas cleaning. By choice of catalyst it may be possible to compress the three stages (1)–(3) above into two, as appears to be done in the Allied process, but such a process requires wet gas cleaning prior to reduction, to prevent catalyst contamination.

Elements such as arsenic, mercury, selenium and tellurium appear in the sulphur product if smelter gas is cleaned only by the dry hot gas method. To meet the specification for sulphur to be used in acid manufacture, it may be necessary to incorporate sulphur purification. Processes based on wet gas cleaning to acid plant standards can produce high purity sulphur in the reduction plant. Direct reduction of sulphur dioxide is a high energy consumer. Sophisticated methods of heat recovery on Allied and Outokumpu plants make use of a large proportion of the fuel consumed. Integration of the direct reduction process with a modern autogenous smelting process results in consumption of purchased fuel comparable with the requirement on a reverberatory plant discharging gases to a stack.

Treatment of dilute gases[29]

The term 'dilute gases' is used in its broadest sense and gases discussed in this section include the following: gases low in sulphur dioxide from primary sulphide treatment process; tail gases from sulphur dioxide treatment processes for producing acid or

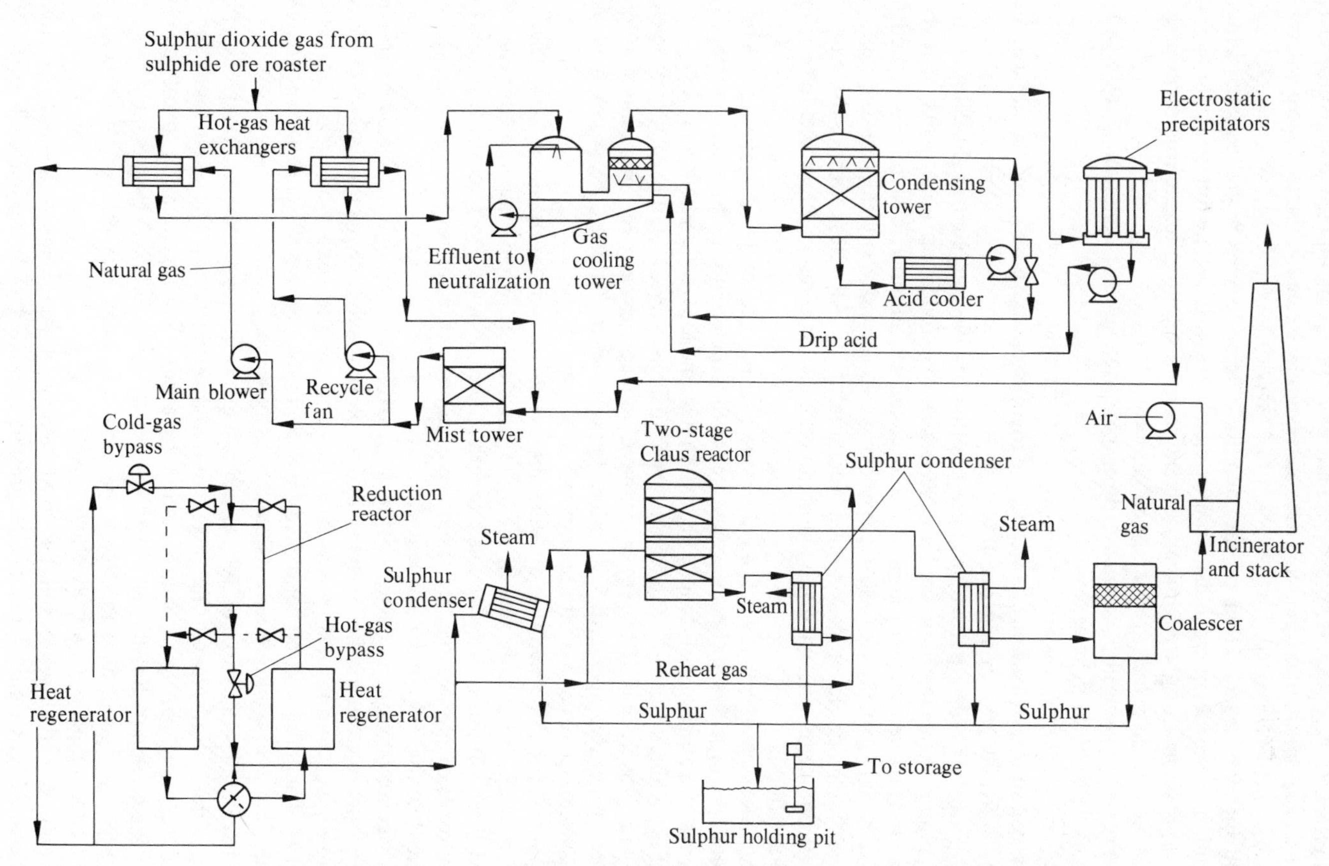

Fig. 15.24. Allied Chemical sulphur dioxide direct reduction process. (*Courtesy* Allied Chemical Co.)

elemental sulphur, gases intermittent in nature or containing unwanted constituents (for feed to a process) such as oxygen. These gases can be treated in the following ways:

1. absorption–regeneration of the sulphur dioxide to give a gas stream richer in sulphur dioxide, in most cases removing oxygen at the same time;
2. improvement of the chemical conversion of a sulphur dioxide treatment process by an add-on process reducing tail gas emissions;
3. scrubbing the weak gas stream to produce solid sulphur-containing waste and tail gas suitable for discharge.

1. Absorption–regeneration. A variety of processes, both dry and wet, have been developed for absorption–regeneration.

Dry processes are of particular interest in that hot discharge gases should give much better plume dispersal than cold moist gases. The following absorbents have been proposed; activated carbon, copper oxide, alkalized alumina, molten salt eutectic of $Na_2SO_3/K_2SO_3/Li_2CO_3$ at 400°C. The last two proposals yield, on regeneration, hydrogen sulphide in a gas stream for treatment. All these proposals are in a development stage; with alkalized alumina, work has been discontinued due to problems of solid attrition.

The processes which have been developed to the prototype commercial stage are generally wet processes. Those receiving considerable attention at present are based on two types of absorbent: organic, e.g., dimethylaniline DMA,[30] and alkaline, e.g., sodium sulphite (Wellman Lord),[31] magnesium oxide suspension,[32] etc. Other processes such as sea water absorption (Boliden) and the Cominco process of ammonium sulphite absorption are also operated. The latter, dating from the 'thirties, is used on acid plant tail gas, but requires a market for ammonium sulphate.

In general it can be said that alkaline absorbents are limited by absorption efficiency to 2·5 per cent sulphur dioxide in the input gas streams, while organic absorbents such as DMA are not attractive with gas strengths under about 3·5 per cent sulphur dioxide, due to losses of organic reagent. The steam requirements for regeneration are less with organic than with alkaline absorbents.

All wet absorption processes suffer from oxidation of sulphur dioxide to trioxide; some absorbent solution has to be vented to prevent sulphate accumulation. Oxidation can take place during absorption or in the process gas stream before it reaches the absorber. The latter can be controlled by limiting the ingress of oxygen, removing dust such as iron oxide, and cooling as quickly as possible.

Incineration of tail gases at 540°C yields a gas with all contained sulphur as sulphur dioxide, which until recently has been emitted to atmosphere. The incinerated gas can alternatively be fed to a sulphur dioxide absorption–regeneration plant and the regenerated sulphur dioxide returned to process. This solution is particularly attractive if absorption–regeneration is already necessary to deal with primary feed gases and volume emit capacity is made available.

Alkaline absorbents are suitable for treatment of acid plant tail gases, and dilute reverberatory gas, and plants are at present operating or installing such equipment.

Stronger gas such as that obtained from Pierce–Smith converters could be absorbed in an organic reagent such as DMA and the regenerated gas supplied to a direct reduction plant. This type of operation is not practised as yet.

2. Improvement of chemical conversion. For acid plants, this can be obtained by double contact, double absorption, as previously discussed. If the feed gas is not suitable for this plant, sulphur dioxide absorption–regeneration or sulphur dioxide scrubbing of gas down to an acceptable exit stack must be practised.

All direct reduction flowsheets involve, as the final reduction step, the Claus reaction between sulphur dioxide and hydrogen sulphide:

$$SO_2 + 2H_2S \rightleftharpoons 3S + 2H_2O$$

Low residual contents of sulphur dioxide and hydrogen sulphide are favoured by low temperature and the use of catalysts. Below the dewpoint of sulphur, condensation of sulphur impairs the catalyst activity. With alumina catalysts the typical Claus reaction temperature is 200–400°C and conversions of 94–96 per cent are obtained. The gas leaving the plant typically contains 1–2 per cent sulphur dioxide, 2 per cent hydrogen sulphide, 0·3 per cent carbon oxysulphide, 0·3 per cent carbon disulphide, 1 per cent carbon monoxide, 2·5 per cent hydrogen, 10 per cent carbon dioxide. Treatment of this gas to yield an acceptable stack gas is a problem being tackled by the natural gas industry.

Lower temperature operation of the Claus reaction can be performed in two ways. Liquid phase processes such as that of the Institut Français du Petrole (IFP)[33] carry out the reaction in an organic solvent-containing catalyst, at a temperature just above the melting point of sulphur. Combinations of a polyalkylene glycol as solvent with alkali metal salts of the carboxylic acids, benzoic acid, salicyclic acid, and nicotinic acid are cited in the literature. Alternatively a solid catalyst bed is operated at low temperature such that it coats with solid sulphur; the sulphur is periodically distilled using heated inert gas. This process (Lurgi Sulphreen)[34] and the IFP process are in operation on natural gas plants; the sulphur emission is reduced substantially, but is similar to that of a single contact acid plant, i.e., 1000–3000 ppm sulphur dioxide, and would not meet US limits. A difficulty in these processes is that residual carbon oxysulphide, carbon disulphide and carbonyl sulphide in Claus kiln tail gas are not removed.

Gaseous sulphides in tail gas can be converted to hydrogen sulphide by use of a cobalt molybdate catalyst at temperatures similar to those of normal Claus operation. Hydrogen sulphide can then be removed to low levels using the Stretford Process.[35] A combination of the catalytic process and Stretford which converts the hydrogen sulphide to sulphur is in an advanced state of development, e.g., the Beavon Process.[36] It is claimed this process can reduce sulphur levels in tail gas to 250 ppm, i.e., within current US emission standards.

3. Scrubbing to solid sulphur-containing waste. Scrubbing the total weak gas from the smelter is the least attractive solution, but may have to be undertaken on old plant producing poor strength gases, where the only alternative may be to close down. Several methods have been proposed for scrubbing sulphur dioxide gas streams with lime or limestone to produce a waste calcium sulphate product. These processes can be categorized into wet or dry processes. Dry processes inject calcium or magnesium oxide into the flue gases, and recover the products in a dry collector such as an electrostatic precipitator or baghouse. However, the efficiency of dry processes is low and they are not attractive for treatment of smelter flue gas.

Wet processes contact the flue gas in a scrubber with a slurry of limestone or lime (which means that raw limestone has to be milled or calcined). The sulphur dioxide is removed as a mixture of calcium sulphite and sulphate. The limestone process has lower capital and operating costs than the lime process, but these advantages are partly offset by a lower efficiency and the larger excess of reagent needed. The limestone process may not be suitable for gases lacking carbon dioxide or oxygen. When limestone is used its reactivity can vary depending on the source, and this can affect the scrubbing efficiency. Both limestone and lime almost invariably contain magnesium in amounts ranging up to several per cent. This makes possible the formation of soluble magnesium sulphate which can give rise to an effluent problem. Scrubbing with lime meets some pollution standards, but problems are encountered with waste disposal. Over three tons of limestone are required for scrubbing one ton of flue gas sulphur, and over four tons of waste gypsum have to be discarded from the operation. Lime and limestone wet scrubbing processes still need engineering development.

Secondary metal processing[37]

Secondary feeds are acquiring increasing importance, because of the need to conserve primary metal resources. Such materials are of two main types:

1. *residual materials*, drosses, fumes and skimmings from both primary and secondary extraction processes;
2. *scrap materials*, which may be
   (a) 'new', i.e., reject material, turnings, borings, stampings, etc., from metal fabrication processes, or
   (b) 'old', i.e., components or fabrications which have reached the end of their service life in fabricated form.

Residual materials are essentially chemical compounds, though frequently admixed with elemental metal. They are reprocessed either by recirculating to a primary or secondary smelting operation or are converted to purified products by the chemical industry. Examples in this category include the following:

1. residues containing lead from solution purification for electrolytic zinc smelting which can be reprocessed by lead smelters;

2. lead and copper-containing drosses which are usually reworked by the primary or secondary copper smelter;
3. zinciferous residues from zinc 'hydros' manufacture which are processed through to zinc chemicals;
4. galvanizers skimmings and ashes which can be treated by primary zinc smelting;
5. drosses from aluminium processing which are converted through to aluminium chemicals.

Air pollution control for the processing of these materials thus follows the procedures outlined in the sections on primary processes.

Scrap materials are the basic raw material for secondary smelters. Since they consist essentially of elemental metal or metals, the refining processes are physical rather than chemical in nature, with the notable exception of removing magnesium from secondary aluminium by chlorination. However, 'old' scrap in its primary form frequently contains a wide variety of metals mixed more or less intimately with inorganic and organic materials such as wood, leather, plastic, rubber, paint, varnish and so on. Since much reprocessing of scrap is carried out in units which are small and unsophisticated by primary smelting standards, and because the crude scrap is frequently both grossly heterogeneous and widely variable in composition, solutions to the complex air-pollution problems are sometimes difficult to define and implement. Thus, although in environmental terms there is a major incentive for scrap recycling, the practical operation of secondary processing plants may result in difficult problems of controlling emissions to the atmosphere.

Pretreatment for scrap feeds

1. Hand-sorting, magnetic and density separations. Scrap in its primary form can vary in composition so widely that few generalizations are possible. At some stage in the reprocessing cycle, not necessarily at the secondary smelter, manual hand-picking is still practised to separate potentially valuable metal-containing material from other worthless constituents. This stage may be followed or preceded by separation of essentially ferrous material by magnetic means, although even after this has been done, additional manual sorting may be necessary to remove items such as 'foreign' screws, inserts, bushings, strappings and the like. More recently, elutriation with water streams has been developed to provide basic separation processes to remove plastic, leather, etc., material from the metal constituents of automobile and consumer-durable scrap by density difference. Density separation can also be effected by forming a fluidized bed of a dense particulate material, e.g., sand, in which scrap separates upwards (organic materials) or downwards (metal). Scope clearly exists to design components so as to facilitate recycling. Wherever feasible, fabrications should be made from materials which are amenable to separation in recycling processes.

2. Oil removal. Contamination of scrap by lubricating and cutting oils is usual, and can be a source of unacceptable emissions of black smoke during subsequent processing, particularly in aluminium and brass reprocessing operations. The smaller sizes of scrap can now be pre-treated thermally in a rotary kiln provided with an after-burner to consume smoke and oil vapours. This process, known as the *Intal* process, is used successfully in many countries, but it is not suitable for treating the larger sizes of scrap. To ensure that such scrap does not generate undesirable emissions during processing, recourse must be made to the rather expensive processes of solvent de-oiling or degreasing or centrifuging. Such processes have their own well-developed technologies of environmental control. If circumstances make this process impracticable, great care must be taken to control the rate of addition of oily scrap to the melting or sweating furnace.

3. Scrap cable. The removal of rubber and plastic insulation by pre-burning is a major source of smoke emission and, with the rapidly growing use of PVC insulations, a troublesome source of hydrochloric acid emissions as well. It has now been a scheduled process under the UK Alkali Acts for more than ten years. Efficient after-burners can reduce smoke emission to satisfactory levels, but reduction of hydrochloric acid requires well-maintained aqueous scrubbing plant. A recent development to overcome this problem is the use of cable-shredding equipment to chop the scrap cable into short lengths from which the plastic insulation can be separated relatively easily by density separation.

4. Cryogenic methods. On cooling to a temperature around −140°C, both organic materials such as rubber and a number of metals become brittle, so that crushing them produces a comminuted material, in which the individual particles are composed of one or another of the original materials, without flattening or bonding together. Although energy is used in the cooling operation, the process can be economical insofar as physically separated fractions can be obtained with less mixture or contamination than is feasible at room temperature or above.

Secondary metal processing techniques

1. Zinc. The relatively low melting and boiling point of zinc permits a two-stage separation–refining process for zinc scrap. The melting stage, known as sweating, is usually carried out in small, sloping-hearth reverberatory type furnaces provided with some means of agitating or stirring the charge, or sometimes in rotary kilns. At the end of a cycle, the molten zinc is separated from unmeltable constituents, which sink to the bottom of the furnace, and is cast for subsequent purification, if required, by distillation or refluxing. Some secondary zinc is processed to zinc dust.

A more sophisticated type of sweating operation involves holding scrap in a basket and raising to closely controlled temperature in a molten salt bath. This yields a liquid metal, which separates downwards out of the salt, and a remaining solid of the other metals still free from oxidation. By arranging for heating to a sequence of temperatures, related to the melting point of the metals or alloys involved, a set of molten metal fractions with minimum intermixture can be obtained.

Emission problems depend greatly on the degree of pre-sorting and the type and homogeneity of the scrap charged. Inadequate pre-sorting leaves a variety of organic plastics to be burned off during the sweating stage, giving smoke emissions usually containing hydrogen chloride. The emission also contains particulates which are predominantly zinc oxide, some zinc and other chlorides (depending on the presence or absence of fluxes in the charge) and particulate impurities derived from other metals in the charge. Emission control systems thus vary with the particular type of scrap being processed, but may comprise systems for smoke suppression, aqueous scrubbers for neutralization of acid emission and bag-plants to remove particulates.

2. Lead. Extensive secondary processing is a feature of the lead industry. Lead scrap (e.g. pipe, sheet or cable sheathing) is usually processed through to soft metal or low alloys. The high proportion of scrap lead-acid batteries which is re-processed provides feed for the alloy lead market.

In the past, battery smelting processes required prior removal of some of the organic components. This was achieved by hand-breaking supplemented by the use of mechanical saws and guillotines. Such arduous labour-intensive methods were replaced by bulk battery processing techniques, e.g. Stolberger, Tonolli, Penarroya, etc., which utilize impact crushing or drum breaking followed by heavy media separation or elutriation to isolate the metal and organic fractions. The heavy media are often made up from a slurry of battery paste fractions.

The bulk processes require careful environmental design to control lead dust emissions from both crushing and dried sludge spillage. Plastic battery cases, which are being increasingly used, are often difficult to crush and satisfactory techniques remain under development.

These difficulties, together with restrictions on the dumping of organic materials containing lead have spurred development of whole battery processes by, for example, Bergsoe and Oerlikon. In the first, the battery is processed in a blast-furnace provided with an incinerator; the carbon content of the battery supplements the furnace coke requirement. In the Oerlikon process, batteries are treated in a co-current kiln where the organic material is burned; some sulphate decomposition necessitates adequate disposal of sulphur oxides.

Fuming of slags produced in lead smelting is important because of the zinc frequently associated with lead in concentrates. The zinc and some residual lead is distilled off and re-collected after oxidation as a fume, which becomes the feed to a plant for zinc metal winning (electrolysis). The slag fuming operation entails production of large volumes of hot combustion products (nitrogen, carbon dioxide, with some sulphur dioxide) with a substantial content of particulate lead and zinc components. The gases pass through waste heat boilers and the fume is collected, usually in bag filters. A similar fuming process is carried out for removing tin from some slags. The tin is customarily fumed as the sulphide; addition of, e.g., pyrites to the feed may be necessary. The gases contain a substantial proportion of sulphur dioxide, insufficient, however, for direct manufacture of sulphuric acid.

Secondary lead is sometimes combined with primary material for refining. Depending on feed materials and product specification, one or other of the conventional pyro-refining routes is employed. If the product specification demands high bismuth elimination, electrolytic refining is used as discussed previously.

3. Copper and its alloys. Substantial proportions of refined copper and copper alloys are obtained by reworking scrap or secondary materials. Pyrometallurgical processes are used. As with primary copper production, final refining, where practised, is electrolytic.

The converters used in primary copper smelting, working on mattes containing iron sulphide, generate surplus heat and additions of scrap copper are often used to control temperature. The converter thus provides a convenient and cheap form of scrap treatment, but often with only moderately efficient gas cleaning. Alternatively hydrometallurgical treatment of scrap is possible, using ammonia leaching, to yield solutions which can be reduced by hydrogen to copper powder (which is a premium product) or subjected to solvent extraction to give feed to a copper-winning cell.

Leaded brasses (about 3 per cent lead) are widely used in the engineering industry, and recycling of scrap waste is a major activity. Such brass scrap feeds frequently contain much swarf and turnings coated with lubricant and cutting oils. Cables and motors invariably contain plastic or rubber insulants, varnishes and lacquers. Scrap pre-treatment to remove these non-metallics, by one or more of the procedures described earlier, is therefore essential to control unnacceptable emissions of black smoke, acid, etc., and to maintain acceptable in-plant working conditions. The production of appreciable quantities of oxide fume, particularly zinc oxide, is inevitably associated with brass melting and casting.

Thus, the reprocessing of brass presents difficult emission control problems, in that hot carbon and metal oxide particles (about 40 per cent of which may be sub-micron) have to be trapped simultaneously. Conventional equipment achieves varying degrees of control and, though increasing success is being obtained with modified equipment, wholly satisfactory solutions have not yet been found.

Bag plants are widely used but the mixture of carbon and zinc oxide particles is spontaneously flammable, and many bag-house fires have occurred. Irrigated bag filters avoid the fire hazard and permit collection of metalliferous fume as sludge; they are currently in the developmental stage, but are showing promise of providing one reliable method of collection.

Low pressure wet scrubbers are inefficient for the collection of the fume, and high pressure venturi scrubbers are considered expensive to operate and to suffer from corrosion and sludge build up. Cyclones are inefficient for trapping the finer particles, but are good for reducing the larger particle emissions and have shown promise experimentally for reducing fume from Intal driers. Wet electrostatic precipitators may also be developed to provide a solution to this problem, but this approach currently suffers from cost and corrosion problems. The impure zinc compounds thus collected are recirculated, either to a primary smelter or to chemicals manufacture.

From the economic and resource viewpoints, growing importance will attach to the reworking of the maximum possible proportion of copper-containing scrap. For this reason it is to be hoped that recent progress in emission control will be maintained, with the object of developing more efficient and reliable systems than those presently based on smoke incineration and cyclones–bag-plants for particulate control.

4. Aluminium. Secondary aluminium is an important fraction of total metal supply, around 20 per cent in the US. Typical of secondary smelting, the industry is characterized by a large number of relatively small plants treating mostly 'new' scrap. These are basically natural gas- or oil-fired reverberatory units, and the emission problems are as indicated earlier (i.e., smoke, acids and particulates) with one major addition, aluminium chloride and its hydrolysis product, hydrochloric acid. Energy for secondary refining consumes only about 5 per cent of that required for primary aluminium production.

Because the charge contains about one per cent magnesium, this has to be removed to prevent off-grade castings when the refined aluminium is cast. This is done by lancing the molten charge with chlorine gas during and after the melting cycle, thus reducing the magnesium to below 0·1 per cent and removing occluded hydrogen but causing emission of corrosive aluminium chloride and hydrochloric acid.

Secondary aluminium processing thus faces the difficult problem of suppressing emissions of corrosive aluminium chloride associated with hydrogen chloride. Conventional gas-cleaning trains have not been very successful in dealing with this problem, but at least two process technologies are now available which deal with the problem at source. One approach, the Derham process, uses proprietary fluxes. It claims a 97 per cent magnesium–chlorine efficiency for the chlorination stage at magnesium levels of less than 0·1 per cent. For a relatively modest installation cost, it claims the following advantages:

1. stoichiometric and hence economic in chlorine consumption,
2. complete air pollution control,
3. improvement in recovery and cycle time,
4. reduced flux consumption
5. clean and gas-free product,
6. simple installation and rugged operation.

The Alcoa fumeless process depends on effecting a stoichiometric chlorination of magnesium in a multi-stage enclosed settler–reactor tank after melting and prior to casting. Efficient gas–liquid contact gives a selective magnesium chlorination reaction (99 per cent efficiency claimed); the magnesium chloride which is separated from the chlorination tank is of saleable grade and is tapped off from time to time as necessary. Its general claims are similar to those made for the Derham process, and so it appears that secondary aluminium smelters now have two technologies to resolve the long-standing pollution problems associated with magnesium removal.

TABLE 15.9

Some typical reactors used in treatment of sulphides

Type	Multi-hearth roaster	Sinter machine	Suspension roaster
Application	Dead roasting pyrites	Roasting zinc-lead concentrates	Roasting zinc concentrates
Reactor geometry	Air in at various levels. Hot gas recirculation. Solid disposed on series of beds.	Air updraught through moving bed of mixed concentrates and recycled crushed - product.	Concentrates/air fed to top of empty column and react in descending cloud. Reaction completed on hearths.
Nature of feed	Concentrates as received; dried on top hearths.	Sized product <5 mm (80%) plus concentrates (15–20%) in moist mixture.	Wet concentrates to drying hearths. Dried concentrate finely ground before use.
Feed distribution	Achieved by raking	Laid with minimum compacting on to hot ignition layer	Dried concentrates delivered via burner in air stream
Inlet gas			
Composition	Cold air and recycled boiler exit gas	Cold air and recycled low SO_2 gas	O_2 enriched air plus recycled product gas
Temperature (°C)			
Pressure	Low pressure		Low pressure
Outlet gas			
Composition	10–11% SO_2, $\simeq$ 5% O_2	6–7% SO_2, $\simeq$10% O_2	7% SO_2, 10% O_2
Temperature (°C)	800	300	1000
Particulates		5–10% of feed	
Gas treatment	Waste heat recovery and acid plant	Acid plant	Waste heat recovery and acid plant
Products	Fine calcine	Porous agglomerate 30–100 mm 0·5–1·0% total S	Fine calcine 2% S
Operation features			
Rate	75 tonnes per day	1000 tonnes per day	350 tonnes per day
Continuity	Continuous	Continuous	Continuous
Restrictions (Advantage)	Maximum temperature restricted to below softening point	(Maximum temperature above softening temperature)	Lead content
Other applications	Zinc and copper concentrates	Lead concentrates	Iron sulphide roasting

Type	Fluidized bed (dead roast)	Fluidized bed (sulphating roast)	Fluidized bed (pellet roast)
Application	Roasting zinc concentrates	Roasting copper concentrates	Roasting nickel sulphide
Reactor geometry	Air fed through grate into single expanded bed of solids. Solid particles well mixed with virtually constant temperature. Heat extraction from bed, or water addition to control temperature.		
Nature of feed	Concentrates or slurry		Pellets
Feed distribution	Via belt slinger or slurry gun		
Inlet gas Composition Temperature (°C) Pressure	 Air Ambient 34 kN m^{-2}	 Air Ambient 17 kN m^{-2}	 O_2 enriched air Ambient 34 kN m^{-2}
Outlet gas Composition Temperature (°C) Particulates Gas treatment	 13% SO_2, 2% O_2 approx. 900–950 50–60% of feed Waste heat recovery plus acid plant	 3–5% SO_2 650–720	 1100
Products	Fine calcine 1·7–2·2% total S (1·1–2·0% S as SO_4^{--})		Granular − 10 + 100 mesh 0·2 S
Operation features Rate Continuity Restrictions (Advantage)	 500 tonnes per day Continuous Temperature restricted to softening point	 240 tonnes per day Continuous Temperature held in appropriate range for sulphate stability	 165 tonnes per day Continuous (Can operate above feed softening point)
Other applications	Pyrites, copper concentrates, nickel sulphide	Cobalt sulphide	Zinc concentrates

TABLE 15.9—continued

Type	Blast furnace (closed top) (Momoda)	Reverberatory furnace (Green feed)	Electric furnace (Green feed)
Application	Copper concentrate matte smelting	Copper concentrate matte smelting	Copper concentrate matte smelting
Reactor geometry	Vertical shaft furnaces with water jackets and tuyères at bottom. Charge fed through central slot at top. Matte and slag are separated in a forehearth.	Burner gases pass over piled solid, underlying liquid matte and slag tapped periodically.	Resistance heating of slag. Piled feed, melted by heat transference from slag. Matte and slag tapped periodically.
Nature of feed	Pugged viscous concentrates, returned lump converter slag, limestone and coke.	Moist or dried concentrates	Moist or dried concentrates.
Feed distribution	Pugged concentrate and lump material alternatively layer charged, pugged material descends in shaft centre, charge passes to walls.	Concentrates through ports in roof. Pulverized coal, natural gas, or oil at burners.	Concentrates through ports in roof. Soderberg electrodes fed with green carbon mixture.
Inlet gas			
Composition	Preheated air and oil injected at tuyères	Air plus O_2 up to 28%	
Temperature (°C)	300–400	200–400	
Pressure	1300 mm W.G. pressure blower	Low pressure fan	

Outlet gas			
Composition	Before dilution 4–6% SO_2, 1% O_2, after dilution 3–4% SO_2, 8% O_2	1·5–2·5% SO_2, 0·25–1·0% O_2	5–6% SO_2, 15–16% O_2
Temperature (°C)	350	1250	500–600
Particulates	Approx 10% of feed	Approx 5% of feed	Dust 3–5% of feed.
Gas treatment	No waste heat recovery, some gas to acid recovery, some to stack.	Waste heat recovery, gas to stack. Not suitable for acid plant or direct reduction.	Waste heat recovery and acid plant.
Products	Matte 35% Cu Slag 0·35–0·45% Cu	Matte 30–45% Cu Slag 0·5–0·7% Cu	Matte 30–45% Cu Slag 0·4–0·6% Cu
Operation features			
Rate	200–250 tonnes per day new concentrates in feed	Up to 1100 tonnes per day dried green feed	Up to 1300 tonnes per day concentrates
Continuity	Continuous	Continuous	Continuous
Restrictions (Advantage)	Limited to small units, high operating cost, (labour and coke).	Substantial fuel requirement. Low SO_2 in off gas.	Electrical energy 400 Kwh/tonne charged. Needs cheap power.
Other applications		Nickel concentrates	Nickel and lead concentrates

TABLE 15.9—continued

Type	Flash smelting Outokumpu	Flash smelting Inco	Cyclons furnace (under development)
Application	Copper concentrates Near autogenous matte smelting	Copper concentrates Near autogeneous matte smelting	Copper concentrates matte smelting
Reactor geometry	Preheated air, dry concentrates and flux sprayed into vertical reaction tower. Molten drops of matte and slag separated in horizontal separator. Off gas passes to vertical offtake. Furnace geometry lends itself to zoning oxidizing-reducing.	Dry concentrates and flux sprayed into horizontal oxygen fed burners at each end of furnace. Drops of matte and slag separated in hearth below burners. Gases pass to central vertical offtake.	Cyclone shaped reaction chamber giving high relative velocities between concentrates and oxidising gases; slag and matte flow down walls, exit with gases at base of cyclone.
Nature of feed	Dry concentrates return dust and flux	Bone dry concentrates and silica flux	Dried concentrates + fine flux
Feed distribution	Concentrate fed to centre of burners, air around periphery. Air-concentrates ratio strictly controlled.	Oxygen-concentrates ratio strictly controlled in burner.	Dry concentrates + fine flux fed through roof of cyclone chamber, preheated air/O_2 fed tangentially.
Inlet gas Composition	Preheated air. Oxygen enrichment can be used. 23–30% O_2	95% O_2 introduced through steam heated lines	Preheated air and/or O_2 can be used.
Temperature (°C)	250–900		
Pressure	Low pressure air		Low pressure air

Outlet gas			
Composition	10–18% SO_2, 0·25–0·5% O_2	80% SO_2	SO_2/O_2 similar to flash smelting
Temperature (°C)	1250–1350	1250	
Particulates	8–9% of feed	3% of feed	2–2·5% of feed
Gas treatment	Waste heat recovered. Gas to acid or direct reduction plant.	No heat recovery practised, but possible. Gas wet cleaned prior to liquid SO_2 production. Gas suitable for acid production or direct reduction.	Waste heat recovery possible, gas suitable for acid plant or direct reduction.
Products	Matte 50–65% Cu (possibly oxidation to white metal 79% Cu). Slag 1·0–1·5% Cu, needs further cleaning treatment.	Matte 47% Cu Slag 0·6% Cu – disposable	Matte and slag mixture flow to settler for separation. Slag requires further cleaning
Operation features			
Rate	1700 tonnes per day concentrates	1500 tonnes per day dry concentrates	High intensity reactor, rate 3 tonnes m^{-3} h^{-1} achieved.
Continuity	Continuous	Continuous	Continuous
Restrictions (Advantage)	Concentrates with high proportion of volatile metals may give waste heat boiler problems.	Need for tonnage oxygen	High heat loss and refractory wear
Other Applications	Nickel concentrates		Mixed copper, lead, zinc concentrates

TABLE 15.9—continued

Type	Pierce-Smith converter	Top-blown rotary converter	Mitsubishi furnace
Application	Conversion Cu matte to metal	Conversion nickel matte to metal	Copper concentrates. Near autogenous smelting to blister copper.
Reactor geometry	Air through tuyères (38–50 mm dia.) bubbles through liquid matte, liquid slag forms top layer.	Rotating 40 rev min^{-1} Kaldo type converter. Air-oxygen-fuel lance blowing onto surface of bath, rotation giving good mixing for slag/metal reactions independent of blowing rate.	Three linked stationary furnaces smelting-slag cleaning-continuous converting. Slag and matte, 65% Cu, pass to electric slag cleaning furnace where slag is separated and discarded; matte underflows to continuous converting furnace.
Nature of feed	Molten matte 30–65% Cu. Agglomerated concentrates or solid Cu scrap can also be fed.	Nickel containing briquettes and pellets from nickel sulphide concentrates, copper/nickel metallics and anode residues.	Concentrate and fine flux dried to >1% moisture, plus return granulated converter slag and flue dust.
Feed distribution	Liquid tipped from ladle into converter mouth, solids fed through mouth.	Agglomerates + fluxes fed through mouth of converter. Converter can be tilted as well as rotated.	Air and solid feed introduced through top jetting consumable lances. Air-concentrate ratio strictly controlled. Similar lances used for smelting and converting furnaces.
Inlet gas			
Composition	Air usual. O_2 enrichment possible to 38%.	Oxygen/air/fuel	Oxygen enriched air
Temperature (°C)	Ambient	Ambient	Ambient
Pressure	103–117 kN m^{-2}		1–2 × 10^2 KNm^{-2}

Outlet gas			
Composition	4–9% SO_2 after dilution	Approx 50% SO_2	$\simeq$12% SO_2 overall
Temperature (°C)	1200 at converter mouth	1380–1600	1150–1200
Particulates		Low dust carryover	Depend on level of impurities
Gas treatment	Waste heat recovery possible, gas suitable for acid production.	Gas suitable for waste heat recovery and acid production.	Waste heat recovered from smelting and converting furnaces. Gas to acid plant or possibly direct reduction plant.
Products	Blister Cu with 0·1–0·2% S and precious metals. Slag with 3–5% Cu requires cleaning	'Fire refined' nickel for feed to carbonyl process.	Blister copper$\simeq$0·5% S from continuous converter. Discard slag$\simeq$0·6% Cu from slag cleaning furnace.
Operation features			
Rate	250–500 tonnes per day	275 tonnes per day fire refined nickel	550 tonnes per day concentrates
Continuity	Batch operation	Batch operation	Continuous
Restrictions (Advantage)			Removal of trace elements, i.e. Pb, Bi, Sb, As, from blister may not be as good as with batch converting. Temperatures kept low $\simeq$1200°C to limit refractory wear.
Other applications	Copper/nickel matte	Copper and lead concentrates	

In-plant conditions

Conditions on pyrometallurgical plant can be hot and dusty, and some of the operations are arduous. Control of this environment to acceptable standards is essential to the health of the operating personnel, and to attract the correct quality of operator into the industry from a diminishing pool of suitable labour. Provision of suitable working conditions must be considered at the early design stage of the plant and suitably integrated. Capital spent on plant environmental control can amount to 10 per cent or more of the plant capital cost.

The main health problems arise from:

1. *dusts* containing heavy metals from materials handling;
2. *fume* (particulate less than 10 μm) produced from processing hot liquid metals and slags;
3. *gas leaks* with dust and fume from process equipment into the working atmosphere.

The aluminium industry is unique among the non-ferrous plants considered, in that maintenance of a suitable in-plant working condition is interrelated to process requirements such as heat removal from pots, type of anode, etc. Usually, in-plant control is separately superimposed on the process by provision of ventilation hoods and gas cleaning equipment at points where required. Volumes of ventilation air are often far in excess of volumes of process gas treated in the plants. The quantity of ventilation air required to capture contaminants in a hood depends upon hood size and shape, position relative to the emission point, and nature and quantity of air contaminant. Some operations can be surrounded by a totally ventilated enclosure, but in others collection hoods remote from the emission source are required. Hood design and location depend on the operating requirements of the particular piece of equipment and are usually a compromise. The velocity of air towards the hood can decline within a distance about equal to the hood diameter to a value only one-tenth of that at the hood face. The contaminants are often released into the atmosphere with considerable velocity but their momentum is quickly dissipated. The aim must be to provide a suitable draught at the null point, the point at which the initial momentum of the contaminant stream has been lost.

Draughts can badly affect capture by hoods located remote from the emission points, and capture velocities of the order of 30 m min^{-1} (100 ft min^{-1}) are minimum. Collection of emissions from hot sources (the usual case in pyrometallurgy) is further complicated by heat transferred to the surrounding air by conduction and convection. This causes a thermal draught which is exacerbated by entrainment of surround air and can cause rising air currents of over 120 m min^{-1} (400 ft min^{-1}). Hood face velocities greater than the velocity of the rising air column are required for capture.

Some examples of different types of hood are as follows:

1. Total enclosure hoods are used in some plants to exhaust emissions of sulphur

dioxide escaping the primary collection system on Pierce–Smith converters. The hoods have doors which are opened during the converter filling (out-of-stack) operation and closed when the converter is blowing to stack. Hoods of this type are also used on lead refining kettles, but the mechanical operations required make it difficult to maintain complete closure; thus, high ventilation volumes are required.

2. Remote hoods are usual near tap holes where fume from slag and matte is collected, and enclosure is not possible due to the mechanical operations required.
3. Another typical example is in collecting the emissions from the charge wells of reverberatory furnaces smelting secondary aluminium.
4. In some applications the whole building forms a ventilated enclosure. This is always the case with primary aluminium pot rooms; recently a copper converter aisle has been designed on these lines.

The threshold limit values (TLVs) shown in Table 15.8 are widely recognized guidlines for the maintenance of acceptable in-plant atmospheres.

Ventilation gases collected in hoods are often scrubbed to remove the pollutants before the gases are emitted to stack. When the fume produced is of micron size, high pressure venturi scrubbers and/or wet and dry electrostatic precipitators are used. Bag filters are used in some applications. Modern high-pressure felt bag filters provide an efficient dry cleaning method for micron sized dust.

Automated process control is tending to reduce the number of jobs which have to be performed on smelters by operators working in hot, dusty and often noisy areas. However, a certain number of manual operations is always necessary and adequate ventilation, properly directed to local 'hot spots', is vital to preserve an adequate working environment. In the future, much more attention will have to be paid to noise suppression.

References

1. *Mineral facts and problems*, US Department of the Interior, Bureau of Mines, Washington DC, 1970 (and other editions); Metalgesellschaft AG. Metalstrahshes, 1965–1975.
2. Pearson, T. G., *The chemical background of the aluminium industry*, Royal Institute of Chemistry Monograph, London, 1955.
3. *U.S.P.*, 3,725,222 (1973).
4. Iversen, R. E., *J. Metals*, **25**(1), 19, 1973.
5. Rossano, A. T., Jr. and M. J. Pilat in *Proceedings of the Second International Clean Air Congress*, p. 701, Academic Press, New York, 1971.
6. *Air pollution by fluorine compounds from primary aluminium smelting*, OECD, Paris, 1973.

7. *Threshold limit values for 1975*, Technical Data, Note 2/75, Health and Safety Executive, UK, H.M. Factory Inspectorate, HMSO, 1976.
8. Cook, C., G. R. Swany and J. W. Colpitts, *APCA Meeting*, 1970, Spokane, Air Pollution Control Assn., Pittsburg, 1970.
9. Nelson, A. H., *Proceedings, BNFMRA* (*now BNF Metals Technology*) *International Conference on Air Pollution and Water Conservation, Basle*, Wantage, Berkshire, 1969.
10. Tyne, F. R. in D. M. Liddell (ed.), *Handbook on non-ferrous metallurgy*, p. 227, McGraw-Hill, London, 1945.
11. Nagano, T. and M. Miimura, *J. Metals*, **21**(7), 76, 1968.
12. Leroy, J. and P. J. Lenoir, IMM Symposium *Advances in extractive metallurgy*, Inst. Mining Metallurgy, London, 1967.
13. Queneau, P., C. E. O'Neill, A. Illis and J. S. Warner, *J. Metals*, **22**(7), 36, 1969; Daniels, R. A. and L. H. Jaquay, *AIME Meeting*, San Francisco, Metallurgical Soc. Amer. Inst. Mining, New York, 1972.
14. *J. Metals*, **7**(6), 742, 1955; Merla, S., C. E. Young and J. W. Matrusch, *AIME Meeting*, San Francisco, 1972.
15. Bryk, P., J. Ryselin, J. Honkasalo and R. Malmstrom, *J. Metal,* **10**(6), 395, 1958; Fuju, T., M. Ando and Y. Fujiwana, *Joint Meeting MMIJ and AIME*, Mining and Met. Inst. Japan, Tokyo, 1972; Mealey, M., *Engineering and Mining J.*, **173**(6), 130, 1972.
16. Herneryd, O., O. A. Sandstrom and A. Norro, *J. Metals*, March **6**(3), 330 1954; Rodolf, D. W. and E. R. Marble, Jr., *J. Metals*, **24**(7), 14, 1972.
17. Themelis, N. J., G. C. McKerrow, P. Tarassoff and G. D. Hallet, *AIME Meeting*, New York, 1971; Themelis, N. J. and G. C. McKerrow, *IMM Meeting*, London, 1971.
18. Suzuki, T. and T. Nagano, *Joint Meeting MMIJ and AIME*, Tokyo, 1972.
19. Worner, H. K., J. O. Reynolds and B. S. Andrews in R. P. Ehrlich, (ed.) *Copper metallurgy*, AIME, New York, 1970; Worner, H. K., *Engineering and Mining Journal*, **072**(8) 64, 1971; Worner, H. K. in *Advances in extractive metallurgy and refining*, IMM, London, 1972; Reynolds, J. O., K. J. Philips, D. E. Fitzgerald and H. K. Worner, *AIME Meeting*, San Francisco, 1972.
20. Kuhn, M. C., N. Arbiter and H. Kling, *Canadian Institute of Metallurgy Meeting*, Edmonton, 1973.
21. Andre, J. A. and R. J. Delvaux, *AIME Meeting*, St Louis, Vol. II, p. 178, 1970; Morijama, E. and Y. Yamamoto, *ibid.*, p. 198; Steinveit, G. *ibid.*, p. 223; Wuthrich, H. R. and A. von Ropenack, *ibid.*, p. 247; DeMichelis, T. and F. Gnesotto, *ibid.*, p. 269; Banes, P. H., R. K. Carpenter and C. E. Paden, *ibid.*, p. 308, 1970.
22. Morgan, S. W. K. and S. E. Woods, *Engineering and Mining J.*, **159**, 95, 1958; Morgan, S. W. K. and J. Lumsden, *AIME Meeting*, San Francisco, 1959; Woods, S. E. and D. A. Temple, *Trans. IMM*, **74**(6), 297, 1964–5; Sellwood, R. M., *AIME Meeting*, St Louis, Vol. II, p. 581, 1970; Bonnemaison, J., M. Defonte and A. Lefaucheux, *ibid.*, p. 619.

23. Beilstein, D. H., *AIME Meeting*, St Louis, Vol. II, p. 702, Gibson, F. W., *ibid.*, p. 738, 1970.
24. Boldt, J R., Jr., in P. Queneau (ed.) *The winning of nickel*, p. 191, Methuen, London, 1967.
25. British Patent 904,982 (5/9/62 to Bayer AG); *Sulphur*, **75**, 34, 1968; Browder, T. J., *Chem. Eng. Prog.*, **67**(5), 45, 1971; Tucker, W. G. and J. R. Burleigh, *Chem. Eng. Prog.*, **67**(5), 57, 1971.
26. Rosenbaum, J. D., D. R. George, L. Crocker, W. I. Nissen, S. L. May and H. R. Beard, *AIME Meeting*, Washington DC, 1971; Sulphur, **101**, 37, 1972.
27. Hunter, W. D., Jr., and A. W. Michener, *Engineering and Mining J.*, **174**(6), 117, 1973; Johnson, C. F., *AIME Meeting*, San Francisco, 1972.
28. Malström, R. and T. Tuominen, *Advances in extractive metallurgy and refining symposium*, London, 1971; *Sulphur*, **50**, 33, 1963.
29. Various authors, IChemE/University of Salford Symposium, *The control of gaseous sulphur compound emission*, Salford, 1973.
30. Fleming, E. P. and T. C. Fitt, *Ind. Eng. Chem.* **42**(11), 2249, 1950.
31. Earl, C. B. and B. H. Potter, Ref. 29.
32. Houston, P. S. and G. R. Koehler, Ref. 29; Shah, I. S., *Chemical Engineering*, **79**(6), 80, 1972.
33. Barthel, Y., S. Franckowiak and Ph. Renault, Ref. 29; Franckowiak, S., P. Bonnifay, R. Dutriau and A. Deschamps, Process Technology International, **17**(10), 785, 1972.
34. Krill, H. Ref. 29.
35. Moyes, A. J. and J. S. Wilkinson, Ref. 29.
36. Beavon, D. K., *Chemical engineering*, **78**(12), 71, 1971; Beavon, D. K., Ref. 29.
37. *EEC Seminar on the Control of Emissions from the Non-ferrous Metallurgical Industries*, Dubrovnik, 1973; *BNFMRA International Conference on Air Pollution and Water Conservation*, Basle, 1969.

16 Inorganic chemical industries

B. C. SPALL

The chemical industry produces a wide variety of solid, liquid and gaseous air pollutants, some of which are damaging to biological life or are objectionable because of odour or appearance. There is concern not only over the nature of these pollutants but also over the amounts involved, as a single chemical plant often represents a very large point source of polluting effluent. Over the last few years this problem has intensified with the very rapid increase in plant capacity that has occurred. For example, a typical single-train ammonia plant now produces 1000–2000 tonnes day^{-1} of ammonia compared with perhaps 200 tonnes day^{-1} only 10 years ago. These increases in output are accompanied by corresponding quantities of gaseous by-products potentially available for discharge, increasing the problems of local dispersion and adding to the total polluting load.

Further problems which acutely affect several sections of the chemical industry are the very obvious nature of the stack emissions and the concern which is always felt over a process which is not understood. This draws a great deal of adverse and often uninformed comment from the general public. The brown plume of nitrogen dioxide from the stack of a nitric acid plant, for example, can be seen over large distances and is often the centre of complaints from local residents. By installing a catalytic decolourizer to convert the nitrogen dioxide to nitric oxide the complaints are silenced, although the total volume of pollutant remains the same. Similarly there is more objection to the emissions from a sulphuric acid plant than from a local power station, because the public realizes that an acid plant produces harmful sulphur oxides but does not realize that much larger quantities of the same materials are produced by most power stations.

Additional problems arise from the grouping of plants at large factory sites, where the aggregate emission of effluents from the site is large, even though the individual plant emissions may be relatively small. In some cases reactions may occur when the gaseous effluents of one plant mix with those from another; e.g., ammonium sulphate fogs produced when ammonia and sulphuric acid plant effluents are mixed is an obvious example. Because of these complexities and the different approaches made to the control of emissions, the problems facing the chemical industry are best considered in relation to individual products.

Sulphuric acid

Sulphuric acid plants emit both sulphur oxides and sulphuric acid mists in their tail gases. A typical 500 tonnes day^{-1} 100 per cent sulphuric acid plant discharges the equivalent of about 6·5 tonnes day^{-1} of sulphur dioxide, which is little more than a 200 MW power station burning heavy fuel oil discharges in one hour.[1] Despite this relatively small contribution to the overall pollution load, sulphuric acid plant emissions have received a great deal of attention over many years, and the legislative authorities continue to set progressively tighter discharge limits.

Emission of sulphur oxides

Figure 16.1 is the flow diagram of a conventional four-stage sulphur burning contact acid plant. Molten sulphur is burnt in a stream of dry air to produce generally a 7–8 per cent strength sulphur dioxide gas. This is converted to sulphur trioxide over a vanadium pentoxide catalyst. The trioxide is absorbed into 98 per cent sulphuric acid in a packed

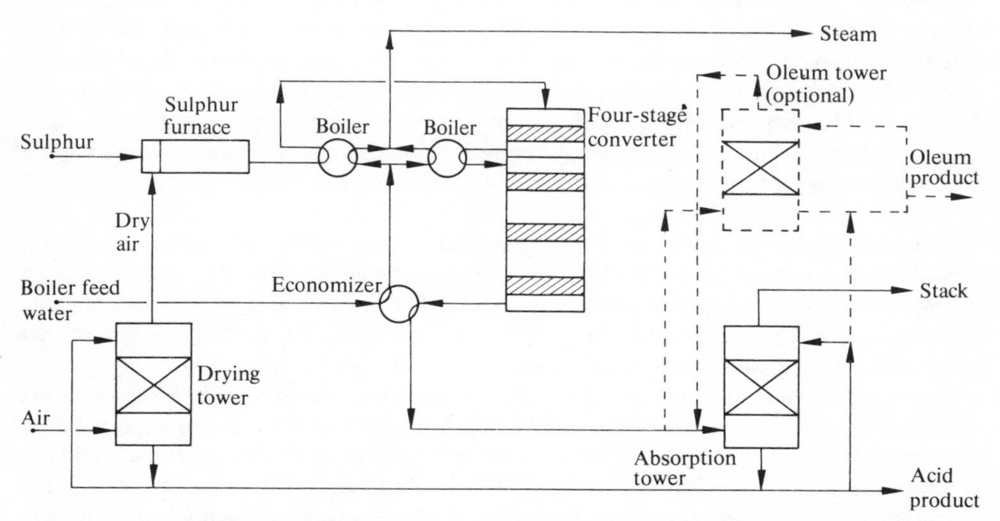

Fig. 16.1. Flow diagram of sulphur burning contact plant for production of sulphuric acid and oleum.

tower operating at high liquid recycle; a second tower is installed when oleum is required. The overall reaction is

$$SO_2 + \tfrac{1}{2}O_2 \rightleftharpoons SO_3 \qquad (16.1)$$

Modern plants operate at a conversion efficiency of 98–98·5 per cent, which corresponds to a level of mixed sulphur dioxide and trioxide in the tail gas of about 2000 ppm. To reduce this further, three broad types of process are available: double absorption, the use of liquid absorbents and the use of solid absorbents.

1. Double absorption. This process depends for its effectiveness on the removal of sulphur trioxide from the reaction gases so that Eq. (16.1) can be shifted to the right. Gases from the second or third stages of conversion are cooled and passed to a primary absorption tower for removal of sulphur trioxide. They are then re-heated and passed over one or two final stages of conversion before entering the secondary absorption tower. A large heat exchanger is required to cool the gases entering the primary absorption tower and reheat those leaving this tower, and serious corrosion problems can occur unless steps are taken to ensure that the gases are not cooled below their dewpoints.[2] The final exit gases contain about 500 ppm of sulphur dioxide and the output of acid is increased by 1–1½ per cent.

For a new plant the additional costs of the heat exchanger and the absorption tower can be offset to some extent by operating the process at a higher sulphur dioxide gas content of about 10–12 per cent, reducing the overall plant size per unit of output. The process can also be added to an existing sulphur burning plant, though here it is difficult to increase sulphur dioxide strength without substantial plant modifications. Further, although there is an increase in the output of acid, the need to cool and reheat the gases at the intermediate stage results in some overall reduction in steam output, which can be a disadvantage. For plants using smelter gas, which is often of varying concentration, sulphur dioxide levels in the gas leaving the primary absorption tower can often fall to the point where reaction over the final catalyst stages cannot be sustained. Temperatures will then drop below the point at which oxidation occurs.

These difficulties reduce the likelihood of the process being applied to many existing plants, but it is being used increasingly for new sulphur burning plants and is favoured by many authorities as a proven method for reducing sulphur dioxide emissions.

2. Liquid absorbents. A large number of liquid absorption processes have been developed for removing sulphur dioxide from gas streams. Most of these have been designed for scrubbing sulphur dioxide containing waste gases from fuel burning installations, but a number can be used for contact acid plant tail gases. Most can reduce the final sulphur dioxide emissions to about 100 ppm.

One process which has been used on the full scale involves scrubbing with ammonia solution to produce an ammonium sulphite–bisulphite solution. By reacting this solution with an acid, nitric or orthophosphoric, sulphur dioxide is recovered for recycle and the appropriate ammonium salt is produced. These processes have application within the fertilizer industry, where the ammonium salts can be disposed of readily. Alternatively, ammonium sulphite can be crystallized from the solution by flash evaporation and the ammonium sulphite then oxidized to ammonium sulphate in a rotary kiln.[3]

Processes based on scrubbing with potassium or sodium sulphite solutions have also been developed and these avoid by-product formation. In the case of potassium sulphite, the bisulphite is formed first and this is cooled to precipitate potassium pyrosulphite, which is separated and then heated with steam to release sulphur dioxide and potassium sulphite for recycle:

$$K_2SO_3 + SO_2 + H_2O \longrightarrow 2KHSO_3$$

$$2KHSO_3 \xrightarrow{\text{cool}} K_2S_2O_5 + H_2O$$

$$K_2S_2O_5 \xrightarrow{\text{heat}} K_2SO_3 + SO_2.$$

The process employing sodium sulphite behaves similarly, but the bisulphite solution can be heated directly to recover sulphur dioxide. Both processes have high capital requirements.

Scrubbing with magnesium oxide slurry causes magnesium sulphite to form, and this can be separated by centrifuge. Calcination of the sulphite allows recovery of sulphur dioxide and the magnesium oxide for recycle.

3. Solid absorbents. Very low sulphur dioxide levels of 15–25 ppm have been claimed for the final tail gas from a 180 tonnes day^{-1} plant using a molecular sieve absorption process[4] in which sulphur dioxide is absorbed into resin at about 40°C. The sulphur dioxide is then stripped from the resin with dry air at a temperature of about 100°C and recycled back to the converters. The stripping air is obtained by heating a side-stream of air from the acid plant drying tower. Any acid mist or sulphur trioxide in the original sulphur dioxide gas stream must be removed by heating and passing through a soda-ash bed, or by passing the gas through a separate resin bed. Laboratory tests suggest a life for the resin bed of about two years, but there is insufficient full-scale operating experience at present to confirm this or to decide on the viability of this new process. If commercially successful, it will set a new standard for emission levels, but in the meantime it is likely that the double absorption process or ammonia scrubbing are likely to be favoured, at least in Europe, for reducing sulphur dioxide emissions.

Mist removal

Sulphuric acid plants can produce persistent mists which are difficult to control. Droplet sizes can vary from 10 μm or more down to well below 1 μm, and the finer the size the more difficult is the removal. The finest mists can pass straight through the absorption tower and up the stack without change.

The origins of these mists are not properly understood, but there are probably three primary causes. The larger particles are almost certainly derived from mechanical breakdown of larger drops in the absorption tower. The finer particles are formed directly from the vapour phase, either when a gas mixture is shock cooled below the dew point of sulphuric acid, or when sulphur trioxide gas reacts with water vapour under conditions where the sulphuric acid dew point is exceeded. High levels of supersaturation lead to the finest mists.

Oleum plants invariably produce fine mists, which are thought to be associated with the relatively low temperatures of 40–60°C prevailing in the primary absorption tower, compared with 70–90°C in a conventional plant.[5] The cooling which occurs in the economizer or the absorption tower of a conventional plant can, however, probably also lead to mist formation under certain conditions. To reduce the possibility of sulphuric acid formation in the vapour phase as far as possible, the air stream entering the plant is always dried before entering the sulphur burner. Even under the best drying conditions, however, there is still sufficient water vapour present in the reaction gases to give a sulphuric acid mist level above that normally considered acceptable to the legislative authorities. This is specified as 'no visible plume'. In the US, the Environmental Protection Agency has specified a mist level on new plants of less than 0·075 kg/te^{-1} 100 per cent sulphuric acid. This corresponds to about 23 mg m^{-3}.

Apart from damp air, the other potential sources of water vapour in the reaction gases are the hydrocarbons present in some sulphurs, and particularly the dark sulphurs produced in Mexico. In the sulphur burner, the hydrocarbons are oxidized and the water vapour produced passes forward with the sulphur dioxide gas; the higher the partial pressures of water vapour and sulphur trioxide, the greater the chance that mist will be formed when the gases are cooled downstream in the plant.

Some sources of sulphur dioxide gas, e.g., hydrogen sulphide or some smelter gases, contain water vapour which cannot readily be removed. Plants producing acid from such raw materials invariably produce mist.

When a conventional contact-acid plant suddenly starts to produce mist it can sometimes be traced to the introduction of water vapour caused perhaps by a steam leak, a drop in drying tower efficiency or a change in the sulphur supply. Alternatively, a relatively small temperature change may have occurred, causing supercooling and mist formation at one particular point in the process. In practice it is often difficult to explain the sudden appearance of mist or to cure it by minor process changes. In these circumstances it is usually necessary to fit a mist eliminator.

1. Coarse mists. For coarse mist of 10 μm and above, simple stainless steel gauze pads mounted in the top of the absorption tower and operating under low pressure drops can be quite successful. These are often made from stainless steel wire, which corrodes rather rapidly, or from Teflon. Venturi scrubbers can also be used, but high pressure drops are required for efficient mist removal, particularly if the droplets are much less than 10 μm in size.[6]

2. Fine mists. Fine mists in the micron and sub-micron range are difficult to remove. Electrostatic precipitators are highly effective, but are expensive and not always easy to maintain in perfect operating order. On the other hand, unlike the fibre mist eliminator, they do not incur a pressure drop penalty.

Fibre mist eliminators are usually made of fibre glass, and are of two types, one operating at a relatively low velocity of 5–10 m min^{-1} and the other at velocities of about 100 m min^{-1}. In both cases pressure drops are about 20 cm water.[5] The low velocity type is highly efficient, removing virtually all particles above 3 μm and up to 99 per cent by weight of those which are smaller. The eliminator itself consists of a series of hollow candles with fibre walls mounted in the top of the absorption tower or in a separate tank alongside; a cross-section through one of the candles is shown in Fig. 16.2. Mist droplets passing through to the inside of the candles are trapped and coalesce within the fibre packing through a combination of impaction and Brownian movement.

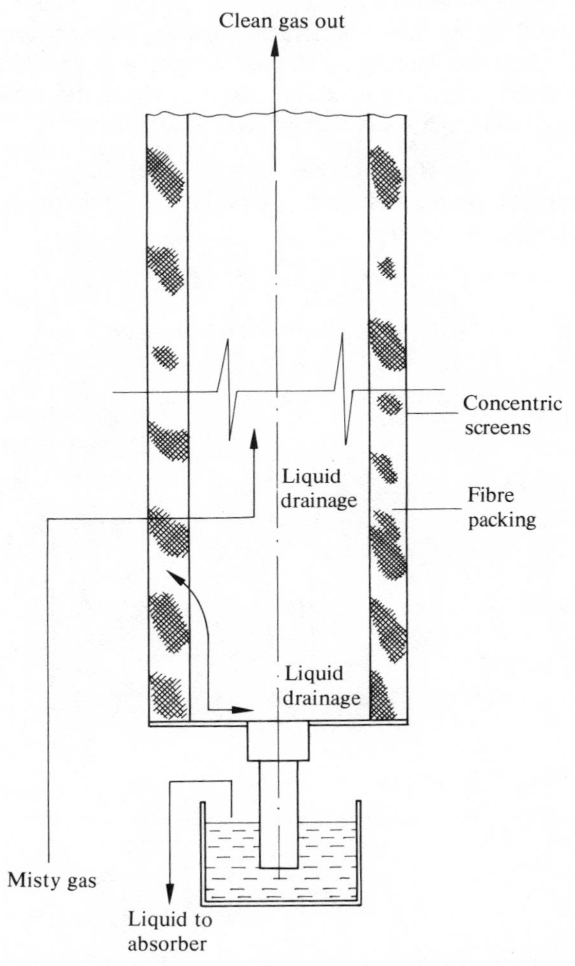

Fig. 16.2. Monsanto high efficiency mist eliminator in production of sulphuric acid and oleum. (Reproduced from J. A. Brink *et al., Chem. Eng. Prog.*, 64, November 1968.)

The liquid acid then drains down through the packing and is discharged back into the absorption tower as shown.

The high velocity eliminators, which are somewhat less efficient at removing the finest particles, depend on impaction for their effect. They are built up from a series of flat fibre pads formed into a polygon and mounted in the absorption tower such that the mist laden gas passes from the outside to the inside of the eliminator.

Nitric acid

Nitric acid is now made almost exclusively by the ammonia oxidation route in plants generally of 500–1000 tonnes day^{-1} nitric acid capacity (100 per cent basis) as shown in Figure 16.3. A mixed gas stream of air and ammonia is passed over a platinum–rhodium gauze catalyst operating at about 900°C, where the ammonia is oxidized primarily to nitric oxide, though some nitrogen is also formed:

$$4NH_3 + 5O_2 \longrightarrow 4NO + 6H_2O. \qquad (16.2)$$

After heat exchange, the cooled gases are passed into the nitric acid absorption tower where the following reactions occur:

$$2NO + O_2 \longrightarrow 2NO_2 \qquad (16.3)$$

$$3NO_2 + H_2O \longrightarrow 2HNO_3 + NO. \qquad (16.4)$$

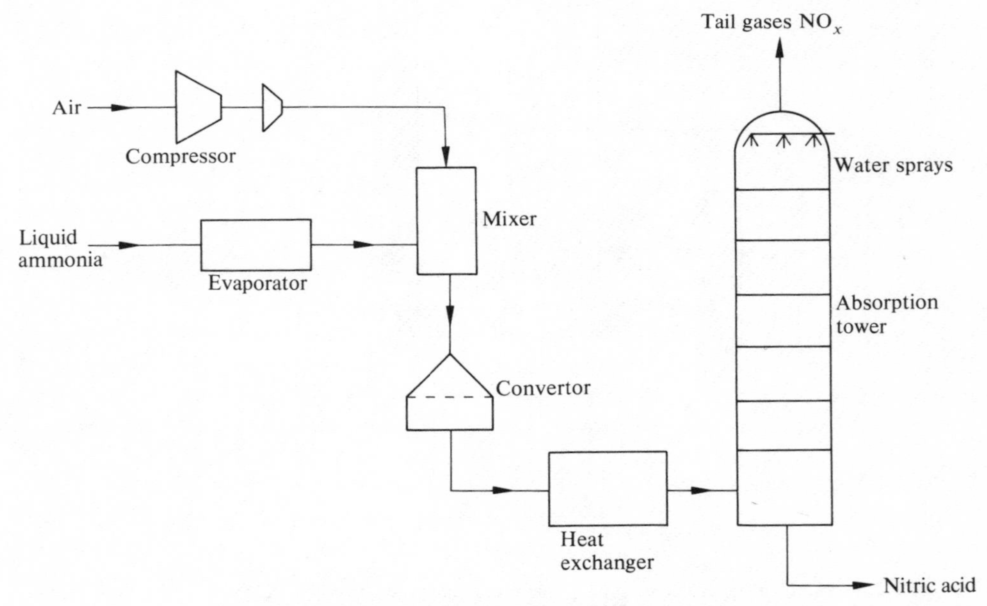

Fig. 16.3. Flow diagram of nitric acid plant.

The gas emerging from the top of the tower contains about 95 per cent nitrogen, 3 per cent oxygen, 1 per cent water vapour and generally 500–5000 ppm of NO_x, that is, a mixture of nitric oxide and nitrogen dioxide, the latter giving the characteristic brown plume. It is the only gaseous discharge from a nitric acid plant and legislative authorities are placing increasingly stringent limits on the permitted levels of NO_x from the new plants. In the UK this is currently about 1000 ppm, but in the US the Environmental Protection Agency have recommended a discharge limit of 1·5 kg nitrogen dioxide (or the equivalent amount of nitric oxide) per 1000 tonnes of 100 per cent nitric acid produced. This corresponds to an NO_x level in the stack gases of less than 200 ppm.

There are four major approaches to the removal of NO_x:

1. extended absorption
2. catalytic reduction
3. liquid absorption processes
4. absorption on solids.

Of these, only the first two have been used at all extensively, (1) being generally favoured in Europe and (2) in the US.

Extended absorption

The recent sharp increase in ammonia prices has meant that raw materials now form a much larger proportion of nitric acid production costs than hitherto. Any loss of NO_x from the process therefore represents a larger financial loss, and irrespective of any legislative requirements there is now a greater incentive to recover NO_x as nitric acid.

The rate-determining step in absorption towers operating at low NO_x emission levels is generally the rate of reoxidation of the nitric oxide to nitrogen dioxide, reaction (16.3). This is favoured at high pressures. On the other hand the ammonia oxidation reaction (16.2) is favoured at low pressures, and selection of the optimum operating pressure for a nitric acid plant has always involved a compromise. Over the last few years, high absorption efficiency has been considered of greater importance, so that design pressures have been raised, and most plant contractors are also now offering dual pressure plants with additional compression before the absorption tower. Pressures of 5 and 10 bar respectively for the oxidation and absorption processes are usual, but processes operating at 10 and 20 bar for the two stages are also offered. By cooling the absorption liquors at the top of the column, the rate of nitrogen dioxide absorption is increased and this provides a further method of reducing NO_x emissions.

By using large towers, high pressures and perhaps cooling, NO_x emissions of 200 ppm or less can be reached on new plants. At least one manufacturer is now offering to add on an additional absorption tower to an existing plant and reduce the existing emission to below 200 ppm. One such system has been installed in the US and another in France.[7]

The capital cost of extended absorption is high, but so is that of the alternative methods, and it is likely to remain an attractive procedure, particularly on high pressure plants, while ammonia costs remain high.

Catalytic reduction

The NO_x and oxygen present in nitric acid plant tail gas can be reduced by mixing with a fuel gas and passing over a catalyst which is normally either palladium, platinum or platinum–rhodium, though other metals have been suggested.[8] If methane is used as fuel, the following reactions occur:

$$CH_4 + 4NO_2 \longrightarrow CO_2 + 2H_2O + 4NO \qquad \Delta H = -138\ \text{kcal} \qquad (16.5)$$

$$CH_4 + 2O_2 \longrightarrow CO_2 + 2H_2O \qquad \Delta H = -182\ \text{kcal} \qquad (16.6)$$

$$CH_4 + 4NO \longrightarrow CO_2 + 2H_2O + 2N_2 \qquad \Delta H = -213\ \text{kcal.} \qquad (16.7)$$

Reaction (16.5) is faster than either (16.6) or (16.7) and can be used to decolorize the tail gas at minimum cost. Many older plants employ this system, which allows for a wide dispersion of the nitric oxide before reoxidation occurs. It does not reduce the overall polluting load measured as nitrogen dioxide, and is being superseded by the full reduction process. The reduction of nitric oxide to nitrogen (reaction (16.7)) is slower than reaction (16.6); hence full abatement of NO_x requires sufficient fuel to remove all of the oxygen present in the tail gas, giving a substantial temperature rise, the reactor gases being expanded through a turbine for power recovery or used to generate steam.

Commercial abatement units use either a honeycomb type of catalyst with a space velocity of about 100 000 h^{-1} or a more robust particulate catalyst with a space velocity of 30 000 h^{-1}. Operating temperatures are usually about 750°C, with 800°C as a maximum, while inlet temperatures are about 500°C for methane and 200°C for hydrogen. The temperature rise across the catalyst depends largely on the oxygen content of the gas and is about 130°C for methane as fuel and 160°C for hydrogen per mole per cent of oxygen in the tail gas. At high oxygen contents, therefore, the maximum catalyst temperature can clearly be exceeded, and a two-stage rather than single-stage process is required, at higher capital cost.

Difficulties have been experienced with the operation of catalytic abaters in the past, particularly with hydrocarbon fuels. Temperature control has been difficult, and sulphur and halogens can poison the catalyst. When a fall off in catalyst performance occurs there is a tendency to raise temperatures and shorten catalyst life further. In general, units using hydrogen as fuel have been more successful than those using hydrocarbons.

Recently a single-stage process with recycle has been developed as shown in Fig. 16.4. Temperature control is much easier than with earlier units, and investment costs are lower than for the two-stage process.

The catalytic abatement process can achieve final NO_x levels of less than 10 ppm, and it also produces large amounts of power which can be used in some of the older

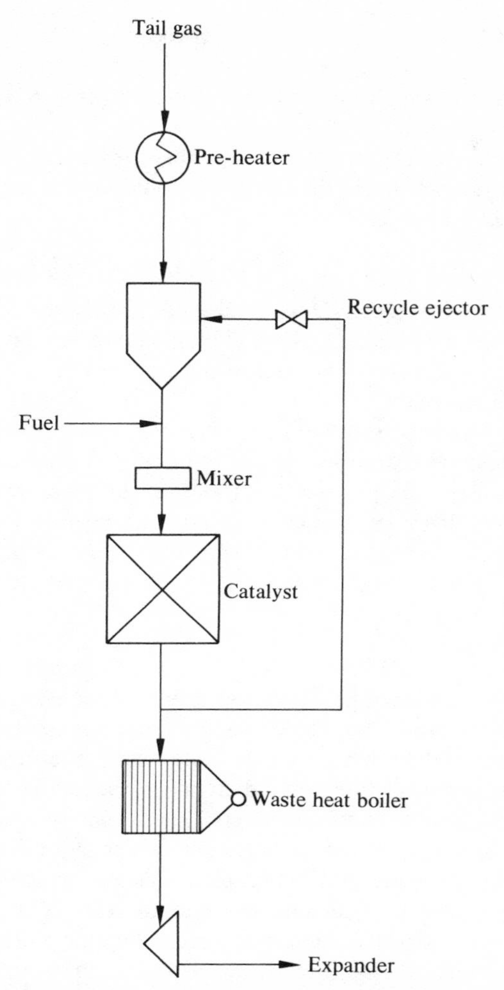

Fig. 16.4. Single-stage catalytic abatement system with recycle to reduce emission of oxides of nitrogen to the atmosphere from nitric acid plant. (Reproduced from W. R. Hatfield, paper to meeting on *Minimizing Pollution from Fertilizer Plants*, Helsinki, August 1974.)

nitric acid plants. Newer high pressure plants operate at high thermal efficiency and need no additional power. There is also some doubt over the future use of hydrocarbon fuels in the process, because they can lead to further aerial pollutants such as cyanides. In this case the process might be restricted to the sole use of hydrogen as fuel.

Selective catalytic reduction has been studied in the past using ammonia as a fuel, and there is a revival of interest in this process. Its chief disadvantage is the possibility of formation of ammonium nitrite, which is very unstable.

Absorption processes

A number of processes have been proposed for removing NO_x from tail gases by absorption in aqueous solution.

Magnesium nitrite is formed when NO_x is absorbed in a mixture of magnesium nitrate, oxide and carbonate. By hydrolysing the solution at a high temperature, nitric oxide can be recovered for recycle.[9]

$$3Mg(NO_2)_2 + 2H_2O \longrightarrow Mg(NO_3)_2 + 2Mg(OH)_2 + 4NO.$$

A French process[10] claims that NO_x can be removed by absorption in an ammonium sulphite–bisulphite mixture obtained by scrubbing sulphuric acid plant tail gases with ammonia. Ammonium sulphate is formed in a complex reaction which also involves residual oxygen in the tail gas.

By scrubbing with an acidic solution of urea, the two nitrogen oxides are reduced to nitrogen with simultaneous decomposition of the urea to carbon dioxide and nitrogen. There is some interest in this process for treating tail gases from medium pressure plants, but a study by TVA[11] suggests that capital and operating costs are higher than for catalytic combustion.

Absorption on solids

The absorption of NO_x on silica gel was used in the Wisconsin pebble bed process for the manufacture of nitric acid[12] but more recently there has been extensive work on the use of molecular sieves.[13] Two beds are used, one being regenerated while the other is absorbing. Water is removed first, and then the nitric oxide which is catalytically oxidized on the absorbent to nitrogen dioxide, and finally any nitrogen dioxide is absorbed. The beds are regenerated with hot air, which is returned together with the desorbed nitrogen dioxide to the nitric acid plant absorption tower, increasing the gas flow by about 25 per cent. A reduction to 10 ppm NO_x in the final effluent gas is claimed, and there is considerable interest in the application of this process on the full scale. It remains to be seen whether the molecular sieve can withstand the large number of absorption–desorption cycles required for practical operation, and whether the batch nature of the process can handle the constant flow of tail gases.

Hydrofluoric acid

Compared with sulphuric acid and nitric acid, hydrofluoric acid is produced on a small scale, the large plants making perhaps 75 tonnes day^{-1}. Virtually all hydrofluoric acid is made by reacting concentrated sulphuric acid with acid grade fluorspar. On one widely used process the reactants are thoroughly mixed and then heated to about 200°C in a rotary kiln, with a hold-up time of perhaps 30–60 min:

$$CaF_2 + H_2SO_4 \longrightarrow 2HF + CaSO_4.$$

In addition to the hydrofluoric acid, the evolved gases also contain some water vapour, sulphur dioxide, carbon dioxide, silicon tetrafluoride and sulphuric acid. The gases are cooled, and scrubbed with sulphuric acid to remove the hydrofluoric acid for eventual recovery. Any non-condensible gases such as sulphur dioxide, carbon dioxide and silicon tetrafluoride which are absorbed in the scrubbing towers are stripped from the sulphuric acid and recycled back into the process.[14]

The silicon tetrafluoride produced in the process arises from the reaction of silica originally present in the fluorspar with the hydrofluoric acid. Silica is a highly undesirable impurity, and acid grade fluorspars are normally required to contain less than 1 per cent of silicon dioxide. In the final water scrubbing tower the silicon tetrafluoride hydrolyses, precipitating silicon dioxide, which can block pipes and pumps. To avoid this, sufficient hydrofluoric acid is left in the gas to redissolve the silicon dioxide and ensure that only fluorosilicic acid is formed in the scrubber. The final tail gases from the plant contain carbon dioxide, sulphur dioxide and traces of hydrofluoric acid, and these can be scrubbed with caustic soda. Water ejectors used to maintain a partial vacuum throughout the system are also effective in removing any final traces of hydrofluoric acid.

Hydrochloric acid

Hydrochloric acid is mainly produced as a by-product from various organic chlorination reactions and from some dehydrochlorination reactions, e.g., the production of vinyl chloride from ethylene dichloride. The production of salt cake, sodium sulphate, by reaction of sodium chloride with sulphuric acid, also provides hydrochloric acid, while some is synthesized directly by burning hydrogen in chlorine:

$$2NaCl + H_2SO_4 \longrightarrow Na_2SO_4 + 2HCl \qquad (16.8)$$

$$H_2 + Cl_2 \longrightarrow 2HCl. \qquad (16.9)$$

The methods used to recover the hydrochloric acid depend on the particular process and whether anhydrous hydrogen chloride or hydrochloric acid are required.

The salt cake route (16.8) provides a 30–60 per cent concentration gas at about 500°C containing residual solids. These are removed in settlement chambers and cyclones.

The hydrochloric acid obtained from organic chlorinations contains some chlorine and small quantities of hydrocarbons, and is evolved at a temperature of perhaps 200°C. The gas stream is scrubbed with solvent, often in several stages, to recover the organic materials. Products from the chlorination of methane, for example, may be scrubbed with a mixture of carbon tetrachloride and trichloromethane from which the hydrocarbons are recovered for recycle.

Synthetic hydrochloric acid is usually produced in water-cooled, brick-lined or graphite combustion chambers. The gas formed is relatively pure, usually contains some excess hydrogen and is at a temperature of perhaps 300°C.

The hydrochloric acid streams from all these processes may be dried if anhydrous hydrogen chloride is wanted; otherwise they are passed into a series of cooler–condensers, where a concentrated solution of hydrochloric acid is formed and the high heat of solution is removed. These cooler–condensers are often of the falling film type. To increase the recovery of hydrochloric acid the process can be operated at elevated pressures. The residual hydrochloric acid is then recovered in packed scrubbing towers irrigated with 20 per cent concentration hydrochloric acid, which is returned to the process. Make-up water fed to this tower can be controlled automatically, depending on the tower temperature and strength of the product acid, ensuring minimum slip of hydrochloric acid in the final tail gas.

Soda ash

Soda ash ($NaCO_3$) is now made almost exclusively by the ammonia–soda process illustrated in Fig. 16.5.

Sodium chloride brine is purified by treatment with process waste gases containing carbon dioxide and ammonia, which precipitate the calcium, magnesium and other heavy metals. The purified brine is then pumped to a tower where it absorbs ammonia, most of which is recycled, with a small quantity of fresh ammonia. The soda ash reaction occurs in two towers operating in series, the first being fed by a weak make-up carbon dioxide gas stream from a lime kiln, while the second is fed by strong gas from the bicarbonate calciner:

$$NH_4OH + CO_2 \longrightarrow NH_4HCO_3$$

$$NH_4HCO_3 + NaCl \longrightarrow NaHCO_3 + NH_4Cl.$$

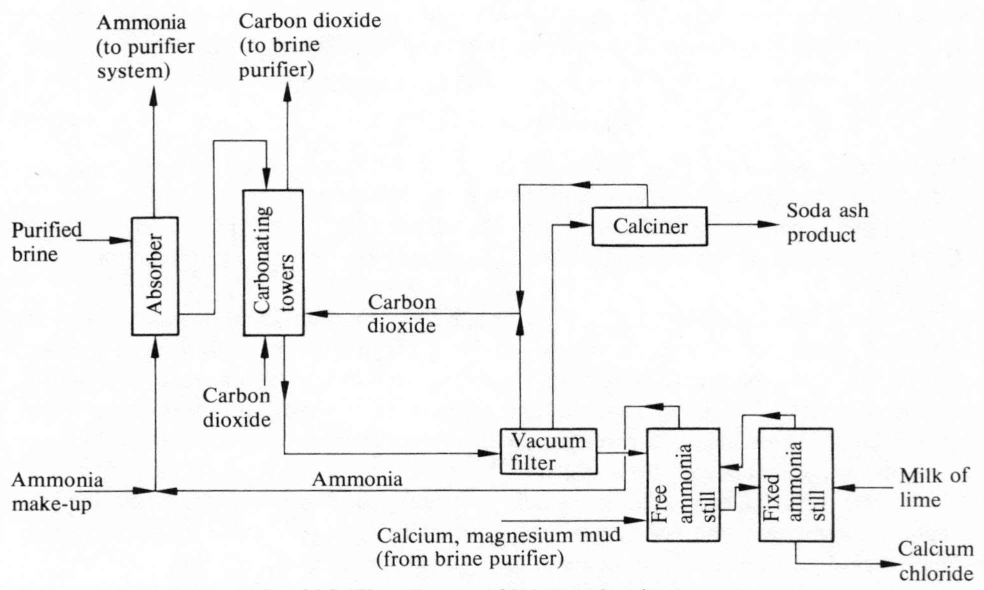

Fig. 16.5. Flow diagram of Solvay soda ash process.

A slurry of sodium bicarbonate forms in the carbonating towers. This is filtered and the recovered solid calcined to give light soda ash, while the filtrate is passed first to one still for separation of free ammonia and then to a second still, where lime from the kiln is added to remove the residual combined ammonia for recycle:

$$2NH_4Cl + Ca(OH)_2 \longrightarrow 2NH_3 + CaCl_2 + 2H_2O.$$

This highly efficient process operates with an ammonia efficiency of about 98·5 per cent, and pollution from ammonia emissions is low. Some dust problems do arise in the exhaust gas streams from rotary driers, the dry solids handling equipment and from the lime kiln. Dust from conveyors, bagging plant and from bulk loading points may be removed in bag filters.

Chlor-alkali production

Chlorine and caustic soda are invariably produced together by the electrolysis of brine either in a diaphragm or mercury cell. The essential features of the diaphragm cell are illustrated in Fig. 16.6. Brine, generally pumped from an adjacent salt mine, is purified by addition of soda ash and caustic soda to precipitate the heavy metals such as calcium, magnesium and iron. The purified saturated brine solution is continuously fed to the anode compartments of the cells, which have graphite electrodes with a diaphragm covering the cathode. The cells are operated at a low d.c. voltage, but high currents are used and chlor-alkali plants are significant base load consumers of

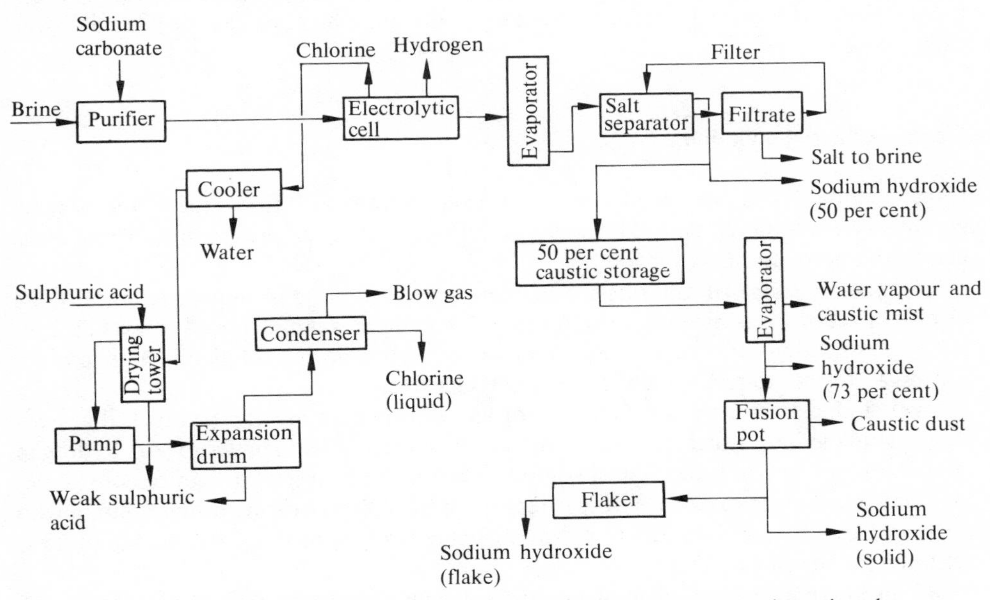

Fig. 16.6. Flow diagram of chlor-alkali plant for production of chlorine and caustic soda.

electricity; 1 tonne of chlorine requires up to 4000 kWh of energy. Wet chlorine gas is evolved from the anode compartment and caustic soda and hydrogen are formed in the cathode compartment; a caustic soda concentration of 10–12 per cent is usually maintained in the electrolyte:

$$2NaCl + 2H_2O \xrightarrow{\text{d.c.}} 2NaOH + H_2 + Cl_2.$$

The caustic solution is evaporated down to 50 per cent strength, causing most of the salt to precipitate out. This is removed and recycled to the cells. The wet chlorine gas is at about 90°C and contains 0·5–1 per cent hydrogen, perhaps 1 per cent of carbon dioxide and some air; it is cooled to remove moisture and then dried by passing it through a sulphuric acid tower. The dry gas is compressed and cooled to provide liquid chlorine. To avoid build up of an explosive hydrogen–chlorine mixture, only about 90 per cent of the chlorine can be liquefied under normal conditions, and the residual or 'blow-off' gas is the chief potential source of air pollution in a chlor-alkali plant.[15]

The mercury cell process differs from the diaphragm process in its use of a moving mercury cathode. Mercury is continuously pumped across the bottom of the gently shelving electrolytic cell and then through the decomposition cell mounted beneath. Hot saturated brine is fed to the cell and electrolysed between the mercury cathode and a series of graphite or platinized titanium anodes held a few millimetres above the mercury surface; chlorine is evolved and the sodium amalgamates with the mercury. The sodium amalgam from the electrolytic cell flows into the lower cell, where it is decomposed with water, forming caustic soda solution and hydrogen. Mercury cell plants provide purer caustic soda, hydrogen and chlorine than diaphragm plants, but the problems of handling hydrogen build-up in the chlorine blow gas are essentially the same.

Control of aerial pollution

Caustic soda dusts are produced from evaporators and from the drum flakers commonly used to provide solid caustic soda from anhydrous caustic melt. These dusts can be removed in wet scrubbers.

The potential loss of chlorine from 'blow gases' is so large that economics alone dictate the need for some recovery process. Unless there is a direct use for this gas, e.g., in the production of hydrochloric acid or of hypochlorite solutions, it is usually processed further to recover additional liquid chlorine.

Overall chlorine recovery efficiencies of 99 per cent or more can be reached in two-stage liquefaction systems, using an air bleed to avoid build-up of hydrogen concentrations beyond the explosive limit. Absorption in water at high pressures can give similar efficiencies; the chlorine is recovered by indirect heating of the solution with hot cell gases followed by direct steam stripping. In this process, the wet chlorine is returned to the cell gas stream.

As an alternative, a number of organic solvents have been claimed as solvents for chlorine in the 'blow gas' and a mixed solvent of carbon tetrachloride and trichloromethane is used commercially. The chlorine is recovered from the solvent.[16]

The tail gases from all these processes invariably contain residual chlorine, and additional steps are usually taken for its recovery. One method is to contact the gas with saturated brine ahead of the cells, decreasing its pH from 10 to 5·6 and reducing the amount of hydrochloric acid needed for brine pH control. The best procedure however is to use caustic soda or lime scrubbing.

Ammonia

Ammonia is synthesized from nitrogen and hydrogen in a high pressure catalytic process and recovered as a liquid from the hot gases by condensation:

$$N_2 + 3H_2 \rightleftharpoons 2NH_3.$$

Ferric oxide catalysts promoted with aluminium oxide and magnesium oxide are generally used at a temperature of about 500°C and pressures in the range 150–400 bar.

The synthesis gases for the process are usually obtained from natural gas or naphtha in a two-stage process. In the first, steam is added and most of the hydrocarbon is converted to carbon monoxide and hydrogen over a nickel catalyst at a temperature of about 800°C and a pressure of 30 bar:

$$CH_4 + H_2O \longrightarrow CO + 3H_2.$$

Air is then added to provide the required amount of nitrogen needed for the ammonia synthesis, and the mixture is passed through a second catalytic reactor in which the nickel content of the catalyst is somewhat lower. Oxidation of the residual hydrocarbon occurs and the temperature rises to 1000–1300°C:

$$2CH_4 + O_2 + N_2 \longrightarrow 2CO + 4H_2 + N_2.$$

The carbon monoxide is converted to dioxide by the shift gas reaction and the carbon dioxide is removed by scrubbing with, for example, potassium carbonate or monoethanolamine solutions, leaving a mixture of hydrogen and nitrogen to feed into the ammonia plant:

$$CO + H_2O \longrightarrow CO_2 + H_2.$$

The carbon dioxide is stripped from the solvent with steam and either vented or recovered for use, for example, in a urea plant.[17]

There are few gaseous emissions from ammonia plants. Carbon dioxide is the major waste material, but this is not considered objectionable and in many plants it is recovered for use in urea synthesis or for sale. Ammonia gas can be lost from leaks, or from pressure release valves on storage vessels, but these losses can be minimized by careful maintenance. In some cases ammonia can be released in gas purged from the system to remove impurities, when it can be removed by scrubbing with water or by burning. In practice, ammonia plants present few, if any, gaseous (as opposed to water) pollution problems.

Ammonium nitrate

Ammonium nitrate is a very important fertilizer ingredient. It can be used alone as a straight nitrogen fertilizer or in combination with phosphate and potassium salts in compound fertilizers. It is also a constituent of many commercial explosives; large tonnages are used in the ANFO blasting agents, which are mixtures of ammonium nitrate with 5–6 per cent of fuel oil.

Manufacture involves first the neutralization of nitric acid with ammonia to give a solution containing from 80–95 per cent ammonium nitrate, followed by evaporation of the solution and solidification of the product, generally by a prilling process. Gaseous effluents are emitted at each stage.

Neutralization

In the liquid phase process for manufacturing ammonium nitrate solution, nitric acid is neutralized in two stages. The first is operated at a relatively low pH to reduce the loss of ammonia carried over in the large volumes of released steam. The second stage is operated at a higher pH, with a correspondingly higher ammonia vapour pressure, but the steam evolution is much lower and the carry over of ammonia is contained within acceptable limits. Ammonium nitrate decomposes rapidly under acidic conditions, and careful control of the pH in the first reactor is essential to avoid a possible hazard. High ammonium nitrate levels and the sensitivity of pH to small changes in ammonia levels, however, cause difficulties, and sophisticated methods of control have been developed to deal with this problem.[18] Steam evolved from the process is condensed to trap out the ammonia and some ammonium nitrate, and this condensate is sewered or can be used, after acidification, in a nitric acid absorption tower.

To avoid the possible hazards associated with handling large volumes of acidic ammonium nitrate solution, the vapour phase neutralization process shown in Fig. 16.7 has been developed. In this process pre-heated nitric acid and an excess of vaporized ammonia are passed through a Raschig ring bed, where virtually instantaneous reaction occurs at a temperature of about 160° C:

$$NH_3 + HNO_3 \rightleftharpoons NH_4NO_3.$$

An ammonium nitrate solution of about 90 per cent concentration is collected in a separator. Ammonia and ammonium nitrate carried forward in the steam from the separator are condensed out as in the liquid phase process.

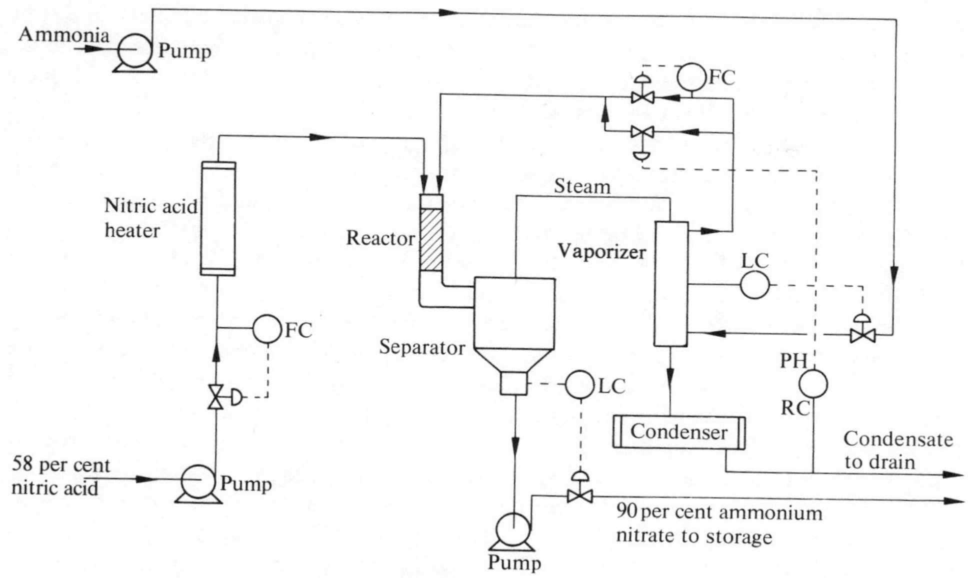

Fig. 16.7. Flow diagram of vapour phase ammonium nitrate process.

Evaporation of the solution

Ammonium nitrate solution is concentrated in falling-film, air swept, or vacuum evaporators at a temperature of about 170°C. Some decomposition to ammonia and nitric acid occurs, but since the nitric acid is more soluble it tends to remain in the solution, reducing its pH and increasing the rate of decomposition. To avoid this and to minimize losses of nitrogen, the pH of the ammonium nitrate solution is maintained at 6–7. Ammonia and nitric acid carried forward in the steam from the evaporator can combine to form a sub-micron ammonium nitrate fume which appears as a blue haze in the exit gases. This fume can be scrubbed out in water, but the pressure drops tend to be unacceptably large and the total losses of nitrogen from evaporators are in any case usually much smaller than those from the solidification stage.

Ammonium nitrate solidification

When ammonium nitrate melt is cooled it solidifies at a temperature of 170°C, with release of latent heat. On further cooling to room temperature, it passes successively through three phase transition points. At about 125°C it converts from form I to form II, at about 80°C, from form II to form III and at 32°C it converts to form IV. The heat evolved during passage through these three transitions is comparable to the latent heat of solidification of the original melt into form I. The commonest method of solidification is by air prilling, which involves the formation of droplets of liquid and their solidification as they fall through a tower. These prilling towers can vary in height from about 40 to over 100 m. With the taller towers, little or no additional product cooling is required, but the shorter towers provide product at perhaps 100°C, which must be cooled further through the transition stages.

For use in explosives, a porous prill able to absorb oil is required, and this is obtained by prilling a 95 per cent strength solution from spray heads or nozzles mounted in the top of the tower; hole size and liquid head, which is usually about 1 m, are adjusted to provide a predominance of droplets in the 1–2 mm region. The droplets solidify and lose some water while passing through the tower, but will contain 2–3 per cent moisture when they reach the bottom. The moist prills are then dried.

Although these porous prills can be used as a fertilizer, a less porous higher density prill of somewhat larger size is preferred. This is obtained by prilling a near anhydrous melt containing about 0·3 per cent of moisture from spray heads with rather larger holes or, in the case of one process, from a spinning perforated basket. With these air prilling processes, it is common to mount the evaporator at the top of the tower, thus adding to the loads on the tower and its foundations and consequently to the costs.

The shower of hot ammonium nitrate droplets at an initial temperature of about 175°C creates a strong updraught in the tower. This is often assisted by fans installed at the base of the tower to promote cooling, and in some designs air velocities can reach 5 m s^{-1}. These massive air flows carry over ammonium nitrate fume, often of sub-micron size, together with some micro prills of about 0·5 mm diameter, which are formed as satellites behind the larger prills with these types of shower heads. For example, a 500 tonnes day^{-1} plant operated at an air velocity of 3 m s^{-1} emits about 1 tonne day^{-1} of ammonium nitrate, much of it as small particles, but at 1 m s^{-1} the larger particles are not emitted and losses fall to less than one-tenth, most of it however in the form of a fine persistent fume.[18] High energy venturi scrubbers can be used to reduce these fume emissions, but costs are high because of the large gas flows and few plants are equipped with these scrubbers. As tighter emission limits are placed on these plants, there is increasing concern about their economics, and interest is now being shown in alternative solidification methods such as granulation and dust prilling, the latter being completely fume-free.

The main features of the dust prilling process are shown in Fig. 16.8. Large, uniformly sized and discrete droplets of anhydrous ammonium nitrate melt are formed in relatively wide bore prilling nozzles operating under a few cm head of liquid. These fall through a cloud of dust into a fluidized dust bed equipped with cooling radiators to remove the latent heat of solidification. Tower heights are only 5–10 m, but during their passage through the dust cloud a solidified shell containing dust forms around the molten droplet whose fall is cushioned when it enters the fluidized dust bed. The process is very compact and air flows are about one-tenth those in an air prilling plant of comparable capacity. Ammonium nitrate fume is produced as in air prilling, but the fume is trapped in the dust which is generally about 50 μm in size. The dust itself is removed in cyclones and bag filters and returned to the bed while the final gaseous effluent is dust and fume-free. The cost of this process is similar to that of air prilling, and it is likely to find increasing use in areas where fume emission cannot be tolerated.[19]

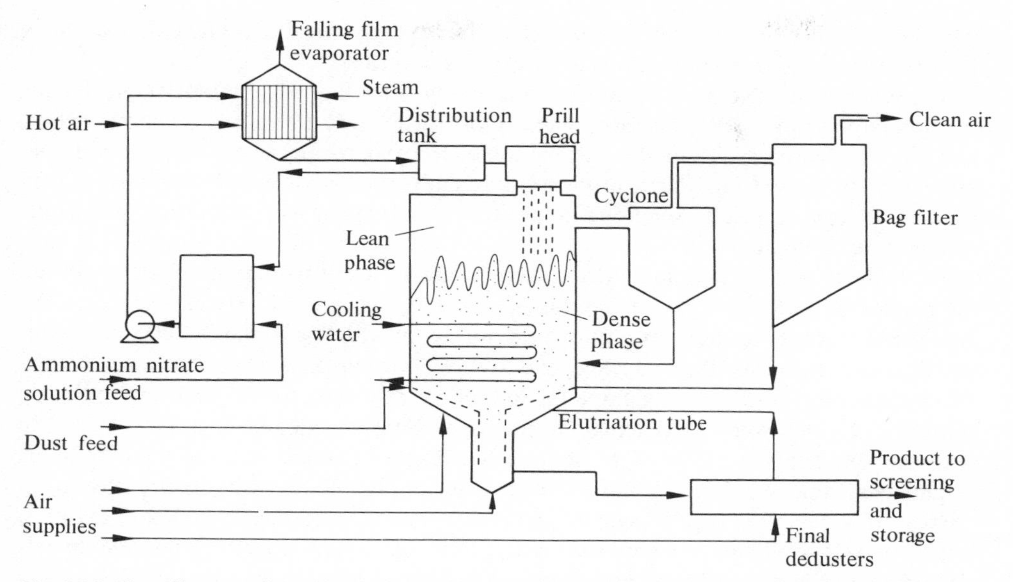

Fig. 16.8. Flow diagram of the Fisons dust prilling process for ammonium nitrate. (Reproduced from G. Drake, *Nitrogen*, November/December 1973.)

Urea

Urea is widely used in tropical and some temperate countries as a nitrogen fertilizer; it also has a number of important industrial uses such as in the manufacture of urea-formaldehyde resins. Modern plants are designed to produce 1000 tonnes day^{-1} or more, and they can emit large quantities of pollutants at the solidification stage.

Manufacture

Urea is manufactured from carbon dioxide and ammonia in a process which involves first the production of ammonium carbamate followed by its decomposition to urea, which is recovered as a solution:

$$CO_2 + 2NH_3 \longrightarrow NH_4CO_2NH_2$$

$$NH_4CO_2NH_2 \rightleftharpoons NH_2{-}CO{-}NH_2 + H_2O.$$

Because they require both ammonia and carbon dioxide, which is available as a by-product from synthesis gas, urea plants are invariably associated with an adjacent ammonia plant. In some modern designs, the ammonia and urea plants are integrated and the carbon dioxide is not separated from the synthesis gas which passes through the ammonia reactor and forward to the urea plant.[20] Urea solution from the plant is generally concentrated further by evaporation to 85–90 per cent strength, and it can then be solidified in vacuum crystallizers. Alternatively the concentrated solution is further evaporated to an anhydrous melt which is air prilled.

The latent heat of solidification of urea is nearly twice that of ammonium nitrate, but it exists in only one allotropic form, hence most of the heat is released on solidification and little is released as the solid is further cooled to room temperature. To cope with this increased latent heat load, urea prills are invariably smaller, 1·0–2·0 mm in size. This ensures the prills are fully solidified before they reach the base of the tower and do not disintegrate on impact.

As in the case of ammonium nitrate there is also a fume problem associated with the air prilling of urea but it is, if anything, even more severe. Urea fume particles are somewhat larger than those of ammonium nitrate, most being in the 2–200 μm region. On the other hand, air flows through the towers are even larger: 500 000 $m^3 h^{-1}$ for a 1000 tones day^{-1} plant with a fume loading in the exit gas of 500–1000 mg m^{-3}. For natural-draught towers, de-dusting systems operating at negligible pressure drops are essential, and simple impinger type de-dusting systems are common. The fume laden air is deflected downwards by a baffle to separate the larger particles before meeting a curtain of water. This type of system can reduce the dust loading to 200–500 mg m^{-3}, but some wet scrubbing systems for natural-draught towers reportedly provide an exit gas containing less than 15 mg m^{-3} urea.

For forced-draught towers, slightly higher pressure drops of about 50 mm water can be used, and a new process which can reduce fume levels below 30 mg m^{-3} has been developed in Japan.[21] The tower exit gases are passed into a void tower equipped with water sprays and a foamed polymer filter as shown in Fig. 16.9. The larger particles are removed in the sprays and dissolved in the solution at the base of the tower. The finer dry fume dust and the mist of water and solution droplets which are formed in the lower

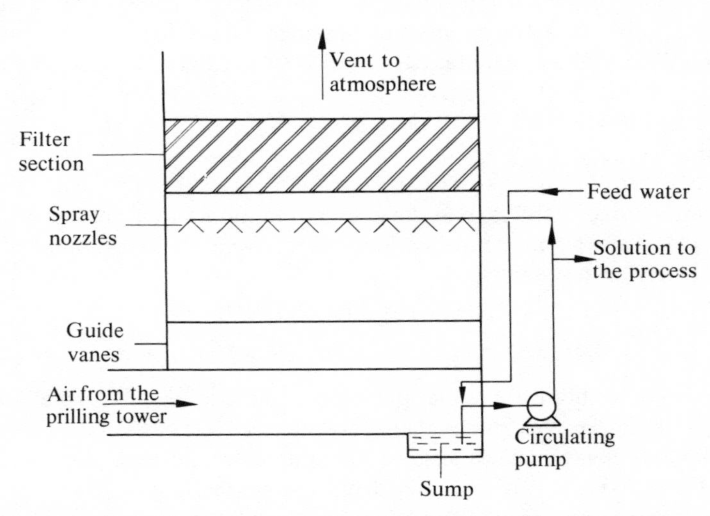

Fig. 16.9. Mitsui Toatsu process for urea fume removal. (Reproduced from K. Akitsune and T. Takae, *Chem. Eng. Prog.* **69** (6), 72, 1973.)

part of the tower are trapped in the polymer filter, where the drops coalesce and drain back to the base of the tower. The back flow of water through the filter dissolves out the urea particles and prevents blockage of the pores which would otherwise occur.

Phosphate fertilizer

The total world production of phosphate fertilizers, expressed as phosphorus pentoxide, is now over 20 million tonnes per annum, virtually all of it derived from phosphate rock. The composition of phosphate rock varies somewhat from one part of the world to another, but the principle deposits in North Africa, Florida and the Kola peninsula in Russia are mainly fluorappatites containing 3–4 per cent fluoride, several per cent of silica and varying amounts of trace metals such as iron and aluminium; the formula approximates to $3Ca_3(PO_4)_2CaF_2$. Evolution of fluorine in one form or another during phosphate rock processing is the main potential source of aerial pollution.

Elemental phosphorus can be recovered from the rock by reduction with carbon in an electric furnace, as described later, but for fertilizers, the more important processes involve attack of the rock by a mineral acid, particularly sulphuric acid, which produces a soluble phosphate, either monocalcium phosphate, $Ca(H_2PO_4)_2$, or phosphoric acid, H_3PO_4. To increase the reaction rate and avoid the effects of insoluble calcium sulphate coating the rock particles, the phosphate rock must first be ground. A closed system is used, with exhaust air drawn through cyclones to classify the particle sizes of the ground rock and then through bag filters to remove dust. Even the best systems release some fine dust and phosphate rock fall-out deposits are common in many fertilizer plants where rock is ground. Further dust arises from rock conveyors associated with the grinding plant and from transfer or discharge points in the system.

Superphosphate production

The earliest, though still widely used, process for partly dissolving the rock involves a solid state reaction with sulphuric acid. The acid and rock are contacted in a pre-mixer, and the semi-solid reacting mass is passed into a closed den where it is held for 30–60 min. Monocalcium phosphate and gypsum are produced as the mixture gradually solidifies. In the extensively used Broadfield Continuous Den, the mixture is propelled forward by the moving sides and base which are constructed of closely fitting wooden slats. At the den exit a rotating knife cuts out the powder superphosphate, which is still reacting and continues to do so for several days. During the reaction perhaps 10–14 per cent of the fluorine present in the rock is released as silicon tetrafluoride, the reaction probably occurring in two stages:

$$CaF_2 + H_2SO_4 \longrightarrow CaSO_4 + 2HF$$

$$4HF + SiO_2 \longrightarrow SiF_4 + 2H_2O.$$

Air containing silicon tetrafluoride at a partial pressure of 2–3 mm of mercury is drawn from the mixer and the den and scrubbed with water in a void tower or sometimes in a venturi scrubber to reduce the final fluorine emission gas level to 108 mg m^{-3} (0·1 gr ft^{-3}) expressed in terms of sulphur trioxide equivalent. The silicon tetrafluoride is hydrolysed to fluorosilicic acid with precipitation of finely divided silica in accordance with the formal equation:

$$3SiF_4 + 2H_2O \longrightarrow 2H_2SiF_6 + SiO_2,$$

though in practice the acid formed has a fluorine-silicon ratio nearer to 5 than 6, due to the presence of some dissolved silicic acid.

The void towers used in most plants have traditionally been built in acid-resistant brick, but rubber lined steel is commonly used at present. A typical tower design is shown in Fig. 16.10. Six passes, both co- and counter-current, are used with make-up water added in the final stage. Two acid sumps are used so that the stronger gas is scrubbed with acid of 15–25 per cent concentration in the first stages, and 10 per cent concentration acid is used for the weaker gas in the final stages. The acid is circulated by glandless, rubber-lined centrifugal pumps and sprayed from simple PVC nozzles to reduce blockages.[22]

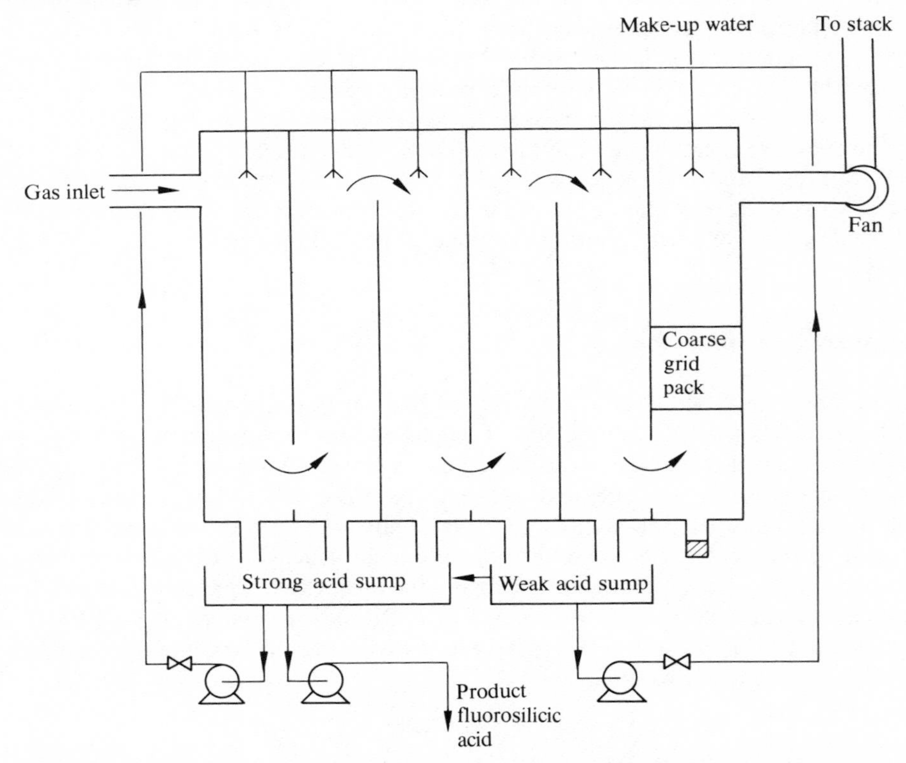

Fig. 16.10. Flow diagram of superphosphate den gas scrubber to reduce discharge of fluorides.

Triple superphosphate is manufactured in much the same way as single superphosphate, with phosphoric acid substituted for sulphuric acid. The reaction occurs, however, at 50–60°C rather than 80°C, and only small quantities of fluorine are emitted. During the curing period superphosphates, and triple superphosphates in particular, continue to emit some fluorine, and it is often necessary to hood and ventilate conveyors handling these materials.

Phosphoric acid

Most phosphoric acid is made by the wet process in which phosphate rock is reacted with sulphuric acid to give a slurry of calcium sulphate in phosphoric acid, which is then filtered. It is essential to provide the best conditions for both rock dissolution and calcium sulphate precipitation. This is generally accomplished in a two- or more-stage process with slurry recycle in which the sulphate and phosphate levels, the temperatures, and slurry densities, etc., are controlled to give optimum conditions.

The well-established process provides acid with 30 per cent phosphorus pentoxide concentration by reaction at a temperature of about 80°C, with gypsum as a by-product. More recently processes which give acid with 50 per cent phosphorus pentoxide concentration have been developed. These operate at a temperature of about 100°C and give hemihydrate as by-product. With any of these processes, the reaction systems must be cooled either by evaporation or by air cooling. The exit gases which contain fluorine, generally in the form of silicon tetrafluoride, must be scrubbed, unless a steam ejector is used to provide the vacuum when the fluorine is removed in the condenser water. Between 10 and 25 per cent of the fluorine present in the rock is evolved at this stage, lower reaction temperatures giving lower fluorine levels. The hot acid slurry from the reactor is filtered either on a moving belt or rotary filter, and these are sometimes hooded and ventilated to remove additional fluorine which is evolved. At this stage 30–50 per cent of the fluorine originally in the rock is present, largely as calcium fluoride, in the calcium sulphate, while the remaining 30–40 per cent is present largely in the filtered acid.

For many purposes, e.g., monoammonium phosphate production, a 50 per cent concentration acid is preferred. This is directly available from the hemihydrate process, but is obtained from a 30 per cent concentration acid by vacuum evaporation, a process which removes most of the fluorine present in the acid. This fluorine which is evolved with the off-gases from the evaporator is trapped in the condenser water from the steam eductor, but there is increasing interest in its recovery and at least two recovery processes are available. In one widely used process the exit gases are scrubbed with a recycled solution of fluorosilicic acid held at the same temperature as the off-gases to prevent condensation of the steam. Under the best conditions, a 93 per cent recovery of the fluorine is claimed at a fluorosilicic acid concentration of 25 per cent maximum.

In many cases much of the reactive silica originally present in the rock is removed as silicon tetrafluoride in the reaction stage, and the 30 per cent filtered acid has a fluorine–silicon ratio of well below 4. During the initial stages of the evaporation, silicone tetrafluoride is given off, but as the silicon level in the acid falls, increasingly the less volatile hydrofluoric acid is driven off. This hydrofluoric acid dissolves any precipitated silicon dioxide present in the scrubbing acid, leaving a colourless solution of fluorosilicic acid, often containing substantial quantities of free hydrofluoric acid; evaporation to 50 per cent concentration phosphorus pentoxide usually provides clear acid, but evaporation to only 40 per cent phosphorus pentoxide concentration does not.

Ammonium and diammonium phosphate

To avoid high transport, storage and handling costs, highly concentrated fertilizers containing few diluting elements are favoured. Ammonium phosphates are much more concentrated sources of phosphorus pentoxide than superphosphates, and are steadily displacing them despite their somewhat higher cost. They are made by reacting ammonia and phosphoric acid.

The neutralization of phosphoric acid by ammonia is highly exothermic, and considerable quantities of steam are evolved. Many processes are operated in 2 stages, giving an intermediate mixed product with a nitrogen–phosphorus ratio of about 1·4, where the solubility of the slurry is highest and hence the viscosity is low:

$$NH_3 + H_3PO_4 \longrightarrow NH_4H_2PO_4$$

$$NH_3 + NH_4H_2PO_4 \longrightarrow (NH_4)_2HPO_4.$$

In the second stage, phosphoric acid or ammonia is added depending on whether the mono- or diammonium salt is wanted. The final product, which is a slurry containing 5–10 per cent moisture, is generally solidified in a granulator either by itself or with other added fertilizer salts as described later.

In one process the neutralization reactor is operated at an elevated pressure at which the temperature is high enough to completely solubilize the monoammonium salt. The solution is sprayed into a short tower, where further quantities of steam are flashed off to give a miniprilled product containing about 7 per cent moisture, known as Minifos.[23]

From all these processes ammonia is evolved with the steam from the reactors, and this must be removed. Packed scrubbers irrigated with phosphoric acid are generally used, and to avoid condensation the acid temperature is controlled to match the partial pressure of the steam in the exit gases; an adequate flow of liquid must be maintained to prevent precipitation of ammonium phosphate which would block the scrubber, the partly neutralized acid being returned to the reactors.

Elemental phosphorus and thermal phosphoric acid

Most elemental phosphorus is produced by the electric furnace process. A sintered mixture of phosphate rock, coke and sand is continuously fed to the electric furnace. Carbon monoxide together with phosphorus vapour and some silicon tetrafluoride and dust is evolved at a temperature of about 400°C, and a calcium silicate slag settles to the bottom of the furnace. The dust is removed in a precipitator and the phosphorus vapour is condensed by spraying the gases with water in a tower held at about 45°C. Liquid phosphorus collects in the bottom of the tower and the spray water is recycled, gradually becoming saturated with fluorosilicates from the absorbed silicon tetrafluoride. To prevent build-up of fluorosilicic acid, which causes corrosion, soda ash and make-up water are added to the spray water system, which is continuously purged. The tail gas leaving the spray tower contains carbon monoxide and this may be burnt as a fuel or used as a raw material, e.g., for phosgene manufacture.

Some of the phosphorus is recovered for direct use, but most is burnt to provide thermal phosphoric acid. This acid is much purer than that obtained from the wet process, and is used for detergent polyphosphates, in the food trade and to some extent, particularly in North America, in liquid fertilizers.

Liquid phosphorus is atomized and burnt to phosphorus pentoxide in a stream of air in graphite-lined or stainless steel furnaces. After cooling, the gases are passed into the hydration tower, where the phosphorus pentoxide is hydrolysed to orthophosphoric acid. The hydrated acid is then passed to a water spray absorption tower where acid of about 50 per cent phosphorus pentoxide concentration is produced.

The main potential air pollutant from a thermal acid plant is phosphoric acid mist. This is produced in large quantities in the absorber tower and can be removed in venturi scrubbers and packed towers. High efficiency mist eliminators of the type used for sulphuric acid mists give much better removal and are preferred.[24]

Traces of lead and arsenic are present in the phosphoric acid and these must be removed if a feed grade acid is required. Sodium bisulphide is usually added to the acid and the precipitated sulphides are filtered off. Periodically hydrogen sulphide is vented from the treated acid and this must be removed by scrubbing with alkali or returning the gas to the phosphorus furnace, where it will be oxidized to sulphur dioxide.

Granular fertilizer

Fertilizers contain one or more of the essential plant food elements, nitrogen, phosphorus and potassium. They can be used in the form of single salts such as ammonium nitrate, ammonium phosphate, potassium chloride, etc., which often supply predominantly one of the nutrients, or as mixtures of the salts containing two or more plant foods. These are called compound or complex fertilizers. Most fertilizer products are manufactured in the form of granules ranging in size from 1 to 4 mm by rotary granulation.

Granulation process

Figure 16.11 illustrates a typical granulation plant. Solid raw materials such as potassium chloride, ammonium sulphate, superphosphates, ammonium phosphate, etc.,

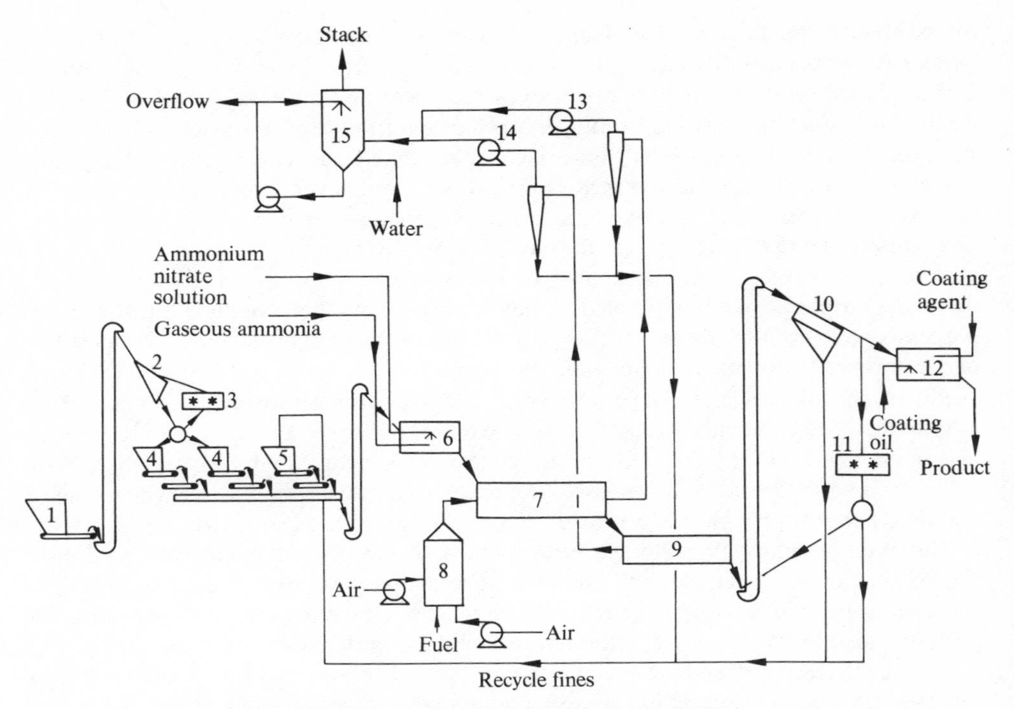

Fig. 16.11. Fisons NPK granulation process for fertilizers.

are screened, crushed and metered into the plant via the weigh belts (4) (numbers on Fig. 16.11). These materials together with recycled fines are fed to a large rotating drum (6), where they are sprayed with hot concentrated ammonium nitrate solution. This and the rolling action of the drum causes the particles to agglomerate to rather wet, weak granules with a very large range of particle sizes. The granules are then dried (7) to remove water and to strengthen them against breakdown during subsequent handling. After cooling (9) the granules are screened (10) to remove both oversize and fine material, leaving the product with a mean particle size of 2–3 mm. The fines and, after crushing (11), the oversize are returned to the process via the hopper and weigh belt (5).

The screened product is coated with a small amount of oil and dust to prevent possible caking, and it is then usually passed into a bulk store. The material is subsequently recovered with a mechanical shovel and packed into plastic sacks of 50 kg (1 cwt). Plants vary in capacity but are generally in the range 10–50 tonnes h^{-1} of product.

The illustrated process, using principally solid intermediates, operates with relatively low amounts of recycled material, the ratio of recycle to product ranging from rather less than 1 : 1 to perhaps 2 : 1. In some variants of the granulation process the intermediates, e.g., ammonium phosphate and ammonium nitrate, are prepared by ammoniating the mixed acids and feeding the slurry mixture into the granulator. These processes usually operate at much higher recycles, perhaps 4–6 : 1, to control the water balance in the process.

Instead of rotating drums the raw materials can be granulated by a set of contra-rotating paddles sometimes known as blungers; tilted rotating pans are also used for granulation.

Granulation plant effluents

Whichever type of granulation process is employed, the gaseous effluent problems are rather similar, differences arising more from the types of raw materials used than from the details of the process.

The discharge gases from the drier and cooler are the principal gaseous effluent streams in most plants. These carry fertilizer dust and a variety of gases which may be acid or alkaline. If diammonium phosphate is present, ammonia may be given off, but most fertilizers, particularly ones containing potassium chloride and superphosphates, give off hydrochloric acid.

The exit gases are passed through cyclone banks, where most of the dust is collected for recycling to the plant with the fines stream, and then through a scrubbing tower (15). These towers are generally void with water sprays, but floating ball scrubbers are used in some newer designs. The wash water may be used only once before discharge or it may be recycled. If it is recycled, a weak fertilizer solution containing perhaps 1 per cent or more of soluble salts and some insoluble residues may be built up, and this is either purged to the sewer or recycled back to the process, as occurs in one modern Finnish fertilizer plant.[25] The final exhaust gases from the scrubbing tower are discharged through the stack.

The major difficulty with recycling scrubber solutions is the additional evaporation load it places on the drier. An alternative approach would be to remove the final traces of dust in bag filters. These would have to be operated at an elevated temperature, otherwise the hygroscopic materials such as urea and ammonium nitrate which may be present would take up moisture from the wet gas, forming sticky deposits which clog the filter.

The other major source of gaseous pollution in a granulation plant, and one which is much more difficult to control, is the dust that arises at every solids handling stage and particularly from the screens, conveyor discharge points and bagging plants. This dust can be controlled, but at great expense, by fitting ventilation hoods to the equipment which discharge through cyclones and bag filters. The hoods and ducts are best made of fibre glass or stainless steel to minimize corrosion pitting.[26]

The dust used to coat the fertilizer can be handled pneumatically or, if bags are used, they can be opened and emptied under properly ventilated conditions.

References

1. Anon., 'Add-on-tail gas treatment increases sulphuric acid plant yields', *Sulphur*, 32, March/April 1975.
2. Anon., 'Sulphuric acid manufacture', *Sulphur*, 56, March/April 1973.
3. Anon., 'Flue gas treatment', *Sulphur*, 49, January/February 1973.
4. Miller, W. C., 'Adsorption cuts', SO_2, NO_x, Hg, *Chem. Eng.*, **80**, 62, August 1973.
5. Brink, J. A., *et al.*, 'Mist eliminators for sulphuric acid plants', *Chem. Eng. Prog.*, **64**, 82, November 1968.
6. Shah, I. S., 'Removing SO_2 and acid mist with venturi scrubbers', *Chem. Eng. Prog.*, **67**, 51, May 1971.
7. Anon., 'Technology newsletter', *Chem. Wk.*, **112**, 36, January 1973.
8. Reitmeier, R. E., 'Removing nitrogen oxides from gas mixtures of varying compositions', *US Patent* 2,924,504, *Chem. Abs.*, **54**, 13581f, 1960.
9. Schmidt, A. and F. Weinrotten, 'Removal of lower oxides of N from gaseous mixtures', *US Patent* 3,034,853, *Chem. Abs.*, **57**, 8195d, 1962.
10. Garlet, R., 'Methods of removing nitrogen oxides from gases', *US Patent* 3,329,478.
11. Faucett, H. L. and C. H. Davis, 'A review of processes for NO_x removal from nitric acid units', Paper read to 164th ACS Meeting, New York, 1972.
12. Foster, E. G. and F. Daniels, 'Recovery of nitrogen oxides by silica gel', *Ind. Eng. Chem.*, **43**, 986, 1951.
13. Anon., 'Acid plants face acid test', *Chem. Wk.*, 44, September 1971.
14. Rogers, W. R. and K. Muller, 'Hydrofluoric acid manufacture', *Chem. Eng. Prog.*, **59**, 85, May 1963.
15. Faith, W. L., *et al.*, *Industrial chemicals*, p. 690, Wiley, New York, 1965.
16. Sommers, H. A., 'The chlor-alkali industry', *Chem. Eng. Prog.*, **61**, 94, March 1965.
17. Reynolds, P. W., 'The manufacture of ammonia', *Fertilizer Soc. Proc.* No. 89, 1965.
18. Carter, R. W. R. and A. G. Roberts, 'The production of ammonium nitrate including safety and handling', *Fertilizer Soc. Proc.* No. 110, 1969.
19. Drake, G., 'Production of ammonium nitrate', *Fertilizer Soc. Proc.* No. 136, 1973.
20. Pagani, G. and U. Zardi, 'Integrated ammonia–urea process offers great reliability', *European Chem. News,* Large Plants Supplement 50, October 1974.
21. Akitsune, K. and T. Takae, 'Abatement of prilling tower effluent', *Chem. Eng. Progr.*, **69**, 72, June 1973.
22. Calver, A. G. and J. D. C. Hemsley, 'The design of void spray towers for silicon tetrafluoride absorption', *Fertilizer Soc. Proc.*, No. 63, 1960.
23. Hemsley, J. D. C., 'A new process for the manufacture of powder mono ammonium phosphate', *Proc. Fert. Industry Round Table, Washington DC,* 109, November 1966.

24. Manufacturing Chemists Association, *Cooperative study project,* Atmospheric Emissions from Thermal Process Phosphoric Acid Manufacture, Washington DC, 1968.
25. Kivela, T., 'Minimizing pollution from phosphate fertilizer plants including captive acid plants', *UNIDO Meeting on Minimizing Fertilizer Plant Pollution, Helsinki,* August 1974.
26. Barber, J. C., 'Pollution control in fertilizer manufacture', *J. Environ. Qual.,* **4**(1), 1, 1975.

17 Manufacture of portland cement

R. J. GATES

Raw materials

Portland cement is formed by burning ground raw materials in a kiln to form a clinker and grinding the clinker with a small addition of gypsum to produce a fine grey powder. The finished product is distinguished chemically by its cement compounds. The calcium, silica, alumina and iron of the raw material are partially sintered to form compounds of calcium and silica, of calcium and alumina, and of calcium, alumina and iron. There is no single formula for Portland cement, but both the proportions of the various chemical compounds and the ratios between them must lie within ranges which are specified similarly but by no means identically by standard authorities in different parts of the world. However, once a suitable chemical balance has been determined for a particular works and proved by the physical quality of cement it produces, it is essential to keep the chemistry constant within very narrow limits, so as to maintain consistent quality of the product. It is common, for example, to regulate the calcium carbonate content of the raw material fed to the kiln to within limits of 0·1–0·2 per cent, close to the repeatability of normal test methods.

The major raw material ingredient of Portland cement is calcium carbonate, which is found naturally as chalk or limestone in varying degrees of purity. Most frequently the contaminants of calcium in the natural stone are the other cement-making elements, and in some cement works sites the calcium and contaminants of the stone are so balanced as to provide a perfect raw material for Portland cement. More often the chalk or limestone is lacking in alumina, and clay or shale has to be added to achieve the required balance. In many places the limestone and clay between them provide suitable proportions and balances of silica and iron, but quite often either or both of these elements have to be added separately, typically as sand or iron oxide. Material otherwise suitable for making Portland cement is sometimes unusable because of other undesirable contaminants, in particular magnesia, which can only be tolerated in very small proportions.

Sulphate-resisting Portland cement is made in exactly the same way as ordinary Portland cement. It differs only in the balance of chemical compounds, and the alumina content is severely restricted to increase the resistance of the cement to sulphate attack.

The main calcium-bearing material is found as soft chalk, as hard limestone and in gradations between these extremes. Chalk is sometimes soft enough to be dug with a bucket wheel excavator, or by a face shovel. Harder forms of chalk and all limestones have to be blasted first before the broken material can be picked up by face or wheeled or tracked shovel. Chalk typically has a moisture content in the ground of about 15 per cent, but it can be as dry as 8 per cent and sometimes as wet as 25 per cent of water content by weight. Limestone is typically much drier, with a water content of 2–8 per cent.

Most cement works are sited where the main raw material component is located, and therefore have their quarry close to them. When a bucket wheel excavator is used the chalk may be conveyed from the quarry face by conveyor belt. Other materials are commonly transported from the quarry face by dump truck to a crushing station, which may be located at any point between the quarry face and the raw material preparation plant. After crushing, wet raw materials, which are difficult to handle in bulk, are passed normally by conveyor belt to the raw plant for processing; but limestone is usually taken, again by conveyor belt, to a store, which is filled and emptied in such a way as to homogenize the stone passing through it. The process is usually highly mechanized and typical modern limestone stores are completely enclosed tent-shaped buildings.

Raw material preparation

To permit the reactions to occur properly in the kiln, the raw materials have to be ground to a fine particle size. For the wet process enough water is added during the grinding process to make a pumpable slurry, the moisture content varying between 30 and 45 per cent of the slurry by weight, depending on the nature of the stone used. For the dry and semi-dry processes the raw material is dried and ground to make a powder containing less than 1 per cent moisture. In the dry process, the powder is fed directly to the kiln. In the semi-dry process, about 10 per cent water is added to the powder, usually in a dish nodulizer, to form marble-size pellets to feed to the kiln. In a modification of the semi-dry process, known as the semi-wet, slurry made as for the wet process is partially dewatered mechanically in a filter press, and the cake so produced is extruded to form worms or sausages as kiln feed.

Wet process raw plant

Soft chalk only needs crushing to remove the large lumps before, with the addition of water, it breaks down easily in a washmill or wash drum to form slurry for the wet process. A washmill is a pit, normally octagonal in plan, which is swept by a set of rotating harrows. The slurry formed in the mill is discharged through grids in some sides of the mill. A wash drum is a rotating cylinder mounted with its axis horizontal. It is fed with chalk and water at one end, and the slurry is discharged through grids in its periphery at the other end. It is fitted with lifters but the grinding is autogenous. The slurry produced by washmill or wash drum is sometimes fine enough for the kiln, but more often its coarse particles have to be passed through a tube mill for further grinding. The tube mill is a horizontal cylinder, often divided into several chambers, and about one-third filled with hard steel balls. For harder material the washmill or wash drum is unsuitable; after crushing, the stone is passed directly to a tube mill for grinding. Screens are often used in the process for control of fineness, the rejects being returned to the mill for further grinding.

When clay is used to complete the raw material blend, it is quarried above or below water by some form of mechanical excavator or dredger. Its sticky nature often demands separate slurrying in a washmill before it is mixed with the chalk or limestone, usually at the start of its milling process. When shale replaces clay as the alumina-bearing material, it may be slightly more difficult to excavate, but it can often be introduced into the main raw material milling process without previous processing.

The milling process has to be carefully controlled to ensure that the slurry produced is of constant quality, and the slurry is subsequently blended and homogenized in a series of tanks or basins, which also provide a reserve store of kiln feed slurry to permit intermittent working of and maintenance of the raw plant.

Semi-wet process raw plant

Wet process slurry is fed to a bank of multi-plate filter presses and dewatered at a pressure of between 12 and 25 atmospheres from a slurry moisture content of 35–40 per cent to a cake moisture of 18–22 per cent by weight. The cake is stored in open silos adjacent to the presses. Screw extractors beneath the silos feed the cake to a kneading machine. This extrudes worms about 15 mm in diameter and 80 mm long, which are fed directly to a semi-dry kiln. So far as the kiln is concerned, the only difference is that the feed is in worms at 20 per cent moisture instead of nodules at 14 per cent moisture. The process has been used when the raw material is too wet and sticky to be suitable for the dryer semi-dry process. It is now attracting new interest as a result of the rapid increase in the cost of energy.

Dry and semi-dry process raw plant

The raw material preparation plant is similar for the dry and semi-dry processes, its function being to produce a finely ground powder with a moisture content of less than 1 per cent. It may have a secondary crusher as its first stage, or the stone may pass direct from the store to the grinding mill in which the material is dried as well as ground. The mills are of two basic forms. The first is a tube mill, similar to that used for wet grinding, except that it is equipped at the feed end with a drying chamber fitted with lifters but without ball charge. Enough hot air is fed to the mill with the raw material to dry it. After leaving the mill the product is graded in a centrifugal separator; the fine powder is carried forward for blending and storage, the coarse is returned to the mill for further grinding. The second type of mill is a roller mill, in which the material is ground by large rollers rotating over a circular track. Hot air is again fed to the mill with the raw material and a separator for grading the product is built in integrally. The drying air leaves both types of mill at about 100°C and passes through a dust collector before being discharged to atmosphere. The powder produced by the raw mill is conveyed in a totally enclosed conveying system first to blending and subsequently to storage silos.

Dry and semi-dry raw material preparation plants differ mainly in the source of drying gas. A dry plant probably relies on the waste gas from the kiln which is at about 350°C, whereas a semi-dry plant is likely to need a separate hot air furnace. After leaving the milling system, the dusty gas in a dry plant is usually returned to the main kiln dust collector, but in a semi-dry raw plant the gas from the mill system is more likely to be cleaned in a separate dust collector.

Kilns

The kiln assembly performs a similar function in all three processes. First the moisture (if any) is boiled away. Secondly the material is calcined, driving off the carbon dioxide (CO_2) from the calcium carbonate. Finally the material is heated to about 1400°C, at which temperature the cement compounds are formed. The resulting clinker is then air cooled, using in part at least air subsequently serving as combustion air in the kiln.

Wet process kilns

A wet process rotary kiln installation is illustrated in Fig. 17.1. It consists basically of a long tube inclined at an angle of about 2° to the horizontal. The largest kilns in the world are 244 m or more in length and 6 m in diameter. A typical modern wet process kiln is about 168 m long and 5 m in diameter, producing some 1500 tonnes of cement clinker a day, but the average productive capacity of kilns in service in Great Britain in 1972 was less than half this amount. The kiln is mounted on tyres and rollers, normally five or six sets over its length, and it is continuously rotated by a gear drive mounted near the mid-point. The slurry which is fed to the kiln at the high end reaches the

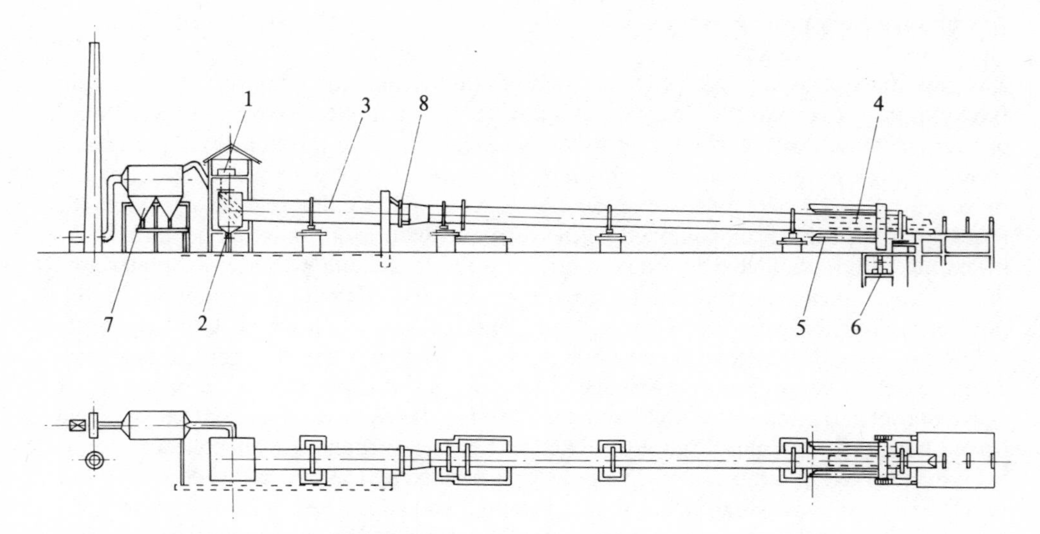

Fig. 17.1. Wet process kiln: (1) slurry feeder; (2) smoke chamber; (3) kiln; (4) firing pipe; (5) clinker coolers (planetary); (6) clinker discharge; (7) electrostatic precipitators; (8) dust return to kiln.

bottom by gravity and the combined effect of the rotation and the slope of the kiln. The passage takes three to four hours, depending on the size of the kiln.

The kiln is lined on the inside with refractory bricks, and it is normal to hang at the upper end, over 15–30 m of the kiln length, an arrangement of chains, which assists the heat exchange process and helps to catch the dust being carried up the kiln by the hot combustion gases.

The kiln is fired by fuel, which may be natural gas, fuel oil, or pulverized coal, through a firing pipe mounted on the kiln's axis at the low, hot end. A supply of primary air is fed with the fuel along the firing pipe; the remaining air for combustion comes from the clinker cooler, where it has already been used to cool the hot clinker as it leaves the kiln. The clinker cooler is located at the hot end of the kiln where the clinker is discharged.

By the time the kiln feed leaves the chain system it is almost dry. As it moves down the kiln it is heated till at about 800°C calcining takes place and the carbon dioxide is driven off. Heating continues as the feed approaches the firing end of the kiln at about 1400°C, by which time partial sintering has taken place and the material has been burnt into cement clinker.

After burning, the clinker immediately leaves the kiln, falling into the clinker cooler. Modern clinker coolers are mainly of two types. The first is a grate cooler, either moving or reciprocating, in which the clinker is cooled by air passing through the grate. Between one-third and one-half of the cooling air is used subsequently for combustion in the kiln; the remainder has to be discharged to atmosphere, preferably through a dust collector. The second type of cooler consists of a number of tubes mounted peripherally round the kiln. The clinker passes into each through holes in the kiln's circumference. In this type of cooler the clinker is cooled partly by the combustion air which passes through them in the reverse direction to the clinker, but as there is no surplus air passing through the cooler, the remainder of the cooling effect is achieved by radiation through the shell of the cooler tubes. Both types of cooler normally cool the clinker to a temperature of between 100° and 200°C.

The waste gases leave a wet kiln normally at a temperature of between 200° and 250°C. After cleaning in the dust collector they pass straight to the chimney stack. Arrangements are often made to return the dust recovered in the dust collector to the kiln. One method of return is through a scoop system fitted to holes in the kiln shell just below the chain system. The dust is returned in enclosed conveyors and elevated into a dust-tight shroud round the kiln from which the scoops pick it up and drop it back into the kiln. An alternative method of returning dust is by pneumatic pump to the firing end of the kiln. Sometimes the dust is blown into the firing pipe itself, but more frequently a separate pipe is used which is generally attached to the firing pipe.

Dry process kilns

A dry process rotary kiln installation is illustrated in Fig. 17.2. Dry kilns were originally very similar to wet kilns, but since 1950 the efficiency of heat exchange at lower temperatures has been much improved by shortening the kiln and replacing the inlet section by a series of cyclones through which the dry powder feed passes in the reverse direction to the combustion gases. The normal arrangement is with four cyclone stages, but some installations retain longer kilns with only one or two cyclone stages. The firing and clinker cooling arrangements in dry kilns are the same as in wet kilns. Over the past few years this has become in the world as a whole the dominant type of kiln for new installations, most of them designed to produce over 2000 tonnes of clinker a day.

The temperature of the waste gas leaving a four-stage pre-heater dry kiln is typically about 350°C. Despite the higher temperature than in the wet process the heat consumption is still 30–50 per cent less, because the feed is a dry powder instead of a slurry, and there is no water to be evaporated in the kiln. The full advantage of the saving in fuel consumption can, however, only be obtained if there is sufficient energy left in the waste gases from the kiln to dry the raw material as it is ground to make the kiln feed meal. The waste gas temperature of 350°C is sufficient to dry raw material having an average moisture content of over 8 per cent, so that wherever the raw material is limestone with a low moisture content, the dry process is able to take full advantage of its fuel economy in the kiln. Where the raw materials are chalk or marl based, additional heat energy may have to be put into the system to assist the kiln waste gas in the raw material drying process.

It is normal for the waste gas leaving a dry process kiln to have two parallel alternative paths to the chimney. The first is through the raw material grinding mill and the second is through a by-pass, usually equipped with a cooling tower because electrostatic precipitators are least effective working on dry gas at about 300°C. To minimize

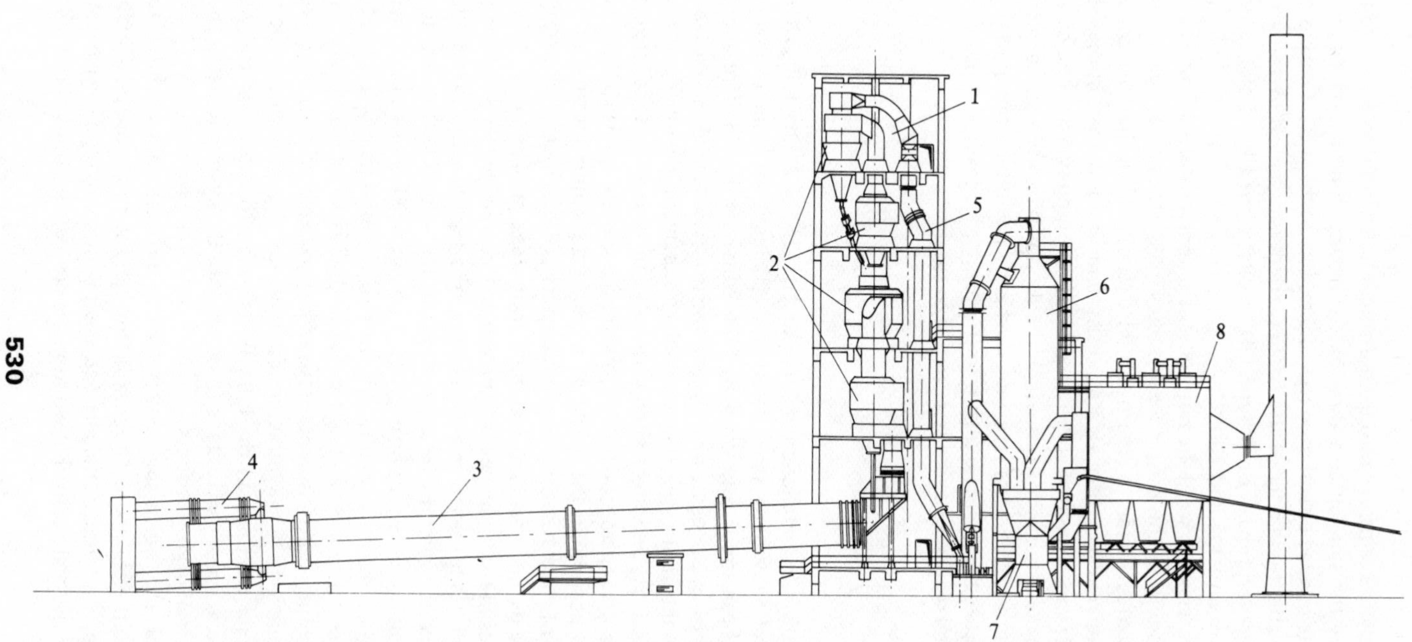

Fig. 17.2. Dry process kiln: (1) raw meal feed; (2) four cyclone stages; (3) kiln; (4) planetary clinker coolers; (5) exhaust gas from cyclones; (6) cooling tower; (7) raw mill; (8) electrostatic precipitator.

the length of hot gas ducting, the dry plant raw material grinding mill is sited close to the pre-heater cyclone tower and the dust collector. In a wet process installation the raw material preparation plant and the kiln are connected only by a slurry pipe and are therefore often at some distance from each other. A recent development of dry process suspension pre-heated kilns, pioneered in Japan, is to install a cyclone or fluid bed calciner combustion chamber between the rotary kiln and the suspension pre-heated cyclones. About half of the required fuel is burned in this combustion chamber, the fuel burned in the rotary kiln itself being correspondingly reduced. This reduces the quantity of combustion product gases passing through the rotary kiln, permitting a smaller diameter rotary kiln to be used than would otherwise be required for a given clinker output. Improved rotary kiln refractory lining life and generally more controlled kiln operation are claimed.

Semi-dry and semi-wet process kilns

In the semi-dry process the prepared dry raw meal powder is fed first to a dish nodulizer, in which it is pelletized with water to form a bed on the grate through which the hot combustion gases can pass freely. In the semi-wet process, the bed on the grate is formed of worms of extruded filter cake. The nodules are fed directly from the pelletizer or the worms directly from the extruder to the kiln system, in which the low temperature part of the heat exchange process takes place on a travelling grate. The combustion gases from the kiln make two downward passes through the grate and the bed of nodules moving along on it. The nodules are dried, heated and partly calcined by the time they leave the grate and fall into the kiln proper, which is short, like the dry process kiln. Clinker cooling and firing arrangements are the same as in the dry and wet processes.

The waste gas leaves the grate at a temperature of 110–130°C. It is cleaned in the dust collector and then passes to the chimney stack. The semi-dry process is almost as economical in its use of heat energy as the dry process, but it provides no surplus energy in its waste gas for drying raw material.

Instrumentation

All parts of the cement making process are well instrumented. Modern raw material preparation plants are often equipped with X-ray analysers which operate directly to maintain consistent chemical quality of the slurry or powder produced by continuously adjusting the proportions of the various raw materials fed to the plant.

It is the instrumentation of the kiln burning process, however, often supported by automatic control loops, which is of importance in the control of air pollution. In operating a kiln the burner operator tries to keep everything as steady and consistent as possible. He is aided by equipment to feed the raw material and fuel very accurately and instruments to record the flow. The temperature at various points in the system is

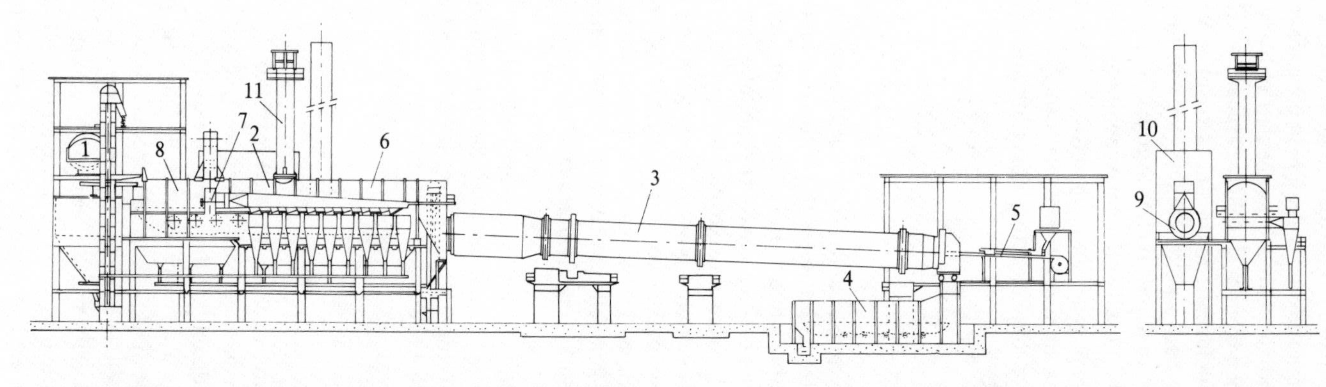

Fig. 17.3. Semi-dry process kiln: (1) dish nodulizer; (2) grate; (3) kiln; (4) clinker cooler (grate type); (5) firing pipe; (6) hot chamber—first gas pass through grate; (7) intermediate fan—gas from below grate in (6) cleaned in cyclone and passed to (8); (8) cool chamber—second gas pass through grate; (9) exhaust fan; (10) electrostatic precipitator; (11) emergency chimney.

recorded. The waste gas leaving the kiln is continuously analysed certainly to determine its oxygen content, and often also its carbon monoxide content. These readings help the operator to ensure that complete combustion of the fuel takes place at all times and that no black smcke or hydrogen sulphide is emitted from the chimney stack. Unsteady conditions do occur in most kilns from time to time; these cause unsteady gas flow conditions to the kiln and cooler dust collectors, which may result in a temporary reduction in their collection efficiency. The material does not always move steadily down the kiln. Deposits, rings, build up on the internal surface of the kiln which affect the flow of both raw material and gas, slowly as they accumulate but sometimes very rapidly when they break down. In such circumstances total control of the burning process is not an easy matter, but with the help of good instrumentation the operator can almost always achieve correct combustion conditions and with them proper control of the gas emission.

Clinker grinding

The clinker leaving the kiln is conveyed either to a store or direct to the mill feed hoppers, which can also be filled from the clinker store.

Clinker is ground with the addition of about 7·5 per cent gypsum to make cement. A tube mill is used, the clinker and gypsum being continuously fed into the mill through a hollow trunnion at one end and delivered through a similar trunnion at the other. The overall feed rate is controlled to ensure that the product has the required fineness of particle size. The long-term strength of concrete, in so far as it depends on the cement, is determined largely by the chemical quality of the clinker, but the manufacturer has to supply cement which provides specified concrete strengths in a given period of time, and the rate of gain of strength is much influenced by the cement fineness. Air is drawn through the mill in the same direction as the clinker and gypsum are moving. It is passed through a dust collector before being discharged to the atmosphere.

Many modern mills are designed for closed circuit grinding in which the cement discharged from the mill is elevated and passed through a centrifugal separator. The fine fraction only is transferred to the clinker silos; the coarse is returned to the mill for further grinding. Modern mills are usually equipped with a system of internal water cooling. Water, atomized by compressed air, is sprayed into the mill through the outlet, and sometimes also the inlet trunnion, to cool the cement as it is being ground.

Cement storage and packing plant

The cement delivered from the cement mill is the final product. It is first conveyed to silos for storage, either by belt conveyor in an enclosed gantry or by totally enclosed pneumatic conveyor. Cement is despatched from the works mainly in bulk, by road or rail tanker. There are two methods of filling tankers, overhead through the manholes in the top of the tanker, and by flexible pipe through an air- and dust-tight fitting in the lower part of the tanker. Typically the silo for overhead loading has its floor suspended at such a height that the tanker can drive underneath. A smaller proportion of cement is despatched in paper sacks which are filled in a packing machine equipped with its own dust collector.

Air pollution and dust collection

Cement making is inevitably a dusty business, as it is much concerned with hot dry powders, or materials containing a proportion of fine dry particles. Dust can escape through the main chimney, through a number of subsidiary venting points, from items of plant, and from conveying systems. The extent to which it is contained depends much on the choice of efficient plant to collect dust wherever air or gas is emitted, and it also depends very much on the overall plant layout, where the aim is to reduce to a minimum the number of transfer points on conveyor systems carrying potentially dusty materials such as dry raw materials, clinker and cement. It is also very much a function of the way in which not only dust arrestment but all plant in the cement works is maintained. There is no evidence to show that cement works' dust is a cause of injury to health, nor is it harmful to vegetation in concentrations normally encountered. Pollution from a cement works is, where it occurs, mainly a matter of inconvenience to those living nearby and of some disfigurement of grass and trees from dust deposited on them. These are important considerations, but it is also important to know the bounds of concern, that air pollution in the cement industry is limited to the effects of non-toxic dust, emitted in very fine particles, in small proportion relative to the gas volume, as a high level dispersion over a wide area from the chimney, and potentially in coarser particles at low level from other plant over an area close to the works.

Kiln gases

The most important potential source of air pollution in cement making is from dust in the kiln combustion gases. Until the 'fifties the wet process was predominant in rotary kilns, which were of modest size, producing 200–400 tonnes of clinker a day. The gas velocity through the kiln was relatively low and the chain system at the wet slurry feed end, which produces a slurry covered screen as the kiln rotates, was effective in catching much of the dust. Nevertheless it was common for 1–3 per cent of the clinker production to be lost in dust emitted through the chimney.

As interest in dust collection increased, attention was first given to cyclones, despite their inherent inefficiency, particularly with very fine particles, and for some years they provided, because of their reliability, the means of catching the highest possible proportion of dust. Dust collection with bag filters has also been tried, particularly in the US. Difficulties have been experienced with blinding of the fabric which can occur with low gas temperature at start-up or sometimes due to a leak of cold air into the collector. Furthermore, unsteadiness in kiln operation can lead to short periods of high exhaust gas temperature, and even with the best glass fibre fabrics there is always a risk of sudden loss of the whole bag installation. Maintenance of rapping gear and problems with the replacement of defective bags have also deterred development of fabric filters of the size required for kiln gas cleaning. Nevertheless, considerable advances have been made during the last few years in fabric types, in over-temperature safeguard systems and in other design aspects. A number of bag-filter systems are operating successfully now, particularly on dry process plants in North America, although for kiln exhaust gas cleaning the electrostatic precipitator is likely to continue to predominate.

Electrostatic precipitators

While bag-filter methods were being improved, electrostatic precipitators were also being installed on cement kilns. At first their reliability, on this as on other duties, was low, and they spent a large proportion of their time under repair, or at least not working properly. Efficiency and reliability have, however, increased significantly, and gradually over the last 20 years electrostatic precipitators have not only established their position as the standard kiln gas dust collector, but their very achievement has encouraged a trend towards more stringent standards of acceptable dust emission from cement works' chimneys.

The electrostatic precipitator has not only changed the standards of acceptable emission, it has made possible a radical change in emphasis in the industry from the wet process to the dry. Until dust collection of its quality was established the advantages, particularly of low fuel consumption, offered by the dry process could not be properly exploited, and competition with the wet process depended on semi-dry kilns, in which the grate loaded with nodules provides a comparable inherent dust collecting capability to the chain system in a wet kiln. Today the dust content of the gas leaving the kiln has no influence on the choice of process. The electrostatic precipitator deals with all normal inlet dust burdens, to produce an outlet burden of as little as 50–200 mg m^{-3}, the precipitator's size depending mainly on the gas flow and the required outlet burden, and to a much smaller degree on the inlet dust burden. For the present, therefore, the best practicable method of cleaning the waste gas from cement kilns of all types is the electrostatic precipitator, whose effectiveness at removing dust from the gas passing through it is a function of the time the gas spends in the electrostatic field, and thus of the precipitator's size.

Basically the reduction in dust burden is at a first approximation in geometric proportion to the number of equal sections in the machine. Thus a precipitator of three sections would reduce the dust burden in the first chamber from 16 to 4 g m^{-3}, in the second chamber from 4 to 1 g m^{-3}, and in the third from 1 to 0·25 g m^{-3}. The addition of a fourth section would reduce the dust to a quarter of 0·25, but the reduction in emission in the extra chamber of only 0·1875 g m^{-3} is hardly won in comparison with the gains in the first two sections. Efficiency of collection expressed as the collected proportion of the inlet dust burden is no criterion of the dust finally emitted, because the inlet dust burden varies widely with different types and sizes of kiln. It is of limited value as a criterion of precipitator performance, because the effectiveness of the precipitator is also extremely dependent on the particle size and resistivity of the dust, and on the temperature and moisture content of the gas. Large particles are easier to catch than small. Precipitation improves with moisture content of the gas. The resistivity of all dust rises with increasing temperature to between 200° and 300°C, above which it falls again. This is less important in wet kilns than in dry kilns, because the resistivity of the dust in a wet kiln never makes collection as difficult as in a dry kiln, in which the resistivity of the dust is so high at 300°C that quite a thin film on the collecting plates causes a phenomenon known as back corona and seriously disturbs the precipitation process.

Wet process. In a wet process installation, the dust entering the precipitator consists of particles from all stages of the burning process, a little raw feed, much calcined material, perhaps a little clinker dust, and the alkalis, potassium and sodium, condensed in the cooling waste gas as both sulphates and chlorides. It is the alkalis which provide the finest particles and are most difficult to catch; their behaviour is not yet fully understood. They do, however, appear to be responsible for some variability in precipitator performance. They also on occasion make the collected dust both fluffy and sticky. In the former condition its bulk density can be less than 500 kg m^{-3}, and it requires a very well-designed collecting hopper to prevent the occurrence of build-up. In the second condition the dust may stick to discharge electrodes, collecting plates, or guide baffles despite the provision of an efficient rapping system. The difficulties may be made worse by the practice of returning dust to the kiln system, thereby causing a circulating load of alkalis. In some wet process kiln precipitators, it is found that disposing of the dust from the outlet section instead of returning it to the kiln removes the problem. It is fortunate that the finest dust is caught last and that the finest dust contains most alkali, so that a large proportion of the undesirable alkali can be removed by disposing of a very small proportion of the dust collected. In a modern large kiln, the dust reaching the precipitator may be as much as 10 per cent of the clinker production, but 90 per cent of this dust can almost always be returned to the kiln without causing a circulating load of alkali detrimental to the operation of the precipitator. This leaves for disposal a quantity of dust only 1 per cent of the clinker production of the kiln. When the raw materials are relatively free of alkali all the dust can be returned to the kiln to be converted to clinker.

Dry process. In the dry process, the dust leaving the kiln assembly, that is to say the last of the pre-heater cyclones, has a different composition from the dust leaving the wet kiln. The proportion of dry raw material is much higher, as the calcined dust leaving the kiln itself is very largely intercepted in the cyclone system and returned to the kiln with the feed material. There is a greater tendency for the alkalis to be returned to the kiln with the feed material as well, so that removal of alkalis from the system by disposing of dust from the last section of the precipitator is not always an effective ploy with raw material of significant alkali content. The alkalis may in any case cause build up of deposit in the cyclones themselves and in time choke the flow of both gas and feed material, although in most dry kilns alkalis are not an operational problem.

When the raw mill and kiln are close coupled, the precipitator has two distinct modes of operation, one when the mill is working, and the other when it is not. When the gas leaving the kiln system passes first to the mill, the dust from the kiln system is absorbed in the raw material in the mill, and the dust in the cool gas, at approximately 100°C, leaving the mill is predominantly from the raw material in the mill. As the gas is cool and damp the precipitator is very efficient, although the dust particles are small and of relatively high resistivity. To put it another way, a modest size of precipitator is adequate to achieve a low dust burden in the final emission. The position changes when the mill is shut down, because the transporting gas is both dry and hot (about 350°C), which increases the resistivity of the dust, and in such conditions the precipitator is relatively inefficient. It is therefore necessary to pass the waste gas first through a vertical cylinder in which water is added through sprays to cool and moisten it. Probably less water is added than in the raw mill and the gas temperature reaching the precipitator is typically about 150°C. The precipitator is then much more efficient, but not quite so effective as when the raw mill is in operation. This is the condition, therefore, that governs the size of precipitator installed, and it follows that when the raw mill is working the final emission of dust is smaller than the standard achieved when the raw mill is shut down.

The dust collected in the precipitator is normally returned in toto to the raw material, either in the storage silo or just before the material is fed to the kiln. In the minority of kilns in which alkalis are a problem, it may be necessary to arrange for a small proportion of gas to be bled from the system between the kiln and the lowest cyclone. It is then very hot, about 1000°C, and has to be cooled by adding air at ambient temperature or by water sprays in a separate cooling tower before being cleaned in a separate precipitator. The alkali-bearing dust collected in this precipitator has to be disposed of separately.

Semi-dry process. In the semi-dry process the dust leaving the kiln proper is largely absorbed in the nodule bed on the grate, and the dust burden in the gas reaching the precipitator is less than in either a large wet kiln or a dry process kiln system. The gas temperature is low, 110°C, and as the gas is slightly moist from the drying out of the nodules, precipitation is normally very effective. Precipitators in the semi-dry process suffer, however, from a very active corrosion process, and it is not possible to use mild steel for the collector plates as in wet and dry process precipitators. Instead the collectors are normally of aluminium alloy and the electrodes must also be specially corrosion-resistant. The mechanism of this corrosion is not yet understood, but it appears to be most violent in the electrostatic field. The dust collected in the precipitator is usually returned to the raw meal, but where alkali problems are encountered it may be necessary to dispose of it.

Blobbing. For most of the time the very fine dust which is emitted from the chimney is dispersed hundreds of feet above ground level, but there are occasions when dust from even the best precipitator installation falls out noticeably. This phenomenon is known generally as 'blobbing', but it takes at least three forms. In the first, blobs of greyish powder are deposited, about 0·5 cm in diameter and conical in shape. In the second and third, the deposit is much whiter, and occurs either as thin snowflakes of much the same diameter as the grey blob, or as small white pinheads. It is difficult to collect enough of the material for chemical analysis, but work is in progress to determine first the nature of the deposits and secondly the mechanism of their formation. Certainly the phenomenon appears to occur most frequently in conditions of near calm, and the fallout is usually concentrated over a small area within a mile of the chimney. It also appears to occur when the precipitator is working at good efficiency, and not, as might be supposed, when the precipitator is in difficulty.

Sulphur dioxide. Only a proportion of the sulphur fed to a cement kiln in the fuel and in the raw material leaves it as sulphur dioxide in the waste gas. Much is combined with the clinker; in a wet process installation some more leaves the system in the dust for disposal. The degree of combination appears to vary from plant to plant, but it is generally accepted that the proportion of sulphur emitted as sulphur dioxide is less than half of that fed to the kiln. In many cases sulphur dioxide emission is very much less. For some dry kilns it has been found that 96–98 per cent of the sulphur leaves the kiln in the clinker when the gas from the kiln is passed through the raw mill, and over 80 per cent when the mill is shut down. In some wet kilns, over 90 per cent absorption of sulphur has been recorded, although in other kilns the sulphur absorption is not nearly so effective. The mechanism of absorption is not yet fully understood, although the subject is under investigation. It is clear, however, that in cement works' emissions dust is the main pollution concern and that the sulphur dioxide emission is only a modest part of the sulphur introduced to the kiln.

Mechanical and electrical condition. Although the operation of precipitators is still attended by some scientific uncertainties, most installations depend for their continuous satisfactory operation on sound design and on regular maintenance. Mechanically there are three major items of concern. The first is electrode breakage, the second is electrode position in the field, and the third is ensuring that the rapping is effective. Electrodes are of two basic types, the plain wire suspending a weight, and the rigid mast, either supporting separate collecting wires, or equipped with pointed fins to establish a satisfactory corona. The plain wires are more subject to breakage as a result of fatigue, spark erosion and possibly corrosion, than the mast electrode. The weight alignment guides may also become blocked with dust, so that the wires are no longer taut and straight. The collecting wires or fins of a mast electrode have to be kept perfectly aligned so that the distance between any point of them and the collecting plates on both sides is the same. In short, one broken electrode will short-circuit the whole electrical section of the precipitator in which it is situated. One bent or twisted electrode can greatly reduce the collecting efficiency of a whole electrical section.

Neither fault is necessarily easy to find, and thorough and regular overhaul is important to minimize undesirable plant stoppages.

Electrically, modern voltage regulators are very effective and reliable, and electrical difficulties are concentrated mainly on insulator failures. The duty is arduous, tracking sometimes occurs along dust deposited on them, and they are on occasion affected by damp when starting up or shutting down.

A precipitator is divided into a number of electrical sections, and a major design consideration is the effect on performance of a fault, electrical or mechanical, which may put a section out of action. It is common practice to arrange precipitators in two banks, so that one can be shut off for service, but the concept does not always work well, as the dampers needed for isolating the defective bank are often either ineffective or time consuming to close and re-open. Furthermore, servicing may take so long that the fall in collection efficiency suffered by operating on one bank only is unacceptable—precipitators are too large and expensive to encourage the installation of full size working and stand-by units. There is now a tendency to fit a single precipitator bank to each kiln and to treat it as an integral part of the kiln unit. Serviceability can be improved by increasing the number of electrical sections from two to three, four, five or six, so that at least one fault can be tolerated without the collecting efficiency suffering seriously.

Reliability of the whole precipitator plant is a matter of the highest importance. If to right a precipitator fault it is necessary to make an unplanned kiln stop, it not only causes a loss of valuable kiln production but it may lead to damage to the kiln's refractory lining. If bricks do fail, their replacement is likely to cost several thousand pounds and several further days' lost production.

None of these very real difficulties detract from the success of electrostatic precipitators in reducing the chimney emission in a modern cement works to a place of concern secondary to low level emission. The success is not, however, complete, and it may still be some years before scientific knowledge about precipitation and kiln dusts and steadily improving mechanical design will ensure continuous trouble-free operation of electrostatic precipitators in every kiln.

In Great Britain the control of emission is in the hands of the Alkali Inspectorate, who base their standards on the 'best practicable means' of dust collections, and in 1967 the Chief Alkali Inspector published a document defining the standards of best practicable means for the cement industry. For emissions from kiln chimneys he recommended a maximum dust burden of 458 mg m^{-3} (0·2 g ft^{-3}) of chimney gas measured at 15 °C wet for small works producing up to 1500 tonnes of clinker a day, and a progressively smaller dust burden at larger works to a figure of 229 mg m^{-3} (0·1 g ft^{-3}) for works producing 3000 or more tonnes of clinker a day. 229 mg m^{-3} is a high standard, which provides only modest scope for future improvement. These standards still apply to existing kilns, but 229 mg m^{-3} is now the standard for all new kilns independent of size. The efficiency of the precipitator required to meet it is often higher than 99·5 per cent, particularly in a large kiln where the dust burden in the gas leaving the kiln is high. The standard of 229 mg m^{-3} of wet gas at 15°C also means that only a very small proportion of the clinker production is permitted to escape as dust via the chimney. In a wet plant it is somewhere about 0·1 per cent, and in a dry plant probably less than half this figure. The exact percentage varies for each kiln, depending on the volume and temperature of chimney gas per unit weight of clinker produced.

As might be expected, recent installations at the beginning of their working life usually do comfortably better than the standard, which is not in the long term meant as a yardstick for the newest plant in trial condition but for all precipitators and all chimneys all the time. The burden has fallen more heavily on older installations, in which dust collection technology and mechanical reliability had not reached the levels expressed in today's plant. Much of this older plant has required radical improvement; some of it, otherwise sound, is having to be renewed years earlier than expected.

In summary it can be said that the Alkali Inspector's best practical means figures for chimney emission represent a high standard of collection. Quite apart from precipitator installations on new kilns, much improvement to and renewal of kiln dust plant on existing kilns has taken place and is continuing. In new works, and in old works, as the level of achievement rises, the emission of dust from the kilns via the chimney is becoming less of a problem, and attention is increasingly turning to the control of low level dust—much greater in quantity than the chimney emission, but happily confined to the area within about 1000 m of the point of emission.

Low level dust

Quarry and raw plant. Cement making is essentially a dusty operation and low level dust is generated at most stages of the process. In the quarry there is no effective means of preventing the generation of dust or of collecting it. Concentration must simply be on good housekeeping, keeping roads clean, swept or damped down, and handling the raw materials as carefully as possible. When wet chalk is crushed it makes very little dust, but dry stone crushing systems have to be sealed and de-dusted, usually through fabric filter systems. Wet process raw material preparation plants make virtually no dust, but dry and semi-dry plants are potentially sources of serious dust emission. The dry plant raw mill is usually associated with the kiln system and the dust carried away from the mill in the drying air is passed to the kiln electrostatic precipitator, in which it is effectively caught and returned to the system. In the semi-dry process the raw mill is not so integrally associated with the kiln system, the drying air usually being separately provided by a furnace. The air leaving the mill is cleaned in a separate dust collector, which may be either of the fabric type or an electrostatic precipitator. A fabric collector is normally more efficient than a precipitator, but the choice between them is far from clear, because the precipitator requires less attention and maintenance. In Great Britain, the Alkali Inspector's standard for all emissions other than from the chimney is 229 mg m^{-3} (0·1 g ft^{-3}).

Both dry and semi-dry raw materials preparation plants produce dry powder which has to be conveyed, first to blending silos, and subsequently to storage silos. The conveying systems are sometimes pneumatic, the powder being carried by air flow along pipelines, but the high power consumption offsets their convenience to some extent. The alternative is the enclosed bucket elevator for lifting, the screw conveyor for horizontal transport and the air slide conveyor giving small downward movement. Both systems are effectively totally enclosed, so it is again a matter of good housekeeping to ensure that they remain at all times free from leakage. At critical points in the system fabric, dust collectors may be installed with venting fans, and the silos are also vented through fabric filters.

Precipitator dust. The disposal of alkali bearing dust from precipitators inevitably creates problems both of dust control during the process of disposal and in the provision of a disposal site. The dust is usually transported from the immediate vicinity of the precipitator in an enclosed conveyor, probably either a screw conveyor or a drag chain, but the disposal site, commonly a worked-out section of quarry, is likely to be some distance away. The most rudimentary solution is to transport it in enclosed trucks, but tipping is bound to cause some dust, even if water sprays can be used during the operation, and its acceptability depends on the location and depth of the disposal quarry. An alternative is to add water to the dust to form a slurry, either in a mixing vessel from which it can be pumped away to the disposal site, or in a tank vehicle which can first be part filled with water and then have dust added to it, the mixing being assisted by air agitation. The material dries out to form a sort of firm sand bank with a crusty surface. The material is changed chemically in the slurrying operation and does not return to a dusty form in natural drying. The settling takes place in lagoons made for the purpose. The water out of which the dust has settled is affected to some extent by the soluble alkalis, and it requires careful control of all the water sources and requirements on a works to ensure that contamination problems are avoided. A third solution is to mix the dust with a small proportion of water either in a pelletizer to make pellets or in a mixer to make a form of damp earth which can be transported by lorry to the disposal area. This solution is hampered by the fact that the dust is often hot and that the water evaporates during the mixing process.

At present a variety of schemes are being tried to deal with the problem of precipitator dust disposal, and as yet no completely satisfactory arrangement has been devised to become the accepted standard.

Clinker coolers. Grate coolers produce a large volume of hot, dry air. The dust burden is often small, and until recently it was normal practice to vent this air through a short stack in the kiln house roof, probably through a cyclone collector. Modern plants are equipped with dust collectors to clean this hot air, but it is not an easy problem. The gas temperature, between 150° and 300°C, may be too hot for normal fabric filters, unless they are sized to carry a large quantity of clean fresh air as well as to cool the mixture. Electrostatic precipitators applied directly are not very efficient because the air is very dry, and the temperature is in the range at which precipitation is least effective. It is necessary therefore to spray water either into the gas stream or more usually into the cooler itself to cool and moisten the air. With this added complication good precipitation is achieved. The water addition must however be carefully controlled to suit variations in kiln output.

A recent development for this duty involves the use of gravel bed filters, which work particularly well on hot dry gas, so that the complication of adding a controlled quantity of water is avoided. Their use is at present limited to a few plants, mostly in Germany.

A more important trend is the increasing use of planetary clinker coolers on new kiln installations in which all the cooling air passes into the kiln and there is no surplus to be cleaned before discharge to the atmosphere. The choice of planetary coolers is often made for this reason more than any technical or process advantage.

Clinker conveying and storage. Probably the most difficult part of the cement making process, so far as control of low level dust emission is concerned, is in the handling and transport of clinker. The clinker leaves the clinker cooler largely in nodular form, but there is always a small proportion of fine gritty dust which is a potential source of emission whenever it has to fall, either through the air or down a chute, when moving from one conveyor to another. The problem is accentuated when the clinker is hot, as the heated air rises sharply, taking the dust with it.

Clinker is often conveyed in totally enclosed drag chain conveyors, but this method is not essential to ensure dust free transport. Vibrating conveyors, bucket and plate conveyors, and rubber belt conveyors convey clinker without making dust over the main body of their length, providing they are protected from wind or strong air currents. Dust is only raised at the points where the clinker is fed on to them and is discharged from them, and it is essential that every conveyor change point is efficiently shrouded and that the shroud is well vented through a fabric dust collector. Most types of fabric collectors are suitable for this purpose (as for most dry material transport dust collecting applications in cement making) but all require maintenance and service to ensure that they are working as they should. There is evidence to show that fabric collectors developed recently using a needle felt envelope cleaned by intermittent pulses from reverse air jets are more effective and very much more reliable than the older form of collector using woven bags cleaned by shaking and rapping. The rapping shortens the life of the bags (it is hard to find a small hole in one of perhaps a hundred bags), and the rapping gear itself requires frequent attention. Whatever the type of fabric dust collector, shrouding of the area to be de-dusted is important (and sometimes very difficult) as the dust collector cannot do its job properly unless the flow of air is always into the shrouded area and through the collector.

The clinker is conveyed from the kiln to the clinker store. Clinker stores were originally open to the air and the clinker was handled by an overhead gantry crane, which made dust whenever clinker was moved. It then became the practice to cover the gantry store, but it was usually provided with some openings to improve working conditions for the crane driver and to enable him to see to do his work. Many gantry stores still exist, and keeping them sealed while making it possible for the crane driver to work efficiently remains a constant problem. For the most part a fair degree of success is obtained.

More recently a tent type of store has been used, either of the ridge or bell tent shape. The clinker is conveyed from the kiln to the top of the totally enclosed store and then allowed to fall freely inside. Much of the clinker is extracted through openings in tunnels constructed beneath the floor of the store. However when the store is nearly empty it is necessary for the clinker to be handled by a mechanical shovel, and similar problems exist in providing acceptable working conditions for the shovel driver as for the crane driver in the gantry store. The situation is particularly difficult when fresh clinker is falling in to the store at the same time as the shovel is working. Design of this type of store is now developing further to ensure that when a shovel is working in the store no fresh clinker is simultaneously falling into it. This can be achieved in two ways; first, by dividing the store into two and feeding clinker into the part of the store in which the shovel is not working, which can thus be kept completely sealed; second, by feeding the majority of the fresh clinker into a separate silo, from which it is conveyed direct to the cement mills. Thus when the kilns are making more clinker than is required to be ground immediately only the surplus is fed to the store, which is then kept completely closed. When the kiln production has to be supplemented from a nearly empty store, the shovel can work in it without being hampered by the dust from fresh falling clinker, and the store can again be practically closed.

A further trend in design of clinker storage, particularly favoured in Continental Europe, is to provide silos for all of the clinker, which are kept permanently closed and are vented through fabric dust collectors mounted on their roofs.

From the clinker store the clinker is conveyed to the hoppers feeding the cement mills. In a gantry store the hoppers are located inside the store and the clinker is fed to the hoppers by overhead crane. Control of emission depends simply on the enclosure of the store. From a tent store or from silos, mechanical conveyors are used to transport the clinker to the mill hoppers, and these are subject to the same need for dust collectors at the change points as the conveyors from the kilns to the store.

Coal. Oil- and gas-fired plants do not suffer from potential dust in handling their fuel, but in coal-fired plants there are problems to be solved both with raw coal and with the pulverized fuel. Some at least of the raw coal is usually stored in the open air. It is nearly always wet enough not to be dusty, even when being handled, but special care may be needed in particularly dry and windy conditions. No devices special to the cement industry are used, and it is largely a matter of good housekeeping.

There are three basic forms of coal-firing systems. In the first the raw coal is fed to the grinding mill through which hot air is passed to dry it. The air is then used to transport the coal to the kiln and as primary combustion air, and both pass immediately to the kiln. The system is totally enclosed, and as all the hot air used passes directly to the kiln there is no problem of dust emission.

In the second scheme the air leaving the mill is again used to transport the pulverized coal, but it is conveyed through a pipe to a cyclone in which most of the coal is separated. The coal passes through a sealing valve into an enclosed fine coal storage hopper; the air is again used immediately as primary air for firing the kiln. It passes through a firing fan between the cyclone and the kiln and the pulverized coal is metered into the firing pipe between the fan and the kiln. The whole system is again totally enclosed; the drying and transport air is again used immediately in the kiln so that there is no problem of dust emission.

In the third scheme, in other ways similar to the second, the drying and transport air is not used as firing air but is discharged to atmosphere through a dust collector. A fabric filter is normally employed. Fresh air is used as primary air for the kiln and the coal feed is metered into it from the fine coal hopper. The system is totally enclosed apart from the air released through the dust collector.

Cement mills. All cement mills have a stream of air passing through them in the same direction as the material being ground. It is needed to keep the cement cool, but it serves a useful purpose at the feed end as it draws into the mill the dust made by the clinker and gypsum falling through the inlet chute. In both open and closed circuit grinding, the air is drawn through the mill by a fan, and between the mill and the fan the dusty air passes through a dust collector.

Both fabric filters and electrostatic precipitators are used on cement mills. When air only is used to cool the cement in the mill and the dusty air to be cleaned is therefore dry, the fabric filter is the clear choice. In most modern mills, however, water sprays are used inside the mill to assist in the cooling and to reduce the air flow required. Thus the dusty air is damp, and electrostatic precipitators can be applied to give very efficient dust collection. They are not, however, normally quite as efficient as fabric dust collectors, and the choice between an electrostatic precipitator, which may be more reliable and require less maintenance, and a fabric collector, which is likely to collect more of the dust, is not at all clear. The efficiency of the precipitator may be impaired at start-up, before the mill is hot enough to require the water sprays to be switched on, unless there is very careful control of the air flow, and care is needed with the fabric filter to ensure that condensation does not take place on shut-down, as this would cause the fabric to become blinded. Both types of collector normally provide reliable and effective treatment.

In closed circuit milling, in which there are ancillary elevators and conveying systems it is usually necessary to provide additional fabric filters at critical points in the transport process.

Cement transport. Cement is transported from the mills to the storage silos either by belt conveyor, which is much more economical in terms of horsepower, or by pneumatic means. There are several types of pumps, but all are totally enclosed, and all convey the cement through steel pipes, so that they do not form a source of dust emission. The conveying air is separated naturally from the cement in the silos, which are equipped with fabric venting filters.

When transport is by belt conveyor, care has first to be taken with design of the feed chute, so that the cement is placed on the belt with as little dust generation as possible. The feed point is also shrouded and vented through a fabric dust collector. The conveyor is in an enclosed gantry and dust is not raised along the main length of the belt. Shrouding and venting through a fabric dust collector is required at the discharge point.

Cement packing and despatch. Dust collectors are fitted to the cement packing machines and at other points inside the cement packing plant. Loading of bags on to the lorries is not therefore a dusty operation, although some dust sticks to the outside of the bags while they are being filled and a small proportion of this is not removed before the bags reach the loading point.

Bulk loading of road cement tankers is carried out sometimes through the lids on top of the tanks, but preferably by a flexible pipe connecting the silo and a special fitting in the back of the tank. The latter method is more easily followed without making dust. The tanker's vent pipe is coupled through a pipe system to a venting dust collector, so that the loading operation is totally enclosed. When top loading is used, a telescopic loading pipe is lowered from above to fit one of the lids on top of the tank. It is difficult always to make a tight seal of the telescopic pipe in the tank opening. The cement loading pipe itself is of smaller diameter than the tank opening. It is surrounded by a second telescopic pipe which fits the tank opening. The annular space between the pipes is used to remove the dusty air displaced during loading. Extraction is through a fabric dust collector. Bulk rail vehicles are loaded similarly through openings in the top.

Buildings. Much cement works plant is enclosed in buildings, erected primarily to shelter the plant and provide dry and warm working conditions for the operators. The buildings also serve to contain whatever small quantity of dust may escape from the plant despite the precautions taken in the installation of local dust collectors. It is not usual for there to be any special means of sealing the buildings, as much of the cement-making process is hot, and natural ventilation is used to keep the working spaces at a satisfactory temperature for the works staff. Kilns are often sited in the open air, with perhaps local shelter for the drive gear, for the firing platform, and for the feed end. A large quantity of heat is radiated from the kiln shell and it would be quite impracticable to enclose them in a sealed building. In fact the main part of the kiln is a sealed cylinder and not therefore liable to dust emission, but it provides an example of the difficulty of applying the principle of sealed buildings as a means of preventing any local dust emission. A move in this direction is possible in the future, but it seems more practicable to concentrate on improving containment of dust in the plant itself, and leave the buildings to continue to perform their present supporting role, most useful when plant is shut down and opened up for overhaul.

Future trends

The control of air pollution is a matter of increasing concern all over the world. The concern is shared by cement manufacturers, responsible government authorities, and by the public, and from it steady improvement will certainly continue. An important element in the control of dust is good housekeeping by works' management and staff, and to this extent effective dust control can be achieved at a moderate cost, because a well-run works is more efficient as well as cleaner than one not so well managed. But much of the plant needed to reduce emissions is very expensive, and its capital and maintenance cost significantly affects the cost of cement production. The manufacturer's concern for improvement is tempered by considerations of cost effectiveness, and he is anxious to assure himself that expensive new plant will operate reliably and acceptably over its planned working life. The government authority, in its concern for improvement, is anxious that all manufacturers should reach a common standard of achievement, but its concern must be tempered with considerations of what is economically as well as technically practicable from year to year. The general public's concern for improvement will be reflected in the end in the price of cement. The trend of improving standards and achievement will advance not only because of the balanced concern of all three parties but also as the result of developing scientific knowledge and of better and more reliable dust collecting plant.

18 The ceramic industries

E. ROWDEN

The great variety of products which the clay industries provide are fired in kilns which vary considerably in design according to the product fired and the firing process used. The main products which are produced by the industries broadly comprise:

1. pottery—domestic earthenware and china ware, sanitary ware, electrical porcelain, glazed wall tiles and teapots;
2. building materials—common building bricks, facing and engineering bricks, roofing tiles, floor tiles and quarries, agricultural land drains, hollow blocks, and salt-glazed and vitrified clay pipes;
3. refractory materials—fireclay and silica bricks and shapes; chrome, magnesite and dolomite high-temperature refractories and insulating refractories.

Each of these products needs to be fired to a definite maximum temperature, which differs for each product, and to a definite firing and cooling time–temperature schedule.

Because of several factors, including the type of fuel used, the special thermal needs and requirements of kiln atmosphere of different products and the presence of certain constituents or impurities in the clay, there are a number of smoke and fume abatement problems which are specific to certain branches of the industry and which require special methods of solution. This was recognized at the time of the passing of the Clean Air Act, 1956. In 1958 certain firing processes presenting special technical difficulties in complying with the Clean Air Act were made registrable under the Alkali, Etc. Works Regulation Act 1906. The definition of registrable ceramic works as later modified are in item 36 of the list of scheduled works as given in chapter 3 of this volume.

The remaining processes, which are not scheduled and must comply with the provisions of the Clean Air Acts 1956 and 1968, include boilers, dryers and air heaters, and all continuous kilns except those which are scheduled.

Pottery industry

Pottery ware is subjected to one or more firings depending on the type of ware. Domestic table ware is subjected to three firings:

1. The biscuit firing in which the clay article after making and having been dried is transformed into a hard durable product by heating according to a predetermined time–temperature schedule to a maximum temperature of 1100–1150°C for earthenware or to 1200–1250°C for bone china.
2. The glost firing in which a coating of powdered glaze, applied to the biscuit ware by dipping in a glaze suspension, is melted on the surface by heating, again according to a schedule, to a maximum temperature of 1050–1100°C. The purpose is to cover the ware with an impermeable coating and make the ware more glossy and attractive.
3. The decorating or enamel firing during which the colours or decoration applied to the biscuit ware or to the surface of the glost ware are fired onto the ware at temperatures of 600–800°C.

Some products are fired only once, for example electrical porcelain at 1250°C and sanitary fireclay at 1200°C, in which case a glaze coating is applied to the dried clay article and the body and glaze are matured by the one firing operation. Other products are twiced fired, wall tiles at 1120°C and 1060°C and sanitary whiteware at 1170°C and 1080°C for biscuit and glost firings respectively. Some extension of single firing has taken place more recently, thus reducing the number of firings to which glazed ware is subjected: some glazed wall tiles, sanitary whiteware, red clay teapots and some earthenware tableware are nowadays produced in this way.

Originally all these products were fired in intermittent kilns charged by hand with coal to a number of fireholes. Figure 18.1 illustrates the type of kiln for firing biscuit and glost ware, typical of the 'bottle' kilns in use until recently in Stoke-on-Trent. Those shown are simple up-draught ovens in which the flames and hot gases from the coal fires pass both upwards to the crown and also through flues under the floor to a central outlet in the floor. Eventually all gases pass through damper-controlled outlets in the crown and into the hovel which serves as a chimney. The biscuit and glost ware were placed in fireclay containers or 'saggers' to protect the ware from the combustion gases, smoke, ash and sulphur oxides. These saggers were stacked in bungs in the oven, one on top of the other.

Decorated ware was not fired in bottle ovens but in smaller, fully muffled coal-fired intermittent kilns. Fully muffled kilns were used to ensure a clean oxidizing atmosphere so essential for the firing of decorated ware. Sanitary fireclay was also fired in

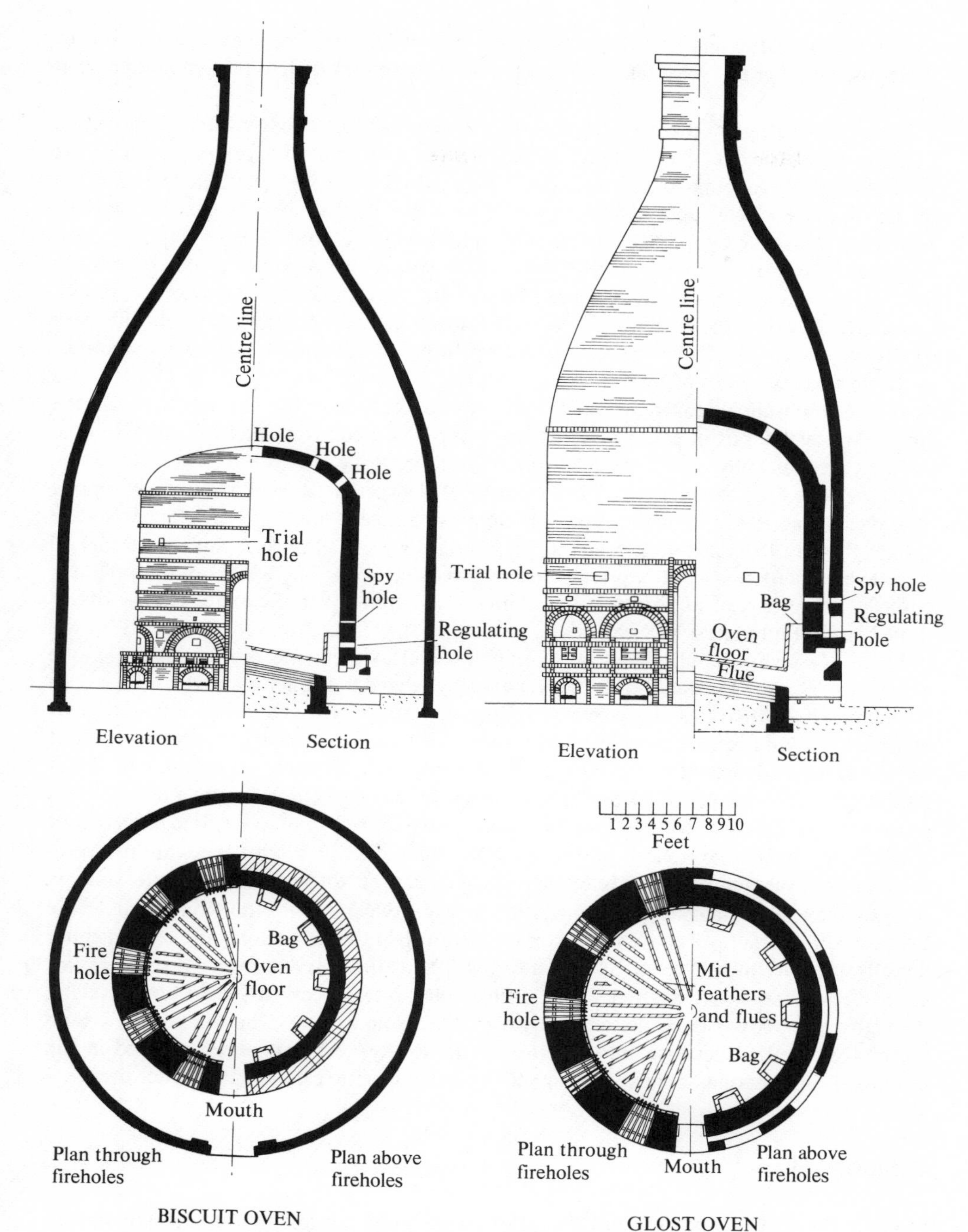

Fig. 18.1. Design of a biscuit and a glost coal-fired intermittent bottle-oven.

rectangular muffle kilns known as 'Scotch' kilns, which had fireholes at each end. In these muffle kilns the ware was placed on shelves made up of fireclay batts supported on props.

Smokeless firing of the coal-fired pottery ovens proved to be virtually impossible. Lengthy emissions of black smoke occurred after each coal charge was given to the fires at intervals ranging from 1 to 3–4 h. Some reduction of the emissions was effected by leaving, after each coal charge, a small air space over the fire doors and, leaving regulating holes open for 30 min or so, and having the four-quarter dampers and, perhaps, the centre damper over the crown holes raised for a period of 15–30 min. A mixture of coal and coke was also tried in an effort to reduce the density of the smoke emissions. Continuous feeding of coal into the top of the fireholes mechanically with overfeed stokers did not produce satisfactory firing conditions, but later conversion to oil firing was utilized to a small extent.

With Scotch muffle kilns, secondary air was admitted over the firebeds to lessen the smoke emissions, but in some cases change-over was made from hand-firing with coal to firing mechanically with underfeed stokers or to oil firing.

These coal-fired intermittent kilns were inefficient, and no real progress in smoke abatement was made until the more efficient car tunnel kilns were developed for the firing of each of the different products and until town gas, electricity and, later, fuel oil were available to be used at a price which allowed fuel costs for firing to be competitive with firing in the coal-fired intermittent kilns. Developments in the increased use of car tunnel kilns commenced well before the passing of the Clean Air Act 1956, both for the purpose of raising the efficiency of production and at the same time reducing smoke emission which the local authority were actively engaged in bringing about.

Figure 18·2 illustrates the principle of the operation of a car tunnel kiln. The kiln consists of a long brick tunnel lined with refractory and of narrow cross-section, about 30–90 m long according to the type of ware being fired, heated over a region at about the centre of its length. Rail tracks are provided through the kiln, on which trucks, running on wheels and with refractory decks on which the ware to be fired is placed or stacked, are pushed at a slow rate through the kiln so as to submit the ware to a predetermined time–temperature schedule. On its passage through the kiln the ware is progressively preheated by the combustion gases travelling from the firing zone to the exhaust fan or stack: fired to maximum temperature in the firing zone and cooled either directly by cold air blown, or drawn, into the kiln at the truck exit end to serve as hot air for combustion of the fuel in the firing zone, or indirectly by cold air travelling through the hollow side walls and crown of the cooling zone. The kilns have a high thermal efficiency, because they operate continuously and heat is recuperated in the pre-heating and cooling zones. Car tunnel kilns may be either muffled or direct fired.

1. Muffled kilns

In these the fuel is burnt in, and the combustion gases confined in, completely closed chambers in the firing and pre-heating zones on each side of the kiln, so as to prevent the

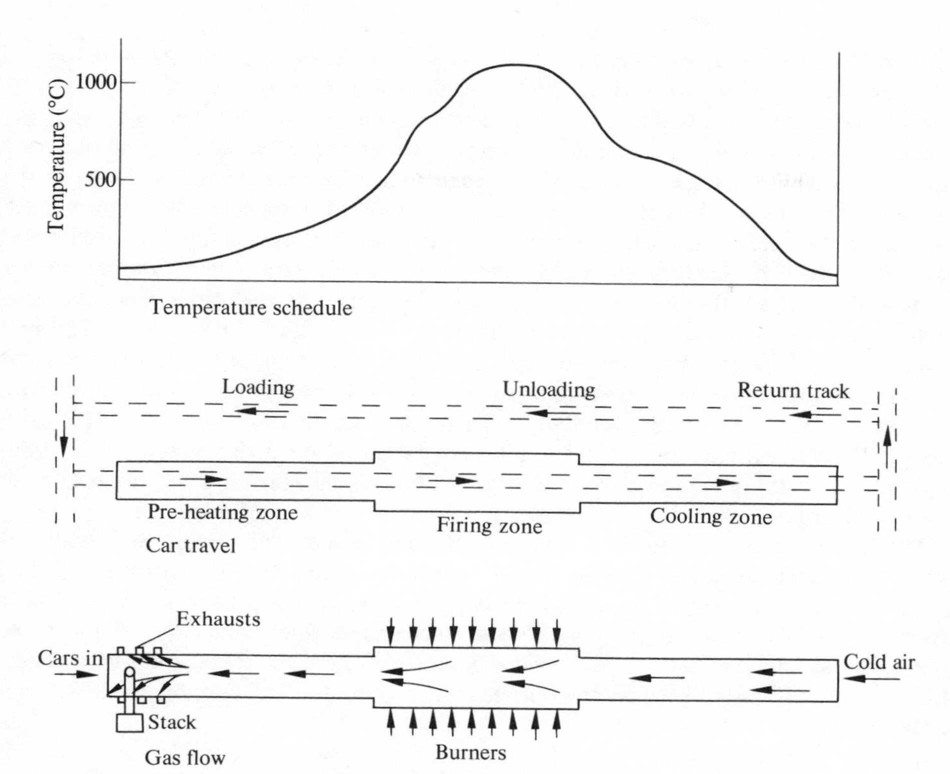

Fig. 18.2. Design of a car tunnel kiln.

gases coming into direct contact with the ware being fired. The ware is open-placed on refractory shelves or batts supported on refractory props or may be placed in special kiln furniture, e.g., cranks.

2. Direct-fired kilns with the ware placed in saggers

With these kilns the ware is placed in saggers to protect it from direct impingement of flames and from the effect of combustion gases, particularly sulphur oxides.

Methods (1) and (2) are necessary when firing with producer gas, fuel oil and also town gas when this fuel is used to fire decorated ware and some glost wares.

3. Direct-fired kiln with the ware open placed

In this case the ware is directly exposed to the heating source. All electric kilns are operated in this way and some gas-fired kilns firing biscuit and some glost wares. This method has been developed because of its higher efficiency. It can only be employed when using clean fuels of low or preferably negligible sulphur content.

As a consequence of kiln and firing developments in the last few decades,[1] the processes in the pottery industry that formerly caused serious pollution by smoke have been so radically changed that they no longer cause serious pollution. Thus, in the Stoke-on-Trent area, where the pottery industry in Great Britain is mainly concentrated, the use of raw coal for pottery firing has been replaced by gaseous fuels, electricity and fuel oil. Coal-fired intermittent kilns have ceased to be used, their number having decreased from 1500–2000 in 1938 to none in 1967. At the same time, the coal consumption fell from about 1 million tonnes in 1938 to 200 000 tonnes in 1959 and zero in 1967. The coal-fired intermittent kilns have been replaced by continuous car tunnel kilns with, in the last ten years or so, modern types of intermittent kilns fired with electricity or gas. The greater number of car tunnel kilns in 1968 were fired with town gas (287) and electricity (90). Electric kilns are favoured for decorative firing, and oil-fired kilns for firing sanitary ware in large muffle tunnel kilns and for firing kilns in outlying districts.

The development of modern intermittent kilns constructed with thin walls of refractory insulating bricks and operated at rapid firing and cooling schedules, assisted in the final replacement of the coal-fired bottle ovens, of which there were 295 still in operation in 1958 when pottery works were scheduled. Such kilns were found to be applicable at small or even medium-sized factories, where the installation of a tunnel kiln was not warranted because the output was too small or too variable or the site not large enough.

These modern rectangular-shaped kilns are of two types, the truck kiln and the top-hat kiln. The truck kilns hold one or more trucks which run on rails both outside and in the kiln. This enables the ware to be loaded and unloaded outside the kiln where and when most convenient.

The top-hat kiln consists of a fixed base on which the ware is placed and a removable cover built of insulating firebrick encased in metal sheeting. The cover which comprises the walls carrying the wire heating elements, or burners, and the roof, can be lifted and lowered by hydraulic rams at each end and locked in the raised position, thus making loading and unloading comparatively simple.

Both the truck and top-hat kilns were initially heated with electric elements in the side walls, but later gas-fired kilns of much bigger capacity were developed. These became popular due to the cheaper price of town gas and freedom from element replacement problems and cost, particularly when firing to high temperatures, such as 1250°C for china biscuit ware. In the tableware industry gas-fired truck kilns with capacities of 7·1 m^3 or more have been brought into use. A battery of these kilns with rail tracks is easy to fit into existing factories to cope with the entire production.

The truck-type kilns have been used to the greater extent to fire tableware, while the top-hat kiln has been used to fire sanitary ware where lack of space or small production precludes the use of tunnel kilns. Figure 18.3 shows a large electric top-hat kiln designed for firing sanitary ware. The top is in the raised position, which gives access to the kiln base for placing the ware on the fixed base. Top-hat kilns are also used for firing large

Fig. 18.3. Electric top-hat kiln firing sanitary ware.

insulators 3–3·5 m high and weighing up to a tonne which can be placed on the base by forklift truck.

Although smokeless combustion has been achieved by change-over in fuel for pottery firing from raw coal to town gas, electricity and fuel oil, further reduction in air pollution has been brought about since 1970 by reduction of the sulphur dioxide emitted. This has been effected by changing to firing with natural gas and liquefied petroleum gases, which are virtually sulphur-free. These two fuels have replaced town gas, which has a small but significant sulphur content, and to some extent fuel oil, which has a high sulphur content. The whole of the gas-fired kilns in the Stoke-on-Trent area have now been converted to firing with natural gas or liquefied petroleum gases.

With processes other than firing, the smoke emission which was particularly heavy at one time from boilers has been curtailed by change-over in operating conditions on the factories. Traditionally steam raised in Lancashire boilers coal-fired by hand provided the ware drying and space heating requirements of the factory and generally was used to drive steam engines to provide the motive power. The use of steam has been reduced drastically by the introduction of electric power, by using gas for some of the drying and space heating, by generating and using the steam more efficiently and by returning condensate. Smoke emission has been lessened not only by reduction of the load but also by the installation of more efficient boilers fired mechanically with coal or with fuel oil.

Generally the virtually complete elimination of smoke emission in the pottery industry has largely resulted from the modernization of the industry, which has led to higher operating efficiency and better working conditions. This has been brought about by closure of a number of the smaller factories, large-scale modification of others and concentration of the industry by mergers into larger groups with more capital available for development and greater technical know-how. The final result is that the pottery industry has now completely discarded its traditional firing practice, which was notoriously smoky, and has replaced it with more efficient and smokeless methods.

Heavy clay and refractories industries

Unlike the pottery industry, the works in the heavy clay and refractories industries are widely distributed throughout the country, and the weight of the products produced is very large. In 1970 approximately 6060 million building bricks of total weight about 20 million tonnes were made; also 720 000 tonnes of sanitary pipes and fittings; about 254 000 tonnes of roofing tiles and floor quarries; 940 000 tonnes of fireclay and siliceous products; 71 000 tonnes of silica bricks and shapes and 258 000 tonnes of basic refractories.

Initially these products were fired with coal, and as in the pottery industry reduction in smoke emission has followed the decrease in use of raw coal. Between 1956 and 1969 the annual consumption of coal decreased from 3 540 000 tonnes to 1 280 000 tonnes, while that of fuel oil increased from 89 000 tonnes to 890 000 tonnes. Reduction in the use of these two fuels has since been made by change to gas firing.

The products of the two industries are all direct fired, with the ware directly exposed to contact with the combustion gases. The main types of kilns[2] used are as follows (the asterisks are explained below):

1. Clamps*
2. Intermittent kilns*
   (a) Up-draught
   (b) Down-draught—round
   —rectangular

3. Continuous kilns
   (a) Annular type
      (i) Longitudinal arch (ring tunnel) kilns
         —Fired from the top of the kiln with coal or oil down amongst the goods, e.g., Hoffmann and zig-zag kilns
         —*Fired with coal (or oil) to grates either from the sides or top of the kiln, e.g., Belgian and related type kilns
      (ii) Transverse arch (chamber) kilns
         —Fired from the top of the kiln with coal or oil down amongst the goods, e.g., Staffordshire kilns
         —*Fired from the top of the kiln with coal or oil to a fender or grate and also down amongst the goods, e.g., Super-Staffordshire kilns.
   (b) Car tunnel type
   Fired from the sides or top of the kiln mechanically with coal, fuel oil or with a gaseous fuel

4. Modern intermittent kilns
   Shuttle and truck kilns fired with gaseous fuel.

Certain of these kilns, marked with an asterisk, when fired with coal or oil present difficulty in complying with the Clean Air Act 1956, because of dark smoke or fume emission. This also applies to any type of kiln, intermittent or continuous, in which a reducing atmosphere is essential during firing.

A large proportion of the products of the industry are fired in continuous kilns without any serious problem with dark smoke emission. For example, over 90 per cent of the building bricks made are fired in continuous kilns with little smoke emission. Apart from the special problem of those products requiring to be fired in a reducing atmosphere to produce certain colour effects, as a broad generalization dark smoke emission is confined to those intermittent and grate-fired continuous kilns which are fired by hand with coal. In other cases the problem is one of fume emission. The most serious problems of the industry at the time of the passing of the Clean Air Act could be summarized as:

1. the firing of intermittent and grate-fired continuous kilns—smoke emission;
2. the firing of blue bricks, tiles and quarries—smoke emission;

3. the firing of salt-glazed pipes and conduits—fume emission;
4. the firing of Fletton building bricks—fume emission.

Since the passing of the Act dark smoke emissions by the industry have been reduced considerably by modification of the method of firing, by change in the type of fuel used and by technological developments. With grate-fired intermittent and continuous kilns hand-fired with coal, application of modified methods of feeding the coal by hand, using mechanical methods of feeding the coal and conversion to oil firing, have led to less emission. Of great importance has been the building of large new works incorporating modern oil-fired continuous car tunnel kilns which, in operation, are smokeless. In the building brick section of the industry there were by 1970 at least 70 modern car tunnel kilns with outputs ranging from 200 000 to 450 000 bricks per week, compared with about 12 such kilns in 1956, mainly coal-fired and of lower output, producing around 150 000 bricks per week. Similarly in the refractories industry there were at least 60 such kilns in operation. In addition, many of the grate-fired Belgian continuous kilns coal-fired by hand which gave dark smoke emissions, have been converted to oil firing or gas firing, thereby eliminating the smoke emissions. With sewer pipes and conduits, 18 oil- or gas-fired car tunnel kilns have been built since 1956

Fig. 18.4. Modern type car tunnel kiln firing facing bricks. (*Courtesy* Gibbons Bros Ltd.)

to fire ceramic glazed or vitrified non-glazed pipes equivalent to over 50 per cent of the annual output, without fume or smoke emission. These kilns, with outputs of 500–1000 tonnes per week, have displaced many coal-fired intermittent kilns firing salt-glazed pipes, thereby eliminating both dark smoke and acid fume emissions.

A further major development was with oil firing, both on intermittent and continuous kilns, particularly the oil slug injection method of firing with oil on top-fired annular kilns (Hoffmann and Staffordshire kilns) and car tunnel kilns. More recently, since 1970, clean gaseous fuels, natural gas and liquefied petroleum gases (butane and propane) of negligible sulphur content, have been increasingly brought into use for firing both intermittent and continuous kilns, displacing coal and fuel oil.[3,4]

Smoke emission from boilers has been virtually eliminated by the change-over to electric power. Whereas previously steam engines were used and exhaust steam was supplied beneath the floors of the drying sheds, now electricity provides the power and waste heat in the form of hot air from the kilns is used to dry the ware in the more efficient chamber or car tunnel dryers. This is sometimes boosted by other sources of heat, e.g., hot air from oil-fired air heaters or the diluted products of combustion from gas or gas-oil burners.

Methods used with different types of kilns and products to reduce smoke or fume emission

Clamps

Clamps, used in south-east England to fire stock bricks, consist of $\frac{1}{2}$–1 million bricks set out in a field in the form of a rectangle about 30 courses high with sides and top covered with burned bricks. The bricks contain added combustible matter such as decayed town refuse, coke breeze or colliery washery slurry. The clamp is provided with layers of coke throughout the base and when completely built is set alight with temporary coal fires at one end. Burning proceeds with little attention for 3–6 weeks. Fumes, which it is not possible to prevent, are emitted as the bricks are burned. These fumes, which have a characteristic smell, are quickly dispersed generally over adjacent fields. The smell has been reduced by replacement of the town refuse, which contains decayed animal and vegetable matter, by coke breeze. The use of clamps is gradually decreasing and methods of firing in continuous kilns have been developed.

Open-top Scotch kilns

The up-draught, or Scotch, intermittent kilns used for firing multi-coloured bricks at small rural works are rectangular in shape and consist of side and end walls only. They have no chimneys. The combustion gases from the fireholes in the walls on each side of the kiln are exhausted from the open top of the kiln after passing upwards through the ware. During the early stages of firing emissions are steamy, but during full firing there is moderately dark smoke from coal-fired kilns and a light to medium haze with some smell from oil-fired kilns. Local pollution is not particularly heavy because the fumes disperse fairly quickly. Progress has been made by converting to stoker and oil-firing and by change-over to down-draught intermittent kilns. In 1973 there were only 24 Scotch kilns in operation, of which 22 had been changed to oil-firing.

Intermittent down-draught kilns

These intermittent or periodic kilns may be either round or rectangular, with fireholes at the periphery or in the side or end walls. The principle of such kilns is for the products of combustion from the fireholes to rise behind the bag walls into the crown of the kiln and then to descend through the setting of goods being fired to openings in the kiln floor through which they pass into flues and then to the chimney. Traditionally these kilns were fired by hand with coal, and as such presented a most serious smoke emission problem.

Methods that have been applied to reduce the smoke emissions include the following:

1. Modification of the coal charging procedure on kilns hand-fired with coal. Such modifications are
   (a) avoiding feeding the fires with large amounts of coal at long intervals;

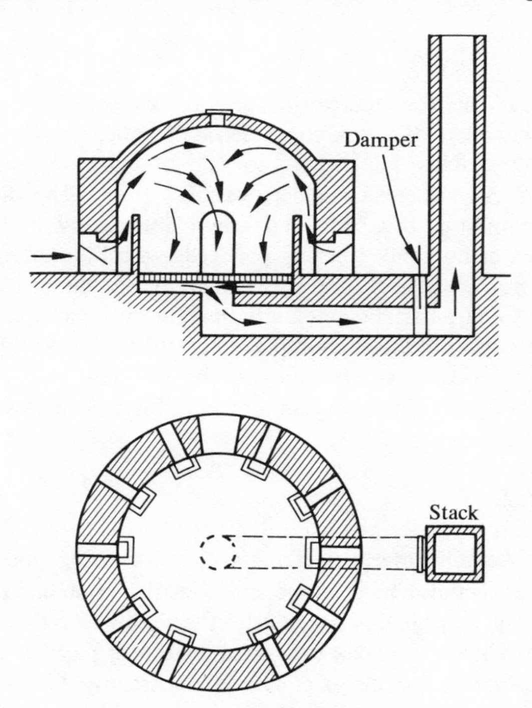

Fig. 18.5. Round down-draught kiln.

(b) adopting the coking method of firing by first pushing the burning fuel down into the firehole and then feeding the fresh charge of coal in at the front top of the firehole;

(c) adopting the alternate firing method, whereby instead of feeding coal to all fireholes the interval between feeds is halved and only alternate fireholes charged with coal. By this means only half the normal amount of coal is charged to the kiln at a time, and when the kiln is above red heat the volatile gases given off from the charged fireholes are burnt in the kiln with air drawn over the top of the partly burnt-down firebeds in the adjacent fireholes;

(d) admitting secondary air over the fires or through secondary air flues to burn the volatile gases given off from the fuel bed;

(e) using a mixture of coal and coke for instance, over the range 200–800°C when difficulty is met in preventing smoke emission;

(f) providing chimneys of adequate height, with damper control, so that sufficient draught is available to draw in the secondary air needed:

2. Installation of mechanical under-feed stokers. The application of mechanical stokers on intermittent kilns allows them to be fired with coal with very little smoke emission except, perhaps, at the time of cleaning fires. Although they have been used to fire a variety of products, their use has not proved entirely satisfactory with reference to uniformity of heating throughout the kiln and the production of a high percentage of best quality products. The number of kilns fired with stokers has gradually declined because of replacement with other methods of firing and disuse of intermittent kilns.
3. Conversion to oil firing. Preference has been to convert hand-fired kilns to oil-firing with atomizing burners rather than to stoker-firing with coal with its problems of ash removal and clinker formation.
4. Conversion to gas firing. The trend finally was firing with gaseous fuels—natural gas, propane and butane—which had now become available and at an economic price.[3,4] Kilns fired with these gaseous fuels are smokeless, and because of their negligible sulphur content the rebuilding or increasing in height of existing kiln chimneys to comply with the conditions as laid down by the Ministry of Housing and Local Government[5] may be avoided, whereas this work is inevitable if the kilns are fired with coal or fuel oil.
5. Replacement of intermittent kilns. The main progress in smoke abatement from intermittent kilns has been through a progressive decrease in the number in use. In 1955 it was estimated that there were approximately 4000–5000 in operation, mainly coal-fired by hand. Table 18.1[6] shows the decline in the total number of intermittent down-draught kilns in operation and in those hand-fired with coal between 1961 and 1973. At the end of 1973 the number of kilns hand-fired with coal was only 36. Finally, under the Alkali Act, there should be none.
6. Raising chimney heights. Although chimney heights of 9–12 m were quite common on intermittent kilns, these were considered far too low for the dispersal of kiln exhaust gases, and the Alkali Inspectorate laid down requirements which had to be met by mid-1970 for intermittent kilns and grate-fired continuous kilns

TABLE 18.1

Number of intermittent down-draught kilns

Year	Coal		Producer-gas fired	Oil-fired	Total
	Hand-fired	Stoker-fired			
1961	2224	297	12	266	2799
1963	1895	312	1	448	2656
1965	1511	274	0	600	2385
1967	1134	231	0	678	2043
1969	657	158	0	619	1434
1971	128	90	0	415	633
1973	36	60	0	328	424

fired by coal or oil, except those kilns firing salt-glazed ware and those kilns firing products under reducing conditions. For kilns, other than those hand-fired with coal, chimney height was to be determined by the sulphur dioxide emission in accordance with the Ministry of Housing and Local Government's publication,[5] with an addition of 10 per cent to compensate for the emission of fluorine compounds and sulphur trioxide. Where hand-firing with coal is adhered to, and is accepted by the Chief Alkali Inspector, the chimney height determined as above is increased by a further 10–25 per cent.

The height of stacks for individual intermittent kilns calculated in this way is approximately 18–23 m or more and required the raising of existing stacks or, where this was not possible, the building of new stacks. The high cost involved encouraged the change-over to firing with clean gaseous fuels of negligible sulphur content, thereby avoiding having to increase the stack height.

Belgian grate-fired continuous kiln

These annular kilns are of the barrel arch or longitudinal arch type as illustrated in Fig. 18.6, without division walls between chambers. Normally they have 22–28 chambers, each about 3·7 m wide and 3·4–3·7 m long. At the beginning of each chamber are grates about 3 m long, extending across the floor, below which is a primary air space or ashpit to which air can be admitted. This space can also serve as the offtake flue from the chamber to a main flue down the centre length of the kiln.

In operation normally there are 3–4 chambers being fired with fuel, 5–8 chambers preheating in front of the fire zone with some connected to the main flue, and 5–8 chambers cooling behind the fire zone with air entering through open wickets at the back of the cooling zone. The whole circuit is progressively moved forward as firing proceeds. Fired goods are withdrawn from chambers behind the cooling zone and fresh goods for firing are placed in chambers in front of the pre-heating zone. When coal

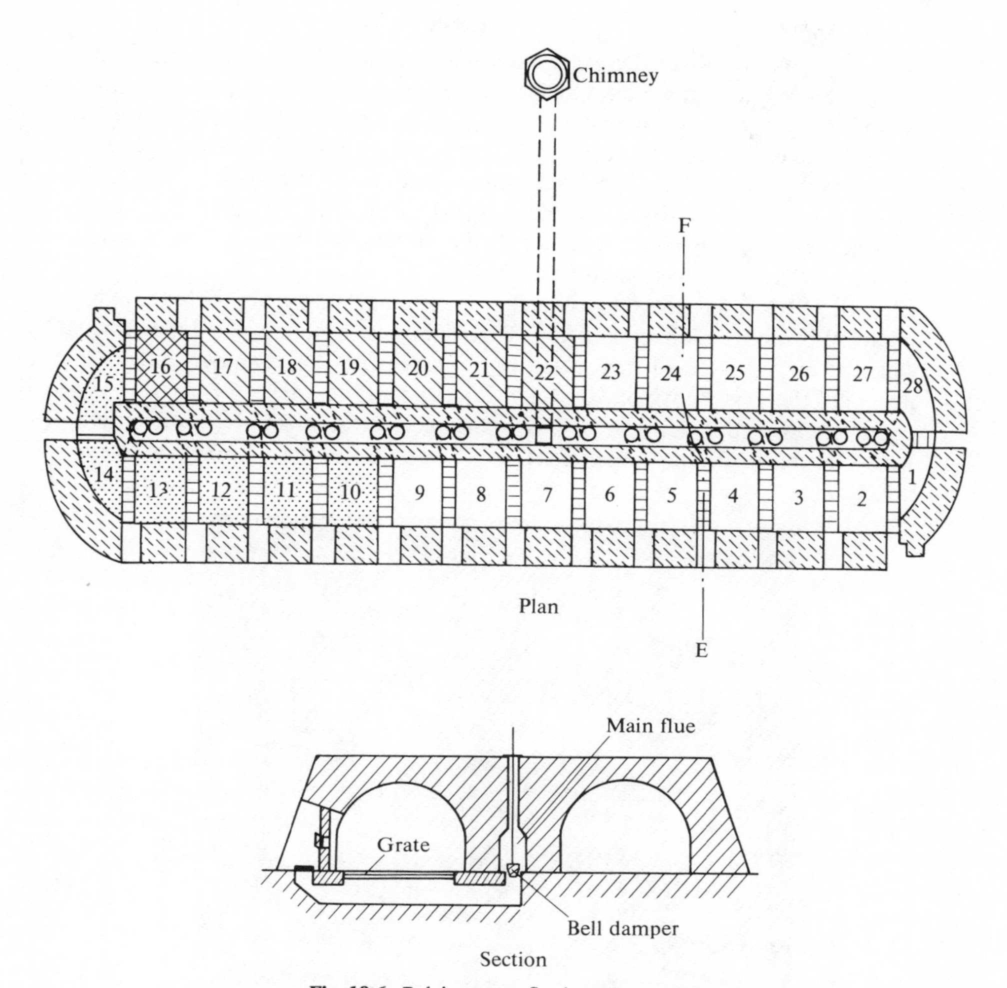

Fig. 18.6. Belgian grate-fired continuous kiln.

fired, the grates are fed with coal by hand from the side of the chamber through a fire door provided in the wicket. Usually 6–9 shovelfuls (30–50 kg) of coal are fed at a time to each of the 3–4 grates.

If the grates are fed regularly every 20–30 min, black smoke emissions can be kept down to 1–3 min duration or less. Methods used or tried to minimize smoke emissions are as follows:

1. Modifying the procedure when hand-firing:
   (a) feeding coal light and often, say every 20 minutes;
   (b) not feeding coal to all 3–4 grates together but at short intervals apart;

(c) using coke to cover the grate when moving the fire zone forward to a fresh grate when a large amount of fuel has to be fed in order to cover the grate and when with coal, smoke emissions are often longer and denser. Alternatively, coke may be fed to the forward grate, a mixture of coal and coke to the second grate and coal to the remaining grates being fired.

2. Firing mechanically with coal. Mechanical firing with sprinkler stokers at the wickets of the grates being fed, and with underfeed stokers has been used at one or two works but not with entire satisfaction.
3. Oil firing. Most progress in reducing smoke emission from Belgian kilns was obtained by conversion to oil firing. The oil impulse system was adopted, whereby slugs of oil are injected to various positions across the chamber above what is normally the grate position, by three injectors in the wicket of each of the 3–4 chambers on fire. Change-over has been made to atomizing oil burners at a few works because of the emission of acid soots from the chimney with oil slug injection.

Fig. 18.7. Oil slug injectors on a Belgian continuous kiln.

4. Gas firing. Recently a number of Belgian kilns have been converted from oil firing to firing with natural gas or butane, using burners at the wickets firing across the chamber above the grate position. These operate smokelessly.

Top-fired annular continuous kilns

These kilns are of two types, the longitudinal arch kilns, of which the Hoffmann kiln is typical, and the transverse arch or chamber continuous kilns of which the Staffordshire kiln is typical.[2] Both are mainly used to fire building bricks.

The Hoffmann kiln is of similar design to the Belgian kiln, except that the fuel is fed through feedholes each fitted with a removable metal lid in the crown of each chamber. Normally the kiln has 16–20 chambers, without division walls between chambers, and according to the size of the chamber has 16, 20 or 25 feedholes per chamber, arranged in rows of 4 or 5 feedholes across the width of the crown.

The transverse arch (Staffordshire) continuous kiln consists of a series of separate chambers built into a common structure, with an equal number of chambers placed back to back, the individual chambers being connected by openings (trace holes) through the inter-chamber walls and the end chambers by flues through the main

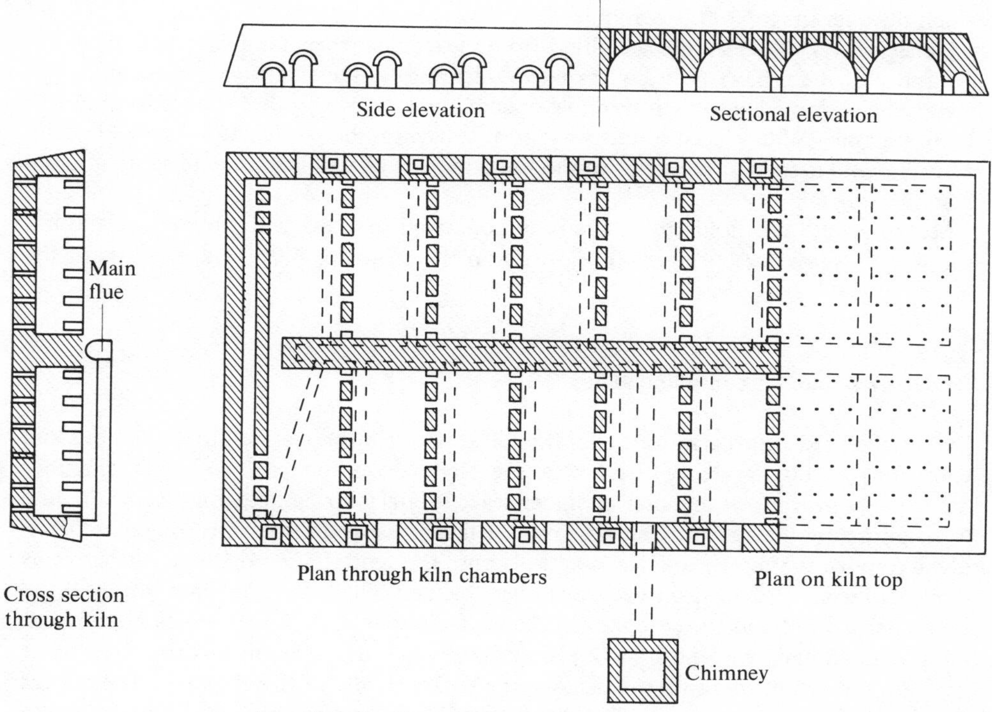

Fig. 18.8. Top-fired chamber continuous kiln.

central wall, so as to provide a continuous circuit. The chambers, maybe 18–20, are provided with 5 rows of feedholes across the chamber crown, the number in each row varying with the size of the chamber.

In both types of kiln coal is fed through numerous feedholes distributed evenly on top of each chamber and falls down and is burnt amongst the goods being fired, special feed-shafts being arranged in the goods below the feedholes. Approximately three chambers are periodically fed with fuel by lifting the metal lid from the feedhole, charging fine coal with a small hand shovel and then replacing the lid.

To avoid smoke emission, a good firing circuit should be maintained on these continuous kilns, with sufficient chambers in front of and behind the fire zone so that the goods are pre-heated to a good red heat before coal is fed, and the air travelling forward as combustion air for the fuel fed is not cold because of the shortness of the cooling zone.

If top-fired kilns are fed regularly with coal to the crown feedholes every 15–20 min, little dark smoke is produced. Feeding heavily at long intervals produces smoke. When this is liable to occur, the installation of mechanical trickle feeders can serve to overcome the emissions. These feeders are designed to feed small quantities of coal, a few grams at a time, every few seconds, one feeder being applied at each feedhole being fed with coal. Each consists of a hopper or cylindrical container holding about 25 kg of coal, from the base of which the coal may be discharged to one side to a downtake which fits over the top of the feedhole.

Generally very little smoke is emitted from these kilns when hand-fired with coal, but a number of them have been converted to oil firing by the slug injection method, one injector being provided to each feedhole or sometimes to every other row of feedholes in the firing zone. With oil fed in this way, soot is likely to be produced if it is fed heavily or to too cold a zone at the front of the fire-zone, which can lead to acid soot emission from the kiln chimney.

More recently top-fired kilns have been converted to firing with butane and natural gas, to give smokeless firing with injectors on the crown or burners at the side of the kiln.

Fender-fired annular continuous kilns

The fender-fired transverse arch kiln (Super-Staffordshire) is a top-fired chamber kiln which has a long grate or fender in the floor down one side of each chamber immediately in front of the inter-chamber wall. Primary air can be admitted to the flue or ashpit below the fender. The first row of feedholes in the crown of the chamber is directly above the fender and the coal falls onto the fender. Two or three further rows of feedholes are provided on each chamber through which the coal falls amongst the bricks being fired and to the chamber floor. The fender row has twice as many feedholes as the other rows and a large proportion of the fuel is fed to this row. The other rows are utilized to bring up the heat on the forward side of the chamber. These kilns often do emit some smoke when hand-fired with coal, because most of the coal is fed to the fender. Smoke emissions have been more or less eliminated by installing mechanical trickle feeders for coal, by converting to oil firing with slug injectors and later to firing with butane, with high pressure gas inspiratory burners or burners with gas and air supplied under pressure on the fender row, and one other row of feedholes on the crown of the chamber.

Firing of blue bricks, roofing tiles and floor quarries

The most serious smoke emission problem of the clay industry, as regards duration and density of the smoke emission, has been that of the firing of blue and brindled bricks, roofing tiles and floor quarries. In order to get the required colour and properties these products are fired in a reducing atmosphere during the late stage of the firing. The bricks produced, Staffordshire blue engineering bricks, are dense, with low porosity, high crushing strength and resistance to acids and abrasion. They are used for foundations to structures, bridges, tunnels, paviors for non-slip floors, stables, sewers, sewage beds, pickling tanks and other acid resisting work.

Generally these products are fired in intermittent kilns which were originally hand-fired with coal. The reducing conditions required were obtained by feeding the fires heavily with coal at approximately 1–2 h intervals and excluding all secondary air. Traditionally the method used was to fill the fireholes with coal at each time of feeding, leaving no air space over the fires, from a kiln temperature of about 500°C upwards. Lengthy black smoke emissions occurred at each coal charge to the end of firing, possibly over a period of 3–4 days.

Improvements with emissions have been obtained when hand-firing with coal. It was found possible to fire these products under oxidizing conditions to temperatures of 950–1000°C, thereby shortening the reducing period, and only then to close the fires with coal to obtain reducing conditions until the finish of firing at 1120–1200°C. It was thus possible to use methods outlined for intermittent kilns to reduce the smoke emissions until the temperature reached 950–1000°C. During the reducing period, tertiary air can be admitted to the flue system after each coal charge, to burn the smoky gases after they have passed through the kiln and carried out their chemical action of reducing the iron oxide in the goods. This air is not operative, however, until a temperature of about 900°C is obtained in the flue, some time after the commencement of reduction. Tertiary air may be admitted as cold air into the main flue at the damper slot or through special tertiary air flues built in the kiln floor or below the main flue, so that the air is pre-heated before it is admitted to the flue system. Application of these procedures has resulted in a decrease in overall smoke emission, but serious emissions do still occur, although they are less in number and of less duration.

Firing with underfeed mechanical stokers and with fuel oil was not generally satisfactory, and smoke emissions although less dense were continuous during part of the reducing period. Difficulties were encountered in getting a good yield of best blue bricks and a satisfactory coloured product. At some works oil-firing has been used to produce 'brindled' bricks.

Investigations were carried out into after-burning of the black smoke produced during the reducing period on kilns hand-fired with coal. The after-burning was effected by means of a heated after-combustion chamber between the kiln and its chimney and by oil burners firing into the main flue, but the method was only partially successful and was considered neither economic nor practicable.

The firing of these products without smoke emission is now being effected almost entirely using gaseous fuels such as natural gas, propane and butane, which from 1969 became available in bulk quantities at works at an economic price.[7] With these gaseous fuels a reducing atmosphere can be obtained without producing smoke. Change-over to gas firing has been made (1) by the conversion of exisiting coal-fired intermittent kilns and (2) by the building of new intermittent kilns of modern design, namely car shuttle kilns.

With existing intermittent kilns the fireholes are fitted with inspirator or nozzle-mix gas burners, and firing procedure and duration of firing are similar to those of coal-fired kilns.

The car shuttle kilns,[8] built of refractory insulation brick, have rail tracks in the base leading in and out of the kiln. They hold several cars on which the bricks to be fired are placed, usually in two stacks about 1 m wide with a space of 0·38–0·45 m between the stacks and into which the burners in the two sides of the kiln fire. At each firing position there are 3–4 burners at different heights above the car deck, so that the kiln is fired evenly from top to bottom. The cars of dry bricks are first pushed into a pre-heater, where they are heated by hot air withdrawn from the cooling kiln, before they are pushed into the kiln for firing immediately after cars of fired and cooled bricks are withdrawn from the kiln, so conserving heat. These kilns have a quick turn round and may be fired twice per week. Firing of blue bricks in them has been effected in under 36 h, with only 10–12 h under reducing conditions, starting at 950–1000°C without smoke emission, and cooling in 48 h.

Firing of sewer pipes and Post Office conduits

In 1956 practically all clay sewer pipes and fittings were fired in intermittent kilns of the round or rectangular down-draught type, coal-fired by hand to the fireholes, and the pipes were salt glazed by feeding common salt to the fires when the kilns had been heated to a temperature in the region of 1100–1150°C. The salting procedure is to throw a shovelful of salt (4·5–5·5 kg) into each of the fireholes in succession and to repeat the process 3–4 times in a period of 4–6 h and then to finish firing the kiln. The salt is volatilized and reacts with the surface of the pipes to form a glaze whilst dense white fumes are emitted from the chimney. As an oxidizing atmosphere is usually needed throughout the firing, smoke emission from hand-fired kilns is not particularly serious, but some kilns have been converted to mechanical under-feed stoker firing with coal or to oil firing.

The main problem in firing salt-glazed ware is one of acid fume emission produced during the salting period. The fumes emitted consist of acid gases and particulate matter, the peak acidities produced immediately after a salt application to the fires exceeding considerably the statutory amount of hydrogen chloride 0·46 g m^{-3} (0·2 gr ft^{-3}) allowed under the Alkali Act for chemical works. It was found that more frequent and smaller applications of salt reduced peak acidity; that there was no suitable alternative to common salt and that scrubbing of the gases would involve a heavy financial burden on the industry. This left dispersal of the fumes from high chimneys or using other glazing techniques. The decision of the Alkali Inspectorate in 1964, which had to be conformed with by mid-1970, was that where salt-glazing was carried out in intermittent kilns, the minimum stack height had to be 36·6 m, but a lesser height might be acceptable for continuous kilns.

In practice, progress has been made by a complete change in the production processes, initially by the introduction of continuous kilns firing salt-glazed ware, followed later by the steady fall in the amount of salt-glazed ware produced as new British Standard specifications for clay pipes permitted ceramic-glazed, with glaze applied by spraying and, later, non-glazed but vitrified pipes to be manufactured. These are fired in car tunnel kilns of large output with negligible smoke and fume emission. Eighteen oil- or gas-fired car tunnel kilns had by 1970 been brought into service to fire

Fig. 18.9. A modern gas-fired car tunnel kiln firing sanitary pipes. (*Courtesy* Gibbons Bros Ltd.)

ceramic-glazed and non-glazed pipes and conduits equivalent, it is estimated, to approximately 55 per cent of the total annual output of pipes.

As a consequence of this change-over from salt glazing, the number of intermittent kilns firing salt-glazed ware has been reduced considerably. It has been reduced still further by concentration of the industry by mergers into fewer and larger firms and by the closure of a number of small works. Whereas in 1964 there were 860 intermittent kilns firing salt-glazed ware, by the end of 1973 there were only 120 such kilns, of which 112 were connected to standard chimneys 36·6 m or more in height for dispersal of acid fume. The 8 kilns connected to sub-standard chimneys must, in the near future, either be connected to standard stacks, be shut down or used to fire unglazed vitrified pipes, possibly with gaseous fuel. Of the 120 kilns, only 21 remain coal-fired by hand, the rest being either gas or oil fired.

Acid soot and fume emission

Smoke emissions from oil-fired continuous kilns are negligible, but in some cases trouble has arisen with the emission of particles of acid soot from the chimney.

The acid soot emission[9,10] is tied up with the firing of continuous kilns, mainly top-fired chamber and side-fired Belgian kilns, with the drip-feed and oil-slug injection methods of firing using medium and heavy fuel oils with sulphur contents of 3–4 per cent, in conjunction with low temperatures in the kiln exhaust system. It can also occur when using coals or firing clays of high sulphur content.

Methods being employed for reducing acid soot emission are as follows:

1. Maintaining efficient combustion to prevent soot formation. This includes the following: regular cleaning and maintenance of the injectors or burners; not feeding oil until the temperature of the goods at the front of the fire zone are at 750–800°C; increasing the temperature in the early part of the fire zone by using atomizing burners, oil gasifier type injectors or liquefied petroleum gas burners and, on Belgian kilns, replacing the oil slug injectors by atomizing burners on the most forward grate or on all grates.
2. Limiting the sulphur contents of the fuel and clay where possible. Methods include change-over to a fuel of lower sulphur content such as gas-oil, liquefied petroleum gas or natural gas. If sulphur is present in the clay possibilities are to change over to an alternative source (but this is not generally possible) or to reject certain seams in the clay pit.
3. Preventing acid deposition by either (a) maintaining satisfactory exhaust gas conditions both with regard to the temperature of the flue gases and the surfaces which they contact, which should be above the acid dewpoint (normally about 150–170°C) and/or (b) injecting very finely divided alkaline powders such as dolomite into the flue gases. These powders serve to neutralize the acid but do not necessarily prevent soots being emitted: they are smaller and less acidic.

A most difficult problem with fume emission occurs with the firing in continuous kilns of certain clays containing compounds of sulphur and fluorine.

With clays with high sulphur content, because of the presence of sulphur trioxide in the flue gases, persistent bluish-white fumes may be emitted which drift for considerable distances from the chimney. This occurs at only a few works, and although a number of methods of treatment have been investigated, so far no practicable solution has been found except that of making use of high stacks to disperse the fumes.

Fume emission from the firing of Fletton bricks has for many years been a major problem, made more so as approximately 45 per cent of the country's annual output of building bricks consists of Flettons. Their production is concentrated in a number of very large works where the bricks are fired by hand with coal in top-fired chamber kilns of large size and output. The lower Oxford clay used to produce these bricks contains organic carbonaceous matter and sulphur and fluorine compounds and, on being fired in the kiln, gives rise to the emission of fluorine compounds and sulphur oxides in the chimney exhaust gases, which also have a typical bad smell. There is no evidence of any public health hazard due to the emissions, but there are complaints about the smell and there is a farming hazard of fluorosis of cattle if grazed on fields adjacent to the works. Problems with acid soot emissions have been treated by insufflation with alkaline powders. So far, otherwise, no practicable means of treating the waste gases has been found and up to the present the conclusion[11] has been that the volume of waste gases is so large and the concentration of the contaminants so low that the only remedy is to use tall chimneys, generally 68·7 m high, to effect adequate dispersion of the gases. At a public inquiry concerning the building of a new brickworks, the conditions laid down were that the chimney should not be less than 91·5 m high and that the gases should leave the chimney at not less than 100°C. Washing of the gases is considered out of the question because of the immense volume of water required, while any treatment that did not remove the odour from the gases would not be worth while.

Investigations into the problem are continuing[6,12,13] with a view to isolating and treating the offensive fraction of the emissions by making it possible to withdraw gases from individual chambers in the forward firing and drying chambers instead of dealing with the stack gases. As a first step, the cycles of emission of sulphur compounds, fluorine compounds, oxides of carbon and hydrocarbons have been determined by heating samples of clay in the laboratory. The compound responsible for the typical 'reek' or odour could not be identified, but the experiments indicated that it is at a maximum at about 500°C and is destroyed at temperatures in excess of 700°C with an adequate supply of air. As the second step, all new kilns are to be constructed with a single steam flue from the separate chambers that will be extended outside the kiln and connected to an outside extraction system, whereby the best practicable means for dealing with the hot gases withdrawn can be investigated. Thus, in the event of a practicable method of treatment being developed, gases can be extracted from any part of the process through a separate ring main. It is possible that existing kilns could be modified similarly. At present the practicalities involved in removing gases from the kiln are being studied.

References

1. Holmes, W. H. 'The pottery industry's contribution to clean air', *Clean Air Conference*, Eastbourne, p. 20, National Society for Clean Air, Brighton, England, 1969.
2. Rowden, E., *The firing of bricks*, Brick Development Association, London.
3. Beech, D. G. and H. W. H. West, *Symposium on gas firing*, British Ceramic Society, Stoke-on-Trent, England, 1971.
4. Harvey, M. R., 'Natural gas firing of kilns—recent developments', *Claycraft*, **44**(12), 7, 1971.
5. Clean Air Act 1956, Memorandum on Chimney Heights (2nd edn.), HMSO, London, 1967.
6. 110th Annual Report on Alkali, Etc., Works 1973, p. 61, HMSO, London, 1974.
7. 109th Annual Report on Alkali, Etc., Works 1972, Appendix VIII, p. 73, Final Report of the Joint Blue Brick Air Pollution Working Party, HMSO, London, 1973.
8. 'Shuttlekiln fires Staffordshire blues', *Claycraft*, **45**(3), 11, 1972.
9. Rowden, E., 'Acid corrosion and smut emission from kiln exhaust gases', *Trans. British Ceramic Soc.*, **68**, 227, 1969.
10. 105th Annual Report on Alkali, Etc., Works 1968, Appendix VI, p. 47, Report and Recommendations of the Ad Hoc Committee on Minimisation of Acid Smut Emissions, HMSO, London, 1969.
11. 102nd Annual Report on Alkali, etc., Works 1965, p. 56–9, HMSO, London, 1966.
12. 106th Annual Report on Alkali, etc., Works 1969, p. 51, HMSO, London, 1970.
13. 107th Annual Report on Alkali, etc., Works 1970, p. 48, HMSO, London, 1971.

19 Lime works

A. PARKER

Lime works have long been detrimental to the environment in several countries by causing objectionable noise and by discharges of smoke and dust, and there have even been complaints in some instances of emissions of sulphur oxides. In Britain the processes are scheduled under the Alkali, Etc. Works Orders under the Alkali, Etc. Works Regulation Act 1906. Item No. 37 of the Orders includes 'Works in which calcium carbonate or calcium magnesium carbonate is burnt through the agency of coal or oil'. Such works have to meet the requirements of the Alkali Inspectorate, who insist on means being used to reduce the emission of air pollutants to the practicable minimum. In his report for the year 1974[1] the Chief Alkali Inspector for England and Wales recorded that the numbers and types of lime kilns registered under the Alkali Act in 1966 and each of the years 1971 to 1974 were as given in Table 19.1.

TABLE 19.1

Numbers and types of lime kilns at registered works in England and Wales in 1966, 1971, 1972, 1973 and 1974

Type of kiln	1966	1971	1972	1973	1974
Coal-fired shaft	112	45	33	30	25
Oil-fired shaft	27	27	18	18	17
Producer gas-fired shaft	9	4	2	1	–
Rotary	NA	NA	9	9	11
Other special kilns	NA	NA	7	7	7

NA means that data are not available.

In his report for 1974 the Chief Alkali Inspector also said:

> The reduction in numbers of registered kilns does not mean that there has been a decline in lime burning. Several kilns have been converted to natural gas firing and some of the new types of kilns have also been designed to use gas. This makes them non-registrable.[2] Lime works were scheduled under the Alkali Act in 1958 because of their black smoke emissions.
>
> At the first opportunity we shall have to consider scheduling all lime works so that all can be properly controlled for dust emissions.

It is also mentioned that rotary kilns that burn dolomite to produce high grade material, some of which is further treated to produce pure magnesium oxide, have recently been equipped with electrical precipitators to reduce dust emission.

What is known as high calcium limestone contains 98 per cent in the UK and 97 per cent in the US of calcium carbonate on a dry basis. The remaining 2–3 per cent of other substances includes magnesium carbonate, aluminium oxide, iron oxide and silica. A limestone containing 30–45 per cent of magnesium carbonate is classed as dolomite.

There are similar problems in other countries in efforts to reduce to the practicable minimum noise and air pollution from lime and dolomite works and in the winning and treatment of other minerals such as roadstone. In the US, where there are limestone deposits in every State, only a small proportion of the deposits is of sufficient purity or quality for industrial use. Interesting and useful reports of experience in the US have been published by the Chemical Committee of the US Pollution Control Association.[3]

Quarrying, crushing and size grading of minerals

Fortunately, most stone quarries are not in densely populated areas. Some, however, are surrounded or adjacent to areas of great natural beauty that are visited by numerous tourists and those on holiday in the district. Examples of such large areas in the UK are the Peak National Park in Derbyshire, the Yorkshire Dales National Park and the Mendip Hills in Somerset. In such circumstances the owners of lime works and quarries are under constant pressure to take all practicable steps to reduce noise and the emissions of smoke and dust from their works.

Where certain types of stone are required in specified shapes and sizes for sculptures or building construction, they are obtained by cutting to the required dimensions and not by blasting. The quantities of dimensioned stone used are small compared with those needed for the production of stone for roads and for use in lime and dolomite works. These latter are obtained by blasting the rock by explosives. In some instances, for example in obtaining stone for roads, the rock is drilled for the insertion of explosive, which is fired and discharges the stone in various sizes. The stone is sorted into ranges of size and as necessary some of it is crushed or otherwise broken down to meet the different requirements.

Limestone quarries are normally developed in a number of benches or lifts, each for example about 20 m (60 ft) in depth, down to the lowest working level. For primary blasting of the limestone, holes are made by drills operated by compressed air. The drilling produces dust and chippings, and the compressed air serves the secondary purpose of removing this from the drilling holes. The dust and chippings are removed from this air by passing it through collectors such as cyclones before the air is discharged to the atmosphere. The series of holes are later filled with explosive and fired in sequence to bring down a length of the quarry face. The stone brought down by the explosives is in pieces of various sizes from large lumps to smaller lumps and particles and dust. Some of the fine dust is unavoidably dispersed in the air.

The excavated limestone is loaded into vehicles which should have vertical exhaust pipes to avoid the exhaust gases stirring up dust from the ground. Each vehicle is loaded to carry 35–50 tonnes of the limestone. It is an advantage in keeping down the dispersion of dust from the ground over which the vehicles travel if the ground is slightly moist. In spells of dry weather this may mean spraying a little water occasionally on the area. The vehicles convey the excavated limestone to a primary crusher.

It is worthy of note that at the lime works scheduled under the Alkali Order in the UK, the official duties of the Alkali Inspectorate do not include the inspection of the excavation of the limestone or its conveyance to the primary crusher. They begin at the crusher.

There are several types of crushing and grinding machines to produce limestone of sizes suitable for the several designs of kiln available for producing quicklime from the limestone. Generally, the machines are appropriately covered and may have a flow of air to carry away dust to be removed by cyclones or bag filters. Such systems not only reduce or prevent the discharge of dust from the machines to the atmosphere but also ensure reasonably good working conditions for the machine operators.

Lime kilns

Over a long period from the 'thirties to the present time many of the kilns have been of the vertical shaft type operated as indicated in the example in Fig. 19.1. These kilns are lined with refractory material. The design shown is about 30 m in height and of an internal diameter of 4–4·5 m in the burning or calcining zone and the cooling zone. There are several designs of shaft kiln, some more complex than that shown. The limestone is charged at intervals, in pieces of approximate, specified size, at or near the top of the kiln. In the top zone the limestone is pre-heated by the gases rising upwards from the burning zone, in which coal or oil was injected at intervals. In this zone the limestone reaches a temperature in the region of 900°C, which drives off carbon dioxide from the limestone ($CaCO_3$) to leave quicklime (CaO). The quicklime descends through the cooling zone where it meets air travelling upwards to be used partly in burning the fuel in the burning zone. The quicklime is discharged at the base of the kiln. For many years coal was the fuel, then in some instances oil or producer gas, but many are now using natural gas.

With coal as the fuel, smoke and other volatile matter are driven off before the coal is fully burning. As a result there are periods after the coal has been charged into the

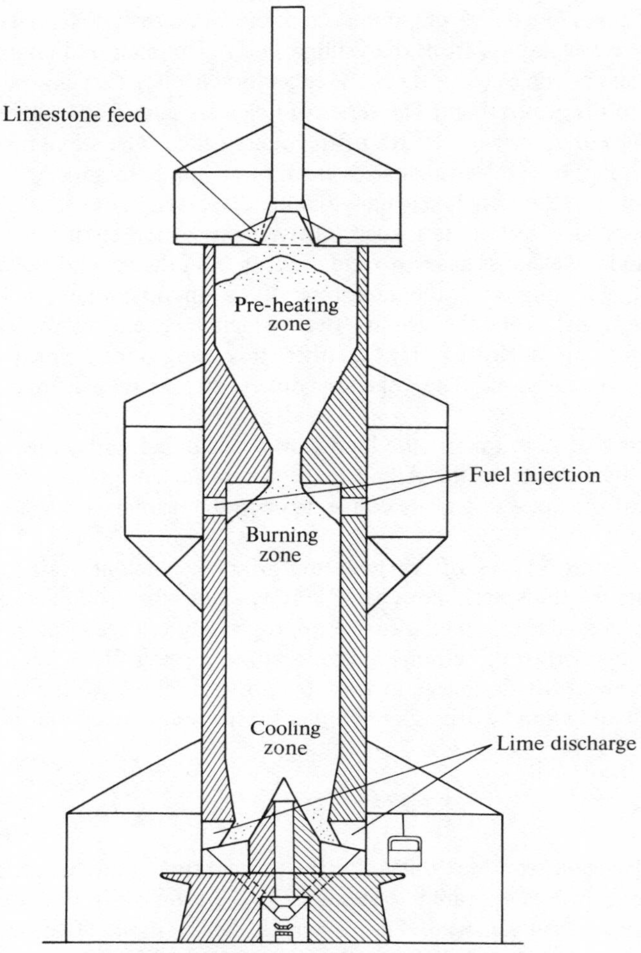

Fig. 19.1. Vertical lime kiln—schematic.

burning zone when there are emissions of dense black smoke from the top or near the top of the kiln. With a works operating a number of shaft kilns burning coal, there is usually always black smoke emission from one or other of the kilns, causing complaints from the people living not far away or visiting the area. The use of oil did not adequately solve the smoke problem. When firing was changed to natural gas, the smoke emission was stopped, but there is still an emission of particles of solid matter in the form of dust in the discharged gases, as from kilns fired with any of the fuels used; the dust is derived from the limestone.

In recent years there have been important developments in the design and use of rotary and rotating hearth kilns in which quicklime is produced by subjecting the

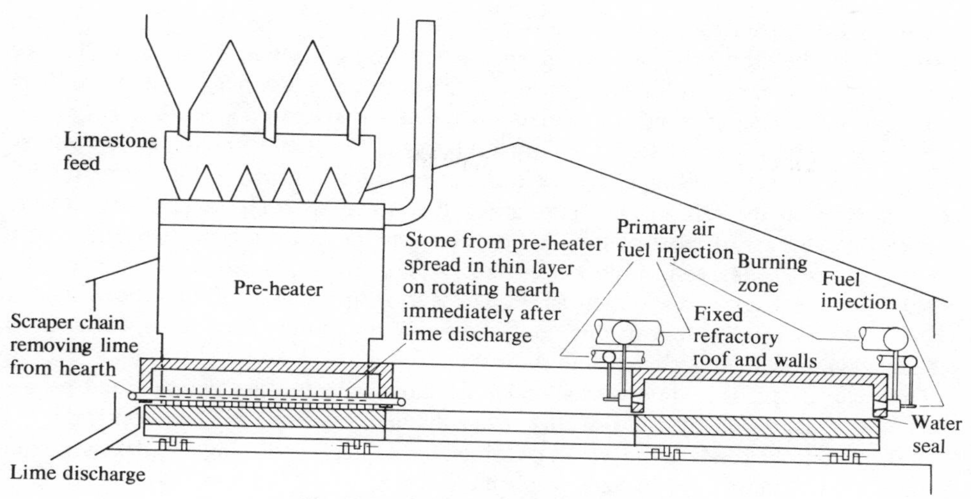

Fig. 19.2. Rotating hearth kiln—schematic.

limestone to relatively high temperatures under carefully controlled conditions. By these methods, quicklime of high quality is obtained. To obtain reasonable thermal efficiency, pre-heaters are installed at the feed end of the kilns to recover heat from the exhaust gases and pre-heat the limestone. At the discharge ends of the kilns heat is recovered from the quicklime product to heat air for combustion. There is a variety of designs of pre-heaters and coolers, which play an important part in determining kiln performance and fuel efficiency. Examples of a rotating hearth kiln and of a rotary kiln are shown in Figs. 19.2 and 19.3 respectively.

Typical rotary lime kilns in use in the UK and the US are inclined tubes of diameter 1·8–3·5 m and length 18–120 m; they are lined with refractory material. They require a more carefully classified and smaller size of limestone than for shaft kilns. The normal

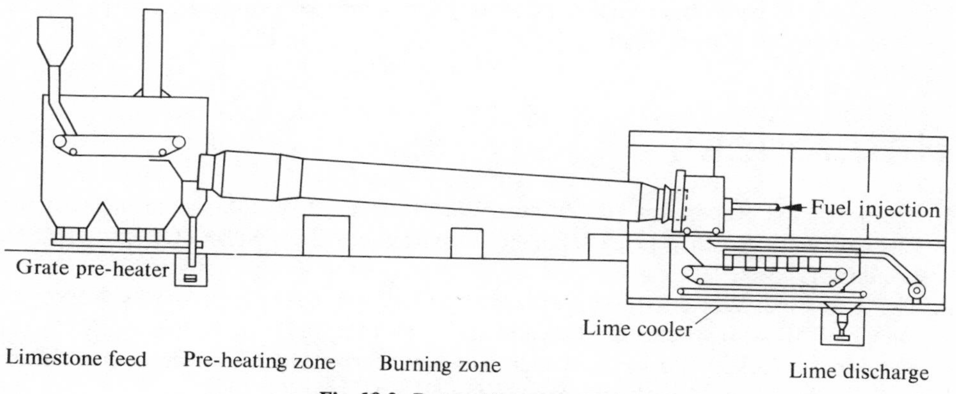

Fig. 19.3. Rotary kiln—schematic.

size of limestone fed to a rotary kiln is about 0·95–4·4 cm and usually 1·9–3·8 cm. There should be little or no trouble from the emission of smoke, whether the fuel be coal, oil or gas. There is, however, dust in the waste gases discharged to atmosphere from rotary and rotating lime kilns unless provision is made to remove the dust or most of the dust from the gases before they are discharged. It is interesting to note that rotary kilns have been used over many years in the production of cement. Such cement kilns are scheduled in the UK in an Order under the Alkali Act. The Alkali Inspectorate require the removal of most of the dust in the waste gases from cement works before the gases are discharged. Modern cement works are equipped with electrostatic precipitators that remove as much as 98 per cent of the dust from the effluent gases. There is the possibility that in due course rotating and rotary lime kilns will also be scheduled in the UK under the Alkali Act.

The main problem in lime works, including those with modern rotating and rotary kilns, is the dispersion by wind of dust from the excavation of the limestone and low level dust from the roads in the works on which the vehicles collecting and delivering the limestone travel; this is not an easy problem to solve.

Hydrated lime

Hydrated lime ($Ca(OH)_2$) is made by adding water to crushed or ground quicklime and thoroughly mixing the quicklime and the water. The reaction of the two causes the emission of a considerable amount of heat and much of the water boils and emits steam carrying some solid matter as dust. Normally the dry hydrated lime is air classified to produce a product in which 95 per cent passes a 200 mesh sieve. There is some production of superfine hydrated lime with particles so small that about 95 per cent will pass through a 325 mesh sieve.

Milk of lime

Milk of lime can be produced either by slaking quicklime with an excess of water or by mixing hydrated lime with water.

References

1. 111th Annual Report on Alkali, etc. Works for 1974, p. 70, Department of the Environment, Scottish Development Department and Welsh Office, HMSO, London, 1975.
2. The Alkali etc. Works Orders 1966 and 1971, 108th Alkali, etc. Works. Report of the Chief Alkali Inspector for England and Wales for 1971, Appendix V p. 66, Item 37, Lime Works, Works in which calcium carbonate or calcium–magnesium carbonate is burnt through the agency of coal or oil. HMSO, 1972.

3. Lewis, C. J. and B. B. Crocker, 'The lime industry's problem of airborne dust', T1–2 Chemical Committee on Air Pollution Problems and Control Measures. Informative Report No. 10. *J. Air Pollution Control Asstn.*, **19**(1), Pittsburg, 1969.
4. Wagstaff, F. J., Paper at a symposium organized by the London and S.E. Branch of the Institution of Chemical Engineers, November, 1972.
5. Hodgson J. M., Series of articles on dust control in quarries. *Quarry Managers' Journal*, May–October, 1967.

20 Incineration of refuse

H. B. JOHNSON and J. M. BURNETT

Pollution is defined in the present chapter as deterioration of the environment due to the construction and operation of a refuse incinerator. Visual pollution due to the intrusion of the plant into the scenery is not dealt with here. Some such intrusion is inevitable, but it can be minimized by suitable choice of site, by satisfactory design of the buildings, by landscaping of the site, by tree planting, etc. In connection with choice of site, it is important to remember that although siting the plant in a hollow can reduce visual pollution it may increase atmospheric pollution.

The other possible sources of pollution from refuse incinerators are atmospheric pollution from the chimneys and the various handling plants, atmospheric and possibly water pollution from the tip for disposal of residues, noise from the plant as a whole and what might be called urban pollution due to traffic bringing refuse to the plant and removing residues for disposal. From the chimneys the visible pollution is likely to be the vapour plume formed in those cases where attemperation of the flue gases prior to passing through dust extraction plant is by water spray or where a wet gas cleaning plant is installed. There are also solid emissions from the chimney, and there is a possibility of smell, particularly when the burning of refuse is combined with the disposal of sludge from sewage treatment plants. Finally there is the possibility of gaseous pollution arising from gases produced in the combustion process. When plants are designed for the burning of domestic refuse and the forms of trade waste delivered at the time the plant is designed, difficulties may arise if other forms of trade waste come to be delivered in future. When incinerators with mechanical grates are used for the burning of carcases, it must be remembered that the residence time for all refuse is the same, and there must therefore be a maximum size of carcase that can be destroyed in this time. From handling plant the possible sources are the blowing about of refuse at the reception point and of dust and clinker from the clinker handling plant. In some cases there may be smell from the handling plant, this pollution usually being a problem internal to the plant. From the residues disposal tip there may be dust pollution, and if the residues contain putrescible matter there may be pollution by smell and by vermin, and possible pollution of water sources. If the plant is required to dispose of carcases, whether of diseased animals (swine fever, fowl pest, etc.) or domestic animals, there may be pollution at the tip if the carcases are not completely destroyed. From the plant as a whole, including the residues disposal tip, there may be noise pollution from vehicles, handling plant, fans, or from the plant for breaking up bulky refuse. Finally there may be pollution from vehicles bringing refuse to the works or taking away clinker. There is pollution from vehicle exhausts, but the main vehicular pollution in the broader sense arises from the traffic problems involved.

The present chapter is concerned mainly with municipal incinerators dealing with the refuse from 100 000 or more people. Plants of this order of size are designed nowadays with environmental conditions very much in mind, and because of their size it is possible to afford substantial sums on prevention of pollution. Smaller plants for small communities, and plants burning industrial or noxious wastes, present different problems which are surveyed in general at the end of the chapter.

Chimneys

Vapour plumes

The gases leaving the furnace of a large municipal incinerator where no heat recovery plant is installed are at a temperature of 1000–1100°C. In the past, where no dust-collecting plant or only a simple water trap was fitted, the gases passed to the chimney at about this temperature without passing through fans, but the chimneys had to be designed to withstand the high temperature. When induced-draught fans and more sophisticated methods of collecting dust are installed, the gas temperature has to be reduced to a level acceptable to the plant. Since this plant is almost certainly built of mild steel, the temperature at entry to the collector must not exceed 275–325°C. For the relatively cheap mechanical collecting plants, such attemperation can be by mixing with cold air, as was first the case at the Polmadie Incinerator at Glasgow, but reduction of the gas temperature of 1100°C to a mixed temperature of 275°C involved an increase in gas mass flow to about 4·7 times the initial flow. Curve (A) on Fig. 20.1 shows the relative mixed volume against temperature to which the gases are cooled.

Where more expensive dust collecting plant is installed, such as electrostatic precipitators, in which the cost is approximately proportional to the volume handled, air attemperation is very expensive and the common practice is to attemperate by spraying water into the gases. Curve (B) in Fig. 20.1 shows that when using spray attemperation the volume of gas at, say, 275°C is only about a third of that when air attemperation is used, and even if town's water has to be used for the spray system the total cost is much less.

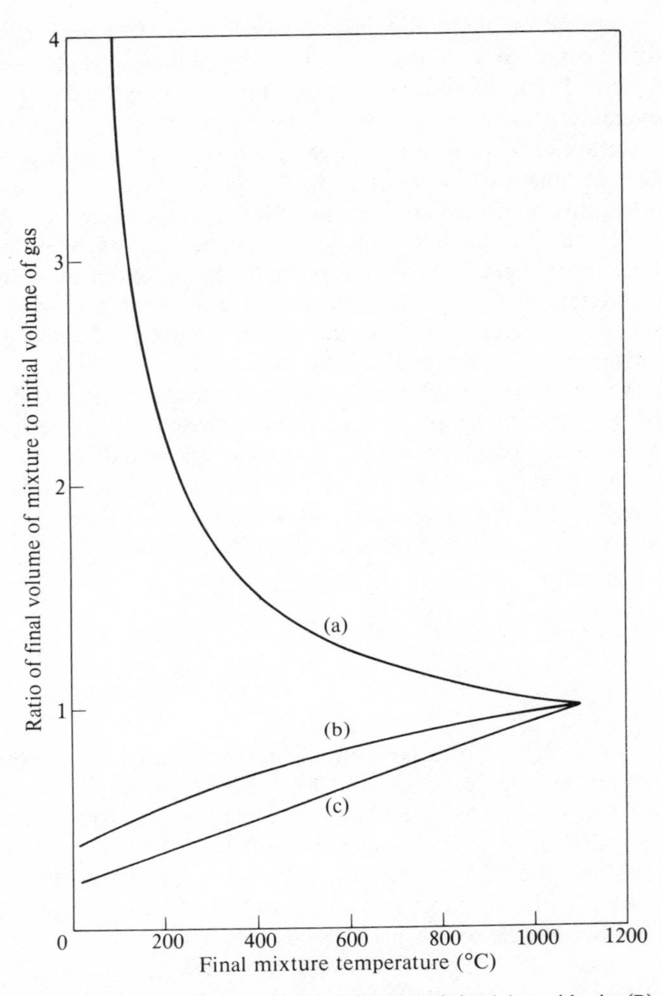

Fig. 20.1. Comparison of three methods of gas attemperation: (A) mixing with air; (B) water spray; (C) absorption of heat in boiler plant.

Figure 20.2 shows, for incinerator gases starting at 1100°C, the relation of dewpoint temperature to the temperature to which the gases are cooled by water spray, indicating that the gases leaving the chimney are highly superheated. However, when mixed with atmospheric air the temperature of the mixture can fall below dewpoint, and Fig. 20.3 shows for various atmospheric dry bulb temperatures and relative humidities the mixing ratios (mass of mixture per unit mass of chimney exit gas) between which there will be a cloud of water particles.

If complete mixing is considered to be the mechanism of visible plume formation, the

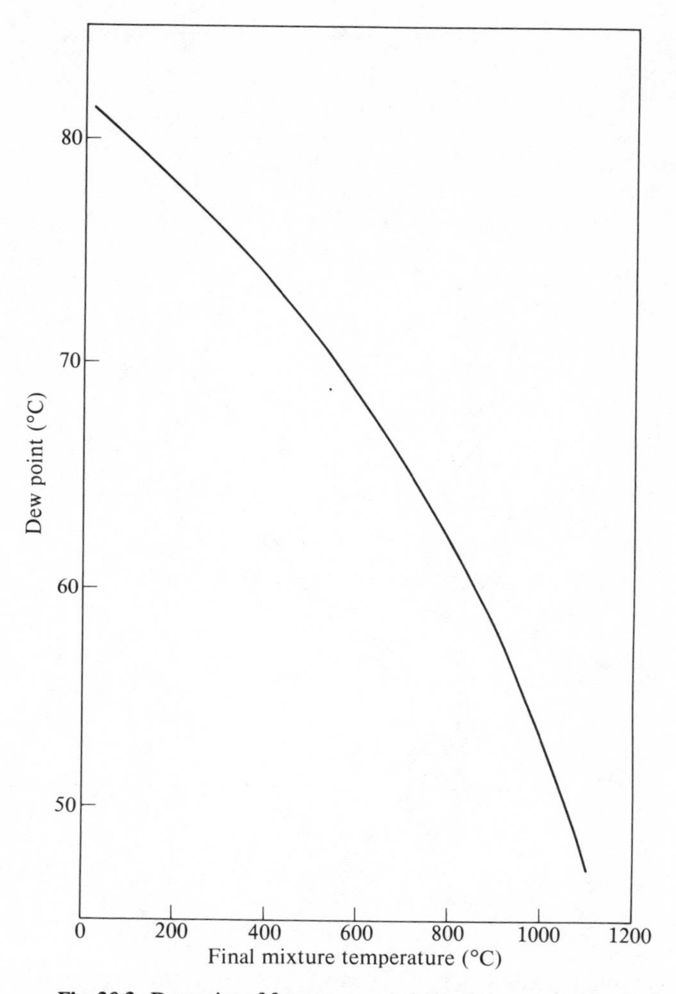

Fig. 20.2. Dewpoint of furnace gases when using spray attemperation.

mixing ratios appear to indicate that the plume would appear at a considerable distance from the chimney. For instance, with a chimney of 5 m diameter a mixing ratio of 100 would mean a plume 50 m in diameter which, at a semi-angle of the plume of 4°, would be 320 m downwind of the chimney. Such a plume would be relatively light in appearance. Experience is that the white vapour begins much nearer the chimney, and thus it appears that mixing alone cannot explain visible plume formation. Other explanations may be

1. loss of heat from the plume by radiation;

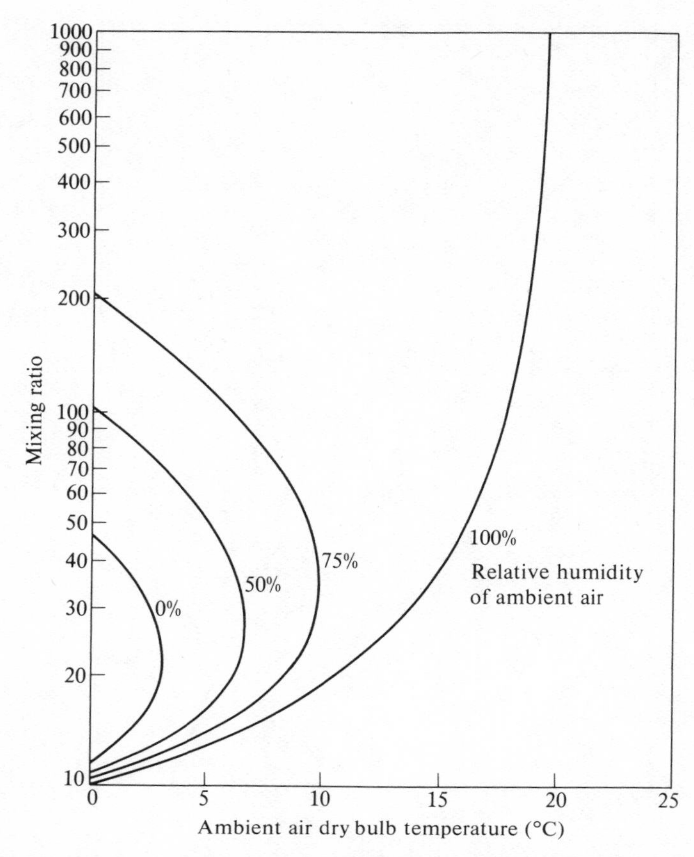

Fig. 20.3. Mixing ratios of air and furnace gas within which vapour is condensed for various dry bulb temperatures and relative humidities.

2. non-uniform mixing in which air, drawn in by injector action of the chimney discharge, causes some parts of the plume vortices to contain a higher proportion of air than others.

Plumes tend, therefore, to appear sooner than indicated above, and to disappear later.

The plume is mainly caused by water vapour, the proportion of steam by volume at the chimney exit being 35–40 per cent, but the water particles dissolve some of the gases in the plume and the dust particles escaping from the collectors act as nuclei for the formation of the water droplets. The general public do not know the nature of the plume and assume that it is 'smoke' and there is experience of public complaint. The point is of particular importance where dust collectors with spray attemperation are added to a plant which had none, for to the public the major change is from a plant with virtually no visible stack emission, except when combustion is bad, to a heavy plume of white vapour. Visual pollution of this kind may be harmless but tends not to be acceptable to the public.

On the assumption that air attemperation is too expensive in dust collecting and draught plant, where the highest standards are required (i.e., electrostatic precipitators) there are two remedies open. First, the hot furnace gases can be passed through a heat exchanger to warm up air which is mixed after the dust collector with the gases leaving the heat exchanger. This reduces the amount of spray attemperation required and increases the dilution. This is discussed below in connection with wet dust collectors. Second, the hot gases can be passed through a boiler and so cooled without adding water or air. The steam produced may be sold if a suitable market exists or condensed in an air-cooled condenser where there is no market or an insufficient one. Some designers have used the steam to drive auxiliaries such as fans by small turbines, but these can only absorb a small proportion of the steam. Power generation must be proved to be worth while and it should be borne in mind that if wet cooling towers are used the problem of the plume reappears. Figure 20.4 shows diagrammatically the arrangement of the furnace, boiler and condenser, and curve (C) in Fig. 20.1 shows the volume of gas against the temperature to which it is cooled, and gives a comparison with the volumes produced by air and by spray attemperation.

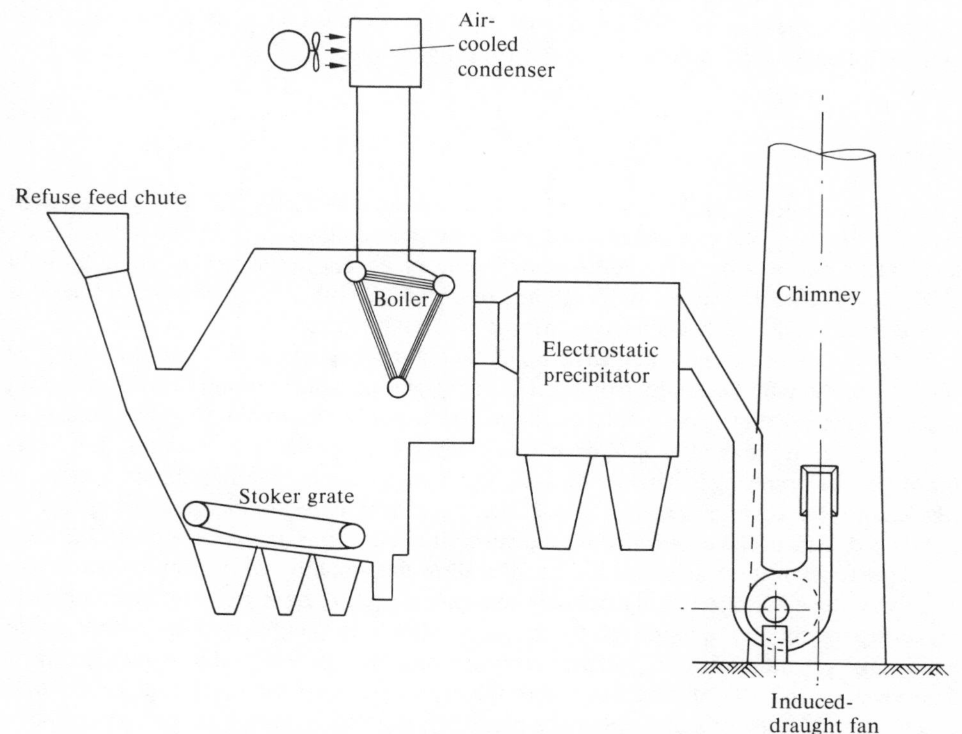

Fig. 20.4. Gas cooling by steam raising and condensing.

Use of the boiler–air-cooled condenser arrangement reduces the gas volume compared with the volume if spray attemperation is used. There is a proportional reduction in the cost of the dust collector, a reduction in the cost of induced-draught fans and fan power, a saving of the water used for attemperation and of pumping power and a saving in the cost of the chimney. There is some saving in maintenance cost of the furnace brickwork if the furnace walls are water cooled. There are extra costs for the boiler, air-cooled condenser and supporting structures, extra fan power for the air-cooled condenser, water-treatment plant for make-up water, etc., and the condenser fans may need to be silenced if the plant is near dwellings. It is likely that on balance the boiler–air-cooled condenser arrangement will lead to a net increase in cost and it is up to the purchaser to decide if the elimination of the plume justifies this extra cost.

In the US, there is a growing use of refuse to replace part of the fuel in coal-fired boilers in power stations. The fact that only 10–20 per cent in heat of the fuel is refuse removes the risk (which has caused much trouble where refuse is the sole fuel) of excessive corrosion and tube fouling. In Britain refuse (after removal of ferrous metals) is replacing part of the coal in some cement works. This process, which is patented, has the advantages that the limestone or chalk in the feed to the kiln absorbs the acid constituents in the furnace gases and that since the residues from burning are acceptable in the cement (just as coal ash is) the process is one of total disposal. Either process eliminates the vapour plume due to spray cooling.

Dust emission

The vapour plume is visible but of no real significance as far as pollution goes, except visually. Black smoke can usually be avoided by providing good combustion conditions and should not be a problem in refuse incinerators, except in the case of some difficult kinds of trade waste such as polyurethane foam. On the other hand dust is a real source of pollution and is always taken seriously in modern incinerators.

There are two ways in which environmental control organizations state the allowable level of particulate pollution from fuel-burning plants. The criterion used is either the mass of dust per unit mass of fuel or the mass of dust per unit volume of gas passing to the chimney, the magnitudes in both cases depending on the nature of the fuel being burnt. In the case of most fossil fuels all the incombustible material stands a more or less equal chance of appearing at the inlet to a dust collector, as in the burning of pulverized fuel. In the case of refuse, however, it is clear that tins, cans, bottles, etc., will not be offered at the collector inlet, and the allowable quantity of dust in relation to the mass of refuse is not so easily defined and may therefore have to differ from place to place depending on the nature of the refuse, particularly the proportion of trade waste. The mass per unit mass of refuse seems unsuitable, therefore, for general control. Moreover it would be difficult to check the composition of the refuse during the time taken to make tests on a dust collecting plant.

The alternative of dust burden per unit volume is subject to abuse if air is allowed in to dilute the gases, and in the US (and occasionally in the UK) this is sometimes prevented by specifying the dustburden per unit volume adjusted to a quoted percentage of carbon dioxide in the gases. This is an entirely artificial adjustment, and the common British practice was much simpler: this was to specify that the rate of burning, the margin of draught plant capacity above normal duty and the performance of the dust collector should be demonstrated in the same series of tests. This ensured that no diluting air could be allowed in without penalty in performance—either by being included in the design, and so leading to high auxiliary costs taken into account in tender assessment, or by taking up the margin and so failing to comply with the specification. Since it is dust burden that the test team is in fact measuring, and since burden measurements are likely to be much better than flow measurements, there was good support for British practice. Probably there should be a range of allowable values of burden depending on the method of attemperation, the allowable value being less with air than with water attemperation.

However, the Report of the second Working Party on Grit and Dust Emissions recommended in 1974 that the dust emitted from an incinerator chimney should be related to the heat release in the furnace. The suggested figures are as follows, and they are based on an assumed calorific value of 5000 Btu/lb, which is greater than has been measured by heat balance on incinerators. Since the calorific value cannot be measured accurately the proposal is entirely artificial.

Heat release (MBtu/h)	Maximum allowable emission of grit and dust (lb/h)
1·0	1·0
2·0	2·0
3·0	3·0
4·0	3·5
5·0	4·0
10·0	6·5
15·0	9·0
20·0	9·6
30·0	10·6
40·0	11·7
50·0	12·8
75·0	19·2
100·0	25·6
200·0	51·1
300·0	76·7
400·0	102
500·0	128
Above 500·0 Add 26 lb/h per 100 MBtu/h	

Intermediate values in the above table should be found by interpolation.

The proposal has not yet (1977) been officially adopted. It is not clear why the Working Party did not propose an allowable emission based solely on mass flow rate of refuse.

The outlet burden from a centrifugal dust collector or a settling chamber depends upon the inlet burden and the grading of the dust. The amount of published information on the performance of mechanical grate incinerators is not large and does not cover all types, but measurements of dust burdens in the gases leaving the furnaces suggest that these may vary from 1000 to 10 000 mg m^{-3} (0·437–4·370 gr ft^{-3}) at NTP dry. There may also be a considerable range of particle size distribution, and Fig. 20.5 shows the probable extent of this range. With such variations it is clear that the most suitable method of control of dust emission depends on individual circumstances.

Figure 20.5 indicates also typical grade efficiencies for a high efficiency cyclone type of centrifugal dust collector. It can be shown that this collector would have an overall efficiency of 66–92 per cent for the range of dusts given. For a situation where an outlet burden of 250 mg m^{-3} (0·11 gr ft^{-3}) at NTP dry was specified, this requirement would thus be met for the coarsest dust provided the collector inlet burden did not

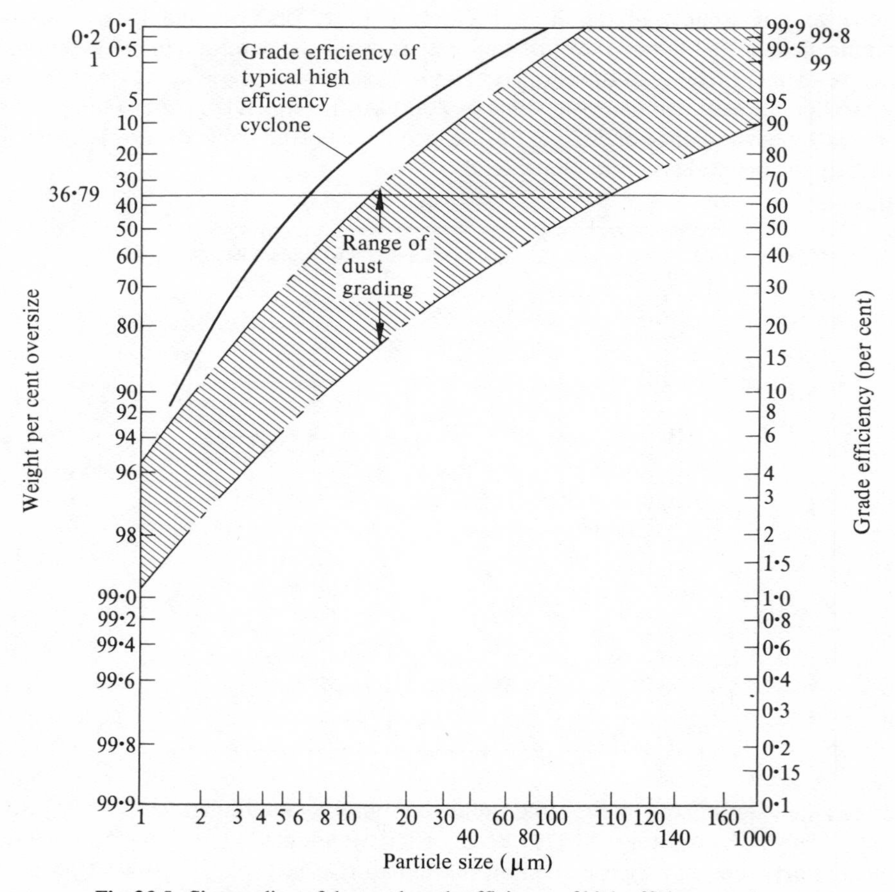

Fig. 20.5. Size grading of dust and grade efficiency of high efficiency cyclone.

exceed 3125 mg m^{-3} (1·366 gr ft^{-3}) at NTP dry, but for the finest dust it is clear that the cyclone type of collector would be inadequate. There is need, therefore, for a great deal more information, notably on inlet burdens and gradings. Account must also be taken of the trend in refuse composition, particularly the proportion of paper and the case for paper recovery, since if paper is salvaged there may be a substantial reduction in dust burden.

The forms of dust collector available for use in refuse incineration plants are the same as those used elsewhere, and are described in another chapter. Incinerators differ from power stations, for example, by the higher temperature and higher water content of the gases, by operating at higher excess air ratios so that fan powers are relatively more important, and by producing part of the dust in flake form (paper char) which presents particular problems in collecting and is readily recognizable as having come from the incinerator.

Where precipitators are used, an important consideration is that if spray attemperation is used, failure of the spray system could lead very quickly to destruction of the precipitator. It is not possible to guard against this by means of a head tank of water, since the spray pressure usually exceeds 100 m of water so that a static tank would be impractical. It would be possible to provide a tank under gas pressure, but the more usual arrangement is to provide a gas temperature control system, which raises an alarm at a fixed temperature and then at a somewhat higher temperature shuts down the induced-draught fans and opens a vent before the precipitator inlet which discharges the gases direct to atmosphere. This means that on a spray failure untreated gas at high temperature would be discharged, but since this would be a rare emergency it should be acceptable. It is important that adequate arrangements be made to protect personnel from danger, and the alarm should be repeated locally to the vent. Access to places in way of the emergency gas path should be subject to a permit-to-work system.

Paper char is not easy to collect by means of a precipitator, since when a flaky material of this kind is shaken from the collecting plates by rapping it is very easily re-entrained. An attempt is frequently made to centrifuge the char out of the gas stream by a sharp bend before the precipitator inlet, but this may well affect seriously the gas distribution in the precipitator. Other possibilities are mechanical collectors before or after the precipitator, or the use of an irrigated precipitator without rapping of the collector plates. Where spray attemperation is used, some of the char is collected in the conditioning towers and attention could well be given to the design of the tower to improve char collection.

Wet dust collectors, or high energy scrubbers, have been used at one incinerator in Britain. The efficiency is high, but since the gases are discharged saturated at a low temperature a dense vapour plume can be formed and the thermal lift or buoyancy of the chimney plume is small, as has been found in power stations with gas-washing plant to remove sulphur dioxide. The formation of the plume was prevented at Govan by passing the gases through a heat exchanger (Fig. 20.6) which reduced the temperature at entry to the washer to 410°C and produced hot air at 430°C. The reduced temperature of the gas resulted in less evaporation in the washer, so that ultimately the gases left the washer saturated at about 50°C, giving a mean chimney temperature of

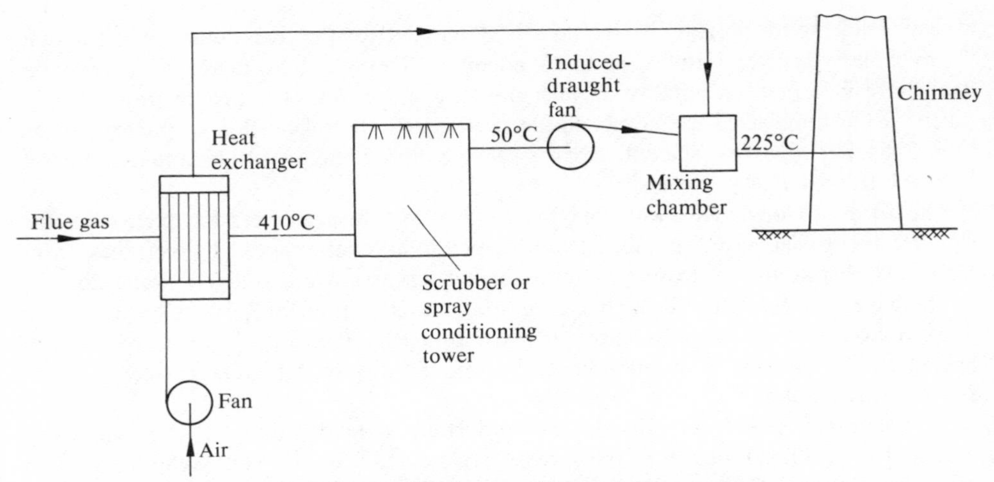

Fig. 20.6. Gas conditioning—avoidance of chimney plume.

225° C. The arrangement has been completely successful in preventing the formation of a plume.

The major trouble with wet collectors in refuse incinerators arises from corrosion. While the sulphur content of household refuse is quite small, there is a growing proportion of chlorine arising mainly from chlorinated plastic materials. These produce hydrochloric acid gas on ignition, and may cause heavy corrosion, which can be overcome by the use of stainless steel. It is necessary to provide pH control of the recirculating water and to neutralize the purge water from the system to prevent possible water pollution. It is necessary also to make frequent inspections of the condition of the plant and to make repairs promptly before any attack is serious. The gas/air heat exchangers at Govan, mentioned above, were designed to avoid low temperatures and corrosion at the cold end, but corrosion protection of the scrubbers involves the use of comparatively expensive materials.

Bag filters have been considered for incinerator gases, but as Table 20.1 shows, the maximum operating temperatures for the various bag materials in common use are such

TABLE 20.1

Maximum operating temperatures for bag filter materials

Material	Maximum temperature, °C
Cotton	85
Wool	25
Nylon	110
Terylene	130
Glass fibre	270

that only glass fibre is suitable for the usual temperature of incinerator gases and the margin of temperature is inadequate even for this. Bag filters are used for control of the air environment within the works.

Mechanical collectors are dealt with in another chapter. It is necessary to add here only that when the chimney gases from an old works have to be improved for the last few years of the life of the plant, simple mechanical collectors of the scroll or shunt-suction type may be adequate. It must be emphasized that efficient collection of combustion solids in the flue gases should not be considered a substitute for efficient combustion.

Gaseous pollution

Little work has been published on gaseous pollution from incinerators. A paper by Barton and Ostle,[1] published in 1966, gave an account of work done in Britain by the National Industrial Fuel Efficiency Service, mainly on incinerators in the Birmingham area. The results are given in Table 20.2, together with measurements made in 1972 in the same area under the supervision of the authors. Table 20.3 gives figures quoted in American references. Table 20.4 gives West German figures and Table 20.5 gives Japanese figures.

These figures may be compared with some limits which have been set in Britain and in Germany, and show that the furnace gases from incinerators contain much less than the allowable limits of concentration, as may be seen by comparing Table 20.2 with

TABLE 20.2

Some constituents of incinerator flue gas by volume[1]

Constituent	Average	Maximum
Nitrous gases, NO and NO_2	100 ppm	200 ppm
Hydrogen sulphide, H_2S	nil	nil
Carbon dioxide, CO_2	6·0–7·0%	11·5%
Carbon monoxide, CO	trace	0·9%
Sulphur dioxide, SO_2	80/90 ppm	192 ppm
Chlorine, Cl_2	trace	2·5 ppm
Hydrogen cyanide, HCN	trace	3·0 ppm
Alcohol, phosgene	nil	nil
Arsine, mercury	nil	nil
Ammonia	nil	nil

Measurements made in the same area in 1972 under the supervision of Merz and McLellan

Constituent	ppm
Hydrochloric acid	150–400
Sulphur dioxide	20–200
Nitrogen oxides	40–65

TABLE 20.3

Composition of incinerator flue gases by volume[2,3]

Constitutent	Furnace gas corrected to 12% CO_2 (ppm or as stated)	
Organic acids	25–133	
Esters	5–137	
Aldehydes	10–82	
Phenols	0·08–1·2	
Sulphur dioxide, SO_2	0·25–1·2	
Nitrous oxides	0·15–1·5	
Ammonia NH_3	0·44–10	
Carbon monoxide, CO	35–400	
Carbon dioxide, CO_2	0·2–1·2	
Carbon dioxide, CO_2	6–10·8%	3·5–12%
Oxygen, O_2	15–9·5%	17–8%
Hydrogen chloride, HCl	90–115	1–330
Sulphur dioxide, SO_2	46–106	0–100
Hydrogen fluoride, HF	trace–0·6	trace–1·1
Nitrogen oxides, NO_2	57–138	107
Hydrocarbons	10	10
Carbon monoxide CO	10	10
Aldehyde	25–58	38

TABLE 20.4

Some constituents of incinerator flue gases by volume[4]

Constituent	Amount present
Sulphur dioxide, SO_2	92–130 ppm (0·122–0·163 gr ft^{-3})
Hydrogen chloride, HCl	172–290 ppm (0·122–0·204 gr ft^{-3})
Nitrous oxide, NO_2	nil
Chlorine, Cl_2	nil

TABLE 20.5

Composition of incinerator flue gases by volume[5]

		HCl (ppm)	SO_2 (ppm)	SO_3 (ppm)	H_2O (%)	CO_2 (%)	O_2 (%)	Temperature (°C)
Furnace outlet	min	300	10	2·0	12·0	9·9	9·2	700
	max	2200	52	6·0	22·0	11·5	10·0	900
	mean	850	28	3·7	17·0	10·6	9·6	800
Dust collector outlet	min	220	49	1·4	18·9	6·3	12·9	190
	max	360	740	1·4	25·4	7·6	14·7	220
	mean	280	60	1·4	22·0	7·0	13·7	200

Table 20.6. The 'presumptive limits' of British practice are those laid down by the Factory Inspectorate.

Examination of these results shows a number of contradictory findings. The UK, West German and Japanese figures for sulphur dioxide all fall within the range 50–200 ppm, but some US figures are very low. For hydrogen chloride, there is good agreement between the West German and Japanese results, with a range of 170–300 ppm. Some US results also fall within this range. However, the Barton and Ostle results make no

TABLE 20.6

Presumptive limits of concentration calculated to NTP which could be applied to gases from a refuse incinerator

1. *Gaseous*	
Total acidity	9·2 g m^{-3} of SO_3 equivalent less than (4·0 gr ft^{-3})
Nitrogen oxides	2·3 g m^{-3} SO_3 equivalent (1·0 gr ft^{-3})
Hydrogen chloride, HCl	0·46 g m^{-3} (0·2 gr ft^{-3})
Chlorine, Cl_2	0·23 g m^{-3} (0·1 gr ft^{-3})
Hydrogen sulphide, H_2S	less than 5 ppm H_2S by volume
Hydrogen fluoride, HF	0·23 g m^{-3} SO_3 equivalent (0·1 gr ft^{-3})
2. *Particulate*	
Arsenic	0·046 g m^{-3} (0·02 gr ft^{-3} as As_2O_3)
Dust (greater than 10 μm)	0·46 g m^{-3} (0·2 gr ft^{-3})
Fume (less than 10 μm)	0·11 g m^{-3} (0·05 gr ft^{-3})

W. German limits	Annual mean	Maximum
Dust emission (units burning more than 20 tonnes per day)	—	0·15 g m^{-3} (0·06 gr ft^{-3})
Non-toxic dusts	0·42 g m^{-2} day (0·6 gr ft^{-2} day)	
Nitrogen oxides	1·0 mg m^{-3} (0·4 × 10^{-3} gr ft^{-3})	2·0 mg m^{-3} (0·03 gr ft^{-3}) ½h in 8 h
Chlorine	0·3 mg m^{-3} (0·12 × 10^{-3} gr ft^{-3})	0·6 mg m^{-3} (0·009 gr ft^{-3}) ½h in 8h
Hydrogen sulphide, H_2S	0·15 mg m^{-3} (0·06 × 10^{-3} gr ft^{-3})	0·3 mg m^{-3} (0·0046 gr ft^{-3}) ½h in 8h
Sulphur dioxide, SO_2	0·40 mg m^{-3} (0·16 × 10^{-3} gr ft^{-3})	0·75 mg m^{-3} (0·011 gr ft^{-3}) ½h in 8h
Hydrogen chloride, HCl	0·70 mg m^{-3}	1·4 mg m^{-3}

mention of the presence of hydrogen chloride, although the 1972 figures from the same area show 150–400 ppm by volume. There is a wide variation in the results for oxides of nitrogen from the nil claimed by West Germany to the 200 ppm found in the UK. Nitric oxide is usually formed in combustion processes, and in the presence of excess air is converted to nitrogen dioxide. Hangerbrauck[6] and others suggest that about 0·9 kg of oxides of nitrogen is produced per tonne of refuse; this gives a concentration of the order of 700 ppm by volume.

Sources of pollution

Sulphur is present in paper, cinders, vegetable matter, rags, rubber and other materials. A typical concentration in refuse is of the order of 0·5 per cent by weight and concentrations of 0·2 per cent are frequently found in the ash. Assuming a 40 per cent ash content of the refuse, these figures give good agreement with the concentration found in flue gas.

Chlorine is present in PVC and in common salt. A typical concentration of plastics in refuse is 1·5 per cent, of which about 20 per cent is PVC. This gives a concentration of 0·18 per cent chlorine in the refuse. If all of this is converted to hydrogen chloride in the flue gas it will give a concentration of less than 50 ppm by volume. Staudinger[7] states that at a temperature of 400–500°C, sodium chloride gives off hydrogen chloride and this could account for the higher concentrations of hydrogen chloride found in flue gases. Further confirmation of the production of hydrogen chloride from common salt is given in a paper by Huck.[8] Road sweepings, which are often burnt with refuse in incinerators, contain a substantial amount of salt in icy weather.

Hydrogen cyanide is formed in the combustion of wool, silk and some plastic materials containing nitrogen. Organic acids such as formic and acetic acid may be produced in burning cellulose. Ammonia can be formed in the combustion of the urea-formaldehyde resins, melamine and nylon. Aldehydes can be produced from combustion of phenol formaldehyde, wood and paper. The production of these and other gases is discussed in a paper by Rasbash.[9]

Smell

The gases from an incinerator leave the furnace at a temperature and after a residence time adequate to ensure that any smell from organic material in the refuse is destroyed. If the chimney is inadequately high or the exit velocity low, so that the gases are brought down to ground level without adequate dilution, there may be some smell from the acids produced in combustion, particularly where spray attemperation is used. These are matters of chimney design and are dealt with in other chapters.

Environment within the works

So far we have been concerned with the protection of the general public outside the plant boundary from pollution arising from the incineration process. It is also necessary to protect those working in the plant, mainly from airborne pollution and noise. The major sources of such pollution within the incinerator works are the handling plant.

Since refuse collection is usually a daytime activity and most refuse arrives at the works between 10.30 am and 4.30 pm on five days a week, while incineration is usually for 24 h per day five days a week, it is necessary to provide storage of refuse. A container system in which the containers are tipped directly into the furnace feed hoppers is regarded as much too expensive, while a possible alternative of airborne carriage of refuse along pipes to an incinerator (which has been done in Scandinavia) is regarded as not practicable except in new towns, because of the difficulty in accommodating the pipes. It is necessary therefore to tip refuse from collection vehicles into a storage pit, and this process is inherently dusty. It is usual to enclose and roof the tipping area, and in a modern works the vehicles are tipped either through curtains or, preferably, through automatically operated doors. When a refuse collection vehicle is reversed into a tipping stall to discharge its load into the refuse storage pit, the associated door is opened and when the vehicle leaves the door is automatically closed. The periods when dust and odours can affect the reception area outside the doors are thus restricted to the short time it takes for vehicles to discharge.

The general practice in incinerator plants is to draw the forced-draught air for combustion from just above the refuse pit, the suction being on the side of the pit remote from the tipping doors. When a tipping door is open, the air velocity is usually adequate to ensure that the dust raised by tipping is drawn inwards to the furnaces and does not significantly pollute the air in the reception area. When the doors are all closed, the combustion air is drawn through balanced louvres located in the doors, and it is essential to ensure by good maintenance that the louvres are free to close, otherwise the louvres by-pass an open door. When the incinerators are operating on low output or when one or both incinerators are not in use, the air flow may be insufficient. To boost the dust extraction facilities provided by the forced-draught fans, separate ventilating fans are provided to be used during such low load conditions. These fans take their suction from the forced-draught fan intake ducting. Thus the air flow across the refuse storage pit takes the same path as when the forced-draught fans are running. The separate fans discharge to atmosphere through dust collecting plant, usually bag filters or mechanical collectors. When a vehicle discharges its load into the pit, high local eddy velocities may be sufficient to overcome locally the air flow through the door and cause the dust to pass into the reception area. To help offset this, the tipping doors should be of the knee type, which when open form a canopy over the discharging vehicles, thus restricting the air flow round and down over the top of vehicles, so that the air velocities local to the vehicle are increased.

With a fuel as varied as refuse, satisfactory combustion can be obtained only if there is adequate mixing of the refuse. In some small plants this is done by passing the whole of the refuse through milling–mixing plant, but this is an expensive process and is adopted mainly because some small grates require it. In the larger plants mixing is obtained by using the grab. Mixing involves spreading of the refuse, so that it is raised 5–8 m and then dropped. This is a dusty process and the dust is extracted by the same system as that from the tipping of the vehicles.

When the grab discharges into the furnace feed chute, however, it is discharged at a higher level, and even if the grab is lowered as far as possible into the chute, much dust is raised. Dust extraction in this area has proved impracticable because in so large a void the air velocity is bound to be low and the dust can only be allowed to settle of its own accord, and is cleared by sweeping or by the use of a vacuum cleaning system, preferably fixed.

Grabbing cranes

The grabbing cranes are directly over the refuse pit and operate in a very dusty atmosphere which would be intolerable for the crane driver. Each crane control cabin or pulpit should be completely sealed and provided with an air conditioning unit which maintains clean air at an acceptable temperature within the control cabin. This is essential for the driver and is desirable for the crane control equipment.

Residuals handling plant

Clinker from incinerator furnace grates is usually quenched by dropping into water and the vapour from the quenching trough or tank and from the subsequent conveyors could produce unpleasant working conditions. From the trough the vapour can readily be sucked into the furnaces, which operate at a negative pressure. The conveyors and any clinker bunker must be adequately ventilated, particularly at manned picking points where large objects have to be removed, to produce acceptable working conditions and minimize condensation and corrosion.

General

The dusty atmosphere of a refuse incinerating works, especially in the crane house, inevitably results in substantial deposits of dust on horizontal surfaces such as ledges and walkways. Ledges should be eliminated as far as possible and this can be achieved most easily by the provision of fire linings which can be designed to give very clean lines, which not only give a pleasing appearance to the building but also reduce dust accumulation.

Noise

So far as is known, there have been no complaints in the UK from the neighbourhood about the noise produced from refuse incinerating plants. Within the plants high values have been measured in certain areas and means are being sought to improve this. Some measured values are as shown in Table 20.7.

TABLE 20.7

Noise levels from refuse incinerating plants

Place	Noise level (db(A))
Near refuse chutes (material falling)	88
Near fans	90
Near bulky refuse reducing plant	90
Near screw conveyors	86
In paper baling house	88

The Greater London Council designed Deephams on certain noise levels and have since produced a memorandum on noise in incineration plants[10] in which plant noise levels are given. Some of these as shown in Table 20.8.

TABLE 20.8

Noise levels from various parts of plants

Source of noise	Noise level (dB(A))
Metal press at 3 m	84–86
Paper press at 3 m	86–88
General plant noise	78–82
Refuse feed chutes	100

Pollution from the ultimate disposal site

The residues from incineration are almost inert, the putrescible matter (by Dusseldorf test) usually being specified as 0·25 per cent by mass, and in most cases being substantially less. The possibility of ground water or aquifer pollution still exists, however, and care should always be taken in the choice of tipping sites that pollution of potable water sources is avoided. This is necessary if only because when tips are open it is difficult to prevent unauthorized tipping and there is a good deal of evidence of clandestine tipping of noxious material on refuse disposal sites.

The other form of pollution arising from the tip is blown dust. It is necessary for the material tipped either to be so conditioned that dust will not blow even if vehicles are driven over the surface or to cover the tip with suitable material.

Traffic pollution

A large incinerator plant burning 1000 tonnes per day has 300 vehicles per day delivering refuse, and 300 empty vehicles, and these are likely to be concentrated in the time between 11 am and 4 pm. It is necessary to remove about 500 tonnes per day of residues, requiring 50 vehicles per day each way, spread over a time interval from 9 am to 6 pm. In addition there are privately owned vehicles delivering trade wastes. The plant must be so sited that this amount of traffic does not cause problems at traffic bottlenecks.

Small incinerators

The discussion above has been concerned mainly with municipal incinerators dealing with the refuse from a substantial number of people—a burning rate of above 5 tonnes h^{-1}, which corresponds to a population of about 100 000 if the incinerator operates for 24 h per day for 5 days a week. Where municipal refuse is disposed of by tipping, and where fuel is expensive, there have been proposals that blocks of flats should be heated by burning the refuse produced by their inhabitants. The amount of heat contained in the refuse from a single flat is of the order of 300 W (1000 Btu/h which is very much less than the space heating requirements of the flat, but might suffice for domestic hot water. Apart from other matters such as handling of the refuse and the clinker, and maintenance and standby requirements, considerations of pollution arise. It is not really practicable to fit such small incinerators with dust collectors of the standard installed in large plants, and in small unattended incinerators it is difficult to ensure that the furnace temperature is high enough at all times to prevent odour. Moreover the chimneys of such small plants cannot be made high enough or the velocity high enough to avoid downdraught and downwash of the gases if the chimneys are on the roof of the flats.

There are many other cases, however, where relatively small incinerators are required. For instance, hospitals need incinerators to dispose of infected or surgical wastes, and many industries produce toxic and other wastes not suitable for disposal without destruction by burning. In plants of this kind, the combustion must be adequate to ensure that the chimney gases are harmless and free from smell and smoke, and may require the burning of supplementary fuel to ensure that the temperature reached and the residence time are adequate. The plants need to be purpose built, their chimneys need to be of adequate height, and control of the plant should be at all times under the charge of a competent person.

References

Note. The Hazardous Wastes Service of the Harwell Laboratory of the UK Atomic Energy Authority publishes monthly abstracts of all literature dealing with refuse disposal. It includes a separate section dealing with incineration.

1. Barton, A. E. and J. Ostle, *Inst. of Public Cleansing Congress*, London, November 1966.
2. Jacobs, M. P., *Proc. Natl. Incinerator Conf. Amer. Soc. Mech. Eng.*, New York, 1964.
3. Miller, P. D., *Corrosion*, 27 January, 1971.
4. Eberhardt, H. and W. Mayer, *Proc. Natl. Incinerator Conf. Amer. Soc. Mech. Eng.*, New York, 1968.
5. Matsumoto, K., *Proc. Natl. Incinerator Conf. Amer. Soc. Mech. Eng.*, New York, 1968.
6. Hangerbrauck, R. P., *J.A.P.C.A.*, 14 July, 1964.
7. Staundinger, J. J. P., *S.C.I. Monograph No 35*.
8. Huck, R., *Brennst Warme-Kraft*, p. 76, 1966.
9. Rasbash, J. D., *J. Plastics Inst.*, January 1967.
10. Greater London Council, *Industrial noise in refuse treatment plants*, Urban Design Bulletin 2, London, September, 1970.
11. Watson, R. H. and J. M. Burnett, 'Municipal refuse as a fuel', *Proc. I. Mech. E.*, **183**(1), 519, 1968–9.
12. Higginson, A. E., 'The analysis of domestic refuse', *Inst. of Public Cleansing Congress*, London, 1966.
13. *Symposium on the prevention of air pollution from incineration*, Filtration Society, Manchester, 1972.
14. Ireland, F. E. and D. H. A. Price, 'Incineration: statutory requirements and environmental aspects, *Brighton Conference*, Inst. of Fuel, 1969.
15. Whitehead, C. and K. Darby, 'Cleaning of gases from the incineration of waste materials', *Brighton Conference*, Inst. of Fuel, 1969.
16. Nonhebel, G., *Gas purification processes*, chs. 18 and 19, Newnes-Butterworth, London, 1972.
17. Watson, R. H. and J. M. Burnett, 'Recent developments and operating experience with British incinerator plant', *ASME National Incinerator Conference*, 1972.
18. Watson, R. H. and J. M. Burnett, 'Incineration as a method of refuse disposal', *Inst. Public Health Eng.*, 1974.
19. Watson, R. H. and P. Knowles, 'Some considerations in the design of refuse incineration plants', *Inst. Public Cleansing*, 1969.
20. 'Incineration—theoretical and practical aspects', *Inst. Chem. Eng. Symposium*, Birmingham, 1972.
21. *Symposium on Solid Wastes and the Environment*, Inst. Public Cleansing, 1973.
22. Snow, D. L., 'Refuse combustion plant monitoring by gas analysis', *Solid Wastes*, **65**, 253, 1975.
23. *Report of the working party on refuse disposal*, British Dept. of the Environment, HMSO, London, 1971.

21 Odours

A. PARKER

During the last two decades there have been considerable reductions in emissions of smoke and other air pollutants from the major industrial processes in operation in the UK, the US, Canada, Western Europe, the USSR, Australia, South Africa, and Japan. In most instances the best practicable means, taking costs into account, have been adopted. This has caused an increasing number of complaints from the inhabitants about the emissions of substances causing objectionable odours, as part of their demand for even further improvement in the quality of the environment. Most of the complaints are justified, some are not really justifiable, and some are incorrect. The number of recorded complaints of odour nuisance received by local authorities in the UK increased from 2101 in 1960 to 2886 in 1965 and 4697 in 1970.[1] This does not mean that the emissions of unpleasant odours have increased during recent years; they have probably decreased.

There are differences of opinion among individuals on whether certain odours are objectionable or pleasant. For example, there are many who object to the odour from coffee grinding and roasting, even on a relatively small scale, while many others consider the odour to be pleasant and enjoyable. It is interesting to note that the materials used for scenting soaps and even for expensive bottles of scents, when in highly concentrated form are very objectionable, but in diluted form are very pleasant and much used.

The increase in complaints about objectionable odours led the Department of the Environment in the UK in 1971 to appoint a Working Party 'To examine the problems of unpleasant odours emitted by offensive and selected other trades and to make recommendations about the best practicable means for their minimization and suppression.' That Working Party, with a Chairman on the staff of the Warren Spring

Laboratory of the Department of Industry, divided much of the work between three Sub-Committees on the following subjects: (1) Animal by-products, (2) Farming and (3) Miscellaneous processes. The Working Party and its Sub-Committees worked quickly and efficiently. The Warren Spring Laboratory published for the Department of the Environment the report of the Working Party in two parts.[1] Part 1 issued in 1974 is an assessment of the problem in Great Britain and part 2 issued in 1975 is on best present practice in odour prevention and abatement. The two parts together cover about 400 pages. The Warren Spring Laboratory is now engaged on intensive investigations with the object of developing practicable methods of eliminating or at least greatly reducing the emissions of objectionable odours. There are so many processes and conditions to be investigated that it will take some time to deal with all the problems; but no doubt the worst odours will be investigated first.

In March 1976, there was a seminar in Birmingham, UK, organized by the National Society for Clean Air, at which fourteen papers on the problems of odour emissions were presented and discussed. Copies of the papers and a report on the discussions can be obtained from the Society.[2]

Measurement of odours

The measurement of the intensity of an objectionable odour and the dilution in air required to remove the objection so far as the population is concerned can only be made by persons whose nose is highly sensitive to the particular odour. The method of odour assessment generally adopted is based on that recommended by the American Society for Testing Materials. A panel of an even number of selected persons, preferably not less than six, who have been tested and found to be good at estimating the relative strengths of a particular odour, is employed. Frequently, the members of the selected panel are young ladies who are non-smokers. They sniff samples of the odorous material drawn from its source and diluted with various amounts of air. Each member of the panel gives an independent opinion on one of the following ratings, nil, very faint, faint, mild, strong, very strong. The dilution is continued until half the members of the panel report no odour and half report very faint odour. The several results obtained in the tests are plotted on a logarithmic scale of concentrations in the diluted samples. The method has been investigated in detail by the Warren Spring Laboratory, where considerable improvements in equipment and methods of sampling have been made. They are described in Part 2 of the Report of the Working Party.[1] With substances of known composition causing the odour the true concentrations in the mixtures tested by the panel can be determined by methods of analysis.

There are instances of effluents causing objectionable odours as a result of constituents of unknown compositions, investigations by chemical and other methods having failed to determine the composition. In some of these instances it is probable that the amount of the substance or substances causing the objectionable odour in the gases and vapours discharged is no more than 1 part in a million parts of the gases.

Masking odours

Materials of a pleasant odour, of undisclosed composition, have been sprayed at various kinds of works near the points of discharging offending odours with the object of masking or disguising the unpleasant odours. The author of this chapter has witnessed trials in chemical works, in works using ethyl and butyl acrylate, which have an unpleasant odour, particularly butyl acrylate, and at sewage works. In these instances the agents masked the unpleasant odours near the points of production and made conditions somewhat more pleasant for some of the operators at the works. But the masking extended for only short distances, not more than 100–125 m in these trials. As a result, complaints of nearby residents continued. With efficiently designed and operated sewage works that are not overloaded and with systems of flow of sewage to the works not allowing sewage to remain in the pipes for so long that it ferments, there should be no objectionable odours from the sewage works.

Works deriving products from maize

There are works in several countries, including the UK, the US and Germany, making several products from maize. The main products of value are glucose and starch, and there is animal feed, known as gluten feed, as a type of by-product. All these works have led to many complaints of the emission of gases with an objectionable odour. Briefly, the usual process begins by soaking or steeping the maize in water from which the glucose and starch are recovered and afterwards purified. The production of these products of value does not lead to an effluent gas with a bad odour.

The process that causes an objectionable odour is that of drying a mixture of maize fibre and residual steep liquor to produce the gluten feed of lower market value than the main products. Investigations have so far failed to determine the composition of the substances causing the bad odour. The usual method of drying the mixture of fibre and steep liquor is by passing the mixture through rotary or flash dryers through which a hot mixture of air and the products of combustion of fuel oil from an oil burner are also passed. The temperature of the mixed gases entering the dryers is not always sufficiently well controlled, though there is a type of automatic control of the rate of burning of oil by the relatively small changes in temperature of the gases leaving the dryer near the position of discharge of the dried gluten feed. There are examples of the temperature of the gases entering the rotary dryers varying between about 500°C and 675°C with an overall average of about 620°C; and a range of temperature of the gases entering a flash dryer of 260–360°C with an overall average of about 320°C. The temperature of the gases leaving the dryers at the position where the dried gluten feed is discharged is usually about 110°C, with fluctuations to a few degrees lower.

There are instances where the flame from the oil burner is so near the inlet of the dryer that there is the probability of scorching some of the fibre near the inlet of the dryer and thus increasing the objectionable odour of the effluent gases. Observations have indicated that the drying of the fibre particles produces a stronger odour than is obtained from the drying of the steep liquor. There is evidence indicating that the strength of the odour would be very greatly reduced and possibly eliminated by drying the mixture of fibre and steep liquor at a much lower temperature, for example by indirect heating with steam. With such an arrangement the rate of drying would be greatly reduced so that larger dryers or more dryers would be required to maintain the same output of gluten feed. Indirect drying of this kind has been suggested but not used on a large scale for gluten feed, though the system is used for drying the more valuable products.

Incineration of the gases and vapours from making gluten feed has been proposed and tried in small pilot plant. Design of incineration equipment has reached an advanced stage, and the erection of full-scale plant for incineration is being seriously considered. It may be that full-scale plant will be in operation in the near future. The temperature of incineration would probably have to be not less than 750°C. Owing to the presence of fine particles of suspended matter in the gases to be incinerated, the use of a catalyst to lower the temperature of incineration greatly would probably not be practicable. There is the probability, however, of utilizing much of the heat in the incinerated gases for drying the mixture of fibre and steep liquor.

Acrylates works

There have been many complaints of objectionable odours from works making acrylates and works using these compounds, for example in the plastics and paint industries. Works making acrylates were first scheduled in 1971 in the UK in the Alkali Works Orders under the Alkali, Etc. Works Regulation Act 1906. Item 58 of the schedule includes 'Works in which acrylates are (a) made; or (b) purified; or (c) made or purified or polymerised.' Acrylates are also included under the Orders in the list of Noxious or Offensive Gases.[3]

In his 110th Annual report for the year 1973[4] the Chief Alkali Inspector for England and Wales says:

> The final emission from a works recovering methacrylate monomer from scrap 'Perspex' is treated in activated carbon filters before discharge.

He also says in the same report:

> There is a large works in the North-East making a wide range of acrylates. The methyl methacrylate plant has operated very satisfactorily since its commissioning nearly two years ago. The same cannot be said for its acrylic esters plant, which has given a considerable amount of trouble. Ethyl acrylate is particularly foul-smelling and even small escapes can be smelt several miles downwind. . . . Investigation showed that the protective foam applied to suppress vapours in an open storage tank, which had not previously given trouble, was being stripped from the surface by gale force winds, leaving the contents open to the atmosphere. Immediate steps were taken to provide a permanent, fixed cover, with the vapour space extracted to the primary air inlet of the adjacent sulphuric acid plant burners. This has been extremely effective and is now a permanent feature.

On the other hand, the Chief Inspector says in his Report for 1974:[5]

> These powerfully smelling compounds are always a potential source of complaint. A large works on Teesside continues to receive complaints despite efficient management and a great endeavour to be a good neighbour. Waste materials are disposed of by incineration, by being dealt with in a spent acid recovery system and by transportation to another part of the country for burial. . . . Progress has been made throughout the industry in gaining steady improvements.

Production of polymers from acrylates

Acrylates, usually purchased from the makers of these chemicals, are used at works making certain kinds of plastics. With the system of transferring acrylates from the road tankers in which the chemicals are delivered to the storage tanks at plastic works, there should be very little emission of acrylate vapours into the atmosphere provided that

1. pipe connections from the road tanker to the storage tanks at the works are made without undue exposure of open ends to the air;
2. the rate of pumping by the pump on the tanker is not such as to cause a pressure in the storage tanks necessitating release of pressure;
3. the tanker and the storage tanks are gas-tight;
4. the pump on the road tanker is in good condition and not leaking.

As there have been some complaints by residents of roads in a village about odours from tankers delivering acrylates to a plastic works near the village, it seems that the means for preventing the escape of vapours from the tankers are not always in good condition or adequately controlled.

One polymer works visited by the author of this chapter frequently received complaints from nearby residents of objectionable odours from the works. The reactor building at that works contained four vertical cylindrical vessels known as a 'slow-add tank', a 'premix tank', a 'reactor' and a 'hold tank'. From the storage tanks the acrylate is pumped into the slow-add tank, which has a pipe line for carrying vapour to the storage tanks. In the premix tank a mixture of soap, surfactant and water is made. The polymerization process is carried out in the reactor vessel, which has an outer jacket through which steam or water can be circulated to heat or cool the reactor and its contents as required. The reactor initially contains nitrogen gas at a pressure of about half an atmosphere. Acrylate and the liquid from the premix tank are run into the reactor at a scheduled rate over a period of 4–5 h and the contents of the reactor are stirred by a turbine agitator. During this period the temperature of the contents is raised to about 70°C and the contents are then cooled to between 45 and 50°C. (At atmospheric pressure the boiling point of methyl acrylate is 80·5°C and that of ethyl acrylate is 99·8°C.) The polymer formed in the reactor is transferred under pressure of nitrogen from the reactor to the hold tank. The evacuated nitrogen, which contains vapour of acrylate, is passed through a vessel containing diesel oil in which acrylate is soluble and then to atmosphere.

Product polymer in the hold tank, which is surrounded by a jacket, is cooled by water passed through the jacket and is tested for content of solid matter. Water is added and mixed in as necessary to give a product of the consistency of thin cream. When the product polymer is of the required consistency as a kind of emulsion, it is passed from the hold tank through an open vibratory filter to remove particles larger than desired and the filtered polymer is pumped into storage tanks ready for transfer to suitable containers for despatch for sale.

The vibratory filter is relatively small and shallow. The vibratory system causes numerous small sprays of the polymer emulsion to rise above the surface of the polymer in the filter. It is intended that the sprays should be directed slightly inwards away from the sides and the top of the filter so that the polymer does not get onto the top and flow down the outside of the filter.

At the time of the visit to the particular works, the filter was not quite horizontal so that the sprays were not all directed away from the sides of the filter. The outside of the filter was covered with dripping polymer and the filter was full to the top rim. Flow to the filter from the hold tank is controlled through a flexible hose. The filter had become clogged to some extent and the inside was being scraped or stirred to remove the material causing the clogging. There was polymer covering a few square metres of the floor near the filter. Some water had been poured onto the polymer on the floor, probably partly to wash it away and partly to make it less dangerous to anyone walking on the floor. Polymer emulsion on a floor makes it very slippery. The filter was close to the hold tank. One half of the underneath of that tank was covered with stalactites of polymer evidently sprayed from the filter. The conclusion in this case was that the troubles, including the emission of much of the undesirable odour, were due to lack of attention to detail and 'bad housekeeping'.

Dried foods for human consumption

The production of some dried foods, such as potato crisps, gives rise to objectionable odours. By drying the material at somewhat lower temperatures, it is probable that the odours that cause complaint could be greatly reduced in intensity or possibly eliminated. To maintain the production of the same amount of dried product per day means having more dryers or larger dryers. In some instances the effluent gases and vapours are discharged into the atmosphere through vents from the roof of the works building in which the dryers are placed. The discharges should be from chimneys of appropriate height to give better dispersion and dilution of the gases and vapours in the air. In deciding on the height of the chimney required, the height of the particular building and nearby buildings should be taken into account.

Man-made fibres

Works for the production of man-made fibres form a branch of the chemical industry. They are generally large works with processes under the control of competent scientists and with a research staff to investigate various problems to improve the products and methods of production and to solve problems such as the reduction or elimination of undesirable odours. They have systems for reducing to the practicable minimum the emissions of odours though these can occur on occasions as a result of temporary breakdown or malfunctioning of plant. Some of the chemicals used in the industry can produce vapours with an objectionable odour. They are largely recovered, and if necessary purified for re-use. In some processes acrylates are used and any excesses are recovered as far as practicable for re-use. In any case the use of acrylates in such works in the UK is scheduled under the Alkali Act and they are subject to inspection by Alkali Inspectors.

Dry cleaning fabrics

Solvents now used in dry cleaning are almost entirely halogenated hydrocarbons of low boiling points and non-flammable; though there is the use of a relatively small amount of white spirit. To minimize the costs of dry cleaning, solvents used are recovered and purified for re-use. For a short period, not more than about 5 min in each hour, air is drawn through the cleaning equipment to remove the last traces of solvent vapour from the fabric that has been cleaned. The air containing a very small proportion of solvent vapour is discharged to the atmosphere through vents above the eaves of the building. These discharges do not ordinarily cause a detectable odour.

Painting industry

There have been complaints in the past of objectionable odours from the use of paints in large quantities, for example in painting numerous motor cars at car factories. Conditions have now greatly improved. They are described in Part 2 of the Report on Odours published in 1975 by the Warren Spring Laboratory[1] and there is useful information in an excellent paper by Morris[6] published by the National Society for Clean Air.

The painting of the steel bodies of cars is carried out usually in three stages. Firstly, the primer coat is applied by electro-deposition from a water-based system in a tank in which the car body is completely immersed. Then a second coat is applied by spray guns and afterwards a third coat also applied by spray guns. The first coat is cured at a temperature in the range 160–180°C, the second at 125–170°C and the third at 125–130°C. The precise temperature of curing at each stage varies with the particular paint used. The principal sources of odour are the vapours from the curing processes, which are done in ovens. As the solvents used in paint production are combustible, it is important to ensure that the concentration of solvent vapour in each oven is well below the lower explosive limit. This is attained by having a good flow of air into each oven and an equivalent exhaust flow of hot fumes. The rates of flow are such that the concentration in the gases is less than one quarter of the lower explosion limit, as required in the UK by the Inspectorate of Factories. This means that the exhaust gases from the curing ovens contain air and vapours with not less than 18 per cent of oxygen by volume.

To avoid odour nuisance, the gases from the curing processes are incinerated in specially designed incinerators, giving good mixing of the air and exhaust fumes for the necessary time at the required temperature in the region of 700°C. Well designed heating burners are available to use natural gas, manufactured town gas, oil or liquefied petroleum gas. Where it is available at a suitable price, natural gas is preferred. Systems of recovery of useful heat from the exhaust gases of the incinerators are available. Incineration of the waste gases before discharge through chimneys of suitable height and at a suitable velocity is the best method of avoiding objectionable odours from painting on a large scale.

Animal feeds from meat and other offal

When animals and fowl are slaughtered to provide food for humans that is sold at butchers' shops, there remains offal, which includes digestive tracts, intestines, horns, bones, feathers and other parts of the animals and fowl. Retail butchers also have fat, bones and offal from final dressing and trimming of meat for customers. The various forms of meat offal are treated at works for the production of food for animals and the recovery of fats and tallow. Offal of the kinds mentioned begins to decompose by the action of organisms such as bacteria as soon as it is taken from the animals. The rate of decomposition or putrefaction increases rapidly[7] at atmospheric temperatures usual in the UK, particularly in the summer. Often the offal is dumped on the floor of one of the buildings of the slaughter house, as a heap later to be conveyed to works where it is processed to give animal feed. In the opinion of the author the offal at the slaughter house should be kept in covered vessels that are cooled by refrigeration or ice and are frequently cleaned. There is often a delay of several days before the treatment works send a vehicle to the slaughter house to collect the offal. The parts of the vehicles holding the consignments should also be cooled, but this is not general practice, so putrefaction continues. As a result the offal has a foul odour when it reaches the treatment works. There may also be some delay before the material is processed. The increasing number of complaints each year during recent years has led to demands for action to eliminate or at least greatly reduce the intensity of the effluents discharged into the air from works producing animal feeds.

At the processing works the offal is broken down into small pieces or particles and fed into a cooker with an automatically operated agitator. The cooker is heated by steam passed through a surrounding jacket and the agitator is also heated by steam. The temperature of the contents of the cooker is raised to about 115°C to give adequate cooking and sterilization but not so high as to cause burning or decomposition of some of the constituents of the offal. After the heating and some cooling of the cooked offal, the gases and vapours are usually passed through a metal type of sieve in a vessel containing a solution of potassium permanganate or a solution of hypochlorite with the object of oxidizing the materials causing odour. Other reagents have been considered or tried for removing or reducing the odour of the gases from the process briefly described, which is a batch process. Some of the equipment for treating the gases with such reagents as permanganate or hypochlorite have not been of good design or not been efficiently operated, and the reduction in the intensity of the odour has not been great. In some cases the gases and vapours have been added to the air supply for combustion of the fuel used for heating the boilers in which steam is raised. This can only be successful in incinerating the gases if the internal walls of the boiler are very hot, and if the fuel is coal the burning coal has become red hot coke. To achieve this condition with coal as the fuel would mean firing the coal in small amounts at small intervals of time, which would not be very convenient in most cases.

Recently there have been installed more automatic processes of supplying offal in a suitable form through a cooker in a continuous stream. Such plants treat much larger quantities of offal per day than are treated in the batch systems.

The above account gives only the broad principles of making animal feed from offal. The problem of greatly reducing the intensity of the objectionable odour from the process has certainly not been solved as a practicable proposition, largely because the problems have not been thoroughly investigated. There should now be detailed investigations by scientists and chemical engineers with previous experience in dealing with problems of air pollution in industries. Such investigations are long overdue and should be given a first order of priority in any programme of investigations of practicable methods of reduction of objectionable odours from various industries.

References

1. Warren Spring Laboratory, Report of Working Party on Odours for UK Department of Environment. *Assessment of the problems in Great Britain*, 1974, Part 1, and *Best present practice in odour prevention and abatement*, Part 2, 1975.
2. *Seminar on Odours by National Society for Clean Air*, Brighton, England, 1976.
3. *108th Annual Report of the Chief Alkali Inspector of England and Wales for the Year 1971*, HMSO, London, 1972.
4. *110th Annual Report of the Chief Alkali Inspector of England and Wales for the Year 1973*, HMSO, London, 1974.
5. *111th Annual Report of the Chief Alkali Inspector of England and Wales for the Year 1974*, HMSO, London, 1975.
6. Morris, J. I., *The incineration of exhaust fumes from solvent based paint stoving ovens.*, National Society for Clean Air, Brighton, England, 1976.
7. Sharples, K., *Animal by-products*, National Society for Clean Air, Brighton, England, 1976.

22 Road Vehicles

R. A. HASLETT

The internal combustion engine has made an important contribution to the high standard of living enjoyed by the developed nations of the World. The raw materials and products of our industries can be transported from mine to factory and from factory to shop, with the minimum of human labour, by vehicles powered by internal combustion engines. It also provides a powerplant for personalized transport, which can be manufactured simply at a price within the reach of the majority of the population.

However, the popularity of the internal combustion engine has introduced environmental problems. Air pollution and its effects are familiar to everyone. It includes a vast array of chemicals and dust. Industries and power stations make a significant contribution, but emissions from engines add to the problem. Furthermore, the exhaust gases of transportation are particularly obnoxious, since they occur at ground level and cannot be vented through high chimneys as with waste gases from stationary sources such as power stations. The main pollutants released by internal combustion engines are as follows:

1. carbon monoxide
2. unburnt hydrocarbons
3. nitrogen oxides
4. sulphur oxides
5. lead compounds
6. smoke
7. particulates
8. odour

Information has been gathered in the US, showing the amounts of these compounds which are released by transportation, in comparison to the contribution of factories (Fig. 22.1). This shows that the internal combustion engine is a primary polluter in terms of carbon monoxide, unburnt hydrocarbons and nitrogen oxides in the US. The situation in Europe is different for three reasons: (1) car engines are smaller in Europe and the mass of air going through the engine is, therefore, reduced, (2) lorries are generally powered by petrol engines in the US, while in Europe diesel engines are almost universal; (3) in Europe there are fewer cars per head of the population. The important constituents of the petrol engine exhaust are carbon monoxide, unburnt hydrocarbons, nitrogen oxides, particulates and lead, while the diesel engine emits unburnt hydrocarbons, nitrogen oxides, sulphur oxides, smoke and smell.

The vehicle population is growing rapidly throughout the world (Fig. 22.2 for the US) and emissions from transportation will increase in proportion, if unchecked. The legislation to reduce automotive emissions, which is being introduced in the US through the 'seventies, is causing a reversal in this upward trend.[1] Since control of emissions is only applied to new vehicles, and existing vehicles continue to emit pollutants until they are scrapped, there is a time lag before the full benefits of the regulations are felt (Fig. 22.3). However, by the middle 'eighties, air pollution from transportation in the US should be below the figure for 1960, in spite of a possible 100 per cent increase in the number of vehicles.

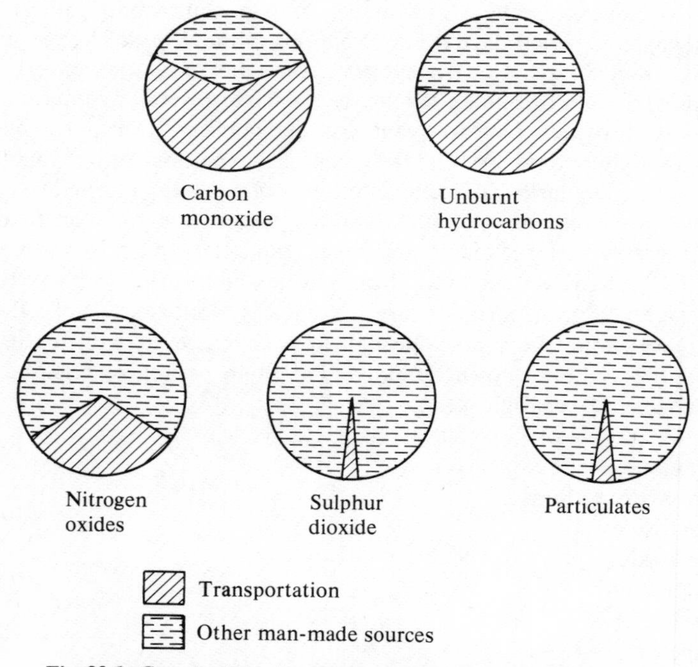

Fig. 22.1. Contribution of transportation to air pollutants in the US.

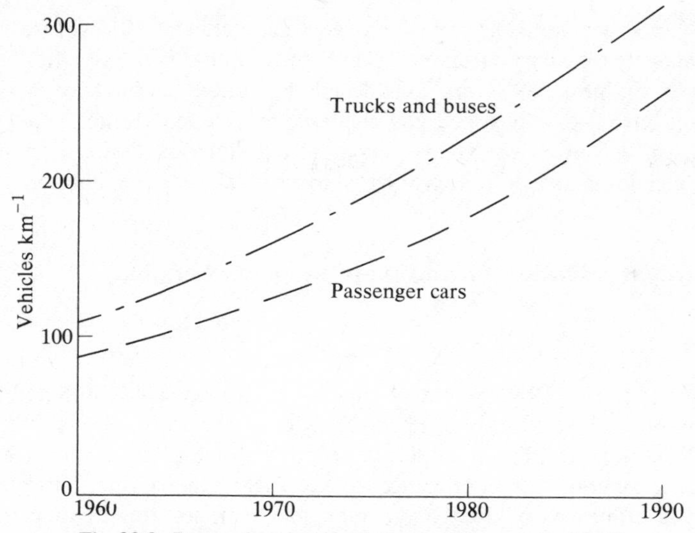

Fig. 22.2. Past and projected vehicles per kilometre in the US.

The list of automotive pollutants can be sub-divided into those which may be a danger to human health, and those which are mainly a nuisance. Compounds (1)–(5) are in the former category, while (6)–(8) are in the latter. Up to the present, the greatest research effort has been directed towards pollutants which may be a health hazard, and

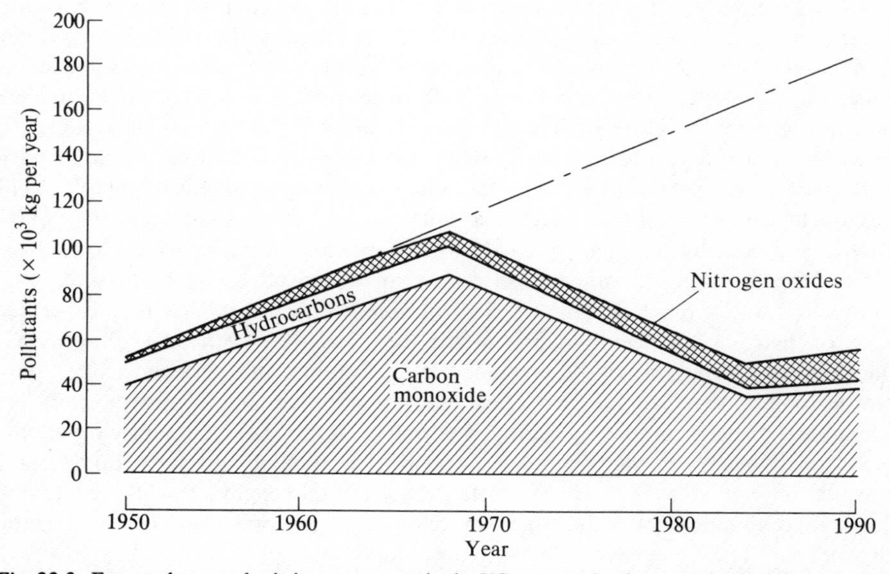

Fig. 22.3. Expected atmospheric improvements in the US as a result of automotive emission controls.

public annoyance was considered to be a secondary concern. However, this attitude is changing, since medical researchers have shown that continual annoyance imposes mental stresses on humans. Before considering the various methods of exhaust emission control which have been developed, the chemical mechanisms which cause the pollutant compounds will be described. Measurement of the emission from internal combustion engines can also introduce certain problems; these will be outlined later in this chapter.

Mechanisms of pollution formation in the petrol engine

Exhaust

The compounds released by the petrol engine which are termed pollutants are carbon monoxide, unburnt hydrocarbons, nitrogen oxides, lead compounds and particulates, and a small amount of sulphur oxides. Carbon monoxide and unburnt hydrocarbons are the result of incomplete combustion of the fuel in the engine. Nitrogen oxides are formed during combustion, due to the high gas temperature. Lead compounds and sulphur oxides are also formed during combustion, due to the inclusion of lead and sulphur in the fuel. Let us now follow the path of the air and petrol through an engine to see how these contaminants arrive in the exhaust. Air, composed mainly of oxygen and nitrogen, is drawn into the engine through the carburettor and throttle. The amount of the air inhaled is controlled by the position of the throttle. There is a pressure drop across the throttle, being ambient on the carburettor side and some lower pressure, termed intake manifold pressure, in the pipes to each cylinder of the engine. The function of the carburettor is to add petrol to the air, such that the ratio of mass of air inhaled to mass of fuel is roughly constant. This constant is about 14, though the ratio can fluctuate between 10 and 17 under some conditions. Petrol is composed of a number of different hydrocarbons derived from hydrogen and carbon in various arrangements. This mixture of hydrocarbons is often simplified in description to the equivalent of a single average component, octene C_8H_{16}. During refining, various additives are mixed with the petrol, the principal one being tetraethyl lead (TEL). This is an anti-detonating compound which prevents the fuel from spontaneously exploding before it is ignited by the spark plug. It is an important additive, as it allows a petrol engine to run at a higher compression ratio than would otherwise be possible. As the compression ratio is raised, engine efficiency improves and the vehicle travels further on a litre of petrol. Additional additives are the scavengers ethylene dibromide and dichloride; their function will be explained later. Sulphur and phosphorus compounds are other components of the fuel. These are not additives but occur naturally in the crude oil and are not completely removed during oil refining. The quantity of sulphur in petrol depends partly on the particular source of the crude oil and partly on the refining processes. The mixture of air and petrol passing the engine throttle contains the following components of importance in relation to air pollution; oxygen, nitrogen, hydrocarbons, TEL, scavengers and sulphur compounds.

The low pressure on the intake manifold side of the throttle causes some of the fuel to vaporize and mix with the air. However, vaporization is not complete and some of the fuel remains in liquid form, running towards the engine cylinder along the bottom of the intake tracts. As the piston descends, the air–fuel mixture and the liquid fuel enter the cylinder through the inlet valve (Fig. 22.4). Since the cylinder is hot, the liquid fuel quickly vaporizes and a homogeneous air and fuel mixture forms after the intake valve closes. During the compression stroke, the pressure and density of the mixture increase and the gas is forced into the crevices, such as between the piston and cylinder above the top piston ring. At the conclusion of the compression stroke, the mixture is ignited by the spark plug and the combustion proceeds. Most of the exhaust pollutants are formed during this combustion period. The amount of carbon monoxide is related to the ratio of air and fuel in the cylinder. If the petrol–air mixture is chemically correct, that is, the ratio of mass of air to mass of fuel is about 14·5 : 1, then the products of combustion are mainly carbon dioxide and water:

$$C_8H_{16} + 12(O_2 + 3{\cdot}79N_2) \longrightarrow 8CO_2 + 8H_2O + 45{\cdot}5N_2.$$

If the mixture is weak, i.e., there is an excess of air, then the products of combustion also contain oxygen:

$$C_8H_{16} + 13(O_2 + 3{\cdot}79N_2) \longrightarrow 8CO_2 + 49{\cdot}3N_2 + O_2 + 8H_2O.$$

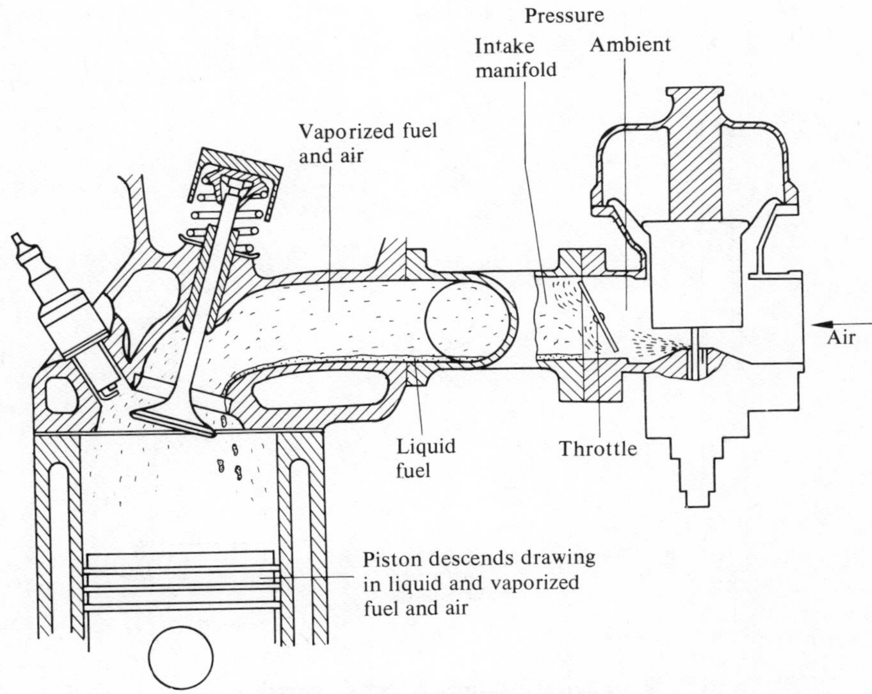

Fig. 22.4. Petrol engine induction stroke.

If the mixture is rich, i.e., there is an excess of fuel, then the products of combustion also contain carbon monoxide, since there is insufficient oxygen to form carbon dioxide from all the carbon. The amount of carbon monoxide measured in the exhaust of an engine at various mixture strengths is shown in Fig. 22.5.

The mixture does not burn spontaneously, but an ignition at the spark plug causes a flame to sweep across the combustion chamber. In front of the flame the mixture is unburnt, while within the flame combustion is intense (Fig. 22.6) and the temperature is high—in the region of 2500°C—so some unusual reactions occur. A small amount of oxygen and nitrogen combine to form nitrogen oxides. The sulphur compounds in the fuel are changed to sulphur oxides. The TEL burns to lead oxide with the air and to lead chloride and bromides with the scavenger. A proportion of the lead compounds is deposited in the combustion chamber. Some of the lead deposits are removed, due to the chemical action of lead chlorides and bromides. A small proportion of the lead enters the lubrication oil and the remainder is carried by the waste gases into the exhaust system in which a small proportion is deposited. The result is that on average from two-thirds to three-quarters of the lead originally in the TEL is discharged in the gases leaving the exhaust system to enter the atmosphere.[2]

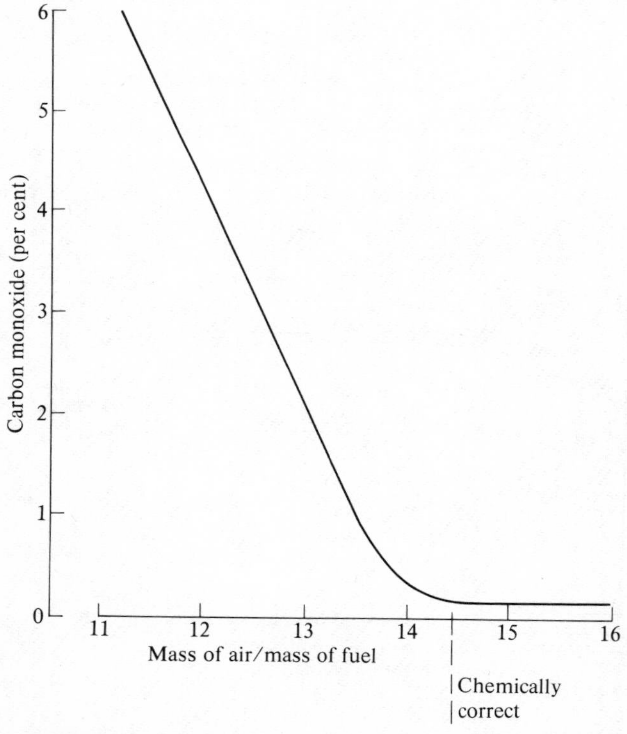

Fig. 22.5. Volume concentration of carbon monoxide in exhaust of spark ignition engine as air–fuel ratio is changed.

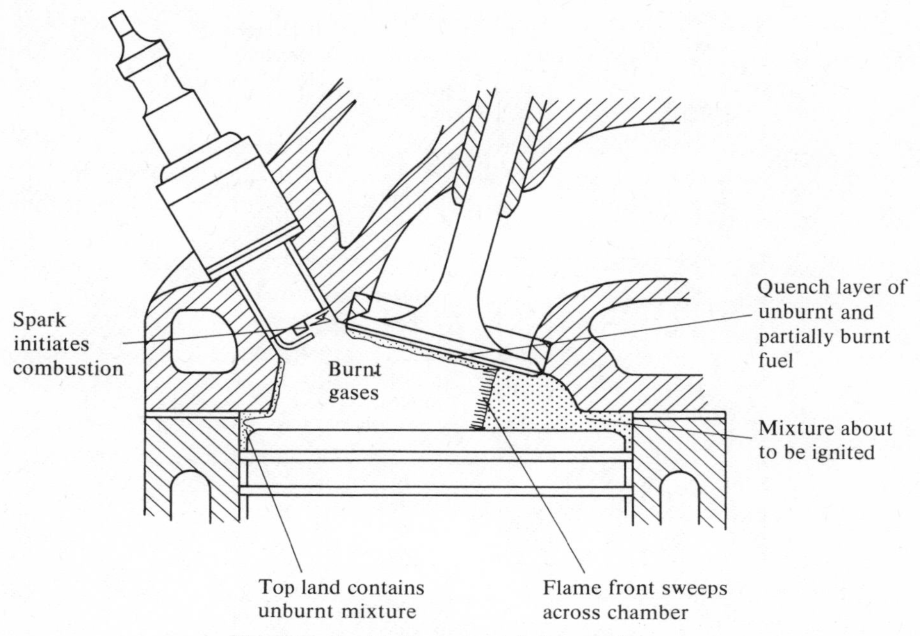

Fig. 22.6. Combustion in spark ignition engine.

The flame sweeps across the combustion chamber initiating all these reactions, but the flame can never quite touch the walls of the chamber. The flame is very hot and as it approaches the walls, it loses heat by radiation to the cooler metal surface and is eventually doused. As a result, a layer of unburnt mixture of air and fuel, less than 0·24 mm thick, remains on the wall of the combustion chamber. In addition, there is a transition layer between this unburnt region and the fully reacted products in the combustion zone, which contains partially oxidized hydrocarbons. Another region to which the combustion does not propagate is the top land, between piston and cylinder, above the top piston ring. Here again a mixture of fuel and air remains.[3] A proportion of the gas in the region is forced into the crankcase due to combustion pressure, but a significant amount still remains.

After combustion, the piston descends and work is extracted from the crankshaft. The exhaust valve opens and gases in the cylinder expand into the exhaust. Then the piston ascends and the remaining gases are displaced into the exhaust pipe. The unburnt fuel and air mixture on the combustion chamber walls is forced into the exhaust and the mixture in the top land, which was compressed during combustion, expands into the main body of the cylinder. The exhaust gases contain mainly nitrogen, carbon dioxide and water vapour, with carbon monoxide if the intake mixture is rich. In addition, there are small proportions of nitrogen oxides, unburnt hydrocarbons, sulphur oxides, lead compounds and particulates. The relative concentrations of these compounds are illustrated in Fig. 22.7.

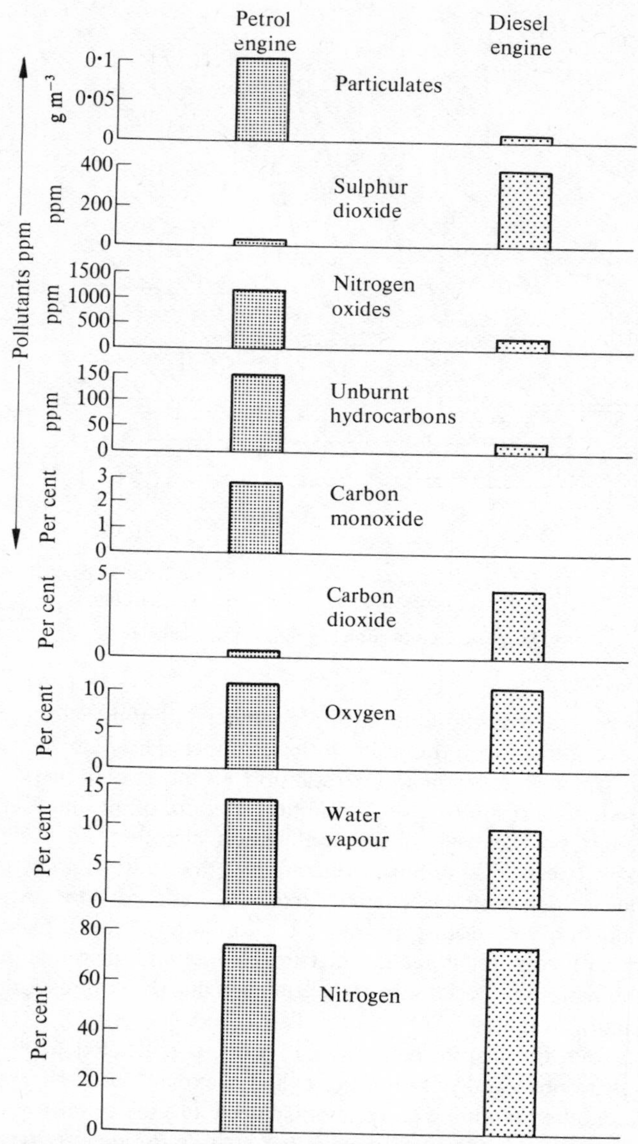

Fig. 22.7. Comparison of exhaust constituents of petrol and diesel engines.

This description of engine operation becomes slightly more complex during vehicle transients, that is, when the throttle is suddenly opened or closed. Consider throttle opening. The fuel flowing along the inlet manifold is partially in liquid form and partially evaporated due to the low pressure. When the throttle is snapped fully open, the pressure rises to ambient conditions and the ratio of evaporated to liquid fuel changes. There is now more liquid fuel in the pipes and less is evaporated. The air and evaporated fuel reach the cylinder before the liquid fuel and the engine will operate very weak for a few revolutions, until the liquid fuel reaches the cylinder. The car driver would feel this as a slight hesitation before the car accelerated. To overcome this, an accelerator pump is incorporated in the carburettor to inject fuel into the induction system as the throttle opens. Therefore, the inlet mixture becomes very rich for a few seconds to prevent vehicle hesitation and this causes a high percentage of carbon monoxide to be emitted from the exhaust. A different effect takes place as the throttle is snapped shut. The liquid fuel in the manifold evaporates almost completely and the intake mixture to the engine becomes very rich. In addition, the intake manifold pressure is low, since the engine is acting like a vacuum pump. These two effects may prevent the mixture being ignited by the spark plug. If this happens, then raw fuel is released to the exhaust and hydrocarbon emissions may instantaneously rise to levels of 8000 ppm (parts per million) in comparison with values of about 200 ppm when the engine is running at a constant throttle position. These vehicle transient effects which cause high emissions are very important. As we shall see later, they govern the methods which can be used to measure emissions.

Other sources

Gases, composed of unburnt fuel and some products of combustion, are forced past the piston rings during combustion and expansion. If these were not vented, the pressure would increase in the crankcase and the oil seals would fail. These gases were allowed to escape into the air until legislation was introduced to curb the practice.

The metal components of a car expand and contract during ambient temperature variations from day to night. In addition, the amount of evaporated fuel in the tank changes. Therefore, to accommodate these changes in volume, the vapour in the fuel tank must be allowed to expand out of the tank, or pressure would build up. This vapour is rich in hydrocarbons and organo-leads (principally TEL). Fuel also vaporizes from

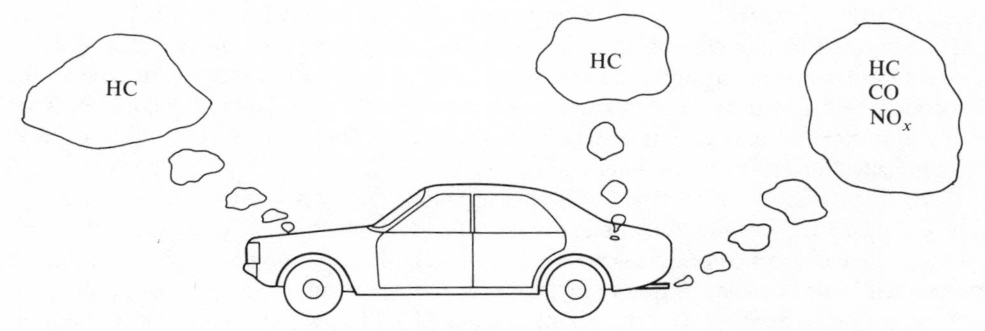

Fig. 22.8. Pollution distribution from car.

the carburettor after a car stops, since the engine and carburettor are still warm. These sources, called evaporative losses, can account for up to 35 per cent of the unburnt hydrocarbons emitted from a vehicle, while the other 65 per cent comes from the exhaust (Fig. 22.8).

Lead is emitted from cars as organic and inorganic compounds. The TEL is an organic lead which is mainly vapour or aerosol, while the inorganic lead from the exhaust is in particulate form. The former are a greater health hazard than the latter.

Mechanism of pollution formation in the diesel engine

The operation of the diesel engine differs from that of the petrol engine in four principle respects:

1. air alone is drawn into the cylinder and compressed and fuel is not added in the induction system;
2. the air entering the engine is not throttled, so the same amount of air is induced in every cycle;
3. the air is highly compressed so that its temperature spontaneously ignites the fuel;
4. the fuel is injected around top dead centre.

There are two main types of diesel engine, depending on the method of achieving the fuel–air mixture in the engine cylinder. In the direct injection type, there is only one combustion chamber (Fig. 22.9). The chamber of the indirect injection diesel is split into two parts and fuel is added to only one chamber. The Comet combustion system (Fig. 22.9), is a popular design for indirect injection engines. Although different amounts of pollutants are produced by these two types of diesels, the mechanisms of formation are similar, so they will be considered together. Let us now follow the movement of air and fuel through the diesel engine.

Air is induced into the cylinder as the piston descends and fills it almost to atmospheric pressure. The inlet valve closes and the piston ascends, compressing the air to a high temperature. The temperature of the air in a diesel engine after compression is about 650°C compared with 350°C at the same instant in a petrol engine. Then diesel fuel is injected from a high-pressure injector. The fuel is more viscous than petrol, since the molecules are larger, and it ignites more easily. An average diesel fuel hydrocarbon is dodecane $C_{12}H_{26}$, and the mixture does not contain any man-made additives. The only important additional component in terms of air pollution is sulphur, which occurs in concentrations 5–20 times that in petrol.

The fuel issuing from the injector is split up into a fine spray and mixes with some of the hot dense air. Locally, the mixture is rich. The fuel which has been injected heats up within a few degrees of crankshaft revolution and ignites spontaneously. Further fuel is injected into this burning region to mix with the remaining air, the amount depending on the power being developed by the engine (Fig. 22.10). The overall mass of air to mass of fuel may be anywhere between 100:1 and 20:1. Since the air–fuel ratio is always

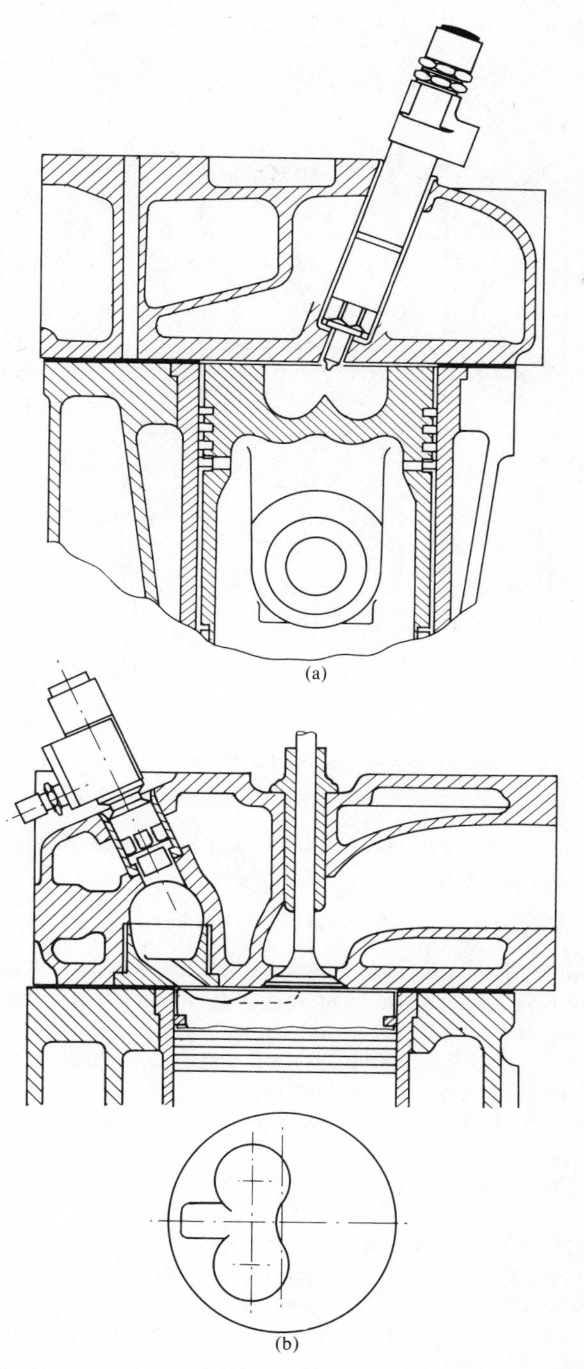

Fig. 22.9. Typical (a) direct injection and (b) indirect injection chambers as used in automotive diesel engines.

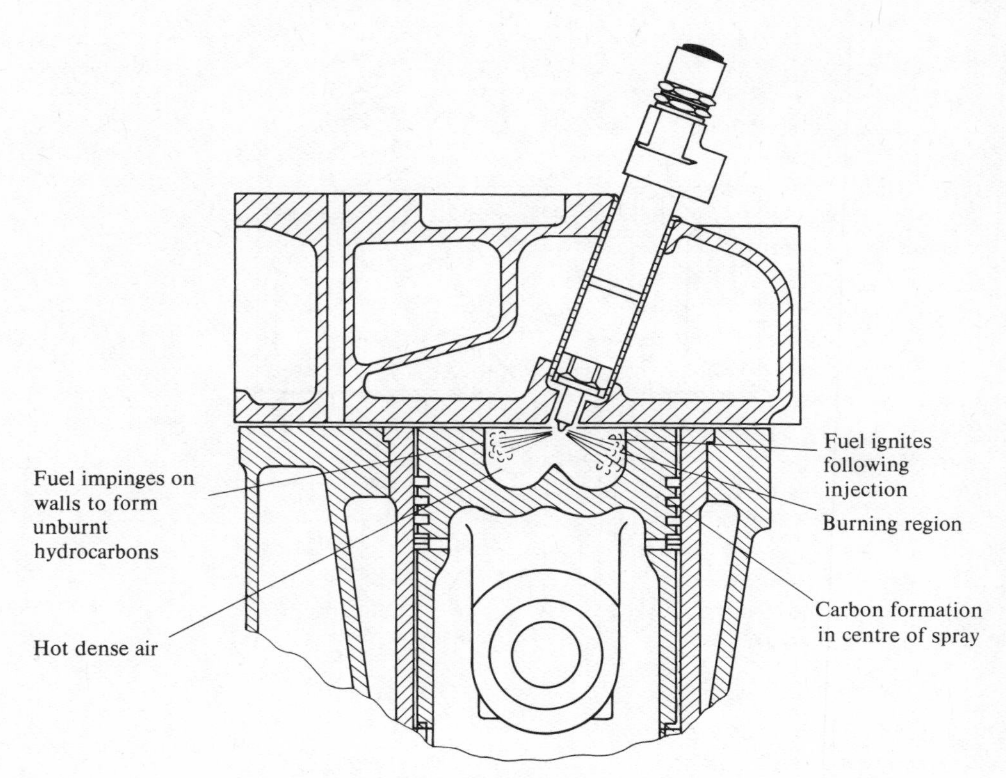

Fig. 22.10. Combustion in a diesel engine.

leaner than chemically correct in a diesel, there is always oxygen remaining after combustion. For example at an air–fuel ratio of 50,

$$C_{12}H_{26} + 50(O_2 + 3{\cdot}79N_2) \rightarrow 12CO_2 + 13H_2O + 31{\cdot}5O_2 + 189{\cdot}5N_2.$$

The concentration of CO formed is very small, usually less than 0·1 per cent by volume in the exhaust gas. The sulphur in the fuel oxidizes during combustion, and some nitrogen oxides are also formed. Like the petrol engine, diesels also produce some unburnt hydrocarbons, but the formation mechanism is slightly different. Some of the injected fuel will impinge on the combustion chamber walls and remain there after combustion is completed. However, a main cause of unburnt hydrocarbons is often secondary injection, or some extra fuel dribbles from the injector during the engine expansion. Gas temperatures during expansion are not high enough to initiate further combustion.

Some of the hydrocarbons are partially oxidized by the air, but do not completely burn to carbon dioxide and water, so they form compounds giving the characteristic diesel engine smells. The reason for the partial burning is probably wall quenching, or partial misfire which stops the reactions before completion. Many of these odour compounds have not yet been chemically identified, so the reactions causing them are unknown.

The air and injected fuel are never completely mixed before combustion in the diesel engine. At high engine load, i.e., when the greatest quantity of fuel is injected, the centre of the combustion zone may be very rich. When combustion occurs in this region, the hydrogen molecules in the fuel are oxidized, but there is insufficient oxygen for the carbon molecules to burn. This effect is called pyrolysis, and the result is soot formation as the carbon particles coagulate.[4] If the region of the combustion is very turbulent, the soot may meet some air from the outskirts of the combustion region and be oxidized. However, this process may not be completed under engine full load conditions, since the combustion cools rapidly as the piston descends and the time available for mixing and chemical reactions is very short.

When expansion is completed, the exhaust valve opens to release the products of combustion. These are mainly nitrogen, oxygen, carbon dioxide and water vapour, with small proportions of the pollutants sulphur oxides, nitrogen oxides, carbon monoxide, soot and unburnt and partially oxidized hydrocarbons. During combustion, some gases leak past the piston rings into the crankcase, but these are not so rich in unburnt fuel as with the petrol engine. As regards vehicle acceleration and deceleration, there are fewer problems than with the petrol engine. The model of combustion which has been outlined does not change significantly during transients and the proportions of pollutants released are similar.

Comparison of petrol and diesel engines

To compare the exhaust gas compositions of typical diesel and petrol engines, a single operating point has been chosen, namely 33 kW (44 bhp). Both engines were running at roughly three-quarters of full load. Figure 22.7 illustrates that the petrol engine produces the greater concentration of carbon monoxide, hydrocarbons, nitrogen oxides and particulates, but less sulphur oxides. For simplicity, soot is considered as a particulate. However, the effect of the exhaust on the environment does not depend only on the concentration of pollutants, but on the mass of pollutants released. The concentration and mass are linked by the formula.

$$\text{Mass of pollutant released per minute} = \text{concentration of pollutant in exhaust} \times \text{exhaust gas mass flow per minute.}$$

In the example chosen, the exhaust mass flow of the petrol engine 26 kg min^{-1} (58 lb min^{-1}) was less than the flow from the diesel engine 37·6 kg min^{-1} (83 lb min^{-1}). Therefore, in terms of mass released, the diesel engine, in comparison with the petrol engine, emits less carbon monoxide, hydrocarbons, nitric oxides and particulates, but more sulphur oxides. The lower particulate emissions of the diesel engine may seem surprising. The exhaust of a diesel lorry is often visible, yet petrol car exhaust is usually clear. As the load on the diesel engine approaches the maximum, the soot formation increases dramatically and the amount shown in Fig. 22.7 may be multiplied by twenty.

On the other hand, the concentration of particulates in the petrol engine exhaust remains roughly constant. In addition, diesel engines are driven at full load much more frequently than petrol engines. Therefore, the particular example chosen for comparison gives a false impression in terms of overall particulate emissions.

The information presented in this comparison contradicts public opinion on vehicle air pollution. The very phrase air pollution conjures up the idea of a smelly sooty diesel lorry, in the minds of the public. In fact, although the smell and smoke are certainly annoying, the exhaust of petrol engines could be more deleterious to public health, particularly when their greater numbers are taken into consideration.

Measurement of vehicle pollution

1. Carbon monoxide, unburnt hydrocarbons and nitrogen oxides

These compounds have been the main targets of exhaust emission regulations throughout the world, because of their potential health hazard. Accurate instruments have been developed to measure the concentration of these compounds in a vehicle exhaust, so that the mass released into the environment can be deduced. However, we must also remember that the total mass of pollutant emitted as a vehicle drives through a town is dependent not only on the steady driving speeds, but also on the accelerations and decelerations. For the diesel engine, this does not raise any serious problems, since these transients can be approximated by certain engine steady speed conditions. Therefore, if an engine is placed on a test bed, with a dynamometer absorbing the load, the emissions can be measured over a range of loads and speeds and weighted according to the frequency with which they occur as a lorry is driven through a town. If the mass flow of the exhaust is recorded, the mass emissions can be calculated from the equation in the last section. This is basically the philosophy behind the CARB (Californian Air Resources Board) test for diesels and the emission result is usually expressed in terms of each unit of power.

The thirteen mode cycle of the CARB test is shown graphically in Fig. 22.11. Each mode lasts ten min and concentration levels of emission of hydrocarbons, carbon monoxide and nitrogen oxides are measured during the last minute of each mode. To obtain the levels in g h^{-1}, the concentration of emissions during each mode is multiplied by a constant and by the mass flow rate of exhaust gases expressed in lb min^{-1}. The constants for hydrocarbons, carbon monoxide and nitrogen oxides are respectively 0·0132, 0·0263 and 0·0432. The average level of emissions during idling is then multiplied by 0·2 and the other levels are each multiplied by 0·08. The sum of these levels indicates the emissions in g h^{-1}. To obtain the final result in g kWh^{-1} this figure is divided by 0·08 times the sum of the output in kW for all modes.

The explanation of pollution formation in the petrol engine described how engine transients contribute disproportionate amounts of carbon monoxide and unburnt hydrocarbons, compared with steady driving. Therefore, any method of measuring pollution from a petrol engine must take account of these transients. It is difficult to

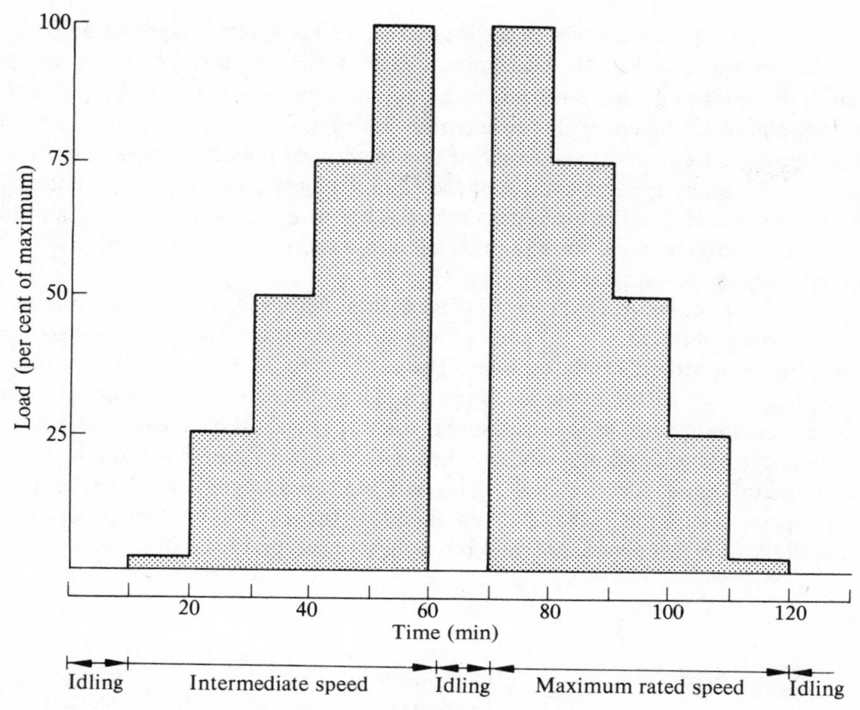

Fig. 22.11. CARB test for diesel engines.

simulate these load and speed conditions continuously on an engine test bed, without very complex control equipment. Therefore, for exhaust emission measurements, vehicles are set on chassis dynamometers which can simulate road tractive resistance and the inertia of the car to be accelerated.

To obtain the mass of exhaust emissions from a vehicle, it is necessary to measure the concentration of the pollutant in the exhaust and the mass flow of air through the engine. Unfortunately, there is no convenient air flow meter which can be easily fitted to the engine air cleaner without affecting the carburation. However, a rather clever idea was adopted by the US authorities to overcome this difficulty, called the constant volume sampling system (CVS).[5] By diluting the exhaust gas with a larger known volume of clean air, measuring the pollutant concentration in the diluted sample, and multiplying by the known volume, the mass of pollutant can be calculated with acceptable accuracy. The principle of this method can be simply explained if the pollutant in the car exhaust is considered as a dye. The exhaust is mixed with a much larger volume of clean air, such that there is only a small relative change in the large volume. The colour of the diluted dye, multiplied by the original clean air volume, is a measure of the total amount of the dye in the exhaust. It is not necesary to store the complete diluted exhaust, as only a small representative sample is analysed. The operation of the CVS equipment is shown in Fig. 22.12, where a positive displacement fan draws in the diluting air and mixes it with the exhaust. A small sample is continuously extracted and retained in a bag for analysis at the end of a test. An additional sample is taken of the ambient air entering the CVS equipment, to form a baseline for measurements. The beauty of this method of pollution measurement is that the car can drive any type of cycle and the sample bag will contain a concentration of pollutant proportional to the total mass released by the car exhaust. The US authorities have devised a driving cycle for exhaust emission tests, which is representative of rush-hour conditions in Los Angeles (Fig. 22.13).

For a Federal emission test, a car is placed on a chassis dynamometer, started from cold and driven through the 23 min of the driving cycle. The continuously diluted exhaust sample is stored in plastic bags. The car is then stopped for 10 min, restarted, and the first $8\frac{1}{2}$ min of the cycle are repeated. This diluted sample is stored in another bag. These emission samples are measured, the total mass of each exhaust pollutant in each bag is calculated, and they are summed with weighting constants to give an overall mass emission figure for the cycle. This is usually expressed in terms of mass of pollutant per mile of the cycle; the test is called the 1975 CVS test procedure. The original 1972 CVS test procedure only incorporated the first 23 min of the test, without

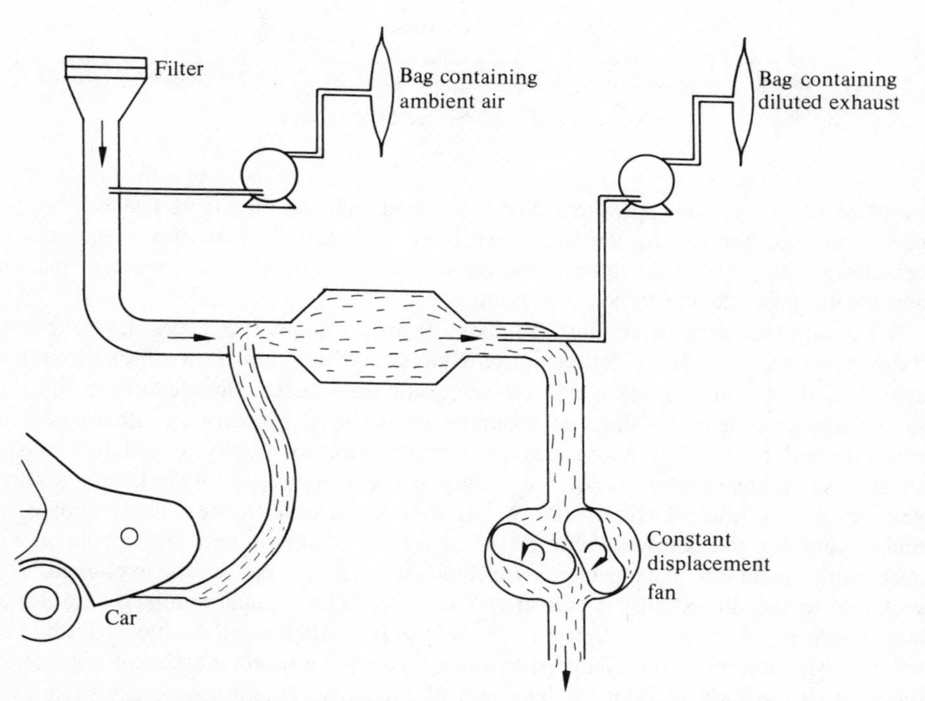

Fig. 22.12. US System for measuring pollution from petrol driven vehicles (constant volume sampling system).

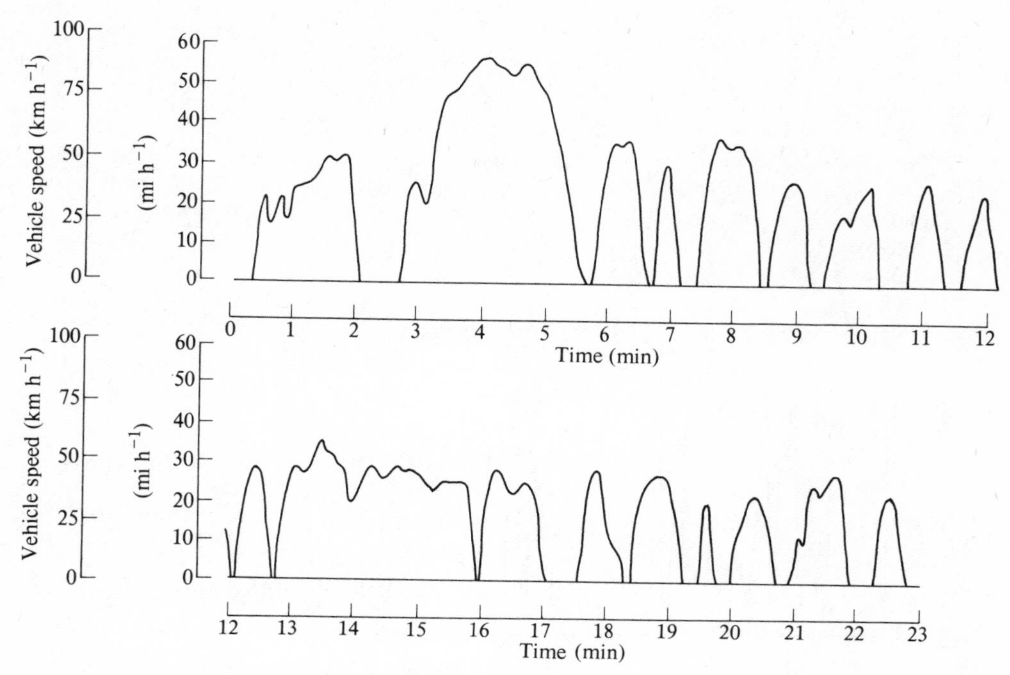

Fig. 22.13. Driving cycle used for exhaust emission tests in US for petrol driven vehicles.

the hot start repeat. Typical results for a car without any emission control equipment would be hydrocarbons 3·7 g km^{-1} (6 g mi^{-1}), carbon monoxide 78 g km^{-1} (125 g mi^{-1}), nitrogen oxides 3·7 g km^{-1} (6 g mi^{-1}).

During development work on the emission control of petrol engines, the raw emissions from the car tailpipe are usually measured before they enter the CVS equipment. This gives the engineer a guide to those modes of vehicle operation which produce the most emissions. Figure 22.14 shows the exhaust constituents during the first 6 min of the US emission test. When the engine is started from cold, the exhaust concentration of carbon monoxide is high, but it decreases when the choke comes off after the first 2 min. Bursts of hydrocarbon emissions occur during accelerations and decelerations. The concentration of nitrogen oxides depends on the throttle opening; when the car is accelerating or driving fast, nitric oxides are high, but they fall to very low levels at idle and during decelerations.

The various authorities in Europe have not followed the same methods of pollution measurement as the US. A simple driving cycle for use on chassis dynamometers has been devised (Fig. 22.15).[6] The car is driven four times through this cycle and all the exhaust is retained in large plastic bags. The mass of unburnt hydrocarbons and that of carbon monoxide released by the car are measured at the conclusion of the test by determining the average concentration of each in the bag, and measuring the volume of

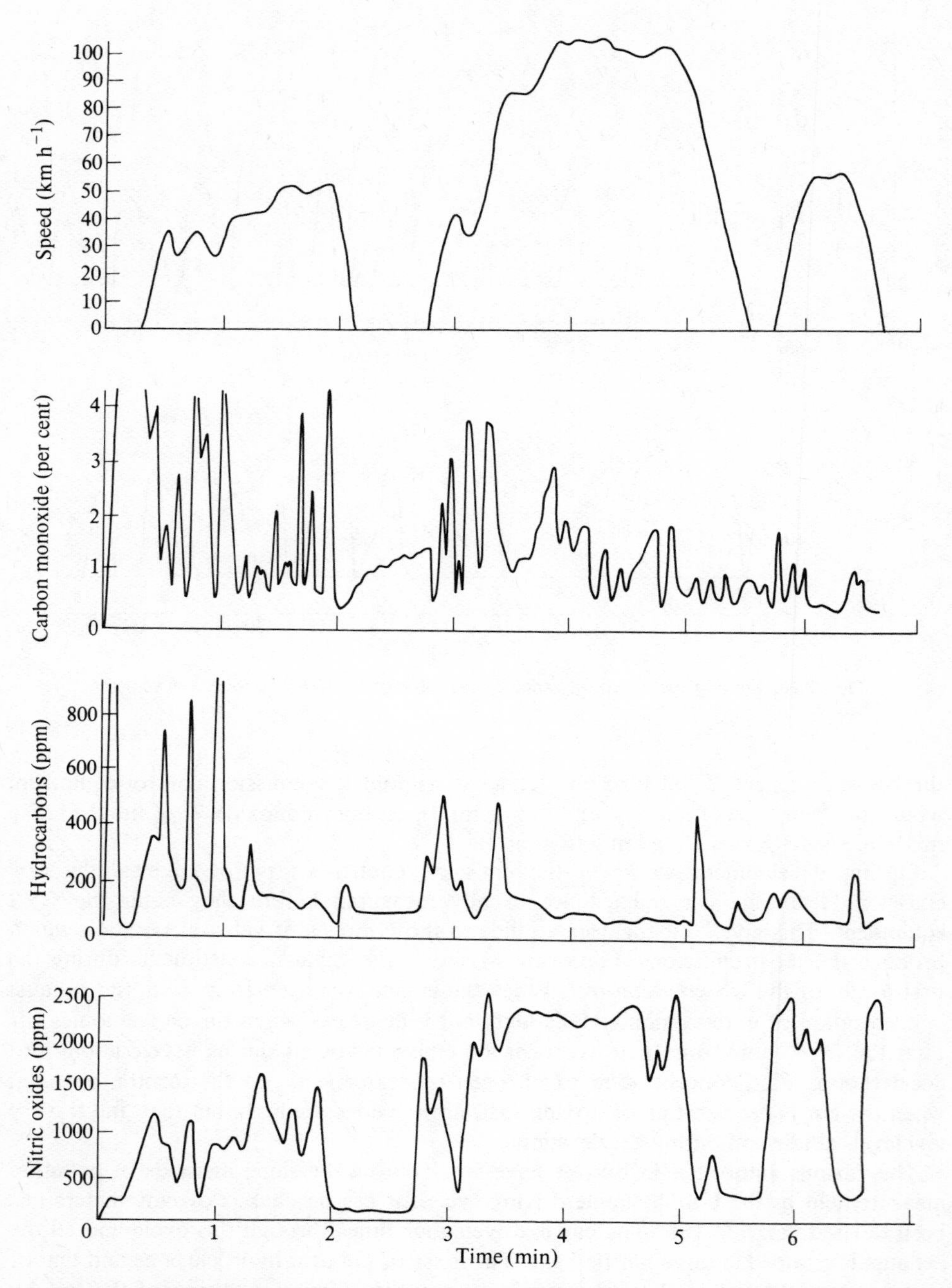

Fig. 22.14. Typical exhaust emission concentration measurements from a petrol driven vehicle on a CVS test.

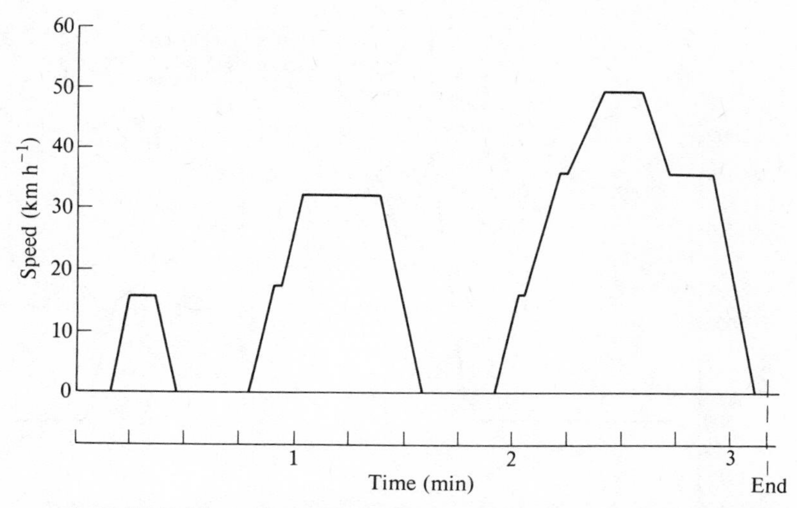

Fig. 22.15. Driving cycle used for ECE emission tests for petrol driven vehicles.

the exhaust with a gas meter. Multiplying these results together yields emission figures in g per test. A typical result for a car from the pre-emission control era is

hydrocarbon 13 g per test, carbon monoxide 270 g per test.

A disadvantage of the ECE procedure is that nitrogen oxides cannot easily be measured.

These descriptions of the various emission tests are by no means complete. Indeed, the regulations governing all of these tests are very complex. In addition to the driving cycles, there are evaporative loss tests in which the vehicle is left in a hot garage for a period of time, so that the amount of fuel which escapes from the car can be monitored. The CARB, CVS and ECE tests which have been described are probably the most important at this time and have been adopted in various countries throughout the world. However, the list is not complete, and various other test methods do exist.

2. Smoke

There is a wide range of techniques for measuring diesel engine smoke, but most are simple steady state test bed procedures. The only one which involves a complete cycle is that of the US Federal authorites (Fig. 22.16). This is conducted on a test bed, and the engine is idled and then accelerated at full rack, without any load. This is followed by a slower acceleration against a load up to 2000 rev min^{-1}, and then a slow increase in dynamometer torque gradually reduces the speed. The latter mode is referred to as 'lugging'.

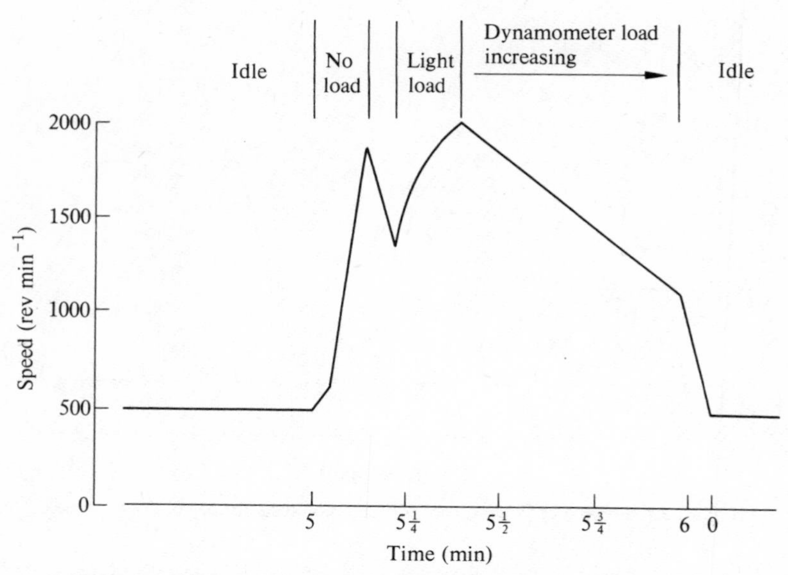

Fig. 22.16. US Federal exhaust smoke certification procedure for diesel engines.

Regulations

The problems arising from vehicle air pollution vary widely throughout the world, depending on vehicle population density, climatic and geographical conditions. Therefore, different emission standards are imposed in various countries. The notorious air conditions in California and Japan have prompted very strict controls, while there are other nations having no controls. We shall not attempt to summarize existing and proposed legislation throughout the world, but only those of Europe and the US, since they are the most important. Air pollution in Japan is also causing concern, but legislation on vehicles is closely following the lead set by the US and will probably develop in the same manner.

The test bed dynamometer cycle for smoke measurements in the US has been described in the previous paragraphs. The maximum acceptable smoke level during this test was 40 per cent opacity in the acceleration mode and 20 per cent opacity in the 'lugging' mode. These limits, set in 1970, were tightened in 1974 to 20 per cent and 15 per cent respectively. The ECE method of smoke measurement is simpler, and consists only of measurements over a range of steady speeds at full load. It is difficult to draw comparisons between these tests because of vast differences in procedure and instrumentation, but approximate conversion factors yield the data in Fig. 22.17. The ECE test also contains a free acceleration smoke measurement, where the engine is operated at full rack, without any load except flywheel inertia. This acts as a simple roadside test. Turbocharged diesel engines have more difficulty attaining these smoke levels than naturally aspirated models.

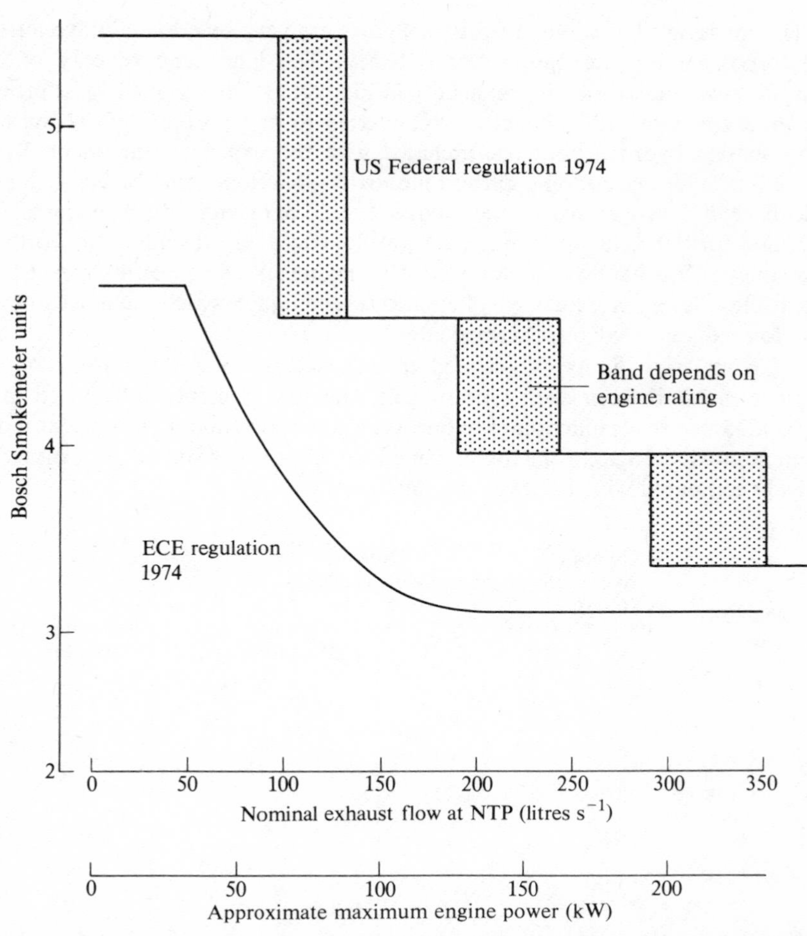

Fig. 22.17. Approximate comparison between legislated smoke limits for naturally aspirated diesel engines.

So far, control of pollutant gases from diesel engines has been exerted only in the US; even there the control was lenient. The limits on the CARB test are shown in Table 22.1.

TABLE 22.1

		Hydrocarbons and nitrogen oxides ($g\ kWh^{-1}$)	Carbon monoxide ($g\ kWh^{-1}$)
Uncontrolled diesel engines	Direct injection	12–17·5	2·5–8
	Indirect injection	7–9·5	2·5–5·5
1973 regulations		21	54
1977 regulations		13·5	33·5

The US emission regulations on petrol engines are very severe, and have introduced difficult problems for the automotive industry. The increasing severity of control through the 'seventies is shown graphically in Fig. 22.18. The results of test procedures before 1972 and after 1975 have been converted to an appropriate 1972 test result. Evaporative loss hydrocarbons are included with the exhaust component. By either 1978 or 1980, hydrocarbon and carbon monoxide emissions must be below 5 per cent of those in 1968. Nitrogen oxides were reduced to 30 per cent of their uncontrolled level in 1977, and further reductions may occur at the end of the decade. It is worth noting that the emission regulations are not related to engine size, so generally smaller engines can reach these levels more easily. The various techniques which have been developed for emission reduction will be described later.

The ECE emission limits are related to car size, so that the same proportional reduction is required from each type of car. The first controls, introduced in 1972, called for a 35 per cent reduction in unburnt hydrocarbons and a 50 per cent reduction in carbon monoxide. In 1975 the limits demanded 45 per cent and 60 per cent reduction respectively compared with 1972 vehicle emissions.

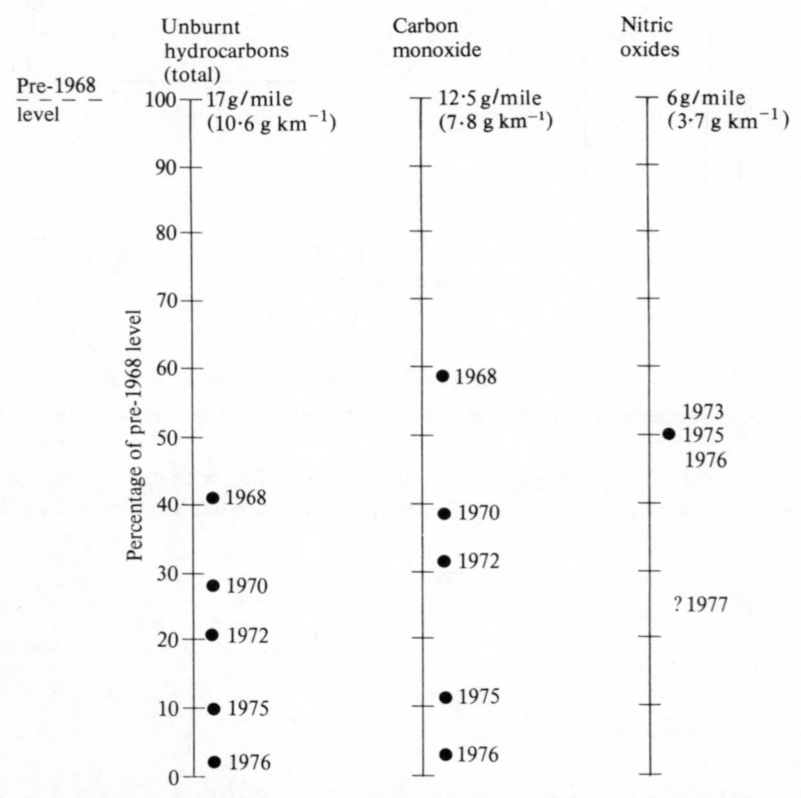

Fig. 22.18. Emission reduction of cars as required by US Federal Authorities.

Other pollutants

In the future, there are likely to be controls on particulate matter from spark ignition engines. Measurement techniques, based on the CVS apparatus, have already been developed. The CVS method is particularly attractive, since the exhaust gas must be mixed with a larger volume of air to complete the formation of particulates. In other words, when car exhaust gas diffuses into the ambient air, many of the small particulates and aerosols coagulate to form dust and droplets. This process must be simulated when measuring particulates. A sample of the diluted exhaust is drawn through a filter, which collects the particulates, and they are weighed at the conclusion of the test. A sieve has been developed for use in car exhaust systems, to collect lead particles before they are emitted into the atmosphere.[7]

Sulphur dioxide emissions cannot easily be controlled in diesel engines, and removal of sulphur from the fuel is the only practical approach. However, de-sulphurization in the refinery is expensive and reduces the overall conversion efficiency of crude oil into diesel. Therefore, some compromise is always required between control of sulphur dioxide emissions and energy conservation.

The process by which the human nose detects smells and its method of differentiating between different types cannot be simulated. These unknowns are reflected in the rather crude techniques which are used for smell measurement. It is usual to set up an odour panel of ten people to act as a jury, to smell the exhaust and rate it on a type and intensity scale. The odour types are burn/smoky, oily, pungent/acid and aldehydic/aromatic. The jury for an odour panel must be chosen and trained carefully to obtain repeatable results. These technical difficulties have dissuaded any attempt to impose legislation on diesel odours. If a suitable instrument which can simulate the human nose is even fully developed, (and there are indications that this may happen in the future), this situation may change.

Methods of pollution control

There is a wide range of techniques which can be applied to the intake system, the combustion chamber, and the exhaust systems of petrol and diesel engines to reduce exhaust emissions; the principal ones are outlined in this section.

Crankcase emissions

A positive crankcase ventilation valve is installed between crankcase and inlet system (Fig. 22.19). The intake manifold vacuum draws the clean air from the air filter through the oil filler cap into the engine crankcase, where it mixes with the blow-by gases. The PVC unit contains a spring-loaded valve which controls the flow of air and blow-by

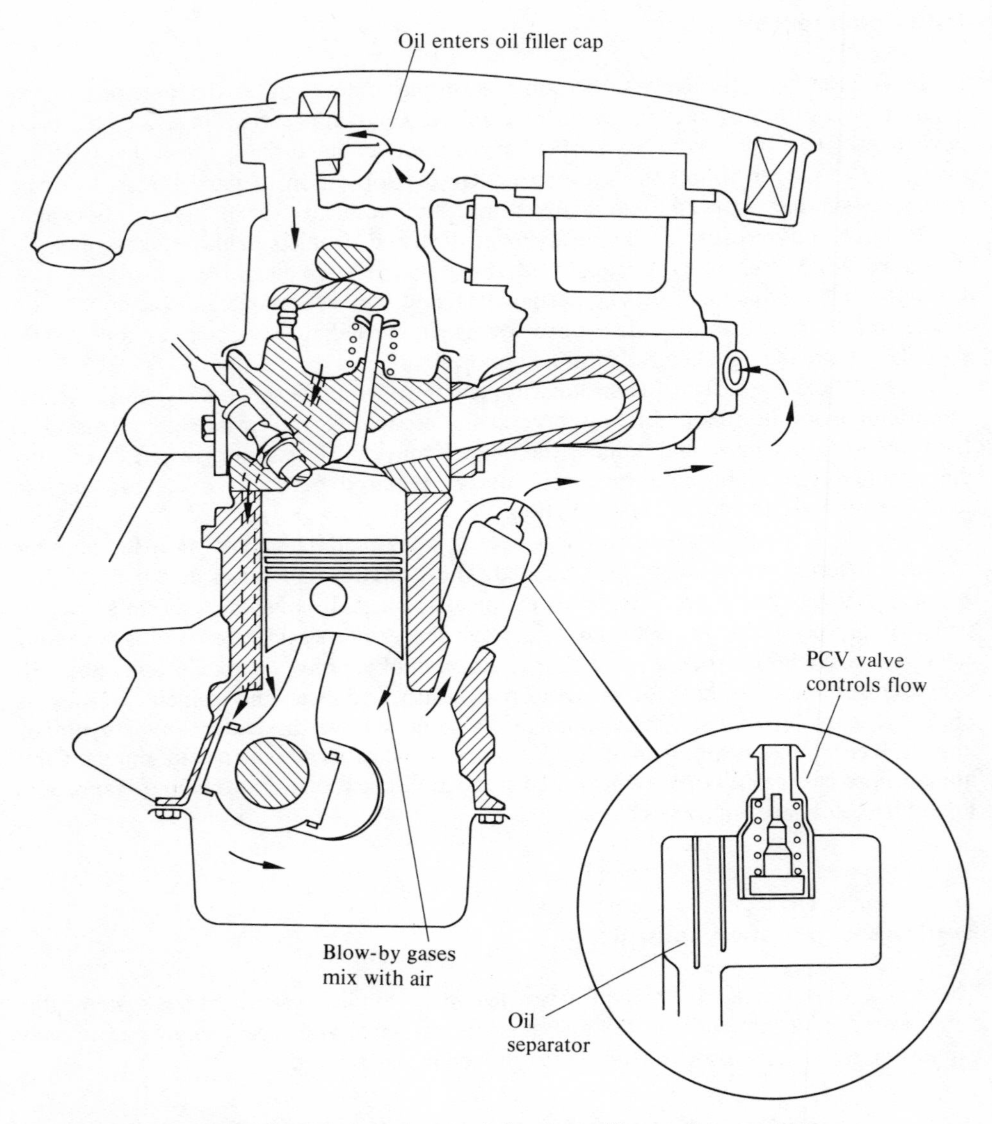

Fig. 22.19. Control of crankcase emissions by positive crankcase ventilation.

gases into the intake system, depending on the manifold depression. Under engine full throttle conditions there is no intake manifold depression, and the blow-by gases flow in the opposite direction through the oil filter cap and into the air filter. Thus, they are still drawn into the engine and burned; 100 per cent control of crankcase emissions can be achieved in this way.

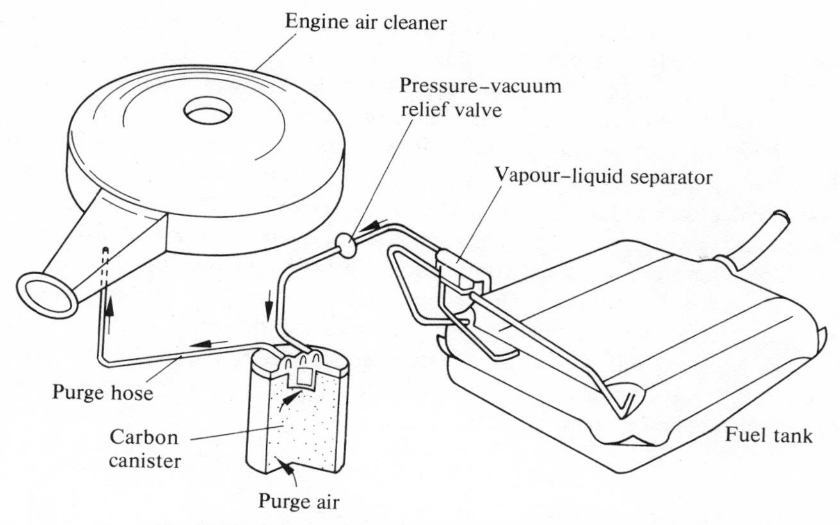

Fig. 22.20. Evaporation control system for petrol engines.

Evaporative losses

The fuel tank vents are connected to a canister containing charcoal (Fig. 22.20). This material has the property that it will adsorb hydrocarbon molecules, and the specific surace area is very large. The surface area of 600 g of charcoal is 69 hectares (170 acres). When the engine is started, purge air is drawn through the charcoal canister and the hydrocarbons desorb into this air. Then the air passes to the engine inlet system and so to the combustion chamber. The pressure–vacuum relief valve in the circuit can act in two different ways, depending on the design. It may release some vapour to the atmosphere in the event of a large pressure build-up in the system, or alternatively it may retain some of the vapour in the fuel tank to prevent the carbon canister being overloaded.

Combustion chamber hydrocarbons

If attention is concentrated on the combustion mechanisms which cause hydrocarbon emissions, then various palliative measures can be applied to the conventional spark ignition engine. It was explained earlier that as the engine decelerated, the high intake manifold vacuum caused the low density incoming charge to be very rich. Under these conditions, the spark could not ignite the mixture and raw fuel passed into the exhaust. This can be avoided if a spring-loaded valve is placed in parallel with the carburettor, such that it opens under high intake manifold vacuum. The valve can draw air and fuel vapour from above the fuel in the carburettor bowl, so that the engine always receives an ignitable mixture. This technique reduces the braking effect of the engine since combustion occurs during every engine cycle.

Another primary source of engine hydrocarbons is the piston top land, which retains an unburnt mixture after combustion, since the flame front does not propagate into this region. There are two alternative philosophies for dealing with this problem: (1) the piston to cylinder liner clearance can be reduced so that the amount of unburnt material is also reduced, or (2) the piston to cylinder liner clearance can be increased so that the flame can ignite the mixture in this region. In either event, reducing the depth of the top land reduces the unburnt hydrocarbons. Hydrocarbon emissions are also related to the surface area of the combustion chamber, since quenched mixture remains on the surface. Therefore, compact combustion chambers release less hydrocarbons.

The source of diesel engine hydrocarbon emissions are not so well defined and prevention techniques can be 'hit or miss'. However, if the volume of the sac at the injector tip is kept as small as possible, hydrocarbons can be reduced. Attention to the design of the fluid dynamic characteristics of injection equipment will prevent secondary fuel injection after combustion.

Carbon monoxide

The amount of carbon monoxide in the exhaust of a petrol engine is related to the mixture strength of the incoming charge. For various reasons which are outside the scope of this text, the mixture must be richer than stoichiometric under idling and full throttle operation. Therefore, some carbon monoxide is always emitted under these conditions. However, the rest of the engine load and speed spectrum can operate with mixtures leaner than stoichiometric. To achieve these engine requirements, modern carburettor design is complex and the manufacturing tolerances are tight.

The heterogeneous mixture of air and fuel in the intake manifold of a multi-cylinder engine can give rise to a phenomenon known as maldistribution. For example, in a four-cylinder engine, the liquid fuel in the manifold may flow preferentially to one cylinder rather than the other three. Thus, three cylinders may be operating leaner than stoichiometric, while the other one emits a large concentration of carbon monoxide. This is detrimental to exhaust emissions, engine power and fuel consumption. If the heterogeneous mixture in the intake can be converted to a homogeneous one, then each cylinder will receive the same mixture strength. This can be achieved by exhaust and water heated hot-spots which vapourize all the liquid fuel below the carburettor. Thus, all the fuel is in a vapour form before it divides between the cylinders. A particularly efficient vaporizer (called the Va-pipe) has been developed, which utilizes a heat pipe to transfer heat from the exhaust to the intake manifold.[8]

The problems of maldistribution and complex carburettor mixture control can be overcome by injection of petrol into each inlet port, beside the valve. This is an expensive solution and is only used on high quality engines at the moment. However, petrol injection has many advantages besides those of exhaust emissions, and its popularity is growing.

Combined carbon monoxide and hydrocarbon control

The exhaust port of a petrol engine is very hot and this fact can be used to advantage. Since carbon monoxide and unburnt hydrocarbons are both the products of incomplete burning, the combustion can be continued if oxygen is mixed with the exhaust gas while the temperature is still high. The air is provided by an air pump, belt driven from the engine crankshaft. This technique, which is called man–air–ox (Fig. 22.21), can give a useful reduction in emissions (Fig. 22.22). The amount of carbon monoxide and unburnt hydrocarbon oxidation can be extended if the exhaust manifold is enlarged and thermally lagged. This prevents heat loss from the exhaust gas and the larger volume allows the exhaust gas to remain in the hot region for a longer period of time. Both of these effects enhance the conditions for oxidation in a thermal reactor (Figs. 22.22 and 22.23).

The oxidation of carbon monoxide and unburnt hydrocarbons with air can be accelerated if a catalyst is present in the exhaust system.[9] The most popular catalysts

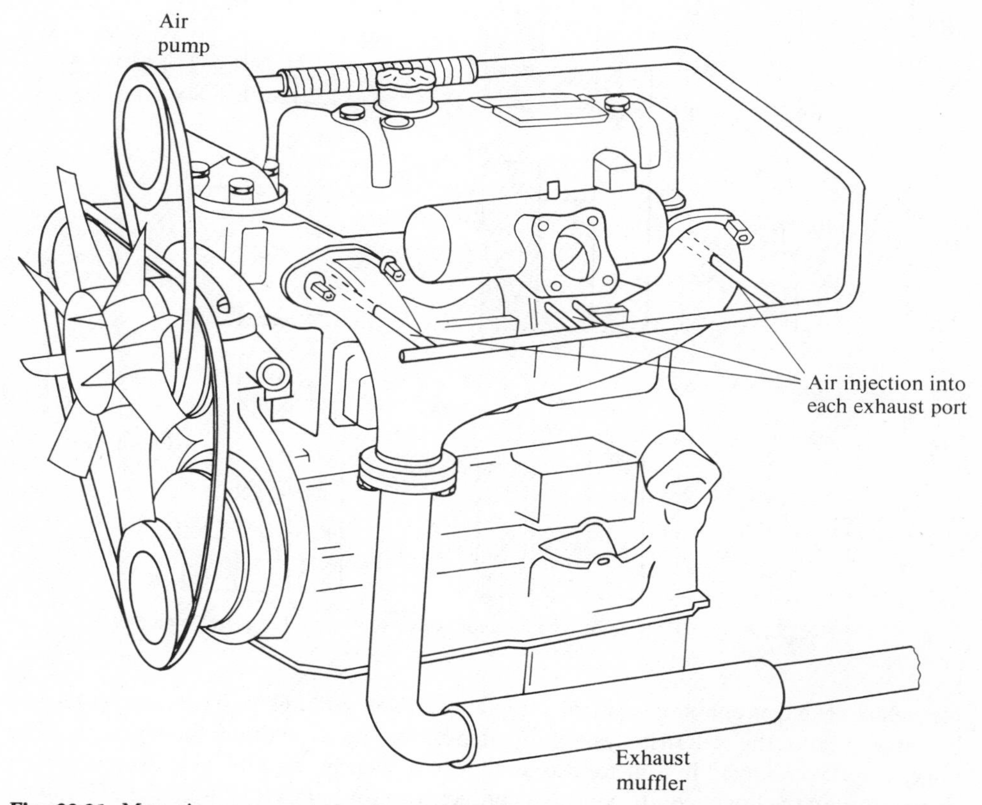

Fig. 22.21. Man–air–ox to control exhaust carbon monoxide and unburnt hydrocarbons for petrol engines.

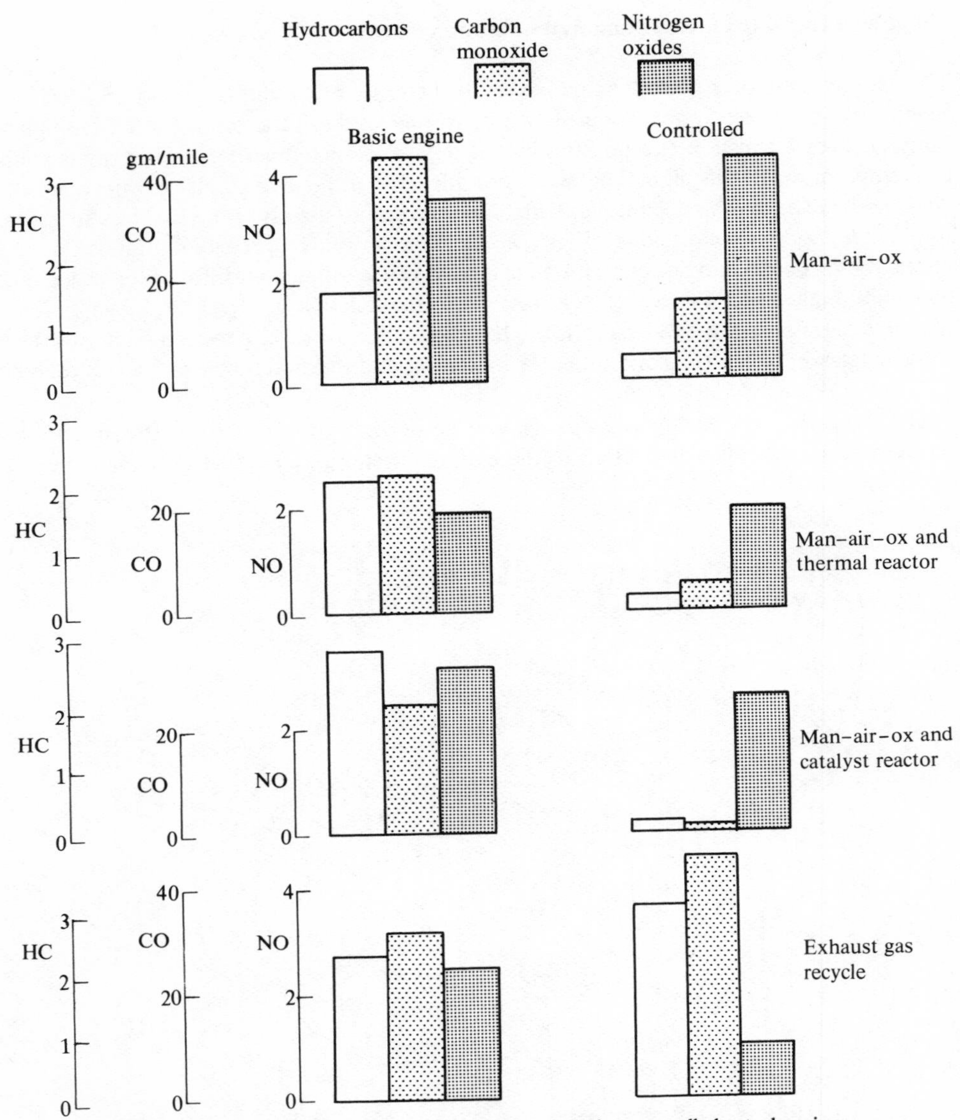

Fig. 22.22. 1975 CVS test results for various emission controlled petrol engines.

for vehicle exhaust emission control are platinum and palladium. The catalyst can be introduced into the exhaust system by deposition on a ceramic honeycomb. The honeycomb is mounted in a metal box similar to a silencer, and the exhaust gas, which flows through the honeycomb, comes in contact with the catalyst (Fig. 22.24). Catalytic reactors of this type can cause a marked reduction in carbon monoxide and unburnt

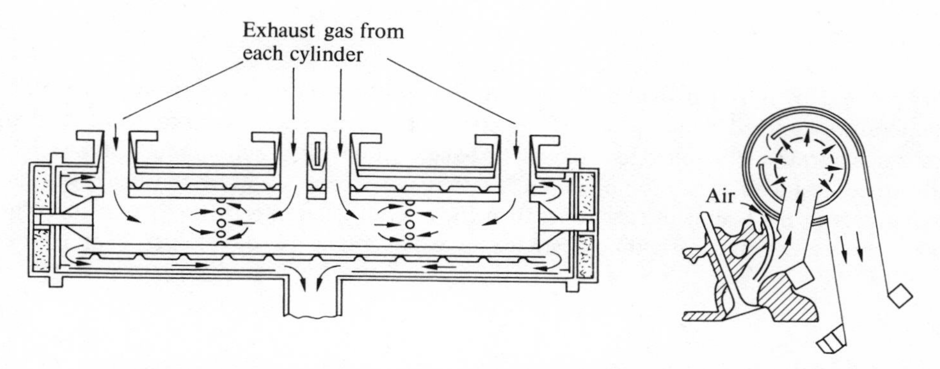

Fig. 22.23. Exhaust reactor to control carbon monoxide and unburnt hydrocarbons.

hydrocarbons (Fig. 22.22). Ceramic pellets, covered with catalyst, can be used in preference to honeycomb supports. Precious metal catalysts are poisoned by lead compounds in the exhaust, so the introduction of catalytic reactors in cars must be coupled with the introduction of an unleaded fuel.

The exhaust gas temperature of diesel engines is lower than that of petrol engines, except at full load. Therefore, the man–air–ox and thermal reactor techniques are unsuitable for emission control of diesel engines. However, some success has been achieved using catalytic reactors on diesel engines, in laboratory experiments.

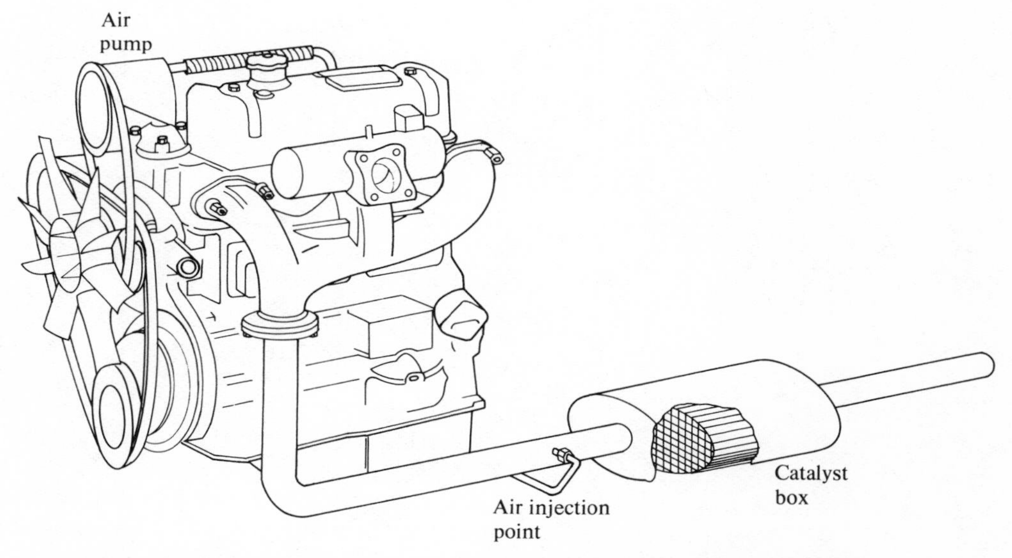

Fig. 22.24. Using a catalyst to control exhaust carbon monoxide and unburnt hydrocarbons.

Nitrogen oxides

The amount of nitrogen oxides formed during combustion depends on the maximum temperature attained by the gases. If the combustion temperature can be reduced, then the concentration of nitrogen oxides is also reduced. When some make-weight substance is added to the combustion, so that it absorbs some of the heat, without taking part in the chemical reactions, then the overall gas temperature is reduced. The most convenient make-weight substance is exhaust gas, since 95 per cent of the

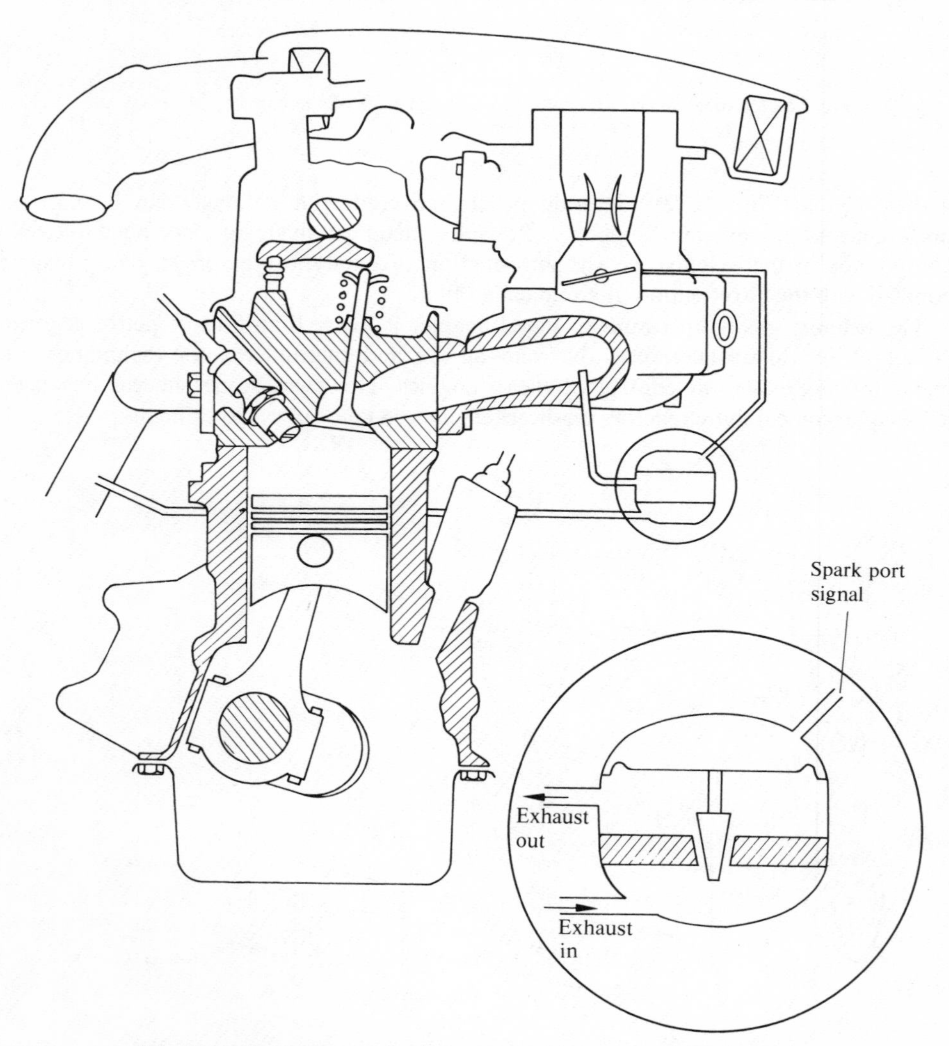

Fig. 22.25. Exhaust gas recirculation to control nitrogen oxide emissions.

composition is carbon dioxide, water vapour and nitrogen. In practice, exhaust gas is extracted from the exhaust and added to the intake system, roughly in proportion to the amount of air flowing through the engine, or the engine power. The signal which controls the cxhaust flow can be provided by the spark port for the distributor (Fig. 22.25), or the pressure depression across the air cleaner. Exhaust gas recirculation controls nitrogen oxides very effectively, since in a petrol engine 11 per cent recycle produced a 50 per cent reduction in nitrogen oxides, Fig. 22.22. However, engine efficiency is reduced and there is an increase in fuel consumption.

Nitrogen oxides can also be reduced by attention to engine combustion. Certain shapes of petrol and diesel combustion chambers tend to quench the combustion and reduce the maximum gas temperature. In addition, the timing of the spark in a petrol engine and the timing of injection in a diesel engine have an effect on gas temperature, and can be modified to reduce nitrogen oxides.[10]. Generally, a reduction in nitrogen oxides from any one type of engine is accompanied by an increase in fuel consumption.

Diesel smoke

Up to the present, the primary method for avoiding smoke has been by careful control of combustion. If very rich zones could be avoided and all the air in the combustion chamber could be utilized, then smoke would not be formed. In practice this ideal can never be achieved, since some air is always contained in inaccessible places such as the piston top land. Fuel additives, based on barium compounds, can reduce smoke by 50 per cent, but have never been sold commercially, due to the increased cost, engine contamination and other marketing problems. Recently, limited success has been achieved in the laboratory with catalysts which burn carbon to carbon dioxide, but this idea has certain technological problems and is a long way from commercial exploitation. Therefore, combustion chamber and injector design to optimize fuel air mixing will remain the principal methods of soot control in the near future.

Other engines

Two strokes

The two-stroke petrol engine, which is popular for utility applications, has a serious exhaust emission problem. Unburnt hydrocarbon emissions are 30–100 times higher than with a conventional four stroke and carbon monoxide emissions are also increased. Conversely, nitrogen oxide emissions are unusually low. The high level of hydrocarbons results from short circuiting of the intake charge from the transfer port to the exhaust port during the induction period and this cannot be avoided in the conventional two-stroke. In addition, it has proved difficult to oxidize these hydrocarbons in the exhaust system. The exhaust gas temperature is lower than in a four-stroke, so man–air–ox does not function. The exhaust gas contains oil droplets, which foul catalysts as they burn on the surface. Therefore catalytic reactors cannot easily be used. Severe exhaust emission legislation has removed the two-stroke engine, as we know it to-day, as a possible powerplant for automotive application in some countries.

Stratified charge

In petrol engines, the petrol–air mixture can only be ignited by a spark plug if the air–fuel ratio is between 8 and 22. It is known that emissions could be decreased considerably if a leaner mixture could be ignited. One method of achieving this is by 'stratifying' the mixture such that the air–fuel ratio at the spark plug is within the required limits, while the rest of the combustion chamber mixture is very lean. If the petrol is injected into the cylinder near the spark plug, just before ignition, it remains close to the plug and can be ignited. This is the principle of the Ford PROCO stratified charge engine.[11] A particularly attractive feature of this engine is that low exhaust emissions (Figs. 22.26 and 22.27) are coupled with a 20 per cent improvement in fuel economy over a conventional petrol engine.

Cylinder injection of petrol requires sophisticated injection equipment, which is expensive to manufacture. However, the stratified charge principle can be utilized in another way. The combustion chamber is divided and a rich mixture is induced beside the spark plug, while the main chamber fills with a very lean mixture. After compression, the rich mixture is ignited by the plug and it expands into the main chamber, firing the lean mixture by a jet ignition process. This principle has been

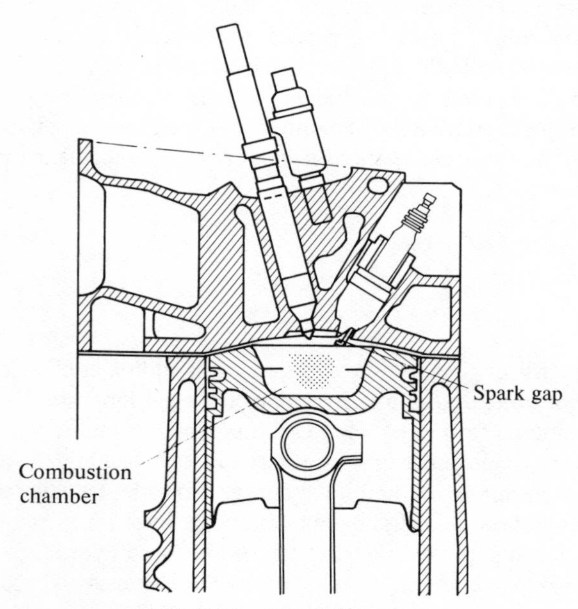

Fig. 22.26. Ford PROCO engine.

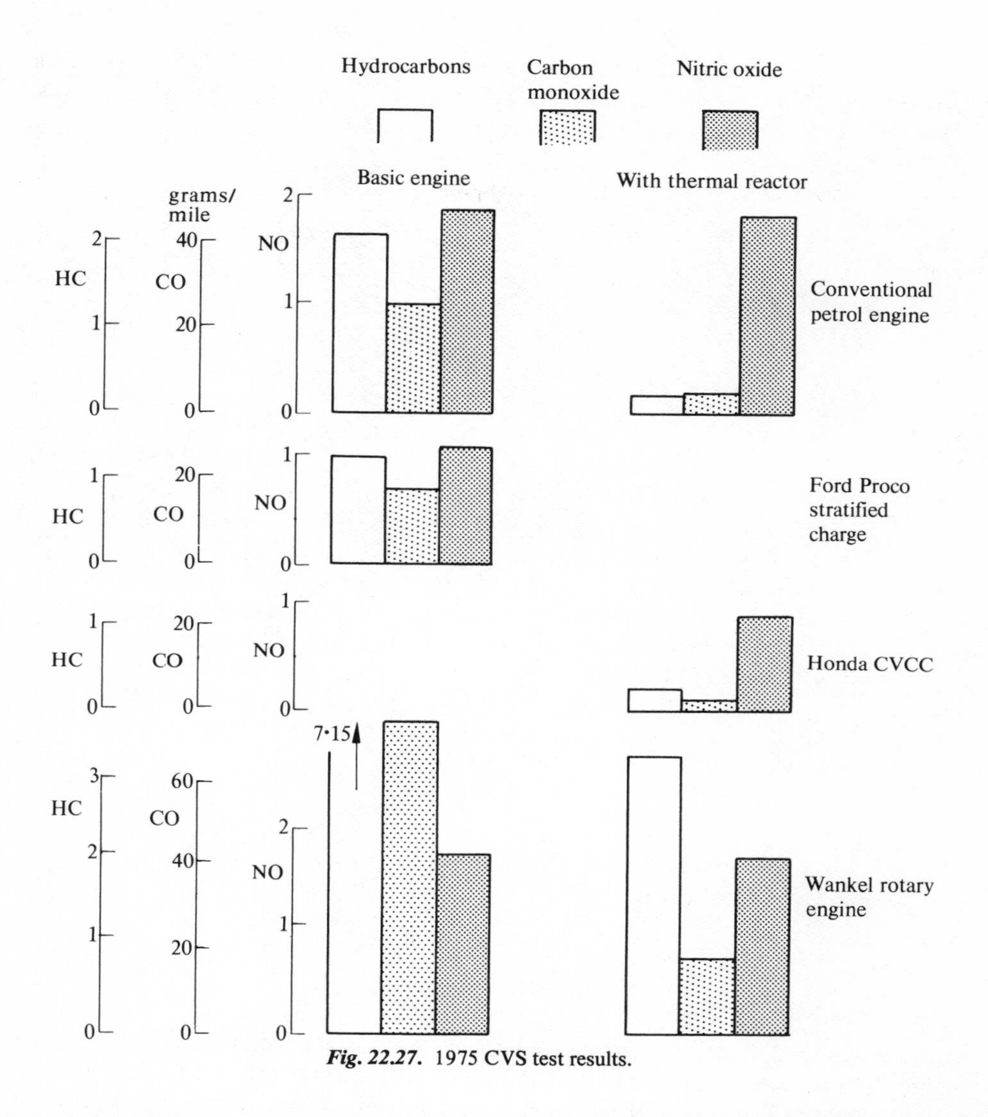

Fig. 22.27. 1975 CVS test results.

utilized by Honda in their CVCC engine (Fig. 22.28). The exhaust manifold of this engine design is lagged to promote hydrocarbon oxidation, and very low emission levels have been achieved (Fig. 22.27). At present, the fuel economy of this engine is worse than a conventional petrol engine.

Wankel engine

The popularity of the Wankel rotary engine has grown recently, due to technological progress and its ability to meet certain exhaust emission criteria. In its basic form,

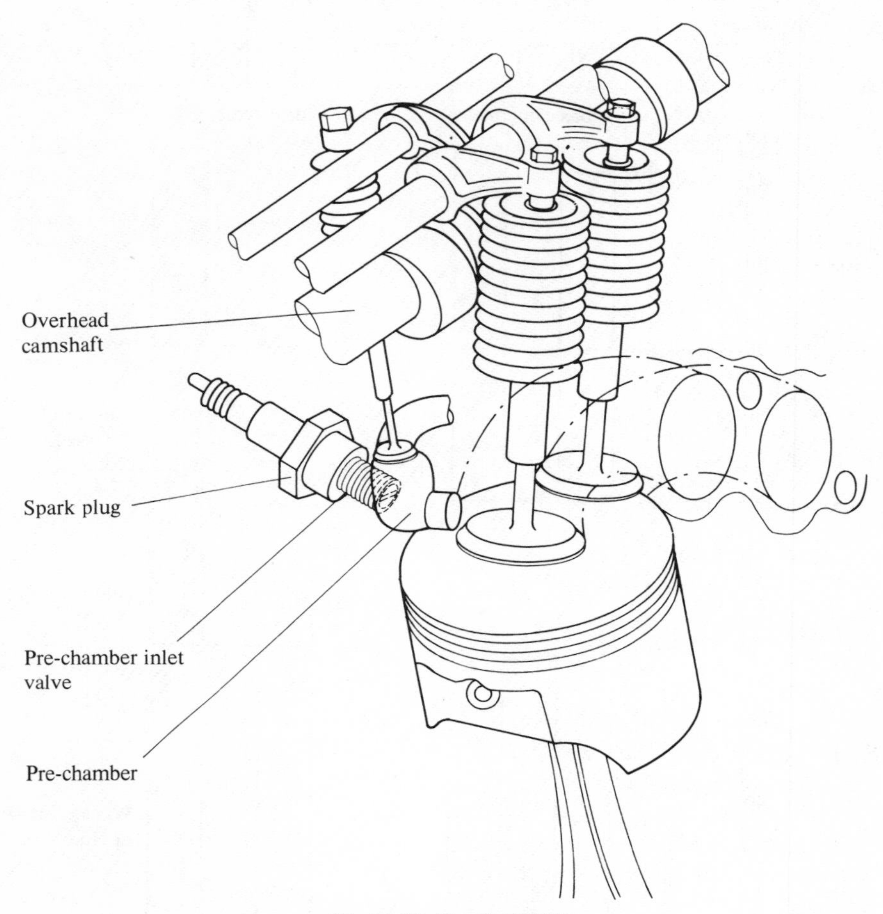

Fig. 22.28. Honda CVCC.

hydrocarbon and carbon monoxide emissions are greater than with reciprocating engines, but nitrogen oxides are lower (Fig. 22.27). Therefore, nitrogen oxide controls are not required. The design of this engine is particularly amenable to emission control by thermal reactors, since the reactor can be made very compact. Hydrocarbon and carbon monoxide emissions can be oxidized effectively and values of 0·21 g km^{-1} for hydrocarbons and 1·55 g km^{-1} for carbon monoxide have been reported, but there is a 15 per cent fuel consumption penalty with Wankel engines, which appears to be an inherent feature that cannot easily be overcome.

Conclusion

Diesel engines are less of a health hazard than petrol engines, although the smoke and odour are certainly a nuisance. A wide range of legislation is being introduced throughout the world to control carbon monoxide, hydrocarbons and nitrogen oxide emissions from petrol engines and smoke from diesels; this trend will continue. In the near future, catalytic reactors will be introduced in the US, where the exhaust emission problem is severe. Simpler measures such as combustion tuning and man–air–ox should fulfil air quality criteria in most other countries in the foreseeable future. Diesel engine emission and smoke targets will probably be achieved by attention to the combustion process. Unfortunately, many techniques for reducing petrol engine emissions involve some increase in fuel consumption. The exception to this trend is the stratified charge engine, and the concept is one of the long term solutions being explored for exhaust emission problems.

Acknowledgements

A general note of this nature could not have been written without the guidance and assistance of a wide range of Ricardo personnel from the different automotive engineering fields. In particular, I should like to mention the Petrol Department, who provided many of the data and illustrations contained in this chapter.

References

1. Myers, P. S. *Automobile emissions—a study in environmental benefits versus technological cost.* Society of Automotive Engineers (SAE) 700182, 1970.
2. (a) *The elimination of vehicle particulate emissions*, Associated Octel Co., OP 72/5, London, December 1972; (b) Campbell, K. and G. Onion, 'Filter exhaust systems for reduction of lead emissions from petrol engines', *Filtration Society's Conference*, London, September 1973.
3. Haskell, W. W. and C. E. Legate, *Exhaust hydrocarbon emissions from gasoline engines—surface phenomena*, SAE 720255, 1972.
4. Broome, D. and I. M. Khan, 'Mechanisms of soot release from combustion of hydrocarbon fuels with particular reference to diesel engines', *I.Mech.E. Symposium on Air Pollution Control in Transport Engines*, London, November 1971.
5. US Federal Register, Vol. 37.
6. *Air pollution by gases from positive ignition engines of motor vehicles*, EEC Directive No. 70/220.
7. *Evaluation of a filter for the removal of lead from the exhausts of petrol engines*, UK Warren Springs Laboratory, Report LR170AP, 1972.
8. Wilson, J. L., 'The vapipe', *First Symposium on Low Pollution Power System Development*, Ann Arbor., October 1973.
9. Haslet, R. A., *A technique for the endurance testing of oxidation catalytic reactors*, SAE740246, 1974.

10. ‘Richardo engine exhaust emission and noise control work’, *Diesel and Gas Turbine Progress*, World Wide Ed. Ricardo and Co, UK, January/February 1973.
11. Simco, A., M. A. Choma and L. L. Repco, *Exhaust emission control by the Ford programmed combustion process—PROCO*, SAE720052, 1972.

INDEX

Printed in Great Britain at The Spottiswoode Ballantyne Press by William Clowes & Sons Limited, London, Colchester and Beccles

ST. LOUIS COMM. COL
AT FLORISSANT VALLEY

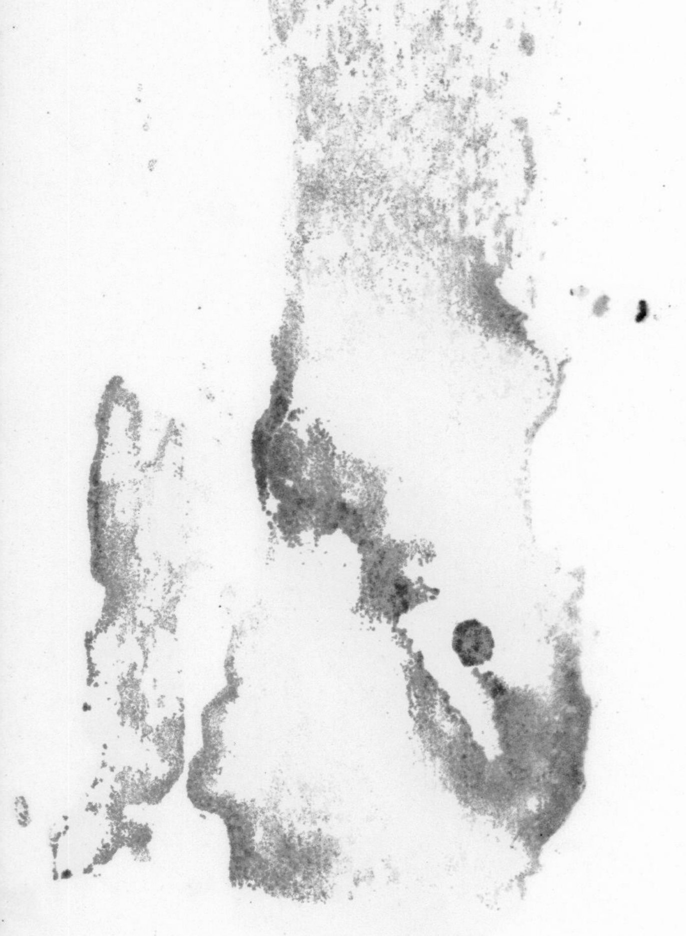